## ACTIVE LEARNING
An active learning focus keeps students engaged as they discover the world of data.

### gUIDED EXERCISES

g **8.27 Gun Control** Historically, the percentage of U.S. residents who support stricter gun control laws has been 52%. A recent Gallup Poll of 1011 people showed 495 in favor of stricter gun control laws. Assume the poll was given to a random sample of people.

**QUESTION** Test the claim that the proportion of those favoring stricter gun control has changed from 0.52. Perform a hypothesis test, using a significance level of 0.05, by following the steps.

**Step 1 ▶ Hypothesize**

$H_0$: The population proportion that supports gun control is 0.52, $p =$ _____.

$H_a$: $p$ _____.

**Step 2 ▶ Prepare**

Choose the one-proportion $z$-test.

### EXERCISES
A wealth of **exercises** appear at the end of each chapter, organized by section and level of difficulty. Challenging exercises are identified with a (*).

- Exercises with a **g** indicate that step-by-step problem-solving help is available in the **Guided Exercises** section at the end of the chapter.

- Exercises marked with a ✳ have accompanying data sets available on the Companion CD-ROM.

### EXAMPLES
**Examples** are abundant and they model solutions to real-world problems. Each example is tied to a **TRY THIS!** exercise, which offers practice solving a similar problem.

### EXAMPLE 11 Stolen Bicycles

According to the website nationalbikeregistry.com, at the campus of UC Berkeley, only 3% of stolen bicycles are returned to owners.

**QUESTIONS** Accepting for the moment that the four characteristics of a binomial experiment are satisfied, write the notation for the probability that exactly 5 bicycles will be returned if 235 are stolen in the course of a year. What is this probability?

**SOLUTION** The number of trials is 235, the probability of success is 0.03, and the number of successes is 5. Therefore, we can write the binomial probability as $b(235, 0.03, 5)$.

The command and output for finding this probability on a TI-83/84 are shown in Figure 6.33. The command binompdf stands for "binomial probability distribution function." The results show us that the probability that exactly 5 bicycles will be returned, assuming that this is in fact a binomial experiment, is about 12.6%.

**TRY THIS!** Exercise 6.59

## INTEGRATED TECHNOLOGY
This text emphasizes statistical software, which enables the use of real data sets and a focus on conceptual learning.

### TECHTIPS
**TechTips** outline the steps for performing calculations using TI-83/84-Plus graphing calculators, Excel,® Minitab,® and StatCrunch.® Whenever a new method or procedure is introduced, a **Tech** refers students to the TechTips section at the end of the chapter.

**TI-83/84**

**NORMAL**

**a. Between Two Values**
1. Press **2ND DISTR** (located below the four arrows on the keypad).
2. Select **2:normalcdf** and press **ENTER**.
3. Enter: (left boundary, right boundary, $\mu$, $\sigma$), in this example 85 ⎬ 115 ⎬ 100 ⎬ 15) and press **ENTER**. (The comma button is above the 7 button.)

Your screen should look like Figure 6A, which shows that the probability that a randomly selected person will have a Wechsler IQ between 85 and 115 is equal to 0.6827.

```
normalcdf(85,115
,100,15)
      .6826894809
```

▲ **FIGURE 6A** TI-83/84 normalcdf (c stands for "cumulative")

### SOFTWARE OUTPUT
Many examples and exercises ask the reader to interpret and critically evaluate **software output**. This builds comfort with various types of software.

**6.16 SAT Scores** Quantitative SAT scores have a population mean of 500 and a population standard deviation of 100 and are approximately Normally distributed. Find the probability that a randomly selected person will have a quantitative SAT score of 650 or less by looking at the StatCrunch output.

# Introductory Statistics: Exploring the World Through Data

# Introductory Statistics: Exploring the World Through Data

## Robert Gould

**University of California, Los Angeles**

## Colleen Ryan

**California Lutheran University**

**PEARSON**

Boston   Columbus   Indianapolis   New York   San Francisco   Upper Saddle River
Amsterdam   Cape Town   Dubai   London   Madrid   Milan   Munich   Paris   Montreal   Toronto
Delhi   Mexico City   Sao Paulo   Sydney   Hong Kong   Seoul   Singapore   Taipei   Tokyo

Editor in Chief: Deirdre Lynch
Acquisitions Editor: Marianne Stepanian
Sponsoring Editor: Christina Lepre
Executive Market Development Manager: Dona Kenly
Executive Director of Development: Carol Trueheart
Editorial Assistant: Sonia Ashraf
Senior Managing Editor: Karen Wernholm
Associate Managing Editor: Tamela Ambush
Senior Production Supervisor: Peggy McMahon
Associate Director of Design, USHE North and West: Andrea Nix
Art Director/Cover Designer: Beth Paquin
Marketing Manager: Erin Lane
Marketing Assistant: Kathleen DeChavez
Senior Author Support/Technology Specialist: Joe Vetere
Executive Manager, Course Production: Peter Silvia
Procurement Manager: Evelyn Beaton
Procurement Specialist: Debbie Rossi
Media Procurement Specialist: Ginny Michaud
Design, Production Coordination, Technical Illustrations, and Composition: Nesbitt Graphics, Inc.
Cover photo: Boardwalk along the salt marsh, Kouchibouguac National Park, New Brunswick,
Canada by Darwin Wiggett/All Canada Photos/Getty Images

Credits and acknowledgments borrowed from other sources and reproduced, with permission, in this text appear on page xxxi.

Many of the designations used by manufacturers and sellers to distinguish their products are claimed as trademarks. Where those designations appear in this book, and Pearson was aware of a trademark claim, the designations have been printed in initial caps or all caps.

**Library of Congress Cataloging-in-Publication Data**
Gould, Robert (Robert Lynn), 1965-
    Introductory statistics : exploring the world through data / Robert
Gould, Colleen N. Ryan. -- Ist ed.
      p. cm.
    ISBN-13: 978-0-321-32215-9
    ISBN-10: 0-321-32215-0
1. Statistics.  I. Ryan, Colleen N. (Colleen Nooter), 1939-  II. Title.
QA276.12.G687 2013
519.5--dc23
                              2011020372

3 4 5 6 7 8 9 10—CKV—15    14 13 12

ISBN-10: 0-321-32215-0
ISBN-13: 978-0-321-32215-9

# Dedication

To my parents and family, my friends, and my colleagues who are also friends. Without their patience and support, this would not have been possible.

—Rob

To my teachers and students, and to my family who have helped me in many different ways.

—Colleen

# About the Authors

## Robert Gould

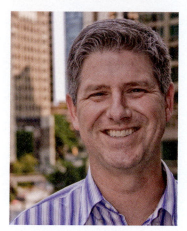

**Robert L. Gould** (Ph.D., University of California, San Diego) is a leader in the statistics education community. He has served as chair of the American Statistical Association's Committee on Teacher Enhancement, has served as chair of the ASA's Statistics Education Section, and served on a panel of co-authors for the *Guidelines for Assessment in Instruction on Statistics Education (GAISE) College Report*. As the associate director of professional development for CAUSE (Consortium for the Advancement of Undergraduate Statistics Education), Rob has worked closely with the American Mathematical Association of Two-Year Colleges (AMATYC) to provide traveling workshops and summer institutes in statistics. For over ten years, he has served as Vice-Chair of Undergraduate Studies at the UCLA Department of Statistics, and he is Director of the UCLA Center for the Teaching of Statistics. In 2009, Rob was elected president of the Southern California Chapter of the American Statistical Association.

In his free time, Rob plays the cello, runs, and is an ardent reader of fiction.

## Colleen Ryan

**Colleen N. Ryan** has taught statistics, chemistry, and physics to diverse community college students for decades. She taught at Oxnard College from 1975 to 2006, where she earned the Teacher of the Year Award. Colleen currently teaches statistics part-time at California Lutheran University. She often designs her own lab activities. Her passion is to discover new ways to make statistical theory practical, easy to understand, and sometimes even fun.

Colleen earned a B.A. in physics from Wellesley College, an M.A.T. in physics from Harvard University, and an M.A. in chemistry from Wellesley College. Her first exposure to statistics was with Frederick Mosteller at Harvard.

In her spare time, Colleen sings with the Oaks Camerata, has been an avid skier, and enjoys time with her family.

# Contents

CHAPTER 14    **Inference for Regression    622**

# Preface

## About This Book

This is a book about data. Data are at the heart of the everyday life of modern citizens, and in particular modern students. Google, Facebook, smartphones, and highways are just a few examples of everyday objects that routinely collect, organize, and transmit data. Students who can think about and reason with data will have an advantage over those who cannot. To become "data literate," students need to develop quantitative literacy, understand how to handle computational technology, understand the role of variability in interpreting data, and learn to communicate clearly. This book teaches these skills.

We realize that students usually do not take this course voluntarily or with great enthusiasm. But we intend to convert them to enthusiasts by showing them that data are everywhere and that everyone can profit both intellectually and financially (and, dare we add, emotionally?) by achieving an understanding of basic statistical concepts. We take special care to accommodate students who have not had a mathematics course in a while and students who may not feel confident in their mathematical abilities. Wherever possible, we build on students' intuition in order to demonstrate that statistics is often just formalized common sense. However, we are sensitive to the fact that students' intuitions can lead to misconceptions, and we proceed carefully to correct common mistakes.

Students might not take another formal course in statistics, but this course is not the end of their relationship with data. Indeed, students will use the concepts and critical thinking skills they learn in this class throughout their lives. The number of careers that require statistical reasoning is growing rapidly—from English majors using digitized texts to analyze works of literature in a larger context, to community activists using statistics to understand patterns of community change. In this class, students will learn how they can use their new-found data skills to expand their world.

## Approach

Our text is concept-based, as opposed to method-based. We do teach useful statistical methods, but we emphasize that applying the method is secondary to understanding the concept.

In the real world, computers do most of the heavy lifting for statisticians. We therefore adopt an approach which frees the instructor from having to teach tedious procedures and leaves more time for teaching deeper understanding of concepts. Accordingly, we present formulas as an aid to understanding the concepts rather than as the focus of study.

We believe students need to learn how to

- determine which statistical procedures are appropriate,
- instruct the software to carry out the procedures, and
- interpret the output.

We understand that students will probably see only one type of statistical software in class. But we believe it is useful for students to compare output from several different sources, so in some examples we ask them to read output from two or more software packages.

One of the authors (Rob Gould) served on a panel of co-authors for the collegiate version of the American Statistical Association-endorsed *Guidelines for Assessment and Instruction in Statistics Education* (*GAISE*). We firmly believe in its main goals and have adopted them in the preparation of this book.

- We emphasize understanding over rote performance of procedures.

- We use real data whenever possible.

- We encourage the use of technology both to develop conceptual understanding and to analyze data.

- We believe strongly that students learn by doing. For this reason, the homework problems offer students both practice in basic procedures and challenges to build conceptual understanding.

## Coverage

The first two-thirds of this book are concept-driven and cover exploratory data analysis and inferential statistics—fundamental concepts that every introductory statistics student should learn. The final third of the book builds on that strong conceptual foundation and is more methods-based. It presents several popular statistical methods and more fully explores methods presented earlier, such as regression and data collection.

Our ordering of topics is guided by the Cycle of Data.

*Chapters 1–4: Exploratory Data Analysis.* The first four chapters cover data collection and summary. Chapter 1 introduces the important topic of data collection and compares and contrasts observational studies with controlled experiments. This chapter also teaches students how to handle raw data so that the data can be uploaded to their statistical software. Chapters 2 and 3 discuss graphical and numerical summaries of single variables based on samples. We emphasize that the purpose is not just to produce a graph or a number but, instead, to explain what those graphs and numbers say about the world. Chapter 4 introduces simple linear regression and presents it as a technique for providing graphical and numerical summaries of relationships between two numerical variables.

We feel strongly that introducing regression early in the text is beneficial in building student understanding of the applicability of statistics to real-world scenarios. After completing the chapters covering data collection and summary, students have acquired the skills and sophistication they need to describe two-variable associations and to generate informal hypotheses. Two-variable associations provide a rich context for class discussion and allow the course to move from fabricated problems (because one-variable analyses are relatively rare in the real world) to real problems that appear frequently in everyday life. We return to regression in Chapter 14, when we discuss statistical inference in the context of regression, which requires quite a bit of machinery. We feel that it would be a shame to delay until the end of the course all the insights that regression without inference can provide.

*Chapters 5–8: Inference.* These chapters teach the fundamental concepts of statistical inference. The main idea is that our data mirror the real world, but imperfectly; although our estimates are uncertain, under the right conditions we can quantify our uncertainty. Verifying that these conditions exist and understanding what happens if they are not satisfied are important themes of these chapters.

*Chapters 9–11: Methods.* Here we return to the themes covered earlier in the text and present them in a new context by introducing additional statistical methods, such as estimating population means, analyzing categorical variables, and analyzing relations between a numerical and a categorical variable. We also introduce multiple comparisons and use them to motivate the need for the statistical method of ANOVA.

*Chapters 12–14: Special Topics.* Students who have covered all topics up to this point will have a solid foundation in statistics. These final chapters build on that foundation and offer more details, as we explore the topics of designing controlled experiments, survey sampling, additional contexts for hypothesis testing, and using regression to make inferences about a population.

In Chapter 12 we provide guidance for reading scientific literature. Even if your schedule does not allow you to cover Chapter 12, we recommend using Section 12.3 to offer students the experience of critically examining real scientific papers.

# Organization

Our preferred order of progressing through the text is reflected in the Contents, but there are some alternative pathways as well.

*10-week Quarter.* The first eight chapters would provide a full, one-quarter course in introductory statistics. If time remains, cover Sections 9.1 and 9.2 as well, so that students can solidify their understanding of confidence intervals and hypothesis tests by revisiting the topic with a new parameter.

*Proportions First.* Ask two statisticians, and you will get three opinions on whether it is best to teach means or proportions first. We have come down on the side of proportions for a variety of reasons. Proportions are much easier to find in popular news media (particularly around election time), so they can more readily be tied to students' everyday lives. Also, the mathematics and statistical theory is simpler; because there's no need to provide a separate estimate for the population standard deviation, inference is based on the Normal distribution, and no further approximations (that is, the *t*-distribution) are required. Hence, we can quickly get to the heart of the matter with fewer technical diversions.

The basic problem here is how to quantify the uncertainty involved in estimating a parameter and how to quantify the probability of making incorrect decisions when posing hypotheses. We cover these ideas in detail in the context of proportions. Students can then more easily learn how these same concepts are applied in the new context of means (and any other parameter they may need to estimate).

*Means First.* Conversely, many people feel that there is time for only one parameter and that this parameter should be the mean. For this alternative presentation, cover Chapters 6, 7, and 9, in that order. On this path, students learn about survey sampling and the terminology of inference (population vs. sample, parameter vs. statistic) and then tackle inference for the mean, including hypothesis testing.

To minimize the coverage of proportions, you might choose to cover Chapters 6, Section 7.1 (which covers in detail the language and framework of statistical inference), and then Chapter 9. Chapters 7 and 8 develop the concepts of statistical inference more slowly than in Chapter 9, but essentially, Chapter 9 develops the same ideas in the context of the mean.

If you present Chapter 9 before Chapters 7 and 8, we recommend that you devote roughly twice as much time to Chapter 9 as you have devoted to previous chapters, because many challenging ideas are explored in this chapter. If you have already covered Chapters 7 and 8 thoroughly, Chapter 9 can be covered more quickly.

# Features

We've incorporated into this book variety of features to aid student learning and to facilitate the use of this text in any classroom.

## Integrating Technology

Modern statistics is inseparable from technology. We have worked to make this textbook accessible for any classroom, regardless of the level of in-class exposure to technology, while still remaining true to the demands of the analysis. We know that students sometimes do not have access to technology when doing homework, so many exercises provide output from software and ask students to interpret and critically evaluate that given output.

Using technology is important because it enables students to handle real data, and real data sets are often large and messy. The following features are designed to guide students.

- **TechTips** outline steps for performing calculations using TI-83/84® graphing calculators, Excel®, Minitab®, and StatCrunch®. We do not want students to get stuck because they don't know how to reproduce the results we show in the book, so whenever a new method or procedure is introduced, an icon, Tech, refers students to the TechTips section at the end of the chapter. Each set of TechTips contains at least one mini-example, so that students are not only learning to use the technology but also practicing data analysis and reinforcing ideas discussed in the text.

- **Check Your Tech** examples help students understand that statistical calculations done by technology do not happen in a vacuum and assure them that they can get the same numerical values by hand. Although we place a higher value on interpreting results and verifying conditions required to apply statistical models, the numerical values are important, too.

- The companion **CD-ROM** includes Technology Tutorial Videos for Excel, Minitab, and the TI83/84, as well as all data sets used in the exposition and exercises. These data are also available at http://www.pearsonhighered.com/mathstatsresources/.

## Guiding Students

- Each chapter opens with a **Theme**. Beginners have difficulty seeing the forest for the trees, so we use a theme to give an overview of the chapter content.

- Each chapter begins by posing a real-world **Case Study**. At the end of the chapter, we show how techniques covered in the chapter helped solve the problem presented in the Case Study.

- **Margin Notes** draw attention to details that enhance student learning and reading comprehension.

     **Caution** notes provide warnings about common mistakes or misconceptions.

     **Looking Back** reminders refer students to earlier coverage of a topic.

     **Details** clarify or expand on a concept.

- **Key Points** highlight essential concepts to draw special attention to them. Understanding these concepts is essential for progress.

- **Snapshots** break down key statistical concepts introduced in the chapter, quickly summarizing each concept or procedure and indicating when and how it should be used.

- An abundance of worked-out **examples** model solutions to real-world problems relevant to students' lives. Each example is tied to an end-of-chapter exercise so that students can practice solving a similar problem and test their understanding. Within the exercise sets, the icon TRY indicates which problems are tied to worked-out examples in that chapter, and the numbers of those examples are indicated.

- The **Chapter Review** that concludes each chapter provides a list of important new terms, student learning objectives, a summary of the concepts and methods discussed, and sources for data, articles, and graphics referred to in the chapter.

## Active Learning

- For each chapter we've included an activity, **Exploring Statistics**, that students are intended to do in class as a group. We have used these activities ourselves, and we have found that they greatly increase student understanding and keep students engaged in class. Detailed instructions are available for instructors in the Instructor's Edition of the text.

- All exercises are located at the end of the chapter. **Section Exercises** are designed to begin with a few basic problems that strengthen recall and assess basic knowledge, followed by mid-level exercises that ask more complex, open-ended questions. **Chapter Review Exercises** provide a comprehensive review of material covered throughout the chapter.

  The exercises emphasize good statistical practice by requiring students to verify conditions, make suitable use of graphics, find numerical values, and interpret their findings in writing. All exercises are paired so that students can check their work on the odd-numbered exercise and then tackle the corresponding even-numbered exercise.

  Challenging exercises, identified with an asterisk (*), ask open-ended questions and sometimes require students to perform a complete statistical analysis. For exercises marked with a ⊛, accompanying data sets are available on the companion CD-ROM included with the book. The answers to all odd-numbered exercises appear in the back of the book.

- Most chapters include select exercises, marked with a **g** within the exercise set, to indicate that problem-solving help is available in the **Guided Exercises** section. If students need support while doing homework they can turn to the Guided Exercises to see a step-by-step approach to solving the problem.

# Acknowledgments

We are grateful for the attention and energy that a large number of people devoted to making this a better book. We extend our gratitude to Elaine Newman (Sonoma State University), Fred J. Rispoli (Dowling College), Delray Schultz (Millersville University), and Catalina Yang (Oxnard Community College), who checked the accuracy of this text and its many exercises. Thanks also to David Chelton, our developmental editor, to Carol Merrigan, who handled production, to Peggy McMahon, senior production project manager, and to Connie Day, our copyeditor. Many thanks to John Norbutas for his technical advice and help with the TechTips. We thank Deirdre Lynch, editor-in-chief, for signing us up and sticking with us, and we are grateful to Dona Kenly for her market development efforts. Special thanks to our acquisitions editor, Marianne Stepanian, whose patience and good advice made it all possible, and to our sponsoring editor, Christina Lepre, who kept us on task and never steered us wrong.

We would also like to extend our sincere thanks for the suggestions and contributions made by the following reviewers, class testers, and focus group attendees.

Arun Agarwal, *Grambling State University*
Anne Albert, *University of Findlay*
Michael Allen, *Glendale Community College*
Eugene Allevato, *Woodbury University*
Dr. Jerry Allison, *Trident Technical College*
Polly Amstutz, *University of Nebraska*
Patricia Anderson, *Southern Adventist University*
MaryAnne Anthony-Smith, *Santa Ana College*
David C. Ashley, *Florida State College at Jacksonville*
Diana Asmus, *Greenville Technical College*
Kathy Autrey, *Northwestern State University of Louisiana*
Wayne Barber, *Chemeketa Community College*
Roxane Barrows, *Hocking College*
Jennifer Beineke, *Western New England College*
Diane Benner, *Harrisburg Area Community College*
Norma Biscula, *University of Maine, Augusta*
KB Boomer, *Bucknell University*
David Bosworth, *Hutchinson Community College*
Diana Boyette, *Seminole Community College*
Elizabeth Paulus Brown, *Waukesha County Technical College*
Leslie Buck, *Suffolk Community College*
RB Campbell, *University of Northern Iowa*
Stephanie Campbell, *Mineral Area College*
Anne Cannon, *Cornell College*
Rao Chaganty, *Old Dominion University*
Carolyn Chapel, *Western Technical College*
Christine Cole, *Moorpark College*
Linda Brant Collins, *University of Chicago*
James A. Condor, *Manatee Community College*
Carolyn Cuff, *Westminster College*

Phyllis Curtiss, *Grand Valley State University*
Monica Dabos, *University of California, Santa Barbara*
Greg Davis, *University of Wisconsin, Green Bay*
Bob Denton, *Orange Coast College*
Julie DePree, *University of New Mexico–Valencia*
Jill DeWitt, *Baker Community College of Muskegon*
Paul Drelles, *West Shore Community College*
Keith Driscoll, *Clayton State University*
Rob Eby, *Blinn College*
Nancy Eschen, *Florida Community College at Jacksonville*
Karen Estes, *St. Petersburg College*
Mariah Evans, *University of Nevada, Reno*
Harshini Fernando, *Purdue University North Central*
Stephanie Fitchett, *University of Northern Colorado*
Elaine B. Fitt, *Bucks County Community College*
Michael Flesch, *Metropolitan Community College*
Melinda Fox, *Ivy Tech Community College, Fairbanks*
Joshua Francis, *Defiance College*
Michael Frankel, *Kennesaw State University*
Heather Gamber, *Lone Star College*
Debbie Garrison, *Valencia Community College, East Campus*
Kim Gilbert, *University of Georgia*
Stephen Gold, *Cypress College*
Nick Gomersall, *Luther College*
Mary Elizabeth Gore, *Community College of Baltimore County–Essex*
Ken Grace, *Anoka Ramsey Community College*
Larry Green, *Lake Tahoe Community College*
Jeffrey Grell, *Baltimore City Community College*

Albert Groccia, *Valencia Community College, Osceola Campus*

David Gurney, *Southeastern Louisiana University*

Chris Hakenkamp, *University of Maryland, College Park*

Melodie Hallet, *San Diego State University*

Donnie Hallstone, *Green River Community College*

Cecil Hallum, *Sam Houston State University*

Josephine Hamer, *Western Connecticut State University*

Mark Harbison, *Sacramento City College*

Beverly J. Hartter, *Oklahoma Wesleyan University*

Laura Heath, *Palm Beach State College*

Greg Henderson, *Hillsborough Community College*

Susan Herring, *Sonoma State University*

Carla Hill, *Marist College*

Michael Huber, *Muhlenberg College*

Kelly Jackson, *Camden County College*

Bridgette Jacob, *Onondaga Community College*

Robert Jernigan, *American University*

Chun Jin, *Central Connecticut State University*

Maryann Justinger, Ed.D., *Erie Community College*

Joseph Karnowski, *Norwalk Community College*

Susitha Karunaratne, *Purdue University North Central*

Mohammed Kazemi, *University of North Carolina–Charlotte*

Robert Keller, *Loras College*

Omar Keshk, *Ohio State University*

Raja Khoury, *Collin County Community College*

Brianna Killian, *Daytona State College*

Yoon G. Kim, *Humboldt State University*

Greg Knofczynski, *Armstrong Atlantic University*

Jeffrey Kollath, *Oregon State University*

Erica Kwiatkowski-Egizio, *Joliet Junior College*

Sister Jean A. *Lanahan, OP, Molloy College*

Katie Larkin, *Lake Tahoe Community College*

Michael LaValle, *Rochester Community College*

Deann Leoni, *Edmonds Community College*

Lenore Lerer, *Bergen Community College*

Quan Li, *Texas A&M University*

Doug Mace, *Kirtland Community College*

Walter H. Mackey, *Owens Community College*

Keith McCoy, *Wilbur Wright College*

Elaine McDonald-Newman, *Sonoma State University*

William McGregor, *Rockland Community College*

Bill Meisel, *Florida State College at Jacksonville*

Bruno Mendes, *University of California, Santa Cruz*

Wendy Miao, *El Camino College*

Robert Mignone, *College of Charleston*

Ashod Minasian, *El Camino College*

Megan Mocko, *University of Florida*

Sumona Mondal, *Clarkson University*

Kathy Mowers, *Owensboro Community and Technical College*

Mary Moyinhan, *Cape Cod Community College*

Junalyn Navarra-Madsen, *Texas Woman's University*

Azarnia Nazanin, *Santa Fe College*

Stacey O. Nicholls, *Anne Arundel Community College*

Helen Noble, *San Diego State University*

Lyn Noble, *Florida State College at Jacksonville*

Keith Oberlander, *Pasadena City College*

Pamela Omer, *Western New England College*

Nabendu Pal, *University of Louisiana at Lafayette*

Irene Palacios, *Grossmont College*

Adam Pennell, *Greensboro College*

Joseph Pick, *Palm Beach State College*

Philip Pickering, *Genesee Community College*

Robin Powell, *Greenville Technical College*

Nicholas Pritchard, *Coastal Carolina University*

Linda Quinn, *Cleveland State University*

William Radulovich, *Florida State College at Jacksonville*

Mumunur Rashid, *Indiana University of PA*

Fred J. Rispoli, *Dowling College*

Nancy Rivers, *Wake Technical Community College*

Corlis Robe, *East Tennesee State University*

Thomas Roe, *South Dakota State University*

Dan Rowe, *Heartland Community College*

Carol Saltsgaver, *University of Illinois–Springfield*

Radha Sankaran, *Passaic County Community College*

Delray Schultz, *Millersville University*

Jenny Shook, *Pennsylvania State University*

Danya Smithers, *Northeast State Technical Community College*

Larry Southard, *Florida Gulf Coast University*

Dianna J. Spence, *North Georgia College & State University*

René Sporer, *Diablo Valley College*

Jeganathan Sriskandarajah, *Madison Area Technical College–Traux*

David Stewart, *Community College of Baltimore County–Cantonsville*

Linda Strauss, *Penn State University*

John Stroyls, *Georgia Southwestern State University*

Joseph Sukta, *Moraine Valley Community College*

Lori Thomas, *Midland College*

Malissa Trent, *Northeast State Technical Community College*

Ruth Trygstad, *Salt Lake Community College*

Gail Tudor, *Husson University*

Manuel T. Uy, *College of Alameda*

Lewis Van Brackle, *Kennesaw State University*

Mahbobeh Vezvaei, *Kent State University*

Joseph Villalobos, *El Camino College*

Barbara Wainwright, *Sailsbury University*

Henry Wakhungu, *Indiana University*

Dottie Walton, *Cuyahoga Community College*

Jen-ting Wang, *SUNY, Oneonta*
Jane West, *Trident Technical College*
Michelle White, *Terra Community College*
Bonnie-Lou Wicklund, *Mount Wachusett Community College*
Sandra Williams, *Front Range Community College*
Rebecca Wong, *West Valley College*
Alan Worley, *South Plains College*

Jane-Marie Wright, *Suffolk Community College*
Haishen Yao, *CUNY, Queensborough Community College*
Lynda Zenati, *Robert Morris Community College*
Yan Zheng-Araujo, *Springfield Community Technical College*
Cathleen Zucco-Teveloff, *Rider University*
Mark A. Zuiker, *Minnesota State University, Mankato*

# Supplements

## Student Resources

*Student Solutions Manual*, by Robert Keller (Loras College), provides detailed, worked-out solutions to all odd-numbered text exercises. (ISBN-13: 978-0-321-75640-4; ISBN-10: 0-321-75640-1).

*Study Cards for Statistics Software* This series of study cards, available for Excel, Minitab, JMP®, SPSS®, R, StatCrunch, and the TI-83/84 graphing calculators, provides students with easy, step-by-step guides to the most common statistics software. Visit www.myPearsonStore.com for more information.

## Instructor Resources

*Instructor's Edition* contains answers to all text exercises, as well as a set of Instructor Resource Pages that offer chapter-by-chapter teaching suggestions and commentary. (ISBN-13: 978-0-321-78027-0; ISBN-10: 0-321-78027-2)

*Instructor's Solutions Manual*, by Robert Keller (Loras College), contains worked-out solutions to all the text exercises. (ISBN-13: 978-0-321-75646-6; ISBN-10: 0-321-75646-0)

*Online Test Bank* (download only), by Jill DeWitt (Baker College of Muskegon), includes three sets of tests for each chapter. The Test Bank is available for download at www.pearsonhighered.com/irc. (ISBN-13: 978-0-321-75639-8; ISBN-10: 0-321-75639-8)

*Instructor Podcasts* are brief audio podcasts from author Robert Gould that focus on the key points of each chapter, helping both new and experienced instructors prepare for class. Available in MyStatLab or at www.pearsonhighered.com/irc. (ISBN-13: 978-0-321-82423-3; ISBN-10: 0-321-82423-7)

*PowerPoint® Lecture Slides,* by Larry Green (Lake Tahoe Community College) provide an outline to use in a lecture setting, presenting definitions, figures, Guided Exercises, Case Studies, and Snapshots from the text. These slides are available within MyStatLab or at www.pearsonhighered.com/irc. (ISBN-13: 978-0-321-78025-6; ISBN-10: 0-321-78025-6)

*Active Learning Questions* Prepared in PowerPoint®, these questions are intended for use with classroom response systems. Several multiple-choice questions are available for each chapter of the book, enabling instructors to assess mastery of material quickly in class. The Active Learning Questions are available to download from within MyStatLab™ or at www.pearsonhighered.com/irc.

## Technology Resources

On the Companion CD-ROM

- **Data sets** from the book are provided in multiple formats.
- **Technology Tutorial Videos.** These brief video clips walk students through common statistical procedures for Minitab, Excel, and the TI-83/84 graphing calculator. They are also available within MyStatLab.

### MyStatLab™ Online Course (access code required)

MyStatLab is a course management systems that delivers **proven results** in helping individual students succeed.

- MyStatLab can be successfully implemented in any environment–lab-based, hybrid, fully online, traditional–and demonstrates the quantifiable difference that integrated usage has on student retention, subsequent success, and overall achievement.

- MyStatLab's comprehensive online gradebook automatically tracks students' results on tests, quizzes, homework, and in the study plan. Instructors can use the gradebook to provide positive feedback or intervene if students have trouble. Gradebook data can be easily exported to a variety of spreadsheet programs, such as Microsoft Excel.

MyStatLab provides **engaging experiences** that personalize, stimulate, and measure learning for each student.

- **Tutorial Exercises with Multimedia Learning Aids:** The homework and practice exercises in MyStatLab align with the exercises in the textbook, and they regenerate algorithmically to give students unlimited opportunity for practice and mastery. Exercises offer immediate helpful feedback, guided solutions, sample problems, animations, videos, and eText clips for extra help at point-of-use.

- **Getting Ready for Statistics:** A library of questions now appears within each MyStatLab course to offer the developmental math topics students need for the course. These can be assigned as a prerequisite to other assignments, if desired.

- **Conceptual Question Library:** In addition to algorithmically regenerated questions that are aligned with your textbook, there is a library of 1,000 Conceptual Questions available in the assessment manager that require students to apply their statistical understanding.

- **StatCrunch:** MyStatLab includes a web-based statistical software, StatCrunch, within the online assessment platform so that students can easily analyze data sets from exercises and the text. In addition, MyStatLab includes access to **www.StatCrunch.com,** a web site where users can access more than 13,000 shared data sets, conduct online surveys, perform complex analyses using the powerful statistical software, and generate compelling reports.

- **Integration of Statistical Software:** Knowing that students often use external statistical software, we make it easy to copy our data sets, both from the ebook and the MyStatLab questions, into software such as StatCrunch, Minitab, Excel, and more. Students have access to a variety of support tools— Technology Instruction Videos, Technology Study Cards, and Manuals for select titles—to learn how to effectively use statistical software.

- **Expert Tutoring:** Although many students describe the whole of MyStatLab as "like having your own personal tutor," students also have access to live tutoring from Pearson. Qualified statistics instructors provide tutoring sessions for students via MyStatLab.

And, MyStatLab comes from a **trusted partner** with educational expertise and an eye on the future.

Knowing that you are using a Pearson product means knowing that you are using quality content. That means that our eTexts are accurate, that our assessment tools work, and that our questions are error-free. And whether you are just getting started with MyStatLab, or have a question along the way, we're here to help you learn about our technologies and how to incorporate them into your course.

To learn more about how MyStatLab combines proven learning applications with powerful assessment, visit **www.mystatlab.com** or contact your Pearson representative.

## MyStatLab™Plus

MyLabsPlus combines effective teaching and learning materials from MyMathLab® and MyStatLab™ with convenient management tools and a dedicated services team. Designed to support growing math and statistics programs, it includes additional features, such as:

- **Batch Enrollment:** Your school can create the login name and password for every student and instructor, so everyone can be ready to start class on the first day. Automation of this process is also possible through integration with your school's Student Information System.

- **Login from Your Campus Portal:** You and your students can link directly from your campus portal into your MyLabsPlus courses. A Pearson service team works with your institution to create a single sign-on experience for instructors and students.

- **Advanced Reporting:** MyLabsPlus's advanced reporting enables instructors to review and analyze students' strengths and weaknesses by tracking their performance on tests, assignments, and tutorials. Administrators can review grades and assignments across all courses on your MyLabsPlus campus for a broad overview of program performance.

- **24/7 Support:** Students and instructors receive 24/7 support, 365 days a year, by phone, email, or online chat.

MyLabsPlus is available to qualified adopters. For more information, visit our website at **www.mylabsplus.com** or contact your Pearson representative.

## MathXL® for Statistics Online Course (access code required)

MathXL® is the homework and assessment engine that runs MyStatLab. (MyStatLab is MathXL plus a learning management system.) With MathXL for Statistics, instructors can:

- Create, edit, and assign online homework and tests using algorithmically generated exercises correlated at the objective level to the textbook.

- Create and assign their own online exercises and import TestGen tests for added flexibility.

- Maintain records of all student work, tracked in MathXL's online gradebook.

With MathXL for Statistics, students can.

- Take chapter tests in MathXL and receive personalized study plans and/or personalized homework assignments based on their test results.

- Use the study plan and/or the homework to link directly to tutorial exercises for the objectives they need to study.

- Students can also access supplemental animations and video clips directly from selected exercises.

- Knowing that students often use external statistical software, we make it easy to copy our data sets, both from the ebook and the MyStatLab questions, into software like StatCrunch, Minitab, Excel and more.

MathXL for Statistics is available to qualified adopters. For more information, visit our website at **www.mathxl .com,** or contact your Pearson representative.

## StatCrunch™

Powerful web-based statistical software that allows users to perform complex analyses, share data sets, and generate compelling reports of their data. The vibrant online community offers more than 13,000 data sets for students to analyze.

- **Collect**. Users can upload their own data to StatCrunch or search a large library of publicly shared data sets, spanning almost any topic of interest. Also, an online survey tool allows users to quickly collect data via web-based surveys.

- **Crunch**. A full range of numerical and graphical methods allow users to analyze and gain insights from any data set. Interactive graphics help users understand statistical concepts, and are available for export to enrich reports with visual representations of data.

- **Communicate**. Reporting options help users create a wide variety of visually-appealing representations of their data.

Full access to StatCrunch is available with a MyStatLab kit, and StatCrunch is available by itself to qualified adopters. For more information, visit our website at www.statcrunch .com or contact your Pearson representative.

## TestGen® (www.pearsoned.com/testgen)

Enables instructors to build, edit, print, and administer tests using a computerized bank of questions developed to cover all the student learning objectives of the text. TestGen is algorithmically based, and instructors can create multiple but equivalent versions of the same question or test with the click of a button. Instructors can also modify Test Bank questions or add new questions. The software and Test Bank are available for download from Pearson Education's online catalog.

# Index of Applications

## SPORTS

## SURVEYS AND OPINION POLLS

## TECHNOLOGY

## TRANSPORTATION

# Credits

# Introductory Statistics: Exploring the World Through Data

# 1 Introduction to Data

**Statistics is the science of data, so we must learn the types of data we will encounter and the methods for collecting data. The method used to collect data is very important because it determines the types of conclusions we can reach and, as you'll learn in later chapters, the types of analyses we can do. By organizing the data we've collected, we can often spot patterns that are not otherwise obvious.**

This book will teach you to examine data to better understand the world around you. If you know how to sift data to find patterns, can communicate the results clearly, and understand whether you can generalize your results to other groups and contexts, you will be able to make better decisions, offer more convincing arguments, and learn things you did not know before. Data are everywhere these days, and making effective use of them is such a crucial task that one prominent economist has proclaimed statistics one of the most important professions of the decade (*McKinsley Quarterly* 2009).

The use of statistics to make decisions and convince others to take action is not new. Some statisticians date the current practice of statistics back to the mid-nineteenth century. One famous example occurred in 1854, when the British were fighting the Russians in the brutal Crimean War. A British newspaper had criticized the military medical facilities, and a young but well-connected nurse, Florence Nightingale, was appointed to study the situation and, if possible, improve it.

Nightingale carefully recorded the numbers of deaths, the causes of the deaths, and the times and dates of the deaths. She organized these data graphically, and these graphs enabled her to see a very important pattern: A large percentage of deaths were due to contagious disease, and many deaths could be prevented by improving sanitary conditions. Within six months, Nightingale had reduced the death rate by half. Eventually she convinced Parliament and military authorities to completely reorganize the medical care they provided. Accordingly, she is credited with inventing modern hospital management.

In modern times, we have equally important questions to answer. Do cell phones cause brain tumors? Are alcoholic drinks healthy in moderation? Which diet works best for losing weight? What percentage of the public is concerned about job security? **Statistics**—the science (and art!) of collecting and analyzing observations to learn about ourselves, our surroundings, and our universe—helps answer questions such as these.

Data are the building blocks of statistics. This chapter introduces some of the basic types of data and explains how we collect them, store them, and organize them. These ideas and skills will provide a basic foundation for your study of the rest of the book.

## CASE STUDY

### Deadly Cell Phones?

In September 2002, Dr. Christopher Newman, a resident of Maryland, sued Motorola, Verizon, and other wireless carriers, accusing them of causing a cancerous brain tumor behind his right ear. As evidence, his lawyers cited a study by Dr. Lennart Hardell. Hardell had studied a large number of people with brain tumors and had found that a greater percentage of them used cell phones than of those who did not have brain tumors (CNN 2002; *New York Times* 2002).

Speculation that cell phones might cause brain cancer began as early as 1993, when (as CNN reports) the interview show *Larry King Live* featured a man who claimed that his wife died because of cancer caused by her heavy cell phone use. However, more recent studies have contradicted Hardell's results, as well as earlier reports about the health risks of heavy cell phone use.

The judge in Dr. Newman's trial was asked to determine whether Hardell's study was compelling enough to support allowing the trial to proceed. Part of this determination involved evaluating the method that Hardell used to collect data. If you were the judge, how would you rule? You will learn the judge's ruling at the end of the chapter. You will also see how the methods used to collect data about important *cause-and-effect relationships*—such as that which Dr. Newman alleged to exist between cell phone use and brain cancer—can affect the conclusions we can draw.

# What Are Data?

The study of statistics rests on two major concepts: variation and data. **Variation** is the more fundamental of these concepts. To illustrate this idea, draw a circle on a piece of paper. Now draw another one, and try to make it look just the same. Now another. Are all three exactly the same? We bet they're not. They might be slightly different sizes, for instance, or slightly different versions of round. This is an example of variation. How can you reduce this variation? Maybe you can get a penny and outline the penny. Try this three times. Does variation still appear? Probably it does, even if you need a magnifying glass to see, say, slight variations in the thickness of the penciled line.

**Data** are observations that you or someone else records. The drawings in Figure 1.1 are data that record our attempts to draw three circles that look the same. Analyzing pictorial data such as these is not easy, so we often try to turn such observations into numbers. How would you measure whether these three circles were the same? Perhaps you would compare diameters or circumferences, or somehow try to measure how and where these circles depart from being perfect circles. Whatever technique you chose, these measurements could also be considered data.

Data are more than just numbers, though. David Moore, a well-known statistician, defined data as "numbers in context." By this he meant that data consist not only of the numbers we record, but also of the story behind the numbers. For example,

$$10.00, 9.88, 9.81, 9.81, 9.75, 9.69, 9.5, 9.44, 9.31$$

are just numbers. But in fact these numbers represent "Weight in pounds of the ten heaviest babies in a sample of babies born in North Carolina in 2004." Now these numbers have a context and have been elevated to data. See how much more interesting data are than numbers?

These data were collected by the state of North Carolina in part to help researchers understand the factors that contribute to low-weight and premature births. If doctors

▶ **FIGURE 1.1** **(a)** Three circles drawn by hand. **(b)** Three circles drawn using a coin. It is clear that the circles drawn by hand show more variability than the circles drawn with the aid of a coin.

(a)

(b)

understand the causes of premature births, they can work to prevent this—perhaps by helping expectant mothers change their behavior, perhaps by medical intervention, and perhaps by a combination of both.

**KEY POINT**   Data are "numbers in context."

The news media regularly collect data through surveys to assess what the public thinks about an issue or a politician. Every few months, approval ratings are taken to see how people think the U.S. president and other politicians are doing. Politicians may say that they don't believe in polls (particularly if they don't like the outcome), but these polls often shape the behavior of the government. This is partly because, when they are done right, polls are an accurate gauge of what the public is thinking.

You collect data. You save pictures of events or people you want to remember. These pictures tell a story that, you hope, others can reconstruct when they flip through your picture book. Perhaps you save records of your spending behavior so that you'll know how much money you have left in your account. Maybe you digitize your music collection on your computer; if so, you have data that can tell us much about your musical tastes. These data are not numbers, but often we can use numbers to describe these things. For example, a computer can tell you how many Rock and how many Classical tracks are on your mp3 player, as well as how long each track lasts.

People collect data about you. Maybe you have a "preferred customer" card for your local market or bookstore. These cards enable the market to track your purchases so that it can offer you coupons that you might be interested in—and thus encourage you to remain a loyal customer. Your school also keeps data on you. Administrators know how many classes you took, when you took them, and what grades you received. Even your car keeps data about you: how many miles you've driven the car and maybe the number of miles since your last fill-up of gas. (Some rental cars record even more data, such as the speed you drove the car at every moment.)

You can see that in our society, data are everywhere. That is one reason why it's important to understand how to work with and analyze data.

In this book you will study the science of data. You will learn (a little) about how data are stored and coded, and how you can organize data to see patterns that otherwise might be invisible. Most important, you'll see how data can sometimes be used to make generalizations about large groups of people or things.

# Classifying and Storing Data

The first step to understanding data is to understand the different types of data you will encounter. As you've seen, data are numbers in context. But that's only part of the story; data are also recorded observations. Your photo from your vacation to Carhenge in Nebraska is data (Figure 1.2). The ultraviolet images streaming from the Earth Observer Satellite system are data (see Figure 1.3 on the next page). These are just two examples of data that are not numbers. Statisticians work hard to help us analyze complex data, such as images and sound files, just as easily as we study numbers. Most of the methods involve recoding the data into numbers. For example, your photos can be digitized in a scanner, converted into a very large set of numbers, and then analyzed. You might have a digital camera that gives you feedback about the quality of a photo you've taken. If so, your camera is not only collecting data but also analyzing it!

▲ **FIGURE 1.2** A photo of Carhenge, Nebraska.

▲ **FIGURE 1.3** Satellites in NASA's Earth Observing Mission record ultraviolet reflections and transmit these data back to earth. Such data are used to construct images of our planet. Earth Observer (http://eos .gsfc.nasa.gov/).

---

🖈 **Details**

**More Grammar**
We're using the word *sample* as a noun—it is an object, a collection of data that we study. Later we'll also use the word *sample* as a verb—that is, to describe an action. For example, we'll sample ice cream cones to measure their weight.

---

| Weight | Gender | Smoke |
|--------|--------|-------|
| 7.69 | F | 0 |
| 0.88 | M | 1 |
| 6.00 | F | 0 |
| 7.19 | F | 0 |
| 8.06 | F | 0 |
| 7.94 | F | 0 |

▲ **TABLE 1.1** Birth Data from North Carolina in 2004

Almost always, our data sets will consist of characteristics of people or things (such as gender and weight). These characteristics are called **variables**. Variables are not "unknowns" like those you studied in algebra. We call these characteristics variables because they have variability: The values of the variable can be different from person to person.

**KEY POINT** ▸ Variables in statistics are different from variables in algebra. In statistics, variables record characteristics of people or things.

When we work with data, they are grouped into a collection, which we call either a **data set** or a **sample**. The word *sample* is important, because it implies that the data we see are just one part of a bigger picture. This "bigger picture" is called a **population**. Think of a population as the Data Set of Everything—it is the data set that contains all of the information about everyone or everything with respect to whatever variable we are studying. Quite often, the population is really what we want to learn about, and we learn about it by studying the data in our sample. However, many times it is enough just to understand and describe the sample. For example, you might collect data from students in your class simply because you want to know about the students in your class, and not because you wish to use this information to learn about all students at your school. Sometimes, data sets are so large they effectively *are* the population, as you'll see in the data from North Carolina that follow.

## Context Is Key

The context is the most important aspect of data, although it is frequently overlooked. Table 1.1 shows a few lines from the data set of births in 2004 in North Carolina (Holcomb 2006).

To understand these data, we need to ask and answer some questions: What are the objects of interest? What variables were measured? How were they measured? What are the units of measurement? Who collected the data? How did they collect the data? Where were the data collected? Why were the data collected?

Many, but not all, of these questions can be answered for these data by reading the information provided on the website that hosts the data. Other times we are not so lucky and must rely on very flimsy supporting documentation. If you collect the data yourself, you should be careful to record this extra supporting information. Or, if you get a chance to talk with the people who collected the data, then you should ask them these questions.

- *What are the objects of interest?* By "objects of interest" we mean who or what was measured. In this case, the object of interest is a baby. Each line in the table represents a newborn baby born in North Carolina in 2004. If we were to see the whole table, we would see a record of every baby born in 2004 in North Carolina.

- *What variables were measured?* For each baby, the state records the weight, the gender, and whether the mother smoked.

- *How were the variables measured?* Unknown. Presumably, most measurements on the baby were taken from a medical care-giver at the time of the birth, but we don't know how or when information about the mother was collected.

- *What are the units of measurement?* Units of measurement are important. The same variable can have different units of measurement. For example, weight could be measured in pounds, in ounces, or in kilograms. For Table 1.1,

   Weight: reported in pounds
   Gender: reported as M for boys and F for girls.
   Smoke: reported as a 1 if the mother smoked during the pregnancy, as a 0 if she did not.

- *Who collected the data?* The government of the state of North Carolina.

- *How did they collect the data?* Data were recorded for *all* births that occurred in hospitals in North Carolina. Later in the chapter you'll see that data can be collected by drawing a random sample of subjects, or by assigning subjects to receive different treatments, as well as through other methods. The exact method used for Table 1.1 is not clear, but the data were probably compiled from publicly available medical records and from reports by the physicians and caregivers.

- *Where were the data collected?* The location that the data were collected often gives us information about who (or what) the study is about. These data were collected in North Carolina and consist of babies born in that state. We should therefore be very wary about generalizing our findings to other states or other countries.

- *Why were the data collected?* Sometimes, data are collected to learn about a larger population. At other times, the goals are limited to learning more about the sample itself. In this case the data consist of all births in North Carolina, and it is most likely that researchers wanted to learn how the health of infants was related to the smoking habits of mothers within this sample.

> **KEY POINT**  The first time you see a data set, ask yourself these questions:
>
> - What are the objects of interest?
> - What variables were measured?
> - How were the variables measured?
> - What are the units of measurement?
> - Who collected the data?
> - How did they collect the data
> - Where were the data collected?
> - Why did they collect the data?

## Two Types of Variables

The variables you'll find in your data set come in two basic types, which can themselves be broken into smaller divisions, as we'll discuss later.

**Numerical variables** describe quantities of the objects of interest. The values will be numbers. The weight of an infant is an example of a numerical variable.

**Categorical variables** describe qualities of the objects of interest. These values will be categories. The gender of an infant is an example of a categorical variable. The possible values are the categories "male" and "female." Eye color of an infant is another example; the categories might be brown, blue, black, and so on. You can often identify categorical variables because their values are *usually* words, phrases, or letters. (We say "usually" because we sometimes use numbers to represent a word or phrase. Stay tuned.)

> **Details**
>
> **Quantitative and Qualitative Data**
> Some statisticians use the word *quantitative* to refer to numerical variables (think "quantity") and *qualitative* to refer to categorical variables (think "quality"). We prefer *numerical* and *categorical*. Both sets of terms are commonly used, and you should be prepared to hear and see both.

## EXAMPLE 1 Crash-Test Results

The data in Table 1.2 on the next page are an excerpt from crash-test dummy studies in which cars are crashed into a wall at 35 miles per hour. Each row of the data set represents the observed characteristics of a single car. This is a small sample of the database, which is available from the National Transportation Safety Administration. The *head injury* variable reflects the risk to the passengers' heads. The higher the number, the greater the risk.

▶ **TABLE 1.2** Crash-Test Results for Cars

| Make | Model | Doors | Weight | Head Injury |
|------|-------|-------|--------|-------------|
| Acura | Integra | 2 | 2350 | 599 |
| Chevrolet | Camaro | 2 | 3070 | 733 |
| Chevrolet | S-10 Blazer 4X4 | 2 | 3518 | 834 |
| Ford | Escort | 2 | 2280 | 551 |
| Ford | Taurus | 4 | 2390 | 480 |
| Hyundai | Excel | 4 | 2200 | 757 |
| Mazda | 626 | 4 | 2590 | 846 |
| Volkswagen | Passat | 4 | 2990 | 1182 |
| Toyota | Tercel | 4 | 2120 | 1138 |

> **Details**
>
> **Categorical Doors**
> Some people might consider *doors* a categorical variable, because nearly all cars have either 2 doors or 4 doors, and for many people, the number of doors designates a certain type of car (small or larger). There's nothing wrong with that.

**QUESTION** For each variable, state whether it is numerical or categorical.

**SOLUTION** The variables *make* and *model* are categorical. Their values are descriptive names. The units of *doors* are, quite simply, the number of doors. The units of *weight* are pounds. The variables *doors* and *weight* are numerical because their values are measured quantities. The units for *head injury* are unclear; head injury is measured using some scale that the researchers developed.

**TRY THIS!** Exercise 1.1

> **! Caution**
>
> **Don't Just Look for Numbers!**
> You can't always tell whether a variable is categorical simply by looking at the data table. You must also consider what the variable represents. Sometimes, researchers code categorical variables with numerical values.

## Coding Categorical Data with Numbers

Sometimes categorical variables are "disguised" as numerical. The smoke variable in the North Carolina data set (Table 1.1) has numbers for its values (0 and 1), but in fact those numbers simply indicate whether or not the mother smoked. Mothers were asked, "Did you smoke?" and if they answered "Yes," the researchers coded this categorical response with a 1. If they answered "No," the response was coded as a 0. These particular numbers represent categories. *Smoke* is a categorical variable.

Coding is used to help both humans and computers understand what the values of a variable represent. For example, a human would understand that a "yes" under the "Smoke" column would mean that the person was a smoker, but to the computer, the word "yes" is just a string of symbols. If instead, we follow a convention where a 1 means "yes" and a 0 means "no," then a human understands that the 1's represent smokers, and a computer can easily add the values together to determine, for example, how many smokers are in the sample.

This approach for coding categorical variables is quite common and useful. If a categorical variable has only two categories, as do *gender* and *smoke*, then it is almost always helpful to code the values with 0 and 1. To help readers know what a "1" means, rename the variable with one its category names. A "1" then means the person belongs to that category, and a 0 means the person belongs to the other category. For example, instead of calling a variable *gender*, we rename it *female*. And then if the baby is a boy we code a 0, and if it's a girl we code a 1.

Sometimes your computer does the coding for you without your needing to know anything about it. So even if you see the words *female* and *male* on your computer, the computer has probably coded these with values of 0 and 1 (or vice versa).

## Storing Your Data

The format in which you record and store your data is very important. Computer programs will require particular formats, and by following a consistent convention, you can be confident that you'll better remember the qualities of your own data set if you need to revisit it months or even years later. Data are often stored in a spreadsheet-like format in which each row represents the object (or person) of interest. Each column represents a variable. In Table 1.3, each row represents a baby. The columns are variables: *Weight*, *Female*, and *Smoke*. All the previous tables use this format, which is sometimes referred to as the **stacked data** format.

When you collect your own data, the stacked format is almost always the best way to record and store your data. One reason is that it allows you to easily record several different variables for each subject. Another reason is that it is the format that most software packages will assume you are using for most analyses. (The exceptions are TI-83/84 and Excel.)

Some technologies, such as the TI calculators, require, or at least accommodate, data stored in a different format, called **unstacked data**. Unstacked data tables are also common in some books and media publications. In this format, each column represents a variable from a different group. For example, one column could represent men's heights, and another column could represent women's heights. The data set, then, is a single variable (*height*) broken into two groups. The groups are determined by a categorical variable. Table 1.4 shows an example of unstacked data, and Figure 1.4 shows the same data in TI-83/84 input format.

By way of contrast, Table 1.5 shows the same data in stacked format.

The great disadvantage of the unstacked format is that it can store only two variables at a time: the variable of interest (for example, height), and a categorical variable that tells us which group the observation belongs in (for example, gender). However, most of the time, we record many variables for each observation. For example, we record a baby's weight, gender, and whether or not the mother smoked. The stacked format enables us to display as many variables as we wish.

## EXAMPLE 2 Downhill Skiers

The accompanying table shows downhill skiing times (seconds) for the top five men and the top five women finalists in the 2010 Winter Olympics in Vancouver. The data are in unstacked form.

(Note that the women raced a different course than the men.)

| Men | Women |
|-----|-------|
| 114.31 | 104.19 |
| 114.38 | 104.75 |
| 114.4 | 105.65 |
| 114.52 | 105.68 |
| 114.64 | 106.17 |

**QUESTION** Create a new data table that displays the data in stacked form.

**SOLUTION** At first it appears that only one numerical variable, time, is displayed. In fact, there is also a categorical variable, gender, which is used to distinguish the two groups. To stack the data, we identify the gender that is associated with each time.

| Weight | Female | Smoke |
|--------|--------|-------|
| 7.69 | 1 | 0 |
| 0.88 | 0 | 1 |
| 6.00 | 1 | 0 |
| 7.19 | 1 | 0 |
| 8.06 | 1 | 0 |
| 7.94 | 1 | 0 |

▲ **TABLE 1.3** Data for Newborns from Table 1.1 Using Coding

| Men's Heights | Women's Heights |
|---------------|-----------------|
| 70 | 59 |
| 68 | 70 |
| 71 | 61 |
| | 62 |

▲ **TABLE 1.4** Data by Groups (Unstacked)

▲ **FIGURE 1.4** TI-83/84 data input screen (unstacked data).

| Height | Gender |
|--------|--------|
| 70 | male |
| 68 | male |
| 71 | male |
| 59 | female |
| 70 | female |
| 61 | female |
| 62 | female |

▲ **TABLE 1.5** The Same Data as in Table 1.4, Shown Here in Stacked Format

| Time (seconds) | Gender |
|---|---|
| 114.31 | M |
| 114.38 | M |
| 114.4 | M |
| 114.52 | M |
| 114.64 | M |
| 104.19 | F |
| 104.75 | F |
| 105.65 | F |
| 105.68 | F |
| 106.17 | F |

**TRY THIS!** Exercise 1.5

# Organizing Categorical Data

Once we have a data set, we next need to organize and display the data in a way that helps us see patterns. This task of organization and display is not easy and we discuss it throughout the entire book. In this section we introduce the topic for the first time, in the context of categorical variables.

With categorical variables, we are usually concerned with knowing how often a particular category occurs in our sample. We then (usually) want to compare how often a category occurs for one group versus another (liberal/conservative, man/woman). To do these comparisons, you need to understand how to calculate percentages and other rates.

A common method for summarizing two potentially related categorical variables is to use a two-way table. **Two-way tables** show how many times each combination of categories occurs. For example, Table 1.6 is a two-way table from the Youth Behavior Risk Survey that shows gender and whether or not the respondent always (or almost always) wears a seat belt when riding in or driving a car. The actual Youth Behavior Risk Survey has over 10,000 respondents, but we are practicing on a small sample from this much larger data set.

The table tells us that 2 people were male and did not always wear a seat belt. Three people were female and did not always wear seat belts. These counts are also called frequencies. A **frequency** is simply the number of times a value is observed in a data set.

Some books and publications present two-way tables as though they displayed the original data collected by the investigators. However, two-way tables do not consist of "raw" data but, rather, are summaries of data sets. For example, the data set that produced Table 1.6 is shown in Table 1.7.

To summarize this table, we simply count how many of the males (a 1 in the Male column) also do not always wear seat belts (a 1 in the Not Always column). We then count how many both are male and always wear seat belts (a 1 in the Male column, a 0 in the Not Always column); how many both are female and don't always wear seat

|  | Male | Female |
|---|---|---|
| Not Always | 2 | 3 |
| Always | 3 | 7 |

▲ **TABLE 1.6** This two-way table shows counts for 15 youths who responded to a survey about wearing seat belts.

belts (a 0 in the Male column, a 1 in the Not Always column); and, finally, how many both are female and always wear a seat belt.

Example 3 illustrates that summarizing the data in a two-way table can make it easy to compare groups.

## EXAMPLE 3 Percentages of Seat Belt Wearers

The 2008 Youth Behavior Risk Survey is a national study that asks American youths about potentially risky behaviors. We show the two-way summary again. All of the people in the table were between 14 and 17 years old. The participants were asked whether they wear a seat belt while driving or riding in a car. The people who said always or almost always were put in the Always group. The people who said sometimes or rarely were put in the Not Always group.

|  | Male | Female |
|---|---|---|
| Not Always | 2 | 3 |
| Always | 3 | 7 |

### QUESTIONS

a. How many men are in this sample? How many women? How many people do not always wear seat belts? How many always wear seat belts?

b. What percent of the sample are men? What percent are women? What percent don't always wear seat belts? What percent always wear seat belts?

c. Are the men in the sample more likely than the women in the sample to take the risk of not wearing a seat belt?

### SOLUTIONS

a. We can count the men by adding the first column: $2 + 3 = 5$ men. Adding the second column gives us the number of women: $3 + 7 = 10$.

   We get the number who do not always wear seat belts by adding the first row: $2 + 3 = 5$ people don't always wear seat belts. Adding the second row gives us the number who always wear seat belts: $3 + 7 = 10$.

b. This question asks us to convert the numbers we found in part (a) to percentages. To do this, we divide the numbers by 15, because there were 15 people in the sample. To convert to percentages, we multiply this proportion by 100%.

   The proportion of men is $5/15 = 0.333$. The percentage is $0.333 \times 100\% = 33.3\%$. The proportion of women must be $100\% - 33.3\% = 66.7\%$ ($10/15 \times 100\% = 66.7\%$).

   The proportion who do not always wear seat belts is $5/15 = 0.333$, or $33.3\%$. The proportion who always wear seat belts is $100\% - 33.3\% = 66.7\%$.

c. You might be tempted to answer this question by counting the number of males who don't always wear seat belts (2 people) and comparing that to the number of females who don't always wear seat belts (3 people). However, this is not a fair comparison because there are more females than males in the sample. Instead, we should look at the percentage of those who don't always wear seat belts in each group. This question should be reworded as follows:

   Is the percentage of males who don't always wear seat belts greater than the percentage of females who don't always wear seat belts?

| Male | Not Always |
|---|---|
| 1 | 1 |
| 1 | 1 |
| 1 | 0 |
| 1 | 0 |
| 1 | 0 |
| 0 | 1 |
| 0 | 1 |
| 0 | 1 |
| 0 | 0 |
| 0 | 0 |
| 0 | 0 |
| 0 | 0 |
| 0 | 0 |
| 0 | 0 |
| 0 | 0 |

▲ **TABLE 1.7** This data set is equivalent to the two-way summary shown in Table 1.6. We highlighted in red those who did not always wear a seat belt (the risk takers).

Because 2 out of 5 males don't always wear seat belts, the percent of males who don't always wear seat belts is $(2/5) \times 100\% = 40\%$.

Because 3 out of 10 females don't always wear seat belts, the percent of females who don't always wear seat belts is $(3/10) \times 100\% = 30\%$.

In fact, females in this sample are less likely than males to engage in this risky behavior. Among all U.S. youth, it is estimated that about 28% of males do not always wear their seat belt, compared to 23% of females.

**TRY THIS!**  Exercise 1.7

---

The calculations in Example 3 took us from frequencies to percentages. Sometimes, we want to go in the other direction. If you know the total number of people in a group, and are given the percentage that meets some qualification, you can figure out *how many* people in the group meet that qualification.

## EXAMPLE 4 Numbers of Seat Belt Wearers

A statistics class has 300 students, and they are asked if they always ride or drive with a seat belt.

**QUESTIONS**

a. Suppose that 30% of the students do not always wear a seat belt. How many students is this?

b. Suppose we know that in another class, 20% of the students do not always wear seat belts, and this is equal to 43 students. How many students are in the class?

**SOLUTIONS**

a. We need to find 30% of 300. When working with percentages, first convert the percentage to its decimal equivalent:

$$30\% \text{ of } 300 = 0.30 \times 300 = 90$$

Therefore, 90 students don't always wear seat belts.

b. The question tells us that 20% of some unknown larger number (call it $y$) must be equal to 43.

$$0.20y = 43$$

Divide both sides by 0.20 and you get

$$y = 215$$

There are 215 total students in the class, and 43 of them don't always wear seat belts.

**TRY THIS!**  Exercise 1.9

---

Sometimes, you may come across data summaries that are missing crucial information. Suppose we wanted to know which team sports are the most dangerous to play. Table 1.8 shows the number of sports-related injuries that were treated in U.S. emergency rooms in 2004 (National Safety Council 2004). (Note that this table is not the table of original data but is, instead, a summary of the original data.)

Wow! It's a dangerous world out there. Which would you conclude is the most dangerous sport? Which is the least dangerous?

| Sport | Injuries |
|---|---|
| Baseball | 178,668 |
| Basketball | 615,546 |
| Bowling | 21,133 |
| Football | 387,948 |
| Ice hockey | 16,435 |
| Soccer | 173,519 |
| Softball | 125,875 |
| Tennis | 19,633 |
| Volleyball | 59,225 |

▲ **TABLE 1.8** Summary of Counts of Sports Injuries

Did you answer that basketball was the most dangerous sport? It does have the most injuries (615,546)—in fact, almost twice as many as in football (387,948). Ice hockey is known for its violence (you've heard the old joke, "I went to a fight and suddenly a hockey match broke out"), but here, it seems to have caused relatively few injuries and looks safe.

The problem with comparing the numbers of injuries for these sports is that the sports have different numbers of participants. Injuries might be more common in basketball simply because more people play basketball. Also, there might be relatively few injuries in ice hockey merely because fewer people play. One important component is missing in Table 1.8, and the lack of this component makes our analysis impossible.

Table 1.9 includes the component missing from Table 1.8: the number of participants in each sport. We can't directly compare the number of injuries from sport to sport, because the numbers of members of the various groups are not the same. This improved table shows us the total membership of each group.

| Sport | Injuries | Participants |
|---|---|---|
| Baseball | 178,668 | 15,600,000 |
| Basketball | 615,546 | 28,900,000 |
| Bowling | 21,133 | 43,900,000 |
| Football | 387,948 | 17,700,000 |
| Ice hockey | 16,435 | 2,100,000 |
| Soccer | 173,519 | 14,500,000 |
| Softball | 125,875 | 13,600,000 |
| Tennis | 19,633 | 11,000,000 |
| Volleyball | 59,225 | 11,500,000 |

◄ **TABLE 1.9** Summary of Counts of Sports Injuries and Numbers of Participants

Which sport is the most dangerous? We now have the information we need to answer this question. Specifically, we can find the percentage of participants injured in each sport. For example, what percent of basketball players were injured? There were 28,900,000 participants and 615,546 were injured, so the percent injured is $(615,546/28,900,000) \times 100\% = 2.1299\%$.

Sometimes, with percentages as small as this, we understand the numbers more easily if we report not a percentage, but "number of events per 1000 objects" or maybe even "per 10,000 objects". We call such numbers **rates**. To get the injury rate per 1000 people, instead of multiplying $(614,546/28,900,000)$ by 100 we multiply by 1000: $(615,546/28,900,000) \times 1000 = 21.299$ injuries per 1000 people.

These results are shown in Table 1.10.

| Sport | Injuries | Participants | Injury/Part | Per Thousand |
|---|---|---|---|---|
| Baseball | 178,668 | 15,600,000 | 0.01145 | 11.45 |
| Basketball | 615,546 | 28,900,000 | 0.02130 | 21.30 |
| Bowling | 21,133 | 43,900,000 | 0.00048 | 0.48 |
| Football | 387,948 | 17,700,000 | 0.02192 | 21.92 |
| Ice hockey | 16,435 | 2,100,000 | 0.00783 | 7.83 |
| Soccer | 173,519 | 14,500,000 | 0.01197 | 11.97 |
| Softball | 125,875 | 13,600,000 | 0.00926 | 9.26 |
| Tennis | 19,633 | 11,000,000 | 0.00178 | 1.78 |
| Volleyball | 59,225 | 11,500,000 | 0.00515 | 5.15 |

◄ **TABLE 1.10** Summary of Rates of Sports Injuries

We see now that football is the most dangerous sport: 21.92 players are injured out of every 1000 players. Basketball is slightly less risky, with 21.30 injuries per 1000 players.

## EXAMPLE 5 Comparing Rates of Stolen Cars

Which model of car has the greatest risk of being stolen? The Highway Loss Data Institute reports that the Cadillac Escalade is the most stolen car; 13.2 Escalades are reported stolen out of every 1000 Escalades on the road. By way of contrast, the Ford Taurus is the least stolen; 0.3 Taurus is reported stolen for every 1000 Tauruses on the road (Insurance Institute for Highway Safety 2006).

**QUESTION** Why does the Highway Loss Data Institute report theft rates rather than the number of each type of car stolen?

**SOLUTION** We need to take into account the fact that some cars are more popular than others. Suppose there were only 10 Escalades stolen all year long, but 350 Tauruses were stolen. It might seem that Tauruses are at a higher risk of being stolen. However, Escalades are expensive compared to Tauruses, so there are probably many more Tauruses on the road available to be stolen. By looking at the *theft rate*, we adjust the total number of cars *of that kind* on the road.

**TRY THIS!** Exercise 1.19

**KEY POINT** In order for us to compare groups, the groups need to be similar. When the data consist of counts, then percentages or rates are often better for comparisons because they take into account possible differences among the sizes of the groups.

## SECTION 1.4

# Collecting Data to Understand Causality

Often, the most important questions in science, business, and everyday life are questions about **causality**. These are usually phrased in the form of "what if" questions. What if I take this medicine; will I get better? What if I change my Facebook profile; will my profile get more hits?

Questions about causality are often in the news. The *Los Angeles Times* reported that many people believe a drink called peanut milk can cure gum disease and slow baldness (Glionna 2006). The BBC News (2010) reported that "Happiness wards off heart disease." Statements such as these are everywhere we turn these days. How do we know whether to believe these claims?

The methods we use to collect data determine the types of conclusions we can make. Only one method of data collection is suitable for making conclusions about causal relationships, but as you'll see, that doesn't stop people from making such conclusions anyway. In this section we talk about three methods commonly used to collect data for questions about causality: anecdotes, observational studies, and controlled experiments.

Most questions about causality can be understood in terms of two variables: the **treatment variable** and the **outcome variable**. (The outcome variable is also sometimes called the **response variable**, because it responds to changes in the treatment.)

We are essentially asking whether the treatment variable causes changes in the outcome variable. For example, the treatment variable might record whether or not a person drinks Peanut Milk, and the outcome variable records whether or not that person's gum disease improved. Or the treatment variable might record whether or not a person is generally happy, and the outcome variable might record whether or not that person suffered from heart disease in a ten-year period.

People who receive the treatment of interest (or have the characteristic of interest) are said to be in the **treatment group**. Those who do not receive that treatment (or do not have that characteristic) are in the **control group**, which is also called the **comparison group.**

## Anecdotes

Peanut milk is a drink invented by Jack Chang, an entrepreneur in San Francisco, California. He noticed that after he drank peanut milk for a few months, he stopped losing hair and his gum disease went away. According to the *Los Angeles Times* (Glionna 2006), another regular drinker of peanut milk says that the beverage caused his cancer to go into remission. Others have reported that drinking the beverage has reduced the severity of their colds, has helped them sleep, and has helped them wake up.

This is exciting stuff! Peanut milk could very well be something we should all be drinking. But can peanut milk really solve such a wide variety of problems? On the face of it, it seems that there's evidence that peanut milk has cured people of illness. The *Los Angeles Times* reports the names of people who claim that it has. However, the truth is that this is simply not enough evidence on the basis of which to judge whether the beverage is helpful, harmful, or without any effect at all.

These testimonials are examples of anecdotes. An **anecdote** is essentially a story that someone tells about her or his own (or a friend's or relative's) experience. Anecdotes are an important type of evidence in criminal justice because eyewitness testimony can carry a great deal of weight in a criminal investigation. However, for answering questions about groups of people with great variability or diversity, anecdotes are essentially worthless.

The primary reason why anecdotes are not useful for reaching conclusions about cause-and-effect relationships is that the most interesting things that we study have so much variety that a single report can't capture the variety of experience. For example, have you ever bought something because a friend recommended it, only to find that after a few weeks it fell apart? If the object was expensive, such as a car, you might have been angry at your friend for recommending such a bad product. But how do you know whose experience was more typical, yours or your friend's? Perhaps the car is in fact a very reliable model, and you just got a lemon.

A very important question to ask when someone claims that a product makes some kind of change is to ask, "Compared to what?" Here the claim is that drinking peanut milk will make you healthier. The question to ask is "Healthier compared to what?" Compared to people who don't drink peanut milk? Compared to people who take medicine for their particular ailment? To answer these questions, we need to examine the health of these other groups of people who do not drink peanut milk.

Anecdotes do not give us a comparison group. We might know that a group of people believe that peanut milk made them feel better, but we don't know how the milk drinkers' experiences compare to those of people who did not drink peanut milk.

 **KEY POINT**  When someone makes a claim about causality, ask, "Compared to what?"

Another reason for not trusting anecdotal evidence is a psychological phenomenon called the placebo effect. People often react to the idea of a treatment, rather than to the treatment itself. A **placebo** is a harmless pill (or sham procedure) that a patient believes is actually an effective treatment. Often, the patient taking the pill feels better, even though the pill actually has no effect whatsoever. In fact, a survey of U.S. physicians published in the *British Medical Journal* found that up to half of physicians prescribe sugar pills—placebos—to manage chronic pain. This psychological wish fulfillment—we feel better because we think we *should* be feeling better—is called the **placebo effect**.

## Observational Studies

The identifying mark of an **observational study** is that the subjects in the study are put into the treatment group or the control group either by their own actions or by the decision of someone else who is not involved in the research study. For example, if we wished to study the effects on health of smoking cigarettes (as many researchers have), then our treatment group would consist of people who had chosen to smoke, and the control group would consist of those who had chosen not to smoke.

Observational studies compare the outcome variable in the treatment group with the outcome variable in the control group. Thus, if many more people are cured of gum disease in the group that drinks peanut milk (treatment) than in the group that does not (control), then we would say that drinking peanut milk is associated with improvement in gum disease; that is, there is an **association** between the two variables. If the group of happy people tend to have less heart disease than the not-happy people, we would say that happiness is associated with improved heart health.

Note that we do not conclude that peanut milk *caused* the improvement in gum disease. In order for us to draw this conclusion, the treatment group and the control group must be very similar in every way except that one group gets the treatment and the other doesn't. For example, if we knew that the group of people who started drinking peanut milk and the group that did not drink peanut milk were alike in every way—both groups have the same overall health, are roughly the same ages, include the same mix of genders and races, education levels, and so on—then if the peanut milk group is healthier after a year, we would be fairly confident in concluding that peanut milk is the reason for their better health.

Unfortunately, in observational studies this goal of having very similar groups is *extremely* difficult to achieve. *Some* characteristic is nearly always different in one group than in the other. This means that the groups may experience different outcomes because of this different characteristic, not because of the treatment. A difference between the two groups that can explain why the outcomes were different is called a **confounding variable**.

For example, early observational studies on the effects of smoking found that a greater percent of smokers than of nonsmokers had lung cancer. However, some scientists argued that genetics was a confounding variable (Fisher 1959). They maintained that the smokers differed genetically from the nonsmokers. This genetic difference made some people more likely to smoke and also more susceptible to lung cancer.

This was a convincing argument for many years. It not only proposed a specific difference between the groups (genetics) but also explained how that difference might come about (genetics makes some people smoke more, perhaps because it tastes better to them or because they have addictive personalities). And the argument also explained why this difference might affect the outcome (the same genetics cause lung cancer). Therefore, the skeptics said, genetics—and not smoking—might be the cause of lung cancer.

Later studies established that the skeptics were wrong about genetics. Some studies compared pairs of identical twins in which one twin smoked and the other did not. These pairs had the same genetic makeup, and still a higher percentage of the smoking twins had cancer than of the nonsmoking twins. Because the treatment and control groups had the same genetics, genetics could not explain why the groups had different cancer rates. When we compare groups in which we force one of the variables to be the same, we say that we are *controlling for* that variable. In these twin studies, the researchers controlled for genetics by comparing people with the same genetic makeup (Kaprio and Koskenvuo 1989).

A drawback of observational studies is that we can never know whether there exists a confounding variable. We can search very hard for it, but just because we don't find a confounding variable does not mean it isn't there. For this reason, we can never make cause-and-effect conclusions from observational studies.

**KEY POINT**
We can never draw cause-and-effect conclusions from observational studies because of potential confounding variables. A single observational study can conclude only that there is an *association* between the treatment variable and the outcome variable.

## EXAMPLE 6 Does Kimchi Cause Gastric Cancer?

In a paper titled "Kimchi and Soybean Pastes Are Risk Factors of Gastric Cancer," researchers noted that gastric cancer rates were ten times higher in Korea and Japan than in the United States. Koreans and Japanese also eat many times more kimchi (a heavily spiced, fermented cabbage dish) than those living in the United States (Nan HongMei et al. 2005).

**QUESTION** On the basis of this information, can we conclude that kimchi causes gastric cancer? If yes, explain why. If no, state a potential confounding variable.

**SOLUTION** No, we cannot. The treatment group consists of Japanese and Korean people, and the control group consists of Americans. Many other factors differ between these two groups. A potential confounding variable is diet, which differs between Koreans and Americans in many ways other than in the amount of kimchi they eat. These differences in diet may also affect cancer rates.

**TRY THIS!** Exercise 1.31

## Controlled Experiments

In order to answer cause-and-effect questions, we need to create a treatment group and a control group that are alike in every way possible, except that one group gets a treatment and the other does not. As you've seen, this cannot be done with observational studies because of confounding variables. In a **controlled experiment**, researchers take control by assigning subjects to the control or treatment group. If this assignment is done correctly, it ensures that the two groups can be nearly alike in every relevant way except whether or not they receive the treatment under investigation.

Well-designed and well-executed controlled experiments are the only means we have for definitively answering questions about causality. However, controlled experiments are difficult to carry out (this is one reason why observational studies are often done instead). Let's look at some of the attributes of a well-designed controlled experiment.

A well-designed controlled experiment has four key features:

- The sample size must be large so that we have opportunities to observe the full range of variability in the humans (or animals or objects) we are studying.

- The subjects of the study must be assigned to the treatment and control groups at random.

- Ideally, the study should be "double-blind," as explained below.

- The study should use a placebo if possible.

These features are all essential in order to ensure that the treatment group and the control group are as similar as possible.

To understand these key design features, imagine that a friend wants to lose weight and has asked your advice about whether he should go on the Atkins diet. (See Dansinger et al. 2005 for an account of a study related to the hypothetical one described here.) The Atkins diet is fairly severe: Eating bread in any form is essentially forbidden. Does this diet work? For a control group, we might compare the Atkins diet to a more traditional diet, such as that advocated by Weight Watchers. The Weight Watchers diet restricts calories and is generally considered a reliable way to lose weight. How do we know whether the Atkins diet (the treatment) works, compared to the Weight Watchers diet (the control)?

## Sample Size

A good controlled experiment designed to determine whether the Atkins diet works should have a large number of people participate in the study. People react to changes in their diet in a variety of ways, and the effects of a diet can vary greatly from person to person. To observe the full range of variability, you therefore need a large number of people.

The question of exactly how many people are required is difficult to answer, and most medical studies hire statisticians to determine the number of participants required. In general, the more the better, and you should be more critical of studies with very few participants. (In Chapter 7 we'll consider how to be more precise about the number of participants required for one type of study.)

## Random Assignment

The next step is to assign people to the treatment and control groups such that the two groups are similar in every way possible. As we saw when we discussed observational studies, letting the participants themselves decide doesn't work, because people who choose the more severe Atkins diet over the more traditional Weight Watchers might differ in other important ways (such as level of motivation) that affect the outcome. Instead, a good controlled experiment uses **random assignment**. One way of doing this is to flip a coin. Heads means the participant goes into the treatment group, tails means the control group (or the other way around—as long as the researchers are consistent). In practice, the randomizing might instead be done with a computer or even with the random number generator on a calculator, but the idea is always the same: No human determines group assignment. Rather, assignment is left to chance.

If both groups have a large enough number of people, random assignment will "balance" the groups. The variation in weights, the mix of ages and genders, and the mix of most variables will be similar in both groups. Note that by "similar" we don't mean exactly the same. We don't expect both groups to have exactly the same percentage of men, for example. Except in rare cases, random variation results in slight differences in the mixes of the groups. But these differences should be small.

If you read about a controlled experiment that does not use random assignment, then there is a very real possibility that the results of the study are invalid. The technical term for what happens with nonrandomized assignment is bias. We say that a study exhibits **bias** when the results are influenced in one particular direction. A researcher who puts the heaviest people in one diet group, for example, is biasing the outcome. It's not always easy or possible to predict what the effects of the bias will be, but the important point is that the bias creates a confounding variable and makes it difficult, or impossible, to determine whether the treatment we're investigating really affects the outcome we're observing.

 **KEY POINT**   **Random assignment** (assignment to treatment groups by a randomization procedure) helps balance the groups to minimize bias. This helps make the groups comparable.

## Blinding

So far, we've recruited a large number of people and randomly assigned half to use the Atkins diet and the other half to use the Weight Watchers diet. In principle, these two groups will be very similar. However, there are still two potential differences. First, we might know who is in which group. This means that when we interact with a participant, we might consciously or unconsciously treat that person differently, depending on which group he or she belongs to. For example, if we believe strongly in the Atkins diet, we might give special advice or encouragement to people on the Atkins diet that we don't give to people on the Weight Watchers diet. If so, then we've biased the study.

To prevent this from happening, researchers should be **blind** to assignment. This means that an independent party—someone who does not regularly see the participants and who does not participate in determining the results of the study—handles the assignment to groups. The researchers do not know who is in which group until the study has ended.

Second, we must consider the participants themselves. If they know they are in the treatment group, they might behave differently than if they know they are in the control group. Perhaps they will work harder at losing weight. Or perhaps they will work less hard, because they might have a false sense of confidence in the diet. Why would we have made the Atkins diet our treatment group, they might reason, if we didn't think it was the better diet? To prevent this from happening, the participants should also not know whether they are in the treatment group or the control group.

When neither the researchers nor the participants know whether they are in the treatment or the control group, we say that the study is **double-blind**. The double-blind format helps prevent the bias that can result if one group acts differently from the other because they know they are being treated differently, or because the researchers treat the groups differently or evaluate them differently because of what the researchers expect.

## Placebos

The treatment and control groups might differ in another way. People often react not just to a particular medical treatment, but also to the *idea* that they are getting medical treatment. This means that patients who receive a pill, a vaccine, or some other form of treatment often feel better even when the treatment in truth does absolutely nothing. Interestingly, this placebo effect also works in the other direction: If told that a certain pill might cause side effects (for example, a rash), some patients experience the side effects even though the pill they were given is just a sugar pill.

To neutralize the placebo effect, it is important that the control group receive attention similar to what the treatment group receives, so that both groups feel they are being treated the same by the researchers. In our diet study, both groups are put on a special diet, so no difference between the groups should arise from the placebo effect.

In our hypothetical study about diets, then, a placebo is not necessary, because we are comparing two diets to each other and are not interested in comparing the dieters to a group of subjects who do not go on any diet at all. But if we were studying whether peanut milk improves baldness, we would require the control group to take a placebo drink so that we could rule out any placebo effect and thus perform a valid comparison between treatment and control.

 **KEY POINT**   The Gold Standard for Experiments:

*Large sample size.* This ensures that the study captures the full range of variation among the population and allows small differences to be noticed.

*Controlled and randomized.* Random assignment of subjects to treatment or control groups to minimize bias.

*Double-blind.* Neither subjects nor researchers know who is in which group.

*Placebo (if appropriate).* This format controls for possible differences between groups that occur simply because subjects think their treatment is effective.

## EXAMPLE 7 Brain Games

Brain training video games, such as Nintendo's Brain Age, claim to improve basic intelligence skills, such as memory. A study published in the journal *Nature* investigated whether playing such games can actually boost intelligence (Owen et al. 2010). The researchers explain that 11,430 people logged onto a Web page and were randomly assigned to one of three groups. Group 1 completed six training tasks that emphasized "reasoning, planning and problem-solving." Group 2 completed games that emphasized a broader range of cognitive skills. Group 3 was a control group and didn't play any of these games; instead, members were prompted to answer "obscure" questions. At the end of six weeks, the participants were compared on several different measures of thinking skills. The results? The control group did just as well as the treatment groups.

**QUESTION** Which features of a well-designed controlled experiment does this study have? Which features are missing?

**SOLUTION** Sample size: The sample size of 11,430 is quite large. Each of the three groups will have about 3800 people.

Randomization: The authors state that patients were randomly assigned to one of the three groups.

Double-blind format: Judging on the basis of this description, there was no double-blind format. It's possible (indeed, it is likely) that the researchers did not know, while analyzing the outcome, to which treatment group individuals had been assigned. But we do not know whether *participants* were aware of the existence of the three different groups and how they differed.

Placebo: The control group participated in a "null" game, in which they simply answered questions. This activity is a type of placebo, because the participants could have thought that this null game was a brain game.

**TRY THIS!** Exercise 1.33

> **! Caution**
>
> **At Random**
> The concept of randomness is used in two different ways in this section. *Random assignment* is used in a controlled experiment. Subjects are randomly assigned to treatment and control groups in order to achieve a balance between groups. This ensures that the groups are comparable to each other and that the only difference between the groups is whether or not they receive the treatment under investigation. *Random selection* occurs when researchers select subjects from some larger group via a random method. We must employ random selection if we wish to extend our results to a larger group.

## Extending the Results

In both observational studies and controlled experiments, researchers are often interested in knowing whether their findings, which are based on a single collection of people or objects, will extend to the world at large.

The researchers in Example 7 concluded that brain games are not effective, but might it just be that the games weren't effective for those people who decided to participate? Maybe if the researchers tested people in another country, for example, the findings would be different.

It is usually not possible to make generalizations to a larger group of people unless the subjects for the study are representative of the larger group. The only way to collect a sample that is representative is to collect the objects we study at random. We will discuss how to collect a random sample, and why we can then make generalizations about people or objects who were not in the sample, in Chapter 7.

Selecting subjects using a random method is quite common in polls and surveys (which you'll also study in Chapter 7), but it is much less common in other types of studies. Most medical studies, for example, are not conducted on people selected randomly, so even when a cause-and-effect relationship emerges between the treatment and the response, it is impossible to say whether this relationship will hold for a larger (or different) group of people. For this reason, medical researchers often work hard to replicate their findings in diverse groups of people.

## Statistics in the News

When reading in a newspaper or blog about a research study that relies on statistical analysis, you should ask yourself several questions to evaluate how much faith you can put in the conclusions reached in the study.

1. *Is this an observational study or a controlled experiment?*

   If it's an observational study, then you can't conclude that the treatment caused the observed outcome.

2. *If the study is a controlled experiment, was there a large sample size? Was randomization used to assign participants to treatment groups? Was the study double-blind? Was there a placebo?*

   See the relevant section of this chapter for a review of the importance of these attributes.

3. *Was the paper published in a peer-reviewed journal? What is the journal's reputation?*

   "Peer-reviewed" means that each paper published in the journal is rigorously evaluated by at least two anonymous researchers familiar with the field. The best journals are very careful about the quality of the research they report on. They have many checkpoints to make sure that the science is as good as it can be. (But remember, this doesn't mean the science is perfect. If you read a medical journal regularly, you'll see much debate from issue to issue about certain results.) Other journals, by contrast, sometimes allow sloppy research results, and you should be very wary of these journals.

4. *Did the study follow people for a long enough time?*

   Some treatments take a long time to work, and some illnesses take a long time to show themselves. For example, many cost-conscious people like to refill water bottles again and again with tap water. Some fear that drinking from the same plastic bottle again and again might lead to cancer. If this is true, it might take a very long time for a person to get cancer from drinking out of the same bottle day after day. So researchers who wish to determine whether drinking water from the same bottle causes cancer should watch people for a very long time.

Often it is hard to get answers to all of these questions from a newspaper article. Fortunately, the Internet has made it much easier to find the original papers, and your college library or local public library will probably have access to many of the most popular journals.

Even when a controlled experiment is well designed, things can still go wrong. One common way in which medical studies go astray is that people don't always do what their doctor tells them to do. Thus, people randomized to the treatment group might not actually take their treatments. Or people randomized to the Atkins diet might switch to Weight Watchers because they don't like the food on the Atkins diet. A good research paper will report on these difficulties and will be honest about the effect on their conclusions.

## EXAMPLE 8 Does City Living Raise Blood Pressure?

The *New York Times* linked to an article that claims higher blood pressure occurs in people who live in urban areas (www.scienceblog.com 2010).

QUESTION   Is this more likely to be an observational study or a controlled experiment? Why? Does this mean that moving to a more urban area will result in an increase in your blood pressure?

SOLUTION  This is most likely an observational study. The treatment variable is whether or not a person lives in an urban area. We are not told that participants were randomly assigned to live in urban or nonurban areas for some period of time, and in fact it is pretty unlikely that such a study could be done. Researchers probably measured blood pressure in people who chose to live in urban or nonurban settings. Because the participants themselves chose where to live, this is an observational study.

Because this is an observational study, we cannot conclude that the *treatment* (living in urban areas) causes the *outcome* (increased blood pressure). This means that simply moving to (or away from) an urban area may not change your blood pressure. One potential confounding variable could be personality, because people who prefer the fast pace of city life might have other personality traits that increase their blood pressure.

TRY THIS!  Exercise 1.39

## EXAMPLE 9 Crohn's Disease

Crohn's disease is a bowel disease that causes cramping abdominal pain, fever, and fatigue. A study reported in the *New England Journal of Medicine* (Columbel et al. 2010) tested two medicines for the disease: injections of infliximab (Inflix) and oral azathioprine (Azath). The participants were randomized into three groups. All groups received an injection and a pill (some were placebos, but still a pill and an injection). One group received Inflix injections alone (with placebo pills), one received Azath pills alone (with placebo injections), and one group received both injections and pills. A good outcome was defined as the disease being in remission after 26 weeks. The accompanying table shows the numbers.

|                  | Combination | Inflix Alone | Azath Alone |
| ---------------- | ----------- | ------------ | ----------- |
| Remission        | 96          | 75           | 51          |
| Not in Remission | 73          | 94           | 119         |

QUESTIONS

a. Compare the percentages in remission for the three treatments. Which treatment was the most effective and which the least effective for this sample?

b. Can we conclude that the combination treatment causes a better outcome? Why or why not?

SOLUTIONS

a. For the combination: 96/169, or 56.8%, success
   For the Inflix alone: 75/169, or 44.4%, success
   For the Azath alone: 51/170, or 30%, success

The combination treatment was the most effective for this sample, and Azath alone was the least effective.

b. Yes, we can conclude that the combination of drugs causes a better outcome than the single drugs. The study was placebo-controlled and randomized. The sample size was reasonably large. Blinding was not mentioned, but at least, thanks to the placebos, the patients did not know what treatment they were getting.

**TRY THIS!**  Exercise 1.41

## CASE STUDY REVISITED

Do cell phones cause cancer? The evidence presented at the beginning of the chapter seems to suggest that they might. But let's consider the nature of the evidence. The man who appeared on *Larry King Live* said that his wife, who had died of brain cancer, talked on her cell phone a great deal. But this is just an anecdote. We can conclude nothing about the effects of cell phones on brain cancer, because we don't know how many other people who talked the same amount of time on their cell phones did not get brain cancer, and we don't know how many people who *did* get brain cancer did not use cell phones at all.

The Hardell study was an observational study. The two groups compared were a group that had brain cancer and a group that did not. Clearly, the researchers could not assign subjects to one of these two groups, so all they could do was observe that one group had a different brain cancer rate than the other. However, many potential confounders might explain this difference. For example, critics of the study point out that Hardell examined only brain cancer survivors, which means that he excluded a large percentage of those with brain cancer. Survivors are different from others, and whether or not people survived cancer determined whether they were in this study or not and affected their responses. (Clearly, it's easier to determine how much time people spend talking on the phone if they are alive and able to answer questions.) Researchers in Australia actually did a controlled experiment with mice. They found that the brain cancer risk for mice that received cell phone levels of radiation was no different from that for mice that did not receive the radiation.

Ruling that the evidence was insufficient to suggest a causal relationship between cell phone use and brain cancer, the judge threw out the lawsuit.

# EXPLORING STATISTICS
## CLASS ACTIVITY

## Collecting a Table of Different Kinds of Data

**ACTIVITY**

As a class, you will suggest possible variables that would help you describe your class to others. You will then collect data and summarize and describe these variables. How would you describe the people in your class as a group? How does your class compare to another class, maybe the class next door? Suggest variables you might collect on individuals that would help you describe your class and make comparisons. (Examples of such variables might include gender, number of credits currently taken, distance the individual lives from campus, and the like.) Be prepared to give your own values for these variables.

Some variables (such as family income, for example) might be too private; inquiring about others might be offensive to some students. You should not feel compelled to provide information that might embarrass you; try to avoid suggesting variables that you think might embarrass others. Table A shows an example of the format for the input

| Gender | Age | Favorite Class | Units | ____ | ____ |
|--------|-----|----------------|-------|------|------|
|        |     |                |       |      |      |

▲ **TABLE A**

**BEFORE THE ACTIVITY**

If someone asked you to describe the students who make up your class, which characteristics would you focus on?

**AFTER THE ACTIVITY**

1. Which variables that were used are numerical? Which are categorical?

2. Write a one-paragraph summary of the data you collected that you think would help someone understand the composition of your class.

# CHAPTER REVIEW

## Key Terms

statistics, *3*
variation, *4*
data, *4*
variables, *6*
data set, *6*
sample, *6*
population, *6*
numerical variable, *7*
categorical variable, *7*

stacked data, *9*
unstacked data, *9*
two-way table, *10*
frequency, *10*
rate, *13*
causality, *14*
treatment variable, *14*
outcome variable
   (or response variable), *14*

treatment group, *15*
control group
   (or comparison group), *15*
anecdotes, *15*
placebo, *15*
placebo effect, *16*
observational study, *16*
association, *16*
confounding variable, *16*

controlled experiment, *17*
random assignment, *18*
bias, *18*
blind, *19*
double-blind, *19*
random selection, *20*

## Learning Objectives

After reading this chapter and doing the assigned homework problems, you should

- Be able to distinguish between numerical and categorical variables and understand methods for coding categorical variables.

- Know how to find and use rates (including percentages) and understand when and why they are more useful than counts for describing and comparing groups.

- Understand when it is possible to infer a cause-and-effect relationship from a research study and when it is not.

- Be able to explain how confounding variables prevent us from inferring causation, and suggest confounding variables that are likely to occur in some situations.

- Be able to distinguish between observational studies and controlled experiments.

## Summary

Statistics is the science (and art) of collecting and analyzing observations (called data) and communicating your discoveries to others. Often, we are interested in learning about a population on the basis of a sample taken from that population. Here are some questions to consider when first examining a data set:

> What are the objects of interest?
> What variables were measured?
> How were they measured?
> What are the units of measurement?
> Who collected the data?
> How did they collect the data?
> Where were the data collected?
> Why were the data collected?

With categorical variables, we are often concerned with comparing rates or frequencies between groups. A two-way table is sometimes a useful summary. Always be sure that you are making valid comparisons by comparing proportions or percentages of groups, or that you are comparing the appropriate rates.

Many studies are focused on questions of causality: If we make a change to one variable, will we see a change in the other? Anecdotes are not useful for answering such questions. Observational studies can be used to determine whether associations exist between treatment and outcome variables, but because of the possibility of confounding variables, observational studies cannot support conclusions about causality. Controlled experiments, if they are well designed, do allow us to draw conclusions about causality.

A well-designed controlled experiment should have the following attributes:

> A large sample size
> Random assignment of subjects to a treatment group or a control group
> A double-blind format
> A placebo

## Sources

BBC News, February 18, 2010. http://news.bbc.co.uk (accessed May 13, 2010).

Britt, C. "U.S. Doctors Prescribe Drugs for Placebo Effects Survey Shows." Bloomberg.com (accessed October 23, 2008).

Brody, J. 2002. Cellphone: A convenience, a hazard or both? *New York Times*, October 1.

cnn.com. September 11, 2002. Cancer study may be used in Motorola suit (accessed September 23, 2002).

Colombel, Sandborn, Reinisch, Mantzaris, Kornbluth, Rachmilewitz, Lichtiger, D'Haens, Diamond, Broussard, Tang, van der Woude, and Rutgeerts. 2010. Infliximab, azathioprine, or combination therapy for Crohn's disease. *New England Journal of Medicine* 362, 1383–1395.

Dansinger, M., J. Gleason, J. Griffith, H. Selker, and E. Schaefer. 2005. Comparison of the Atkins, Ornish, Weight Watchers, and Zone diets for weight loss and heart disease risk reduction: A randomized trial. *JAMA* 293(1), January 5.

Decmick, B. 2006. Koreans' kimchi adulation, with a side of skepticism. *Los Angeles Times*, May 21.

Fisher, R. 1959. *Smoking: The cancer controversy*. Edinburgh, UK: Oliver and Boyd.

Glionna, J. 2006. Word of mouth spreading peanut milk. *Los Angeles Times*, May 17.

Holcomb, J. 2006. http://www2.irss.unc.edu/ncvital/index.html (accessed May 17, 2006).

Insurance Highway Safety Institute. 2006. http://www.iihs.org/news/2006/hldi_news_060706.pdf (accessed July 12, 2006).

Kaprio, J., and M. Koskenvuo. 1989. Twins, smoking and mortality: A 12-year prospective study of smoking-discordant twin pairs. *Social Science and Medicine* 29(9), 1083–1089.

*McKinsley Quarterly*. 2009. Hal Varian on how the Web challenges managers. Business Technology Office, January 2009.

Nan HongMei, Park JinWoo, Song YoungJin, Yun HyoYung, Park JooSeung, Hyun TaiSun, Youn SeiJin, Kim YongDae, Kang JongWon, and Kim Heon. 2005.

Kimchi and soybean pastes are risk factors of gastric cancer. *World Journal of Gastroenterology*, June 2005.

National Safety Council. 2004. *Injury Facts*, 2004 edition. Itasca, Illinois.

Owen, A., et al. 2010. Letter: Putting brain training to the test. *Nature* advance online publication, April 20, www.nature.com (accessed May 15, 2010).

www.scienceblog.com. 2010. Higher blood pressure found in people living in urban areas (accessed May 2010).

## SECTION EXERCISES

## SECTION 1.2

*The data in Table 1A were collected from one of the authors' statistics classes. The column heads give the variable, and each of the other rows represents a student in the class. Refer to this table for Exercises 1.1–1.4, 1.11, and 1.12.*

| Male | Age | Eye Color | Shoe Size | Height (inches) | Weight (pounds) | Number of Siblings | College Units This Term | Handedness |
|------|-----|-----------|-----------|-----------------|-----------------|--------------------|-------------------------|------------|
| 1 | 20 | Brown | 9.5 | 71 | 170 | 1 | 16 | Right |
| 0 | 19 | Blue | 8 | 66 | 135 | 1 | 13 | Right |
| 0 | 42 | Brown | 7.5 | 63 | 130 | 3 | 5 | Right |
| 0 | 19 | Brown | 8.5 | 65 | 150 | 0 | 15 | Left |
| 1 | 21 | Brown | 11 | 70 | 185 | 5 | 19.5 | Right |
| 0 | 20 | Hazel | 5.5 | 60 | 105 | 2 | 11.5 | Right |
| 1 | 21 | Blue | 12 | 76 | 210 | 2 | 9.5 | Right |
| 0 | 21 | Brown | 10 | 70 | 140 | 0 | 8 | Left |
| 0 | 32 | Brown | 8 | 64 | 165 | 1 | 13.5 | Right |
| 1 | 23 | Brown | 7.5 | 63 | 145 | 6 | 12 | Right |
| 0 | 21 | Brown | 6.5 | 61.5 | 110 | 4 | 14 | Right |

▲ TABLE 1A

TRY **1.1 (Example 1)** Are the following variables, from Table 1A, numerical or categorical? Explain.

 a. Handedness

 b. Age

**1.2** Are the following variables, from Table 1A, numerical or categorical? Explain.

 a. Shoe size

 b. Eye color

**1.3 Coding** Explain why the variable Male, in Table 1A, is categorical, even though its values are numbers. Often, it does not make sense, or is not even possible, to add the values of a categorical variable. Does it make sense for Male? If so, what does the sum represent?

**1.4 Coding** Students with fewer than 12 units in the current term are considered part-time. Create a new categorical variable that classifies each student in Table 1A as full-time (12 or more units) or part-time. Call this variable Full. Report the values in a column in the same order as those in the table.

TRY **1.5 Brain Size (Example 2)** In 1991, researchers conducted a study on brain size as measured by pixels in a magnetic resonance imagery (MRI) scan. The numbers are in hundreds of thousands of pixels. The data table provides the sizes of the brains and the gender. (Source: www.lib.stat.cmu.edu/DASL)

 a. Is the format of the data set stacked or unstacked?

 b. Explain the coding. What do 1 and 0 represent?

 c. If you answered "stacked" in part a, then unstack the data into two columns labeled Male and Female. If you answered "unstacked," then stack the data into one column; choose an appropriate name for the stacked variable.

| Brain | Male |
|-------|------|
| 9.4 | 1 |
| 9.5 | 0 |
| 9.5 | 1 |
| 9.5 | 1 |
| 9.5 | 0 |
| 9.7 | 1 |
| 9.9 | 0 |

**1.6 Students' Ages** The accompanying table gives ages for some of the students in two statistics classes. One class met at noon and the other at 5 p.m.

a. Is the format for the data set stacked or unstacked?

b. If you answered "stacked," then unstack the data set. If you answered "unstacked," then stack and code the data set.

| 5 p.m. | Noon |
|--------|------|
| 31 | 24 |
| 34 | 18 |
| 46 | 21 |
| 47 | 20 |
| 50 | 20 |

## SECTION 1.3

**TRY 1.7 Changing Answers (Example 3)** On a multiple-choice exam, if you think you have made a mistake, is it better to stick with your first choice, or to erase and change your answer? One of the authors instructed her students to change the answer if the second answer they got seemed better than the first (Change!). Her colleague told his students that their first answer was usually the best and they should avoid changing answers (Don't Change!). After both classes had taken several multiple-choice exams, the exams were studied to see whether answers were erased and replaced with different answers. The results are shown in the accompanying table.

a. What is the percentage of answers that were changed for the students who were told not to change their answers?

b. What is the percentage of answers that were changed for the students who were told to change their answers?

c. Compare the answers to parts a and b, and state whether you think the instruction had any effect on whether students changed their answers.

| | | Instruction | |
|---|---|---|---|
| **Behavior** | | Don't Change! | Change! |
| | Changed | 189 | 124 |
| | Unchanged | 29,428 | 14,389 |

**1.8 Changing Answers** Read the information in Exercise 1.7. The table summarizes the outcomes based on the instructions given by the teacher.

a. For the teacher who said Don't Change!, what percentage of the changes were from wrong to right? What percentage of the changes were from right to wrong? Did the results of the changes raise students' grades or lower them?

b. For the teacher who said Change!, what percentage of the changes were from wrong to right? What percentage of the changes were from right to wrong? Did the results of the changes raise students' grades or lower them?

c. On the basis of these data, what advice would you give to a student who asked you whether it's better to stick with your first hunch or to change your answer?

| | Don't Change! | Change! |
|---|---|---|
| Wrong to right | 91 | 86 |
| Right to wrong | 48 | 24 |
| Wrong to wrong | 50 | 14 |

**TRY 1.9 Finding and Using Percentages (Example 4)**

a. A statistics class is made up of 15 men and 23 women. What percentage of the class is male?

b. A different class has 234 students, and 64.1% of them are men. How many men are in the class?

c. A different class is made up of 40% women and has 20 women in it. What is the total number of students in the class?

**1.10 Finding and Using Percentages**

a. A hospital employs 346 nurses and 35% of them are male. How many male nurses are there?

b. An engineering firm employs 178 engineers and 112 of them are male. What percentage of these engineers are *female*?

c. A large law firm is made up of 65% male lawyers, or 169 male lawyers. What is the total number of lawyers at the firm?

**1.11 Women** Find the frequency, proportion, and percentage of women in Table 1A on page 26.

**1.12 Right-Handed People** Find the frequency, proportion, and percentage of right-handed people in Table 1A on page 26.

**1.13 Population Prediction** The *2009 World Almanac and Book of Facts* predicted that the United States will have an elderly population (65 and older) of 88,547,000 in the year 2050 and that this will be 20.2% of the population. What is the total predicted U.S. population in 2050?

**1.14 2007 Population** The *2009 World Almanac and Book of Facts* reported that in 2007 there were 12,608,000 people age 16 or older who had a "go outside the home" disability and that this was 5.5% of the U.S. population (of this age group). These are people who cannot go outside the home without help. How large was the total population (of this age group) in 2007?

**g 1.15 Living with AIDS** The table gives the number of people living with AIDS in 2007 in the five states with the largest number of cases, as well as the District of Columbia, as reported by the

U.S. Centers for Disease Control and Prevention (CDC). It also shows the population of those regions at that time from the U.S. Census Bureau.

Find the number of people living with AIDS per thousand residents in each region, and rank the six regions from highest rate (rank 1) to lowest rate (rank 6). Compare these rankings (of rates) with the ranks of total number of cases. If you moved to one of these regions and met 50 random people, in which region would you be most likely to meet at least one person living with AIDS? In which of these regions would you be least likely to meet at least one person with AIDS? *See page 33 for guidance.*

| State | AIDS | Population |
|---|---|---|
| New York | 75,253 | 19,297,729 |
| California | 65,582 | 36,553,215 |
| Florida | 48,059 | 18,251,243 |
| Texas | 34,940 | 23,904,380 |
| Pennsylvania | 19,236 | 12,432,792 |
| District of Columbia | 8,895 | 588,292 |

**1.16 Population Density** The accompanying table gives the population of the six U.S. states with the largest populations in 2008 and the area of these states. (Source: www.infoplease.com)

| State | Population | Area (square miles) |
|---|---|---|
| Pennsylvania | 12,448,279 | 44,817 |
| Illinois | 12,901,563 | 55,584 |
| Florida | 18,328,340 | 53,927 |
| New York | 19,490,297 | 47,214 |
| Texas | 24,326,974 | 261,797 |
| California | 36,756,666 | 155,959 |

a. Determine and report the rankings of the population density by dividing each population by the number of square miles to get the population density (in people per square mile). Use rank 1 for the highest density.

b. If you wanted to live in the state (of these six) with the lowest population density, which would you choose?

c. It you wanted to live in the state (of these six) with the highest population density, which would you choose?

**1.17 Marriage Rates** The number of married people in the United States and the total number of adults in the United States (in millions) are provided in the accompanying table for several years.

Find the percentage of people married in each of the given years, and describe the trend over time. (Source: *2009 World Almanac and Book of Facts*)

| Year | Married | Total |
|---|---|---|
| 1990 | 112.6 | 191.8 |
| 1997 | 116.8 | 207.2 |
| 2000 | 120.2 | 213.8 |
| 2007 | 129.9 | 235.8 |

**1.18 Elderly Population** The number of people in the United States age 65 or older and the total number of people living in the United States (in thousands) are reported in the accompanying table for several years. Find what percentage of population is elderly for each year, and describe the trend of these percentages. (Source: *2009 World Almanac and Book of Facts*)

| Year | Elderly | Population |
|---|---|---|
| 1900 | 3,080 | 75,100 |
| 1920 | 4,933 | 105,000 |
| 1940 | 9,019 | 133,000 |
| 1960 | 16,560 | 180,000 |
| 1980 | 25,549 | 226,000 |
| 2000 | 34,992 | 282,000 |

TRY *1.19 Course Enrollment Rates (Example 5)** Two sections of statistics are offered, the first at 8 a.m. and the second at 10 a.m. The 8 a.m. section has 25 women, and the 10 a.m. section has 15 women. A student claims this is evidence that women prefer earlier statistics classes than men do. What information is missing that might contradict this claim?

* **1.20 Pedestrian Fatalities** In 2008, the National Highway Traffic Safety Administration reported that the number of pedestrian fatalities in Miami-Dade County, Florida, was 65 and that the number in Hillsborough County, Florida, was 45. Can we conclude that pedestrians are safer in Hillsborough County? Why or why not?

## SECTION 1.4

*For Exercises 1.21–1.28, indicate whether the study is an observational study or a controlled experiment.*

**1.21** A student watched picnickers with a large cooler of soft drinks to see whether teenagers were less likely than adults to choose diet soft drinks over regular soft drinks.

**1.22** The medical records of people who live near high-power lines are compared with the medical records of people who do not live near high power lines to see whether people living near high-power lines are more likely to have cancer.

**1.23** A researcher wonders whether the order in which people taste beverages influences their preference. Students are told by the researcher to drink from two unmarked cups. The researcher assigns one group of students to drink Coke first and then Pepsi. The researcher assigns the other group to drink Pepsi first and then Coke. The subjects are asked whether they prefer the first or the second beverage.

**1.24** Patients with Alzheimer's disease are randomly divided into two groups. One group is given a new drug, and the other is given a placebo. After six months they are given a memory test to see whether the new drug fights Alzheimer's better than a placebo.

**1.25** A group of boys is randomly divided into two groups. One group watches violent cartoons for one hour, and the other group watches cartoons without violence for one hour. The boys are then observed to see how many violent actions they take in the next two hours, and the two groups are compared.

**1.26** A local public school encourages, but does not require, students to wear uniforms. The principal of the school compares the grade point averages of students at this school who wear uniforms with the GPAs of those who do not wear uniforms to determine whether those wearing uniforms tend to have higher GPAs.

**1.27** A researcher was interested in the effect of exercise on academic performance in elementary school children. She went to the recess area of an elementary school and identified some students who were exercising vigorously and some who were not. The researcher then compared the grades of the exercisers with the grades of those who did not exercise.

**1.28** A researcher was interested in the effect of exercise on memory. She randomly assigned half of a group of students to run up a stairway three times and the other half to rest for an equivalent amount of time. Each student was then asked to memorize a series of random digits. She compared the numbers of digits remembered for the two groups.

**1.29 Effects of Tutoring on Math Grades** A group of educators want to determine how effective tutoring is in raising students' grades in a math class, so they arrange free tutoring for those who want it. Then they compare final exam grades for the group that took advantage of the tutoring and the group that did not. Suppose the group participating in the tutoring tended to receive higher grades on the exam. Does that show that the tutoring worked? If not, explain why not and suggest a confounding variable.

**1.30 Treating Depression** A doctor who believes strongly that antidepressants work better than "talk therapy" tests depressed patients by treating half of them with antidepressants and the other half with talk therapy. After six months the patients are evaluated on a scale of 1 to 5, with 5 indicating the greatest improvement.

a. The doctor is concerned that if his most severely depressed patients do not receive the antidepressants, they will get much worse. He therefore decides that the most severe patients will be assigned to receive the antidepressants. Explain why this will affect his ability to determine which approach works best.

b. What advice would you give the doctor to improve his study?

c. The doctor asks you whether it is acceptable for him to know which treatment each patient receives and to evaluate them himself at the end of the study to rate their improvement. Explain why this practice will affect his ability to determine which approach works best.

d. What improvements to the plan in part c would you recommend?

**TRY 1.31 Treating Heart Disease (Example 6)** A study reported by Hannon et al. looked at people treated for heart disease and reported lower death rates for those who received a coronary bypass (CABG) compared to those who received a stent. A stent is a device to keep blood vessels open. The beginning of the abstract is given below.

*Methods* We identified patients with multivessel disease who received drug-eluting stents or underwent CABG in New York State between October 1, 2003, and December 31, 2004, and we compared adverse outcomes (death, death or myocardial infarction, or repeat revascularization) through December 31, 2005, after adjustment for differences in baseline risk factors among the patients.

The adverse outcomes were greater with stenting. Was this an observational study or a controlled experiment? How do you know? Can we say that the use of CABG causes a better success rate? Why or why not? (Source: E. H. Hannan et al., Drug-eluting stents vs. coronary-artery bypass grafting in multivessel coronary disease, *New England Journal of Medicine*, vol. 358: 331-341, #4, January 24, 2008)

**1.32 Effects of Exercise on Aging** In a study reported on HealthNews.com, Dr. Yonas E. Geda and colleagues at Mayo Clinic in Rochester, Minnesota, analyzed data on 1324 individuals without dementia who completed a questionnaire about physical activity between 2006 and 2008 as part of the Mayo Clinic Study of Aging. The participants had an average age of 80 and were classified as having either normal cognition (1126) or mild cognitive impairment (MCI; 198). Those who reported performing moderate exercise such as yoga, aerobics, or brisk walking during midlife were 39% less likely to develop MCI, and moderate exercise later in life was associated with a 32% reduction.

Was this more likely to have been a controlled experiment or an observational study, and how do you know? Can we say that exercise reduces cognitive impairment? Why or why not?

**TRY 1.33 Vitamin C Study (Example 7)** A pair of college students have decided to test vitamin C to see whether it prevents colds. They recruit 500 students with a sign-up sheet, containing a numbered list. The first half of those on the sheet (Numbers 1–250) are asked to take 500 mg of vitamin C per day, and the second half are told *not* to use vitamin C. At the end of the school year, participants are asked how many colds they had. How would you improve this study, and why?

**1.34 Weight Loss Study** A group of overweight people are asked to participate in a weight loss program. Participants are allowed to choose whether they want to go on a vegetarian diet or follow a traditional low-calorie diet that includes some meat. Half of the people choose the vegetarian diet, and half choose to be in the control group and continue to eat meat. Suppose that there is greater weight loss in the vegetarian group.

a. Suggest a plausible confounding variable that would prevent us from concluding that the weight loss was due to the lack of meat in the diet. Explain why it is a confounding variable?

b. Explain a better way to do the experiment that is likely to remove the influence of confounding variables.

**1.35 Are Optimists Healthier?** In 2004 "On Health," the *Consumer Reports* medical newsletter, reported a study on the effects of optimism on health. Researchers studied the emotional styles of 334 subjects and then squirted a cold virus in the noses of these subjects. Those who scored high on "energy, happiness, and relaxation" were significantly less likely to develop colds. Is this study evidence that you can reduce your chance of catching colds by training yourself to be more relaxed and happy? Explain your answer.

**1.36 Reducing Migraines** A 2004 article in *The Nutrition Reporter* stated that melatonin supplements were effective in preventing migraine headaches. The article quoted a Brazilian doctor who used melatonin for 3 months to treat 34 patients with a history of two to eight migraine headaches per month. Thirty-two patients completed the study, and 25 patients reported experiencing at least a 50% reduction in headaches. None of the patients had an increase in headaches. What is the major flaw in this research design that might prevent us from concluding that melatonin helps prevent migraine headaches?

**1.37 Flu Vaccine** In the fall of 2004, there was a shortage in flu vaccine in the United States after it was discovered that vaccines from one of the manufacturers were contaminated. The *New England Journal of Medicine* reported on a study that was done to see whether a smaller dose of the vaccine could be used successfully. If that were the case, then a small amount of vaccine could be divided into more flu shots. In this study, the usual amount of vaccine was injected into half the patients, and the other half of the patients had only a small amount of vaccine injected. The response was measured by looking at the production of antibodies (more antibodies generally result in less risk of getting the flu). In the end, the lower dose of vaccine was just as effective as a higher dose for those under 65 years old. What more do we need to know to be able to conclude that the lower dose of vaccine was equally effective at preventing the flu for those under 65? (Source: Beishe et al., Serum antibody responses after intradermal vaccination against influenza, *New England Journal of Medicine*, 2004)

**1.38 Effect of Confederates on Compliance** A study was conducted to see whether participants would ignore a sign that said, "Elevator may stick between floors. Use the stairs." The study was done at a university dorm on the ground floor of a three-level building. Those who used the stairs were said to be compliant, and those who used the elevator were said to be noncompliant. There were three possible situations, two of which involved confederates. A confederate is a person who is secretly working with the experimenter. In the first situation, there was no confederate. In the second situation, there was a compliant confederate (one who used the stairs), and in the third situation, there was a noncompliant confederate (one who used the elevator). The subjects tended to imitate the confederates. What more do you need to know about the study to determine whether the presence or absence of a confederate causes a change in the compliance of subjects? (Source: Wogalter, et. al, (1987), reported in Shaffer & Merrens (2001), Research Stories in Introductory Psychology, Allyn and Bacon: Boston.)

TRY **1.39 Vitamin C and Allergies (Example 8)** The March 2005 issue of *The Nutrition Reporter* reported on the effects of vitamin C in breast milk on preventing allergies in infants. Researchers analyzed levels of vitamin C in the breast milk of some new mothers who had decided whether to take vitamin C during their pregnancy. The highest levels of vitamin C were associated with a 30 percent lower risk of allergies in the infants. Was this an observational study or a controlled experiment? On the basis of this study, can you conclude that vitamin C lowers the risk of allergies in infants? Why or why not?

**1.40 Childhood Exposure to Tobacco Smoke and Emphysema** A headline on the website e! Science News in December 2009 stated, "Exposure to tobacco smoke in childhood home [is] associated with early emphysema in adulthood."

Is this likely to have been a controlled experiment or an observational study? How do you know? Is it possible to conclude, on the basis of this study, that exposure to tobacco smoke for children causes early emphysema in adulthood?

TRY **1.41 Effects of Light Exposure (Example 9)** A study carried out by Baturin and colleagues looked at the effects of light on female mice. Fifty mice were randomly assigned to a regimen of 12 hours of light and 12 hours of dark (LD), while another fifty mice were assigned to 24 hours of light (LL). Researchers observed the mice for two years, beginning when the mice were two months old. Four of the LD mice and 14 of the LL mice developed tumors. The accompanying table summarizes the data. (Source: Baturin et al., The effect of light regimen and melatonin on the development of spontaneous mammary tumors in mice, *Neuroendocrinology Letters*, 2001)

|  | LD | LL |
|---|---|---|
| Tumors | 4 | 14 |
| No tumors | 46 | 36 |

a. Determine the percentage of mice that developed tumors from each group (LL and LD). Compare them and comment.

b. Was this a controlled experiment or an observational study? How do you know?

c. Can we conclude that light for 24 hours a day causes an increase in tumors in mice? Why or why not?

**1.42 Preschool Attendance and High School Graduation Rates** The Perry preschool project was started in the early 1960s by David Weikart in Ypsilanti, Michigan. In this project, 123 African American children were randomly assigned to one of two groups: One group enrolled in the Perry preschool, and one group did not enroll in preschool. Follow-up studies were done on the children after 27 years and after 40 years. The accompanying table shows whether the students graduated from high school or not. The treatment variable is attendance at preschool (or not) and the response variable is graduation from high school (or not). (Source: Schweinhart, L.J. et al., The High/Scope Perry Preschool Study through age 40. *Journal of Experimental Criminology*, (1)1, April, 2005)

|  | Preschool | No Preschool |
|---|---|---|
| Grad HS | 37 | 29 |
| No Grad | 20 | 35 |

a. What percentage of those who went to preschool graduated from high school?

b. What percentage of those who did not go to preschool graduated from high school?

c. Is the better graduation rate from those who attended preschool or those who did not attend preschool?

d. Can you conclude that attending Perry preschool affects graduation rates? Explain.

# CHAPTER REVIEW EXERCISES

**1.43 Obesity and Marital Status** A 2009 study analyzed data from the National Longitudinal Study of Adolescent Health. Participants were studied into adulthood. Each study participant was categorized as to whether they were obese or not and whether they were dating, cohabiting, or married. Obesity was defined as having a Body Mass Index of 30 or more. (Source: The and Larsen, Entry into romantic partnership is associated with obesity, *Obesity*, vol. 17, no. 7: 1441–1447, 2009)

|  | Dating | Cohabiting | Married |
|---|---|---|---|
| Obese | 81 | 103 | 147 |
| Not Obese | 359 | 326 | 277 |
|  | 440 | 429 | 424 |

a. What percentage of those who were dating were obese?

b. What percentage of those who were cohabiting were obese?

c. What percentage of those who were married were obese?

d. Which group had the highest rate of obesity? Does this imply that marital status causes obesity? Why or why not? If not, can you name a confounding variable?

**1.44 Are Cell Phones More Dangerous Than Cigarettes?** Fox News reported on a study in 2008, headed by Dr. Vini Khurana, that suggested that there is some evidence that using cell phone head sets for 10 years or more doubles the risk of brain cancer. He said that three billion people now use cell phones, but only one billion smoke and that smoking kills about five million each year. He concluded that cell phones are more dangerous than cigarettes, and so he suggested that people should avoid using cell phones. (Source: www.independent.co.uk)

a. Is this study more likely to be a controlled experiment or an observational study? Explain.

b. Do you think this study shows that using cell phones causes brain cancer? Also, does it show that cell phones are more dangerous than cigarettes? Why or why not?

**1.45 Probation** A statistics student conducted a study on young male and female criminals 15 years of age and under who were on probation. The purpose of the study was to see whether there was an association between type of crime and gender. The subjects of the study lived in Ventura County, California. Violent crimes involve physical contact such as hitting or fighting. Nonviolent crimes are vandalism, robbery, or verbal assault. The raw data are shown in the accompanying table; v stands for violent, n for nonviolent, b for boy, and g for girl.

a. Make a two-way table that summarizes the data. Label the columns (across the top) Boy and Girl. Label the rows Violent and Nonviolent.

b. Find the percentage of girls on probation for violent crimes and the percentage of boys on probation for violent crimes, and compare them.

c. Are the boys or the girls more likely to be on probation for violent crimes?

| Gen | Viol? | Gen | Viol? | Gen | Viol? |
|---|---|---|---|---|---|
| b | n | b | n | g | n |
| b | n | b | n | g | n |
| b | n | b | n | g | n |
| b | n | b | n | g | v |
| b | n | b | v | g | v |
| b | n | b | v | g | v |
| b | n | b | v | g | v |
| b | n | b | v | g | v |
| b | n | b | v | g | v |
| b | n | b | v | g | v |
| b | n | b | v | g | v |
| b | n | b | v | g | v |
| b | n | b | v | g | v |
| b | n | b | v | g | v |
| b | n | g | n |  |  |

**1.46 Scorpion Antivenom** A study was done on children (6 months to 18 years of age) who had (nonlethal) scorpion stings. Each child was randomly assigned to receive an experimental antivenom or a placebo. Good results were no symptoms after four hours. Make a summary of the data in the form of a two-way table. Label the columns Antivenom and Placebo. Label the rows Better and Not Better. Compare the percentage better for the antivenom group and the placebo group. (Source: Boyer, Leslie V. et al., Antivenom for critically ill children with neurotoxicity from scorpion stings, *New England Journal of Medicine*, vol. 360: 2090–2098, no. 20, May 14, 2009)

| Antivenom | Better | Antivenom | Better |
|---|---|---|---|
| 1 | 1 | 0 | 0 |
| 1 | 1 | 0 | 0 |
| 1 | 1 | 0 | 0 |
| 1 | 0 | 0 | 1 |
| 1 | 1 | 0 | 0 |
| 1 | 1 | 0 | 0 |
| 1 | 1 | 1 | 1 |
| 0 | 0 |  |  |

\* **1.47 Activated Charcoal** A man posting on an Internet message board claimed that his wife was sick for more than 6 months with many problems, including nausea, confusion, and night sweats. Doctors were unable to help her. However, when she started taking "activated charcoal," her condition began to improve. The posting on the message board also explained that his wife had possibly ingested herbicides and might have been poisoned by these "toxins."

Describe the design of a controlled experiment to determine whether activated charcoal can cure people suffering from what we will call "toxin disease," assuming that such a disease exists. Assume that you have access to 200 people suffering from this condition. Your description of your experiment should address all of the major features of a controlled experiment.

\* **1.48 Allergy Experiment** A post on an Internet message board reported that a person suffering from allergies took grapeseed extract and had a remarkable decrease in allergy attacks.

Describe the design of a controlled experiment to determine whether grapeseed extract can relieve the symptoms of allergy sufferers. Assume that you have access to 200 people suffering from allergies. Your description of your experiment should address all of the major features of a controlled experiment.

**1.49 Clinical Trials** In a study of more than 2000 British patients with heart failure, patients were asked whether they would be willing to participate in a clinical trial. After 55 months, those who said they would be willing to participate were about half as likely to have died as those who were not willing to participate. Does this imply that if you have heart failure, you can increase your chance of living by being willing to participate in clinical trials? Why or why not? (Source: Clark et al., Is taking part in clinical trials good for your health? A cohort study, *European Journal of Heart Failure*, vol. 11: 1078–1083, 2009)

**1.50 Recidivism Rate** In May 2010 *Time* magazine published an article about an innovative Norwegian prison, Halden Fengsel, which provides prisoners with dorm-like accommodations, flat-screen televisions, and mini-fridges. The staff does not carry guns. Within two years of release from Halden Fengsel, the rate of recidivism (being imprisoned again) is 20%. In the United States that rate is between 50% and 60%. Is this sufficient evidence to conclude that if the United States changed its prisons to be more like Halden Fengsel, the recidivism rate in the United States would go down? Explain.

**1.51 Death Row and Head Trauma** A study conducted by Lewis et al. in 1986 looked at 14 juveniles awaiting execution. They found that 57% (8 of the 14) had had a serious brain injury. Can we conclude that head trauma causes bad behavior later in life? What primary factor is not present here that should be present in both observational studies and controlled experiments? (Source: Psychiatric, neurological, and psychoeducational characteristics of 15 death row inmates in the United States, *American Journal of Psychiatry*, vol. 143: 838–845. 1986)

**1.52 Brief Exercise and Diabetes** As part of a study, sixteen young men performed high-intensity exercise that totaled only 15 minutes in a two-week period. At the end of two weeks, several (but not all) tests for diabetes, such as an insulin sensitivity test, showed improvement. Do these results indicate that brief, high-intensity exercise causes an improvement in markers for diabetes? What essential component of both controlled experiments and observational studies is missing from this study? (Source: Babraj et al., Extremely short duration high intensity interval training substantially improves insulin action in young healthy males, *BMC Endocrine Disorders*, vol. 9: 3doi: 10.1186/1472-6823-9-3, January 2009)

# gUIDED EXERCISES

**g 1.15 Living with AIDS** The accompanying table gives the numbers of people living with AIDS in 2007 in the five states with the largest numbers of cases, as well as the District of Columbia, as reported by the U.S. Centers for Disease Control and Prevention (CDC). It also shows the population of those regions at that time, from the U.S. Census Bureau.

**Question** Find the number of people living with AIDS per thousand in each region, and rank the six regions from highest rate (rank 1) to lowest rate (rank 6). Compare these rankings (of rates) with the ranks of total number of cases. If you moved to one of these regions and met 50 random people, in which region would you be most likely to meet at least one person living with AIDS? In which of these regions would you be least likely to meet at least one person living with AIDS?

**Step 1 ▶ Figure out the populations of the remaining regions in thousands, and add them to the table.**

**Step 2 ▶ For each region, divide the number of people living with AIDS by the population in thousands, and fill in column 6.**

**Step 3 ▶ Enter the ranks for the rates of AIDS patients per 1000 population, using 1 for the largest value and 6 for the smallest.**

**Step 4 ▶ Are the ranks for the rates the same as the ranks for the numbers of cases? If not, describe at least one difference.**

**Step 5 ▶ Finally, if you moved to one of these regions and met 50 random people, in which region would you be most likely to meet at least one person living with AIDS? In which region would you be least likely to meet at least one person living with AIDS?**

| State | AIDS | Rank Cases | Population | Population (thousands) | AIDS per 1000 population | Rank Rate |
|---|---|---|---|---|---|---|
| New York | 75,253 | 1 | 19,297,729 | 19,298 | 3.90 | 2 |
| California | 65,582 | 2 | 36,553,215 | 36,553 | | |
| Florida | 48,059 | 3 | 18,251,243 | | | |
| Texas | 34,940 | 4 | 23,904,380 | | | |
| Pennsylvania | 19,236 | 5 | 12,432,792 | | | |
| District of Columbia | 8,895 | 6 | 588,292 | | | |

# 2 Picturing Variation with Graphs

## THEME

Any collection of data exhibits variation. The most important tool for organizing this variation is called the distribution of the sample, and visualizing this distribution is the first step in every statistical investigation. We can learn much about a numerical variable by focusing on three components of the distribution: the shape, the center, and the variability, or horizontal spread. Examining a graph of a distribution can lead us to deeper understanding of the situation that produced the data.

One of the major concepts of statistics is that although individual events are hard to predict, large numbers of events usually exhibit predictable patterns. The search for patterns is a key theme in science and business. An important first step in this search is to identify and visualize the key features of your data.

Using graphics to see patterns and identify important trends or features is not new. One of the earliest statistical graphs dates back to 1786, when a Scottish engineer named William Playfair published a paper examining whether there was a relationship between the price of wheat and wages. To help answer this question, Playfair produced a graph (shown in Figure 2.1) that is believed to be the first of its kind. This graph became the prototype of two of the most commonly used tools in statistics: the bar chart and the histogram.

Graphics such as these can be extraordinarily powerful ways of organizing data, detecting patterns and trends, and communicating findings. The graphs that we use have changed somewhat since Playfair's day, but graphics are of fundamental importance to analyzing data. The first step in any statistical analysis

▲ **FIGURE 2.1** Playfair's chart explores a possible relationship between wages and the price of wheat. (Source: Playfair 1786)

is to make a picture of some kind in order to check our intuition against the data. If our intuitions are wrong, it could very well be because the world works differently than we thought. Thus, by making and examining a display of data (as illustrated in this chapter's Case Study), we gain some insight into how the world works.

In Chapter 1 we discussed some of the methods used to collect data. In this chapter we'll cover some of the basic graphics used in analyzing the data we collect. Then, in Chapter 3, we'll comment more precisely on measuring and comparing key features of our data.

## CASE STUDY

## Student-to-Teacher Ratio at Colleges

Are private four-year colleges better than public four-year colleges? That depends on what you mean by "better." One measure of quality that many people find useful (and there are many other ways to measure quality) is the student-to-teacher ratio: the number of students enrolled divided by the number of teachers. For schools with small student-to-teacher ratios, we expect class sizes to be small; students can get extra attention in a small class.

The data in Table 2.1 on the next page were collected from some schools that award four-year degrees. The data are for the 2004–2005 academic year; 85 private colleges and 57 state-supported (public) colleges were sampled. Each ratio was rounded to the nearest whole number for simplicity. For example, the first private

| 85 Private Colleges | | | | | | | | |
|---|---|---|---|---|---|---|---|---|
| 10 | 20 | 24 | 10 | 11 | 12 | 20 | 14 | 9 |
| 16 | 10 | 11 | 13 | 15 | 32 | 14 | 7 | 7 |
| 10 | 13 | 11 | 10 | 10 | 17 | 11 | 11 | 21 |
| 15 | 12 | 15 | 38 | 10 | 14 | 8 | 13 | 11 |
| 16 | 5 | 17 | 14 | 9 | 11 | 10 | 16 | 55 |
| 11 | 13 | 13 | 16 | 9 | 8 | 11 | 16 | |
| 14 | 11 | 16 | 9 | 14 | 32 | 10 | 15 | |
| 12 | 30 | 10 | 10 | 9 | 10 | 12 | 24 | |
| 16 | 10 | 12 | 11 | 6 | 15 | 13 | 11 | |
| 12 | 10 | 11 | 15 | 11 | 8 | 11 | 11 | |

| 57 Public Colleges | | | | | |
|---|---|---|---|---|---|
| 15 | 20 | 22 | 18 | 1 | 23 |
| 20 | 20 | 21 | 20 | 29 | 21 |
| 15 | 25 | 20 | 13 | 23 | 16 |
| 15 | 15 | 16 | 16 | 18 | 17 |
| 20 | 20 | 21 | 15 | 16 | 13 |
| 15 | 16 | 18 | 16 | 21 | 17 |
| 28 | 16 | 18 | 11 | 17 | 17 |
| 12 | 15 | 17 | 20 | 20 | |
| 17 | 9 | 15 | 12 | 22 | |
| 13 | 14 | 17 | 20 | 24 | |

▲ **TABLE 2.1** Ratio of Students to Teachers at Private and Public Colleges
(Source: *2006 World Almanac and Book of Facts*)

college listed has a student-to-teacher ratio of 10, which means that there are about 10 students for every teacher. What differences do you expect between the two groups? What similarities do you anticipate?

It is nearly impossible to compare the two groups without imposing some kind of organization on the data. In this chapter you will see several ways in which we can graphically organize groups of data like these so that we can compare the two types of colleges. At the end of this chapter, you'll see what the right graphical summaries can tell us about how these types of colleges compare.

# Visualizing Variation in Numerical Data

One of the most important conceptual tools in statistics and data analysis is the distribution. The **distribution of a sample** of data is simply a way of organizing the data. Think of the distribution of the sample as a list that records (1) the values that were observed and (2) the frequencies of these values. **Frequency** is another word for *count*, of how many times the value occurred in the collection of data.

 **KEY POINT** The distribution of a sample is one of the central organizational concepts of data analysis. The distribution organizes data by recording all of the values observed in a sample, as well as how many times each value was observed.

Distributions are important because they capture much of the information we need in order to make comparisons between groups, examine data for errors, and learn about real-world processes. Distributions allow us to examine the variation of the data in our sample, and from this variation we can often learn more about the world.

The first step of almost every statistical investigation is to visualize the distribution of the sample. By creating an appropriate graphic, we can see patterns that might otherwise escape our notice.

For example, here are some raw data from the National Collegiate Athletic Association (NCAA), available online. This set of data shows the number of goals scored per game for Division III women soccer players for all games in the 2009 season. (Division III schools are colleges or universities that are not allowed to offer scholarships to athletes.) To make the data set smaller, we show only first-year students.

0.727, 0.737, 0.737, 0.737, 0.750, 0.750, 0.750, 0.762, 0.765, 0.765, 0.789, 0.789, 0.813, 0.833, 0.875, 0.889, 0.909, 0.941, 0.947, 1.053, 1.091, 1.158, 1.294, 1.400

This list includes only the values. A distribution lists the values and also the frequencies. The distribution of this sample is shown in Table 2.2.

It's hard to see patterns when the distribution is presented as a table. A picture makes it easier for us to answer questions such as "What's the typical number of goals per game scored by a player?" and "Is 1.4 goals per game an unusually high number?" Data are also available for male soccer players and for other divisions and classes. A picture would make it easier to compare the numbers of goals for different groups. For example, in a season, do men typically score more goals per game or fewer goals per game than women?

When examining distributions, we use a two-step process:

1. See it.

2. Summarize it.

In this section we explain how to visualize the distribution. In the next section, we discuss the characteristics you should look for to help you summarize it.

All of the methods we use for visualizing distributions are based on the same idea: Make some sort of mark that indicates how many times each value occurred in our data set. In this way, we get a picture of the sample distribution so that we can see at a glance which values occurred and how often.

Two very useful methods for visualizing distributions of numerical variables are dotplots and histograms. Dotplots are simpler; histograms are more commonly used and perhaps more useful.

## Dotplots

In constructing a **dotplot**, we simply put a dot above a number line where each value occurs. We can get a sense of frequency by seeing how high the dots stack up. Figure 2.2 shows a dotplot for the number of goals for first-year female soccer players.

Tech

With this simple picture, we can see more than we could from Table 2.2. We can see from this dotplot that most women scored an average of less than 1 goal per game, but a few scored more. Also, we can see that the woman who scored 1.4 goals per game was exceptional within this group. Not only is 1.4 the largest value in this data set, but it stands apart from most of the others by a large gap.

| Value | Frequency |
|-------|-----------|
| 0.727 | 1 |
| 0.737 | 3 |
| 0.750 | 3 |
| 0.762 | 1 |
| 0.765 | 2 |
| 0.789 | 2 |
| 0.813 | 1 |
| 0.833 | 1 |
| 0.875 | 1 |
| 0.889 | 1 |
| 0.909 | 1 |
| 0.941 | 1 |
| 0.947 | 1 |
| 1.053 | 1 |
| 1.091 | 1 |
| 1.158 | 1 |
| 1.294 | 1 |
| 1.400 | 1 |

▲ **TABLE 2.2** Distribution of the number of goals per game scored by first-year women soccer players in NCAA Division III in 2009.

### Details

**Making Dotplots**
Dotplots are easy to make with pen and paper, but don't worry too much about recording values to great accuracy. The purpose of a plot like this is to help us see the overall shape of the distribution, not to record details about individual observations.

◄ **FIGURE 2.2** Dotplot of number of goals per game scored by first-year women soccer players in NCAA Division III, 2009. Each dot represents a soccer player. Note that the horizontal axis begins at 0.7.

## SNAPSHOT  THE DOTPLOT

| | |
|---|---|
| **WHAT IS IT?** ▶ | A graphical summary. |
| **WHAT DOES IT DO?** ▶ | Shows a picture of the distribution of a numerical variable. |
| **HOW DOES IT DO IT?** ▶ | Each observation is represented by a dot on a number line. |
| **HOW IS IT USED?** ▶ | To see patterns in samples of data with variation. |

## Histograms

While dotplots have one dot for each observation in the data set, **histograms** produce a smoother graphic by grouping observations into intervals, called bins. These groups are formed by dividing the number line into bins of equal width and then counting how many observations fall into each bin. To represent the bins, histograms display vertical bars, where the height of each bar is proportional to the number of observations inside that bin.

For example, with the goals scored in the dotplot in Figure 2.2, we could create a series of bins that go from 0.7 to 0.8, 0.8 to 0.9, 0.9 to 1.0, and so on. Twelve women scored between 0.7 and 0.8 goals per game, so the first bar has a height of 12. The second bin contains four observations and consequently has a height of 4. The finished graph is shown in Figure 2.3. (Note that some statisticians use the word *interval* in place of *bin*. You might even see another word that means the same thing: *class*.) Figure 2.3 shows, among other things, that three women scored between 0.9 and 1.0 goals per game, two scored between 1.0 and 1.1 goals per game, and so on.

▶ **FIGURE 2.3** Histogram of goals per game for female first-year soccer players in NCAA Division III, 2009. The first bar, for example, tells us that 12 players scored between 0.7 and 0.8 goals per game during the season.

Making a histogram requires paying attention to quite a few details. For example, we need to decide on a rule for what to do if an observation lands exactly on the boundary of two bins. In which bin would we place an observation of 0.8 goals per game? A common rule is to decide always to put "boundary" observations in the bin on the right, but we could just as well decide always to put them in the bin on the left. The important point is to be consistent. The graphs here use the right-hand rule and put boundary values in the bin to the right.

Figure 2.4 shows two more histograms of the same data. Even though they display the same data, they look very different—both from each other and from Figure 2.3. Why?

Changing the width of the bins in a histogram changes its shape. Figure 2.3 has bins with width of 0.1. In contrast, Figure 2.4a has much smaller bins, and Figure 2.4b has wider bins. Note that when we use small bins, we get a spiky histogram. When we use wider bins, the histogram gets less spiky. Using wide bins hides more detail. If you chose very wide bins, you would have no details at all. You would see just one big rectangle!

How large should the bins be? Too small and you see too much detail (as in Figure 2.4a). Too large and you don't see enough (as in Figure 2.4b). Most computer software will automatically make a good choice. Our software package (Minitab, in this case) automatically chose the binwidth of 0.1 that you see in Figure 2.3. Still,

◄ **FIGURE 2.4** Two more histograms of goals scored per game, the same data as in Figure 2.3. **(a)** This histogram has narrow bins and is spiky. **(b)** This histogram has wide bins and offers less detail.

if you can, you should try different sizes to see how different choices change your impression of the distribution of the sample. Fortunately, most statistical software packages make it quite easy to change the bin width.

## Relative Frequency Histograms

A variation on the histogram (and statisticians, of course, love variation) is to change the units of the vertical axis from frequencies to relative frequencies. A **relative frequency** is simply a proportion. So instead of reporting that the first bin had 12 observations in it, we would report that the proportion of observations in the first bin was $12/24 = 0.50$. We divide by 24 because there were a total of 24 observations in the data set. Figure 2.5 is the same as the first histogram shown for the distribution of goals per game (Figure 2.3); however, Figure 2.5 reports relative frequencies, and Figure 2.3 reports frequencies.

Using relative frequencies does not change the shape of the graph; it just communicates different information to the viewer. Rather than answering the question "How many players scored between 0.7 and 0.8 goals per game?" (12 players), it now answers the question "What *proportion* of players scored between 0.7 and 0.8 goals per game?" (0.50).

◄ **FIGURE 2.5** Relative frequency histogram of goals scored per game by first-year women soccer players in NCAA Division III, 2009.

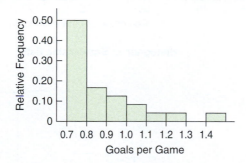

## SNAPSHOT   THE HISTOGRAM

| | |
|---|---|
| **WHAT IS IT?** ► | A graphical summary for numerical data. |
| **WHAT DOES IT DO?** ► | Shows a picture of the distribution of a numerical variable. |
| **HOW DOES IT DO IT?** ► | Observations are grouped into bins, and bars are drawn to show how many observations (or what proportion of observations) lie in each bin. |
| **HOW IS IT USED?** ► | By smoothing over details, histograms help our eyes pick up on more important, large-scale patterns. Be aware that making the bins wider hides detail, and making the bins smaller can show too much detail. The vertical axis can display frequency, relative frequency, or percents. |

## EXAMPLE **1** Visualizing Bar Exam Pass Rates at Law Schools

In order to become a lawyer, you must pass your state's bar exam. When you are choosing a law school, it therefore makes great sense to choose one that has a high percentage of graduates passing the bar exam. The Internet Legal Research Group website provides the pass rate for 181 law schools in the United States in 2006 (Internet Research Legal Group 2008).

**QUESTION** Is 80% a good pass rate for a law school?

**SOLUTION** This is a subjective question. An 80% pass rate might be good enough for a prospective student, but another way of looking at this is to determine whether there are many schools that do better than 80%, whether 80% is a typical pass rate, or whether there are very few schools that do so well. Questions such as these can often be answered by considering the distribution of our sample of data, so our first step is to choose an appropriate graphical presentation of the distribution.

Either a dotplot or a histogram would show us the distribution. Figure 2.6a shows a dotplot generated by Minitab. It's somewhat difficult (but not impossible) to read a dotplot with 181 observations; there's too much detail to allow us to answer broad questions such as these. It is easier to use a histogram, as in Figure 2.6b. In this histogram, each bar has a width of 10 percentage points, and the y-axis tells us how many law schools had a pass rate within those limits.

► **FIGURE 2.6 (a)** A dotplot for the bar-passing rate for 181 law schools. Each dot represents a law school, and the dot's location indicates the bar-passing rate for that school. **(b)** A histogram shows the same data as in part (a), except that the details have been smoothed.

**(a)**

**(b)**

We see that a little more than 50 law schools had a pass rate between 80% and 90%. Another 30 or so had a pass rate over 90%. Adding these together, about 80 schools had a pass rate higher than 80%, and this is $80/181 = 0.4420$ or about 44% of the schools. This tells us that a pass rate of 80% is not all that unusual; about 44% of all law schools do this well or better.

**TRY THIS!** Exercise 2.5

## Stemplots

Stemplots, which are also called stem-and-leaf plots, are useful for visualizing numerical variables when you don't have access to technology and the data set is not large. Stemplots are also useful if you want to be able to easily see the actual values of the data.

To make a **stemplot**, divide each observation into a "stem" and a "leaf." The **leaf** is the last digit in the observation. The **stem** contains all the digits that precede the leaf. For the number 60, the *6* is the stem and the *0* is the leaf. For the number 632, the *63* is the stem and the *2* is the leaf. For the number 65.4, the *65* is the stem and the *4* is the leaf.

A stem-and-leaf plot can help us understand data such as drinking behaviors. Alcohol is a big problem at many colleges and universities. For this reason, a collection of college students who said that they drink alcohol were asked how many alcoholic drinks they had consumed in the last seven days. Their answers were

1, 1, 1, 1, 1, 2, 2, 2, 3, 3, 3, 3, 3, 4, 5, 5, 5, 6, 6, 6, 8, 10, 10, 15, 17, 20, 25, 30, 30, 40

For one-digit numbers, imagine a 0 at the front. The observation of 1 drink becomes 01, the observation of 2 drinks becomes 02, and so on. Then each observation is just two digits; the first digit is the stem, and the last digit is the leaf. Figure 2.7 shows a stemplot of these data.

If you rotate a stemplot 90 degrees counterclockwise, it looks not too different from a histogram. Unlike histograms, stemplots display the actual values of the data. With a histogram, you know only that the values fall somewhere within an interval.

Stemplots are often organized with the leaves in order from lowest to highest, which makes it easier to locate particular values, as in Figure 2.7. This is not necessary, but it makes the plot easier to use.

From the stemplot, we see that most students drink a moderate amount in a week but that a few drink quite a bit. Forty drinks per week is almost six per day, which qualifies as problem drinking by some physicians' definitions.

Figure 2.8 shows a stemplot of some exam scores. Note the empty stems at 4 and 5, which show that there were no exam grades between 40 and 59. Most of the scores are between 60 and 100, but one student scored very low relative to the rest of the class.

**Tech**

| Stem | Leaves |
|------|--------|
| 0 | 11111222333334555668 |
| 1 | 0057 |
| 2 | 05 |
| 3 | 00 |
| 4 | 0 |

▲ **FIGURE 2.7** A stemplot for alcoholic drinks consumed by college students. Each digit on the right (the leaves) represents a student. Together, the stem and the leaf indicate the number of drinks for one student who is a drinker.

| Stem | Leaves |
|------|--------|
| 3 | 8 |
| 4 | |
| 5 | |
| 6 | 0257 |
| 7 | 00145559 |
| 8 | 0023 |
| 9 | 0025568 |
| 10 | 00 |

▲ **FIGURE 2.8** A stemplot for exam grades. Two students had scores of 100, and no students scored in the 40's or 50's.

### SNAPSHOT  THE STEMPLOT

| | |
|---|---|
| **WHAT IS IT?** ▸ | A graphical summary for numerical data. |
| **WHAT DOES IT DO?** ▸ | Shows a picture of the distribution of a numerical variable. |
| **HOW DOES IT DO IT?** ▸ | Numbers are divided into leaves (the last digit) and stems (the preceding digits). Stems are written in a vertical column, and associated leaves are "attached." |
| **HOW IS IT USED?** ▸ | In very much the same way as a histogram. It is often useful when technology is not available and the data set is not large. |

## SECTION 2.2

# Summarizing Important Features of a Numerical Distribution

When examining a distribution, pay attention to describing the shape of the distribution, the **typical value (center)** of the distribution, and the **variability (spread)** in the distribution. The typical value is subjective, but the location of the center of a distribution often gives us an idea which values are typical for this variable. The variability is reflected in the amount of horizontal spread the distribution has.

> **KEY POINT**  When examining distributions of numerical data, pay attention to the shape, center, and horizontal spread.

Figure 2.9 compares distributions for two groups. You've already seen histogram (a)—it's the histogram for the goals scored per game by first-year women soccer players in Division III. Histogram (b) shows the same goals per game for first-year male soccer players in Division III. How do these two distributions compare?

▶ **FIGURE 2.9** Distributions of the goals scored per game for **(a)** first-year women and **(b)** first-year men in Division III soccer in 2009.

1. *Shape.* Are there any interesting or unusual features about the distributions? Are the shapes very different? (If so, this might be evidence that men play the game differently than women.)

2. *Center.* What is the typical value of each distribution? Is the typical number of goals scored per game different for men than for women?

3. *Spread.* The horizontal spread presents the variation in goals per game for each group. How do the amounts of variation compare? If one group has low variation, it suggests that the soccer skills of the members are pretty much the same. Lots of variability might mean that there is a wider variety of skill levels.

Let's consider these three aspects of a distribution one at a time.

## Shape

You should look for three basic characteristics of a distribution's shape:

1. Is the distribution symmetric or skewed?

2. How many mounds appear? One? Two? None? Many?

3. Are unusually large or small values present?

### Symmetric or Skewed?

A symmetric distribution is one in which the left-hand side of the graph is roughly a mirror image of the right-hand side. The idealized distributions in Figure 2.10 show several possibilities. Figure 2.10a is a **symmetric distribution** with one mound. (Statisticians often describe a distribution with this particular shape as a **bell-shaped distribution**. Bell-shaped distributions play a major role in statistics, as you will see throughout this book.)

> **Details**
>
> **Vague Words**
>
> For now, we have left the three ideas *shape*, *center*, and *spread* deliberately vague. In Chapter 3 you'll learn ways of measuring more precisely where the center is and how much spread exists. However, the first task in a data analysis is to examine a distribution to informally evaluate the shape, center, and spread.

▶ **FIGURE 2.10** Sketches of **(a)** a symmetric distribution and **(b)** a right-skewed distribution.

Figure 2.10b represents a nonsymmetric distribution with a skewed shape that has one mound. Note that it has a "tail" that extends out to the right (toward larger values). Because the tail goes to the right, we call it a **right-skewed distribution**. This is a typical shape for the distribution of a variable in which most values are relatively small but that also has a few very large values. If the tail goes to the left, it is a **left-skewed distribution**, where most values are relatively large but that also has a few very small values.

Figure 2.11 shows a histogram of 123 college women's heights. How would you describe the shape of this distribution? This is a good real-life example of a symmetric distribution. Note that it is not perfectly symmetric, but you will never see "perfect" in real-life data.

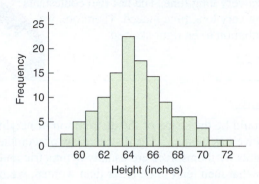

◄ **FIGURE 2.11** Histogram of heights of women. (Source: Brian Joiner in Tufte 1983)

Suppose we asked a sample of people how many hours of TV they watched in a typical week. Would you expect a histogram of these data to be bell-shaped? Probably not. The smallest possible value for this data set would be 0, and most people would probably cluster near a common value. However, a few people probably watch quite a bit more TV than most other people. Figure 2.12 shows the actual histogram. We've added an arrow to emphasize that the tail of this distribution points to the right; this is a right-skewed distribution.

▲ **FIGURE 2.12** This data set on TV hours viewed per week is skewed to the right (Source: Minitab Program)

▲ **FIGURE 2.13** This data set on test scores is skewed to the left. (Source: Ryan 2006)

Figure 2.13 shows a left-skewed distribution of test scores. This is the sort of distribution that you should hope your next exam will have. Most people scored pretty high, so the few people who scored low create a tail on the left-hand side. A very difficult test, one in which most people scored very low and only a few did well, would be right-skewed.

Another circumstance in which we often see skewed distributions is when we collect data on people's income. When we graph the distribution of incomes of a large group of people, we quite often see a right-skewed distribution. You can't make less than 0 dollars per year. Most people make a moderate amount of money, but there is no upper limit to how much a person can make, and a few people in any large sample will make a very large amount of money.

Example 2 shows that you can often make an educated guess as to the shape of a distribution even without collecting data.

## EXAMPLE 2 Roller Coaster Endurance

A morning radio show is sponsoring a contest in which contestants compete to win a car. About 40 contestants are put on a roller coaster, and whoever stays on it the longest wins. Suppose we make a histogram of the amount of time the contestants stay on (measured in hours or even days).

**QUESTION** What shape do we expect the histogram to have and why?

**SOLUTION** Probably most people will drop out relatively soon, but a few will last for a very long time. The last two contestants will probably stay for a very long time indeed. Therefore, we would expect the distribution to be right-skewed.

**TRY THIS!** Exercise 2.9

## How Many Mounds?

What do you think would be the shape of the distribution of heights if we included men in our sample as well as women? The distributions of women's heights by themselves and men's heights by themselves are usually symmetric and have one mound. But because we know that men tend to be taller than women, we might expect a histogram that combines men's and women's heights to have two mounds.

The statistical term for a one-mound distribution is a **unimodal distribution**, and a two-mound distribution is called a **bimodal distribution**. Figure 2.14a shows a bimodal distribution. A **multimodal distribution** has more than two modes. The modes do not have to be the same height (in fact, they rarely are). Figure 2.14b is perhaps the more typical bimodal distribution.

▶ **FIGURE 2.14** Idealized bimodal distributions. **(a)** Modes of roughly equal height. **(b)** Modes that differ in height.

These sketches are idealizations. In real life you won't see distributions this neat. You will have to make a decision about whether a histogram is close enough to be called symmetric or whether it has one mound, two mounds, no mounds, or many mounds. The existence of multiple mounds is sometimes a sign that two very different groups have been combined into a single collection (such as combining men's heights with women's heights). When you see multimodal distributions, you may want to look back at the original data and see whether you can examine the groups separately, if separate groups exist. At the very least, whenever you see more than one mound, you should ask yourself, "Could these data be from different groups?"

## EXAMPLE 3 Two Marathons, Merged

Data were collected on the finishing times for two different marathons. One marathon consisted of a small number of elite runners: the 2004 Olympic Games. The other marathon included a large number of amateur runners: a marathon in Ontario, Canada.

**QUESTION** What shape would you expect the distribution of this sample to have?

**SOLUTION** We expect the shape to be bimodal. The elite runners would tend to have faster finishing times, so we expect one mound on the left for the Olympic runners and another, taller mound (because there were more amateur runners) on the right. Figure 2.15 is a histogram of the data.

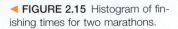

There appears to be one mound centered at about 9,000 seconds (2.5 hours) and another centered at about 14,000 seconds (about 3.9 hours).

**TRY THIS!** Exercise 2.11

When we view a histogram, our understanding of the shape of a distribution is affected by the width of the bins. Figure 2.15 reveals the bimodality of the distribution partly because the width of the bins is such that we see the right level of detail. If we had made the bins too big, we would have got less detail and might not have seen the bimodal structure. Figure 2.16 shows what would happen. Experienced data analysts usually start with the bin width the computer chooses and then also examine histograms made with slightly wider and slightly narrower bins. This lets them see whether any interesting structure emerges.

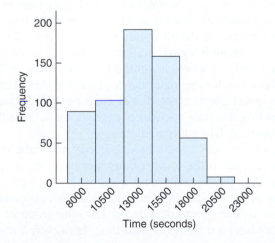

◀ **FIGURE 2.16** Another histogram of the same running times as in Figure 2.15. Here, the bins are wider and "wash out" detail, so the distribution no longer looks bimodal, even though it should.

## Do Extreme Values Occur?

Sometimes when you make a histogram or dotplot, you see extremely large or extremely small observations. When this happens, you should report these values and, if necessary, take action (although *reporting* is often action enough). Extreme values can appear when an error is made in entering data. For example, we asked students in one of our classes to record their weights in pounds. Figure 2.17 shows the distribution. The student who wrote 1200 clearly made a mistake. He or she probably meant to write 120.

Extreme values such as these are called **outliers**. The term *outlier* has no precise definition. Sometimes you may think an observation is an outlier, but another person might disagree, and this is fine. However, if there is no gap between the observation in question and the bulk of the histogram, then the observation probably should not be considered an outlier. Outliers are points that don't fit the pattern of the rest of the data, so if a large percentage of the observations are extreme, it might not be accurate to label them as outliers. After all, if lots of points don't fit a pattern, maybe you aren't seeing the right pattern!

► **FIGURE 2.17** Histogram of weights with an extreme value. (Source: Ryan 2005)

**KEY POINT** — Outliers are values so large or small that they do not fit into the pattern of the distribution. There is no precise definition of the term *outlier*. Outliers can be caused by mistakes in data entry, but genuine outliers are sometimes unusually interesting observations.

Sometimes outliers result from mistakes, and sometimes they do not. The important thing is to make note of them when they appear and to investigate further, if you can. You'll see in Chapter 3 that the presence of outliers can affect how we measure center and spread.

## Center

> ⚠ **Caution**
>
> **Multimodal Centers**
> Be careful about giving a single typical value for a multimodal or bimodal distribution. These distributions sometimes indicate the existence of different and diverse groups combined into the same data set. It may be better to report a different typical value for each group, if possible.

An important question to ask about any data set is "What is the typical value?" The typical value is the one in the center, but we use the word *center* here in a deliberately vague way. We might not all agree on precisely where the center of a graph is. But the idea here is to get a rough impression so that we can make comparisons later. For example, judging on the basis of the histogram shown in Figure 2.9, the center for the women soccer players is about 0.8 goals per game. Thus we could say that the typical first-year woman soccer player scored about 0.8 goals per game in 2009. In contrast, the center of the distribution for the male soccer players is about 0.55. It would seem that the typical male soccer player scores fewer goals per game than the typical female player, perhaps indicating that men's soccer is stronger on defense than women's soccer, at least among Division III first-year players.

If the distribution is bimodal or multimodal, it may not make sense to seek a "typical" value for a data set. If the data set combines two very different groups, then it might be more useful to find separate typical values for each group. What is the typical finishing time of the runners in Figure 2.15? There is no single typical time, because there are two distinct groups of runners. The elite group have their typical time, and the amateurs have a different typical time. However, it *does* make sense to ask about the typical test score for the student scores in Figure 2.13, because there is only one mound and only one group of students.

## EXAMPLE **4** Typical Bar-Passing Rate for Law Schools

Examine the distribution of bar-passing rates for law schools in the United States (see Figure 2.6b, which is repeated here for convenience).

**QUESTION** What is a typical bar-passing rate for a law school?

**SOLUTION** We show Figure 2.6b again. The center is somewhere in the range of 70% to 80%, so we would say that the typical bar-passing rate is between 70% and 80%.

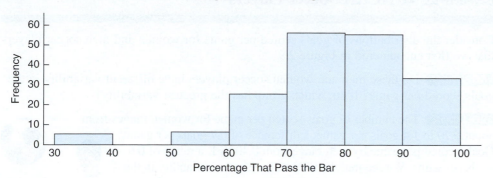

**Histogram of Bar-Passing Rate**

◄ **FIGURE 2.6b (repeated)**
Percentage passing the bar exam from different law schools.

**TRY THIS!** Exercise 2.13

## Variability

The third important feature to consider when examining a distribution is the amount of variation in the data. If all of the values of a numerical variable are the same, then the histogram (or dotplot) will be skinny. On the other hand, if there is great variety, the histogram will be spread out, thus displaying greater variability.

Here's a very simple example. Figure 2.18a shows a family of four people who are all very similar in height. Note that the histogram of these heights (Figure 2.18b) is quite skinny; in fact, it is just a single bar!

**(a)**

**(b)**    **Family One, Heights**

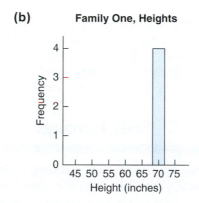

▲ **FIGURE 2.18** Family One with a Small Variation in Heights

Figure 2.19a, on the other hand, shows a family that exhibits large variation in height. The histogram for this family (Figure 2.19b) is more spread out.

**(a)**

**(b)**    **Family Two, Heights**

▲ **FIGURE 2.19** Family Two with a Large Variation in Height

## EXAMPLE 5 NCAA Soccer Players

Consider the distributions of goals scored per game for women and men soccer players that we first encountered in Figure 2.9.

**QUESTION** Do these men and women soccer players have different variability in the goals scored per game? If so, which group has the greatest variability?

**SOLUTION** The number of goals scored per game for women ranges from about 0.70 to 1.4 goals per game; a difference of 0.7 goals per game. Goals scored per game by men range from 0.40 to 1, a range of 0.6 goals per game. We see that the women have more variability in the number of goals scored per game. One interpretation of this difference in spread is that women's soccer has a wider range of skill levels. This is also supported by the shape of the distribution, which suggests that most women score in the neighborhood of about 0.80 goals per game, a few as many as 1 goal per game, but a very few do quite a bit better. Men, on the other hand, are much more similar to each other, at least in terms of the number of goals scored per game. In Chapter 3 you'll see some more precise ways of measuring the spread of a numerical variable.

**TRY THIS!** Exercise 2.15

## Describing Distributions

When you are asked to describe a distribution or to compare two distributions, your description should include the center of the distribution (What is the typical value?), the spread (How much variability is there?), and the shape. If the shape is bimodal, it might not be appropriate to mention the center, and you should instead identify the approximate location of the mounds. You should also mention any unusual features, such as extreme values.

 ## EXAMPLE 6 Body Piercings

How common are body piercings among college students? How many piercings does a student typically have? One statistics professor asked a large class of students to report (anonymously) the number of piercings they possessed.

**QUESTION** Describe the distribution of body piercings for students in a statistics class.

**SOLUTION** The first step is to "see it" by creating an appropriate graphic of the distribution of the sample. Figure 2.20 shows a histogram for these data. To summarize this distribution, we examine the shape, center, and spread and comment on any unusual features.

▶ **FIGURE 2.20** Numbers of piercings of students in a statistics class.

The distribution of piercings is right-skewed, as we might expect, given that many people will have 0, 1, or 2 piercings, but a few people are likely to have more. The typical number of piercings (the center of the distribution) seems to be about 1, although a majority of students have none. The number of piercings ranges from 0 to about 10. An interesting feature of this distribution is that it appears to be multimodal. There are three peaks: at 0, 2, and 6, which are even numbers. This makes sense, because piercings often come in pairs. But why is there no peak at 4? (The authors do not know.) What do you think the shape of the distribution would look like if it included only the men? Only the women?

**TRY THIS!** Exercise 2.17

# SECTION 2.3

# Visualizing Variation in Categorical Variables

When visualizing data, we treat categorical variables in much the same way as numerical variables. We visualize the distribution of the categorical variable by displaying the values (categories) of the variable and the number of times each value occurs.

To illustrate, consider the Statistics Department at UCLA. UCLA offers an introductory statistics course every summer, and it needs to understand what sorts of students are interested in this class. In particular, understanding whether the summer students are mostly first-year students (eager to complete their general education requirements) or seniors (who put off the class as long as they could) can help the department better plan its course offerings.

Table 2.3 shows data from a sample of students in an introductory course offered during the 2009 summer term at UCLA. The "unknown" students are probably not enrolled in any university (adult students taking the course for business reasons or high school students taking the class to get a head start).

*Class* is a categorical variable. Table 2.4 on the next page summarizes the distribution of this variable by showing us all of the values in our sample and the frequency with which each value appears. Note that we added a row for first-year students.

Two commonly used graphs to display the distribution of a sample of categorical data are bar charts and pie charts. Bar charts look, at first glance, very similar to histograms, but they have several important differences, as you will see.

## Bar Charts

A **bar graph** (also called a **bar chart**) shows a bar for each observed category. The height of the bar is proportional to the frequency of that category. Figure 2.21a on the next page shows a bar graph for the UCLA statistics class data. The vertical axis measures frequency. We see that the sample has one graduate student and four juniors. We could also display relative frequency if we wished (Figure 2.21b). The shape does not change; only the numbers on the vertical axis change.

Note that there are no first-year students in the sample. We might expect this of a summer course, because entering students are unlikely to take courses in the summer before they begin college, and students who have completed a year of college are generally no longer first-year students (assuming they have completed enough units).

### Bar Charts vs. Histograms

Bar charts and histograms look a lot alike, but they have some very important differences.

- In a bar chart, it sometimes doesn't matter in which order you place the bars. Quite often, the categories of a categorical variable have no natural order. If they

| Student ID | Class |
|:---:|:---|
| 1 | Senior |
| 2 | Junior |
| 3 | Unknown |
| 4 | Unknown |
| 5 | Senior |
| 6 | Graduate |
| 7 | Senior |
| 8 | Senior |
| 9 | Unknown |
| 10 | Unknown |
| 11 | Sophomore |
| 12 | Junior |
| 13 | Junior |
| 14 | Sophomore |
| 15 | Unknown |
| 16 | Senior |
| 17 | Unknown |
| 18 | Unknown |
| 19 | Sophomore |
| 20 | Junior |

▲ **TABLE 2.3** Identification of Classes for Students in Statistics

| Class | Frequency |
|---|---|
| unknown | 7 |
| First-year student | 0 |
| Sophomore | 3 |
| Junior | 4 |
| Senior | 5 |
| Graduate | 1 |
| **Total** | **20** |

▲ **TABLE 2.4** Summary of Classes for Students in Statistics

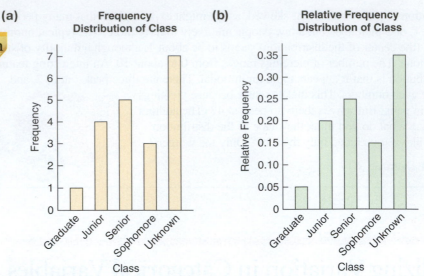

▲ **FIGURE 2.21** **(a)** Bar chart showing numbers of students in each class enrolled in an introductory statistics section. The largest "class" is the group made up of seven unknowns. First-year students are not shown because there were none in the data set. **(b)** The same information as shown part (a), but now with relative frequencies. The unknowns are about 0.35 (35%) of the sample.

do have a natural order, you might want to sort them in that order. For example, in Figure 2.22a we've sorted the categories into a fairly natural order, from "Unknown" to "Graduate." In Figure 2.22b we've sorted them from most frequent to least frequent. Either choice is acceptable.

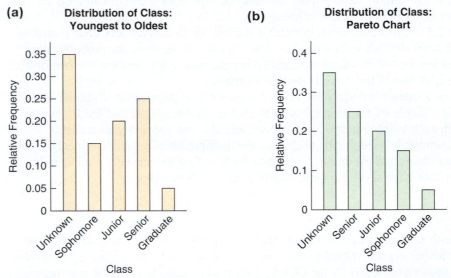

▲ **FIGURE 2.22** **(a)** Bar chart of classes using natural order. **(b)** Pareto chart of the same data. Categories are ordered with the largest frequency on the left and arranged so the frequencies decrease to the right.

Figure 2.22b is interesting because it shows more clearly that the most populated category consists of "Unknown." This result suggests that there might be a large demand for this summer course outside of the university. Bar charts that are sorted from most frequent to least frequent are called **Pareto charts**. (These charts were invented by the Italian economist and sociologist Vilfredo Pareto (1848–1943).) They are often an extremely informative way of organizing a display of categorical data.

- Another difference between histograms and bar charts is that in a bar chart, it doesn't matter how wide or narrow the bars are. The widths of the bars have no meaning.

- A final important difference is that a bar chart has gaps between the bars. This indicates that it is impossible to have observations between the categories. In a histogram, a gap indicates that no values were observed in the interval represented by the gap.

## SNAPSHOT  THE BAR CHART

| | |
|---|---|
| **WHAT IS IT?** ▶ | A graphical summary for categorical data. |
| **WHAT DOES IT DO?** ▶ | Shows a picture of the distribution of a categorical variable. |
| **HOW DOES IT DO IT?** ▶ | Each category is represented by a bar. The height of the bar is proportional to the number of times that category occurs in the data set. |
| **HOW IS IT USED?** ▶ | To see patterns of variation in categorical data. The categories can be presented in order of most frequent to least frequent, or they can be arranged in another meaningful order. |

## Pie Charts

Pie charts are another popular format for displaying relative frequencies of data. A **pie chart** looks, as you would expect, like a pie. The pie is sliced into several pieces, and each piece represents a category of the variable. The area of the piece is proportional to the relative frequency of that category. The largest piece in the pie in Figure 2.23 belongs to the category "Unknown" and takes up about 35% of the total pie.

Some software will label each slice of the pie with the percentage occupied. This isn't always necessary, however, because a primary purpose of the pie chart is to help us judge how frequently categories occur relative to one another. For example, the pie chart in Figure 2.23 shows us that "Unknown" occupies a fairly substantial portion of the whole data set. Also, labeling each slice gets cumbersome and produces cluttered graphs if there are many categories.

Although pie charts are very common (we bet that you've seen them before), they are not commonly used by statisticians or in scientific settings. One reason for this is that the human eye has a difficult time judging how much area is taken up by the wedge-shaped slices of the pie chart. Thus, in Figure 2.23, the "Sophomore" slice looks only slightly smaller than the "Junior" slice. But you can see from the bar chart (Figure 2.22) that they're actually quite different. Pie charts are also extremely difficult to use to compare the distribution of a variable across two different groups (such as comparing males and females as well as which classes the students come from).

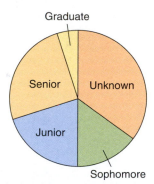

▲ **FIGURE 2.23** Pie chart showing the distribution of the categorical variable *Class* in a statistics course.

## SNAPSHOT  THE PIE CHART

| | |
|---|---|
| **WHAT IS IT?** ▶ | A graphical summary for categorical data. |
| **WHAT DOES IT DO?** ▶ | Shows the proportion of observations that belong to each category. |
| **HOW DOES IT DO IT?** ▶ | Each category is represented by a wedge in the pie. The area of the wedge is proportional to the relative frequency of that category. |
| **HOW IS IT USED?** ▶ | To understand which categories are most frequent and which are least frequent. Sometimes it is useful to label each wedge with the proportion of occurrence. |

# Summarizing Categorical Distributions

The concepts of *shape*, *center*, and *spread* that we used to summarize numerical distributions sometimes don't make sense for categorical distributions, because we can often order the categories any way we please. The center and shape would be different for every ordering of categories. However, we can still talk about typical outcomes and the variability in the sample.

## The Mode

When describing a distribution of a categorical variable, pay attention to which category occurs most often. This value, the one with the tallest bar in the bar chart, can sometimes be considered the "typical" outcome. There might be a tie for first place, and that's okay. It just means there's not as much variability in the sample. (Read on to see what we mean by that.)

The category that occurs most often is called the **mode**. This meaning of the word *mode* is similar to its meaning when we use it with numerical variables. However, one big difference between categorical and numerical variables is that we call a categorical variable bimodal only if two categories are nearly tied for most frequent outcomes. (The two bars don't need to be exactly the same height, but they should be very close.) Similarly, a categorical variable's distribution is multimodal if more than two categories all have roughly the tallest bars. For a numerical variable, the heights of the mounds do not need to be the same height for the distribution to be multimodal.

For an example of a mode, as part of their Internet and American Life Project, the Pew Research Center interviewed 2054 Americans in 2002, 2006, and 2007 to ask them how they used various electronic devices. One question they asked was whether it would be "very hard to give up" their cell phone, or Internet service, or television, or landline phone, or email service, or a wireless email device such as a BlackBerry. Figure 2.24 shows the results in 2002.

▶ **FIGURE 2.24** Percents of respondents who reported, in 2002, that they would find it very hard to give up a cell phone, access to the Internet, television, a landline phone, email service, or a wireless email device in 2002. Note that the percents do not add up to 100% because people could find it difficult to give up more than one of these devices (and some might even claim that every one of them would be hard to live without).

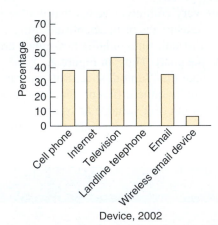

Device, 2002

We see that many people (about 60%) would find it very hard to give up their landline telephones. The landline phone is the mode of this distribution. Very few people in 2002 said they could not live without a BlackBerry (wireless email) device, whereas email, a cell phone, and access to the Internet were all considered about equally dispensable.

## EXAMPLE 7 Electronic Devices: What Can't You Live Without?

In 2007, the Pew survey asked a new group of people whether they would find it very hard to give up their electronic devices. The bar chart in Figure 2.25 shows the distribution of responses.

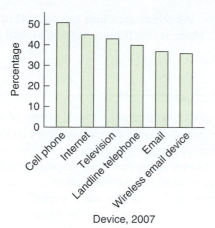

◄ **FIGURE 2.25** Percents of respondents who reported, in 2007, that they would find it very hard to give up a cell phone, access to the Internet, television, a landline phone, email service, or a wireless email device in 2007. Compare to Figure 2.24, where the data for 2002 are given.

**QUESTION** What is the typical response? How would you compare the responses in 2007 to to those in 2002?

**SOLUTION** The mode is cell phone, which 50% of the sample would find hard to give up. Between 2002 and 2007, the mode has changed from landline phone to cell phone.

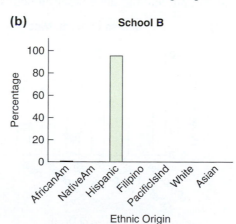

**TRY THIS!** Exercise 2.31

## Variability

When thinking about the variability of a categorical distribution, it is sometimes useful to think of the word *diversity*. If the distribution has a lot of diversity (many observations in many different categories), then variability is high. If the distribution has only a little diversity (many of the observations fall into the same category), then variability is low.

For example, Figure 2.26 shows bar charts of the ethnic composition of two schools in the Los Angeles City School System. Which school has the greater variability in ethnicity?

School A (Figure 2.26a) has much more diversity. School B (Figure 2.26b) consists almost entirely of a single ethnic group, with very small numbers of the other groups.

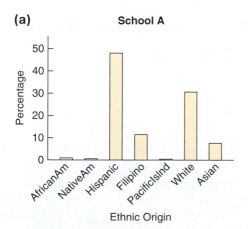

◄ **FIGURE 2.26** Percents of students at two Los Angeles elementary schools who are identified with several ethnic groups. Which school has more ethnic variability?

## EXAMPLE 8 Electronic Devices: Growing Toward Consensus?

Compare the responses to the Pew study in 2002 (Figure 2.24) with the responses in 2007 (Figure 2.25).

**QUESTION** In which year is more variability apparent? Explain.

**SOLUTION** In 2002, the bar chart shows that the landline phone was the clear favorite: there was little variability in the sample. But in 2007, while *cell phone* is, strictly speaking, the most frequent category, the others are close behind. All of these devices seem popular with the respondents. There is much more variability and diversity in choices in 2007 than in 2002.

**TRY THIS!** Exercise 2.33

You might be surprised by the answer to Example 8 because the bars in the 2007 bar chart are nearly all the same height, so it looks like things don't change much from category to category. However, variability is not just about the heights of the bars; variability is measured by how many different categories have responses in them. The most diverse distribution for categorical data would be represented by a bar chart in which all categories had bars of exactly the same height. The least variability would be represented by one in which there was only one bar, and all the other categories were empty.

## Describing Distributions of Categorical Variables

When describing a distribution of categorical data, you should mention the mode (or modes) and say something about the variability. Example 9 illustrates what we mean.

**KEY POINT** When summarizing graphs of categorical data, report the mode or modes and describe the variability (diversity).

### EXAMPLE 9 Causes of Death

According to some experts, about 51.5% of babies currently born in the United States are male. But among people between 100 and 104 years old, there are four times as many women as men (U.S. Census Bureau, 2000). How does this happen? One possibility is that the percent of boys born has changed over time. Another possibility is presented in the two bar charts in Figure 2.27. These bar charts show the numbers of deaths per 100,000 people in one year, for people aged 15–24 years, for both males and females.

**QUESTION** Compare the distributions depicted in Figure 2.27. Note that the categories are put into the same order on both graphs.

▶ FIGURE 2.27 The number of deaths per 100,000 males **(a)** and females **(b)** for people 15–24 years old in a one-year period.

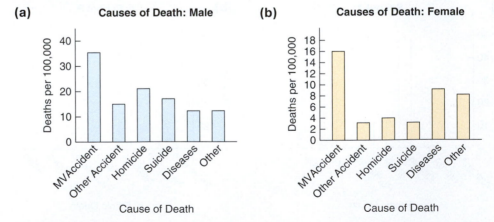

**SOLUTION** First, note that the histograms are not on the same scale, as you can see by comparing the values on the *y*-axes. This presentation is typical of most software and is sometimes desirable, because otherwise, one of the bar charts might have such small bars that we couldn't easily discern differences.

Although motor vehicle accidents are the mode for both groups, males show a consistently high death rate for all other causes of death, whereas females have relatively low death rates in the categories for other accident, homicide, and suicide. In other words, the cause of death for females is less variable than that for males. It is also worth noting that the death rates are higher for males in every category. For example, roughly 16 out of every 100,000 females died in a motor vehicle accident in one year, while roughly 35 out of every 100,000 males died in car accidents in the same year.

**TRY THIS!** Exercise 2.37

We can also make graphics that help us compare two distributions of categorical variables. When comparing two groups across a categorical variable, it is often useful to put the bars side-by-side, as in Figure 2.28. This graph makes it easier to compare rates of death for each cause. The much higher death rate for males is made clear.

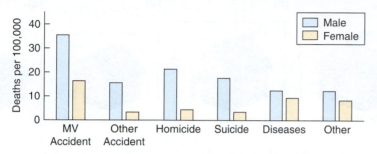

◄ **FIGURE 2.28** Death rates of males and females, graphed side by side.

# SECTION 2.5

# Interpreting Graphs

The first step in every investigation of data is to make an appropriate graph. Many analyses of data begin with visualizing the distribution of a variable, and this requires that you know whether the variable is numerical or categorical. When interpreting these graphics, you should pay attention to the center, spread, and shape.

Often, you will come across graphics produced by other people, and these can take extra care to interpret. In this section, we'll warn you about some potential pitfalls when interpreting graphs and show you some unusual visualizations of data.

## Misleading Graphs

A well-designed statistical graphic can help us discover patterns and trends and can communicate these patterns clearly to others. However, our eyes can play tricks on us, and manipulative people can take advantage of this to use graphs to give false impressions.

The most common trick—one that is particularly effective with bar charts—is to change the scale of the vertical axis so that it does not start at the origin (0). Figure 2.29a shows the number of violent crimes per year in the U.S. as reported by the FBI (http://www.fbi.gov). The graphic seems to indicate a dramatic drop in crime in 2009. Notice that the vertical axis starts at about 1,240,000 (1.24 million) crimes.

Because the origin begins at 1.24 million and not at 0, the bars are all shorter than they should be. This creates a much more dramatic apparent decline in crime, since the shortest bar is roughly one-tenth the height of the tallest bar. What does this chart look like if we make the bars the correct height? Figure 2.29b shows the same data, but to the correct scale. It's still clear that there has been a decline, but the decline doesn't look nearly so dramatic now, does it? Why not?

The reason is that when the origin is correctly set to 0, as in Figure 2.29b, it is clear that violent crimes were only slightly lower in 2009 than in 2008. The actual decline from 2008 to 2009 was about 5 percentage points. The largest percentage decline in several years, and certainly good news. But not as dramatically good as the first graph erroneously suggests.

**(a)**

**(b)**

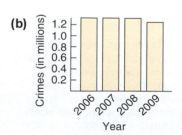

▲ **FIGURE 2.29** **(a)** This bar chart shows a dramatic decline in the number of violent crimes in 2009. The origin for the vertical axis begins at about 1.24 million, not at 0. **(b)** This bar chart reports the same data as part (a), but here the vertical axis begins at the origin (0).

Most of the misleading graphics you will run across exploit a similar theme: Our eye tends to compare the relative sizes of objects. Many newspapers and magazines like to use pictures to represent data. For example, the plot in Figure 2.30 attempts to illustrate the number of homes sold for some past years, with the size of each house representing the number of homes sold that year. Such graphics can be very misleading, because the pictures are often not to scale with the actual numbers.

In Figure 2.30 the *heights* of the homes are indeed proportional to the sales numbers, but our eye reacts to the *areas* instead. The smallest house is 69% as tall as the largest house (because 4.9 is 69% of 7.1), but the area of the smallest house is only about 48% of that of the largest house, so our tendency to react to area rather than to height exaggerates the difference.

**(a)**      **(b)**      **(c)**      **(d)**

▶ **FIGURE 2.30** Deceptive graphs: Image **(a)** represents 7.1 million homes sold in 2005, image **(b)** represents 6.5 million homes sold in 2006, image **(c)** represents 5.8 million homes sold in 2007, and image **(d)** represents 4.9 million homes sold in 2008. (Source: *L.A. Times*, April 30, 2008)

## The Future of Statistical Graphics

The Internet allows for a great variety of graphical displays of data that take us beyond simple visualizations of distributions. Many statisticians, computer scientists, and designers are experimenting with new ways to visualize data. Most exciting is the rise of interactive displays. The *State of the Union Visualization*, for example (http://stateoftheunion .onetwothree.net), makes it possible to compare the content of State of the Union speeches. Every U.S. president delivers a State of the Union address to Congress near the beginning of each year. This interactive graphic enables users to compare words from different speeches and "drill down" to learn details about particular words or speeches.

For example, Figure 2.31 is based on the State of the Union address that President Barack Obama delivered on January 27, 2010. The largest words are the words that appear most frequently in the speech. The words that appear to the left are words that typically appear earlier in the speech, and the words located to the right typically appear later. Thus we see that near the beginning, President Obama talked about the economy, using words such as *wall*, *street*, *recovery*, and *businesses*. Toward the end, there was talk about more general things, and we see *families*, *values*, and *deficit*. The word *Americans* is more

▶ **FIGURE 2.31** President Obama's 2010 State of the Union speech, visualized as a "word cloud." This array shows us the most commonly used words, approximately where these words occurred in the speech, how often they occurred, and how unusual they are compared to the content of other State of the Union speeches. (Courtesy of Brad Borevitz, http://onetwothree.net. Used with permission.)

or less in the middle, which suggests that it was used frequently throughout the speech. The vertical position of a word indicates how unusual it is compared to the content of other State of the Union addresses. Words that appear near the top, such as *businesses* and *jobs*, are words that are particular to this speech, compared to all other State of the Union addresses. Words near the bottom, such as *kids, talk,* and *cuts*, are words that occur relatively frequently throughout all State of the Union addresses.

Figure 2.31 is packed with information. On the bottom, in red, you see dots that represent the grade level required to read and understand the speech. Each dot represents a different speech, and the speeches are sorted from first (George Washington) to most recent (Barack Obama). Near the bottom right we see that the grade level of the speech is 9.0, or about ninth grade, and that this is considerably lower than it was 100 years ago but consistent with more recent speeches. The white "bars" represent the numbers of words in the State of the Union addresses. This speech contains 7068 words, which is fairly typical of the last few years. The tallest spike represents Jimmy Carter's speech in 1981, which had 33,613 words and a grade level of 15.3.

## CASE STUDY REVISITED

# Student-to-Teacher Ratio at Colleges

How do public colleges compare to private colleges in student-to-teacher ratios? The data were presented at the beginning of the chapter. This list of raw data makes it hard to see patterns, so it is very difficult to compare groups. But because the student-to-teacher ratio is a numerical variable, we can display these distributions as two histograms to enable us to make comparisons between public and private schools.

In Figure 2.32, the student-to-teacher ratio for the private schools is shown in the left histogram, and that for state schools is shown in the right histogram. The distributions are quite different. The typical student-to-teacher ratio for private schools is around 10 or 11 students per teacher, while for state schools it's around 16 or 20. Although both schools have quite a bit of spread in these ratios, there may be a little less spread—a little less variation—in the state schools. The private schools' distribution is right-skewed. The state schools' distribution appears bimodal (at least with these bins), with mounds centered at about 16 and 20 students per teacher. It makes us wonder whether there are two kinds of state colleges.

▲ **FIGURE 2.32** Histograms of the student-to-teacher ratio at samples of private colleges and state colleges.

By the way, many of the schools with the smallest student-to-teacher ratios deal with specialties such as health or design. In case you are wondering about the state college with the smallest student-to-teacher ratio in Figure 2.32b, it is the University of Colorado at Denver Health Science Center. At medical facilities like this one, the physicians are often listed as faculty. The private school with the 54-to-1 ratio is Western Governors University at Salt Lake City, Utah. It is an online university offering accredited courses in distance learning.

# EXPLORING STATISTICS
## CLASS ACTIVITY

### Personal Distance

How much personal distance do people require when they're using an automatic teller machine?

| GOALS | MATERIALS |
|---|---|
| In this activity, you will learn to make graphs of sample distributions in order to answer questions about data in comparing two groups of students. | Meter stick (or tape measure). |

**ACTIVITY**

Work in groups of three students. Each group must have a meter stick. The first person stands (preferably in front of a wall) and imagines that she or he is at an ATM getting cash. The second student stands behind the first. The first student tells the second student how far back he or she must stand for the first student to be just barely comfortable, saying, for example, "Move back a little, now move forward just a tiny bit," and so on. When that distance is set, the third student measures the distance between the heel of the first person's right shoe to the toe of the second person's right shoe. That will be called the "personal distance."

For each student in your group, record the gender and personal distance. Your instructor will help you pool your data with the rest of the class.

*Note:* Be respectful of other people's personal space. Do not make physical contact with other students during this activity.

**BEFORE THE ACTIVITY**

1. Do you think men and women will have different personal distances? Will the larger distances be specified by the men or the women?

2. Which group will have distances that are more spread out?

3. What will be the shape of the distributions?

**AFTER THE ACTIVITY**

Do men and women have different personal distances? Create appropriate graphics to compare personal distances of men and women to answer this question. Then describe these differences.

# CHAPTER REVIEW

## Key Terms

distribution of a sample, *36*
frequency, *36*
dotplot, *37*
histogram, *38*
relative frequency, *39*
stemplot (or stem-and-leaf plot), *41*

leaf, *41*
stem, *41*
typical value (center), *41*
variability, *41*
symmetric distribution, *42*
bell-shaped distribution, *42*

right-skewed distribution, *43*
left-skewed distribution, *43*
unimodal distribution, *44*
bimodal distribution, *44*
multimodal distribution, *44*
outlier, *45*

bar graph (bar chart), *49*
Pareto chart, *50*
pie chart, *51*
mode (in categorical variables), *52*

## Learning Objectives

After reading this chapter and doing the assigned homework problems, you should

- Understand that a distribution of a sample of data displays a variable's values and the frequencies (or relative frequencies) of those values.

- Know how to make graphs of distributions of numerical and categorical variables and how to interpret the graphs in context.

- Be able to compare centers and spreads of distributions of samples informally.

## Summary

The first step in any statistical investigation is to make plots of the distributions of the data in your data set. You should identify whether the variables are numerical or categorical so that you can choose an appropriate graphical representation.

If the variable is numerical, you can make a dotplot, histogram, or stemplot. Pay attention to the shape (Is it skewed or symmetric? Is it unimodal or multimodal?), to the center (What is a typical outcome?), and to the spread (How much variability is present?). You should also look for unusual features, such as outliers.

Be aware that many of these terms are deliberately vague. You might think a particular observation is an outlier, but another

person might not agree. That's okay, because the purpose isn't to determine whether such points are "outliers" but to indicate whether further investigation is needed. An outlier might, for example, be caused by a typing error someone made when entering the data.

If you see a bimodal or multimodal distribution, ask yourself whether the data might contain two or more groups combined into the single graph.

If the variable is categorical, you can make a bar chart, a Pareto chart (a bar chart with categories ordered from most frequent to least frequent), or a pie chart. Pay attention to the mode (or modes) and to the variability.

## Sources

Internet Research Legal Group. 2008. http://www.ilrg.com/rankings/law/index.php/1/desc/LSATLow/2006

National Safety Council. 2002. *Injury Facts*, 2002 edition. Itasca, Illinois.

Playfair, W. 1786. Commercial and political atlas: Representing, by copperplate charts, the progress of the commerce, revenues, expenditure, and debts of England, during the whole of the eighteenth century. London: Corry. Reprinted in H. Wainer and I. Spence (eds.), *The commercial and political atlas and statistical breviary*, New York: Cambridge University Press, 2005. ISBN 0-521-85554-3. http://www.math.yorku.ca/SCS/Gallery/milestone/refs.html#Playfair:1786

Tufte, E. 1983. *The visual display of quantitative information*, Graphics Press: Cheshire, Connecticut. 1st ed., p. 141. Photo by Brian Joiner.

Minitab 15 Statistical Software (2007). [Computer software]. State College, PA: Minitab, Inc. (www.minitab.com)

*World Almanac and Book of Facts*. 2006. World Almanac.

U.S. Census Bureau. 2000. http://ceic.mt.gov/C2000/SF12000/Pyramid/pptab00.htm

# SECTION EXERCISES

## SECTIONS 2.1 AND 2.2

**2.1 Body Mass Index** The dotplot shows body mass index (BMI) for 134 people according to the National Health and Nutrition Examination Survey (NHANES) in 2010, as reported in *USA Today*.

a. A BMI of more than 40 is considered morbidly obese. Report the number of morbidly obese shown in the dotplot.

b. Report the percentage of people who are morbidly obese. Compare this with an estimate from 2005 that 3% of people in the United States at that time were morbidly obese.

**2.2 Cholesterol Levels** The dotplot shows the cholesterol level of 93 adults from the 2010 NHANES data.

a. A total cholesterol level of 240 mg/dL (milligrams per deciliter) or more is considered unhealthy. Report the number of people in this group with unhealthy cholesterol levels.

b. Knowing there are a total of 93 people in this sample, report the percentage of people with unhealthy total cholesterol levels. How does this compare with an estimate from 2010 that 18% of people in the United States had unhealthy cholesterol levels?

**2.3 Ages of CEOs** The histogram shows frequencies for the ages of 25 CEOs listed at Forbes.com. Convert this histogram to one showing relative frequencies by relabeling the vertical axis with the appropriate relative frequencies. You may just report the new labels for the vertical axis because that is the only thing that changes.

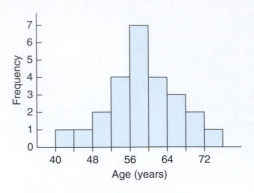

**2.4 Private College Tuition** The histogram shows the distribution of tuition at 85 private colleges, as reported in the *2006 World Almanac and Book of Facts*. Approximately how many colleges charged $30,000 or more in tuition?

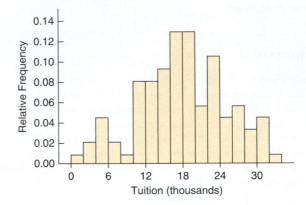

**TRY  2.5 Televisions (Example 1)** The histogram shows the distribution of the number of televisions in the homes of 90 community college students.

a. According to the histogram, about how many homes do not have a television?

b. How many televisions are in the homes that have the most televisions?

c. How many homes have three televisions?

d. How many homes have six or more televisions?

e. What proportion of homes have six or more televisions?

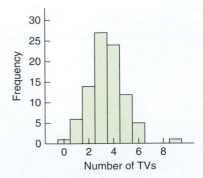

**2.6 Exercise Hours** The histogram shows the distribution of self-reported number of hours of exercise per week for 50 community college students. This graph uses a right-hand rule: If someone exercised for exactly 4 hours, they would be in the third bin, the bin to the right of 4.

Exercise Hours per Week

a. According to the histogram, there are two possible values for the maximum number of hours of exercise. What are they?

b. How many people exercised 0 or 1 hour (less than 2 hours)?

c. How many people exercised 10 or more hours?

d. What proportion of people exercised 10 or more hours?

**2.7 Rental Prices** Have you ever wondered whether you could afford to move to another city? The dotplot shows rental prices (dollars per month) in three cities in December 2009 for units with at least one bedroom and one bathroom. (Source: Realtor.com)

Dollars per Month

a. Center: Which city typically has the highest rents?

b. Spread: Which city's distribution has the smallest variation?

c. Shape: Which city's distribution is the least skewed?

**2.8 Rental Prices** The dotplot shows rental prices (dollars per month) in three cities in December 2009 for units with at least one bedroom and one bathroom. (Source: Realtor.com)

Dollars per Month

a. Center: Which city typically has the highest rental prices?

b. Which city's distribution has the greatest variation?

c. Describe the shape of the distribution of prices in Houston.

TRY **2.9 CEO Salaries (Example 2)** Predict the shape of the distribution of the salaries of 25 chief executive officers (CEOs).

A typical value is about 50 million per year, but there is an outlier at about 200 million.

**2.10 Cigarettes** A physician asks all of his patients to report the approximate number of cigarettes they smoke in a day. Predict the shape of the distribution of number of cigarettes smoked per day.

TRY **2.11 Armspans (Example 3)** According to the ancient Roman architect Vitruvius, a person's armspan (the distance from fingertip to fingertip with the arms stretched wide) is approximately equal to his or her height. For example, people 5 feet tall tend to have an armspan of 5 feet. Explain, then, why the distribution of armspans for a class containing roughly equal numbers of men and women might be bimodal.

★ **2.12 Tuition** The distribution of in-state annual tuition for all colleges and universities in the United States is bimodal. What is one possible reason for this bimodality?

TRY **2.13 Ages of CEOs (Example 4)** From the histogram in Exercise 2.3, approximately what is a typical age of a CEO in this sample?

**2.14 Private College Tuition** From the histogram shown in Exercise 2.4, approximately what is a typical tuition, in 2005, for a private college?

TRY **2.15 Commute Times (Example 5)** Use the histograms to compare the times spent commuting for community college students who drive to school in a car with the times spent by those who take the bus. Which group typically has the longer commute time? Which group has the more variable commute time?

Time for Commute by Car (minutes)

Time for Commute by Bus (minutes)

**2.16 Spending on Clothes** The histograms show the distribution of the estimated number of dollars per month spent on clothes for college women (left) and college men (right).

a. Compare and describe the shape of the distributions.

b. Which group tends to spend more?

c. Which group has more variation in its expenses?

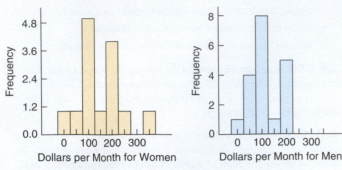

Dollars per Month for Women    Dollars per Month for Men

**2.17 Education (Example 6)** In 2008, the General Social Survey (GSS), a national survey conducted nearly every year, reported the number of years of formal education for 2018 people. The histogram shows the distribution of data.

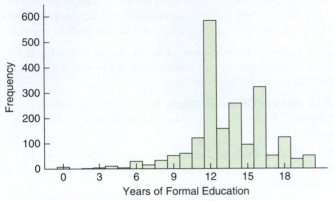

a. Describe and interpret the distribution of years of formal education. Mention any unusual features.

b. Assuming that those with 16 years of education completed a bachelor's degree, estimate how many of the people in this sample got a bachelor's degree or higher.

c. The sample includes 2018 people. What percentage of people in this sample have a bachelor's degree or higher? How does this compare with Wikipedia's estimate that 27% have a bachelor's degree?

**2.18 Siblings** The histogram shows the distribution of the numbers of siblings (brothers and sisters) for 2000 adults surveyed in the 2008 General Social Survey.

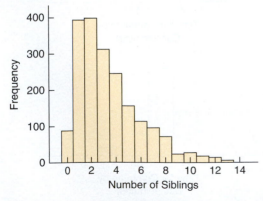

a. Describe the shape of the distribution.

b. What is the typical number of siblings, approximately?

c. About how many people in this survey have no siblings?

d. What percentage of the 2000 adults surveyed have no siblings?

**\*2.19 Years of Education** The GSS asked people how many years of education they had and how many years their mothers had.

If people who responded to the survey (respondents) completed high school but had no further education, they reported 12 years of education. If they stopped after a bachelor's degree, they reported 16 years. There were 2018 people who answered the question about their own education and 1780 who answered the question about the education of their mothers. Compare the distribution of years of education of the respondents with the distribution of years of education of their mothers.

**Education of Respondent**

**Education of Respondent's Mother**

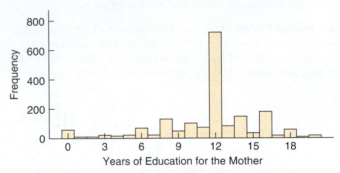

**\*2.20 Hours Worked** In the 2008 General Social Survey, 636 male paid employees and 537 female paid employees were asked how many hours they worked in the last week. (Those who said, "I don't know" or "I don't work" were not included in the data set.) Compare the distributions of hours of work for the men and the women.

**2.21 Matching Histograms** Match each of the following histograms to the correct situation.

1. The age of death of a sample of 19 typical women in the U.S.
2. The yearly tuition (according to the 2006 *World Almanac and Book of Facts*) for 142 colleges, 85 of which are private and 57 of which are state-supported.
3. The outcomes of rolling a fair die (with six sides) 5000 times.

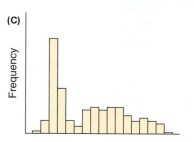

(A)   (B)

(C)

**2.22 Matching Histograms** Match each of the following histograms to the correct situation.

1. Test scores for students on an easy test.
2. The numbers of hours of television watched by a large, typical group of Americans.
3. The heights of a large, typical group of adults in the U.S.

(A)   (B)

(C)

**g 2.23 Eating Out and Jobs** College student Jacqueline Loya asked students who had full-time jobs and students who had part-time jobs how many times they went out to eat in the last month.

Write a brief comparison of the distributions of the two groups. Include appropriate graphics. *See page 71 for guidance.*

Full-time: 5, 3, 4, 4, 4, 2, 1, 5, 6, 5, 6, 3, 3, 2, 4,
5, 2, 3, 7, 5, 5, 1, 4, 6, 7
Part-time: 1, 1, 5, 1, 4, 2, 2, 3, 3, 2, 3, 2, 4,
2, 1, 2, 3, 2, 1, 3, 3, 2, 4, 2, 1

**2.24 Comparing Weights of Baseball and Soccer Players** College students Edward Lara and Anthony Dugas recorded the self-reported weights (in pounds) of some community college male baseball players and male soccer players. Write a brief comparison of the distributions of weight for the two groups. Include appropriate graphics. For one approach, look at the guidance given for exercise 2.23.

**Baseball Players' Weights**

| 190 | 205 | 230 | 198 | 195 |
|-----|-----|-----|-----|-----|
| 200 | 185 | 195 | 180 | 182 |
| 187 | 177 | 169 | 182 | 193 |
| 181 | 207 | 186 | 193 | 190 |
| 192 | 225 | 210 | 200 | 186 |

**Soccer Players' Weights**

| 165 | 189 | 167 | 173 | 184 |
|-----|-----|-----|-----|-----|
| 190 | 170 | 190 | 158 | 174 |
| 185 | 182 | 185 | 150 | 190 |
| 187 | 172 | 156 | 172 | 156 |
| 183 | 180 | 168 | 180 | 163 |

**2.25 Textbook Prices** The table shows prices of 50 college textbooks in a community college bookstore, rounded to the nearest dollar. Make an appropriate graph of the distribution of the data, and describe the distribution.

| 76 | 19 | 83 | 45 | 88 | 70 | 62 | 84 | 85 | 87 |
|----|----|----|----|----|----|----|----|----|----|
| 86 | 37 | 88 | 45 | 75 | 83 | 126 | 56 | 30 | 33 |
| 26 | 88 | 30 | 30 | 25 | 89 | 32 | 48 | 66 | 47 |
| 115 | 36 | 30 | 60 | 36 | 140 | 47 | 82 | 138 | 50 |
| 126 | 66 | 45 | 107 | 112 | 12 | 97 | 96 | 78 | 60 |

**2.26 SAT scores** The table shows a random sample of 50 quantitative SAT scores of first-year students admitted at a university. Make an appropriate graph of the distribution of the data, and describe the distribution.

| | | | | | | | | | |
|---|---|---|---|---|---|---|---|---|---|
| 649 | 557 | 734 | 653 | 652 | 538 | 674 | 705 | 729 | 737 |
| 672 | 583 | 729 | 677 | 618 | 662 | 692 | 692 | 672 | 624 |
| 669 | 529 | 609 | 526 | 665 | 724 | 557 | 647 | 719 | 593 |
| 624 | 611 | 490 | 556 | 630 | 602 | 573 | 575 | 665 | 620 |
| 629 | 593 | 665 | 635 | 700 | 665 | 677 | 653 | 796 | 601 |

**2.27 Animal Longevity** The accompanying table shows the average lifespan for some mammals in years, according to infoplease.com. Graph these average lifespans and describe the distribution. What is a typical lifespan? Identify the three outliers and report their lifespans. If you were to include humans in this graph, where would the data point be? Humans average about 75 years.

**2.28 Animal Gestation Periods** The accompanying table also shows the gestation period (in days) for some animals. The gestation period is the length of pregnancy. Graph the gestation period and describe the distribution. If there are any outliers, identify the animal(s) and give their gestation periods. If you were to include humans in this graph where would the data point be? The human gestation period is about 266 days.

| Animal | Gestation (days) | Lifespan (years) | Animal | Gestation (days) | Lifespan (years) |
|---|---|---|---|---|---|
| Baboon | 187 | 12 | Hippo | 238 | 41 |
| Bear, grizzly | 225 | 25 | Horse | 330 | 20 |
| | | | Leopard | 98 | 12 |
| Beaver | 105 | 5 | Lion | 100 | 15 |
| Bison | 285 | 15 | Monkey, rhesus | 166 | 15 |
| Camel | 406 | 12 | | | |
| Cat, domestic | 63 | 12 | Moose | 240 | 12 |
| Chimp | 230 | 20 | Pig, domestic | 112 | 10 |
| Cow | 284 | 15 | Puma | 90 | 12 |
| Deer | 201 | 8 | Rhino, black | 450 | 15 |
| Dog, domestic | 61 | 12 | | | |
| Elephant, African | 660 | 35 | Sea Lion, Cal | 350 | 12 |
| | | | Sheep | 154 | 12 |
| Elephant, Asian | 645 | 40 | Squirrel, gray | 44 | 10 |
| Elk | 250 | 16 | Tiger | 105 | 16 |
| Fox, red | 52 | 7 | Wolf, maned | 63 | 5 |
| Giraffe | 457 | 10 | | | |
| Goat | 151 | 8 | Zebra, Grant's | 365 | 15 |
| Gorilla | 258 | 20 | | | |

**2.29 Matching** Match each description with the correct histogram of the data.

1. Heights of students in a large UCLA statistics class that contains about equal numbers of men and women.

2. Numbers of hours of sleep the previous night in the same large statistics class.

3. Numbers of driving accidents for students in a large university in the U.S.

**(A)**

**(B)**

**(C)**

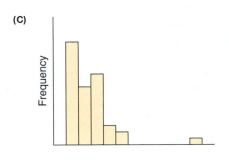

**2.30 Matching** Match each description with the correct histogram of the data.

1. Quantitative SAT scores for 1000 students who have been accepted to a private university.

2. Weights of over 500 adults, about half of whom are men and half of whom are women.

3. Ages of all 39 students in a community college statistics class that is made up of full-time students and meets during the morning.

(A)

(B)

(C)

## SECTIONS 2.3 AND 2.4

TRY **2.31 Changing Multiple-Choice Answers When Told *Not* to Do So (Example 7)** One of the authors wanted to determine the effect of changing answers on multiple-choice tests. She studied the tests given by another professor, who had told his students before their exams that if they had doubts about an answer they had written, they would be better off *not changing* their initial answer. The author went through the exams to look for erasures, which indicate that the first choice was changed. In these tests, there is only one correct answer for each question. Do the data support the view that students should not change their initial choice of an answer?

**2.32 Changing Multiple-Choice Answers When Told to Do So** One of the authors wanted to determine the effect of changing answers on multiple-choice tests. She had advised her students that if they had changed their minds about a previous answer, they should replace their first choice with their new choice. By looking for erasures on the exam, she was able to count the number of changed answers that went from wrong to right, from right to wrong, and from wrong to wrong. The results are shown in the bar chart.

a. Do the data support her view that it is better to replace your initial choice with the revised choice?

b. Compare this bar chart with the one in Exercise 2.31. Does changing answers generally tend to lead to higher or to lower grades?

TRY **2.33 Search Engines (Example 8)** The bar chart shows market share for the search engines used in China. (Source: Data from *Wall Street Journal*, April 27, 2010)

a. Report the mode for both time periods. What changes do you see from 2009 to 2010?

b. In which time period, 2009 (fourth quarter) or 2010 (first quarter) was there more variability in the market share for search engines used for computers in China?

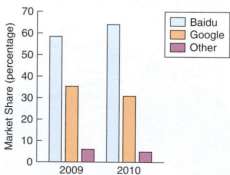

**2.34 Browsers** The bar chart shows the market share of browsers according to w3schools, an online Web development school. Write a sentence or two describing changes in the market share of these browsers from 2007 to 2010. (Source: Data from www.w3schools.com)

**Market Share: Browsers**

**2.35 Political Party Affiliation: Men** The 2008 General Social Survey (GSS) asked its respondents to report their political party affiliation. The graphs show the results for 926 men.

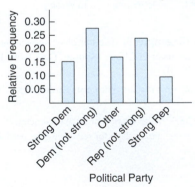

**Bar Chart of Political Party for Men**

**Pie Chart of Political Party for Men**

a. Which political affiliation has the most men? Is this easier to determine with the bar chart or with the pie chart? Why?

b. Which political affiliation has the second largest number of men?

**2.36 Political Party Affiliation: Women** The 2008 GSS asked its participants to report their political party affiliation. The graphs show the distribution of political party affiliation for 1084 women.

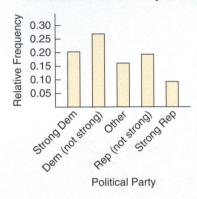

**Bar Chart of Political Party for Women**

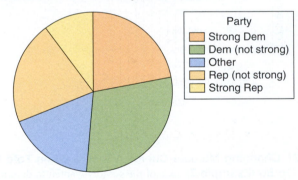

**Pie Chart of Political Party for Women**

a. Which political affiliation has the most women? Is this easier to see from the bar chart or the pie chart?

b. Which political affiliation has the second largest number of women?

c. Some people believe that women tend to lean more towards liberal political positions (like Democrats). By comparing the shape of the distribution of political affiliation for women to that for men (Exercise 2.35), do you see evidence of this? Explain.

TRY **2.37 Age by Year (Example 9)** The bar chart shows the projected percentage of U.S. residents in different age categories by year, according to the 2009 *World Almanac and Book of Facts*.

a. Comment on the predicted changes from 2010 through to 2030. Which age groups are predicted to become larger, which are predicted to become smaller, and which are predicted to stay roughly the same?

b. Comment on the effect this might have on Social Security, a government program that collects money from those currently working and gives it to retired people.

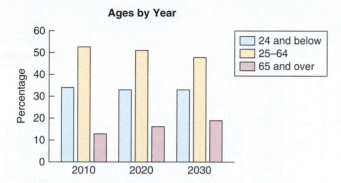

Ages by Year

**2.38 Retail Car Sales** With gas prices rising, as they did between 1985 and 2007, you might expect people to move toward buying smaller cars. Compare the types of cars sold in 1985, 2000, and 2007 as shown in the figure. (Source: 2009 *World Almanac and Book of Facts*.)

a. Which type of car sold the most in all three years?

b. What is the trend for small cars? Has a higher or a lower percentage of small cars sold in more recent years?

c. What is the trend for large cars?

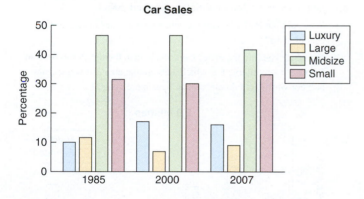

Car Sales

**2.39 Majors** The table gives information on college majors at Wellesley College, a women's college outside of Boston. Sketch an appropriate graph of the distribution, and comment on its important features.

| Major | Percentage |
|---|---|
| Humanities | 30% |
| Social Science | 38% |
| Math and Science | 15% |
| Interdisciplinary | 17% |

**2.40 Adoptions** The table gives information on the top five countries from which U.S. residents adopted children in 2007. Sketch an appropriate graph of the distribution, and comment on its important features. (Source: 2009 *World Almanac and Book of Facts*)

| Country | Number |
|---|---|
| China | 5453 |
| Guatemala | 4728 |
| Russia | 2310 |
| Ethiopia | 1255 |
| South Korea | 939 |

## CHAPTER REVIEW EXERCISES

**2.41 Television** The table shows the first few entries for the number of hours of television viewed per week for some fifth-grade students, stacked and coded, where 1 represents a girl and 0 represents a boy.

What would be appropriate graphs to compare the distributions of hours of TV watched per week for boys and girls if you had all the data? Explain.

| TV | Girl |
|----|------|
| 3  | 1    |
| 8  | 1    |
| 11 | 1    |
| 12 | 1    |
| 7  | 0    |
| 5  | 0    |
| 4  | 0    |

**2.42 Job** The table shows the job categories for some employees at a business. What type of graph(s) would be appropriate to compare the distribution of jobs for men and for women if you had all of the data? Explain.

| Male | Job |
|------|-----|
| 1    | Custodial  |
| 0    | Clerical   |
| 0    | Managerial |
| 0    | Clerical   |
| 1    | Managerial |

**2.43 Hormone Replacement Therapy** The use of the drug Prempro, a combination of two female hormones that many women take after menopause, is called hormone replacement therapy (HRT). In July 2002, a medical article reported the results of a study that was done to determine the effects of Prempro on many diseases. (Source: Women's Writing Group, Risks and benefits of estrogen plus progestin in healthy postmenopausal women, *JAMA*, 2002).

The study was placebo-controlled, randomized, and double-blind. From studies like these, it is possible to make statements about cause and effect. The figure shows comparisons of disease rates in the study.

a. For which diseases was the disease rate higher for those who took HRT? And for which diseases was the rate lower for those who took HRT?

b. Why do you suppose we compare the rate per 10,000 women (per year), rather than just reporting the numbers of women observed who get the disease?

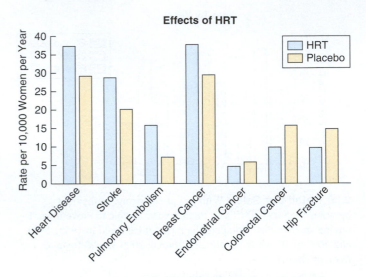

**2.44 ER Visits for Injuries** The graph shows the rates of visits to the ER for injuries by gender and by age. Note that we are concerned with the rate per 100 people of that age and gender in the population. (Source: National Safety Council 2004)

a. Why does the National Safety Council give us rates instead of numbers of visits?

b. For which ages are the males more likely than the females to have an ER visit for an injury? For which ages are the men and women similar? For which ages do the women have more visits?

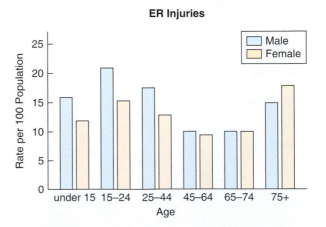

**2.45 Hormone Replacement Therapy Again** The bar chart shows a comparison of breast cancer rates for those who took HRT and those who took a placebo. Explain why the graph is deceptive, and indicate what could be done to make it less so.

**Bar Chart of Rate of Breast Cancer**

**2.46 Holding Breath** A group of students held their breath as long as possible and recorded the times in seconds. The times went from a low of 25 seconds to a high of 90 seconds, as you can see in the stemplot. Suggest improvements to the histogram below generated by Excel, assuming that what is wanted is a histogram of the data (not a bar chart).

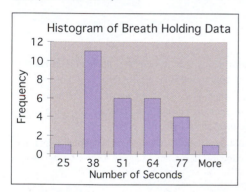

Histogram of Breath Holding Data

| 2 | 588 |
| 3 | 00035678 |
| 4 | 0033557 |
| 5 | 3357 |
| 6 | 005559 |
| 7 | |
| 8 | |
| 9 | 0 |

**2.47 Gender**

a. An elementary school class includes 12 girls and 8 boys. Which type of graph would be appropriate to show the distribution of gender in the class: a pie chart or a histogram?

b. In a school district, there are 45 elementary school classrooms, each containing 20 students. Which type of graph would be appropriate to show the distribution of the number of boys in the classrooms: a pie chart or a histogram?

**2.48 Handedness**

a. An elementary school class is made up of 3 children who are left-handed and 17 who are right-handed. The objects of study are the children, and the variable of interest is whether the child is right- or left-handed. Which type of graph would be appropriate to show the distribution of handedness in the class: a pie chart or a histogram?

b. In a school district, there are 45 elementary school classrooms, each containing 20 students. The unit of study is a classroom, and the variable is the number of right-handed students in each class. Which type of graph would be appropriate to show the distribution of the number of right-handed students in the classrooms: a pie chart or a histogram?

\* **2.49 Global Temperatures** The histograms show the average global temperature per year for two 26-year ranges in degrees Fahrenheit. The range for 1880 to 1905 is on the top, and the range for 1980 to 2005 is on the bottom. Compare the two histograms for the two time periods, and explain what they show. Also estimate the difference between the centers. That is, about how much does the typical global temperature for the 1980–2005 time period differ from that for the 1880–1905 period? (Source: Goddard Institute for Space Studies, NASA, 2009)

**1880–1905**

**1980–2005**

\* **2.50 Employment after Law School** Accredited law schools were ranked from 1 for the best (Harvard) down to number 181 by the Internet Research Legal Group (ILRG). When you decide on a law school to attend, one of the things you might be interested in is whether, after graduation, you will be able to get a job for which your law degree is required. We split the group of 181 law schools in half, with the top-ranked schools in one group and the lower ranked schools in the other. The histograms show the distribution of the percentages of graduates who, 9 months after graduating, have obtained jobs that require a law degree. Compare the two histograms.

**Top 90 Law Schools**

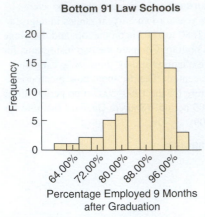

**Bottom 91 Law Schools**

Percentage Employed 9 Months after Graduation

**2.51** Create a dotplot that has at least 10 observations and is right-skewed.

**2.52** Create a dotplot that has at least 10 observations and does not have skew.

**\* 2.53 Home Prices** How do home prices compare in West Los Angeles (WLA) and Midtown (MT) Los Angeles? Using appropriate graphics, write a few sentences comparing the distribution of home prices. The data are sampled from a real estate website, and prices are in hundreds of thousands of dollars. The data are on the disk that comes with the book. (Thus 10 represents $1,000,000.)

**\* 2.54 Traffic Cameras** College students Jeannette Mujica, Ricardo Ceja Zarate, and Jessica Cerda conducted a survey in Oxnard, California, of the number of cars going through a yellow light at intersections with and without traffic cameras that are used to automatically send notification of fines to drivers who run red lights. The cameras were very noticeable to drivers. The amount of traffic was constant throughout the study period (the afternoon commute.) The data record the number of cars that crossed the intersection during a yellow light for each light cycle. A small excerpt of the data is shown in the table; see the disk for all the data. What differences, if any, do you see between intersections with cameras and those without? Use an appropriate graphical summary, and write a comparison of the distributions.

| # Cars | Cam |
|--------|--------|
| 1 | Cam |
| 2 | Cam |
| 1 | No Cam |
| 3 | No Cam |

**\* 2.55 Forbes 100 Companies** The 100 companies with the highest net incomes (for 2008) were obtained from Forbes.com. First predict the shape of a histogram of the data, and explain your prediction. Then make a histogram of the data (on the disk) and describe it. Mention the names of any companies that are outliers.

**\* 2.56 Ideal Weight** Thirty-nine students (26 women and 13 men) reported their ideal weight (in most cases, not their current weight). The tables show the data.

| Women | | | | | |
|-----|-----|-----|-----|-----|-----|
| 110 | 115 | 123 | 130 | 105 | 110 |
| 130 | 125 | 120 | 115 | 120 | 120 |
| 120 | 110 | 120 | 150 | 110 | 130 |
| 120 | 118 | 120 | 135 | 130 | 135 |
| 90 | 110 | | | | |

| Men | | | | | |
|-----|-----|-----|-----|-----|-----|
| 160 | 130 | 220 | 175 | 190 | 190 |
| 135 | 170 | 165 | 170 | 185 | 155 |
| 160 | | | | | |

a. Explain why the distribution of ideal weights is likely to be bimodal if men and women are both included in the sample.

b. Make a histogram combining the ideal weights of men and women. Use the default histogram provided by your software. Report the bin width and describe the distribution.

c. Vary the number of bins, and print out a second histogram. Report the bin width and describe this histogram. Compare the two histograms.

# gUIDED EXERCISES

**g 2.23 Eating Out and Jobs** College student Jacqueline Loya asked students who had full-time jobs and students who had part-time jobs how many times they went out to eat in the last month.

Full-time: 5, 3, 4, 4, 4, 2, 1, 5, 6, 5, 6, 3, 3, 2, 4, 5, 2, 3, 7, 5, 5, 1, 4, 6, 7

Part-time: 1, 1, 5, 1, 4, 2, 2, 3, 3, 2, 3, 2, 4, 2, 1, 2, 3, 2, 1, 3, 3, 2, 4, 2, 1

**Question** Compare the two groups by following the steps below. Include appropriate graphics.

**Step 1 ▶ Create graphs.**
Make dotplots (or histograms) using the same axis with one set of data above the other, as shown in the figure.

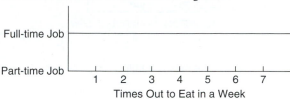

**Step 2 ▶ Examine shape.**
What is the shape of the data for those with part-time jobs? What is the shape of the data for those with full-time jobs?

**Step 3 ▶ Examine center.**
Which group tends to go out to eat more often, those with full-time or those with part-time jobs?

**Step 4 ▶ Examine variation.**
Which group has a wider spread of data?

**Step 5 ▶ Check for outliers.**
Were there any numbers separated from the other numbers as shown in the dotplots? (In other words, were there any gaps in the dotplots?)

**Step 6 ▶ Summarize.**
Finally, in one or more sentences, compare the shape, center, and variation (and mention outliers if there were any).

# TechTips

## General Instructions for All Technology

**EXAMPLE A ▶** Use the following ages to make a histogram:

$$7, 11, 10, 10, 16, 13, 19, 22, 42$$

### Columns
Data sets are generally put into columns, not rows. The columns may be called variables, lists, or the like. Figure 2A shows a column of data.

▲ **FIGURE 2A** TI-83/84
Data Screen

### TI-83/84

**Resetting the Calculator (Clearing the Memory)**
If you turn off the TI-83/84, it does not reset the calculator. All the previous data and choices remain. If you are having trouble and want to start over, you can reset the calculator.

1. **2nd Mem** (with the + sign)
2. **7** for **Reset**
3. **1** for **All RAM**
4. **2** for **Reset**

It will say **RAM cleared** if it has been done successfully.

**Entering Data into the Lists**
1. Press **STAT**, and select **EDIT** (by pressing **ENTER** when **EDIT** is highlighted).
2. If you find that there are data already in **L1** (or any list you want to use), you have three options:
   a. Clear the entire list by using the arrow keys to highlight the **L1** label and then pressing **CLEAR** and then **ENTER**. Do not press **DELETE**, because then you will no longer have an **L1**. If you delete **L1**, to get it back you can **Reset** the calculator.

b. Delete the individual entries by highlighting the top data entry, then pressing **DELETE** several times until all the data are erased. (The numbers will scroll up.)

c. Overwrite the existing data. CAUTION: Be sure to **DELETE** data in any cells not overwritten.

3. Type the numbers from the example into **L1** (List1). After typing each number, you may press **ENTER** or use ▼ (the arrow down on the keypad). Double-check your entries before proceeding.

### Histogram

1. Press **2nd**, **STATPLOT** (which is in the upper left corner of the keypad).

2. If more than the first plot is **On**, press **4** (for **PlotsOff**) and **ENTER** to turn them all off.

3. Press **2nd**, **STATPLOT**, and **1** (for **Plot 1**).

4. Turn on **Plot1** by pressing **ENTER** when **On** is flashing; see Figure 2B.

▲ **FIGURE 2B** TI-83/84

5. Use the arrows on the keypad to get to the histogram icon (highlighted in Figure 2B) and press **ENTER** to choose it. Use the down arrow to get to Xlist and press **2nd** and **1** for **L1**. The settings shown in Figure 2B will lead to a histogram of the data in List 1 (L1).

6. Press **ZOOM** and **9** (**ZoomStat**) to create the graph.

7. To see the numbers shown in Figure 2C, press **TRACE** (in the top row on the keypad) and move around using the group of four arrows on the keypad to see other numbers.

Figure 2C shows a histogram of the numbers in the example.

▲ **FIGURE 2C** TI-83/84 Histogram

The TI-83/84 cannot make stemplots, dotplots, or bar charts.

# Downloading Numerical Data from a Computer into a TI-84

Before you can use your computer with your TI-84, you must install (on your computer) the software program, TI Connect™, and a driver for the calculator. You need do this only once. If you have done steps 1–3 and are ready to download the data, start with Step 4.

1. Downloading and saving the TI Connect™ setup program. Insert the CD that came with the TI-84 into the computer disk drive and follow the on screen instructions. This will copy the setup file into the Downloads folder in your hard drive. (If you no longer have the CD, the programs can be downloaded free from the TI website, www.TI.com.)

2. Installing TI Connect™. Click on the globe in the lower left corner of your desktop and in the **Search** box, type **Downloads**. When you get to the **Downloads**, double click on **TIConnect** to begin the installation.

Follow the on screen instructions. This should also create a TI Connect icon on your desktop screen.

3. Installing the calculator driver on the computer. To do this, connect the TI-84 to the computer using the USB cable that came with the calculator; then follow the on screen instructions.

This completes the installation and the CD may now be removed from the computer

4. Double click the TI Connect icon on your desktop screen. If it doesn't exist, Click **Start**, **Programs**, **TI Tools**, and **TI-Connect**.

5. Click on **TI-Data Editor**.

6. Refer to Figure 2D. Click on the icon for the white sheet of paper. The arrow points to it. This will give you a white column to use for the data. If you have more than one column of data click on the piece of paper until you have the correct number of columns for your use. Figure 2D shows two columns.

▲ **FIGURE 2D** TI Data Editor

7. Copy a column or more than one column of *numerical* data from your computer. You may use Excel. Minitab, or any spreadsheet for the source of the column(s) of numbers but do not include any labels or words. (If there are any letters or special characters in the column, they will not transfer and may show up as zeroes.) Then click on the 0 in the first cell of the column (see Figure 2D). The cell *must* become colored (blue, as shown in the figure) before pasting. If it is not colored, click out of the column (for example on the gray area to the right of the column) and then click in the first cell of the first column again. When that cell is blue, paste the column(s) of numbers into the TI-84 data editor. Alternatively you can just type numbers in the column on the data editor.

8. While your cursor is in the column you want to name, choose **File** and **Properties** from the **Data Editor**. Then refer to Figure 2E: Check a list number like **L1** as shown (or if you want a name, it cannot be more than 8 characters) and click **OK**. Then go back to any other column and do the same but pick a different name such as **L2**.

9. Connect the TI-84 to the computer with the cable and turn on the TI-84 calculator.

10. Refer to Figure 2D. To paste the column(s) of data click **Actions, Send All Lists**.

11. When you get the **Warning**, click **Replace or Replace All** to overwrite the old data in the lists.

▲ **FIGURE 2E** Variable Properties

Look in your calculator lists (**STAT**, **EDIT**) to see the data there. If it is not there, check the cable connection, check that the TI-84 is turned on and start again with step 6.

Caution: While the data is transferring you will not be able to use the calculator; it is thinking.

---

### Entering the Data

When you open Minitab, you will see a blank spreadsheet for entering data. Type the data from the example into **C1**, column 1. Be sure your first number is put into Row 1, not above Row 1 in the label region. Be sure to enter only numbers. (If you want a label for the column, type it in the label region *above* the numbers.) You may also paste in data from the computer clipboard. Double-check your entries before proceeding.

### All Minitab Graphs

After making the graph, double click on what you want to change, such as labels.

### Histogram

1. Click **Graph > Histogram**
2. Leave the default option **Simple** and click **OK**.
3. Double click **C1** (or the name for the column) and click **OK**. (Another way to get **C1** in the big box is to click **C1** and click **Select**.)
4. After obtaining the histogram, if you want different bins (intervals), double click on the *x*-axis and look for **Binning**.

Figure 2F shows a Minitab histogram of the ages.

### Stemplot
Click **Graph > Stem-and-Leaf**

### Dotplot
Click **Graph > Dotplot**

### Bar Chart
Click **Graph > Bar Chart**

▲ **FIGURE 2F** Minitab Histogram

When you open Excel, you will see a blank spreadsheet for entering data. But first, click the **Data** tab (top of screen, middle); you should see **Data Analysis** just below and on the far right. If you do not see it, you will need to load the Data Analysis Toolpak (instructions below). Now click the **Add-Ins** tab; you should see the XLSTAT icon ▶ just below and on the far left. If you do not see it, you should install XLSTAT (instructions below). You will need both of these add-ins in order to perform all of the statistical operations described in this textbook

### Data Analysis Toolpak

1. Click **File** > **Options** > **Add-Ins**.
2. In **Manage** Box, select **Excel Add-Ins**, and click **Go**.
3. In **Add-ins available** box, check **Analysis Toolpak** and click **OK**.

### XLSTAT

1. Close Excel.
2. Download XLSTAT from www.myPearsonstore.com.
3. Install XLSTAT.
4. Open Excel, click **Add-Ins tab**, click the XLSTAT icon ▶.

You need to install the Data Analysis Toolpak only once. The Data Analysis tab should be available now every time Excel is opened. For XLSTAT, however, step 4 above may need to run each time Excel is opened (if you expect to run XLSTAT routines).

### Entering Data

See Figure 2G. Enter the data from the example into column A with **Ages** in cell A1. Double check your entries before proceeding.

|    | A    |
|----|------|
| 1  | Ages |
| 2  | 7    |
| 3  | 11   |
| 4  | 10   |
| 5  | 10   |
| 6  | 16   |
| 7  | 13   |
| 8  | 19   |
| 9  | 22   |
| 10 | 42   |

▲ **FIGURE 2G** Excel Data Screen

### Histogram

1. Click **Add-Ins**, **XLSTAT**, **Describing data**, and **Histograms**.
2. When the box under **Data** is activated, drag your cursor over the column containing the data including the label **Ages**.
3. Click **OK** and **Continue**.

Figure 2H shows the histogram.

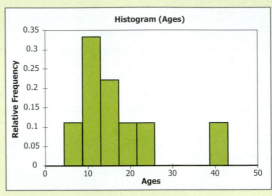

▲ **FIGURE 2H** XLSTAT Histogram

### Dotplot and Stemplot

1. Click **Add-Ins**, **XLSTAT**, **Visualizing data**, and **Univariate plots**.
2. When the box under **Quantitative Data** is activated, drag your cursor over the column containing the data including the label **Ages**.

#### Dotplot

3. Click the **Options, Charts, Charts(1)** and select **Scattergrams** and **Horizontal**.
4. Click **OK** and **Continue**.

#### Stemplot

3. Click **Options, Charts, Charts (1)** and select **Stem-and-leaf plots**.
4. Click **OK** and **Continue**.

### Bar Chart

After typing a summary of your data in table form (including labels), drag your cursor over the table to select it, click **Insert**, **Column** (in the **Charts** group), and select the first option.

For Help: After logging in, click on **Help** or **Resources** and **Watch StatCrunch Video Tutorials on YouTube**.

### Entering Data

1. Click **Open StatCrunch** and you will see a spreadsheet as shown in Figure 2I.
2. Enter the data from the example into the column labeled var1.
3. If you want labels on the columns, click on the variable label, such as **var1**, and backspace to remove the old label and type the new label. Double-check your entries before proceeding.

### Pasting Data

1. If you want to paste data from your computer clipboard, click **Paste** (on the left side of Figure 2I), which will open a screen that says **Load Data from Paste**.
2. Click in the **Input** box and then paste your data.
3. Note that it will use the first line as variable name(s) unless you uncheck it. Scroll down to the very bottom and click **Load Data**.

### Histogram

1. Click **Graphics > Histogram**
2. Under **Select Columns**, click the variable you want a histogram for.
3. Click **Create Graph!**
4. To copy the graph for pasting into a document for submission, click **Options** and **Copy** and then paste it into a document.

Figure 2J shows the StatCrunch histogram of the ages.

### Stemplot

Click **Graphics > Stem-and-leaf**

### Dotplot

Click **Graphics > Dotplot**

### Bar Chart

Click **Graphics > Chart > Columns**

▲ **FIGURE 2I** StatCrunch Data Table

▲ **FIGURE 2J** StatCrunch Histogram

# 3

# Numerical Summaries of Center and Variation

## THEME

The complexity of numerical distributions can often be captured with a small number of summaries. In many cases, two numbers—one to measure the typical value and one to measure the variability—are all we need to summarize a set of data and make comparisons between groups. We can use many different ways of measuring what is typical and also of measuring variability; which method is best to use depends on the shape of the distribution.

Google the phrase *average American* and you'll find lots of entertaining facts. For example, the average American lives within 3 miles of a McDonald's, showers 10.4 minutes a day, and prefers smooth peanut butter to chunky. The average American woman is 5'4" tall and weighs 140 pounds, while the average American female fashion model is considerably taller and weighs much less: 5'11" and 117 pounds.

Whether or not these descriptions are correct, they are attempting to describe a typical American. The reason for doing this is to try to understand a little better what Americans are like, or perhaps to compare one group (American women) to another (female fashion models).

These summaries can seem odd, because we all know that people are too complex to be summarized with a single number. Characteristics such as weight, distance from a McDonald's, and length of a shower vary quite a bit from person to person. If we're describing the "typical" American, shouldn't we also describe how much variation exists among Americans? If we're making comparisons between groups, as we do in this chapter's Case Study, how do we do it in a way that takes into account the fact that individuals may vary considerably?

In Chapter 2, we talked about looking at graphs of distributions of data to get an intuitive, informal sense of the typical value (the center) and the amount of variation (the spread). In this chapter, we explore ways of making these intuitive concepts more precise by assigning numbers to them. We will see how this step makes it much easier to compare and interpret sets of data, for both symmetric and skewed distributions. These measures are important tools that we will use throughout the course.

## CASE STUDY

### Living in a Risky World

Our perception of how risky an activity is plays a role in our decision making. If you think that flying is very risky, for example, you will be more willing to put up with a long drive to get where you want to go. A team of UCLA psychologists were interested in understanding how people perceive risk and whether a simple reporting technique was enough to detect differences in perceptions between groups. The researchers asked over 500 subjects to consider various activities and rate them in terms of how risky they thought the activities were. The ratings were on a scale of 0 (no risk) to 100 (greatest possible risk). For example, subjects were asked to assign a value to the following activities: "Use a household appliance" and "Receive a diagnostic X-ray every 6 months." One question of interest to the researchers was whether men and women would assign different risk levels to these activities (Carlstrom et al. 2000).

Figure 3.1a on the next page shows a histogram for the perceived risk of using a household appliance, and Figure 3.1b shows a histogram for the risk of twice-annual X-rays. (Women are represented in the left panels, men in the right.) What differences do you see between the genders? How would you quantify these differences? In this chapter, you'll learn several techniques for answering these questions. And at the end of the chapter, we'll use these techniques to compare perceived risk between men and women.

► **FIGURE 3.1** Histograms showing the distributions of perceived risk by gender, for two activities. **(a)** Perceived risk of using everyday household appliances. **(b)** Perceived risk of receiving a diagnostic X-ray twice a year. The left panel in each part is for women and the right panel for men.

**(a)**

**(b)**

# Summaries for Symmetric Distributions

In Chapter 2 you learned that we can characterize the typical value of a distribution by the center of the distribution, and the variability in that distribution by the horizontal spread. We left these concepts somewhat vague, but now our goal is to quantify them—to measure these concepts with numbers. However, assigning a number to the center and spread of a distribution is not all that straightforward. Statisticians have different ways of thinking about both center and spread, and these different ways of thinking play different roles, depending on the context of the data.

In this chapter, the two different ways in which we will think about the concept of center are (1) center as the balancing point (or center of mass), and (2) center as the halfway point. In this section we introduce the idea of the center as a balancing point, useful for symmetric distributions. Then in Section 3.3, we introduce the idea of the center as the halfway point, useful for skewed distributions. Each of these approaches results in a different measure, and our choice also affects the method we use to measure variability.

## The Center as Balancing Point: The Mean

The most commonly used measure of center is the mean. The **mean** of a collection of data is the arithmetic average. The mean can be thought of as the balancing point of a distribution of data, and when the distribution is symmetric, the mean closely matches our concept of the "typical value."

### Visualizing the Mean

Different groups of statistics students often have different backgrounds and levels of experience. Some instructors collect student data to help them understand whether one class of students might be very different from another. For example, if one class is offered earlier in the morning, or later in the evening, it might attract a different composition of students. An instructor at Peoria Junior College in Illinois collected data from two classes, including the students' ACT scores. (ACT is a standardized national college entrance exam.) Figure 3.2 shows the distribution of self-reported ACT scores for one statistics class.

 **Looking Back**

**Symmetric Distributions**
Recall that symmetric distributions are those for which the left-hand side of the graph of the distribution is roughly a mirror image of the right-hand side.

◄ **FIGURE 3.2** The distribution of ACT scores for one class of Statistics students. The mean is indicated here with a fulcrum. The mean is at the point on the dotplot that would balance if the points were placed on a seesaw. (Source: StatCrunch, Statistical_Data_499)

The balancing point for this distribution is roughly in the middle, because the distribution is fairly symmetric. The mean ACT for these students is calculated to be 24.6. You'll see how to do this calculation shortly. At the moment, rather than worrying about how to get the precise number, note that 24.6 is about the point where the distribution would balance if it were on a seesaw.

When the distribution of the data is more or less symmetric, the balancing point is roughly in the center, as in Figure 3.2. However, when the distribution is not symmetric, as in Figure 3.3, the balancing point is off-center and the average may not match what our intuition tells us is the center.

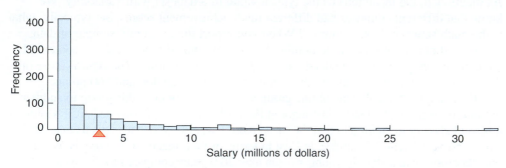

◄ **FIGURE 3.3** Salaries of Professional Baseball Players in 2010

Figure 3.3 shows a very skewed distribution: the salaries of professional baseball players in 2010. Salaries are in millions of dollars. (The highest paid player was Alex Rodriguez at $33 million. The lowest salary was $400,000, which 58 players received. The first bin includes over 400 players who earned $1 million dollars or less.) Where would you place the center of this distribution? Because the distribution is skewed to the right, the balancing point is fairly high; the mean is at 3.2 million dollars. However, when you consider that almost 70% of the players made less than this amount, you might not think that 3.2 million is what the "typical" player made. In other words, in this case the mean might not represent our idea of a typical salary, even for this group of famously well-paid professionals.

## EXAMPLE 1 Math Scores

Figure 3.4 shows the distribution of math achievement scores for 46 countries as determined by the National Assessment of Education Progress. The scores are meant to measure the accomplishment of each country's eighth grade students with respect to mathematical achievement.

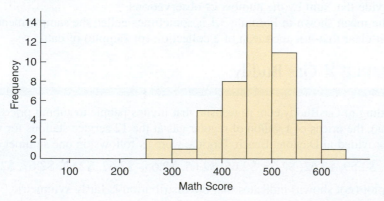

◄ **FIGURE 3.4** Distribution of international math scores for 46 countries. The scores measure the mean level of mathematical achievement for a country's eighth grade students.

**QUESTION** Based on the histogram, approximately what value do you think is the mean math achievement score?

**SOLUTION** The mean score is at the point where the histogram would balance if it were placed on a seesaw. The distribution looks roughly symmetric, and the balancing point looks to be somewhere between 400 and 500 points.

**CONCLUSION** The mean international math achievement score looks to be about 450 points. By the way, the U.S. score was 504, and the highest three scores were from Singapore (highest at 605), South Korea, and Hong Kong. This tells us that the typical score, among the 46 countries, on the international math achievement test was between 400 and 500 points.

**TRY THIS!** Exercise 3.3

## The Mean in Context

As mentioned, the mean tells us the typical value in a data set with variability. We know that different countries had different math achievement scores, but typically, what is the math score of these countries? When you report the mean of a sample of data, you should not simply report the number but, rather, should report it in the context of the data so that your reader understands what you are measuring: The typical score on the international math test among these countries is between 400 and 500 points.

Knowing the typical value of one group allows us to compare this group to another group. For example, that same instructor at Peoria Junior College recorded ACT scores for a second classroom and found that the mean there was also 24.6. This tells us that these two classrooms were comparable, at least in terms of their mean ACT score. However, one classroom had a slightly younger mean age: 26.9 years compared to 27.2 years.

 **KEY POINT** The mean of a collection of data is located at the "balancing point" of a distribution of data. The mean is one representation of the "typical" value of a variable.

## Calculating the Mean for Small Data Sets

The mean is used so often in statistics that it has its own symbol: $\bar{x}$, which is pronounced "x-bar." To calculate the mean, find the (arithmetic) **average** of the numbers; that is, simply add up all the numbers and divide that sum by the number of observations. Formula 3.1 shows you how to calculate the mean, or average.

$$\text{Formula 3.1:}\quad \text{Mean} = \bar{x} = \frac{\sum x}{n}$$

The symbol $\sum$ is the Greek capital sigma, or capital S, which stands for *summation*. The x that comes after $\sum$ represents the value of a single observation. Therefore, $\sum x$ means that you should add all the values. The letter n represents the number of observations. Therefore, this equation tells us to add the values of all the observations and divide that sum by the number of observations.

The mean shown in Formula 3.1 is sometimes called the sample mean in order to make it clear that it is the mean of a collection (or sample) of data.

## EXAMPLE 2 Gas Buddy

According to GasBuddy.com (a website that invites people to submit prices at local gas stations), the prices of 1 gallon of regular gas at the 12 service stations for which data were provided in Daytona Beach, Florida, were as follows on one summer day in 2010:

$2.59, $2.59, $2.62, $2.65, $2.65, $2.65, $2.65, $2.67, $2.67, $2.67, $2.69, $2.69

A dotplot (not shown) indicates that the distribution is fairly symmetric.

**QUESTION** Find the mean price of a gallon of gas for these service stations. Explain what the value of the mean signifies in this context (in other words, interpret the mean).

**SOLUTION** Add the 12 numbers together to get $31.79. We have 12 observations, so we divide $31.79 by 12 to get $2.649.

$$\bar{x} = \frac{2.59 + 2.59 + 2.62 + 2.65 + 2.65 + 2.65 + 2.65 + 2.67 + 2.67 + 2.67 + 2.69 + 2.69}{12}$$

$$= \frac{31.79}{12} = 2.649$$

**CONCLUSION** The typical price of 1 gallon of gas at these gas stations in Daytona Beach, Florida, is $2.65 on this particular day.

**TRY THIS!** Exercise 3.5a

## Calculating the Mean for Larger Data Sets

Formula 3.1 tells you how to compute the mean if you have a small set of numbers that you can easily type into a calculator. But for large data sets, you are better off using a computer or a statistical calculator. That is true for most of the calculations in this book, in fact. For this reason, we will often just display what you would see on your computer or calculator and describe in the TechTips section the exact steps used to get the solution.

For example, the histogram in Figure 3.5 shows the distribution of the amount of particulate matter, or smog, in the air in 333 cities in the United States in 2008, as reported by the Environmental Protection Agency. (The units are micrograms of particles per cubic meter.) When inhaled, these particles can affect the heart and lungs, so you would prefer your city to have a fairly low amount of particulate matter. (The EPA says that levels over 15 micrograms per cubic meter are unsafe.) Looking at the histogram, you can estimate the mean value using the fact that, since the distribution is fairly symmetric, the average will be about in the middle (around 11 micrograms per cubic meter). If you were given the list of 333 values, you could find the mean using Formula 3.1. But you'll find it easier to use the pre-programmed routines of your calculator or software. Example 3 demonstrates how to do this.

◄ **FIGURE 3.5** Levels of particulate matter for 333 U.S. cities in 2008. Because the distribution is fairly symmetric, the balancing point is roughly in the middle at about 11 micrograms per cubic meter.

## ✳ EXAMPLE **3** Mean Smog Levels

We used four different statistical software programs to find the mean particulate level for the 333 cities. The data were uploaded into StatCrunch, Minitab, Excel, and the TI-83/84 calculator.

**Tech**

**QUESTION** For each of the computer outputs shown in Figure 3.6, find the mean particulate matter. Interpret the mean.

▶ FIGURE 3.6a  StatCrunch output.

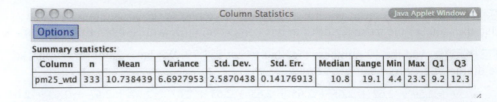

**Summary statistics:**

| Column | n | Mean | Variance | Std. Dev. | Std. Err. | Median | Range | Min | Max | Q1 | Q3 |
|--------|---|------|----------|-----------|-----------|--------|-------|-----|-----|----|----|
| pm25_wtd | 333 | 10.738439 | 6.6927953 | 2.5870438 | 0.14176913 | 10.8 | 19.1 | 4.4 | 23.5 | 9.2 | 12.3 |

▶ FIGURE 3.6b  Minitab output.

### Minitab Descriptive Statistics: pm25_wtd

```
Variable    N    N*    Mean   SE Mean   StDev   Minimum      Q1   Median       Q3
pm25_wtd   333   213  10.738    0.142   2.587     4.400   9.200   10.800   12.300

Variable   Maximum
pm25_wtd    23.500
```

▶ FIGURE 3.6c  TI-83/84 calculator output.

```
1-Var Stats
x̄=10.73843844
Σx=3575.9
Σx²=40621.59
Sx=2.587043708
σx=2.583156337
↓n=333
```

```
1-Var Stats
↑n=333
minX=4.4
Q₁=9.2
Med=10.8
Q₃=12.3
maxX=23.5
```

▶ FIGURE 3.6d  Excel output.

| Column1 | |
|---------|---|
| Mean | 10.73843844 |
| Standard Error | 0.141769122 |
| Median | 10.8 |
| Mode | 11.4 |
| Standard Deviation | 2.587043708 |
| Sample Variance | 6.692795145 |
| Kurtosis | 2.494762112 |
| Skewness | 0.542367196 |
| Range | 19.1 |
| Minimum | 4.4 |
| Maximum | 23.5 |
| Sum | 3575.9 |
| Count | 333 |

**SOLUTION**  We have already seen, from Figure 3.5, that the distribution is close to symmetric, so the mean is a useful measure of center. With a large data set like this, it makes sense to use a computer to find the mean. Minitab and many other statistical software packages produce a whole slew of statistics with a single command, and your job is to choose the correct value.

**CONCLUSION**  The software outputs give us a mean of 10.7 micrograms per cubic meter. We interpret this to mean that the typical level of particulate matter for these cities is 10.7 micrograms per cubic meter. For StatCrunch, Minitab, and Excel the mean is easy to find; it is labeled clearly. For the TI output, the mean is labeled as $\bar{x}$, "x-bar."

**TRY THIS!**  Exercise 3.9

**KEY POINT** ▶ In this textbook, pay more attention to how to apply and interpret statistics than to individual formulas. Your calculator or computer will nearly always find the correct values without your having to know the formula. However, you need to tell the calculator *what* to compute, you need to make sure the computation is meaningful, and you need to be able to explain what the result tells you about the data.

## SNAPSHOT  THE MEAN OF A SAMPLE

| | |
|---|---|
| **WHAT IS IT?** ▶ | A numerical summary. |
| **WHAT DOES IT DO?** ▶ | Measures the center of a sample distribution. |
| **HOW DOES IT DO IT?** ▶ | The mean identifies the "balancing point" of the distribution, which is the arithmetic average of the values. |
| **HOW IS IT USED?** ▶ | The mean represents the typical value in a set of data when the distribution is roughly symmetric. |

## Measuring Variation with the Standard Deviation

The mean amount of particulate matter in the air, 10.7 micrograms per cubic meter, does not tell the whole story. Just because the mean of the 333 cities is at a safe level (below 15 micrograms per cubic meter) does not imply that any particular city—yours for example—is at a healthful level. Are most cities close to the mean level of 10.7? Or do cities tend to have levels of particulate matter far from 10.7? Values in a data set vary, and this variation is measured informally by the horizontal spread of the distribution of data. A measure of variability, coupled with a measure of center, helps us understand whether most observations are close to the typical value or far from it.

### Visualizing the Standard Deviation

The histograms in Figure 3.7 show daily high temperatures in degrees Fahrenheit recorded over one recent year at two locations in the United States. The histogram shown in Figure 3.7a records data collected in Provo, Utah, a city far from the ocean and at an elevation of 4500 feet.

The histogram shown in Figure 3.7b records data collected in San Francisco, California, which sits on the Pacific coast and is famously chilly in the summer. (Mark Twain is alleged to have said, "The coldest winter I ever spent was a summer in San Francisco.")

▲ **FIGURE 3.7** Distributions of daily high temperatures in two cities: **(a)** Provo, Utah; **(b)** San Francisco. Both cities have about the same mean temperature, although the variation in temperatures is much greater in Provo than in San Francisco.

The distributions of temperatures in these cities are similar in several ways. Both temperature distributions are fairly symmetric. You can see from the histograms that both cities have about the same mean temperature. For San Francisco the mean daily high temperature was about 65 degrees; for Provo it was about 67 degrees. But note the difference in the spread of temperatures!

One effect of the ocean on coastal climate is to moderate the temperature: The highs are not that high, and the lows not too low. This suggests that a coastal city should have less spread in the distribution of temperatures. How can we measure this spread, which we can see informally from the histograms?

Note that in San Francisco, most days were fairly close to the mean temperature of 65 degrees; rarely was it more than 10 degrees warmer or cooler than 65. (In other words, rarely was it colder than 55 or warmer than 75 degrees.) Provo, on the other hand, had quite a few days that were more than 10 degrees warmer or cooler than average.

The **standard deviation** is a number that measures how far away the typical observation is from the mean. Distributions such as San Francisco's temperatures have smaller standard deviations because more observations are fairly close to the mean. Distributions such as Provo's have larger standard deviations because more observations are farther from the mean.

As you'll soon see, for most distributions, a majority of observations are within one standard deviation of the mean value.

**KEY POINT**    The standard deviation should be thought of as the typical distance of the observations from their mean.

## EXAMPLE 4 Comparing Standard Deviations from Histograms

Each of the three graphs in Figure 3.8 shows a histogram for a distribution of the same number of observations, and all the distributions have a mean value of about 3.5.

**QUESTION**  Which distribution has the largest standard deviation, and why?

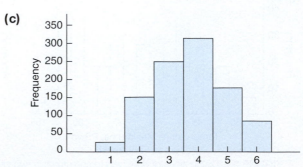

▶ **FIGURE 3.8** All three of these histograms have the same mean, but each has a different standard deviation.

**SOLUTION** All three groups have the same minimum and maximum values. However, the data shown in Figure 3.8a have the largest standard deviation. Why? The standard deviation measures how widely spread the points are from the mean. Note that the histogram in Figure 3.8a has the greatest number of observations farthest from the mean (at the values of 1 and 6). Figure 3.8c has the smallest standard deviation because so many of the data are near the center, close to the mean, which we can see because the taller bars in the center show us that there are more observations there.

**CONCLUSION** Figure 3.8a has the largest standard deviation, and Figure 3.8c has the smallest standard deviation.

**TRY THIS!** Exercise 3.13

## The Standard Deviation in Context

The standard deviation is somewhat more abstract, and harder to understand, than the mean. Fortunately, in a symmetric, unimodal distribution, a handy rule of thumb helps make this measure of spread more comprehensible. In these distributions, the majority of the observations (in fact, about two-thirds of them) are less than one standard deviation from the mean.

For temperatures in San Francisco (see Figure 3.7), the standard deviation is about 8 degrees, and the mean is 65 degrees. This tells us that in San Francisco, on a majority of days, the high temperature is within 8 degrees of the mean temperature of 65 degrees—that is, usually it is no colder than $65 - 8 = 57$ degrees and no warmer than $65 + 8 = 73$ degrees. In Provo, the standard deviation is substantially greater: 21 degrees. On a typical day in Provo, the high temperature is within 21 degrees of mean. Provo has quite a bit more variability in temperature.

## EXAMPLE 5 Standard Deviation of Smog Levels

The mean particulate matter in the 333 cities graphed in Figure 3.5 is 10.7 micrograms per cubic meter, and the standard deviation is 2.6 micrograms per cubic meter.

**QUESTION** Find the level of particulate matter one standard deviation above the mean and one standard deviation below the mean. Keeping in mind that the EPA says that levels over 15 micrograms per cubic meter are unsafe, what can we conclude about the air quality of most of the cities in this sample?

**SOLUTION** The typical city has a level of 10.7 micrograms per cubic meter, and because the distribution is unimodal and (roughly) symmetric, most cities have levels within 2.6 micrograms per cubic meter of this value. In other words, most cities have levels of particulate matter between

$$10.7 - 2.6 = 8.1 \text{ micrograms per cubic meter and}$$
$$10.7 + 2.6 = 13.3 \text{ micrograms per cubic meter}$$

**CONCLUSION** Because the value of 13.3 (one standard deviation above the mean) is lower than 15, most cities are below the safety limit. (The three cities reporting the highest levels of particulate matter were Phoenix, Arizona; Visalia, California; and Hilo, Hawaii. The three cities reporting the lowest levels of particulate matter were Cheyenne, Wyoming; Santa Fe, New Mexico; and Dickinson, North Dakota.)

As you'll soon see, we can say even more about this example. In a few pages you'll learn about the Empirical Rule, which tells us that about 95% of all cities should be within two standard deviations of the mean particulate level.

**TRY THIS!** Exercise 3.15

## Calculating the Standard Deviation

The formula for the standard deviation is somewhat more complicated than that for the mean, and a bit more work is necessary to calculate it. A calculator or computer is pretty much required for all but the smallest data sets. Just as the mean of a sample has its own symbol, the standard deviation of a sample of data is represented by the letter $s$.

$$\textbf{Formula 3.2:} \quad \text{Standard deviation} = s = \sqrt{\frac{\sum (x - \bar{x})^2}{n - 1}}$$

Think of this formula as a set of instructions. Essentially, the instructions say that we need to first calculate how far away each observation is from the mean. This distance, including the positive or negative sign, $(x - \bar{x})$, is called a **deviation**. We square these deviations so that they are all positive numbers, and then we essentially find the average. (If we had divided by $n$, and not $n - 1$, it would be the average. We'll discuss in Chapter 9 why we divide by $n - 1$ and not $n$.) Finally, we take the square root, which means that we're working with the same units as the original data, not with squared units.

## EXAMPLE 6 A Gallon of Gas

| $x$ | $x - \bar{x}$ | $(x - \bar{x})^2$ |
|------|------|------|
| 2.59 | −0.06 | 0.0036 |
| 2.59 | −0.06 | 0.0036 |
| 2.62 | −0.03 | 0.0009 |
| 2.65 | 0 | 0 |
| 2.65 | 0 | 0 |
| 2.65 | 0 | 0 |
| 2.65 | 0 | 0 |
| 2.67 | 0.02 | 0.0004 |
| 2.67 | 0.02 | 0.0004 |
| 2.67 | 0.02 | 0.0004 |
| 2.69 | 0.04 | 0.0016 |
| 2.69 | 0.04 | 0.0016 |

▲ **TABLE 3.1**

From the website Gasbuddy.com, we collected the prices of a gallon of regular gas at all 12 reported gas stations in Daytona Beach, Florida, for one day in June 2010.

$2.59, $2.59, $2.62, $2.65, $2.65, $2.65, $2.65, $2.67, $2.67, $2.67, $2.69, $2.69

**QUESTION** Find the standard deviation for the prices. Explain what this value means in the context of the data.

**SOLUTION** We show this result two ways. The first way is by hand, which illustrates how to apply Formula 3.2. The second way uses a statistical calculator.

The first step is to find the mean. We did this earlier in Example 2, using Formula 3.1, which gave us a mean value of $2.65. We substitute this value for $\bar{x}$ in Formula 3.2.

Table 3.1 shows the first two steps. First we find the deviations (in column 2). Next we square each deviation (in column 3). The numbers are sorted so you can more easily compare the differences.

The sum of the squared deviations—the sum of column 3—is 0.0125. Dividing this by 11 (because $n - 1 = 12 - 1 = 11$), we get 0.001136364. The last step is to take the square root of this. The result is our standard deviation:

$$s = \sqrt{0.001136364} = 0.03371$$

To recap:

$$s = \sqrt{\frac{\sum(x - \bar{x})^2}{n - 1}} = \sqrt{\frac{0.0125}{12 - 1}} = \sqrt{\frac{0.0125}{11}} = \sqrt{0.001136364}$$

$$= 0.03371, \text{ or about 3 cents}$$

When doing these calculations, your final result will be more accurate if you do not round any of the intermediate results. For this reason, it is far easier, and more accurate, to use a statistical calculator or statistical software to find the standard deviation. Figure 3.9 shows the standard deviation as $Sx = 0.0336987546$, which we round to about 0.03 dollars (or, if you prefer, 3 cents). Note that the value reported by the calculator in Figure 3.9 is not exactly the same as the figure we obtained by hand. In part, this is because we used an approximate value for the mean in our hand calculations.

**Tech**

```
1-Var Stats
 x̄=2.649166667
 Σx=31.79
 Σx²=84.2295
 Sx=.0336987546
 σx=.0322641011
↓n=12
```

▲ **FIGURE 3.9** The standard deviation is denoted "Sx" in the TI calculator output.

**CONCLUSION** The standard deviation is about 3 cents, or $0.03. Therefore, at most of these gas stations, the price of a gallon of gas is within 3 cents of $2.65.

**TRY THIS!** Exercise 3.17

One reason why we suggest using statistical software rather than the formulas we present is that we nearly always look at data using several different statistics and approaches. We nearly always begin by making a graph of the distribution. Usually the next step is to calculate a measure of the center and then a measure of the spread. It does not make sense to have to enter the data again every time you want to examine them; it is much better to enter them once and use the functions on your calculator (or software).

**SNAPSHOT** THE STANDARD DEVIATION OF A SAMPLE

| | |
|---|---|
| **WHAT IS IT?** ▸ | A numerical summary. |
| **WHAT DOES IT DO?** ▸ | Measures the spread of a distribution. |
| **HOW DOES IT DO IT?** ▸ | It measures the typical distance of the observations from the mean. |
| **HOW IS IT USED?** ▸ | To measure the amount of variability in a sample when the distribution is fairly symmetric. |

## Variance, a Close Relative of the Standard Deviation

Another way of measuring spread—a way that is closely related to the standard deviation—is the variance. The **variance** is simply the standard deviation squared, and it is represented symbolically by $s^2$.

**Formula 3.3:**    $\text{Variance} = s^2 = \dfrac{\sum(x - \bar{x})^2}{n - 1}$

In Example 5, the standard deviation of the concentration of particulate matter in the cities in our sample was 2.6 micrograms per cubic meter. The variance is therefore $2.6 \times 2.6 = 6.76$ micrograms squared per cubic meter squared. The standard deviation in daily high temperatures in Provo is 21 degrees, so the variance is $21 \times 21 = 441$ degrees squared.

For most applications, the standard deviation is preferred over the variance. One reason is that the units for the variance are always squared (degrees squared in the last paragraph), which implies that the units used to measure spread are different from the units used to measure center. The standard deviation, on the other hand, has the same units as the mean. In Chapter 11 you will see that the variance is useful for some analyses.

SECTION 3.2

# What's Unusual? The Empirical Rule and z-Scores

Finding the standard deviation and the mean is a useful way to compare different samples and to compare observations from one sample with those in another sample.

## The Empirical Rule

The **Empirical Rule** is a rough guideline, a rule of thumb, that helps us understand how the standard deviation measures variability. This rule says that if the distribution is unimodal and symmetric, then

- Approximately 68% of the observations (roughly two-thirds) will be within one standard deviation of the mean.

- Approximately 95% of the observations will be within two standard deviations of the mean.

- Nearly all the observations will be within three standard deviations of the mean.

When we say that about 68% of the observations are within one standard deviation of the mean, we mean that if we count the observations that are between the mean minus one standard deviation and the mean plus one standard deviation, we will have counted about 68% of the total observations.

The Empirical Rule is illustrated in Figure 3.10 in the context of the data on particulate matter in 333 U.S. cities, introduced in Example 3. Suppose we did not have access to the actual data and knew only that the distribution is unimodal and symmetric, that the mean particulate matter is 10.7 micrograms per cubic meter, and that the standard deviation is 2.6 micrograms per cubic meter. The Empirical Rule predicts that about 68% of the cities will fall between 8.1 micrograms per cubic meter $(10.7 - 2.6 = 8.1)$ and 13.3 micrograms per cubic meter $(10.7 + 2.6 = 13.3)$.

The Empirical Rule predicts that about 95% of the cities will fall within two standard deviations of the mean, which means that about 95% of the cities will be between 5.5 and 15.9 micrograms per cubic meter $(10.7 - (2 \times 2.6) = 5.5$ and $10.7 + (2 \times 2.6) = 15.9)$. Finally, nearly all cities, according to the Empirical Rule, will be between 2.9 and 18.5 micrograms per cubic meter. This is illustrated in Figure 3.10.

▶ **FIGURE 3.10** The Empirical Rule predicts how many observations we will see within one standard deviation of the mean (68%), within two standard deviations of the mean (95%), and within three standard deviations of the mean (almost all).

**KEY POINT**  In a relatively large collection of observations, if the distribution is unimodal and roughly symmetric, then about 68% of the observations are within one standard deviation of the mean; about 95% are within two standard deviations of the mean, and almost all observations are within three standard deviations of the mean. Not all unimodal, symmetric distributions are the same, so your actual outcomes might differ from these values, but the Empirical Rule works well enough in a surprisingly large number of situations.

## EXAMPLE 7 Comparing the Empirical Rule to Actual Smog Levels

Because the distribution of levels of particular matter (PM) in 333 U.S. cities is roughly unimodal and symmetric, the Empirical Rule predicts that about 68% of the cities will have particulate matter levels between 8.1 and 13.3 micrograms per cubic meter, about 95% of the cities will have PM levels between 5.5 and 15.9 micrograms per cubic meter, and nearly all will have PM levels between 2.9 and 18.5 micrograms per cubic meter.

QUESTION  Figure 3.11 shows the actual histograms for the distribution of PM levels for these 333 cities. The location of the mean is indicated, as well as the boundaries for points within one standard deviation of the mean (a), within two standard deviations of the mean (b), and within three standard deviations of the mean (c). Using these figures, compare the actual number of cities that fall within each boundary to the number predicted by the Empirical Rule.

SOLUTION  From Figure 3.11a, by counting the heights of the bars between the two boundaries, we find that there are about $20 + 50 + 52 + 60 + 48 = 230$ cities that actually lie between these two boundaries. (No need to count very precisely; we're after approximate numbers here.) The Empirical Rule predicts that about 68% of

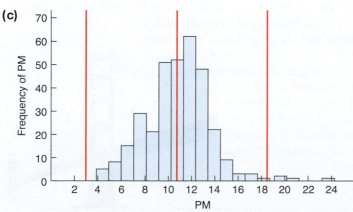

▲ **FIGURE 3.11** **(a)** Boundaries are placed one standard deviation below the mean (8.1) and one standard deviation above the mean (13.3). The middle line in all three figures indicates the mean (10.7). **(b)** The Empirical Rule predicts that about 95% of observations will fall between these two boundaries (5.5 and 15.9 micrograms per cubic meter). **(c)** The Empirical Rule predicts that nearly all observations will fall between these two boundaries (2.9 and 18.5 micrograms per cubic meter).

the cities, or $0.68 \times 333 = 226$ cities, will fall between these two boundaries. The Empirical Rule is pretty accurate in this case.

From Figure 3.11b, we count that there are about $15 + 30 + 20 + 50 + 52 + 60 + 48 + 20 + 8 + 2 = 305$ cities within two standard deviations of the mean. The Empirical Rule predicts that about 95% of the cities, or $0.95 \times 333 = 316$ cities, will fall between these two boundaries. So again, the Empirical Rule is not too far off.

Finally, from Figure 3.11c, we clearly see that all but a few cities (about 4 or 5), are within three standard deviations of the mean. We summarize in a table.

**CONCLUSION**

| Interval | Empirical Rule Prediction | Actual Number |
|---|---|---|
| within 1SD of mean | 226 | 230 |
| within 2SD of mean | 316 | 305 |
| within 3SD of mean | nearly all | all but 5 |

**TRY THIS!** Exercise 3.21

# EXAMPLE 8 Temperatures in San Francisco

We have seen that the mean daily high temperature in San Francisco is 65 degrees Fahrenheit and that the standard deviation is 8 degrees.

**QUESTION** Using the Empirical Rule, decide whether it is unusual in San Francisco to have a day when the maximum temperature is colder than 49 degrees.

**SOLUTION** One way of answering this question is to find out about how many days in the year we would expect the high temperature to be colder than 49 degrees. The Empirical Rule says that about 95% of the days will have temperatures within two standard deviations of the average—that is, within $2 \times 8 = 16$ degrees of 65 degrees. Therefore, on most days the temperature will be between $65 - 16 = 49$ degrees and $65 + 16 = 81$ degrees. Because only (approximately) 5% of the days are outside this range, we know that days either warmer or cooler than this are rare. According to the Empirical Rule, which assumes a roughly symmetric distribution, about half the days outside the range—2.5%—will be colder, and half will be warmer. This shows that about 2.5% of 365 days (roughly 9 or 10 days) should have a maximum temperature colder than 49 degrees.

**CONCLUSION** A maximum temperature colder than 49 degrees is fairly unusual for San Francisco. According to the Empirical Rule, only about 2.5% of the days should have maximum temperatures below 49 degrees, or about 9 or 10 days out of a 365-day year. Of course, the Empirical Rule is only a guide. In this case, we can compare with the data to see just how often these cooler days occurred.

Your concept of "unusual" might be different from ours. The main idea is that "unusual" is rare, and in selecting two standard deviations, we chose to define temperatures that occur 2.5% of the time or less as rare and therefore unusual. You might very reasonably set a different standard for what you wish to consider unusual.

**TRY THIS!** Exercise 3.23

## *z*-Scores: Measuring Distance from Average

The question "How unusual is this?" is perhaps the statistician's favorite question. (It is just as popular as "Compared to what?") Answering this question is complicated because the answer depends on the units of measurement. Eighty-four is a big value if we are measuring a man's height in inches, but it is a small value if we are measuring his weight in pounds. Unless we know the units of measurement and the objects being measured, we can't judge whether a value is big or small.

One way around this problem is to change the units to standard units. **Standard units** measure a value relative to the sample rather than with respect to some absolute measure. A measurement converted to standard units is called a **z-score**.

### Visualizing *z*-Scores

Specifically, a standard unit measures how many standard deviations away an observation is from the mean. In other words, it measures a distance, but instead of measuring in feet or miles, it counts the number of standard deviations. A measurement with a *z*-score of 1.0 is one standard deviation above the mean. A measurement with a *z*-score of −1.4 is 1.4 standard deviations below the mean.

Figure 3.12 shows a dotplot of the heights (in inches) of 247 adult men. The average height is 70 inches, and the standard deviation is 3 inches (after rounding). Below the dotplot is a ruler that marks off how far from average each observation is, measured in terms of standard deviations. The average height of 70 inches is 0 because 70 is zero standard deviations away from the mean. The height of 76 is marked as 2 because it is two standard deviations above mean. The height of 67 is marked as −1 because it is one standard deviation *below* mean.

◀ **FIGURE 3.12** A dotplot of heights of 247 men marked with *z*-scores as well as heights in inches.

We would say that a man from this sample who is 73 inches tall has a *z*-score of 1.0 standard unit. A man who is 67 inches tall has a *z*-score of −1.0 standard unit.

### Using *z*-Scores in Context

*z*-Scores allow us to compare observations in one group with those in another, even if the two groups are measured in different units or under different conditions. For instance, some students might choose their math class on the basis of which professor they think is an easier grader. So if one student gets a 65 on an exam in a hard class, how do we compare his score to another student who gets a 75 in an easy class? If we converted to standard units, we would know how far above (or below) the average each test score was, so we could compare these students' performances.

## EXAMPLE **9** Exam Scores

Maria scored 80 out of 100 on her first stats exam in a course and 85 out of 100 on her second stats exam. On the first exam, the mean was 70 and the standard deviation was 10. On the second exam, the mean was 80 and the standard deviation was 5.

**QUESTION** On which exam did Maria perform better when compared to the whole class?

**SOLUTION** On the first exam, Maria is 10 points above the mean (from 80 − 70). Because the standard deviation is 10 points, she is one standard deviation above the mean. In other words, her z-score for the first exam is 1.0.

On the second exam, she is 5 points above average (from 85 − 80). Because the standard deviation is 5 points, she is one standard deviation above the mean. In other words, her z-score is again 1.0.

**CONCLUSION** The second exam was a little easier; on average, students scored higher and there was less variability in the scores. But Maria scored one standard deviation above average on both exams, so she did equally well on both when compared to the whole class.

**TRY THIS!** Exercise 3.27

## Calculating the z-Score

It's straightforward to convert to z-scores when the result is a whole number, as in the last few examples. More generally, to convert a value to its z-score, first subtract the mean. Then divide by the standard deviation. This simple recipe is summarized in Formula 3.4.

$$\text{Formula 3.4:} \quad z = \frac{x - \bar{x}}{s}$$

Let's apply this to the data shown in Figure 3.12. What is the z-score of a man who is 75 inches tall? Remember that the mean height is 70 inches and the standard deviation is 3 inches. Formula 3.4 says first to subtract the mean height.

$$75 - 70 = 5 \text{ inches}$$

Next divide by the standard deviation:

$$5/3 = 1.67 \text{ (rounding off to two decimal digits.)}$$

$$z = \frac{x - \bar{x}}{s} = \frac{75 - 70}{3} = \frac{5}{3} = 1.67$$

This person has a z-score of 1.67. In other words, we would say that he is 1.67 standard deviations taller than average.

## EXAMPLE 10 Daily Temperatures

The mean daily high temperature in San Francisco is 65 degrees F, and the standard deviation is 8 degrees. On one day, the high temperature is 49 degrees.

**QUESTION** What is this temperature in standard units? Assuming that the Empirical Rule applies, does this seem unusual?

**SOLUTION**

$$z = \frac{x - \bar{x}}{s} = \frac{49 - 65}{8} = \frac{-16}{8} = -2.00$$

**CONCLUSION** This is an unusually cold day. From the Empirical Rule, we know that 95% of z-scores are between −2 and 2 standard units, so it is fairly unusual to have a day as cold as or colder than this one.

**TRY THIS!** Exercise 3.29

**SNAP**SHOT  **THE z-SCORE**

| | | |
|---|---|---|
| **WHAT IS IT?** | ► | A standardized observation. |
| **WHAT DOES IT DO?** | ► | Converts a measurement into standard units. |
| **HOW DOES IT DO IT?** | ► | By measuring how many standard deviations away a value is from the sample mean. |
| **HOW IS IT USED?** | ► | To compare values from different groups, such as two exam scores from different exams, or to compare values measured in different units, such as inches and pounds. |

# Summaries for Skewed Distributions

As you saw earlier, for a skewed distribution, the center of balance is not a good way of measuring a "typical" value. Another concept of center, which is to think of the center as being the location of the *middle* of a distribution, works better in these situations. You saw one example of this in Figure 3.3, which showed that the mean baseball player salary was quite a bit higher than what a majority of the players actually earned. Figure 3.13 shows another example of a strongly right-skewed distribution. This is the distribution of incomes of 788 New York State residents, drawn from the 2000 U.S. census. The mean income of $27,198 is marked with a triangle. However, note that the mean doesn't seem to match up very closely with what we think of as typical. The mean seems to be too high to be typical. In fact, over half (about 55%) of residents earn less than this mean amount.

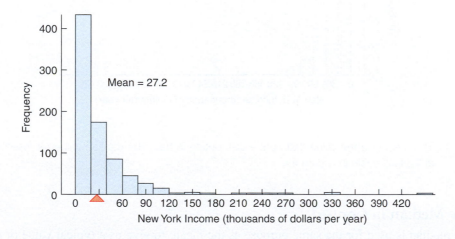

◄ **FIGURE 3.13** The distribution of annual incomes for a collection of New York State residents is right-skewed. Thus the mean is somewhat greater than what many people would consider a typical income. The location of the mean is shown with a triangle.

In skewed distributions, the mean can often be a poor measure of "typical." Instead of using the mean and standard deviation in these cases, we measure the center and spread differently.

## The Center as the Middle: The Median

The median provides an alternative way of determining a typical observation. The **median** of a sample of data is the value that would be right in the middle if you were to sort the data from smallest to largest. The median cuts a distribution down the middle, so about 50% of the observations are below it and about 50% are above it.

## Visualizing the Median

One of the authors found herself at the grocery store, trying to decide whether to buy ham or turkey for sandwiches. Which is more healthful? Figure 3.14 shows a dotplot of the percentage of fat for each of ten types of sliced ham. The vertical line marks the location of the median at 23.5 percent. Note that five observations lie below the median and five lie above it. The median cuts the distribution exactly in half. In Example 12, you'll see how the median percentage of fat in sliced turkey compares to that in sliced ham.

► **FIGURE 3.14** A dotplot of percentage fat from ham has a median of 23.5%. This means that half the observations are below 23.5% and half are above it.

Finding the median for the distribution of incomes of the 778 New York State residents, shown again in Figure 3.15, is slightly more complicated because we have many more observations. The median (shown with the red vertical line) cuts the total area of the histogram in half. The median is at $15,800, and roughly 50% of the observations are below this value and about 50% are above it.

► **FIGURE 3.15** The distribution of incomes of New York State residents, with the median indicated by a vertical line. The median has about 50% of the observations above it and about 50% below it.

**KEY POINT**    The median is the value that cuts a distribution in half. The median value represents a "typical" value in a data set.

## The Median in Context

The median is used for the same purpose as the mean: to give us a typical value of a set of data. Knowing the typical value of one group helps us to compare it to another. For example, as we've seen, the typical median income of this sample of New York State residents is $15,800. How does the typical income in New York compare to the typical income in Florida? A representative sample of 1000 Florida residents (where many New Yorkers go to retire) has a median income of $18,000, which is higher than the median income in New York (Figure 3.16).

The typical person in our data set of New York incomes makes less than the typical person in our Florida data set, as measured by the median. Because the median for Florida is $18,000, we know that more than half of Florida residents in the sample make more than the median New York salary of $15,800.

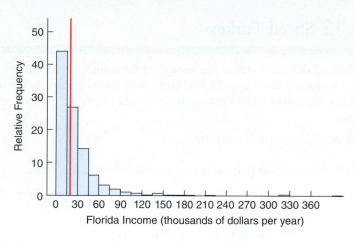

◀ **FIGURE 3.16** Distribution of incomes of a selection of Florida residents (in thousands of dollars). The median income is $18,000.

The median is often reported in news stories when the discussion involves variables with distributions that are skewed. For example, you may hear reports of "median housing costs" and "median salaries."

## Calculating the Median

To calculate the value of the median, follow these steps:

1. Sort the data from smallest to largest.

2. If the set contains an odd number of observed values, the median is the middle observed value.

3. If the set contains an even number of observed values, the median is the average of the two middle observed values. This places the median precisely halfway between the two middle values.

## EXAMPLE 11 Twelve Gas Stations

The prices of a gallon of regular gas at 12 Daytona Beach gas stations in June 2010 (see Example 6) were

$2.65, $2.67, $2.65, $2.65, $2.59, $2.65, $2.69, $2.67, $2.67, $2.59, $2.62, $2.69

**QUESTION** Find the median price for a gallon of gas and interpret the value.

**SOLUTION** First we sort the data from smallest to largest.

2.59, 2.59, 2.62, 2.65, 2.65, 2.65, 2.65, 2.67, 2.67, 2.67, 2.69, 2.69

Because the data set contains an even number of observations (12), the median is the average of the two middle observations, the sixth and seventh: 2.65 and 2.65.

2.59, 2.59, 2.62, 2.65, 2.65, 2.65, 2.65, 2.67, 2.67, 2.67, 2.69, 2.69

                              Med

**CONCLUSION** The median is $2.65, which is the typical price of a gallon of gas at these 12 gas stations. (Note: because the distribution of gas prices at these stations is roughly symmetric, the mean and median prices are very similar. We'll explore this more in a few pages.)

**TRY THIS!** Exercise 3.35a

Example 12 demonstrates how to find the median in a sample with an odd number of observations.

### EXAMPLE 12 Sliced Turkey

Figure 3.14 showed that the median percentage of fat from the various brands of sliced ham for sale at a grocery store was 23.5%. How does this compare to the median percentage of fat for the turkey? Here are the percentages of fat for the available brands of sliced turkey:

14, 10, 20, 20, 40, 20, 10, 10, 20, 50, 10

**QUESTION**  Find the median percentage of fat and interpret the value.

**SOLUTION**  The data are sorted and displayed below. Because we have 11 observations, the median is the middle observation, 20.

10  10  10  10  14  20  20  20  20  40  50

                            Med

**CONCLUSION**  The median for the turkey is 20% fat. Thus the typical percentage of fat for these types of sliced turkey is 20%. This is (slightly) less than that for the typical sliced ham, which has 23.5% fat. Figure 3.17 provides TI-83/84 output that confirms our calculation.

**TRY THIS!**  Exercise 3.37a

```
1-Var Stats
↑n=11
 minX=10
 Q₁=10
 Med=20
 Q₃=20
 maxX=50
```

▲ **FIGURE 3.17** Some TI-83/84 output for the percentage of fat in the turkey.

## SNAPSHOT    THE MEDIAN OF A SAMPLE

**WHAT IS IT?**  ▶  A numerical summary.

**WHAT DOES IT DO?**  ▶  Measures the center of a distribution.

**HOW DOES IT DO IT?**  ▶  It is the value that has roughly the same number of observations above it and below it.

**HOW IS IT USED?**  ▶  To measure the typical value in a data set, particularly when the distribution is skewed.

## Measuring Variability with the Interquartile Range

The standard deviation measures how spread out observations are with respect to the mean. But if we don't use the mean, then it doesn't make sense to use the standard deviation. When a distribution is skewed and you are using the median to measure the center, an appropriate measure of variation is called the interquartile range. The **interquartile range (IQR)** tells us, roughly, how much space the middle 50% of the data occupy.

### Visualizing the IQR

To find the IQR, we cut the distribution into four parts with roughly equal numbers of observations. The distance taken up by the middle two parts is the interquartile range.

The dotplot in Figure 3.18 shows the distribution of weights for a class of introductory statistics students. The vertical lines slice the distribution into four parts so that each part has about 25% of the observations. The IQR is the distance between the first "slice" (at about 121 pounds) and the third slice (at about 160 pounds). This is an interval of 39 pounds (160 − 121 = 39).

◀ **FIGURE 3.18** The distribution of weights (in pounds) of students in a class is divided into four sections so that each section has roughly 25% of the observations. The IQR is the distance between the outer vertical lines (at 121 pounds and 160 pounds).

Figure 3.19 shows distributions for the same students, but this time the weights are separated by gender. The vertical lines are located so that about 25% of the data are below the leftmost line, and 25% are above the rightmost line. This means that about half of the data lie between these two boundaries. The distance between these boundaries is the IQR. You can see that the IQR for the males, about 38 pounds, is much larger than the IQR for the females, which is about 20 pounds. The females have less variability in their weights.

◀ **FIGURE 3.19** The dotplot of Figure 3.18 with weights separated by gender (the women on top). The men have a much larger interquartile range than the women.

The IQR focuses only on the middle 50% of the data. We can change values outside of this range without affecting the IQR. Figure 3.20 shows the men's weights, but this time we've changed one of the men's weights to be very small. The IQR is still the same as in Figure 3.19.

◀ **FIGURE 3.20** The men's weights are given with a fictitious point (in red) below 100. Moving an extremely large (or small) value does not change the interquartile range.

## The Interquartile Range in Context

The IQR for the incomes of the New Yorkers in the data set previously shown in Figure 3.15 is $33,200, as shown in Figure 3.21 at the top of the next page. This tells us that the middle 50% of people in our data set had incomes that varied by as much as $33,200. Compare this to the incomes from Florida, which have an IQR of $26,400. There is less variability among the Floridians; they are more similar (at least in terms of their incomes).

An IQR of $33,200 for New Yorkers seems like a pretty large spread. However, considered in the context of the entire distribution (see Figure 3.21 on the next page), which includes incomes near $0 and as large as nearly $400,000, the IQR looks fairly small. The reason is that lots of people (half of our data set) have incomes in this fairly narrow interval.

**(a)**
New York Income

**(b)**
Florida Income

▲ FIGURE 3.21  **(a)** The distribution of incomes for New Yorkers. **(b)** The distribution of income for Floridians. In both figures, vertical bars are drawn to divide the distribution into areas with about 25% of the observations. The IQR is the distance between the outer vertical lines, and it is wider for the New York incomes.

## Calculating the Interquartile Range

Calculating the interquartile range involves two steps. First, you must determine where to "cut" the distribution. These points are called the **quartiles**, because they cut the distribution into quarters (fourths). The **first quartile (Q1)** has roughly one-fourth, or 25%, of the observations at or below it. The **second quartile (Q2)** has about 50% at or below it; actually, Q2 is just another name for the median. The **third quartile (Q3)** has about 75% of the observations at or below it. The second step is the easiest: To find the interquartile range you simply find the interval between Q3 and Q1—that is, Q3 − Q1.

To find the quartiles:

- First find the median, which is also called the second quartile, Q2. The median cuts the data into two regions.

- The first quartile (Q1) is the median of the lower half of the sorted data. (Do not include the median observation in the lower half if you started with an odd number of observations.)

- The third quartile (Q3) is the median of the upper half of the sorted data. (Again, do not include the median itself if your full set of data has an odd number of observations.)

**Formula 3.5:**   Interquartile Range = Q3 − Q1

## EXAMPLE **13** Heights of Children

A group of eight children have the following heights (in inches):

48.0, 48.0, 53.0, 53.5, 54.0, 60.0, 62.0, and 71.0

They are shown in Figure 3.22.

**QUESTION**  Find the interquartile range for the distribution of the children's heights.

**SOLUTION**  As before, we first explain how to do the calculations by hand and then show the output of technology.

First we find Q2 (the median). Note that the data are sorted and there are four observed values below the median and four observed values above the median.

| Duncan | Charlie | Grant | Aidan | | Sophia | Seamus | Cathy | Drew |
|--------|---------|-------|-------|--|--------|--------|-------|------|
| 48 | 48 | 53 | 53.5 | | 54 | 60 | 62 | 71 |

Q2
53.75

To find Q1, examine the numbers below the median and find the median of them, as shown.

| Duncan | Charlie | | Grant | Aidan |
|--------|---------|--|-------|-------|
| 48 | 48 | | 53 | 53.5 |

Q1
50.50

**(a)**

**(b)**

▲ **FIGURE 3.22** **(a)** Eight children sorted by height. **(b)** A dotplot of the heights of the children. The median and quartiles are marked with vertical lines. Note that two dots, or 25% of the data, appear in each of the four regions.

To find Q3, examine the numbers above the median and find the median of them.

|  | Sophia | Seamus |  |  | Cathy | Drew |
|--|--------|--------|--|--|-------|------|
|  | 54 | 60 |  |  | 62 | 71 |

$$\underset{61.00}{\text{Q3}}$$

Together, these values are

| Duncan | Charlie |  | Grant | Aidan |  | Sophia | Seamus |  | Cathy | Drew |
|--------|---------|--|-------|-------|--|--------|--------|--|-------|------|
| 48 | 48 |  | 53 | 53.5 |  | 54 | 60 |  | 62 | 71 |

$$\underset{50.50}{\text{Q1}} \qquad \underset{53.75}{\text{Q2}} \qquad \underset{61.00}{\text{Q3}}$$

Here's how we calculated the values:

$$Q1 = \frac{48 + 53}{2} = \frac{101}{2} = 50.50 \qquad \text{(Halfway between Charlie and Grant)}$$

$$Q2 = \frac{53.5 + 54}{2} = \frac{107.5}{2} = 53.75 \qquad \text{(Halfway between Aidan and Sophia)}$$

$$Q3 = \frac{60 + 62}{2} = \frac{122}{2} = 61.00 \qquad \text{(Halfway between Seamus and Cathy)}$$

The last step is to subtract:

$$IQR = Q3 - Q1 = 61.00 - 50.50 = 10.50$$

Figure 3.22b shows the location of Q1, Q2, and Q3. Note that 25% of the data (two observations) lie in each of the four regions created by the vertical lines.

Figure 3.23 shows the TI-83/84 output. The TI-83/84 does not calculate the IQR directly; you must subtract Q3 – Q1 yourself. The IQR is $61 - 50.5 = 10.5$, which is the same as the IQR done by hand above.

**CONCLUSION**  The interquartile range of the heights of the eight children is 10.5 inches.

**TRY THIS!**  Exercise 3.39b

```
1-Var Stats
↑n=8
 minX=48
 Q₁=50.5
 Med=53.75
 Q₃=61
 maxX=71
▮
```

▲ **FIGURE 3.23** Some output of a TI-83/84 for eight children's heights.

**Tech**

## Finding the Range, Another Measure of Variability

Another measure of variability is similar to the IQR but much simpler. The **range** is the distance spanned by the entire data set. It is very simple to calculate: the largest value minus the smallest value.

**Formula 3.6:**  Range = maximum − minimum

For the heights of the eight children (Example 13), the range is $71.0 - 48.0 = 23.0$ inches.

The range is useful for a quick measurement of variability because it's very easy to calculate. However, because it depends on only two observations—the largest and the smallest—it is very sensitive to any peculiarities in the data. For example, if someone makes a mistake when entering the data and enters 710 inches instead of 71 inches, then the range will be very wrong. The IQR, on the other hand, depends on many observations and is therefore more reliable.

## SNAPSHOT   THE INTERQUARTILE RANGE

| | |
|---|---|
| **WHAT IS IT?** ▶ | A numerical summary. |
| **WHAT DOES IT DO?** ▶ | It measures the spread of the distribution of a data set. |
| **HOW DOES IT DO IT?** ▶ | It computes the distance taken up by the middle half of the sorted data. |
| **HOW IS IT USED?** ▶ | To measure the variability in a sample, particularly when the distribution is skewed. |

## SECTION 3.4

# Comparing Measures of Center

Which should you choose, the mean (accompanied by the standard deviation) or the median (with the IQR)? These pairs of measures have different properties, so you need to choose the pair that's best for the data you're considering. Our primary goal is to choose a value that is a good representative of the typical values of the distribution. (In later chapters, you will learn about other criteria for choosing a mean or median.)

## Look at the Shape First

This decision begins with a picture. The shape of the distribution will determine which measures are best for communicating the typical value and the variability in the distribution.

###  EXAMPLE **14** MP3 Song Lengths

One of the authors created a data set of the songs on his mp3 player. He wants to describe the distribution of song lengths.

QUESTION  What measures should he use for the center and spread: the mean (250.2 seconds) and standard deviation (152.0 seconds) or the median (226 seconds) and interquartile range (117 seconds)? Interpret the appropriate measures. Refer to the histogram in Figure 3.24.

SOLUTION  Before looking at the histogram, you should think about what shape you expect the graph to be. No song can be shorter than 0 seconds. Most songs on the radio are around 4 minutes (240 seconds), with some a little longer and some a little shorter. However, a few songs, particularly classical tracks, are quite long, so we might expect the distribution to be right-skewed. This suggests that the median and IQR are the best measures to use.

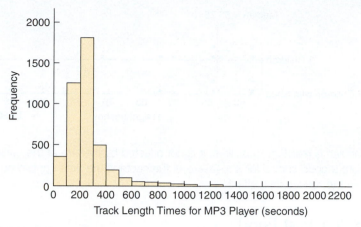

◀ **FIGURE 3.24** The distribution of lengths of songs (in seconds) on the mp3 player of one of the authors.

Figure 3.24 confirms that, as we predicted, the distribution is right-skewed, so the median and interquartile range would be the best measures to use.

**CONCLUSION** The median length is 226 seconds (roughly 3 minutes and 46 seconds), and the interquartile range is 117 seconds (close to 2 minutes). In other words, the typical track on the author's mp3 player is about 4 minutes, but there's quite a bit of variability, with the middle 50% of the tracks differing by about 2 minutes.

**TRY THIS!** Exercise 3.47

Sometimes, you don't have the data themselves, so you can't make a picture. If so, then you must deduce a shape for the distribution that is reasonable and choose the measure of center on the basis of this deduction.

When a distribution is right-skewed, as it is with the mp3 song lengths, the mean is generally larger than the median. You can see this in Figure 3.24; the right tail means the balancing point must be to the right of the median. With the same reasoning, we can see that in a left-skewed distribution, the mean is generally less than the median. In a symmetric distribution, the mean and median are approximately the same.

**KEY POINT**    In a symmetric distribution, the mean and median are approximately the same. In a right-skewed distribution, the mean tends to be greater than the median, and in a left-skewed distribution, the mean tends to be less than the median.

## The Effect of Outliers

Even when a distribution is mostly symmetric, the presence of one or more outliers can have a large effect on the mean. Usually, the median is a more representative measure of center when an outlier is present.

The average height of the eight children from Example 13 was 56.2 inches. Imagine that we replace the tallest child (who is 71 inches tall) with basketball player Shaquille O'Neal, whose height is 85 inches. Our altered data set is now

48, 48, 53, 53.5, 54, 60, 62, and 85 inches

In order to keep the balance of the data, the mean has to shift higher. The mean of this new data set is 57.9 inches—over 1 inch higher. The median of the new data set, however, is the same: 53.75 inches, as shown in Figure 3.25.

When outliers are present, the median is a good choice for a measure of center. In technical terms, we say that the median is **resistant to outliers**; it is not affected by the size of an outlier and does not change even if a particular outlier is replaced by an even more extreme value.

**🔄 Looking Back**

**Outliers**
Recall from Chapter 2 that an outlier is an extremely large or small observation relative to the bulk of the data.

▶ **FIGURE 3.25** The effect of changing the tallest child's height into Shaquille O'Neal's height. Note that the mean (shown with triangles) changes, but the median (the vertical line) stays the same.

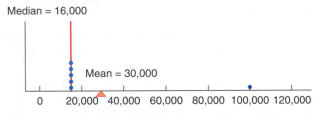

Median = 53.75

8 Children

7 Children with Shaq

Height (inches)              Shaq

**KEY POINT** The median is resistant to outliers; it is not affected by the size of the outliers. It is therefore a good choice for a measure of the center if the data contain outliers and you want to reduce their effect.

## EXAMPLE 15 Fast Food

A (very small) fast-food restaurant has five employees, all of whom work full-time for $7 per hour. Each employee's annual income is about $16,000 per year. The owner, on the other hand, makes $100,000 per year.

**QUESTION** Find both the mean and the median. Which would you use to represent the typical income at this business—the mean or the median?

**SOLUTION** Figure 3.26 shows a dotplot of the data. The mean income is $30,000, and the median is $16,000. Nearly all of the employees earned less than the mean amount!

▶ **FIGURE 3.26** Dotplot of salaries for five employees and their boss at a fast-food company.

Median = 16,000

Mean = 30,000

0    20,000    40,000    60,000    80,000    100,000    120,000

**CONCLUSION** Given the choice between mean and median, the median is better at showing the typical income.

Why are the mean and median so different? Because the owner's salary of $100,000 is an outlier.

**TRY THIS!** Exercises 3.59

## Many Modes: Summarizing Center and Spread

What should you do if the distribution is bimodal or has several modes? Unfortunately, the answer is "It's complicated."

You learned in Chapter 2 that multiple modes in a graphical display of a distribution sometimes indicate that the data set combines different groups. For example, perhaps a data set containing heights includes both men and women. The distribution could very well be bimodal, because we're combining two groups of people who differ markedly in terms of their heights. In this case, and in many other contexts, it is more useful to separate the groups and report summary measures separately for each group. If we know which observations belong to the men and which to the women, then we can separate the data and compute the mean height for men separately from the mean height for women.

For example, Figure 3.27 shows a histogram of the finishing times of female marathon runners. The most noticeable feature of this distribution is that there appear to be two modes. When confronted with this situation, a natural question to ask is "Are two different groups of runners represented in this data set?"

 **Looking Back**

**Bimodality**
Recall from Chapter 2 that we said that a mode was a major mound in a graph (such as a histogram) of a single numerical variable. A bimodal distribution has two major mounds.

 **Caution**

**Don't Get Modes from Computer Output**
Avoid using your computer to calculate the modes. For data sets with many numerical observations, many software programs give meaningless values for the modes. For the data shown in Figure 3.27, for example, most software packages would report the location of five different modes, none of them corresponding to the high points on the graph.

◄ **FIGURE 3.27** Marathon times reported for two groups: amateur and Olympic athletes. Note the two modes. Only women runners were included.

As it turns out, the answer is "yes." Table 3.2 shows the first few lines of the data set. From this, we see that the times belong to runners from two different events. One event was the 2008 Olympics, which includes the best marathoners in the world. The other event was an amateur marathon in Canada (London, Ontario) in 2009. The "L" in Table 3.2 means the event took place in London, Ontario, and the "O" means it took place during the Olympics.

Figure 3.28 shows the data separately for each of these events. We could now compute measures for center and spread separately for the Olympic and amateur events. However, you will sometimes find yourself in situations where a bimodal distribution occurs but it *does* make sense to compute a single measure of center. We can't give you advice for what to do in all situations. Our best advice is always to ask, "Does my summary make sense?"

| Time (seconds) | Event |
|---|---|
| 12,981 | L |
| 12,590 | L |
| 13,722 | O |
| 11,905 | O |

▲ **TABLE 3.2** First four lines of data set of 2008 marathon times for women.

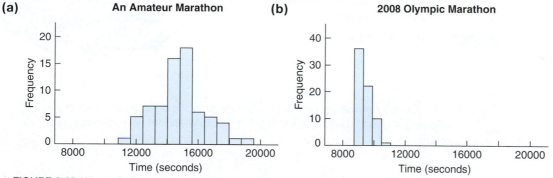

▲ **FIGURE 3.28** Women's times for a marathon, separated into two groups: **(a)** amateur athletes and **(b)** Olympic athletes.

## Comparing Two Groups with Different-Shaped Distributions

Note that the times for the amateur runners in Figure 3.28a are fairly symmetric, which suggests we should use the mean to report the typical finishing time. But the Olympic times are right-skewed, which suggests we should compute the median. Which do you think is the better measure for comparing these two groups, the mean or the median?

When comparing two distributions, you should always use the same measures of center and spread for both distributions. Otherwise, the comparison is not valid.

## EXAMPLE 16 Marathon Times, Revisited

In Figure 3.27 we lumped all of the marathon runners' finishing times into one group. But in fact, our data set had a variable that told us which time belonged to an Olympic runner and which to an amateur runner, so we could separate the data into groups. Figure 3.28 shows the same data, separated by groups.

**! Caution**

**Average**

A guy comes across a statistician who is standing with one foot in a pot of boiling water and one foot frozen in a block of ice. "Isn't that painful?" the man asks. "Maybe so," says the statistician, "but on average it feels just right." Beware of applying the mean to situations where it will not provide a "typical" measure!

**(a)**   An Amateur Marathon

**(b)**   2008 Olympic Marathon

▲ **FIGURE 3.29** Women's times for a marathon, separated into two groups: **(a)** amateur athletes and **(b)** Olympic athletes. Medians are shown for each group.

**QUESTION**   Typically, which group has the fastest finishing times?

**SOLUTION**   The distribution of Olympic runners is right-skewed, so the median would be the best measure. Although the distribution of amateur runners is relatively symmetric, we report the median because we want to compare the typical running time of the amateurs to the typical running time of the Olympic runners. The median of the women Olympic runners is 9291 seconds (about 2.6 hours), and the median for the amateur women runners is 14,780 seconds (about 4.1 hours). Figure 3.29 shows the location of each group's median running time.

**CONCLUSION**   The typical woman Olympic runner finishes the marathon considerably faster: a median time of 9291 seconds (about 2.6 hours) compared to 14,780 seconds (about 4.1 hours) for the amateur athlete.

**TRY THIS!**   Exercise 3.61

> **KEY POINT**   When you are comparing groups, if any one group is strongly skewed or has outliers, it is usually best to compare the medians and interquartile ranges for all groups.

## SECTION 3.5

# Using Boxplots for Displaying Summaries

**Boxplots** are a useful graphical tool for visualizing a distribution, especially when comparing two or more groups of data. A boxplot shows us the distribution divided into fourths. The left edge of the box is at the first quartile (Q1) and the right edge is at the third quartile (Q3). Thus the middle 50% of the sorted observations lie inside the box. Therefore, the length of the box is proportional to the IQR.

A vertical line inside the box marks the location of the median. Horizontal lines, called whiskers, extend from the ends of the box to the smallest and largest values, or nearly so. (We'll explain soon.) Thus the entire length of the boxplot spans most, or all, of the range of the data.

Figure 3.30 compares a dotplot (with the quartiles marked with vertical lines) and a boxplot for the price of gas at stations in Daytona, Florida, as discussed in Examples 6 and 11.

Unlike many of the graphics used to visualize data, boxplots are relatively easy to draw by hand, assuming that you've already found the quartiles. Still, most of the time you will use software or a graphing calculator to draw the boxplot. Most software packages produce a variation of the boxplot that helps identify observations that are extremely large or small compared to the bulk of the data.

These extreme observations are called potential outliers. Potential outliers are different from the outliers we discussed in Chapter 2, because sometimes points that look extreme in a boxplot are not that extreme when shown in a histogram or dotplot. They are called *potential* outliers because you should consult a histogram or dotplot of the distribution before deciding whether the observation is too extreme to fit the pattern of the distribution. (Remember, whether or not an observation is an outlier is a subjective decision.)

**Potential outliers** are identified by this rule: They are observations that are a distance of more than 1.5 interquartile ranges below the first quartile (the left edge of a horizontal box) and above the third quartile (the right edge).

To allow us to see these potential outliers, the whiskers are drawn from the edge of each box to the most extreme observation that is not a potential outlier. This implies that before we can draw the whiskers, we must identify any potential outliers.

**(a)**

**(b)**

**▲ FIGURE 3.30 (a)** A dotplot with Q1, Q2, and Q3 indicated and **(b)** a boxplot for the price of regular, unleaded gas at stations in Daytona Beach, Florida.

 **KEY POINT** Whiskers in a boxplot extend to the most extreme values that are not potential outliers. Potential outliers are points that are more than 1.5 IQRs from the edges of the box.

Figure 3.31 is a boxplot of temperatures in San Francisco (see Examples 8 and 10). From the boxplot, we can see that

$$IQR = 70 - 59 = 11$$

$$1.5 \times IQR = 1.5 \times 11 = 16.5$$

$$\text{Right limit} = 70 + 16.5 = 86.5$$

$$\text{Left limit} = 59 - 16.5 = 42.5$$

Any points below 42.5 or above 86.5 would be potential outliers.

**◀ FIGURE 3.31** Boxplot of maximum daily temperatures in San Francisco.

The whiskers go to the most extreme values that are not potential outliers. On the left side of the box, observations smaller than 42.5 would be potential outliers. However, there are no observations that small. The smallest is 49, so the whisker extends to 49.

On the right, several values in the data set are larger than 86.5. The whisker extends to the largest temperature that is less than (or equal to) 86.5, and the larger values are shown in Figure 3.31 with dots. These represent days that were unusually warm in San Francisco. Figure 3.32 shows a boxplot made with a TI-83/84.

**Tech**

**▲ FIGURE 3.32** TI-83/84 output for a boxplot of San Francisco temperatures.

## Investigating Potential Outliers

What do you do with potential outliers? The first step is always to investigate. A potential outlier might not be an outlier at all. Or a potential outlier might tell an interesting story, or it might be the result of an error in entering data.

Figure 3.33a is a boxplot of the NAEP International math scores for 42 countries (International math scores 2007). One country (South Africa, as it turns out) is flagged as a potential outlier. However, if we examine a histogram, shown in Figure 3.33b, we see that this outlier is really not that extreme. Most people would not consider South Africa to be an outlier in this distribution, because it is not separated from the bulk of the distribution in the histogram.

▶ **FIGURE 3.33 (a)** Distribution of International Math Scores for eighth grade achievement. The boxplot indicates a potential outlier. **(b)** The histogram of the distribution of math scores shows that although South Africa's score of 264 might be the lowest, it is not that much lower than the bulk of the data.

Figure 3.34 shows a boxplot and histogram for the fuel economy (in city driving) of the 2010 model sedans from Ford, Toyota, and GM, in miles per gallon, as listed on their websites. Two potential outliers appear, which are far enough from the bulk of the distribution as shown in the histogram that many people would consider them real outliers. These outliers turn out to be hybrid cars: the Ford Fusion and the Toyota Prius. These hybrids run on both electricity and gasoline, so they have much better fuel economy.

▶ **FIGURE 3.34** Distribution of fuel economies for cars from three manufacturers. **(a)** The boxplot identifies two potential outliers. **(b)** The histogram confirms that these points are indeed more extreme than the bulk of the data.

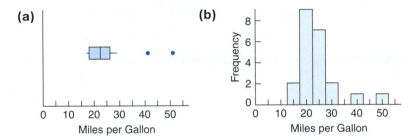

## Horizontal or Vertical?

Boxplots do not have to be horizontal. Many software packages (such as Minitab) provide you with the option of making vertical boxplots. (See Figure 3.35a). Which direction you choose is not important. Try both to see which is more readable.

▶ **FIGURE 3.35 (a)** Default output of Minitab for a boxplot of the San Francisco temperatures. **(b)** Minitab boxplot with the horizontal orientation.

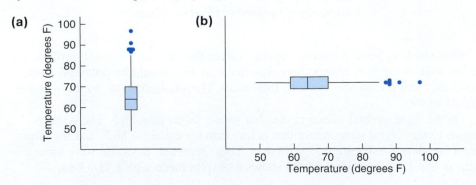

## Using Boxplots to Compare Distributions

Boxplots are often a very effective way of comparing two or more distributions. How do temperatures in Provo compare with those in San Francisco? Figure 3.36 shows boxplots of daily maximum temperatures for San Francisco and Provo. At a glance, we can see how these two distributions differ and how they are similar. Both cities have similar typical temperatures (the median temperatures are about the same). Both distributions are fairly symmetric (because the median is in the center of the box, and the boxplots are themselves fairly symmetric). However, the amount of variation in daily temperatures is much greater in Provo than in San Francisco. We can see this easily because the box is wider for Provo's temperatures.

◀ FIGURE 3.36 Boxplots of daily maximum temperatures in Provo and San Francisco emphasize the difference in variability of temperature in the two cities.

Also note that although both cities do have days that reach about 100 degrees, these days are unusual in San Francisco—they're flagged as potential outliers—but merely fall in the upper 25% for Provo.

## Things to Watch For with Boxplots

Boxplots are best used only for unimodal distributions because they hide bimodality (or any multimodality). For example, Figure 3.37a repeats the histogram of marathon running times for two groups of women runners: amateurs and Olympians. The distribution is clearly bimodal. However, the boxplot in part (b) doesn't show us the bimodality. Boxplots can give the misleading impression that a bimodal distribution is really unimodal.

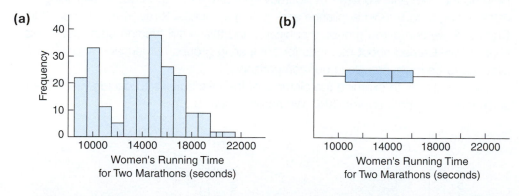

◀ FIGURE 3.37 (a) Histogram of finishing times (in seconds) for two groups of women marathon runners: Olympic athletes and amateurs. The graph is bimodal because the elite athletes tend to run faster, and therefore there is a mode around 10,000 seconds (about 2.8 hours) and another mode around 15,000 seconds (about 4.2 hours). (b) The boxplot hides this interesting feature.

Boxplots should *not* be used for very small data sets. It takes at least five numbers to make a boxplot, so if your data set has fewer than five observations, you can't make a boxplot.

## Finding the Five-Number Summary

Boxplots are not really pictures of a distribution in the way that histograms are. Instead, boxplots help us visualize the location of various summary statistics. The boxplot is a visualization of a numerical summary called the **five-number summary**. These five numbers are

the minimum, Q1, the median, Q3, and the maximum

For example, for daily maximum temperatures in San Francisco, the five-number summary is

49, 59, 64, 70, 97

as you can see in Figure 3.31.

Note that a boxplot is not just a picture of the five-number summary. Boxplots always show the maximum and minimum values, but sometimes they also show us potential outliers.

## SNAPSHOT    THE BOXPLOT

**WHAT IS IT?** ▶ A graphical summary.

**WHAT DOES IT DO?** ▶ Provides a visual display of numerical summaries of a distribution of numerical data.

**HOW DOES IT DO IT?** ▶ The box stretches from the first quartile to the third quartile, and a vertical line indicates the median. Whiskers extend to the largest and smallest values that are not potential outliers, and potential outliers are indicated with special marks.

**HOW IS IT USED?** ▶ Boxplots are useful for comparing distributions of different groups of data.

## CASE STUDY REVISITED

# Living in a Risky World

How do the men and women of this study compare when it comes to assigning risk to using a household appliance and getting an annual X-ray at the doctor's? In Chapter 2 we compared groups graphically, and this is still the first step. But in this chapter, we learned about methods for comparing groups numerically, and this will allow us to be more precise in our comparison.

Our first step is to examine the pictures of the distributions to decide which measures would be most appropriate. (We repeat Figure 3.1.)

▶ **FIGURE 3.1a (repeated)**
Histograms showing the distributions of perceived risk of using appliances. The women's data are in the left graph, and the men's data are in the right graph.

## Risk of Appliances

The histograms for the perceived risk from using appliances (Figure 3.1a) do not show large differences between men and women. We can see that both distributions are right-skewed, and both appear to have roughly the same typical value, although the women's typical value might be slightly higher than the men's. Because the distribution is right-skewed, we compute the median and IQR to compare the two groups. (Table 3.3 summarizes these comparisons.) The men assigned a median risk of 10 to using household appliances, and the women assigned a median risk of 15. We see that, first impressions aside, these women tend to feel that using appliances is a riskier activity than do these men. Also, more differences in opinion occurred among these women than among these men. The IQR was 25 for women and 20 for men. Thus the middle 50% of the women varied by as much as 25 points in how risky they saw this activity; there was less variability for the men.

| | Risk of Appliances | |
| --- | --- | --- |
| | **Median** | **IQR** |
| Men | 10 | 20 |
| Women | 15 | 25 |

| | Risk of X-rays | |
| --- | --- | --- |
| | **Mean** | **SD** |
| Men | 46.8 | 20.0 |
| Women | 47.8 | 20.8 |

▲ **TABLE 3.3** Comparison of Perceived Risks for Men and Women

## Risk of X-rays

Both distributions for the perceived risk level of X-rays (Figure 3.1b) were fairly symmetric, so it makes sense to compare the two groups using the mean and standard deviation. The mean risk level for men was 46.8 and for women 47.8. Typically, men and women feel about the same as far as the risk associated with X-rays. The standard deviations are about the same, too: men have a standard deviation of 20.0 and women 20.8. From the Empirical Rule, we know that a majority (about two-thirds) of men in this sample rated the risk level between 26.8 and 66.8. The majority of women rated it between 27.0 and 68.6.

◄ **FIGURE 3.1b (repeated)** Histograms showing the distributions of perceived risk of X-rays. The women's data are shown on the left and the men's data are shown on the right.

The comparisons are summarized in Table 3.3.

## Does Reaction Distance Depend on Gender?

**GOALS**

Apply the concepts introduced in this chapter to data you have collected in order to compare two groups.

**MATERIALS**

A meter stick or stiff ruler for each group.

**ACTIVITY**

Work in groups of two or three. One person holds the meter stick vertically, with one hand near the top of the stick, so that the 0-centimeter mark is at the bottom. The other person then positions his or her thumb and index finger about 5 cm apart (2 inches apart) on opposite sides of the meter stick at the bottom. Now the first person drops the meter stick without warning, and the other person catches it. Record the location of the middle of the thumb of the catcher. This is the distance the stick traveled and is called the reaction distance, which is related to reaction time. A student who records a small distance has a fast reaction time, and a student with a larger distance has a slower reaction time. Now switch tasks. Each person should try catching the meter stick twice, and the better (shorter) distance should be reported for each person. Then record the gender of each catcher. Your instructor will collect your data and combine the class results.

**BEFORE THE ACTIVITY**

1 Imagine that your class has collected data and you have 25 men and 25 women. Sketch the shape of the distribution you expect to see for the men and the distribution you expect to see for the women. Explain why you chose the shape you did.

2. What do you think would be a reasonable value for the typical reaction distance for the women? Do you think it will be different from the typical reaction distance for the men?

**AFTER THE ACTIVITY**

1. Now that you have actual data, how do the shapes of the distributions for men and women compare to the sketches you made before you collected data?

2. What measures of center and spread are appropriate for comparing men and women's reaction distances? Why?

3. How do the actual typical reaction distances compare to the values you predicted?

4. Using the data collected from the class, write a short paragraph (a couple of sentences) comparing the reaction distances of men and women. You should also talk about what group you could extend your findings to, and why. For example, do your findings apply to all men and women? Or do they apply only to college students?

## CHAPTER REVIEW

### Key Terms

mean, *78*
average, *80*
standard deviation, *84*
deviation, *86*
variance, *87*

Empirical Rule, *88*
standard units, *91*
z-score, *91*
median, *93*

interquartile range (IQR), *96*
first quartile (Q1), *98*
second quartile (Q2), *98*
third quartile (Q3), *98*

resistant to outliers, *101*
boxplot, *104*
potential outlier, *105*
five-number summary, *107*

### Learning Objectives

After reading this chapter and doing the assigned homework problems, you should

- Understand how measures of center and spread are used to describe characteristics of real-life samples of data.

- Understand when it is appropriate to use the mean and standard deviation and when it is better to use the median and interquartile range.

- Understand the mean as the balancing point of the distribution of a sample of data and the median as the point that has roughly 50% of the distribution below it.

- Be able to write comparisons between samples of data in context.

### Summary

The first step in any statistical investigation is to make a picture of the distribution of a numerical variable using a dotplot or histogram. Before computing any summary statistics, you must examine a graph of the distribution to determine the shape and whether or not there are outliers. As noted in Chapter 2, you should report the shape, center, and variability of every distribution.

If the shape of the distribution is symmetric and there are no outliers, you can describe the center and spread by using either the median with the interquartile range or the mean with the standard deviation, although it is customary to use the mean and standard deviation.

If the shape is skewed or there are outliers, you should use the median with the interquartile range.

If the distribution is multimodal, try to determine whether the data consist of separate groups; if so, you should analyze these groups separately. Otherwise, see whether you can justify using a single measure of center and spread.

If you are comparing two distributions and one of the distributions is skewed or has outliers, then it is usually best to compare the median and interquartile ranges for both groups.

The choices are summarized in Table 3.4.

Converting observations to z-scores changes the units to standard units, and this allows us to compare individual observations from different groups.

### Formulas

**Formula 3.1:**   Mean $= \bar{x} = \dfrac{\sum x}{n}$

The mean is the measure of center best used if the distribution is symmetric.

**Formula 3.2:**   Standard deviation $= s = \sqrt{\dfrac{\sum(x - \bar{x})^2}{n - 1}}$

The standard deviation is the measure of variability best used if the distribution is symmetric.

| Shape | Summaries of Center and Variability |
|---|---|
| If distribution is bimodal or multimodal | Try to separate groups, but if you cannot, decide whether you can justify using a single measure of center. If a single measure will not work, then report the approximate locations of the modes. |
| If any group's distribution is strongly skewed or has outliers | Use medians and interquartile ranges for all groups. |
| If all groups' distributions are roughly symmetric | Use means and standard deviations for all groups. |

▲ **TABLE 3.4** Preferred measures to report when summarizing data or comparing two or more groups.

**Formula 3.3:**   Variance $= s^2 = \dfrac{\sum(x - \bar{x})^2}{n - 1}$

The variance is another measure of variability used if the distribution is symmetric.

**Formula 3.4:**   $z = \dfrac{x - \bar{x}}{s}$

This formula converts an observation to standard units.

**Formula 3.5:**   Interquartile range $= Q3 - Q1$

The interquartile range is the measure of variability best used if the distribution is skewed.

**Formula 3.6:**   Range $=$ maximum $-$ minimum

The range is a crude measure of variability.

## Sources

Carlstrom, L., J. Woodward, P. Arthur, and G. Christina. 2000. Evaluating the simplified conjoint expected risk model: Comparing the use of objective and subjective information. *Risk Analysis* 20(3).

Baseball salaries. 2010. http://www.bizofbaseball.com/index.php?option= com_wrapper&view=wrapper&Itemid=17 (accessed June 13, 2010).

EPA. Particulate matter. http://www.epa.gov/airtrends/factbook.html

Gasoline mileage. 2010. http://www.gm.com and www.fordvehicles.com (accessed June 2010).

International math scores. 2007. http://www.edweek.org/ew/ articles/2007/05/02/35air.h26.html

Provo temperatures. 2007. http://www.pgjr.alpine.k12.ut.us/science/james/ provo2006.html (accessed August 2007).

San Francisco temperatures. 2007. http://169.237.140.1/calludt.cgi/ WXDATAREPORT (accessed August 2007).

## SECTION EXERCISES

### SECTION 3.1

**3.1 Earnings** A sociologist says, "Typically, men in the United States still earn more than women." What does this statement mean? (Pick the best choice.)

a. All men make more than all women in the United States.

b. All U.S. women's salaries are less varied than all men's salaries.

c. The center of the distribution of salaries for U.S. men is greater than the center for women.

d. The highest paid people in the United States are men.

**3.2 Houses** A real estate agent claims that all things being equal, houses with swimming pools tend to sell for less than those without swimming pools. What does this statement mean? (Pick the best choice.)

a. There are fewer homes with swimming pools than without.

b. The typical price for homes with pools is smaller than the typical price for homes without pools.

c. There's more variability in the price of homes with pools than in the price of those without.

d. The most expensive houses sold do not have pools.

**TRY** **3.3 Age of CEOs (Example 1)** The histogram shows the ages of 25 CEOs listed at Forbes.com. Based on the distribution, what is the approximate mean age of the CEOs in this data set? Write a sentence in context (using words in the question) interpreting the estimated mean. The typical CEO is about _____ years old.

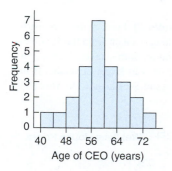

**3.4 Siblings** The histogram shows the number of siblings (brothers and sisters) for 2000 adults surveyed for the 2008 General Social Survey. Based on the distribution, what is the approximate mean number of siblings for these 2000 people? Write a sentence in context interpreting the estimated mean. You may want to refer to the model in Exercise 3.3. (Source: www.norc.uchicago.edu)

**TRY** **3.5 Paid Days (Example 2)** This list represents the numbers of paid vacation days required by law for different countries. (Source: *2009 World Almanac and Book of Facts*)

| United States | 0 |
|---|---|
| Australia | 20 |
| Italy | 20 |
| France | 30 |
| Germany | 24 |
| Canada | 10 |

a. Find the mean, rounding to the nearest tenth of a day. Interpret the mean in this context.

b. Find the standard deviation, rounding to the nearest tenth of a day. Interpret the standard deviation in context.

c. Which number of days is farthest from the mean and therefore contributes most to the standard deviation?

**3.6 Children of First Ladies** This list represents the number of children for the first six First Ladies of the United States. (Source: *2009 World Almanac and Book of Facts*)

| Martha Washington | 0 |
|---|---|
| Abigail Adams | 5 |
| Martha Jefferson | 6 |
| Dolley Madison | 0 |
| Elizabeth Monroe | 2 |
| Louisa Adams | 4 |

a. Find the mean number of children, rounding to the nearest tenth. Interpret the mean in this context.

b. According to eh.net/encyclopedia, women living around 1800 tended to have between 7 and 8 children. How does the mean of these first ladies compare to that?

c. Which of the first ladies listed here had the number of children that is farthest from the mean and therefore contributes most to the standard deviation?

d. Find the standard deviation, rounding to the nearest tenth.

**3.7 Ages of Presidents at Inauguration** At their inauguration, the ages of the first six presidents of the United States were 57, 61, 57, 57, 58, and 57. (Source: *2009 World Almanac and Book of Facts*)

a. Find the mean age at inauguration, rounding to the nearest tenth. The mean of the most recent six presidents (Carter to Obama) was 55.3. Did the first six presidents tend to be a bit *older* or a bit *younger* than the most recent six presidents were?

b. Find the standard deviation of the ages at inauguration, rounding to the nearest tenth. The standard deviation of the most recent six presidents was 9.3 years. Did the first six tend to have *more* or *less* variation than the most recent presidents did?

**3.8 Ages of Chief Justices at Installation** At their installations as chief justice of the United States, the first six Chief Justices were 44, 56, 51, 45, 59, and 56 years old, respectively. (Source: *2009 World Almanac and Book of Facts*)

a. Find and interpret (report in context) the mean age at installation, rounding to the nearest tenth. The mean age at installation of the six most recent chief justices was 62. Did the first six tend to be *older* or *younger* at installation than the most recent six chief justices were?

b. Find the standard deviation of the ages, rounding to the nearest tenth. The standard deviation of the most recent six chief justices was 10.1. Did the first six tend to have *more* or *less* variation than the most recent six chief justices did?

**TRY 3.9 Areas of States (Example 3)** The Minitab output gives some numerical summaries for the areas of the 50 states and Washington D.C. (in square miles). The regions are categorized as those east (e) of the Mississippi River and those west (w) of the river. (Source: *2009 World Almanac and Book of Facts*)

```
Minitab Statistics: Area square miles

Variable  e or w    N    Mean  StDev  Minimum
Area      e        27   35520  24704       68
          w        24  117845 124646    10932

Variable  e or w   Q1  Median     Q3  Maximum
Area      e       9615  40411  53821    96705
          w      70103  84239 113146   656424
```

a. Compare the mean areas by completing this sentence: The mean of the eastern regions is _____ square miles and the mean of the western regions is _____ square miles, showing that the _____ regions tend to have smaller areas.

b. Compare the standard deviations (StDev) of the areas by completing this sentence: The standard deviations of the eastern regions is _____ square miles and the standard deviation of the western regions is _____ square miles, showing that areas of the _____ regions tend to have more variation.

c. Does the TI-83/84 output given represent the eastern or western regions?

**3.10 State Populations** The StatCrunch output shows some statistics for the populations (in thousands) of the 50 states in 2010. The states are categorized by those east (e) of the Mississippi River and those west (w) of the river. (Source: www.TheGreenPapers.com)

**Summary statistics for 2010 / 1000:**

Group by: Mississippi

| Mississippi | n | Mean | Variance | Std. Dev. |
|---|---|---|---|---|
| e | 26 | 6871.2188 | 2.637322E7 | 5135.4863 |
| w | 24 | 5323.1333 | 6.86806E7 | 8287.376 |

a. Compare the means in a sentence or two. The means reported are in thousands of people; round to the nearest thousand. See exercise 3.9 for one possible model of a comparison.

b. Compare the standard deviations in a sentence or two. The standard deviations reported are in thousands of people; round to the nearest thousand.

**3.11 Surfing** College students and surfers Rex Robinson and Sandy Hudson collected data on the self-reported number of days surfed in a month for 30 longboard surfers and 30 shortboard surfers.

Longboard: 4, 9, 8, 4, 8, 8, 7, 9, 6, 7, 10, 12, 12, 10, 14, 12, 15, 13, 10, 11, 19, 19, 14, 11, 16, 19, 20, 22, 20, 22

Shortboard: 6, 4, 4, 6, 8, 8, 7, 9, 4, 7, 8, 5, 9, 8, 4, 15, 12, 10, 11, 12, 12, 11, 14, 10, 11, 13, 15, 10, 20, 20

a. Compare the means in a sentence or two.

b. Compare the standard deviations in a sentence or two.

**3.12 Eating Out** College student Jacqueline Loya asked a group of 50 employed students how many times they went out to eat in the last week. Half of the students had full-time jobs, and half had part-time jobs.

Full-time jobs: 5, 3, 4, 4, 4, 2, 1, 5, 6, 5, 6, 3, 3, 2, 4, 5, 2, 3, 7, 5, 5, 1, 4, 6, 7

Part-time jobs: 1, 1, 5, 1, 4, 2, 2, 3, 3, 2, 3, 2, 4, 2, 1,2, 3, 2, 1, 3, 3, 2, 4, 2, 1

a. Compare the means in a sentence or two.

b. Compare the standard deviations.

**TRY 3.13 Real Estate Prices (Example 4)** Look at the two histograms, created from 2009 real estate data taken from the *Ventura County Star*, and decide whether you think the standard deviation of home prices in Agoura, California (A), was larger or smaller than the standard deviation of home prices in Westlake, California (B). Explain.

**3.14 Dice** The histograms contain data with a range of 1 to 6. Which group would have the larger standard deviation, group A or group B? Why?

(A) Outcome of a Single Die

(B) Average of Two Dice

**TRY** **3.15 Birth Weights (Example 5)** The mean birth weight for U.S. children born at full term (after 40 weeks) is 3462 grams (about 7.6 pounds). Suppose the standard deviation is 500 grams and the shape of the distribution is symmetric and unimodal. (Source: www.babycenter.com)

a. According to the Empirical Rule, what is the range of birth weights (in grams) of U.S.-born children from one standard deviation below the mean to one standard deviation above the mean?

b. Is a birth weight of 2800 grams (about 6.2 pounds) more than one standard deviation below the mean?

**3.16 Birth Length** The mean birth length for U.S. children born at full term (after 40 weeks) is 52.2 cm (about 20.6 inches). Suppose the standard deviation is 2.5 cm and the distributions are unimodal and symmetric. (Source: www.babycenter.com)

a. According to the Empirical Rule, what is the range of birth lengths (in centimeters) of U.S.-born children from one standard deviation below the mean to one standard deviation above the mean?

b. Is a birth length of 54 cm more than one standard deviation above the mean?

**TRY** **3.17 Children's Ages (Example 6)** Mrs. Johnson's children are 2, 2, 3, and 5 years of age.

a. Calculate the standard deviation of their current ages.

b. Without doing any calculation, indicate whether the standard deviation of the ages in 20 years will be larger, smaller, or the same as the standard deviation of their current ages. Check your answer by calculating the standard deviation of the ages in 20 years. Explain how adding 20 to each number affects the standard deviation.

c. Find the mean of the children at their current ages.

d. Without doing any calculation, indicate whether the mean age in 20 years will be larger, smaller, or about the same as the mean of the current ages. Confirm your answer and describe how adding 20 to each number affects the mean.

**3.18 Pay Rate in Different Currencies** The pay rates for 3 people were the following numbers of dollars per hour: 8, 10.5, and 7. In Mexican pesos (if the exchange rate is 10 pesos per dollar), these pay rates are 80, 105, and 70 pesos per hour.

a. Compare the mean in dollars with the mean in pesos; don't forget the units. Explain what multiplying each number in a data set by 10 does to the mean.

b. Compare the standard deviations in pesos and in dollars. Explain what multiplying each number in a data set by 10 does to the standard deviation.

**3.19 Olympics** In the most recent summer Olympics, do you think the standard deviation of the running times for all men who ran the 100-meter race would be larger or smaller than the standard deviation of the running times for the men's marathon? Explain.

**3.20 Weights** Suppose you have a data set with the weights of all members of a high school soccer team and all members of a high school academic decathlon team (a team of students selected because they often answer quiz questions correctly). Which team do you think would have a larger standard deviation of weights? Explain.

## SECTION 3.2

**TRY** **3.21 Violent Crime: West (Example 7)** In 2007, the mean rate of violent crime (per 100,000 people) for the 24 states west of the Mississippi River was 406. The standard deviation was 177. Assume that the distribution of violent crime rates is approximately unimodal and symmetric. *See page 125 for guidance.* (Source: www.worldatlas.com)

a. Between what two values would you expect to find about 95% of the rates?

b. Between what two values would you expect to find about 68% of the violent crime rates?

c. If a western state had a violent crime rate of 584 crimes per 100,000 people, would you consider this unusual? Explain.

d. If a western state had a violent crime rate of 30 crimes per 100,000 people, would you consider this unusual? Explain.

**3.22 Violent Crime: East** In 2007, the mean rate of violent crime (per 100,000 people) for the 26 states east of the Mississippi River was 409; the standard deviation was 193. Assume the distribution of violent crime rates is approximately unimodal and symmetric. (Source: www.worldatlas.com)

a. Between what two values would you expect to find about 95% of the rates?

b. Between what two values would you expect to find about 68% of the rates?

c. If an eastern state had a violent crime rate of 896 crimes per 100,000 people, would you consider this unusual? Explain.

d. If an eastern state had a violent crime rate of 503 crimes per 100,000 people, would you consider this unusual? Explain.

**TRY** **3.23 Property Crime (Example 8)** In 2007, the mean property crime rate (per 100,000 people) for the 24 states west of the Mississippi River was 3331; the standard deviation was 729. Assume the distribution of crime rates is unimodal and symmetric. (Source: www.worldatlas.com)

a. What percentage of western states would you expect to have property crime rates between 2602 and 4060?

b. What percentage of western states would you expect to have property crime rates between 1873 and 4789?

c. If someone guessed that the property crime rate in one state west of the Mississippi River was 9000, would you agree that that number was consistent with this data set?

**3.24 Property Crime** In 2007, the mean property crime rate (per 100,000 people) for the 26 states east of the Mississippi River was 3009; the standard deviation was 732. Assume the distribution of property crime rates is approximately unimodal and symmetric. (Source: www.worldatlas.com)

a. Approximately what percentage of eastern states would you expect to have property crime rates between 1545 and 4473?

b. Approximately what percentage of eastern states would you expect to have property crime rates between 2277 and 3741?

c. If someone guessed that the property crime rate in one state east of the Mississippi River was 500, would you agree that number was consistent with this data set?

**3.25 Heights and z-Scores** The dotplot shows heights of college women; the mean is 64 inches (5 feet 4 inches) and the standard deviation is 3 inches.

Height (inches) and Standard Units

a. What is the z-score for a height of 58 inches (4 feet 10 inches)?

b. What is the height of a woman with a z-score of 1?

**3.26 Heights** Refer to the dotplot in the previous question.

a. What is the height of a woman with a z-score of −1?

b. What is the z-score for a woman who is 70 inches tall (5 feet 10 inches)?

**TRY** **3.27 SAT and ACT Scores (Example 9)** Quantitative SAT scores have a mean of 500 and a standard deviation of 100, while ACT scores have a mean of 21 and a standard deviation of 5. Assuming both types of scores have distributions that are unimodal and symmetric, which is more unusual: a quantitative SAT score of 750 or an ACT score of 28? Show your work.

**3.28 Children's Heights** Mrs. Diaz has two children: a three-year-old boy 43 inches tall and a ten-year-old girl 57 inches tall. Three-year-old boys have a mean height of 38 inches and a standard deviation of 2 inches, and ten-year-old girls have a mean height of 54.5 inches and a standard deviation of 2.5 inches. Assume the distributions of boys' and girls' heights are unimodal and symmetric. Which of Mrs. Diaz's children is more unusually tall for his or her age and gender? Explain, showing any calculations you perform. (Source: www.kidsgrowth.com)

**TRY** **3.29 Low-Birth-Weight Babies (Example 10)** Babies born weighing 2500 grams (about 5.5 pounds) or less are called low-birth-weight babies, and this condition sometimes indicates health problems for the infant. The mean birth weight for U.S.-born children is about 3462 grams (about 7.6 pounds). The mean birth weight for babies born one month early is 2622 grams. Suppose both standard deviations are 500 grams. Also assume that the distribution of birth weights is roughly unimodal and symmetric. (Source: www.babycenter.com)

a. Find the standardized score (z-score), relative to all U.S. births, for a baby with a birth weight of 2500 grams.

b. Find the standardized score for a birth weight of 2500 grams for a child born one month early, using 2622 as the mean.

c. For which group is a birth weight of 2500 grams more common? Explain what that implies. Unusual z-scores are far from 0.

**3.30 Birth Lengths** Babies born after 40 weeks gestation have a mean length of 52.2 centimeters (about 20.6 inches). Babies born one month early have a mean length of 47.4 cm. Assume both standard deviations are 2.5 cm and the distributions are unimodal and symmetric. (Source: www.babycenter.com)

a. Find the standardized score (z-score), relative to all U.S. births, for a baby with a birth length of 45 cm.

b. Find the standardized score of a birth length of 45 cm for babies born one month early, using 47.4 as the mean.

c. For which group is a birth length of 45 cm more common? Explain what that means.

**3.31 Exam Scores** An exam has a mean of 70 and a standard deviation of 10. What exam score corresponds to a z-score of 1.5?

**3.32 Boys' Heights** Three-year-old boys in the United States have a mean height of 38 inches and a standard deviation of 2 inches. How tall is a three-year-old boy with a z-score of −1.0? (Source: www.kidsgrowth.com)

## SECTION 3.3

**3.33** Name two measures of the center of a distribution, and state the conditions under which each is preferred for describing the typical value of a single data set.

**3.34** Name two measures of the variation of a distribution, and state the conditions under which each measure is preferred for measuring the variability of a single data set.

**TRY** **3.35 Pixar Animated Movies (Example 11)** The ten top-grossing Pixar Animated movies for the U.S. box office up to June 2010 are shown in the table on the next page, in millions of dollars. (Source: www.pixar.com)

a. Arrange the gross income from smallest (on the left) to largest (showing the arrangement), and find the median by averaging the two middle numbers. Interpret the median in context.

b. Using the sorted data, find Q1 and Q3. Then find the interquartile range and interpret it in context.

| Movie | $ Millions |
|-------|-----------|
| *Toy Story* | 192 |
| *A Bug's Life* | 163 |
| *Toy Story 2* | 246 |
| *Monsters, Inc.* | 256 |
| *Finding Nemo* | 340 |
| *The Incredibles* | 261 |
| *Cars* | 244 |
| *Ratatouille* | 206 |
| *WALL-E* | 224 |
| *Up* | 293 |

**3.36 DreamWorks Animated Movies** The ten top-grossing DreamWorks animated movies for the U.S. box office up to June 2010 are shown in the table, in millions of dollars. (Source: www.the-top-tens.com)

| Movie | $ Millions |
|-------|-----------|
| *Shrek 2* | 441 |
| *Shrek the Third* | 323 |
| *Shrek* | 268 |
| *Kung Fu Panda* | 215 |
| *Monsters vs. Aliens* | 198 |
| *Madagascar* | 194 |
| *Madagascar: Escape 2 Africa* | 180 |
| *Shark Tale* | 161 |
| *Over the Hedge* | 155 |
| *Bee Movie* | 127 |

a. Find and interpret the median box office dollars for the ten top grossing DreamWorks animated movies.

b. Find and interpret the interquartile range for these movies.

**TRY 3.37 Gas Taxes (Example 12)** The gasoline taxes (in cents per gallon) are given for each state in the plains, as reported in the June 2010 issue of the *AARP Bulletin*. These are state taxes and do not include federal tax.

| State | Tax |
|-------|-----|
| Iowa | 22.0 |
| Kansas | 25.0 |
| Minnesota | 27.2 |
| Missouri | 17.3 |
| Nebraska | 27.7 |
| North Dakota | 23.0 |
| South Dakota | 24.0 |

a. Find the median. Show the data sorted from smallest (on the left) to largest (on the right) and indicate the location of the median.

b. Indicate the locations of the first quartile (Q1) and the third quartile (Q3). Then find the interquartile range.

**3.38 Gas Taxes** The gasoline taxes (in cents per gallon) are given for each state in the far west, as reported in the June 2010 issue of the *AARP Bulletin*. These are state taxes and do not include federal tax. (Of all 50 states, California has the highest gas tax, and Alaska has the lowest.)

| State | Tax |
|-------|-----|
| Alaska | 8.0 |
| California | 48.6 |
| Hawaii | 45.1 |
| Nevada | 33.1 |
| Oregon | 25.0 |
| Washington | 37.5 |

a. Find and interpret the median gas tax for these states. Between which two states is the median rate?

b. Find, Q1, Q3, and the interquartile range (IQR) for these states. Interpret the IQR.

**TRY 3.39 Surfing, Again (Example 13)** College students and surfers Rex Robinson and Sandy Hudson collected data on the self-reported numbers of days surfed in a month for 30 longboard surfers and 30 shortboard surfers.

Longboard: 4, 9, 8, 4, 8, 8, 7, 9, 6, 7, 10, 12, 12, 10, 14, 12, 15, 13, 10, 11, 19, 19, 14, 11, 16, 19, 20, 22, 20, 22

Shortboard: 6, 4, 4, 6, 8, 8, 7, 9, 4, 7, 8, 5, 9, 8, 4, 15, 12, 10, 11, 12, 12, 11, 14, 10, 11, 13, 15, 10, 20, 20

a. Compare the typical number of days surfing for these two groups by placing the correct numbers in the blanks in the following sentence: The median for the longboards was _____ days, and the median for the shortboards was _____ days, showing that those with _____ boards typically surfed more days in this month.

b. Compare the interquartile ranges by placing the correct numbers in the following sentence: The interquartile range for the longboards was _____ days, and the interquartile range for the shortboards was _____ days, showing more variation in the days surfed this month for the _____ boards.

**3.40 Eating Out, Again** College student Jacqueline Loya asked 50 employed students how many times they went out to eat last week. Half of the students had full-time jobs and half had part-time jobs.

Full-time: 5, 3, 4, 4, 4, 2, 1, 5, 6, 5, 6, 3, 3, 2, 4, 5, 2, 3, 7, 5, 5, 1, 4, 6, 7

Part-time: 1, 1, 5, 1, 4, 2, 2, 3, 3, 2, 3, 2, 4, 2, 1, 2, 3, 2, 1, 3, 3, 2, 4, 2, 1

a. Using the median values, write a sentence comparing the typical numbers of times the two groups ate out. See exercise 3.39 for one possible model.

b. Using the interquartile ranges, write a sentence comparing the variability of these two groups.

# SECTION 3.4

### 3.41 Outliers

a. In your own words, describe to someone who has knows only a little statistics how to recognize when an observation is an outlier. What action(s) should be taken with an outlier?

b. Which measure of the center (mean or median) is more resistant to outliers, and what does "resistant to outliers" mean?

**3.42 Center and Variation** When you are comparing two sets of data, and one set is strongly skewed and the other is symmetric, which measures of the center and variation should you choose for the comparison?

**3.43 An Error** A dieter recorded the number of calories he consumed at lunch for one week. As you can see, a mistake was made on one entry. The calories are listed in increasing order:

331, 374, 387, 392, 405, 4200

When the error is corrected by removing the extra 0, will the median calories change? Will the mean? Explain without doing any calculations.

**3.44 Baseball Strike** In 1994, major league baseball players went on strike. At the time, the average salary was $1,049,589, and the median salary was $337,500. If you were representing the owners, which summary would you use to convince the public that a strike was not needed? If you were a player, which would you use? Why was there such a large discrepancy between the mean and median salaries? Explain. (Source: www.usatoday.com)

**3.45 Students' Ages** Here are the ages of some students in a statistics class: 17, 19, 35, 18, 18, 20, 27, 25, 41, 21, 19, 19, 45, and 19. The teacher's age is 66 and should be included as one of the ages when you do the calculations below. The figure shows a histogram of the data.

a. Describe the distribution of ages by giving the shape, the numerical value for an appropriate measure of the center, and the numerical value for an appropriate measure of spread, as well as mentioning any outliers.

b. Make a rough sketch (or copy) of the histogram, and mark the approximate locations of the mean and of the median. Why are they not at the same location?

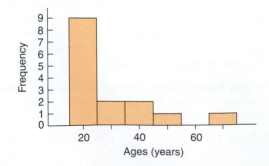

**3.46 House Prices** The figure, which is from data taken from the *Ventura County Star*, shows a histogram of house prices in Thousand Oaks, California, in 2009. The location of the mean and median are marked with letters. Which is the location of the mean, A or B? Explain why the mean and median are not the same.

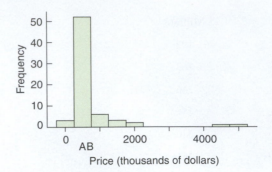

TRY **3.47 Study Hours (Example 14)** A group of 50 statistics students, 25 men and 25 women, reported the number of hours per week spent studying statistics.

a. Refer to the histograms. Which measure of the center should be compared: the means or the medians? Why?

b. Compare the distributions in context using appropriate measures. (Don't forget to mention outliers, if appropriate). Refer to the Minitab output for the summary statistics.

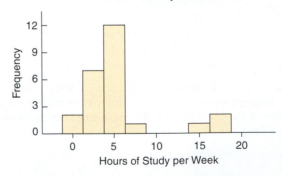

**Minitab Statistics: Men, Women**

| Variable | N | Mean | StDev | Minimum | Q1 | Median | Q3 | Maximum |
|----------|-----|------|-------|---------|------|--------|------|---------|
| Men | 25 | 5.20 | 4.378 | 1.00 | 2.50 | 4.00 | 5.50 | 17.00 |
| Women | 25 | 7.52 | 3.787 | 2.00 | 5.00 | 7.00 | 9.50 | 20.00 |

**3.48 Driving Accidents** College student Sandy Hudson asked a group of college students the total number of traffic accidents they had been in as drivers. The histograms are shown, and the table displays some descriptive statistics.

**Men's Driving Accidents**

**Women's Driving Accidents**

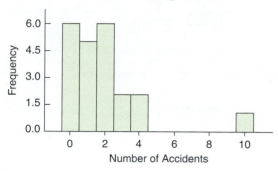

**Minitab Statistics: Men, Women**

| Variable | N | Mean | StDev | Min | Q1 | Median | Q3 | Max |
|---|---|---|---|---|---|---|---|---|
| Men | 20 | 0.850 | 0.988 | 0.00 | 0.00 | 0.505 | 2.00 | 3.00 |
| Women | 22 | 1.864 | 2.210 | 0.00 | 0.00 | 1.50 | 2.25 | 10.00 |

a. Refer to the histograms. If we wish to compare the typical numbers of accidents for these men and women, should we compare the means or the medians? Why?

b. Write a sentence or two comparing the distributions of numbers of accidents for men and women in context.

# SECTION 3.5

**3.49 Regional Rates of Millionaires** The boxplot shows the number of millionaires by state per 1000 residents for the United States, as reported by *Fortune Magazine* in March 2007. Assume all distributions are unimodal.

a. List the regions from lowest to highest in terms of the median rate of millionaires in that region.

b. Which region has the largest interquartile range?

c. Which region has the smallest interquartile range?

d. Which region has potential outliers?

*e. Why is a rate of about 24 millionaires per 1000 people in the South a potential outlier, while the rate of about 24.5 in the Northeast is not a potential outlier?

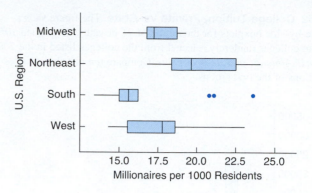

**3.50 Regional Gas Taxes** Each state in the United States can charge its own tax on gasoline. The boxplots show the gas tax (in cents per gallon) for the 50 states by region. Assume all distributions are unimodal. (Source: *AARP Bulletin,* June 2010)

a. Compare the medians. List the regions from smallest to largest in terms of the median amount of gas tax.

b. Compare the interquartile ranges. List the regions from smallest to largest interquartile range.

c. Which region had the largest range? Is it the same as the region with the largest interquartile range?

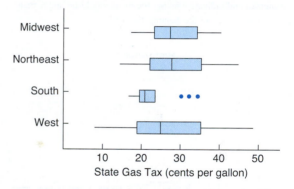

**3.51 City Temperatures** The boxplot shows temperatures for six cities. Each city's boxplot was made from 12 temperatures: the average monthly temperature over a period of years. Which city tends to be warmest? Which city has the most variation in temperatures? Compare the temperatures of the cities by interpreting the boxplots. If temperature were the only factor to consider, which city you would choose to live in, and why? (Source: *2009 World Almanac and Book of Facts*)

**3.52 College Tuition, Private vs. State** The figure shows side-by-side boxplots for the tuition of 71 private colleges and 105 state colleges randomly selected from the colleges listed in the *2006 World Almanac and Book of Facts*. Compare the distribution of tuitions of the two groups.

### Histogram 3

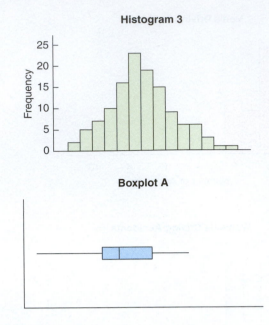

**3.53 Matching Boxplots and Histograms**

a. Match each of the boxplots (A, B, and C) with the corresponding histogram (1, 2, or 3). Explain your reasoning.

b. For each histogram, label the shape (right-skewed, left-skewed, or symmetric) and indicate whether the mean would be larger than, smaller than, or about the same as the median.

### Boxplot A

### Histogram 1

### Boxplot B

### Histogram 2

### Boxplot C

**3.54 Matching Boxplots and Histograms** Match each of the histograms (X, Y, and Z) with the corresponding boxplot (C, M, or P). Explain your reasoning.

X

Y

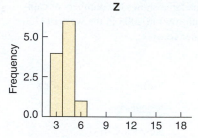

Z

One way in which a state can have a high death rate is to have many deaths. What is the other reason why a state might have a high death rate?

```
1-Var Stats
↑n=25
 minX=16
 Q₁=17.5
 Med=20
 Q₃=23
 maxX=33
```

**3.56 Educated Residents** The U.S. Census Bureau reported the percentage of residents 25-years-old and older in each state who held a Bachelor's degree. We provide data for the states in the western U.S. in 2004. Make a boxplot of the data and describe it. If there are any potential outliers, explain what they show, including the name of the state(s).

| | |
|---|---|
| Alaska | 17.3 |
| Arizona | 17.2 |
| California | 21.2 |
| Colorado | 29.3 |
| Hawaii | 17.8 |
| Idaho | 17.4 |
| Montana | 18.3 |
| Nevada | 16.0 |
| New Mexico | 14.9 |
| Oregon | 17.6 |
| Utah | 21.2 |
| Washington | 20.2 |
| Wyoming | 16.3 |

**\*3.55 Driving Deaths** Have you ever wondered whether driving is more dangerous in some states than in others? The data on the disk give the automobile fatality rates (deaths per 100,000 residents) in 2005 for the 25 states with the highest death rates. The TI-83/84 output shows some descriptive statistics for this data set. (Source: www.nhtsa.dot.gov)

a. Make a boxplot of the data, and report the median and interquartile range. If you find potential outlier(s), identify the state(s).

b. There are two variables involved in the calculation of the death rate for a state (deaths per 100,000 residents):

$$\text{Death rate} = \frac{\text{deaths}}{\text{population}}(100{,}000)$$

# CHAPTER REVIEW EXERCISES

**3.57 Marathon Times by Gender** The figure shows side-by-side boxplots of marathon times for the 2009 Forest City Marathon. Who tended to run faster, the men or the women? Make appropriate comparisons without reporting numbers. The men's distribution and the women's distribution were both unimodal. (Remember, smaller times mean faster runners.)

**3.58 Family Income** The boxplots show median family income in 2008 for the 50 U.S. states and Washington, D.C. by region, according to the U.S. Census Bureau. The distributions were unimodal.

a. Compare the medians and interquartile ranges without reporting numbers. List the regions from smallest (on the left) to largest median income.

b. Explain why the medians are more appropriate for comparisons than the means.

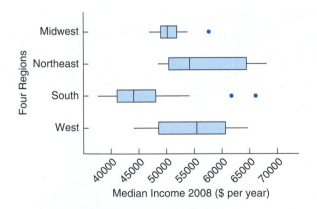

**TRY 3.59 Death Row: South (Example 15)** The table shows the numbers of capital prisoners (prisoners on death row) in 2005 in the southern U.S. states. (Source: www.statemaster.com)

a. Find the median number of prisoners and interpret (using a sentence in context).

b. Find the interquartile range (showing Q3 and Q1 in the process) to measure the variability in the number of prisoners.

c. What is the mean number of prisoners?

d. Why is the mean so much larger than the median?

e. Why is it better to report the median, instead of the mean, as a typical measure?

| State | CapPris | State | CapPris |
|---|---|---|---|
| Alabama | 193 | North Carolina | 166 |
| Arkansas | 36 | Oklahoma | 84 |
| Florida | 374 | South Carolina | 62 |
| Georgia | 105 | Tennessee | 102 |
| Kentucky | 40 | Texas | 391 |
| Louisiana | 86 | Virginia | 20 |
| Maryland | 6 | West Virginia | 0 |
| Mississippi | 69 | | |

**3.60 Death Row: West** The table shows the numbers of capital prisoners (prisoners on death row) in 2005 in the western U.S. states. (Source: www.statemaster.com)

| State | CapPris | State | CapPris |
|---|---|---|---|
| Alaska | 0 | Nevada | 82 |
| Arizona | 110 | New Mexico | 2 |
| California | 656 | Oregon | 32 |
| Colorado | 2 | Utah | 9 |
| Hawaii | 0 | Washington | 9 |
| Idaho | 18 | Wyoming | 2 |
| Montana | 2 | | |

a. Find the median.

b. Find the interquartile range (showing Q3 and Q1 in the process).

c. Find the mean number of capital prisoners.

d. Why is the mean so much larger than the median?

**TRY g 3.61 Head Circumference (Example 16)** Following are head circumferences, in centimeters, for some men and women in a statistics class.

Men: 58, 60, 62.5, 63, 59.5, 59, 60, 57, 55

Women: 63, 55, 54.5, 53.5, 53, 58.5, 56, 54.5, 55, 56, 56, 54, 56, 53, 51

Compare the circumferences of the men's and the women's heads. Start with histograms to determine shape; then compare appropriate measures of center and spread, and mention any outliers. *See page 125 for guidance.*

**3.62 Heights of Sons and Dads** The data on the disk give the heights of 18 male college students and their fathers, in inches.

a. Make histograms and describe the shapes of the two data sets from the histograms.

b. Fill in the following table to compare descriptive statistics.

|  | Mean | Median | Standard deviation | Interquartile range |
|---|---|---|---|---|
| **Sons** | _____ | _____ | _____ | _____ |
| **Dads** | _____ | _____ | _____ | _____ |

c. Compare the heights of the sons and their dads, using the means and standard deviations.

d. Compare the heights of the sons and their dads, using the medians and interquartile ranges.

e. Which pair of statistics is more appropriate for comparing these samples: the mean and standard deviation or the median and interquartile range? Explain.

**3.63 Final Exam Grades** The data that follow are final exam grades for two sections of statistics students at a community college. One class met twice a week relatively late in the day; the other class met four times a week at 11 a.m. Both classes had the same instructor and covered the same content. Is there evidence that the performances of the classes differed? Answer by making appropriate plots (including side-by-side boxplots) and reporting and comparing appropriate summary statistics. Explain why you chose the summary statistics that you used. Be sure to comment on the shape of the distributions, the center, and the spread, and be sure to mention any unusual features you observe.

11 a.m. grades: 100, 100, 93, 76, 86, 72.5, 82, 63, 59.5, 53, 79.5, 67, 48, 42.5, 39

5 p.m. grades: 100, 98, 95, 91.5, 104.5, 94, 86, 84.5, 73, 92.5, 86.5, 73.5, 87, 72.5, 82, 68.5, 64.5, 90.75, 66.5

**3.64 Speeding Tickets** College students Diane Glover and Esmeralda Olguin asked 25 men and 25 women how many speeding tickets they had received in the last three years.

Men: 14 men said they had 0 tickets, 9 said they had 1 ticket, one had 2 tickets, and 1 had 5 tickets.

Women: 18 said they had 0 tickets, 6 said they had 1 ticket, and 1 said she had 2 tickets.

Is there evidence that the men and women differed? Answer by making appropriate plots and comparing appropriate summary statistics. Be sure to comment on the shape of the distributions and to mention any unusual features you observe.

**3.65 Heights** The graph shows the heights for a large group of adults. Describe the distribution and explain what might cause this shape. (Source: www.amstat.org)

*3.66 **Marathon Times** The histogram of marathon times includes data for men and women and also for both an Olympic marathon and an amateur marathon. Greater values indicate slower runners. (Sources: www.forestcityroadraces.com and www.runnersworld.com)

a. Describe the shape of the distribution.

b. What are two different possible reasons for the two modes?

c. Knowing that there are usually fewer women who run marathons than men and that more people ran in the amateur marathon than the Olympic marathon, look at the size of the mounds and try to decide which of the reasons stated in part b is likely to cause this. Explain.

*3.67 **Law School GPAs** The histogram shows the minimum GPA of accepted students at 181 nationally ranked law schools in 2006. (Source: www.ilrg.com)

a. State an approximate value for the mean GPA.

b. Here is a proposed method for finding an approximation to the standard deviation based on a histogram: Find the approximate range and divide by 6. Use this method to find an approximate standard deviation for the GPAs. For comparison, the true standard deviation is 0.22.

c. Explain why dividing the range by 6 should produce a reasonable approximation of the standard deviation for symmetric, unimodal distributions.

**3.68 Ideal Family** In 2008, the General Social Survey asked respondents how many children they felt would be in an "ideal" family. The histogram contains the data from 1730 people who responded to the survey.

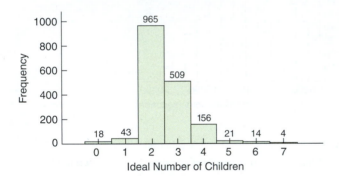

a. Approximately what is the mean ideal number of children? Explain how you chose this value.

b. What is the approximate value for the median ideal number of children? Describe how you chose this value.

c. Find the mean by completing the work that is started below:

$$\bar{x} = \frac{18(0) + 43(1) + 965(2) + \cdots}{1730}$$

d. Explain how the method in part c is related to the usual method of finding the mean, which has all the raw numbers given, without frequencies.

e. Which is more appropriate to report for these data, the mean or the median? Why?

**3.69–3.72** *Construct two sets of numbers with at least five numbers in each set (showing them as dotplots) with the following characteristics:*

**3.69** The means are the same, but the standard deviation of one of the sets is larger than that of the other. Report the mean and both standard deviations.

**3.70** The means are different, but the standard deviations are the same. Report both means and the standard deviation.

**3.71** The mean of set A is larger than that of set B, but the median of set B is larger than the median of set A. Label each dotplot with its mean and median in the correct place.

**\*3.72** The standard deviation of set A is larger, but the interquartile range of set B is larger. Report both standard deviations and interquartile ranges.

**\*3.73 Population Density** Data were recorded for each of the 50 U.S. states: the state, its population (in 2010), its area (in square miles), and region in which it is located (Northeast, Midwest, South, or West). For each state, find the population density: the number of people divided by the number of square miles. Write a few sentences comparing the distribution of population densities in the four regions. Support your description with appropriate graphs. (Sources: U.S. Census Bureau and *2009 World Almanac and Book of Facts*)

**\*3.74 Population Increase** Data were recorded for each of the 50 U.S. states: the state, its population in 2000, its population in 2010, and the region in which it is located (Northeast, Midwest, South, or West). Find the percentage population increase for each state by applying the following formula:

$$\frac{\text{pop}_{2010} - \text{pop}_{2000}}{\text{pop}_{2000}} \times 100\%$$

Write a few sentences comparing the distribution of percentage population increases for the four regions. Support your description with appropriate graphs. (Source: U.S. Census Bureau)

**3.75 Gas Barrels** Data were recorded on the number of barrels of oil per person that were used for transportation in each state (in 2008), and the region where the state is located (Northeast, Midwest, South, or West). Write a few sentences comparing the four regions. Include boxplots. (Source: www.eia.doe.gov)

**3.76 Sales Tax** Data were recorded on the sales tax as a percentage of purchase for each of the 50 states and the state's region (Northeast, Midwest, South, or West). Write a few sentences comparing the four regions. Include boxplots. (Source: www.taxadmin.org)

# gUIDED EXERCISES

**g 3.21 Violent Crime: West** In 2007, the mean rate of violent crime (per 100,000 people) for the 24 states west of the Mississippi River was 406, and the standard deviation was 177. Assume the distribution of violent crime rates is approximately unimodal and symmetric. The Empirical Rule says 95% of data should lie within two standard deviations of the mean and 68% of the data should lie within one standard deviation of the mean with unimodal and symmetric data.

## Questions
Answer these questions by following the numbered steps.

a. Between which two values would you expect to find about 95% of the violent crime rates?

b. Between which two values would you expect to find about 68% of the violent crime rates?

c. If a western state had a violent crime rate of 584 crimes per 100,000 people, would you consider this unusual? Explain.

d. If a western state had a violent crime rate of 30 crimes per 100,000 people, would you consider this unusual? Explain.

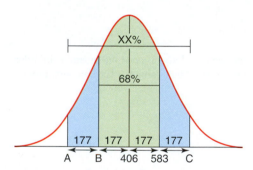

The green area is 68% of the area under the curve. The green and blue areas together shade XX% of the area. The numbers without percentage signs are crime rates, with a mean of 406 and a standard deviation of 177.

**Step 1 ▶ Percentage**
Reproduce Figure A, which is a sketch of the distribution of the crime rates. What percentage of data should occur within two standard deviations of the mean? Include this number in the figure, where it now says XX.

**Step 2 ▶ Why 583?**
How was the number 583, shown on the sketch, obtained?

**Step 3 ▶ A, B, and C**
Fill in numbers for the crime rates for areas A, B, and C.

**Step 4 ▶ Boundaries for 95%**
Read the answer from your graph:

a. Between which two values would you expect to find about 95% of the rates?

**Step 5 ▶ Boundaries for 68%**
Read the answer from your graph:

b. Between which two values would you expect to find about 68% of the violent crime rates?

**Step 6 ▶ Unusual?**

c. If a western state had a violent crime rate of 584 crimes per 100,000 people, would you consider this unusual? Many people would consider any numbers outside your boundaries of A and C to be unusual because such values occur in 5% of the states or fewer.

**Step 7 ▶ Unusual?**

d. If a western state had a violent crime rate of 30 crimes per 100,000 people, would you consider this unusual? Explain.

---

**g 3.61 Head Circumference** The head circumferences in centimeters for some men and women in a statistics class are given.

Men: 58, 60, 62.5, 63, 59.5, 59, 60, 57, 55

Women: 63, 55, 54.5, 53.5, 53, 58.5, 56, 54.5, 55, 56, 56, 54, 56, 53, 51

## Question
Compare the circumferences of the men's and women's heads by following the numbered steps.

**Step 1 ▶ Histograms**
Make histograms of the two sets of data separately. (You may use the same horizontal axes—if you want to—so that you can see the comparison easily.)

**Step 2 ▶ Shapes**
Report the shapes of the two data sets.

**Step 3 ▶ Measures to Compare**
If either data set is skewed or has an outlier (or more than one), you should compare medians and interquartile ranges for *both* groups. If both data sets are roughly symmetric, you should compare means and standard deviations. Which measures should be compared with these two data sets?

**Step 4 ▶ Compare Centers**
Compare the centers (means or medians) in the following sentence: The _____ (mean or median) head circumference for the men was _____ cm, and the _____ (mean or median) head circumference for the women was _____ cm. This shows that the typical head circumference was larger for the_____.

**Step 5 ▶ Compare Variations**
Compare the variations in the following sentence: The _____ (standard deviation or interquartile range) for the head circumferences for the men was _____ cm, and the _____ (standard deviation or interquartile range) for the women was _____ cm. This shows that the _____ tended to have more variation, as measured by the _____ (standard deviation or interquartile range).

**Step 6 ▶ Outliers**
Report any outliers and state which group(s) they belong to.

**Step 7 ▶ Final Comparison**
Finally, in a sentence or two, make a complete comparison of head circumferences for the men and the women.

## CHECK YOUR TECH

# Finding the Standard Deviation of Vacation Days for Several Countries

 The Minitab output shown gives the mean and standard deviation of the data set. (Source: *2009 World Almanac and Book of Facts*)

**Descriptive Statistics: Days**

```
Variable   N    Mean   StDev
Days       6   30.00   10.35
```

Minitab Output

| Country | Mean Days |
|---|---|
| United States | 13 |
| Japan | 25 |
| Italy | 42 |
| France | 37 |
| Germany | 35 |
| U. K. | 28 |

The table at the left reports the mean number of vacation days per year for several countries.

$$\text{Standard deviation} = s = \sqrt{\frac{\sum(x - \bar{x})^2}{n - 1}}$$

**QUESTION** Find the standard deviation of vacation days (by following the numbered steps) and verify that it is the same as StDev given in the Minitab output.

**SOLUTION**

**Step 1 ▶ Mean**
From the standard deviation formula above, you can see that you will need the mean, $\bar{x}$, in order to calculate the standard deviation. Check that the mean is 30 by adding the six numbers, reporting the sum, and dividing by 6.

**Step 2 ▶ Table**
Fill in the table below.

| $x$ | $x - \bar{x}$ | $(x - \bar{x})^2$ |
|---|---|---|
| 13 | $13 - 30 = -17$ | $(-17)^2 = 289$ |
| 25 | | |
| 42 | $42 - 30 = 12$ | $12^2 = 144$ |
| 37 | | |
| 35 | | |
| 28 | | |

**Step 3 ▶ Sum of Squares**
Add the numbers in the last column of your table to get $\sum(x - \bar{x})^2$

**Step 4 ▶ Variance**
Divide your answer to step 3 by $n - 1$, which is $6 - 1$, or 5, to get the following:

$$\frac{\sum(x - \bar{x})^2}{n - 1}$$

**Step 5 ▶ Standard Deviation**
Finally, to get the standard deviation, take the square root of your answer to step 4. Show a long version of your answer, and then round it to hundredths and compare it with the StDev in the Minitab output.

# Making a Boxplot of the Area of Western States

The area of western states is given in the table (in thousands of square miles). Verify that the boxplot given in Figure A is correct by following the numbered steps. (Source: *2009 World Almanac and Book of Facts*)

0  100  200  300  400  500  600  700
Area of Western States (thousands of square miles)
▲ **FIGURE A** Minitab Output

```
1-Var Stats
↑n=13
 minX=11
 Q₁=84.5
 Med=104
 Q₃=134.5
 maxX=656
```

▲ **FIGURE B** TI-83/84 Output

| State | Area |
|---|---|
| Alaska | 656 |
| Arizona | 114 |
| California | 164 |
| Colorado | 104 |
| Hawaii | 11 |
| Idaho | 84 |
| Montana | 147 |
| Nevada | 111 |
| New Mexico | 122 |
| Oregon | 98 |
| Utah | 85 |
| Washington | 71 |
| Wyoming | 98 |

**QUESTION**    Make a boxplot by hand, using graph paper, a ruler, and a pencil by following the numbered steps. You may use the TI-83/84 output reported in Figure B.

**SOLUTION**

### Step 1 ▶ The Axis
Draw a horizontal line (using a ruler or other straight edge) across the page and label it in equal intervals up to a bit above the highest area (700), and down to a bit below the lowest area (0). Below the numbers, add a label telling what the numbers are.

### Step 2 ▶ The Box
Draw vertical lines at the positions of Q1 and Q3 and join them with horizontal lines to make a box. The height of the box is arbitrary.

### Step 3 ▶ The Median
Put a vertical line representing the median in the correct position inside the boxplot. Which state is at the median?

### Step 4 ▶ The Interquartile Range
Find and report the interquartile range, Q3 − Q1.

### Step 5 ▶ Lower Limit and Potential Lower Outliers
Find and report the lower limit by finding

$$Q1 - 1.5 \, IQR$$

If there are any points lower than that limit, make separate marks showing they are potential outliers.

### Step 6 ▶ Lower Whisker
Draw the lower whisker (horizontal line) from the box to the lowest point that is not a potential outlier. Which state is at the left end of the lower whisker?

### Step 7 ▶ Upper Limit and Potential Upper Outliers
Now find and report the upper limit by finding

$$Q3 + 1.5 \, IQR$$

If there are any points higher than that, make separate marks showing they are potential outliers. Report the name of the state that is a potential outlier.

### Step 8 ▶ Upper Whisker
Draw the upper whisker to the largest point that is not a potential outlier. Report this point and report the name of the state that the point represents.

### Step 9 ▶ Title
Give your graph an informative title, different from the one given.

# TechTips

## Example

Analyze the data given by finding descriptive statistics and making boxplots. The tables give calories per ounce for sliced ham and turkey. Table 3A shows unstacked data, and Table 3B shows stacked data. We coded the meat types with numerical values (1 for ham and 2 for turkey) but you could also use descriptive terms, such as "ham" or "turkey."

| Ham | Turkey |
|-----|--------|
| 21 | 35 |
| 25 | 25 |
| 35 | 25 |
| 35 | 25 |
| 25 | 25 |
| 30 | 25 |
| 30 | 29 |
| 35 | 29 |
| 40 | 23 |
| 30 | 50 |
|    | 25 |

▲ TABLE 3A

| Cal | Meat |
|-----|------|
| 21 | 1 |
| 25 | 1 |
| 35 | 1 |
| 35 | 1 |
| 25 | 1 |
| 30 | 1 |
| 30 | 1 |
| 35 | 1 |
| 40 | 1 |
| 30 | 1 |
| 35 | 2 |
| 25 | 2 |
| 25 | 2 |
| 25 | 2 |
| 25 | 2 |
| 25 | 2 |
| 29 | 2 |
| 29 | 2 |
| 23 | 2 |
| 50 | 2 |
| 25 | 2 |

▲ TABLE 3B

### TI-83/84

Enter the unstacked data (Table 3A) into **L1** and **L2**.

#### For Descriptive Comparisons of Two Groups

Follow the steps twice, first for **L1** and then for **L2**.

#### Finding One-Variable Statistics

1. Press **STAT**, choose **CALC** (by using the right arrow on the keypad), and choose **1** (for **1-Var Stats**)
2. Specify **L1** (or the list containing the data) by pressing **2ND**, **1**, and **ENTER**.
3. Output: On your calculator, you will need to scroll down using the down arrow on the keypad to see all of the output.

### Making Boxplots

1. Press **2ND**, **STATPLOT**, **4** (**PlotsOff**), and **ENTER** to turn the Plots Off. This will prevent you from seeing old plots as well as the new ones.
3. Press **2ND**, **STATPLOT**, and **1**.
4. Refer to Figure 3A. Turn on **Plot1** by pressing **ENTER** when **On** is flashing. (**Off** will no longer be highlighted.)
5. Use the arrows on the keypad to locate the boxplot in the lower left corner of the six graphs (as shown highlighted in Figure 3A), and press **ENTER**. (If you accidentally choose the other boxplot—the one in the middle—there will never be any separate marks for potential outliers.)

▲ FIGURE 3A TI-83/84 Plot Dialogue Screen

6. Use the down arrow on the keypad to get to the **XList**. Choose **L1** by pressing **2ND** and **1**.
7. Press **ZOOM** and **9** (**Zoomstat**) to make the graph.
8. Press **TRACE** and move around with the arrows on the keypad to see the numerical labels.

### Making Side-by-Side Boxplots

For side-by-side boxplots, turn on a second boxplot (**Plot2**) for data in a separate list, such as **L2**. Then, when you choose **ZOOM** and **9**, you should see both boxplots. Press **TRACE** to see numbers.

Figure 3B shows side-by-side boxplots with the boxplot from the turkey data on the bottom and the boxplot for the ham data on the top.

▲ FIGURE 3B TI-83/84 Boxplots

### MINITAB

#### For Comparisons of Two Groups

Upload the data from the disk or manually enter the data given. You may use unstacked data entered into two different columns (Table 3A), or you may use stacked data and use labels such as Ham and Turkey or codes such as 1 and 2 (Table 3B).

#### Finding Descriptive Statistics: One-Column Data or *Unstacked* Data in Two or More Columns

1. **Stat > Basic Statistics > Display Descriptive Statistics**.
2. Double click on the column(s) containing the data to put it (them) in the **Variables** box.
3. Ignore the **By variables** box.
4. (Optional) Click on **Statistics**; you can choose what you want to add, such as the interquartile range.
5. Click **OK**.

## Finding Descriptive Statistics: *Stacked and Coded*

1. **Stat > Basic Statistics > Display Descriptive Statistics**.
2. See Figure 3C: Double click on the column(s) containing the stack of data to put it in the **Variables** box.

▲ **FIGURE 3C** Minitab Descriptive Statistics Dialog Box

3. When the **By variables** box is activated (by clicking in it), double click the column containing the categorical labels or code.
4. (Optional) Click on **Statistics**; you can choose what you want to add, such as the interquartile range. You can also make boxpots by clicking **Graphs**.
5. Click **OK**.

## Making Boxplots

1. **Graph > Boxplot**.
2. For a single boxplot, choose **One Y**, **Simple** and click **OK**. See Figure 3D.
3. Double click the label for the column(s) containing the data and click **OK**.
4. For side-by-side boxplots
   a. If the data are unstacked, choose **Multiple Y's Simple**, shown in Figure 3D. Then double click both labels for the columns and click **OK**.
   b. If the data are stacked, choose **One Y With Groups** (the top right in Figure 3D). Then see Figure 3E and double click on the label for the stack (such as Calories) and then click in the **Categorical variables ...** box and double click the label for codes or words defining the groups such as **Meat**.
5. Labeling and transposing the boxplots. If you want to change the labeling, double click on what you want to change after the boxplot(s) are made. To change the orientation of the boxplot to horizontal, double click on the *x*-axis and select **Transpose value and category scales**.

▲ **FIGURE 3D** Minitab First Boxplot Dialog Box

▲ **FIGURE 3E** Minitab Dialog Screen for Boxplots: One Y with Groups

Figure 3F Shows Minitab boxplots of the ham and turkey data, without transposition.

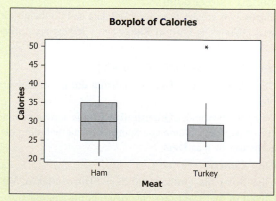

▲ **FIGURE 3F** Minitab Boxplots

## Entering Data

Input the unstacked data from the disk or enter it manually. You may have labels in the first row such as **Ham** and **Turkey**.

## Finding Descriptive Statistics

1. Click **Data**, **Data Analysis,** and **Descriptive Statistics**.
2. See Figure 3G: In the dialogue screen, for the **Input Range** highlight the cells containing the data (one column only) and then Click **Summary Statistics**. If you include the label in the top cell such as A1 in the **Input Range**, you need to check **Labels in First Row**. Click **OK**.

▲ **FIGURE 3G** Excel Dialog Box for Descriptive Statistics

## Comparing Two Groups

To compare two groups, use unstacked data and do the above analysis twice. (Choosing **Output Range** and selecting appropriate cells on the same sheet makes comparisons easier.)

## Boxplots (Requires XLSTAT Add-in)

1. Click **Add-ins**, **XLSTAT**, **Visualizing data**, and **Univariate plots.**
2. When the box under **Quantitative data** is activated, drag your cursor over the column containing the data including the label at the top such as **Ham**.
3. Click **Charts.**
4. Click **Box plots, Outliers** and choose **Horizontal** (or **Vertical**).
5. Click **OK** and **Continue**. See step 6 in the side-by-side instructions.

## Side by side Boxplots

Use unstacked data with labels in the top row.

1. Click **Add-ins**, **XLSTAT**, **Visualizing data**, and **Univariate plots**.
2. When the box under **Quantitative data** is activated, drag your cursor over all the columns containing the unstacked data including labels at the top such as **Ham** and **Turkey**.
3. Click **Charts**
4. Click **Box plots, Group plots, Outliers,** and choose **Horizontal** (or **Vertical**).
5. Click **OK** and **Continue**.
6. When you see the small labels **Turkey** and **Ham**, you may drag them to where you want them and you can increase the font size.

Figure 3H shows boxplots for the ham and turkey data. The red crosses give the location of the two means.

▲ **FIGURE 3H** XLSTAT Boxplots

## For Comparisons of Two Groups

Upload the data from the disk or enter the data manually. You may use stacked or unstacked data. Use the data from the example.

## Finding Summary Statistics (*stacked* or *unstacked* data)

1. **Stat > Summary Stats > Columns**
2. Refer to Figure 3I. If the data are stacked put the stack (here, Calories) in the big box on the right and put the code (here, Meat) in the rectangle labeled **Group by**:

   If the data are unstacked put both lists into the large box on the right.
3. Click **Next** for additional options such as extra statistics to be computed.
4. Click **Calculate** to compute the summary statistics.

▲ FIGURE 3I StatCrunch Dialog Screen for Summary Statistics

## Making Boxplots (stacked or unstacked)

1. **Graphics > Boxplot**
2. *Unstacked* data: Select the columns to be displayed. A boxplot for each column will be included in a single graph.
3. *Stacked* and coded data: Refer to Figure 3J:

▲ FIGURE 3J StatCrunch Dialog Screen for Boxplots

Put the one column with all the data into the big upper box and then select the column of codes or categories to put in the **Group by** small rectangle.

4. Click **Next** and check **Use fences to identify outliers** to make sure that the outliers show up as separate marks. You may also check **Draw boxes horizontally** if that is what you want.
5. Click **Create Graph!**
6. To copy your graph, click **Options** and **Copy** and paste it into a document.

Figure 3K Shows boxplots of the ham and turkey data; the top box comes from the turkey data (calories per ounce).

▲ FIGURE 3K StatCrunch Boxplots

# 4

# Regression Analysis: Exploring Associations between Variables

## THEME

Relationships between two numerical variables can be described in several ways, but as always, the first step is to understand what is happening visually. When it fits the data, the linear model can help us understand how average values of one variable vary with respect to another variable, and we can use this knowledge to make predictions about one of the variables on the basis of the other.

The website www.Zillow.com can estimate the market value of any home. You need merely type in an address. Zillow estimates the value of a home even if the home has not been on the market for many years. How can this site come up with an estimate for something that is not for sale? The answer is that Zillow takes advantage of associations between the value and other easily observed variables—size of the home, selling price of nearby homes, number of bedrooms, and so on.

What role does genetics play in determining basic physical characteristics, such as height? This question fascinated nineteenth-century statistician Francis Galton (1822–1911). He examined the heights of thousands of father-son pairs to determine the nature of the relationship between these heights. If a father is 6 inches taller than average, how much taller than average will his son be? How certain can we be of the answer? Will there be much variability? If there's a lot of variability, then perhaps factors other than the father's genetic material play a role in determining height.

Associations between variables can be used to predict as yet unseen observations. You might think that estimating the value of a piece of real estate and understanding the role of genetics in determining height are unrelated. However, both take advantage of associations between two numerical variables. They use a technique called regression, invented by Galton, to analyze these associations.

As in previous chapters, graphs play a major role in revealing patterns in data, and graphs become even more important when we have two variables, not just one. For this reason, we'll start by using graphs to visualize associations between two numerical variables, and then we'll talk about quantifying these relationships.

## CASE STUDY

## Catching Meter Thieves

Parking meters are an important source of revenue for many cities. Collecting the money from these meters is no small task, particularly in a large city. In the 1970s, New York City collected the money from its meters using several different private contractors and also some of its own employees. In 1978, city officials became suspicious that employees from one of the private contractors, Brink's Inc., were stealing some of the money. Several employees were later convicted of this theft, but the city wanted its money back, so it sued Brink's.

But how could the city tell how much money had been stolen? Fortunately, the city had collected data. For each month, it knew how much money its own employees had collected from parking meters and how much the private contractors (excluding Brink's) had collected. If there was a relationship between how much the city collected and how much the honest private contractors had collected, then that information could be used to predict about how much the city should have received from Brink's. City officials could then compare the predicted amount with the amount they actually collected from Brink's. (*Source*: De Groot et al., 1986)

At the end of this chapter, you'll see how a technique called linear regression can take advantage of patterns in the data to estimate how much money had been stolen.

# Visualizing Variability with a Scatterplot

At what age do men and women first marry? How does this vary among the 50 states in the United States? These are questions about the relationship between two variables: age at marriage for women, and age at marriage for men. The primary tool for examining two-variable relationships, when both variables are numerical, is the **scatterplot**. In a scatterplot, each point represents one observation. The location of the point depends on the values of the two variables. For example, we might expect that states where men marry later would also have women marrying at a later age. Figure 4.1 shows a scatterplot of these data, culled from the 2008 U.S. Census. Each point represents a state (and one point represents Washington, D.C.) and shows us the typical age at which men and women marry in that state. The two points in the lower left corner represent Idaho and Utah, where the typical woman first marries at about 23 years of age, the typical man at about age 25. The point in the upper right corner represents Washington, D.C.

▶ **FIGURE 4.1** A scatterplot of typical marrying ages for men and women in the United States. The points represent the 50 states and the District of Columbia.

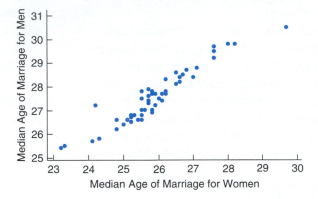

When examining histograms (or other pictures of distributions for a single variable), we look for center, spread, and shape. When studying scatterplots, we look for **trend** (which is like center), **strength** (which is like spread), and **shape** (which is like, well, shape). Let's take a closer look at these characteristics.

## Recognizing Trend

The trend of an association is the general tendency of the scatterplot as you scan from left to right. Usually trends are either increasing (uphill, /) or decreasing (downhill, \), but other possibilities exist. Increasing trends are called **positive associations** (or **positive trends**) and decreasing trends are **negative associations** (or **negative trends**).

Figure 4.2 shows examples of positive and negative trends. Figure 4.2a reveals a positive trend between the age of a used car and the miles it was driven (mileage). The positive trend matches our common sense: We expect older cars to have been driven farther, because generally, the longer a car is owned, the more miles it travels. Figure 4.2b shows a negative trend—the birthrate of a country against that country's literacy rate. The negative trend suggests that countries with higher literacy rates tend to have a lower rate of childbirth.

▲ **FIGURE 4.2** Scatterplots with **(a)** positive trend and **(b)** negative trend. (Source: United Nations [UN], Statistics Division, http://unstats.un.org)

Sometimes, the absence of a trend can be interesting. For example, running a marathon requires considerable training and endurance, and we might expect that the speed at which a person runs a marathon is related to his or her age. But Figure 4.3a shows no trend at all between the ages of runners in a marathon (in Forest City, Canada) and their times. The lack of a trend means that no matter what age group we examine, the runners have about the same times as any other age group, and so we conclude that at least for this group of elite runners, age is not associated with running speed in the marathon.

Figure 4.3b shows simulated data of an association between two variables that cannot be easily characterized as positive or negative—for smaller *x*-values the trend is negative (\), and for larger *x*-values it is positive (/).

▲ **FIGURE 4.3** Scatterplots with **(a)** no trend and **(b)** a changing trend. (Sources: (a) Forest City Marathon, http://www.forestcityroadraces.com/; (b) simulated data)

## Seeing Strength of Association

Weak associations result in a large amount of scatter in the scatterplot. A large amount of scatter means that points have a great deal of spread in the vertical direction. This vertical spread makes it somewhat harder to detect a trend. Strong associations have little vertical variation.

Figure 4.4 on the next page allows us to compare the strengths of two associations. Figure 4.4a shows the association between height and weight for a sample of active adults. Figure 4.4b involves the same group of adults, but this time we examine the association between waist size and weight. Which association is stronger?

The association between waist size and weight is the stronger one (Figure 4.4b). To see this, in Figure 4.4a, consider the data for people who are 65 inches tall. Their

▲ **FIGURE 4.4** Part **(a)** This graph shows a relatively weaker association. **(b)** This graph shows a stronger association, because the points have less vertical spread. (Heinz et al. 2003)

weights vary anywhere from about 120 pounds to 230 pounds, a range of 110 pounds. If you were using height to predict weight, you could be off by quite a bit. Compare this with the data in Figure 4.4b for people with a waist size of 30 inches. Their weights vary from about 120 pounds to 160 pounds, only a 40-pound range. The association between waist size and weight is stronger than that between height and weight because there is less vertical spread, so tighter predictions can be made. If you had to guess someone's weight, and could ask only one question before guessing, you'd do a better job if you asked about the person's waist size than about his or her height.

Labeling a trend as strong, very strong, or weak is a subjective judgment. Different statisticians might have different opinions. Later in this section, we'll see how we can measure strength with a number.

## Identifying Shape

The simplest shape for a trend is **linear**. Fortunately, linear trends are quite common in nature. Linear trends always increase (or decrease) at the same rate. They are called linear because the trend can be summarized with a straight line. Scatterplots of linear trends often look roughly football-shaped, as shown in Figure 4.4a, particularly if there is some scatter and there are a large number of observations. Figure 4.5 shows a linear trend from data provided by Google Trends. Each point represents a week between January 2006 and December 2009. The numbers measure the frequency of Google searches for the term *vampire* or *zombie*. Figure 4.5 shows that a positive, linear association exists between the number of searches for *vampire* and the number for *zombie*. We've added a straight line to the scatterplot to highlight the linear trend.

▶ **FIGURE 4.5** A line has been inserted to emphasize the linear trend.

Not all trends are linear; in fact, a great variety of shapes can occur. But don't worry about that for now: All we want to do is classify trends as either linear or not linear.

Figure 4.6 shows the relationship between levels of the pollutant ozone in the air (measured in parts per million, or ppm) and air temperature (degrees Fahrenheit) in Upland, California, near Los Angeles, over the course of a year. The trend is fairly flat at first and then becomes steeper. For temperatures less than about 55 degrees, ozone levels do not vary all that much. However, higher temperatures (above 55 degrees) are associated with much greater ozone levels. The curved line superimposed on the graph shows the nonlinear trend.

◄ **FIGURE 4.6** Ozone (ppm) is associated with temperature (degrees Fahrenheit) in a nonlinear way. (Source: Breiman, L. From the R earth package. See Faraway 2005.)

Nonlinear trends are more difficult to summarize than linear trends. This book does not cover nonlinear trends. Although our focus is on linear trends, it is very important that you first examine a scatterplot to be sure that the trend is linear. If you apply the techniques in this chapter to a nonlinear trend, you might reach disastrously incorrect conclusions!

**KEY POINT** ➤ When examining associations, look for the trend, the strength of the trend, and the shape of the trend.

## Writing Clear Descriptions of Associations

Good communication skills are vital for success in general, and being able to clearly describe patterns in data is an important goal of this book. Here are some tips to help you describe two-variable associations.

- A written description should always include (1) trend, (2) shape, and (3) strength (not necessarily in that order) and should explain what all of these mean in (4) the context of the data. You should also mention any observations that do not fit the general trend.

Example 1 demonstrates how to write a clear, precise description of an association between numerical variables.

## EXAMPLE 1 Age and Mileage of Used Cars

Figure 4.2a on page 135 displays an association between the age and mileage of a sample of used cars.

QUESTION Describe the association.

SOLUTION The association between the age and mileage of used cars is positive and linear. This means that older cars tend to have greater mileage. The association is moderately strong; some scatter is present, but not enough to hide the shape of the relationship. There is one exceptional point: One car is only about 6 years old but has been driven many miles.

TRY THIS! Exercise 4.7

The description in Example 1 is good because it mentions trend (a "positive" association), shape ("linear"), and strength ("moderately strong") and does so in context ("older cars tend to have greater mileage").

- It is very important that your descriptions be precise. For example, it would be wrong to say that older cars have greater mileage. This statement is not true of every car in the data set. The one exceptional car (upper left corner of the plot) is relatively new (about 6 years old) but has a very high mileage (about 250,000 miles). Some older cars have relatively few miles on them. To be precise, you could say older cars *tend* to have higher mileage. The word *tend* indicates that you are describing a trend that has variability, so the trend you describe is not true of all individuals but instead is a characteristic of the entire group.

- When writing a description of a relationship, you should also mention unusual features, such as outliers, small clusters of points, or anything else that does not seem to be part of the general pattern. Figure 4.7 includes an outlier. These data are from a statistics class in which students reported their weights and heights. One student wrote the wrong height.

► **FIGURE 4.7** A fairly strong, positive association between height and weight for a statistics class in 2006. One student reported the wrong height. (Source: R. Gould, UCLA, Department of Statistics)

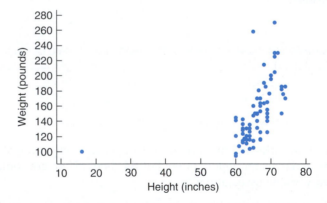

# SECTION 4.2

# Measuring Strength of Association with Correlation

The **correlation coefficient** is a number that measures the strength of the linear association between two numerical variables—for example, the relationship between people's heights and weights. We can't emphasize enough that the correlation coefficient *makes sense only if the trend is linear and if both variables are numerical.*

The correlation coefficient, represented by the letter $r$, is always a number between $-1$ and $+1$. Both the value and the sign (positive or negative) of $r$ have information we can use. If the value of $r$ is close to $-1$ or $+1$, then the association is very strong; if $r$ is close to 0, the association is weak. If the value of the correlation coefficient is positive, then the trend is positive; if the value is negative, the trend is negative.

## Visualizing the Correlation Coefficient

Figure 4.8 presents a series of scatterplots that show associations of gradually decreasing strength. The strongest linear association appears in Figure 4.8a; the points fall exactly along a line. Because the trend is positive and perfectly linear, the correlation coefficient is equal to 1.

The next scatterplot, Figure 4.8b, is slightly weaker. The points are more spread out vertically. We can see a linear trend, but the points do not fall exactly along a line. The trend is still positive, so the correlation coefficient is also positive. However, the value of the correlation coefficient is less than 1 (it is 0.98).

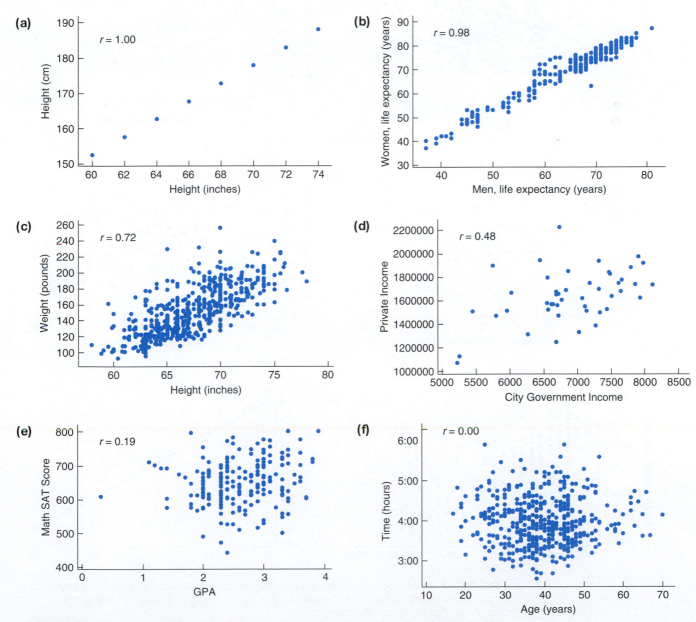

▲ **FIGURE 4.8** Scatterplots with gradually decreasing positive correlation coefficients. (Sources: (a) simulated data; (b) www.overpopulation.com; (c) Heinz, 2003; (d) Bentow and Afshartous; (e) Fathom™ Sample Document, Educational Testing Service [ETS] validation study; (f) Forest City Marathon.)

The remaining scatterplots show weaker and weaker associations, and their correlation coefficients gradually decrease. The last scatterplot, Figure 4.8f, has no association at all between the two variables, and the correlation coefficient has a value of 0.00.

The next set of scatterplots (Figure 4.9) starts with that same marathon data (having a correlation of 0.00), and the negative correlations gradually get stronger. The last figure has a correlation of −1.00.

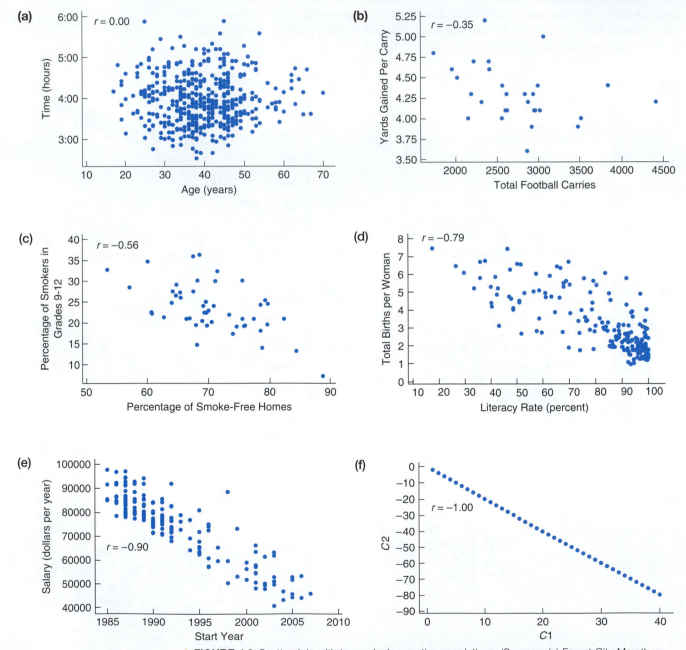

▲ **FIGURE 4.9** Scatterplots with increasingly negative correlations. (Sources: (a) Forest City Marathon; (b) http://wikipedia.org; (c) Centers for Disease Control; (d) UN Statistics; (e) Minitab Student 12 file "Salary," adjusted for inflation; (f) simulated data.)

## The Correlation Coefficient in Context

The correlation between the number of Google searches for *zombie* and *vampire* is $r = 0.924$. If we are told, or already know, that the association between these variables is linear, then we know that the trend is positive and strong. The fact that the correlation is close to 1 means that there is not much scatter in the scatterplot.

College admission offices sometimes report correlations between students' Scholastic Aptitude Test (SAT) scores and their first-year GPA. If the association is linear and the correlation is high, this justifies using the SAT to make admissions decisions, because a high correlation would indicate a strong association between SAT scores and academic performance. A positive correlation means that students who score above average on the SAT tend to get above-average grades. Conversely, those who score below average on the SAT tend to get below-average grades. Note that we're careful to say "tend to." Certainly, some students with low SAT scores do very well, and some with high SAT scores struggle to pass their classes. The correlation coefficient does not tell us about individual students; it tells us about the overall trend.

## More Context: Correlation Does Not Mean Causation!

Quite often, you'll hear someone use the correlation coefficient to support a claim of cause and effect. For example, one of the authors once read that a politician wanted to close liquor stores in a city because there was a positive correlation between the number of liquor stores in a neighborhood and the amount of crime.

As you learned in Chapter 1, we can't form cause-and-effect conclusions from observational studies. If your data came from an observational study, it doesn't matter how strong the correlation is. Even a correlation near 1 is not enough to conclude that changing one variable (closing down liquor stores) will lead to a change in the other variable (crime rate).

A positive correlation also exists between the number of blankets sold in Canada per week and the number of brush fires in Australia per week. Are brush fires in Australia caused by cold Canadians? Probably not. The correlation is likely to be the result of weather. When it is winter in Canada, people buy blankets. When winter is happening in Canada, summer is happening in Australia (which is located in the southern hemisphere), and summer is brush-fire season.

What, then, can we conclude from the fact that the number of liquor stores in a neighborhood is positively correlated with the crime rate in that neighborhood? Only that neighborhoods with a higher-than-average number of liquor stores typically (but not always) have a higher-than-average crime rate.

If you learn nothing else from this book, remember this: No matter how tempting, do *not* conclude that a cause-and-effect relationship between two variables exists just because there is a correlation, no matter how close to $+1$ or $-1$ that correlation might be!

 **KEY POINT**    Correlation does not imply causation.

## Finding the Correlation Coefficient

The correlation coefficient is best determined through the use of technology. We calculate a correlation coefficient by first converting each observation to a *z*-score, using the appropriate variable's mean and standard deviation. For example, to find the correlation coefficient that measures the strength of the linear relation between weight and height, we first convert each person's weight and height to *z*-scores. The next step is to multiply the observations' *z*-scores together. If both are positive or both negative—meaning that both *z*-scores are above average or both are below average—then the product is a positive number. In a strong positive association, most of these products are positive values. In a strong negative association, however, observations above

 **Looking Back**

**z-scores**
Recall that z-scores show how many standard deviations a measurement is from the mean. To find a z-score from a measurement, first subtract the mean and then divide the difference by the standard deviation.

average on one variable tend to be below average on the other variable. In this case, one $z$-score is negative and one positive, so the product is negative. Thus, in a strong negative association, most $z$-score products are negative.

To find the correlation coefficient, add the products of $z$-scores together and divide by $n - 1$ (where $n$ is the number of observed pairs in the sample). In mathematical terms, we get

$$\text{Formula 4.1: } r = \frac{\sum z_x z_y}{n - 1}$$

The following example illustrates how to use Formula 4.1 in a calculation.

 ## EXAMPLE **2** Heights and Weights of Six Women

Figure 4.10a shows the scatterplot for heights and weights of six women.

**QUESTION** Using the data provided, find the correlation coefficient between heights (inches) and weights (pounds) for these six women.

| Heights | 61 | 62 | 63 | 64 | 66 | 68 |
|---------|-----|-----|-----|-----|-----|-----|
| Weights | 104 | 110 | 141 | 125 | 170 | 160 |

▶ **FIGURE 4.10a** Scatterplot showing heights and weights of six women.

**SOLUTION** Before proceeding, we verify that the conditions hold. Figure 4.10a suggests that a straight line is an acceptable model; a straight line through the data might summarize the trend, although this is hard to see with so few points.

Next, we calculate the correlation coefficient. Ordinarily, we use technology to do this, and Figure 4.10b shows the output from StatCrunch, which gives us the value $r = 0.88093364$.

**Tech**

▶ **FIGURE 4.10b** StatCrunch, like all statistical software, lets you calculate the correlation between any two columns you choose.

**Data set from:**

| StatCrunch | Data | Stat | Graphics | Help |

| Row | Height | Weight | var3 |
|-----|--------|--------|------|
| | 61 | 104 | |
| | 62 | 110 | |
| | 63 | 141 | |
| | 64 | 125 | |
| | 66 | 170 | |
| 6 | 68 | 160 | |
| 7 | | | |
| 8 | | | |

**Correlation**

**Options**

Correlation between Height and Weight is:
0.88093364

Because the sample size is small, we confirm this output using Formula 4.1. It is helpful to go through the steps of this calculation to better understand how the correlation coefficient measures linear relationships between variables.

The first step is to calculate average values of height and weight and then determine the standard deviation for each.

$$\text{For the height: } \bar{x} = 64 \text{ and } s_x = 2.608$$

$$\text{For the weight: } \bar{y} = 135 \text{ and } s_y = 26.73$$

Next we convert all of the points to pairs of $z$-scores and multiply them together. For example, for the woman who is 68 inches tall and weighs 160 pounds,

$$z_x = \frac{x - \bar{x}}{s} = \frac{68 - 64}{2.608} = \frac{4}{2.608} = 1.53$$

$$z_y = \frac{y - \bar{y}}{s} = \frac{160 - 135}{26.73} = \frac{25}{26.73} = 0.94$$

The product is

$$z_x \times z_y = 1.53 \times 0.94 = 1.44$$

Note that this product is positive and shows up in the upper right quadrant in Figure 4.10c.

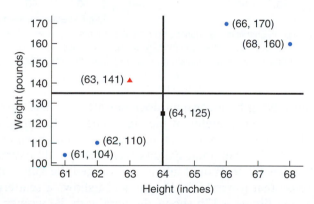

◀ **FIGURE 4.10c** The same scatterplot as in 4.10a, but with the plot divided into quadrants based on average height and weight. Points represented with blue circles contribute a positive value to the correlation coefficient (positive times positive is positive or negative times negative equals a positive). The red triangle represents an observation that contributes negatively (a negative $z$-score times a positive $z$-score is negative), and the black square contributes nothing because one of the $z$-scores is 0.

Figure 4.10c can help you visualize the rest of the process. The two blue circles in the upper right portion represent observations that are above average in both variables, so both $z$-scores are positive. The two blue circles in the lower left region represent observations that are below average in both variables; the products of the two negative $z$-scores are positive, so they add to the correlation. The red triangle has a positive $z$-score for weight (it is above average) but a negative $z$-score for height, so the product is negative. The black square is a point that makes no contribution to the correlation coefficient. This person is of average height, so her $z$-score for height is 0.

The correlation between height and weight for these six women comes out to be about 0.881.

CONCLUSION  The correlation coefficient for the linear association of weights and heights of these six women is $r = 0.881$. Thus, there is a strong positive correlation between height and weight for these women. Taller women tend to weigh more.

TRY THIS!  Exercise 4.19a

## Understanding the Correlation Coefficient

The correlation coefficient has a few features you should know about when interpreting a value of $r$ or deciding whether you should compute the value.

- Changing the order of the variables does not change $r$. Note that in the equation for $r$ it doesn't matter which variable is called $x$ and which is called $y$. In practice, this means that if the correlation between life expectancy for men and women is 0.977, then the correlation between life expectancy for women and men is also 0.977. This makes sense because the correlation measures the strength of the linear relationship between $x$ and $y$, and that strength will be the same no matter which variable gets plotted on the horizontal axis and which on the vertical.

Figure 4.11a and b have the same correlation; we've just swapped axes.

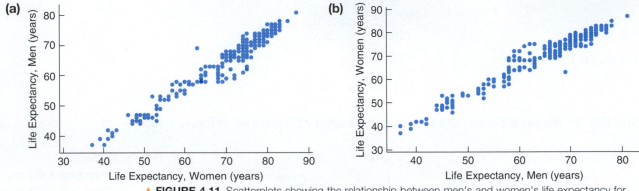

▲ **FIGURE 4.11** Scatterplots showing the relationship between men's and women's life expectancy for various countries. **(a)** Women's life expectancy is plotted on the $x$-axis. **(b)** Men's life expectancy is plotted on the $x$-axis. (Sources: http://www.overpopulation.com and Fathom™ Sample Documents)

- Adding or multiplying by a constant does not affect $r$. The correlation between the heights and weights of the six women in Example 2 was 0.881. What would happen if all six women in the sample had been asked to wear 3-inch platform heels when their heights were measured? Everyone would have been 3 inches taller. Would this have changed the value of $r$? Intuitively, you should sense that it wouldn't. Figure 4.12a shows a scatterplot of the original data, and Figure 4.12b shows the data with the women in 3-inch heels.

▲ **FIGURE 4.12 (a)** A repeat of the scatterplot of height and weight for six women. **(b)** The same women in 3-inch heels. The correlation remains the same.

We haven't changed the strength of the relationship. All we've done is shift the points on the scatterplot 3 inches to the right. But shifting the points doesn't change the relationship between height and weight. We can verify that the correlation is unchanged by looking at the formula. The heights will have the same $z$-scores both before and after the women put on the shoes; since everyone "grows" by the same amount, everyone is still the same number of standard deviations away from the average, which also "grows" by 3 inches. As another example, if science found a way to add 5 years to the life expectancy of men in all countries in the world, the correlation between life expectancies for men and women would still be the same.

More generally, we can add a constant (a fixed value) to all of the values of one variable, or of both variables, and not affect the correlation coefficient.

For the very same reason, we can multiply either or both variables by positive constants without changing $r$. For example, to convert the women's heights from inches to feet, we multiply their heights by 1/12. Doing this does not change how strong the association is; it merely changes the units we're using to measure height. Because the strength of the association does not change, the correlation coefficient does not change.

- The correlation coefficient is unitless. Height is measured in inches and weight in pounds, but $r$ has no units because the $z$-scores have no units. This means that we will get the same value for correlation whether we measure height in inches, meters, or fathoms.

- Linear, linear, linear. We've said it before, but we'll say it again: We're talking only about linear relationships here. The correlation can be misleading if you do not have a linear relationship. Figure 4.13a through d illustrate the fact that different nonlinear patterns can have the same correlation. All of these graphs have $r = 0.817$, but the graphs have very different shapes. The take-home message is that the correlation alone does not tell us much about the shape of a graph. We must also know that the relationship is linear to make sense of the correlation.

Remember: *Always* make a graph of your data. If the trend is nonlinear, the correlation (and, as you'll see in the next section, other statistics) can be very misleading.

> **! Caution**
>
> **Correlation Coefficient and Linearity**
> A value of $r$ close to 1 or $-1$ does *not* tell you that the relationship is linear. You must check visually; otherwise, your interpretation of the correlation coefficient might be wrong.

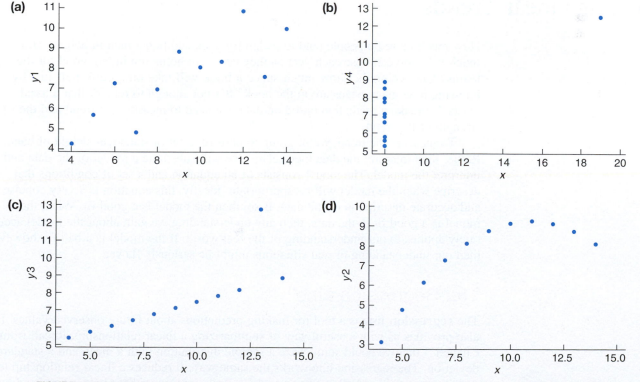

▲ **FIGURE 4.13 (a–d)** Four scatterplots with the same correlation of 0.817 have very different shapes. The correlation coefficient is meaningful only if the trend is linear. (Source: Anscombe, F. Anscombe's Quartet)

> **KEY POINT**
> The correlation coefficient does not tell you whether an association is linear. However, if you already know that the association is linear, then the correlation coefficient tells you how strong the association is.

## SNAPSHOT  THE CORRELATION COEFFICIENT

**WHAT IS IT?** ▶ Correlation coefficient.

**WHAT DOES IT DO?** ▶ Measures the strength of a linear association.

**HOW DOES IT DO IT?** ▶ By comparing *z*-scores of the two variables. The products of the two *z*-scores for each point are averaged.

**HOW IS IT USED?** ▶ The sign tells us whether the trend is positive (+) or negative (−). The value tells us the strength. If the value is close to 1 or −1, then the points are tightly clustered about a line; if the value is close to 0, then there is no linear association.

*Note:* The correlation coefficient can be interpreted only with linear associations.

## SECTION 4.3

# Modeling Linear Trends

How much more do people tend to weigh for each additional inch in height? How much value do cars lose each year as they age? Are home run hitters good for their teams? Can we predict how much space a book will take on a bookshelf just by knowing how many pages are in the book? It's not enough to remark that a trend exists. To make a prediction based on data, we need to measure the trend and the strength of the trend.

To measure the trend, we're going to perform a bit of statistical sleight of hand. Rather than interpret the data themselves, we will substitute a model of the data and interpret the model. The model consists of an equation and a set of conditions that describe when the model will be appropriate. Ideally, this equation is a very concise and accurate description of the data; if so, then the model is a good fit. When the model is a good fit to the data, then any understanding we gain about the model accurately applies to our understanding of the real world. If the model is a bad fit, however, then our understanding of real situations might be seriously flawed.

## The Regression Line

The **regression line** is a tool for making predictions about future observed values. It also provides us with a useful way of summarizing a linear relationship. Recall from Chapter 3 that we could summarize a sample distribution with a mean and a standard deviation. The regression line works the same way: It reduces a linear relationship to its bare essentials and allows us to analyze a relationship without being distracted by small details.

## Review: Equation of a Line

The regression line is given by an equation for a straight line. Recall from algebra that equations for straight lines contain an **intercept** and a **slope**. The equation for a straight line is

$$y = mx + b$$

The letter $m$ represents the slope, which tells how steep the line is, and the letter $b$ represents the $y$-intercept, which is the value of $y$ when $x = 0$.

Statisticians write the equation of a line slightly differently and put the intercept first; they use the letter $a$ for the intercept and $b$ for the slope and write

$$y = a + bx$$

We often use the names of variables in place of $x$ and $y$ to emphasize that the regression line is a model about two real-world variables. We will sometimes write the word *predicted* in front of the $y$-variable to emphasize that the line consists of predictions for the $y$-variable, not actual values. A few examples should make this clear.

## Visualizing the Regression Line

Can we know how wide a book is on the basis of the number of pages in the book? A student took a random sample of books from his shelf, measured the width of the spine (in millimeters, mm), and recorded the number of pages. Figure 4.14 illustrates how the regression line captures the basic trend of a linear association between width of the book and the number of pages for this sample. The equation for this line is

$$\text{Predicted Width} = 6.22 + 0.0366 \text{ Pages}$$

◄ **FIGURE 4.14** The regression line summarizes the relationship between the width of the book and the number of pages for a small sample of books. (Source: Onaga, E. 2005, UCLA, Department of Statistics)

In baseball, two numbers used to measure how good a batter is are the number of runs batted in (RBI) and the number of home runs. (A home run occurs when the batter rounds all three bases with one hit and scores a run. An RBI occurs when a player is already on base, and a batter hits the ball far enough for the on-base runner to score a run.) Some baseball fans believe that players who hit a lot of home runs might be exciting to watch but are not that good for the team. (Some believe that home run hitters tend to strike out more often; this takes away scoring opportunities for the team.) Figure 4.15 on the next page shows the relationship between the number of home runs and RBIs for the best home run hitters in the major leagues in the 2008 season. The association seems fairly linear, and the regression line can be used to predict how many RBIs a player will score, given the number of home runs hit. The data suggest that players who score a large number of home runs also tend to score a large number of points through RBIs.

$$\text{Predicted RBIs} = 30.46 + 2.16 \text{ Home Runs}$$

▶ **FIGURE 4.15** The regression line summarizes the relationship between RBIs and home runs for the top home run hitters in the 2008 major league baseball season. (Source: www.baseball1.com)

## Regression in Context

Suppose you have a 10-year-old car and want to estimate how much it is worth. One of the more important uses of the regression line is to make predictions about what *y*-values can be obtained for particular *x*-values. Figure 4.16 suggests that the relationship between age and value is linear, and the regression line that summarizes this relationship is

$$\text{Predicted Value} = 21375 - 1215 \text{ Age}$$

We can use this equation to predict approximately how much a 10-year-old car is worth:

$$\text{Predicted Value} = 21375 - 1215 \times 10$$
$$= 27375 - 12150$$
$$= 9225$$

▶ **FIGURE 4.16** The regression line summarizes the relationship between the value of a car, according to the Kelley Blue Book, and the car's age for a small sample of students' cars. (Source: C. Ryan 2006)

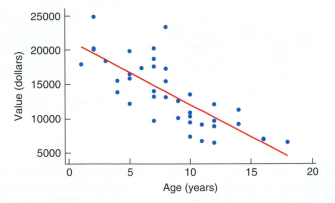

The regression line predicts that a 10-year-old car will be valued at about $9225. As we know, many factors other than age affect the value of a car, and perhaps with more information we might make a better prediction. However, if the only thing we know about a car is its age, this may be the best prediction we can get. It is also important to keep in mind that this sample comes from one particular group of students in one particular statistics class and is not representative of all used cars on the market in the United States.

Using the regression line to make predictions requires certain assumptions. We'll go into more detail about these assumptions later, but for now, just use common sense. This predicted value of $9225 is useful only if our data are valid. For instance, if all the cars in our data set are Toyotas, and our 10-year-old car is a Chevrolet, then the prediction is probably not useful.

## EXAMPLE 3 Book Width

A college instructor with far too many books on his shelf is wondering whether he has room for one more. He has about 20 mm of space left on his shelf, and he can tell from the online bookstore that the book he wants has 598 pages. The regression line is

$$\text{Predicted Width} = 6.22 + 0.0366 \, \text{Pages}$$

QUESTION  Will the book fit on his shelf?

SOLUTION  Assuming that the data used to fit this regression line are representative of all books, we would predict the width of the book corresponding to 598 pages to be

$$\text{Predicted Width} = 6.22 + 0.0366 \times 598$$
$$= 6.22 + 21.8868$$
$$= 28.1068 \, \text{mm}$$

CONCLUSION  The book is predicted to be 28 mm wide. Even though the actual book width is likely to differ somewhat from 28 mm, it seems that the book will probably not fit on the shelf.

TRY THIS!  Exercise 4.21

Common sense tells us that not all books with 598 pages are exactly 28 mm wide. There is a lot of variation in the width of a book for a given number of pages. In Chapter 14, we'll see how to place upper and lower limits on this predicted width so that we can take into account the uncertainty caused by the variation in books.

## Finding the Regression Line

In almost every case, we'll use technology to find the regression line. However, it is important to know how the technology works, and to be able to calculate the equation when we don't have access to the full data set.

To understand how technology finds the regression line, imagine trying to draw a line through the scatterplots in Figures 4.14 through 4.16 to best capture the linear trend. We could have drawn almost any line, and some of them would have looked like pretty good summaries of the trend. What makes the regression line special? How do we find the intercept and slope of the regression line?

The regression line is chosen because it is the line that comes closest to most of the points. More precisely, the square of the vertical distances between the points and the line, on average, is bigger for any other line we might draw than for the regression

---

📌 **Details**

**Least Squares**
The regression line is also called the least squares line because it is chosen so that the sum of the squares of the differences between the observed $y$-value, $y$, and the value predicted by the line, $\hat{y}$, is as small as possible. Mathematically, this means that the slope and intercept are chosen so that $\Sigma(y - \hat{y})^2$ is as small as possible.

line. Figure 4.17 shows these vertical distances with black lines. The regression line is sometimes called the "best fit" line because, in this sense, it provides the best fit to the data.

**◄ FIGURE 4.17** The "best fit" regression line, showing the vertical distance between each observation and the line. For any other line, the average of the squared distances is larger.

To find this best fit line, we need to find the slope and the intercept. The slope, $b$, of the regression line is the ratio of the standard deviations of the two variables, multiplied by the correlation coefficient:

**Formula 4.2a:** The slope, $b = r\dfrac{s_y}{s_x}$

Once we have found the slope, we can find the intercept. Finding the intercept, $a$, requires that we first find the means of the variables, $\bar{x}$ and $\bar{y}$:

**Formula 4.2b:** The intercept, $a = \bar{y} - b\bar{x}$

Now put these quantities into the equation of a line, and the regression line is given by

**Formula 4.2c:** The regression line, Predicted $y = a + bx$

##  EXAMPLE 4 SAT Scores and GPAs

A college in the northeastern United States used SAT scores to help decide which applicants to admit. To determine whether the SAT was useful in predicting success, the college examined the relationship between the SAT scores and first-year GPAs of admitted students. Figure 4.18a shows a scatterplot of the math SAT scores and first-year GPAs for a random sample of 200 students. The scatterplot suggests a weak, positive linear association: Students with higher math SAT scores tend to get higher first-year GPAs. The average math SAT score of this sample was 649.5 with standard deviation 66.3. The average GPA was 2.63 with standard deviation 0.58. The correlation between GPA and math SAT score was 0.19.

**QUESTIONS**

a. Find the equation of the regression line that best summarizes this association. Note that the *x*-variable is the math SAT score and the *y*-variable is the GPA.

b. Using the equation, find the predicted GPA score for a person in this group with a math SAT score of 650.

◀ **FIGURE 4.18a** Students with higher math SAT scores tend to have higher first-year GPAs, but this linear association is fairly weak. (Source: ETS validation study of an unnamed college and Fathom™ Sample Documents, 2007, Key Curriculum Press)

**SOLUTIONS**

a. With access to the full data set on the text's companion CD-ROM, we can use technology to find (and plot) the regression line. Still, when summary statistics are provided (means and standard deviations of the two variables as well as their correlation), it is not time-consuming to use Formula 4.2 to find the regression line.

Figure 4.18b shows StatCrunch output for the regression line. (StatCrunch provides quite a bit more information than we will use in this chapter.)

According to technology, the equation of the regression line is

$$\text{Predicted GPA} = 1.53 + 0.0017 \text{ Math}$$

 Tech

◀ **FIGURE 4.18b** StatCrunch output for the regression line to predict GPA from math SAT score. The structure of the output is typical for many statistical software packages.

**Simple Linear Regression**  Java Applet Window

**Options**

**Simple linear regression results:**
Dependent Variable: gpa
Independent Variable: math
gpa = 1.5264336 + 0.001699023 math
Sample size: 200
R (correlation coefficient) = 0.1942
R-sq = 0.037730377
Estimate of error standard deviation: 0.57071334

**Parameter estimates:**

| Parameter | Estimate | Std. Err. | DF | T-Stat | P-Value |
|---|---|---|---|---|---|
| Intercept | 1.5264336 | 0.39811763 | 198 | 3.8341272 | 0.0002 |
| Slope | 0.001699023 | 6.097748E-4 | 198 | 2.786312 | 0.0058 |

**Analysis of variance table for regression model:**

| Source | DF | SS | MS | F-stat | P-value |
|---|---|---|---|---|---|
| Model | 1 | 2.5286899 | 2.5286899 | 7.7635355 | 0.0058 |
| Error | 198 | 64.49131 | 0.3257137 | | |

<- Back    Next ->

We now check this calculation by hand, using Formula 4.2.
We are given that

For math SAT scores: $\bar{x} = 649.5$, $s_x = 66.3$

For GPAs: $\bar{y} = 2.63$, $s_y = 0.580$

and $r = 0.194$

First we must find the slope:

$$b = r\frac{s_y}{s_x} = 0.194 \times \frac{0.580}{66.3} = 0.001697$$

Now we can use the slope to find the intercept:

$$a = \bar{y} - b\bar{x} = 2.63 - 0.001697 \times 649.5 = 2.63 - 1.102202 = 1.53$$

Rounding off:

Predicted GPA = 1.53 + 0.0017 math

b. Predicted GPA = 1.53 + 0.0017 math

$$= 1.53 + 0.0017 \times 650$$
$$= 1.53 + 1.105$$
$$= 2.635$$

CONCLUSION

We would expect someone with a math SAT score of 650 to have a GPA of about 2.64.

TRY THIS!   Exercise 4.23

---

Different software packages present the intercept and slope differently. Therefore, you need to learn how to read the output of the software you are using. Example 5 shows output from several packages.

## EXAMPLE 5 Technology Output for Regression

Figure 4.19 shows outputs from Minitab, StatCrunch, the TI-83/84, and Excel for finding the regression equation in Example 4 for GPA and math SAT scores.

QUESTION   For each software package, explain how to find the equation of the regression line from the given output.

CONCLUSION   Figure 4.19a: Minitab gives us a simple equation directly:
GPA = 1.53 + 0.00170 Math.
  However, the more statistically correct format would be

Predicted GPA = 1.53 + 0.00170 Math

▶ FIGURE 4.19a Minitab output.

> **Regression Analysis: GPA versus Math**
> The regression equation is
> GPA = 1.53 + 0.00170 Math

Figure 4.19b: StatCrunch gives the equation directly near the top, but it also lists the intercept and slope separately in the table near the bottom.

**Simple linear regression results:**
Dependent Variable: gpa
Independent Variable: math
gpa = 1.5264336 + 0.001699023 math
Sample size: 200
R (correlation coefficient) = 0.1942
R-sq = 0.037730377
Estimate of error standard deviation: 0.57071334

**Parameter estimates:**

| Parameter | Estimate | Std. Err. | DF | T–Stat | P–Value |
|-----------|----------|-----------|-----|--------|---------|
| Intercept | 1.5264336 | 0.39811763 | 198 | 3.8341272 | 0.0002 |
| Slope | 0.001699023 | 6.097748E-4 | 198 | 2.786312 | 0.0058 |

**Analysis of variance table for regression model:**

| Source | DF | SS | MS | F–stat | P–value |
|--------|-----|-----|-----|--------|---------|
| Model | 1 | 2.5286899 | 2.5286899 | 7.7635355 | 0.0058 |
| Error | 198 | 64.49131 | 0.3257137 | | |
| Total | 199 | 67.02 | | | |

▲ **FIGURE 4.19b** StatCrunch output.

```
LinReg
 y=a+bx
 a=1.526433632
 b=.0016990229
 r²=.0377303763
 r=.1942430855
```

▲ **FIGURE 4.19c** TI-83/84 output.

| | Coefficients | Standard Error | t Stat | P-valve | Lower 95% | Upper 95% | Lower 95.0% | Upper 95.0% |
|---|---|---|---|---|---|---|---|---|
| Intercept | 1.526434 | 0.398118 | 3.834127 | 0.000169 | 0.741339 | 2.311528 | 0.741339 | 2.311528 |
| Variable 1 | 0.001699 | 0.00061 | 2.786312 | 0.00585 | 0.000497 | 0.002902 | 0.000497 | 0.002902 |

▲ **FIGURE 4.19d** Excel output.

Figure 4.19c: TI-83/84 gives us the coefficients to put together. The "a" value is the intercept, and the "b" value is the slope. If the diagnostics are on (use the CATALOG button), the TI-83/84 also gives the correlation.

Figure 4.19d: Excel shows the coefficients in the column labeled "Coefficients." The intercept is in the first row, labeled "Intercept," and the slope is in the row labeled "Variable 1."

**TRY THIS!** Exercise 4.27

## Interpreting the Regression Line

An important use of the regression line is to make predictions about the value of $y$ that we will see for a given value of $x$. However, the regression line provides more information than just a predicted $y$-value. The regression line can also tell us about the rate of change of the mean value of $y$ with respect to $x$ and can help us understand the underlying theory behind cause-and-effect relationships.

### Choosing $x$ and $y$: Order Matters

In Section 4.2, you saw that the correlation coefficient is the same no matter which variable you choose for $x$ and which you choose for $y$. With regression, however, order matters.

Consider the collection of data about book widths. We used it earlier to predict the width of a book, given the number of pages. The equation of the regression line for this prediction problem (shown in Figure 4.20a) is

$$\text{Predicted Width} = 6.22 + 0.037 \text{ Pages}$$

But what if we instead wanted to predict how many *Pages* there are in a book on the basis of the width of the book?

To do this, we would switch the order of the variables, and use *Pages* as our *y*-variable and *Width* as our *x*-variable. Then the slope is calculated to be 19.6 (Figure 4.20b).

It is tempting to think that because we are flipping the graph over when we switch *x* and *y*, we can just flip the slope over to get the new slope. If this were true, then we could find the new slope simply by calculating 1/(old slope). However, that approach doesn't work. That would give us a slope of

$$\frac{1}{0.037} = 27.0, \text{ which is not the same as the correct value of } 19.6.$$

How, then, do we know which variable goes where?

We use the variable plotted on the horizontal axis to make predictions about the variable plotted on the vertical axis. For this reason, the *x*-variable is called the **explanatory variable**, the **predict*or* variable**, or the **independent variable**. The *y*-variable is called the **response variable**, the **predict*ed* variable**, or the **dependent variable**. These names reflect the different roles played by the *x*- and *y*-variables in regression. Which variable is which depends on what the regression line will be used to predict.

You'll see many pairs of terms used for the *x*- and *y*-variables in regression. Some are shown in Table 4.1.

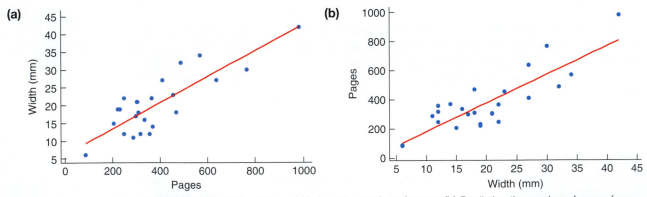

▲ **FIGURE 4.20** (a) Predicting width from the number of pages. (b) Predicting the number of pages from width.

▶ **TABLE 4.1** Terms used for the *x*- and *y*-variables.

| x-Variable | y-Variable |
| --- | --- |
| Predictor variable | Predicted variable |
| Explanatory variable | Response variable |
| Independent variable | Dependent variable |

## EXAMPLE **6** Bedridden

It is hard to measure the height of people who are bedridden, and for many medical reasons it is often important to know a bedridden patient's height. However, it is not so difficult to measure the length of the ulna (the bone that runs from the elbow to the wrist). Data collected on non-bedridden people shows a strong linear association between ulnar length and height.

**QUESTION** When making a scatterplot to predict height from ulnar length, which variable should be plotted on the x-axis and which on the y-axis?

**CONCLUSION** We are measuring ulnar length to predict a person's height. Therefore, ulnar length is the predictor (independent variable) and is plotted on the x-axis, while height is the response (dependent variable) and is plotted on the y-axis.

**TRY THIS!** Exercise 4.35

## The Regression Line Is a Line of Averages

Figure 4.21 shows a histogram of the weights of a sample of 507 active people. What is the typical weight?

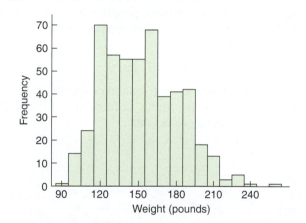

◄ **FIGURE 4.21** A histogram of the weights of 507 active people.

One way of answering this question is to calculate the mean of the sample. The distribution of weights is a little right-skewed but not terribly so, and so the mean will probably give us a good idea of what is typical. The average weight of this group is 152.5 pounds.

Now, we know that an association exists between height and weight and that shorter people tend to weigh less. If we know someone's height, then, what weight would we guess for that person? Surely not the average weight of the whole group! We can do better than that. For instance, what's the typical weight of someone 66 inches tall? To answer this, it makes sense to look only at the weights of those people in our sample who are about 66 inches tall. To make sure we have enough 66-inch-tall people in our sample, let's include everyone who is *approximately* 66 inches tall. So let's look at a slice of people whose height is between 65.5 inches and 66.5 inches. We come up with 47 people. Some of their weights are

130, 150, 142, 149, 118, 189 . . .

Figure 4.22a shows this slice, which is centered at 66 inches.

The mean of these numbers is 141.5 pounds. We put a special symbol on the plot to record this point—a triangle at the point (66, 141.5), shown in Figure 4.22b.

▲ **FIGURE 4.22a** Heights and weights with a slice at 66 inches.

▲ **FIGURE 4.22b** Heights and weights with the average at 66 inches, which is 141.5 pounds.

The reason for marking this point is that if we wanted to predict the weights of those who were 66 inches tall, one good answer would be 141.5 pounds.

What if we wanted to predict the weight of someone who is 70 inches tall? We could take a slice of the sample and look at those people who are between 69.5 inches and 70.5 inches tall. Typically, they're heavier than the people who are about 66 inches tall. Here are some of their weights:

189, 197, 151, 207, 157 . . .

Their mean weight is 173.0 pounds. Let's put another special triangle symbol at (70, 173.0) to record this. We can continue in this fashion, and Figure 4.22c shows where the mean weights are for a few more heights.

▶ **FIGURE 4.22c** Heights and weights with more means marked.

Note that the means fall (nearly) on a straight line. What could be the equation of this line? Figure 4.22d shows the regression line superimposed on the scatterplot with the means. They're nearly identical.

In theory, these means should lie exactly on the regression line. However, when working with real data, we often find that the theory doesn't always fit perfectly.

The series of graphs in Figure 4.22 illustrates a fundamental feature of the regression line: It is a line of means. You plug in a value for *x*, and the regression line "predicts" the mean *y*-value for everyone in the population with that *x*-value.

◀ **FIGURE 4.22d** Heights and weights with means and a straight line.

**KEY POINT** When the trend is linear, the regression line connects the points that represent the mean value of $y$ for each value of $x$.

## EXAMPLE 7 Funny Car Race Finishing Times

A statistics student collected data on National Hot Rod Association driver John Force. The data consist of a sample of Funny Car races. The drivers race in several different trials and change speed frequently. One question of interest is whether the fastest speed (in miles per hour, mph) driven during a trial is associated with the time to finish that trial (in seconds). If a driver's fastest speed cannot be maintained and the driver is forced to slow down, then it might be best to avoid going that fast. An examination of the scatterplot Fig. 4.23 of finishing time (response variable) versus fastest speed (predictor) for a sample of trials for John Force shows a reasonably linear association.

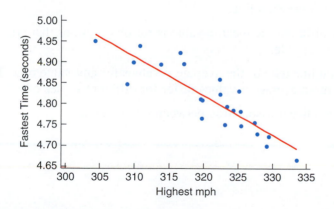

◀ **FIGURE 4.23** A scatterplot for the funny car race. (Source: Hettinga, J. 2005, UCLA, Department of Statistics)

The regression line is

Predicted Time = 7.84 − 0.0094 (Highest mph)

**QUESTION** What is the predicted mean finishing time when the driver's fastest speed is 318 mph?

**SOLUTION** We predict by estimating the mean finishing time when the speed is 318 mph. To find this, plug 318 into the regression line:

Predicted Time = 7.84 − 0.0094 × 318 = 4.85 seconds

**CONCLUSION**  We predict that the finishing time will be 4.85 seconds.

**TRY THIS!**  Exercise 4.37

### Interpreting the Slope

The slope tells us how to compare the mean *y*-values for objects that are 1 unit apart on the *x*-variable. For example, how different is the mean finishing time when the driver's fastest speed is 1 mph faster? The slope in Example 7 tells us that the means differ by a slim 0.0094 second. (Of course, races can be won or lost by such small differences.) What if the fastest time is 10 mph faster? Then the mean finishing time is $10 \times 0.0094 = 0.094$ second faster. We can see that in a typical race, if the driver wants to lose 1 full second from racing time, he or she must go over 100 mph faster, which is clearly not possible.

We should pay attention to whether the slope is 0 or very close to 0. When the slope is 0, we say that no (linear) relationship exists between *x* and *y*. The reason is that a 0 slope means that no matter what value of *x* you consider, the predicted value of *y* is always the same. The slope is 0 whenever the correlation coefficient is 0.

For example, the slope for runners' ages and their marathon times (see Figure 4.3a) is very close to 0 (0.000014 second per year of age).

## SNAPSHOT  THE SLOPE

**WHAT IS IT?** ▶ The slope of a regression line.

**WHAT DOES IT DO?** ▶ Tells us how different the mean *y*-value is for observations that are 1 unit apart on the *x*-variable.

**HOW DOES IT DO IT?** ▶ The regression line tells us the average *y*-values for any value of *x*. The slope tells us how these average *y*-values differ for different values of *x*.

**HOW IS IT USED?** ▶ To measure the linear association between two variables.

## EXAMPLE **8** Math and Critical Reading SAT Scores

A regression analysis to study the relationship between SAT math and SAT critical reading scores resulted in this regression model:

$$\text{Predicted Critical Reading} = 398.81 + 0.3030\ \text{Math}$$

This model was based on a sample of 200 students.

The scatterplot in Figure 4.24 shows a linear relationship with a weak positive trend.

◀ **FIGURE 4.24** The regression line shows a weak positive trend between critical reading SAT score and math SAT score.

**QUESTION** Interpret the slope of this regression line.

**SOLUTION** The slope tells us that students who score 10 points higher on the math SAT had an average critical reading SAT score that was

$$10 \times 0.3030 = 3.03 \text{ points higher}$$

**CONCLUSION** Students who score 10 points higher on the math SAT score, on average, 3.03 points higher on critical reading.

**TRY THIS!** Exercise 4.55a

## Interpreting the Intercept

The intercept tells us the predicted mean $y$-value when the $x$-value is 0. Quite often, this is not terribly helpful. Sometimes it is ridiculous. For example, the regression line to predict weight, given someone's height, tells us that if a person is 0 inches tall, then his or her predicted weight is negative 231.5 pounds!

Before interpreting the intercept, ask yourself whether it makes sense to talk about the $x$-variable taking on a 0 value. For the SAT data, you might think it makes sense to talk about getting a 0 on the math SAT. However, the lowest possible score on the SAT math portion is 200, so it is not possible to get a score of 0. (One lesson statisticians learn early is that you must know something about the data you analyze—knowing only the numbers is not enough!)

## EXAMPLE 9 Predicting Funny Car Races

What can we learn about the relationship between the fastest speed of a funny car driver and the finishing time in a race by looking at the intercept and slope of the regression line? The regression model is

$$\text{Predicted Time} = 7.84 - 0.0094 \text{ (Highest mph)}$$

**QUESTION** Interpret the intercept and slope.

**CONCLUSION** The intercept is not meaningful. It tells us that if the fastest the car moves is 0 mph, the car will finish the race in 7.84 seconds. Of course, we know that if the car goes 0 mph, it will never finish the race!

The slope is meaningful. The negative sign indicates that there is a negative trend: The faster the maximum speed during the race, the shorter, on average, the driver's time. It also shows that each additional mile per hour of maximum speed shaves off, on average, 0.0094 second of time.

> TRY THIS! Exercise 4.59

## SNAPSHOT THE INTERCEPT

| | | |
|---|---|---|
| **WHAT IS IT?** | ▶ | The intercept of a regression line. |
| **WHAT DOES IT DO?** | ▶ | Tells us the average $y$-value for all observations that have a zero $x$-value. |
| **HOW DOES IT DO IT?** | ▶ | The regression line is the best fit to a linear association, and the intercept is the best prediction of the $y$-value when the $x$-value is 0. |
| **HOW IS IT USED?** | ▶ | It is not always useful. Often, it doesn't make sense for the $x$-variable to assume the value 0. |

## EXAMPLE 10 Age and Value of Cars

Figure 4.25 shows the relationship between age and Kelley Blue Book values for a sample of cars. These cars were owned by students in one of the author's statistics class.

▶ **FIGURE 4.25** The age of a car and its value for a chosen sample.

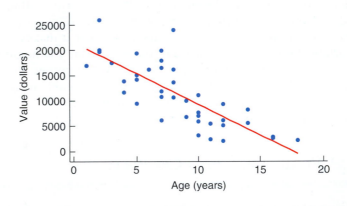

The regression line is

$$\text{Predicted Value} = 21375 - 1215 \text{ Age}$$

> QUESTION   Interpret the slope and intercept.

> CONCLUSION   The intercept estimates that the average value of a new car (0 years old) owned by students in this class is \$21,375. The slope tells us that, on average, cars lost \$1215 in value each year.

> TRY THIS! Exercise 4.61

# Evaluating the Linear Model

Regression is a very powerful tool for understanding linear relationships between numerical variables. However, we need to be aware of several potential interpretation pitfalls so that we can avoid them. We will also discuss methods for determining just how well the regression model fits the data.

## Pitfalls to Avoid

You can avoid most pitfalls by simply making a graph of your data and examining it closely before you begin interpreting your linear model. This section will offer some advice for sidestepping a few subtle complications that might arise.

### Don't Fit Linear Models to Nonlinear Associations

Regression models are useful only for linear associations. If the association is not linear, a regression model can be misleading and deceiving. For this reason, before you fit a regression model, you should always make a scatterplot to verify that the association seems linear. Chapter 14 will provide a slightly more sophisticated (and discerning) method for checking linearity.

Figure 4.26 shows an example of a bad regression model. The association between mortality rates for several industrialized countries (deaths per 1000 people) and wine consumption (liters per person per year) is nonlinear. The regression model is

$$\text{Predicted Mortality} = 7.69 - 0.0761 \, (\text{Wine Consumption})$$

but it provides a poor fit. The regression model suggests that countries with middle values of wine consumption should have higher mortality rates than they actually do.

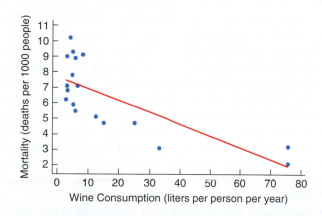

◀ **FIGURE 4.26** The straight-line regression model is a poor fit to this nonlinear relationship. (Source: Leger et al. 1979)

### Correlation Is Not Causation

One important goal of science and business is to discover relationships between variables such that if you make a change to one variable, you know the other variable will change in a reliably predictable way. This is what is meant by "*x* causes *y*": Make a change in *x*, and a change in *y* will usually follow. For example, the distance it takes to stop your car depends on how fast you were traveling when you first applied the brakes (among other things); the amount of memory an mp3 file takes on your hard drive depends on the length of the song; and the size of your phone bill depends on how many minutes you talked. In these cases, a strong causal relationship exists between two variables, and if you were to collect data and make a scatterplot, you would see an association between the variables.

In statistics, however, we are often faced with the reverse situation, in which we see an association between two variables and wonder whether there is a cause-and-effect relationship. The correlation coefficient for the association could be quite strong, but as we saw earlier, correlation does not mean cause and effect. A strong correlation or a good-fitting regression line is not evidence of a cause-and-effect relationship.

> **KEY POINT** An association between two variables is not sufficient evidence to conclude that a cause-and-effect relationship exists between the variables, no matter how strong the correlation or how well the regression line fits the data.

Be particularly careful about drawing cause-and-effect conclusions when interpreting the slope of a regression line. For example, for the SAT data,

$$\text{Predicted Critical Reading} = 398.81 + 0.3030 \text{ Math}$$

Even if this regression line fits the association very well, it does not give us sufficient evidence to conclude that if you improve your math score by 10 points, your critical reading score will go up by 3.03 points. As you learned in Chapter 1, because these data were not collected from a controlled experiment, the presence of confounding factors could prevent you from making a causal interpretation.

Beware of the algebra trap. In algebra, you were taught to interpret the slope to mean that "as $x$ increases by 1 unit, $y$ goes up by $b$ units." However, quite often with data, the phrase "as $x$ increases" doesn't make sense. When looking at the height and weight data, where $x$ is height and $y$ is weight, to say "$x$ increases" means that people are growing taller! This is not accurate. It is much more accurate to interpret the slope as making comparisons between groups. For example, when comparing people of a certain height with those who are 1 inch taller, you can see that the taller individuals tend to weigh, on average, $b$ pounds more.

When can we conclude that an association between two variables means a cause-and-effect relationship is present? Strictly speaking, never from an observational study and only when the data were collected from a controlled experiment. (Even in a controlled experiment, care must be taken that the experiment was designed correctly.) However, for many important questions, controlled experiments are not possible. In these cases, we can sometimes make conclusions about causality after a number of observational studies have been collected and examined, and if there is a strong theoretical explanation for why the variables should be related in a cause-and-effect fashion. For instance, it took many years of observational studies to conclude that smoking causes lung cancer, including studies that compared twins—one twin who smoked and one who did not—and numerous controlled experiments on lab animals.

## Beware of Outliers

Recall that when calculating sample means, we were warned that outliers can have a big effect. Because the regression line is a line of means, you might think that outliers would have a big effect on the regression line. And you'd be right. You should always check a scatterplot before performing a regression analysis to be sure there are no outliers.

The graphs in Figure 4.27 illustrate this effect. Both graphs in Figure 4.27 show crime rates (number of reported crimes per 100,000 people) versus population density (people per square mile) in 2008. Figure 4.27a includes all 50 states and the District of Columbia. The District is an outlier and has a strong influence on the regression line. Figure 4.27b excludes the District of Columbia.

Without the District of Columbia, the slope of the regression line actually changes sign! Thus, whether you conclude that a positive or a negative association occurs between crime and population density depends on whether this one city is included.

▲ **FIGURE 4.27** Crime rates and population densities in the states. Part (a) includes Washington, D.C. and part (b) does not. (Source: *The world almanac and Book of Facts 2009*)

These types of observations are called **influential points** because their presence or absence has a big effect on conclusions. When you have influential points in your data, it is good practice to try the regression and correlation with and without these points (as we did) and to comment on the difference.

## Regressions of Aggregate Data

Researchers sometimes do regression analysis based on what we call **aggregate data**. Aggregate data are those for which each plotted point is from a summary of a group of individuals. For example, in a study to examine the relationship between SAT math and critical reading scores, we might use the *mean* of each of the 50 states rather than the scores of individual students. The regression line provides a summary of linear associations between two variables. The mean provides a summary of the center of a distribution of a variable. What happens if we have a collection of means of two variables and we do a regression with these?

This is a legitimate activity, but you need to proceed with caution. For example, Figure 4.28 shows scatterplots of critical reading and math SAT scores. However, in Figure 4.28a, each point represents an individual: an SAT score for each first-year student enrolled in one northeastern university for a fall semester. The scatterplot in Figure 4.28b seems to show a much stronger relationship, because in this case each point is an aggregate. Specifically, each point represents a state in the United States: the mean SAT score for all students in the state taking SAT tests in 2005.

> **🔄 Looking Back**
>
> **Outliers**
> You learned about outliers for one variable in Chapter 3. Outliers are values that lie a great distance from the bulk of the data. In two-variable associations, outliers are points that do not fit the trend or are far from the bulk of the data.

▲ **FIGURE 4.28** Critical reading and math SAT scores. **(a)** Scores for individuals. **(b)** Means for states.

We can still interpret Figure 4.28b, as long as we're careful to remember that we are talking about states, not people. We *can* say a strong correlation exists between a state's mean math SAT score and a state's mean critical reading SAT score. We *cannot* say that there is a strong correlation between individual students' math and critical reading scores.

## Don't Extrapolate!

**Extrapolation** means that we use the regression line to make predictions beyond the range of our data. This practice can be dangerous, because although the association may have a linear shape for the range we're observing, that might not be true over a larger range. This means that our predictions might be wrong outside the range of observed *x*-values.

Figure 4.29a shows a graph of height versus age for children between 2 and 9 years old from a large national study. We've superimposed the regression line on the graph (Figure 4.29a), and it looks like a pretty good fit. However, although the regression model provides a good fit for children ages 2 through 9, it fails when we use the same model to predict heights for older children.

The regression line for the data shown in Figure 4.29a is

$$\text{Predicted Height} = 31.78 + 2.45 \text{ Age}$$

However, we observed only children between the ages of 2 and 9. Can we use this line to predict the height of a 20-year-old?

The regression model predicts that the mean height of 20-year-olds is 80.78 inches:

$$\text{Predicted Height} = 31.78 + 2.45 \text{ Age} = 31.78 + 2.45 \times 20 = 80.78: \text{ nearly 7 feet!}$$

We can see from Figure 4.29b that the regression model provides a poor fit if we include people over the age of 9. Beyond that age, the trend is no longer linear, so we get bad predictions from the model.

▲ **FIGURE 4.29** **(a)** Ages and heights for children between 2 and 9 years old. **(b)** Heights included for people up to age 21 years. The straight line is not valid in the upper range of ages. (Source: National Health and Nutrition Examination Survey, Centers for Disease Control)

It is often tempting to use the regression model beyond the range of data used to build the model. Don't. Unless you are very confident that the linear trend continues without change beyond the range of observed data, you must collect new data to cover the new range of values.

**KEY POINT**  Don't extrapolate!

### The Origin of the Word *Regression* (Regression toward the Mean)

The first definition in the *Oxford Dictionary* says that regression is a "backward movement or return to a previous state." So why are we are using it to describe predictions of one numerical variable (such as value of a car) from another numerical variable (such as age of the car)?

The term *regression* was coined by Francis Galton, who used the regression model to study genetic relationships. He noticed that even though taller-than-average fathers tended to have taller-than-average sons, the sons were somewhat closer to average than the fathers were. Also, shorter-than-average fathers tended to have sons who were closer to the average than their fathers. He called this phenomenon regression toward mediocrity, but later it came to be known as **regression toward the mean**.

You can see how regression toward the mean works by examining the formula for the slope of the regression line:

$$b = r\frac{s_y}{s_x}$$

This formula tells us that fathers who are one standard deviation taller than average ($s_x$ inches above average) have sons who are not one standard deviation taller than average ($s_y$) but are instead $r$ times $s_y$ inches taller than average. Because $r$ is a number between $-1$ and 1, $r$ times $s_y$ is usually smaller than $s_y$. Thus the "rise" will be less than the "run" in terms of standard deviations.

The *Sports Illustrated* jinx is an example of regression toward the mean. According to the jinx, athletes who make the cover of *Sports Illustrated* end up having a really bad year after appearing. Some professional athletes have refused to appear on the cover of *Sports Illustrated*. (Once, the editors published a picture of a black cat in that place of honor, because no athlete would agree to grace the cover.) However, if an athlete's performance in the first year is several standard deviations above average, the second year is likely to be closer to average. This is an example of regression toward the mean. For a star athlete, closer to average can seem disastrous.

## The Coefficient of Determination, $r^2$, Measures Goodness of Fit

If we are convinced that the association we are examining is linear, then the regression line provides the best numerical summary of the relationship. But how good is "best"? The correlation coefficient, which measures the strength of linear relationships, can also be used to measure how well the regression line summarizes the data.

The **coefficient of determination** is simply the correlation coefficient squared: $r^2$. In fact, this statistic is often called *r-squared*. Usually, when reporting *r*-squared, we multiply by 100% to convert it to a percentage. Because $r$ is always between $-1$ and 1, *r*-squared is always between 0% and 100%. A value of 100% means the relationship is perfectly linear and the regression line perfectly predicts the observations. A value of 0% means there is no linear relationship and the regression line does a very poor job.

For example, when we predicted the width of a book from the number of pages in the book we found, the correlation between these variables to be $r = 0.9202$. So the coefficient of determination is $0.9202^2 = 0.8468$, which we report as 84.7%.

What does this value of 84.7% mean? A useful interpretation of *r*-squared is that it measures how much of the variation in the response variable is explained by the explanatory variable. For example, 84.7% of the variation in book widths was explained by the number of pages. What does this mean?

Figure 4.30 shows a scatterplot (simulated data) with a constant value for $y$ ($y = 6240$) no matter what the $x$-value is. You can see that there is no variation in $y$, so there is also nothing to explain.

▶ **FIGURE 4.30** Because $y$ has no variation, there is nothing to explain.

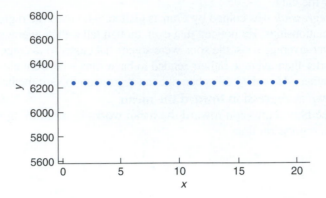

Figure 4.31 shows the height in inches and in centimeters for several people. Here, variation in the $y$-variable does occur. Height as measured in centimeters varies from about 150 cm to about 190 cm. However, the points are perfectly linear and have a correlation of 1.000. That means that if you are given an $x$-value (a person's height in inches), then you know the $y$-value (the person's height in centimeters) precisely. Thus all of the variation in $y$ is explained by the regression model. In this case, the coefficient of determination is 100%; all variation in $y$ is perfectly explained by the best-fit line.

Real data are messier. Figure 4.32 shows a plot of the age and value of some cars. The regression line has been superimposed to remind us that there is, in fact, a linear trend and that the regression line does capture it. The regression model explains some of the variation in $y$, but as we can see, it's not perfect; plugging the value of $x$ into the regression line gives us an imperfect prediction of what $y$ will be. In fact, for these data, $r = -0.778$, so we've explained $(-0.778)^2 = 0.605$, or about 60.5%, of the variation in $y$ with this regression line.

The practical implication of $r$-squared is that it helps determine which explanatory variable would be best for making predictions about the response variable. For example,

▲ **FIGURE 4.31** Heights of people in inches and in centimeters (cm) with a correlation of 1. The coefficient of determination is 100%.

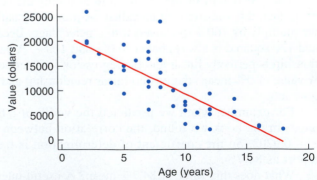

▲ **FIGURE 4.32** Age and value of cars; the correlation is $-0.778$, and the coefficient of determination is 60.5%.

is waist size or height a better predictor of weight? We can see the answer to this question from the scatterplots in Figure 4.33, which show that the linear relationship is stronger (has less scatter) for waist size.

▲ **FIGURE 4.33** Scatterplots of height vs. weight **(a)** and waist size vs. weight **(b)**. Waist size has the larger coefficient of determination.

The *r*-squared for predicting weight from height is 51.4% (Figure 4.33a), and the *r*-squared for predicting weight from waist size is 81.7% (Figure 4.33b). We can explain more of the variation in these people's weights by using their waist sizes than by using their heights, and therefore, we can make better (more precise) predictions using waist size as the predictor.

**KEY POINT**  If the association is linear, the larger the coefficient of determination (*r*-squared), the smaller the variation or scatter about the regression line, and the more accurate the predictions tend to be.

## SNAPSHOT  R-SQUARED

| | |
|---|---|
| **WHAT IS IT?** ▶ | *r*-squared or coefficient of determination. |
| **WHAT DOES IT DO?** ▶ | Measures how well the data fit the linear model. |
| **HOW DOES IT DO IT?** ▶ | If the points are tightly clustered around the regression line, $r^2$ has a value near 1, or 100%. If the points are more spread out, it has a value closer to 0. If there is no linear trend, and the points are a formless blob, it has a value near 0. |
| **HOW IS IT USED?** ▶ | Works only with linear associations. Large values of *r*-squared tell us that predicted *y*-values are likely to be close to actual values. The coefficient of determination shows us the percentage of variation that can be explained by the regression line. |

## CASE STUDY REVISITED

Brinks was contracted by New York City to collect money from parking meters in some months, but some Brinks employees had been stealing some of the money. The city knew how much money its own employees had collected from meters each month and how much money honest private contractors had collected. It used this relation to predict the amount that Brinks should have collected.

The first step in this analysis was to determine the relation between the amount of money collected by the city itself and that collected by the honest contractors. Figure 4.34 shows a positive (though somewhat weak) linear association. In months in which the city collected more money, the (honest) private contractors also tended to collect more money.

▶ **FIGURE 4.34** Regression Plot without Brinks. The association between amounts of meter income collected from honest private contracts (vertical axis) and amounts collected by the city government. The trend is positive, linear, and somewhat weak.

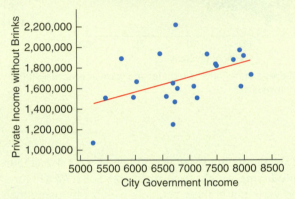

The regression line is

$$\text{Predicted Contractor Collection} = 688497 + 145.5\,(\text{City Income})$$

This model is useful for predicting how much Brinks should have collected in any given month. For example, in one particular month during which Brinks was the contractor, the city collected $7,016. The regression line predicts that if Brinks were like the other private contractors, it should have collected about

$$688497 + 145.5 \times 7016 = \$1,709,325$$

However, Brinks collected only $1,330,143—about $400,000 too little. This is just one month, but as Figure 4.35 shows, Brinks was consistently low in its collections. Figure 4.35 shows data both from Brinks and from the other private contractors. The Brinks data points are shown with squares, and the other contractors' data points are shown with circles. You can see that the regression line for the Brinks collection (the dotted line) is below the regression line for the other contractors, indicating that the mean amount collected by Brinks was consistently lower than the mean amount collected by the other contractors.

▶ **FIGURE 4.35** Scatterplot of Income from Parking Meters. The dashed red line represents the regression line for the money that Brinks collected. The solid blue line represents the money they should have collected, judging on the basis of the amount collected by the city. Notice that the predicted means for Brinks tend to be lower than those of the other contractors.

For each month that Brinks had worked for the city, the city used the regression line to compute the amount Brinks owed, by finding the difference between the amount Brinks would presumably have collected if there had been no stealing and the amount they actually collected.

# EXPLORING STATISTICS
## CLASS ACTIVITY

## Guessing the Age of Famous People

| GOALS | MATERIALS |
|---|---|
| In this activity, you will learn how to interpret the slope and intercept of a regression line, using data you collect in class. | Graph paper and a calculator or computer. |

**ACTIVITY**

Your instructor will give you a list of names of famous people. Beside each name, write your guess of the person's age, in years. Even if you don't know the age or don't know who the person is, give your best guess. If you work in a group, your group should discuss the guessed ages and record the best guess of the group. After you've finished, your instructor will give you a list of the actual ages of these people.

To examine the relationship between the actual ages and the ages you guessed, make a scatterplot with actual age on the $x$-axis and guessed age on the $y$-axis. Use technology to find the equation for the regression line and insert the line in the graph. Calculate the correlation.

**BEFORE THE ACTIVITY**

1. Suppose you guessed every age correctly. What would be the equation of the regression line? What would be the correlation?

2. What correlation do you think you will actually get? Why?

3. Suppose you consistently guess that people are older than they actually are? How will the intercept of the regression line compare with your intercept in Question 1? How about the slope?

**AFTER THE ACTIVITY**

1. How would you describe the association between the ages you guessed and the actual ages?

2. Is a regression line appropriate for your data? Explain why or why not.

3. What does it mean if a point falls above your regression line? Below your line?

4. What is the intercept of your line? What is the slope? Interpret the slope and intercept. Explain what these tell you about your ability to guess the ages of these people.

## CHAPTER REVIEW

## Key Terms

Scatterplot, *134*
Trend, *134*
Strength, *134*
Shape, *134*
Positive association (positive trend), *134*

Negative association (negative trend), *134*
Linear, *136*
Correlation coefficient, *138*
Regression line, *146*
Intercept, *147*

Slope, *147*
Explanatory variable, predictor variable, independent variable, *154*
Response variable, predicted variable, dependent variable, *154*

Influential point, *163*
Aggregate data, *163*
Extrapolation, *164*
Regression toward the mean, *165*
Coefficient of determination, $r^2$, *r*-squared, *165*

## Learning Objectives

After reading this chapter and doing the assigned homework problems, you should

- Be able to write a concise and accurate description of an association between two numerical variables based on a scatterplot.

- Understand how to use a regression line to summarize a linear association between two numerical variables.

- Interpret the intercept and slope of a regression line in context and know how to use the regression line to predict mean values of the response variable.

- Critically evaluate a regression model.

## Summary

The first step in looking at the relationship between two numerical variables is to make a scatterplot and learn as much about the association as you can. Examine trend, strength, and shape. Look for outliers.

Regression lines and correlation coefficients can be interpreted meaningfully only if the association is linear. A correlation coefficient close to 1 or −1 does *not* mean the association is linear. When the association is linear, the correlation coefficient can be calculated by Formula 4.1:

**Formula 4.1:** $r = \dfrac{\sum z_x z_y}{n - 1}$

A regression line can summarize the relationship in much the same way as a mean and standard deviation can summarize a distribution. Interpret the slope and intercept, and use the regression line to make predictions about the mean *y*-value for any given value of *x*.

The regression line is given by

**Formula 4.2c:** Predicted $y = a + bx$

where the slope, *b*, is

**Formula 4.2a:** $b = r\dfrac{s_y}{s_x}$

and the intercept, *a*, is

**Formula 4.2b:** $a = \bar{y} - b\bar{x}$

When interpreting a regression analysis, be careful:

Don't extrapolate.
Don't make cause-and-effect conclusions if the data are observational.
Beware of outliers, which may (or may not) strongly affect the regression line.
Proceed with caution when dealing with aggregated data.

## Source

Afshartous, D., Bentow, S., "Statistical Analysis of Brink's Data," www .stat.ucla.edu/cases

Faraway, J. J. 2005. *Extending the linear model with R*. Boca Raton, FL: CRC Press.

Heinz, G., L. Peterson, R. Johnson, and C. Kerk. 2003. Exploring relationships in body dimensions. *Journal of Statistics Education* 11(2). http://www.amstat.org/publications/jse/

Joyce, C. A., ed. *The 2009 World Almanac and Book of Facts*. New York: World Almanac.

Leger, A., A. Cochrane, and F. Moore. 1979. Factors associated with cardiac mortality in developed countries with particular reference to the consumption of wine. *The Lancet* 313(8124): 1017–1020.

# SECTION EXERCISES

## SECTION 4.1

**4.1 Predicting Land Value** Both figures concern the assessed value of land (with homes on the land), and both use the same data set.

a. Which do you think has a stronger relationship with value of the land—the number of acres of land or the number of rooms in the homes? Why?

b. If you were trying to predict the value of a parcel of land in this area (on which there is a home), would you be able to make a better prediction by knowing the acreage or the number of rooms in the house? Explain.

(Source: Minitab File, Student 12, "Assess.")

**(A)**

**(B)**

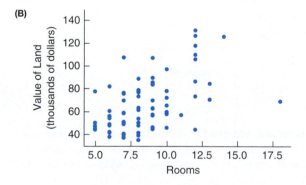

**4.2 Predicting Total Value of Property** Both figures concern the total assessed value of properties that include homes, and both use the same data set.

a. Which do you think has a stronger relationship with value of the property—the number of square feet in the home (shown in part B of the figure) or the number of fireplaces in the home (shown in part A of the figure)? Why?

b. If you were trying to predict the value of a property in this area (where there is a home), would you be able to make a better prediction by knowing the number of square feet or the number of fireplaces? Explain.

**(A)**

**(B)**

**4.3 Car Value and Age of Student** The figure shows the age of students and the value of their cars according to the Kelley Blue Book. Does it show an increasing trend, a decreasing trend, or very little trend? Explain.

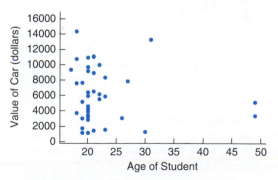

**4.4 Shoe Size and GPA** The figure shows a scatterplot of shoe size and GPA for some college students. Does it show an increasing trend, a decreasing trend, or no trend? Is there a strong relationship?

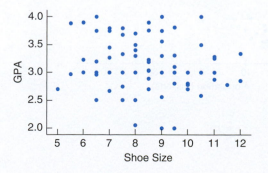

**4.5 Weight Loss** The scatterplot shows the actual weight and desired weight change of some students. Thus, if they weighed 220 and wanted to weigh 190, the desired weight change would be negative 30.

Explain what you see. In particular, what does it mean that the trend is negative?

**4.6 Comparing Salaries** The figure shows the salary and year of first employment for some professors at a college. Explain, in context, what the negative trend shows. Who makes the most and who makes the least?
(Source: Minitab, Student 12, "Salary.")

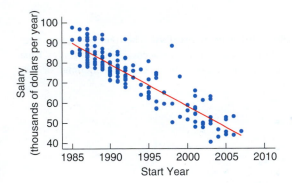

TRY **4.7 Population and Motor Vehicle Fatalities (Example 1)** The figure shows the number of motor vehicle fatalities and the population of the 50 states and Washington, D.C., for 2006, from the U.S. Census Bureau. Explain what the trend shows.

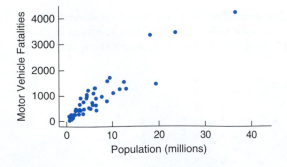

**4.8 BAs and Median Income** The scatterplot shows data from the 50 states taken from the U.S. Census—the percentage of the population (25 years or older) with a college degree or higher and the median family income. Describe and interpret the trend. (The outlier is Colorado.)

**4.9 Value of Homes and Acreage** The figure shows the total assessed value of some properties (the total value of the home and the land) and the acreage that comes with the homes. Describe what you see. Is the trend positive or negative? What does that mean?

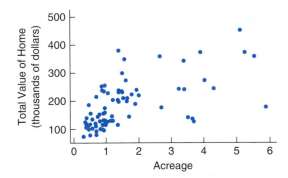

**4.10 Height and Weight for Women** The figure shows a scatterplot of the heights and weights of some women taking statistics. Describe what you see. Is the trend positive, negative, or near zero? Explain.

# SECTION 4.2

**4.11 Children's Ages and Heights** The figure shows information about the ages and heights of several children. Why would it not make sense to find the correlation or to perform linear regression with this data set? Explain.

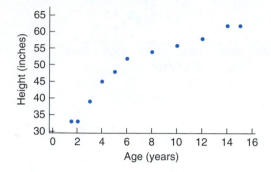

**4.12 Blackjack Tips** The figure shows the amount of money won by people playing blackjack and the amount of tips they gave to the dealer (who was a statistics student), in dollars.

Would it make sense to find a correlation for this data set? Explain.

**4.13 SAT Scores and Percentage Taking SAT** The figure shows data from The College Board on the percentage of students who take the SAT and the average score on the quantitative part for the 50 states and Washington, D.C., in 2009.

a. Does the graph show an increasing or a decreasing trend? Explain what that means in the context of these data.

b. Do the points seem to follow a straight line or a curve?

c. Is it appropriate to find the correlation for these data? Why or why not?

**4.14 Ages of Women Who Give Birth** The figure shows a scatterplot of birthrate (live births per 1000 women) and age of the mother in the United States. Would it make sense to find the correlation for this data set? Explain. According to this graph, at approximately what age is the highest fertility rate?
(Source: Helmut T. Wendel and Christopher S. Wendel (Editors): *Vital Statistics of the United States: Births, Life Expectancy, Deaths, and Selected Health Data.* Second Edition, 2006, Bernan Press.)

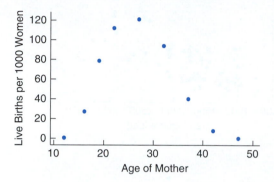

**4.15 Do Older Students Have Higher GPAs?** On the basis of the scatterplot, do you think that the correlation coefficient between age and GPA for this figure is positive, negative, or near zero?

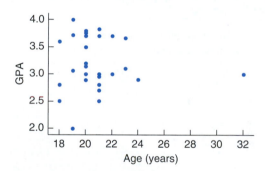

**4.16 Handspans** Refer to the figure. Is the correlation coefficient between the handspan of the dominant hand and that of the nondominant hand positive, negative, or near zero?

**4.17 Matching** Pick the letter of the graph that goes with each numerical value listed below for the correlation. Correlations:

0.767 _____

0.299 _____

−0.980 _____

**(A)**

**(B)**

**(C)**

**4.18 Matching** Pick the letter of the graph that goes with each numerical value listed below for the correlation. Correlations:

−0.903 _____

0.374 _____

0.777 _____

**(A)**

**(B)**

**(C)**

TRY **4.19 Cost of Flights (Example 2)** The table for part a shows approximate distances between selected cities and the approximate cost of flights between those cities in November 2009.

a. Calculate the correlation of the numbers shown in the part a table by using a computer or statistical calculator.

| Cost | Miles |
|------|-------|
| 180  | 960   |
| 400  | 3100  |
| 270  | 2000  |
| 100  | 430   |
| 443  | 3000  |

b. The table for part b shows the same information, except that the distance was converted to kilometers by multiplying the number of miles by 1.609. What happens to the correlation when numbers are multiplied by a constant?

| Cost | Kilometers |
|------|------------|
| 180  | 1545       |
| 400  | 4988       |
| 270  | 3218       |
| 100  | 692        |
| 443  | 4827       |

c. Suppose the Federal Aviation Administration (FAA) adds a tax to each flight. Fifty dollars is added to every flight, no matter how long it is. The table for part c shows the new data. What happens to the correlation when a constant is added to each number?

| Cost | Miles |
|------|-------|
| 230 | 960 |
| 450 | 3100 |
| 320 | 2000 |
| 150 | 430 |
| 493 | 3000 |

**4.20 Trash** The table shows the number of people living in a house and the weight of trash (in pounds) at the curb just before trash pickup.

| People | Trash (pounds) |
|--------|----------------|
| 2 | 18 |
| 3 | 33 |
| 6 | 93 |
| 1 | 23 |
| 7 | 83 |

a. Find the correlation between these numbers by using a computer or a statistical calculator.

b. Suppose some of the weight was from the container (each container weighs 3 pounds). Subtract 3 pounds from each weight and find the new correlation with the number of people. What happens to the correlation when a constant is added (we added negative 3) to each number?

c. Suppose each house contained exactly twice the number of people, but the weight of the trash was the same. What happens to the correlation when numbers are multiplied by a constant?

## SECTION 4.3

TRY **4.21 Are Men Paid More Than Women? (Example 3)** The scatterplot shows the median annual pay for college-educated men and women in the 50 states in 2005, according to the U.S. Census Bureau. The correlation is 0.834. The regression equation is above the graph.

a. Find a rough estimate (by using the scatterplot) of median pay for women in a state that has a median pay of about $60,000 for men.

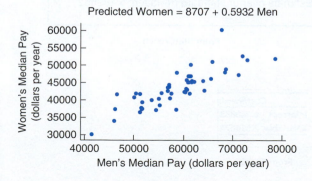

b. Use the regression equation above the graph to get a more precise estimate of the median pay for women in a state that has a median pay for men of $60,000.

**4.22 Number of Births and Population** The figure shows the number of births and the populations for the 50 states in 2005, according to the U.S. Census Bureau. The correlation is 0.99.

a. Find a rough estimate (by using the scatterplot) of number of births in a state with a population of about 10 million.

b. Use the regression equation above the graph to get a more precise estimate of the number of births for a state with a population of 10 million.

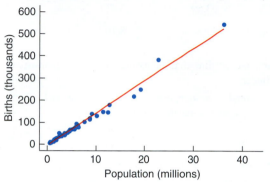

**4.23–4.26** Refer to the following table for Exercises 4.23–4.26. It provides various body measurements for some college women. The height is self-reported in inches. All other measurements are in centimeters.

Head is the circumference of the head.

HandL is the length of the hand.

FootL is the length of the foot.

HandW is the width of the hand with the fingers spread wide.

Arm is the armspan with the arms spread wide.

| Ht | Head | HandL | FootL | HandW | Armspan |
|------|------|-------|-------|-------|---------|
| 62.0 | 56.0 | 17.0 | 23.0 | 20.0 | 159.5 |
| 65.0 | 54.0 | 17.5 | 23.5 | 18.0 | 161.0 |
| 66.0 | 55.0 | 18.0 | 24.0 | 16.5 | 161.5 |
| 63.0 | 58.0 | 18.5 | 23.5 | 19.5 | 160.5 |
| 68.0 | 55.0 | 19.0 | 25.0 | 18.0 | 173.0 |
| 56.0 | 54.5 | 16.5 | 22.0 | 17.0 | 142.0 |
| 61.0 | 50.0 | 16.0 | 22.0 | 18.0 | 155.0 |
| 67.0 | 53.5 | 18.0 | 23.5 | 15.0 | 165.0 |
| 66.0 | 55.5 | 16.0 | 21.0 | 20.0 | 166.0 |
| 64.0 | 57.0 | 19.0 | 24.5 | 20.0 | 163.0 |
| 61.5 | 55.0 | 19.0 | 24.5 | 21.0 | 152.0 |

TRY * **4.23 Height and Armspan for Women (Example 4)** TI-83/84 output from a linear model for predicting armspan (in cm) from height (in inches) is given in the figure. Summary statistics are also provided.

```
LinReg
y=a+bx
a=16.80271378
b=2.249707146
r²=.8987441215
r=.9480211609
```

|            | Mean   | Standard Deviation |
|------------|--------|--------------------|
| Height, $x$  | 63.59  | 3.41               |
| Armspan, $y$ | 159.86 | 8.10               |

To do parts a–c, assume that the association between armspan and height is linear.

a. Report the regression equation, using the words "Height" and "Armspan," not $x$ and $y$, using the output given.

b. Verify the slope by using the formula $b = r\dfrac{s_y}{s_x}$.

c. Verify the $y$-intercept using $a = \bar{y} - b\bar{x}$.

d. Using the regression equation, predict the armspan (in cm) for someone 64 inches tall.

* **4.24 Hand and Foot Length for Women** Refer to the data shown in the table for Exercises 4.23–4.26. Some computer output is shown in the figure. The output is for a linear model used to predict foot length from hand length.

Assume the trend is linear. Summary statistics for these data are shown below.

|          | Mean   | Standard Deviation |
|----------|--------|--------------------|
| Hand, $x$  | 17.682 | 1.168              |
| Foot, $y$  | 23.318 | 1.230              |

```
The regression equation is
Y = 5.67 + 0.998 X
Pearson correlation of HandL and FootL = 0.948
```

a. Report the regression equation, using the words "Hand" and "Foot," not $x$ and $y$.

b. Verify the slope by using the formula $b = r\dfrac{s_y}{s_x}$.

c. Verify the $y$-intercept by using the formula $a = \bar{y} - b\bar{x}$.

d. Using the regression equation, predict the foot length (in cm) for someone who has a hand length of 18 cm.

**4.25 Hand Width and Armspan for Women** Refer to the data shown in the table for Exercises 4.23–4.26. Use hand width as the predictor and armspan as the response.

a. Make a scatterplot of the data.

b. Explain why linear regression is probably not appropriate for these variables.

**4.26 Head Circumference and Hand Width for Women** Refer to the data shown in the table for Exercises 4.23–4.26. Use head circumference as $x$ and hand width as $y$.

a. Make a scatterplot of the data.

b. Explain why linear regression is probably not appropriate for these variables.

TRY **4.27 Height and Armspan for Men (Example 5)** Measurements were made for a sample of adult men. A regression line was fit to predict the men's armspan from their height. The output from several different statistical technologies is provided. The scatterplot confirms that the association between armspan and height is linear.

a. Report the equation for predicting armspan from height. Use words such as armspan, not just $x$ and $y$.

b. Report the slope and intercept from each technology, using all the digits given.

(A)

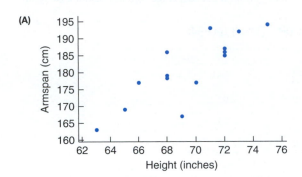

(B)
```
The regression equation is
Armspan = 6.2 + 2.51 Height
```
**From Minitab**

(C)
**Simple linear regression results:**
Dependent Variable: Armspan
Independent Variable: Height
Armspan = 6.2408333 + 2.514674 Height
Sample size: 15
R (correlation coefficient) = 0.8681
R-sq = 0.7535989
Estimate of error standard deviation: 5.409662

**From StatCrunch**

(D)

|           | Coefficients |
|-----------|--------------|
| Intercept | 6.240833     |
| X Variable | 2.514674    |

**From Excel**

(E)
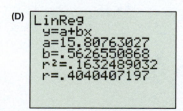

```
LinReg
y=a+bx
a=6.240833333
b=2.514673913
r²=.7535988997
r=.8681007428
```

**From TI-83/84**

**4.28 Hand Length and Foot Length for Men** Measurements were made for a sample of adult men. Assume that the association between their hand length and foot length is linear. Output for predicting foot length from hand length is provided from several different statistical technologies.

a. Report the equation for predicting foot length from hand length. Use words like foot or foot length, not just *x* and *y*.

b. Report the slope and intercept from each technology, using all the digits given.

(A)

```
The regression equation is
FootL = 15.8 + 0.563 HandL
```

**From Minitab**

(B)

**Simple linear regression results:**

Dependent Variable: FootL

Independent Variable: HandL

FootL = 15.807631 + 0.5626551 HandL

Sample size: 17

R (correlation coefficient) = 0.404

R-sq = 0.1632489

Estimate of error standard deviation: 1.6642156

**From StatCrunch**

(C)

| | Coefficients |
|---|---|
| Intercept | 15.80763 |
| X Variable | 0.562655 |

**From Excel**

(D)
```
LinReg
y=a+bx
a=15.80763027
b=.5626550868
r²=.1632489032
r=.4040407197
```

**From TI-83/84**

**4.29 Height and Head Circumference for Men** Explain what makes this scatterplot hard to interpret. What should have been done differently?

**4.30 Elbow-to-Wrist and Knee-to-Ankle Measurements** The scatterplot shows measurements (in centimeters) for the distance from elbow to wrist and from knee to ankle for some college men. Explain what makes this scatterplot difficult to interpret.

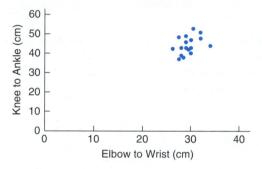

**4.31 Comparing Correlation for Armspan and Height** The correlation between height and armspan in a sample of adult women was found (in Exercise 4.23) to be $r = 0.948$. The correlation between armspan and height in a sample of adult men was found (in Exercise 4.27) to be $r = 0.868$. Which association—the association between height and armspan for women, or the association between height and armspan for men—is stronger? Explain.

* **4.32 Age and Weight for Men and Women** The scatterplot shows a solid blue line for predicting weight from age of men; the dotted red line is for predicting weight from age of women. The data were collected from a large statistics class.

a. Which line is higher and what does that mean?

b. Which line has a steeper slope and what does that mean?

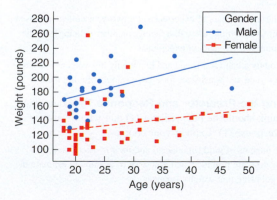

**4.33 Social Security Number and Age** The figure shows a scatterplot of the last two digits of some students' Social Security numbers and their ages.

a. If a regression line were drawn on this graph, would it have a positive slope, a negative slope, or a slope near 0?

b. Give an estimate of the numerical value of the correlation between age and Social Security number.

c. Explain what this graph tells us about the relation between Social Security number and age.

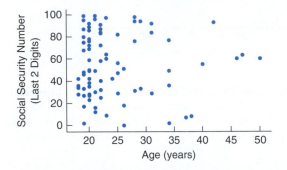

**4.34 Seesaw** The figure shows a scatterplot of the height of the left seat of a seesaw and the height of the right seat of the same seesaw. Estimate the numerical value of the correlation and explain the reason for your estimate.

**TRY** **4.35 Choosing the Predictor and Response (Example 6)** Pick out which variable you think should be the predictor ($x$) and which variable should be the response ($y$). Explain your choices.

a. You collect data on the number of gallons of gas it takes to fill up the tank after driving a certain number of miles. You wish to know how many miles you've driven based on the number of gallons it took to fill up the tank.

b. Data on salaries and years of experience at a two-year college are used in a law suit to determine whether a faculty member is being paid the correct amount for her years of experience.

c. You wish to buy a belt for a friend and know only his weight. You have data on the weight and waist sizes for a large sample of adult men.

**4.36 Choosing the Predictor and Response** Pick out which variable you think should be the predictor ($x$) and which variable should be the response ($y$). Explain your choices.

a. Weights of nuggets of gold (in ounces) and their market value over the last few days are provided, and you wish to use this to estimate the value of a gold ring that weighs 4 ounces.

b. You have data collected on the amount of time since chlorine was added to the public swimming pool and the concentration of chlorine still in the pool. (Chlorine evaporates over time.) Chlorine was added to the pool at 8 A.M., and you wish to know what the concentration is now, at 3 P.M.

c. You have data on the circumference of oak trees (measured 12 inches from the ground) and their age (in years). An oak tree in the park has a circumference of 36 inches, and you wish to know approximately how old it is.

**TRY** **4.37 Percentage of Smoke-Free Homes and Percentage of High School Students Who Smoke (Example 7)** The figure shows a scatterplot with the regression line. The data are for the 50 states. The predictor is the percentage of smoke-free homes. The response is the percentage of high school students who smoke. The data came from the Centers for Disease Control and Prevention (CDC).

a. Explain what the trend shows.

b. Use the regression equation to predict the percentage of students in high school who smoke, assuming that there are 70% smoke-free homes in the state. Use 70 not 0.70.

Predicted Pct. Smokers = 56.32 − 0.464 (Pct. Smoke-free)

**4.38 Effect of Adult Smoking on High School Student Smoking** The figure shows a fitted line plot. The data are for the 50 states. The predictor is the percentage of adults who smoke. The response is the percentage of high school students who smoke. (The point in the lower left is Utah.)

a. Explain what the trend shows.

b. Use the regression equation to predict the percentage of high school students who smoke, assuming that 25% of adults in the state smoke. Use 25, not 0.25.

Predicted Pct. Smokers = −0.838 + 1.124 (Pct. Adult Smoke)

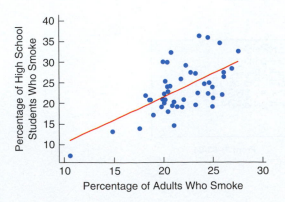

**4.39 Drivers' Deaths and Ages** The figure shows a graph of the death rate in automobile accidents and the age of the driver. The numbers came from the Insurance Information Institute.

a. Explain what the graph tells us about drivers at different ages; state which ages show the safest drivers and which show the most dangerous drivers.

b. Explain why it would not be appropriate to use these data for linear regression.

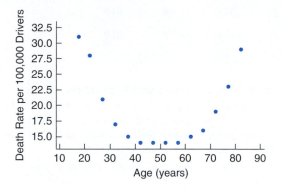

**4.40 Do Women Tend to Live Longer Than Men?** The figure shows life expectancy versus age for males and females in the United States in 2009, up to the age of 119. Females are represented by the blue circles and males by the red squares. These figures were reported by the CDC.

a. Find your own age on the graph and estimate your life expectancy from the appropriate graph.

b. Would it make sense to find the best straight line for this graph? Why or why not?

c. Is it reasonable to predict the life expectancy for a person who is 120 from the regression line for these data? Why or why not?

d. Explain what it means that nearly all of the blue circles (for women) are above the red squares (for men). (Above the age of 100, the red squares cover the blue circles because both are in the same place.)

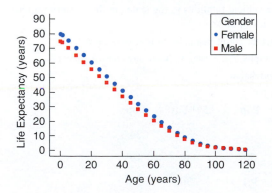

**g 4.41 Does the Cost of a Flight Depend on the Distance?**
The table gives the round-trip fare for flights between Boston and some other cities on major airlines. These were the lowest prices shown on www.Sidestep.com on January 21, 2010. The airlines varied. (The travel dates were all the same, in early February.) *See page 188 for guidance.*

**Round-trip Air Fares from Boston**

| City | Cost | Miles |
|---|---|---|
| New York | 179 | 206 |
| Chicago | 179 | 963 |
| San Francisco | 397 | 3095 |
| Denver | 274 | 1940 |
| Los Angeles | 314 | 2979 |
| Washington, D.C. | 99 | 429 |
| Atlanta | 317 | 1037 |
| Seattle | 443 | 2976 |
| Dallas | 404 | 1748 |
| St. Louis | 334 | 1141 |

How much would it cost, on average, to fly 500 miles? To answer this question, perform a complete regression analysis, including a scatterplot with a regression line.

**4.42 English in California Schools** This problem concerns the vote of the Ventura County cities on Proposition 227 in June of 1998. Proposition 227 in California stated that children should be taught primarily in English and abolished bilingual education. The statewide vote showed that 61% of the voters were in favor of this proposition. Reported in the table are the percentages of voters in each city who voted for Proposition 227 and the percentages of students in those cities with limited English, according to the *Ventura County Star*.

**Vote on Proposition 227 (English in schools)**

| City | %Yes | %Students with limited English |
|---|---|---|
| Thousand Oaks | 71.5 | 7.4 |
| Simi Valley | 71.0 | 6.0 |
| Camarillo | 68.3 | 7.8 |
| Moorpark | 67.8 | 16.8 |
| Ventura | 61.7 | 11.9 |
| Port Hueneme | 61.0 | 42.5 |
| Ojai | 57.4 | 6.5 |
| Santa Paula | 49.9 | 16.6 |
| Fillmore | 48.4 | 36.1 |
| Oxnard | 47.9 | 40.4 |

Assume that the association between the percentage of voters who favor the initiative and the percentage of students with limited English is linear enough to proceed. Find the regression equation for predicting percent Yes from percent with limited English, and report it. Interpret the sign of the slope clearly, in words that someone who had not studied statistics could understand. Pretend you are writing a conclusion for a newspaper article.

### 4.43 Do States with Higher Populations Have More Millionaires?
The table gives the number of millionaires (in thousands) and the population (in hundreds of thousands) for the states in the northeastern region of the United States in 2008. The numbers of millionaires come from *Forbes Magazine* in March of 2007.

a. Without doing any calculations, predict whether the correlation and slope will be positive or negative. Explain your prediction.

b. Make a scatterplot with the population (in hundreds of thousands) on the *x*-axis and the number of millionaires (in thousands) on the *y*-axis. Was your prediction correct?

c. Find the numerical value for the correlation.

d. Find the value of the slope and explain what it means in context. Be careful with the units.

e. Explain why giving and interpreting the value for the intercept does not make sense in this situation.

| State | Millionaires | Population |
|---|---|---|
| Connecticut | 86 | 35 |
| Delaware | 18 | 8 |
| Maine | 22 | 13 |
| Massachusetts | 141 | 64 |
| New Hampshire | 26 | 13 |
| New Jersey | 207 | 87 |
| New York | 368 | 193 |
| Pennsylvania | 228 | 124 |
| Rhode Island | 20 | 11 |
| Vermont | 11 | 6 |

### *4.44 Semesters and Units
The table shows the self-reported number of semesters completed and the number of units completed for 15 students at a community college. All units were counted, but attending summer school was not included.

a. Make a scatterplot with the number of semesters on the *x*-axis and the number of units on the *y*-axis. Does one point stand out as unusual? Explain why it is unusual. (At most colleges, full-time students take between 12 and 18 units per semester.)

   Finish each part *two ways*, with and without the unusual point, and comment on the differences.

b. Find the numerical values for the correlation between semesters and units.

c. Find the two equations for the two regression lines.

d. Insert the lines. Use technology if possible.

e. Report the slopes and intercepts of the regression lines and explain what they show. If the intercepts are not appropriate to report, explain why.

| Sems | Units | Sems | Units |
|---|---|---|---|
| 2 | 21.0 | 3 | 30.0 |
| 4 | 130.0 | 4 | 60.0 |
| 5 | 50.0 | 3 | 45.0 |
| 7 | 112.0 | 5 | 70.0 |
| 3 | 45.5 | 3 | 32.0 |
| 3 | 32.0 | 8 | 70.0 |
| 8 | 140.0 | 6 | 60.0 |
| 0 | 0.0 | | |

### 4.45 Motor Vehicle Fatalities and Population
The table gives the population (in hundreds of thousands of people) and the number of motor vehicle fatalities in the states in the northeastern region of the United States in 2008, according to the National Highway Traffic Safety Administration (NHTSA).

a. Do you expect the correlation to be positive or negative? Why?

b. Make a scatterplot of the data.

c. Report the correlation.

d. Report the equation of the regression line and insert the line in the scatterplot, or use technology to insert the line in your plot.

e. Report the slope of the regression line and explain what it shows.

| MVFat | Pop | MVFat | Pop |
|---|---|---|---|
| 322 | 35 | 773 | 87 |
| 47 | 8 | 1522 | 193 |
| 216 | 13 | 1614 | 124 |
| 459 | 64 | 84 | 11 |
| 127 | 13 | 78 | 6 |

### 4.46 Motor Vehicle Fatalities and Population
The table gives the population (in hundreds of thousands of people) and the number of motor vehicle fatalities in the states in the western region of the United States in 2008.

a. Do you expect the correlation to be positive or negative? Why?

b. Make a scatterplot of the data. Are there any unusual points?

c. Report the correlation.

| MVFat | Pop | MVFat | Pop |
|---|---|---|---|
| 87 | 7 | 381 | 24 |
| 1117 | 50 | 449 | 19 |
| 4078 | 361 | 436 | 36 |
| 742 | 47 | 328 | 25 |
| 119 | 13 | 659 | 63 |
| 264 | 14 | 176 | 5 |
| 270 | 9 | | |

d. Report the equation of the regression line and insert the line in the scatterplot.

e. Report the slope of the regression line and explain in context what it shows.

(Source: www.encarta.msn.com and www-fars.nhtsa.dot.gov)

## SECTION 4.4

**4.47** Answer the questions using complete sentences.

a. What is an influential point?

b. It has been noted that people who go to church frequently tend to have lower blood pressure than people who don't go to church. Does this mean you can lower your blood pressure by going to church? Why or why not? Explain.

**4.48** Answer the questions, using complete sentences.

a. What is extrapolation and why is it a bad idea in regression analysis?

b. How is the coefficient of determination related to the correlation, and what does the coefficient of determination show?

*c. When testing the IQ of a group of adults (aged 25 to 50), an investigator noticed that the correlation between IQ and age was negative. Does this show that IQ goes down as we get older? Why or why not? Explain.

**4.49** If there is a positive correlation between number of years studying math and shoe size (for children), does that prove that larger shoes cause more studying of math, or vice versa? Can you think of a hidden variable that might be influencing both of the other variables?

**4.50** Suppose that the growth rate of children looks like a straight line if the height of a child is observed at the ages of 24 months, 28 months, 32 months, and 36 months. If you use the regression obtained from these ages and predict the height of the child at 21 years, you might find that the predicted height is 20 feet. What is wrong with the prediction and the process used?

**4.51 Coefficient of Determination** If the correlation between height and weight of a large group of people is 0.67, find the coefficient of determination (as a percent) and explain what it means. Assume that height is the predictor and weight is the response, and assume that the association between height and weight is linear.

**4.52 Coefficient of Determination** Does a correlation of −0.70 or +0.50 give a larger coefficient of determination? We say that the linear relationship having the larger coefficient of determination is more strongly correlated. Which of the values shows a stronger correlation?

* **4.53 Decrease in Cholesterol** A doctor is studying cholesterol readings in his patients. After reviewing the cholesterol readings, he calls the patients with the highest cholesterol readings (the top 5% of readings in his office) and asks them to come back to discuss cholesterol-lowering methods. When he tests these patients a second time, the average cholesterol readings tended to have gone down somewhat. Explain what statistical phenomenon might have been partly responsible for this lowering of the readings.

* **4.54 Test Scores** Suppose that students who scored much lower than the mean on their first statistics test were given special tutoring in the subject. Suppose that they tended to show some improvement on the next test. Explain what might cause the rise in grades other than the tutoring program itself.

**TRY** **4.55 Salary and Year of Employment (Example 8)** The equation for the regression line relating the salary and the year first employed is given above the figure.

a. Report the slope and explain what it means.

b. Either interpret the intercept (4,255,000) or explain why it is not appropriate to interpret the intercept.

Predicted Salary (K) = 4255 − 2.099 Start Year

**4.56 MPG: Highway and City** The figure shows the relationship between the number of miles per gallon on the highway and in the city for some cars.

a. Report the slope and explain what it means.

b. Either interpret the intercept (7.792) or explain why it is not appropriate to interpret the intercept.

Predicted Highway Mpg = 7.792 + 0.9478 City Mpg

**4.57 Cost of Turkeys** The table on the next page shows a list of the weights and prices of some turkeys at different supermarkets.

a. Make a scatterplot with weight on the x-axis and cost on the y-axis. If using computer statistical technology, include the regression line on your scatterplot.

b. Find the numerical value for the correlation between weight and price. Explain what the sign of the correlation shows.

c. Report the equation of the best straight line, using weight as the predictor (x) and cost as the response (y).

d. Insert the line on the scatterplot if you did not do it in part a.

e. Report the slope and intercept of the regression line and explain what they show. If the intercept is not appropriate to report, explain why.

f. Add a new point to your data: a 30-pound turkey that is free. Give the new value for *r* and the new regression equation. Explain what the negative correlation implies. What happened?

| Weight (pounds) | Price |
|:---:|:---:|
| 12.3 | $17.10 |
| 18.5 | $23.87 |
| 20.1 | $26.73 |
| 16.7 | $19.87 |
| 15.6 | $23.24 |
| 10.2 | $ 9.08 |

### 4.58 Iraq Casualties and Population of Hometowns The

figures show the number of Iraq casualties through October 2009 and the population of some hometowns from which the servicemen or servicewomen came, according to the *Los Angeles Times*. Comment on the difference in graphs and in the coefficient of determination between the graph that included L.A. (a) and the one that did not include L.A. (b). L.A. is the point with a population of nearly 4 million.

**(A)**    Predicted Casualties = 1.980 + 0.00641 Population (K)

**(Includes LA)**

**(B)**    Predicted Casualties = 2.995 + 0.00309 Population (K)

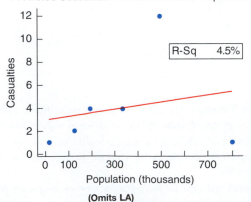

**(Omits LA)**

### TRY 4.59 Teachers' Pay and Costs of Education (Example 9)

The figure shows a scatterplot with a regression line for teachers' average pay and the expenditure per pupil for each state for public schooling in 2007, according to *The 2009 World Almanac and Book of Facts*.

a. From the graph, is the correlation between teachers' average pay and the expenditure per pupil positive or negative?

Predicted Expenditure = −1424 + 0.2315 (Average Pay)

b. Interpret the slope.

c. Interpret the intercept or explain why it should not be interpreted.

### 4.60 Teachers' Pay The figure shows a scatterplot with a

regression line for the average teacher's pay and the percentage of students graduating from high school for each state in 2007, according to *The 2009 World Almanac and Book of Facts*. On the basis of the graph, do you think the correlation is positive, negative, or near 0? Explain what this means.

### TRY 4.61 Does Having a Job Affect Students' Grades?

**(Example 10)** Grades on a political science test and the number of hours of paid work in the week before the test were studied. The instructor was trying to predict the grade on a test from the hours of work. The figure shows a scatterplot and the regression line for these data.

a. By looking at the plot and the line (without doing any calculations), state whether the correlation is positive or negative and explain your prediction.

b. Interpret the slope.

c. Interpret the intercept.

**4.62 Weight of Trash and Household Size** Data that included information on the weight of the trash (in pounds) on the street one week and the number of people who live in the house were collected. The figure shows a scatterplot with the regression line.

a. Is the trend positive or negative? What does that mean?

b. Now calculate the correlation between the weight of trash and the number of people. (Use R-Sq from the figure and take the square root of it.)

c. Report the slope. For each additional person in the house, there are, on average, how many additional pounds of trash?

d. Either report the intercept or explain why it is not appropriate to interpret it.

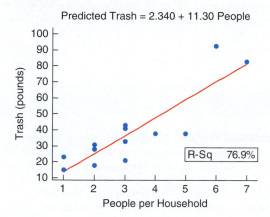

Predicted Trash = 2.340 + 11.30 People

**4.63 Education of Fathers and Mothers** The data shown in the table are the number of years of formal education of the fathers and mothers of a sample of 29 statistics students at a small community college in an area with many recent immigrants. (The means are both about 8, and the standard deviations are both about 4.6.) The scatterplot (not shown) suggests a linear trend.

| Father | Mother | Father | Mother |
|--------|--------|--------|--------|
| 3 | 3 | 4 | 7 |
| 11 | 13 | 16 | 13 |
| 3 | 4 | 13 | 3 |
| 12 | 12 | 11 | 10 |
| 8 | 16 | 6 | 6 |
| 1 | 2 | 3 | 2 |
| 8 | 8 | 12 | 14 |
| 6 | 4 | 14 | 12 |
| 12 | 8 | 12 | 12 |
| 0 | 10 | 12 | 13 |
| 8 | 12 | 12 | 18 |
| 5 | 5 | 13 | 8 |
| 8 | 8 | 0 | 0 |
| 5 | 5 | 3 | 5 |
| 12 | 6 | | |

a. Find and report the regression equation for predicting the mother's years of education from the father's. Then find the predicted number of years for the mother if the father has 12 years of education, and find the predicted number of years for a mother if the father has 4 years of education.

b. Find and report the regression equation for predicting the father's years of education from the mother's. Then find the predicted number of years for the father if the mother has 12 years of education, and find the predicted number of years for the father if the mother has 4 years of education.

c. What phenomenon from the chapter does this demonstrate? Explain.

* **4.64 Heights of Fathers and Sons** The table shows some data on a sample of heights of fathers and their sons.

You may want to use the computer or a statistics calculator to verify that

$$\bar{x}_{\text{father}} = 69.233 \quad s_{\text{father}} = 2.82$$

$$\bar{y}_{\text{son}} = 69.033 \quad s_{\text{son}} = 2.84$$

$$r = 0.762$$

The scatterplot (not shown) suggests a linear trend.

a. Find and report the regression equation for predicting the son's height from the father's height. Then predict the height of a son with a father 74 inches tall. Also predict the height of a son of a father who is 65 inches tall.

b. Find and report the regression equation for predicting the father's height from the son's height. Then predict a father's height from that of a son who is 74 inches tall and also predict a father's height from that of a son who is 65 inches tall.

c. What phenomenon does this show?

| Father's Height | Son's Height |
|-----------------|--------------|
| 75 | 74 |
| 72.5 | 71 |
| 72 | 71 |
| 71 | 73 |
| 71 | 68.5 |
| 70 | 70 |
| 69 | 69 |
| 69 | 66.5 |
| 69 | 72 |
| 68.5 | 66.5 |
| 67.5 | 65.5 |
| 67.5 | 70 |
| 67 | 67 |
| 65.5 | 64.5 |
| 64 | 67 |

g * **4.65 Test Scores** Assume that in a political science class, the teacher gives a midterm exam and a final exam. Assume that the association between midtern and final scores is linear. The summary statistics have been simplified for clarity.

> Midterm:  Mean = 75,  Standard deviation = 10
>
> Final:  Mean = 75,  Standard deviation = 10
>
> Also, $r = 0.7$ and $n = 20$.

According to the regression equation, for a student who gets a 95 on the midterm, what is the predicted final exam grade? What phenomenon from the chapter does this demonstrate? Explain. *See page 189 for guidance.*

* **4.66 Test Scores** Assume that in a sociology class, the teacher gives a midterm exam and a final exam. Assume that the association between midterm and final scores is linear. Here are the summary statistics:

> Midterm:  Mean = 72,  Standard deviation = 8
>
> Final:  Mean = 72,  Standard deviation = 8
>
> Also, $r = 0.75$ and $n = 28$.

a. Find and report the equation of the regression line to predict the final exam score from the midterm score.

b. For a student who gets 55 on the midterm, predict the final exam score.

c. Your answer to part b should be higher than 55. Why?

d. For a student who gets a 100 on the midterm, without doing any calculations, state whether the predicted value would be higher, lower, or the same as 100.

# CHAPTER REVIEW EXERCISES

* **4.67 Heights and Weights of People** The table shows the height and weight of some people. The figure shows that the association is linear enough to proceed.

| Height (inches) | Weight (pounds) |
|---|---|
| 60 | 105 |
| 66 | 140 |
| 72 | 185 |
| 70 | 145 |
| 63 | 120 |

a. Calculate the correlation, and find and report the equation of the regression line, using height as the predictor and weight as the response.

b. Change the height to centimeters by multiplying each height in inches by 2.54. Find the weight in kilograms by dividing the weight in pounds by 2.205. Retain at least six digits in each number so there will be no errors due to rounding.

c. Calculate the correlation between height in centimeters and weight in kilograms, and compare it with the correlation for the heights in inches and the weights in pounds.

d. Find the equation of the regression line for predicting weight from height, using height in cm and weight in kg. Is the equation for weight (in pounds) and height (in inches) the same or different from the equation for weight (in kg) and height (in cm)?

* **4.68 Heights and Weights of Men** The table shows the heights (in inches) and weights (in pounds) of 14 college men. The figure shows that the association is linear enough to proceed.

| Height (inches) | Weight (pounds) | Height (inches) | Weight (pounds) |
|---|---|---|---|
| 68 | 205 | 70 | 200 |
| 68 | 168 | 69 | 175 |
| 74 | 230 | 72 | 210 |
| 68 | 190 | 72 | 205 |
| 67 | 185 | 72 | 185 |
| 69 | 190 | 71 | 200 |
| 68 | 165 | 73 | 195 |

a. Find the equation for the regression line with weight (in pounds) as the response and height (in inches) as the predictor. Report the slope and intercept of the regression line and explain what they show. If the intercept is not appropriate to report, explain why.

b. Find the correlation between weight (in pounds) and height (in inches).

c. Find the coefficient of determination and interpret.

d. If you changed each height to centimeters by multiplying heights in inches by 2.54, what would the new correlation be? Explain.

e. Find the equation with weight (in pounds) as the response and height (in cm) as the predictor, and interpret the slope.

f. Summarize what you found: Does changing units change the correlation? Does changing units change the regression equation?

**4.69 House Price and Square Feet** The figure gives a scatterplot, using data taken from *The Ventura County Star* (February, 2006), of the number of square feet in some houses and their list price for sale.

a. Choose the correct correlation from these choices: $+1.00, 0, -1.00, -0.75, +0.75$.

b. The equation of the regression line is given above the graph. Report the slope and intercept of the regression line and explain what they show in this context. If the intercept is not appropriate to report, explain why.

c. By looking at the graph or using the equation, predict the average cost of a house that is about 1300 square feet.

d. Find the coefficient of determination from the correlation that you chose in part a and explain what it means in the context of the problem.

Predicted Price = 223.5 + 0.174 (Square Feet)

**4.70 Heights and Weights of First and Second Graders** Some statistics students recorded the heights and weights of a sample of first and second graders. Refer to the figure.

a. Choose the correct correlation coefficient from these choices: $+1.00, 0, -1.00, -0.84, +0.84$.

b. The equation is given above the graph. Report the slope and intercept of the regression line and explain what they show in this context. If the intercept is not appropriate to report, explain why.

c. By looking at the graph or using the equation of the regression line, predict the weight for a first or second grader who is 54 inches tall.

d. Find the coefficient of determination from the correlation that you chose in part a and explain what it means in the context of the problem.

Predicted Weight = −83.05 + 2.993 Height

**4.71 Hours of Study and of TV Viewing** The number of hours of study and the number of hours of TV viewing per week were recorded for some community college students.

a. Make a scatterplot of the data, using hours of study as the predictor.

b. Describe the association, if any, between hours of study and hours of TV viewing.

c. Is the correlation strong and positive, strong and negative, or near zero? Do not calculate.

**4.72 Salary and Age** The age and salary (dollars per hour) were recorded for some college students with jobs.

a. Make a scatterplot of the data, using age as the predictor.

b. Find the equation of the regression line.

c. Report the slope and intercept of the regression line and explain what they show. If the intercept is not appropriate to report, explain why not.

**4.73 Age and Weight for Women** The figure shows a scatterplot and regression line for the age and weight of some women in a statistics class.

a. Find the correlation by taking the positive square root of R-Sq shown on the graph and attaching the proper sign.

b. Using the equation given, find the predicted weight for a woman 35 years old.

*c. Report the slope and explain what it means. This study did not follow the same people over time.

d. If it is appropriate, report the intercept and explain what it means. Otherwise, explain why it is not appropriate to report.

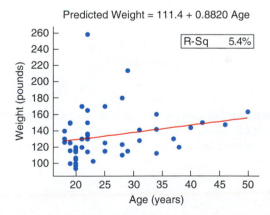

Predicted Weight = 111.4 + 0.8820 Age

R-Sq    5.4%

**4.74 Age and Weight for Men** The figure on the next page shows a scatterplot and regression line for the age and weight of some men in a statistics class.

a. Find the correlation by taking the positive square root of R-Sq shown on the graph and attaching the proper sign.

b. Using the equation given, find the predicted weight for a man 35 years old.

*c. Report the slope and explain what it means. (Note that this study did not follow the same people over time.)

d. Either report the intercept and explain what it shows or explain why it should not be reported.

Predicted Weight = 132.9 + 2.040 Age

**4.75** The figure shows a scatterplot of the educational level of twins. Describe the scatterplot. Explain the trend and mention any unusual points. (Source: www.stat.ucla.edu)

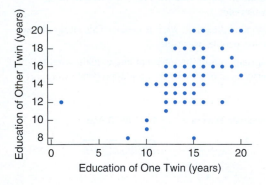

**4.76 Wages and Education** The figure shows a scatterplot of the wages and educational level of some people. Describe what you see. Explain the trend and mention any unusual points. (Source: www.stat.ucla.edu)

**4.77 Do Students Taking More Units Study More Hours?** The figure shows the number of units that students were enrolled in and the number of hours (per week) that they reported

studying. Do you think there is a positive trend, a negative trend, or no noticeable trend? Explain what this means about the students.

**4.78 Hours of Exercise and Hours of Homework** The scatterplot shows the number of hours of exercise per week and the number of hours of homework per week for some students. Explain what it shows.

**4.79 Tree Heights** Loggers gathered information about some trees. The diameter is in inches, the height is in feet, and the volume of the wood is in cubic feet. Loggers are interested in whether they can estimate the volume of the tree given any single dimension. Which is the better predictor of volume: the diameter or the height?

**4.80 Salary and Education** Does education pay? The salary per year in dollars, the number of years employed (YrsEm), and number of years of education after high school (Educ) for the employees of a company were recorded. Determine whether number of years employed or number of years of education after high school is a better predictor of salary. Explain your thinking. (Source: Minitab File)

**4.81 Film Budgets and Grosses** Movie studios spend much effort trying to predict how much money their movies will make. One possible predictor is the amount of money spent on the production of the movie. In the table you can see the budget and the amount of money made worldwide for the 13 movies with the highest profit. The budget and gross are in millions of dollars. Make a scatterplot and comment on what you see. If appropriate, find, report, and interpret the regression line. If it is not appropriate to do so, explain why. (Source: www.the-numbers.com)

| Movie | Budget | Gross |
|-------|--------|-------|
| Titanic | 200.0 | 1848.8 |
| The Lord of the Rings: The Return of the King | 94.0 | 1133.0 |
| Jurassic Park | 63.0 | 920.1 |
| Star Wars Ep. IV: A New Hope | 11.0 | 797.9 |
| Shrek 2 | 70.0 | 915.3 |
| ET: The Extra-Terrestrial | 10.5 | 792.9 |
| Pirates of the Caribbean: Dead Man's Chest | 150.0 | 1065.7 |
| The Lord of the Rings: The Two Towers | 94.0 | 926.3 |
| Harry Potter and the Sorcerer's Stone | 125.0 | 976.5 |
| Star Wars Ep. I: The Phantom Menace | 115.0 | 924.3 |
| Harry Potter and the Chamber of Secrets | 100.0 | 879.0 |
| Finding Nemo | 94.0 | 866.6 |
| Avatar | 237.0 | 2212.0 |

**4.82 Gas Mileage of Cars** The table gives the number of miles per gallon in the city and on the highway for the cars reported to have the best gasoline mileage in 2008, according to CNN. Make a scatterplot, using the city mileage as the predictor. Find the equation of the regression line for predicting the number of miles per gallon (mpg) on the highway from the number of miles per gallon in the city. Use the equation to predict the highway mileage from a city mileage of 60 mpg. Also find the coefficient of determination and explain what it means.

| | City | Highway |
|---|------|---------|
| Honda Insight | 60 | 66 |
| Toyota Prius | 60 | 51 |
| Honda Civic Hybrid | 48 | 47 |
| Volkswagen Diesel | 38 | 46 |
| Honda Civic | 36 | 44 |
| Toyota Echo | 35 | 43 |
| Toyota Scion | 32 | 38 |
| Toyota Corolla | 32 | 40 |
| Toyota Celica | 29 | 36 |

For 4.83–4.86 show your points in a rough scatterplot and give the coordinates of the points.

* **4.83** Construct a small set of numbers with at least three points with a perfect positive correlation of 1.00.

* **4.84** Construct a small set of numbers with at least three points with a perfect negative correlation of $-1.00$.

* **4.85** Construct a set of numbers (with at least three points) with a strong negative correlation. Then add one point (an influential point) that changes the correlation to positive. Report the data and give the correlation of each set.

* **4.86** Construct a set of numbers (with at least three points) with a strong positive correlation. Then add one point (an influential point) that changes the correlation to negative. Report the data and give the correlation of each set.

# gUIDED EXERCISES

g **4.41 Does the Cost of a Flight Depend on the Distance?**   The table gives the round-trip fare for flights between Boston and some other cities on major airlines. These were the lowest prices shown on www.Sidestep.com on January 21, 2010. The airlines varied. (The travel dates were all the same, in early February.)

How much would it cost, on average, to fly 500 miles? To answer this question, perform a complete regression analysis, including a scatterplot with a regression line, following the steps below.

**Round-trip
Air Fares from Boston**

| City | Cost | Miles |
|------|------|-------|
| NY | 179 | 206 |
| Chicago | 179 | 963 |
| San Fran | 397 | 3095 |
| Denver | 274 | 1940 |
| LA | 314 | 2979 |
| WashDC | 99 | 429 |
| Atlanta | 317 | 1037 |
| Seattle | 443 | 2976 |
| Dallas | 404 | 1748 |
| St. Louis | 334 | 1141 |

### Step 1 ▶ Make a scatterplot
Be sure that miles is the *x*-variable and cost is the *y*-variable. You may want to include the regression line with your scatterplot if you have that option with your technology; refer to Step 4.

### Step 2 ▶ Is the linear model appropriate?
In this case the answer is yes, because there is a linear trend. It is hard to see with so few points, but a strong curvature is not present and the cost tends to increase as the miles increase.

### Step 3 ▶ Obtain the equation
When finding the regression equation, be sure that you use miles as the *x*-variable and cost as the *y*-variable. For example, if you are using the TI-83/84, the predictor (which you probably put into List 2 because the second column of numbers has the miles) has to be entered first and the response (in List 1) is entered second after the comma. See the TI-83/84 figure.

### Step 4 ▶ Add the regression line to the scatterplot
If your technology will make a plot with a line, do so. Refer to the TechTips, which begin on page 191, to see how to do this.

If your technology will not draw the line, you can choose two *x*-values (distances), find the corresponding *y*-values (cost), plot these, and then draw the line to connect them. For example,

```
LinReg(a+bx) L₂,
L₁
```

to choose the first point on the line, choose an *x*-value such as $x = 200$ and follow these additional steps:

- **First Predicted Point**
  Pick an arbitrary, fairly small distance such as 200 miles (not less than the smallest distance given) and substitute into the equation you got to find the predicted cost:

$$\text{Predicted Cost} = 162.60 + 0.0795 \text{ Miles}$$
$$= 162.60 + 0.0795(200)$$
$$= 162.60 + 15.90$$
$$= 178.5$$

  A predicted point is (200, 178.5), so a flight of about 200 miles should cost about $178.50. Put the point on the graph with a symbol you will remember.

- **Second Point**
  To get a second point, pick an arbitrary large value for distance such as 3000 (not larger than the largest distance) and substitute it into the equation to find the cost. Put the point on the graph using the same symbol you chose before.

- **The Line**
  The regression line will be a straight line between these two predicted points (use a ruler to make the line). Extend the line out to the edges of the data, to the left to about 100 miles, and to the right to about 3000 miles.

### Step 5 ▶ Interpret the slope and intercept in context

Predicted   Y   =   a   +   b   X
Predicted Cost = 162.60 + 0.0795 Miles

The slope is *b* (which is the multiplier for *Miles*), which is 0.0795, and the intercept is *a,* which is the first number, $162.60.

Fill in the blanks that follow.

For the slope: For every additional mile, on average, the price goes up by ____ dollars.
For the intercept: A trip with zero miles should cost about ____ dollars. Explain why interpreting the *y*-intercept like this is questionable.

### Step 6 ▶ Answer the question by using the regression equation
How much would it cost to fly 500 miles?

**g 4.65 Test Scores**    Assume that in a political science class, the teacher gives a midterm exam and a final exam. Assume that the association between midterm and final scores is linear. The summary statistics have been simplified for clarity.

| | | |
|---|---|---|
| Midterm: | Mean = 75, | Standard deviation = 10 |
| Final: | Mean = 75, | Standard deviation = 10 |

Also, $r = 0.7$ and $n = 20$.

For a student who gets 95 on the midterm, what is the predicted final exam grade? Assume the graph is linear.

**Step 1** ▶ Find the equation of the line to predict the final exam score from the midterm score. Standard form: $y = a + bx$

a. First find the slope: $b = r\left(\dfrac{s_{final}}{s_{midterm}}\right)$

b. Then find the $y$-intercept, $a$, from the equation

$$a = \bar{y} - b\bar{x}$$

c. Write out the following equation:

$$\text{Predicted } y = a + bx$$

However, use "Predicted Final" instead of "Predicted $y$" and "Midterm" in place of $x$.

**Step 2** ▶ Use the equation to predict the final exam score for a student who gets 95 on the midterm.

**Step 3** ▶ Your predicted final exam grade should be less than 95. Why?

## CHECK YOUR TECH

## Verifying Minitab Output for Correlation and Regression

| Husband | Wife |
|---------|------|
| 20 | 20 |
| 30 | 30 |
| 40 | 25 |

Data are given on the right for the ages at which husbands and wives married. The data were simulated to make the calculations easier.

We used Minitab to find the correlation and regression equation given in Figure A, using the husband's age ($x$) to predict his wife's age ($y$).

▶ FIGURE A

```
Pearson correlation of Husband and Wife = 0.500
The regression equation is
Wife = 17.5 + 0.250 Husband
```

From Figure B: for the husband, $\bar{x} = 30$ and $s_x = 10$; for the wife, $\bar{y} = 25$ and $s_y = 5$.

$$r = \frac{\sum z_x z_y}{n-1} \text{ and } z_x = \frac{x - \bar{x}}{s_x} \text{ and } z_y = \frac{y - \bar{y}}{s_y}$$

▶ FIGURE B

```
Variable  N   Mean   StDev
Husband   3   30.00  10.00
Wife      3   25.00   5.00
```

**QUESTION**   Using the summary statistics provided in Figure B, verify the output given in Figure A by following the steps below.

**SOLUTION**

**Step 1** ▶ Fill in the missing numbers (indicated by the blanks) in the table.

| $x$ | $x - \bar{x}$ | $z_x$ | $y$ | $y - \bar{y}$ | $z_y$ | $z_x z_y$ |
|-----|---------------|-------|-----|---------------|-------|-----------|
| 20 | $20 - 30 = -10$ | $-10/10 = -1$ | 20 | $20 - 25 = -5$ | $-5/5 = -1$ | $(-1) \times (-1) = 1$ |
| 30 | $30 - 30 = 0$ | $0/10 = \_$ | 30 | $30 - \_ = \_$ | $5/5 = \_$ | $0 \times (\_) = \_$ |
| 40 | $40 - 30 = 10$ | $\_/\_ = \_$ | 25 | $\_ - \_ = \_$ | $0/\_ = \_$ | $\_ \times (\_1) = \_$ |

**Step 2** ▶ Add the last column to get $\sum z_x z_y = \_$.

**Step 3** ▶ Find the correlation and check it with the output.

$$r = \frac{\sum z_x z_y}{n-1} = \frac{\_}{3-1} = \frac{\_}{2} = \_$$

**Step 4** ▶ Find the slope: $b = r\dfrac{s_y}{s_x} = r\dfrac{5}{10} = \_$.

**Step 5** ▶ Find the $y$-intercept: $a = \bar{y} - b\bar{x} = 25 - 30b = \_$.

**Step 6** ▶ Finally, put together the equation:

$$y = a + bx$$

Predicted Wife $= a + b$ Husband

and check the equation with the Minitab output. The equations should match.

# TechTips

## General Instructions for All Technology

Upload data from the disk or website, or enter data manually using two columns of equal length. Refer to TechTips in Chapter 2 for a review of entering data. Each row represents a single observation, and each column represents a variable. All technologies will use the example that follows.

**EXAMPLE** ▶Analyze the six points in the data table with a scatterplot, correlation, and regression. Use heights (in inches) as the x-variable and weight (in pounds) as the y-variable.

| Height | Weight |
|--------|--------|
| 61 | 104 |
| 62 | 110 |
| 63 | 141 |
| 64 | 125 |
| 66 | 170 |
| 68 | 160 |

---

### TI-83/84

#### Making a Scatterplot

1. Press **2ND**, **STATPLOT** (which is the button above **2ND**), **4**, and **ENTER**, to turn off plots made previously.
2. Press **2ND**, **STATPLOT**, and **1** (for Plot1).
3. Refer to Figure 4A: Turn on **Plot1** by pressing **ENTER** when **On** is highlighted.

▲ **FIGURE 4A** TI-83/84 Plot1 Dialogue Screen

4. Use the arrows on the keypad to get to the scatterplot (upper left of the six plots) and press **ENTER** when the scatterplot is highlighted. Be careful with the **Xlist** and **Ylist**. To get **L1**, press **2ND** and **1**. To get **L2**, press **2ND** and **2**.
5. Press **ZOOM** and **9** (**Zoomstat**) to create the graph.
6. Press **TRACE** to see the coordinates of the points, and use the arrows on the keypad to go to other points. Your output will look like Figure 4B, but without the line.

▲ **FIGURE 4B** TI-83/84 Plot with Line

7. To get the output with the line in it, shown in Figure 4B: **STAT**, **CALC**, 8:LinReg(a+bx), **L1ʼ L2ʼ Y1** (You get the **Y1** by pressing **VARS**, **Y-VARS**, I: Function, I:Y1, **ENTER**.) The comma button is on the keypad above the 7 button.
8. Press **ZOOM** and **9**.
9. Press **TRACE** to see the numbers, and use the arrows on the keypad to get to other number.

#### Finding the Correlation and Regression Equation Coefficients

Before finding the correlation, you must turn the diagnostics on, as shown here.

Press **2ND**, **CATALOG**, and scroll down to **DiagnosticOn** and press **ENTER** twice. The diagnostics will stay on unless you **Reset** your calculator or change the batteries.

1. Press **STAT**, choose **CALC**, and **8** (for LinReg (a + bx)).
2. Press **2ND L1** (or whichever list is X, the predictor), press **ʼ** (comma: the button above the 7), press **2ND L2** (or whichever list is Y, the response), and press **ENTER**.

Figure 4C shows the output.

▲ **FIGURE 4C** TI-83/84 Output

## Making a Scatterplot

1. **Graph > Scatterplot**
2. Leave the default **Simple** and click **OK**.
3. Double click the column containing the weights so that it goes under the **Y Variables**. Then double click the column containing the heights so that it goes under the **X Variables**.
4. Click **OK**. After the graph is made, you can edit the labels by clicking on them.

## Finding the Correlation

1. **Stat > Basic Statistics > Correlation**
2. Double click both the predictor column and the response column (in either order).
3. Click **OK**. You will get 0.881.

## Finding the Regression Equation Coefficients

1. **Stat > Regression > Regression**
2. Put in the **Response** (y) and **Predictor** (x) columns.
3. Click **OK**. You may need to scroll up to see the regression equation. It will be easier to understand if you have put in

labels for the columns, such as "Height" and "Weight." You will get: Weight = −443 + 9.03 Height.

## To Display the Regression Line on a Scatterplot

1. **Stat > Regression > Fitted Line Plot**
2. Double click the **Response (y)** column and then double click the **Predictor (x)** column.
3. Click **OK**. Figure 4D shows the fitted line plot.

▲ **FIGURE 4D** Minitab Fitted Line Plot

## Making a Scatterplot

1. Select (highlight) the two columns containing the data, with the predictor column to the *left* of the response column. You may include the labels at the top or not include them.
2. Click **Insert**, in **Charts** click the picture of a scatterplot, and click the upper left option shown here:

3. Click **OK**.
4. Note that the lower left corner of the chart is not at the origin, (0, 0). If you want to zoom in or out on the data by changing the minimum value for an axis, right-click on the axis numbers, click **Format axis**, in **Axis Options** change the **Minimum** to **Fixed**, and put in the desired value. You may want to do this twice: once for the *x*-axis and once for the *y*-axis. Then click **Close**.
5. When the chart is active (click on it), **Chart Tools** are shown at the top of screen, right of center. Click **Layout** (not **Page Layout**), then **Axis Titles**, and **Chart Title** to add appropriate labels. After the labels are added, you can click on them to change the spelling or add words. Delete the labels on the right-hand side, such as **Series 1**, if you see any.

## Finding the Correlation

1. Click on **Data**, click on **Data Analysis**, select **Correlation**, and click **OK**.

2. For the **Input Range**, select (highlight) both columns of data (if you have highlighted the labels as well as the numbers, you must also click on the **Labels in first row**).
3. Click **OK**. You will get 0.880934.

(Alternatively, just click the $f_x$ button, for **category** choose **statistical**, select **CORREL**, click **OK**, and highlight the two columns containing the numbers, one at a time. The correlation will show up on the dialogue screen, and you do *not* have to click **OK**.)

## Finding the Coefficients of the Regression Equation

1. Click on **Data**, **Data Analysis**, **Regression**, and **OK**.
2. For the **Input Y Range**, select the column of numbers (not words) that represents the response or dependent variable. For the **Input X Range**, select the column of numbers that represents the predictor or independent variable.
3. Click **OK**.

A large summary of the model will be displayed. Look under **Coefficients** at the bottom. For the **Intercept** and the slope (next to **XVariable1**), see Figure 4E, which means the regression line is

$$y = -442.9 + 9.03x$$

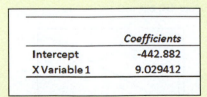

|  | Coefficients |
|---|---|
| Intercept | -442.882 |
| X Variable 1 | 9.029412 |

▲ **FIGURE 4E** Excel Regression Output

## To Display the Regression Line on a Scatterplot

4. After making the scatterplot, under **Chart Tools** click **Design**. In the **Chart Layouts** group, click the triangle to the right of **Quick Layout**. Choose Layout 9 (the option in the lower right portion, which shows a line in it and also *fx*).

   Refer to Figure 4F.

▲ **FIGURE 4F** Excel Fitted Line Plot with Equation

---

STATCRUNCH

### Making a Scatterplot

1. **Graphics > Scatterplot**
2. Select an **X variable** and a **Y variable** for the plot.
3. Click **Create Graph!** to construct the plot.
4. To copy the graph, click **Options** and **Copy**.

### Finding the Correlation and Coefficients for the Equation

1. **Stat > Regression > Simple Linear**
2. Select the **X variable** and **Y variable** for the regression.
3. Click **Calculate** to view the equation and numbers, which are shown in Figure 4G.

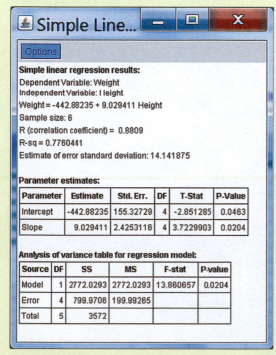

▲ **FIGURE 4G** StatCrunch Regression Output

### Plotting the Regression Line on a Scatterplot

1. **Click Stat > Regression > Simple Linear**
2. Select your columns for X and Y and click **Next**.
3. Don't select any **Options** and click **Next** again. Do this step twice.
4. Check **Plot the fitted line**.
5. Click **Calculate**. When you see the numbers in the output, enlarge it and click **Next** at the very bottom to see the Fitted line plot. Refer to Figure 4H.
6. To copy the graph, click **Options** and **Copy**.

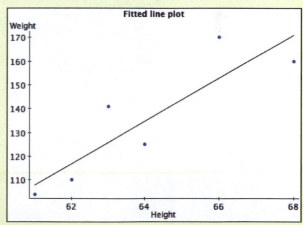

▲ **FIGURE 4H** StatCrunch Fitted Line Plot

# 5 Modeling Variation with Probability

# THEME

Probabilities are long-run relative frequencies used to describe processes where the outcome depends on chance, such as flipping a coin or rolling a die. Theoretical probabilities are based on specific assumptions (usually based on a theory) about the chance process. Empirical probabilities are based on observation of the actual process. The two types of probabilities are closely related, and we need both to analyze samples of data in the real world.

In 1971, the United States was fighting the Vietnam War and drafting men to serve in the military. To determine who was chosen, government officials wrote the days of the year (January 1, January 2, and so on) on capsules. The capsules were placed in a large container and mixed up. They were then drawn out one at a time. The first date chosen was assigned the rank 1, the second date was assigned the rank 2, and so on. Men were drafted on the basis of their birthday. Those whose birthday had rank 1 were drafted first, then those whose birthday had rank 2, and so on until the officials had enough men.

Although the officials thought that this method was random, some fairly convincing evidence indicates that it was not (*Journal of Statistics Education*). Figure 5.1a shows boxplots with the actual ranks for each month. Figure 5.1b shows what boxplots might have looked like if the lottery had been truly random. In Figure 5.1b, each month has roughly the same rank. However, in Figure 5.1a, a few months had notably lower ranks than the other months. Bad news if you were born in December—you were more likely to be called up first.

What went wrong? The capsules, after having dates written on them, were clustered together by month as they were put into the tumbler. But the capsules weren't mixed up enough to break up these clusters. The mixing wasn't adequate to create a truly random mix.

It's not easy to generate true randomness, and humans have a hard time recognizing random events when they see them. Probability gives us a tool for understanding randomness. Probability helps us answer the question "How often does something like this happen by chance?" By answering that question, we create an important link between our data and the real world. In previous chapters, you learned how to organize, display, and summarize data to see patterns and trends. Probability is a vital tool because it gives us the ability to generalize our understanding of data to the larger world. In this chapter, we'll explore issues of randomness and probability: What is randomness? How do we measure it? And how do we use it?

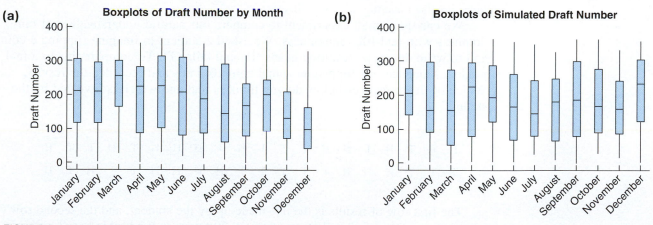

▲ **FIGURE 5.1** Boxplots of **(a)** actual Vietnam draft numbers by month, and **(b)** what might have happened if the draft had really been random.

# CASE STUDY

## SIDS or Murder?

In November 2000, Sally Clark was convicted in England of killing her two children. Her children had died several years apart, and the initial diagnosis was sudden infant death syndrome (SIDS). SIDS is the sudden, unexplained death of an infant under 1 year of age. It is estimated that in the United States, about 5500 infants die of SIDS every year. Although some risk factors have been identified (most are related to the mother's health during pregnancy), the cause or causes are as yet unknown.

Clark was convicted of murder on the basis of the expert testimony of Sir Roy Meadow, a prominent British physician who was an expert on SIDS. Dr. Meadow quoted a published statistic: The probability of an infant dying of SIDS is 1/8543. That is, in the U.K., one child in every 8543 dies of SIDS. Dr. Meadow concluded that the probability of two children in the same family dying of SIDS was therefore (1/8543) × (1/8543), or about 1 in 73 million. The jury agreed that this event—two children in the same family dying of SIDS—was so unlikely that they convicted Ms. Clark of murder, and she was sent to prison.

But then in 2003, Sally Clark's conviction was overturned and she was released from prison. Dr. Meadow was accused of professional misconduct. Why did Dr. Meadow multiply the two probabilities together? Why was the verdict overturned? We will answer these questions at the end of the chapter.

SECTION 5.1

# What Is Randomness?

What exactly is randomness? The *Oxford English Dictionary* says *random* means "having no definite aim or purpose." You probably use the word to describe events that seem to happen with no purpose at all. Sometimes the word is used to describe things that have no apparent pattern. However, in statistics the word *random* has a more precise meaning, as you will see.

We can do a small experiment to compare our natural understanding of randomness with real-life randomness. We asked a student to imagine flipping a coin and to write down the resulting heads (H) and tails (T). To compare this to a real random sequence, we flipped an actual coin 20 times and recorded the results. Which of the sequences below do you think is real and which is invented by the student?

| H | H | T | H | T | H | H | T | T | T | H | T | H | H | H | H | T | H | T | T | H |
|---|---|---|---|---|---|---|---|---|---|---|---|---|---|---|---|---|---|---|---|---|
| T | T | T | H | H | H | T | T | T | T | T | H | H | H | H | T | T | H | T | H |

The first row of results is the one made up by the student, and the second row records the results of actually tossing a coin 20 times. Is it possible to tell by comparing the two sequences? Not always, but this time the student did something that most people do in this situation. Most of the time, he wrote very short "streaks." (A streak is any sequence of consecutive heads or tails. TT is a streak of two tails.) Only once did he write as many as three heads or three tails in a row. It's as if, after writing three heads in a row, he thought, "Wait! If I put a fourth, no one will believe it is random."

However, the real, truly random sequence had one streak of five tails (beginning at the seventh flip) and another streak with four heads. These long streaks are examples of the way chance creates things that look like a pattern or structure, but really are not.

## You Try It

Use a computer to flip a coin 20 times for you. Go to http://www.socr.ucla.edu.
  Click on the **Experiments** tab and select **Coin Sample Experiment** from the dropdown menu that will appear. Set $n$ to be 20 (by moving the slider to the left of $n = 10$), and set $p$ to be 0.50. Press the **Play** (▶) button, and count the resulting streaks. Do this a few times, and keep track of the longest streak you get each time. How long does the longest streak have to be before you think it is unusually long?

 **KEY POINT** — People are not good at identifying truly random samples or random experiments, so we need to rely on outside mechanisms such as coin flips or random number tables.

Real randomness is hard to achieve without help from a computer or some other randomizing device. If a computer is not available to generate random numbers, another useful approach is to use a random number table. (An example is provided in Appendix A of this book.) A random number table provides a sequence of digits from 0 to 9 in a random order. Here, **random** means that no predictable pattern occurs and that no digit is more likely to appear than any other. (Of course, if you use the same table often enough, it might seem predictable to you, but to an outsider, it will not seem predictable.)

For example, if we are doing a controlled experiment, we might assign each subject in our study a random number from this table as the subjects come into our office. The odd-numbered subjects would then go into the Control group, and the even-numbered would go into the Treatment group.

To use a random number table to simulate coin flipping, we assign a number to each outcome. For example, suppose we let even numbers (0, 2, 4, 6, 8) represent tails and let odd numbers (1, 3, 5, 7, 9) represent heads. Now choose any row of the table you wish and any column. For example, we arbitrarily chose column 11 and row 30 because that's the date of the day we wrote this paragraph (November 30, or 11/30). Read off the next 20 digits, but translate them into "heads" and "tails." What's the longest streak you get?

| Line | | | | | | |
|------|-------|-------|-------|-------|-------|-------|
| 28 | 31498 | 85304 | 22393 | 21634 | 34560 | 77404 |
| 29 | 93074 | 27086 | 62559 | 86590 | 18420 | 33290 |
| 30 | 90549 | 53094 | 76282 | 53105 | 45531 | 90061 |
| 31 | 11373 | 96871 | 38157 | 98368 | 39536 | 08079 |
| 32 | 52022 | 59093 | 30647 | 33241 | 16027 | 70336 |

◄ **TABLE 5.1** Lines 28–32 (indicated at the left side of the table) from the random number table in Appendix A. The red 7 in the 11th column and the 30th line, or row, is our starting point.

## EXAMPLE 1  Simulating Randomness

Let's play a game. You roll a six-sided die until the first 6 appears. You win one point for every roll. You pass the die to your opponent, who also rolls until the first 6 appears. Whoever has the most points, wins.

**QUESTION**  Simulate rolling the die until a 6 appears. Use Table 5.1, and start at the very first entry of line 28 in Table 5.1. How many rolls did it take?

**SOLUTION** We'll let the digits 1 through 6 represent the outcome shown on the die. We'll ignore the digits 7, 8, 9, and 0. Starting at line 28, we get these random digits:

3, 1, 4, (ignore, ignore, ignore) 5, 3, (ignore), 4, 2, 2, 3,
    (ignore), 3, 2, 1, 6

**CONCLUSION** We rolled the die 12 times before the first 6 appeared.

**TRY THIS!** Exercise 5.1

Computers and calculators have random number generators that come close to true randomness. For instance, the Internet game you played (to simulate flipping the coin 20 times) uses a random number generator. Computer-generated random numbers are sometimes called pseudo random numbers, because they are generated on the basis of a seed value—a number that starts the random sequence. If you input the same seed number, you will always see the same sequence of pseudo-random numbers. However, it is not possible to predict what number will come next in the sequence, and in this sense the generated numbers are considered random. For most practical work (and certainly for everything we cover in this book), these pseudo-random numbers are as random as we need. (You should be aware, though, that not all statistics packages produce equally convincing sequences of random numbers.)

Less technological ways exist for generating random outcomes, such as actually flipping a coin. When you play a card game, you carefully shuffle the cards to mix them up. In the games of Mah Jong and tile rummy, players create randomness by scrambling the tiles on the tabletop. In many board games, players either roll dice or spin a spinner to determine how far they will move their game pieces. In raffles, tickets are put into a basket, the basket is spun, and then a ticket is drawn by someone who does so without looking into the basket.

Such physical randomizations must be done with care. A good con artist can flip a coin so that it always comes up heads. A child learns quickly that a gentle tap of a spinner can move the spinner to the desired number. A deck of cards needs to be shuffled at least seven times in order for the result to be considered random (as mathematician and magician Persi Diaconnis proved). Many things that we think are random might not be. This is the lesson the government learned from its flawed Vietnam draft lottery, which was described in the chapter introduction. Quite often, statisticians are employed to check whether random processes, such as the way winners are selected for state lotteries, are truly random.

## Empirical and Theoretical Probabilities

**Probability** is used to measure how often random events occur. When we say, "The probability of getting heads when you flip a coin is 50%," we mean that flipping a coin results in heads about half the time (assuming that the outcome of the flips is random). This is sometimes called the "frequentist" approach to probability, because in it, probabilities are defined as relative frequencies.

We will examine two kinds of probabilities. These two kinds of probabilities are connected, as we will see later in this chapter. **Theoretical probabilities** are long-run relative frequencies. The theoretical probability is the relative frequency at which an event happens after *infinitely* many repetitions. When we say that a coin has a 50% probability of coming up heads, we mean that if it were possible to flip the coin *infinitely* many times, then exactly 50% of the flips would be heads.

Because we can't do *anything* infinitely many times, finding theoretical probabilities requires relying on theory. (That's why they're called theoretical!) As an example, a coin has two sides. My theory is that either side is equally likely to come

**Looking Back**

**Relative Frequencies**
Relative frequencies, introduced in Chapter 2, are the same as proportions.

up when I flip a coin. I don't know for a fact that this is true; I *assume* it is true. And so I conclude that the probability of seeing heads is 50%. However, I haven't actually done an experiment. I have simply reasoned on the basis of a theory about how coins behave.

**Empirical probabilities,** on the other hand, are short-run relative frequencies based on an experiment. I toss a coin 10 times and get 6 heads. My empirical probability of getting heads is therefore $6/10 = 0.6$, or 60%.

Empirical and theoretical probabilities have some striking differences. Chief among them is that theoretical probabilities are always the same value; if we all agree on the theory, then we all agree on the value for the theoretical probabilities. Empirical probabilities, however, change with every experiment. Suppose I flip the exact same coin 10 times again, and this time I get 3 heads. Now my empirical probability of heads is 0.3. Empirical probabilities are themselves random and vary from experiment to experiment.

 **KEY POINT** Empirical probabilities tell us how often an event occurred in an actual set of experiments. Theoretical probabilities are based on theory and tell us how many times an event would occur if an experiment were repeated *infinitely* many times.

## Why Do We Need Both?

Theoretical probabilities are very abstract things. They measure what proportion of the time events would occur if an experiment were repeated *infinitely many times*. That's a very long time—forever, in fact. This means it is impossible to carry out an experiment that will provide the exact value of a theoretical probability; to do so would literally take forever. However, it turns out that we can use empirical probabilities to *estimate* and to *test* theoretical probabilities.

Why do we need to estimate a theoretical probability? Sometimes, it is just too difficult to compute a theoretical probability. Development of the mathematics of probability began in the 1600s, and we can now compute probabilities for quite complex events. But this does not mean it's easy! Sometimes, it may be adequate to get a good approximate value using an experiment that allows us to estimate how often a certain event might happen if we could repeat the experiment infinitely many times. On other occasions, the event for which we need a probability may be too complex for theory, and if we really need to know the probability, then running an experiment is the only way of finding a useful value.

Why do we need to test? We might not trust the theoretical probability value. For example, it might be based on assumptions that we are not sure are true. If done well, an experiment that produces an empirical probability can be used to verify or refute a theoretically derived value. In fact, much of the rest of this book will develop this theme.

If you do the "Let's Make a Deal" activity described at the end of this chapter, you will find empirical probabilities of winning for each of the two strategies offered to contestants: stay or switch. These empirical probabilities allow you to estimate the probability of winning, and they also allow you to test whether your own ideas about which strategy is best are correct.

**Simulations** are experiments used to produce empirical probabilities, because the investigators hope that these experiments simulate the situation they are examining. As you will see in Section 5.4, a mathematical discovery called the Law of Large Numbers tells us that if the relationship between the situation and our experiment is strong, then our empirical probability will be a good estimate for the true theoretical probability.

# Finding Theoretical Probabilities

We can use empirical probabilities as an estimate of theoretical probabilities, and if we do our experiment enough times, the estimate can be pretty good. But how good is pretty good? To understand how to use empirical probabilities to estimate and to verify theoretical probabilities, we first need to learn how to calculate theoretical probabilities.

## Facts about Theoretical Probabilities

Probabilities are always numbers between 0 and 1 (including 0 and 1). Probabilities can be expressed as fractions, decimals, or percents: 1/2, 0.50, and 50% are all used to represent the probability that a coin comes up heads.

Some values have special meanings. If the probability of an event happening is 0, then that event never happens. If you purchase a lottery ticket after all the prizes have been given out, the probability is 0 that you will win one of those prizes. If the probability of an event happening is 1, then that event always happens. If you flip a coin, the probability of a coin landing heads or tails is 1.

Another useful property to remember is that the probability that an event will *not* happen is 1 minus the probability that it will happen. If there is a 0.90 chance that it will rain, then there is a $1 - 0.9 = 0.10$ chance that it will not. If there is a 1/6 chance of rolling a "1" on a die, then there is a $1 - (1/6) = 5/6$ probability that you will not get a 1.

We call such a "not event" a **complement**. The complement of the event "it rains today" is the event "it does not rain today." The complement of the event "coin lands heads" is "coin lands tails." The complement of the event "a die lands with a 1 on top" is "a die lands with a 2, 3, 4, 5, or 6 on top"; in other words, the die lands with a number that is *not* 1 on top.

Events are usually represented by uppercase letters: A, B, C, and so on. For example, we might let A represent the event "it rains tomorrow." Then the notation P(A) means "the probability that it will rain tomorrow." In sentence form, the notation P(A) = 0.50 translates into English as "The probability that it will rain tomorrow is 0.50, or 50%."

A common misinterpretation of probability is to think that large probabilities mean that the event will certainly happen. For example, suppose your local weather reporter predicts a 90% chance of rain tomorrow. Tomorrow, however, it doesn't rain. Was the weather reporter wrong? Not necessarily. When the weather reporter says there is a 90% chance of rain, it means that on 10% of the days like tomorrow it does not rain. Thus a 90% chance of rain means that on 90% of all days just like tomorrow, it rains, but on 10% of those days it does not.

---

**Summary of Probability Rules**

**Rule 1:** A probability is always a number from 0 to 1 (or 0% to 100%) inclusive (which means 0 and 1 are allowed). It may be expressed as a *fraction*, a *decimal*, or a *percent*.

In symbols: For any event A,

$$0 \leq P(A) \leq 1$$

**Rule 2:** The probability that an event will not occur is 1 minus the probability that the event will occur. In symbols, for any event A,

$$P(A \text{ does } not \text{ occur}) = 1 - P(A \text{ does occur})$$

The symbol $A^c$ is used to represent the complement of A. With this notation, we can write Rule 2 as

$$P(A^c) = 1 - P(A)$$

## Finding Theoretical Probabilities with Equally Likely Outcomes

In some situations, all of the possible outcomes of a random experiment occur with the same frequency. We call these situations "equally likely outcomes." For example, when you flip a coin, heads and tails are equally likely. When you roll a die, 1, 2, 3, 4, 5, and 6 are all equally likely.

When we are dealing with equally likely outcomes, it is sometimes helpful to list all of the possible outcomes. A list that contains *all* possible (and equally likely) outcomes is called the **sample space**. We often represent the sample space with the letter S. An **event** is any collection of outcomes in the sample space. For example, the sample space S for rolling a die is the numbers 1, 2, 3, 4, 5, 6. The event "get an even number on the die" consists of the even outcomes in the sample space: 2, 4, and 6.

When the outcomes are equally likely, the probability that a particular event occurs is just the number of outcomes that make up that event, divided by the total number of equally likely outcomes in the sample space. In other words, it is the number of outcomes resulting in the event divided by the number of outcomes in the sample space.

**Summary of Probability Rules**

**Rule 3:**

$$\text{Probability of A} = P(A) = \frac{\text{Number of outcomes in A}}{\text{Number of all possible outcomes}}$$

This is true *only* for equally likely outcomes.

For example, suppose 30 people are in your class, and one person will be selected at random by a raffle to win a prize. What is the probability that you will win? The sample space is the list of the names of the 30 people. The event A is the event that contains only one outcome: your name. The probability that you win is 1/30, because there is only 1 way for you to win and there are 30 different ways that this raffle can turn out. We write this using mathematical notation as follows:

$$P(\text{you win prize}) = 1/30$$

We can be even more compact:
   Let *A* represent the event that you win the raffle. Then

$$P(A) = 1/30.$$

One consequence of Rule 3 is that the probability that *something* in the sample space will occur is 1. In symbols, $P(S) = 1$. This is because

$$P(S) = \frac{\text{Number of outcomes in S}}{\text{Number of outcomes in S}} = 1$$

## EXAMPLE 2 Ten Dice in a Bowl

Reach into a bowl that contains 5 red dice, 3 green dice, and 2 white dice (Figure 5.2). But assume that, unlike what you see in Figure 5.2, the dice have been well mixed.

▶ **FIGURE 5.2** Ten dice in a bowl.

**QUESTION** What is the probability of picking (a) a red die? (b) a green die? (c) a white die?

**SOLUTIONS** The bowl contains 10 dice, so we have ten possible outcomes. All are equally likely (assuming all the dice are equal in size, they are mixed up within the bowl, and we do not peek when choosing).

a. Five dice are red, so the probability of picking a red die is 5/10, 1/2, 0.50, or 50%. That is,

$$P(\text{red die}) = 1/2, \text{ or } 50\%$$

b. Three dice are green, so the probability of picking a green die is 3/10, or 30%. That is,

$$P(\text{green die}) = 3/10, \text{ or } 30\%.$$

c. Two dice are white, so the probability of picking a white die is 2/10, 1/5, or 20%. That is,

$$P(\text{white die}) = 1/5, \text{ or } 20\%.$$

Note that the probabilities add up to 1, or 100%, as they must.

**TRY THIS!** Exercise 5.11

Example 3 shows that it is important to make sure the outcomes in your sample space are equally likely.

## EXAMPLE 3 Adding Two Dice

Roll two dice and add the outcomes. Assume each side of each die is equally likely to appear face up when rolled. Event A is the event that the sum of the two dice is 7.

**QUESTION** What is the probability of event A? In other words, find P(A).

**SOLUTION** This problem is harder because it takes some work to list all of the equally likely outcomes, which are shown in Table 5.2.

| Die 1 | 1 | 1 | 1 | 1 | 1 | 1 | 2 | 2 | 2 | 2 | 2 | 2 |
|-------|---|---|---|---|---|---|---|---|---|---|---|---|
| Die 2 | 1 | 2 | 3 | 4 | 5 | 6 | 1 | 2 | 3 | 4 | 5 | 6 |

| Die 1 | 3 | 3 | 3 | 3 | 3 | 3 | 4 | 4 | 4 | 4 | 4 | 4 |
|-------|---|---|---|---|---|---|---|---|---|---|---|---|
| Die 2 | 1 | 2 | 3 | 4 | 5 | 6 | 1 | 2 | 3 | 4 | 5 | 6 |

| Die 1 | 5 | 5 | 5 | 5 | 5 | 5 | 6 | 6 | 6 | 6 | 6 | 6 |
|-------|---|---|---|---|---|---|---|---|---|---|---|---|
| Die 2 | 1 | 2 | 3 | 4 | 5 | 6 | 1 | 2 | 3 | 4 | 5 | 6 |

◀ **TABLE 5.2** Possible outcomes for two six-sided dice.

Table 5.2 lists 36 possible equally likely outcomes. Here are the outcomes in event A:

(1, 6), (2, 5), (3, 4), (4, 3), (5, 2), (6, 1)

There are six outcomes for which the dice add to 7.

**CONCLUSION** The probability of rolling a sum of 7 is 6/36, or 1/6.

**TRY THIS!** Exercise 5.15

It is important to make sure that the outcomes in the sample space are equally likely. A common mistake when solving Example 3 is listing all the possible *sums*, instead of listing all the equally likely outcomes of the two dice. If we made that mistake here, our list of sums would look like this:

$$2, 3, 4, 5, 6, 7, 8, 9, 10, 11, 12$$

This list has 11 sums, and only 1 of them is a 7, so we would incorrectly conclude that the probability of getting a sum of 7 is 1/11.

Why didn't we get the correct answer of 1/6? The reason is that the outcomes we listed—2, 3, 4, 5, 6, 7, 8, 9, 10, 11, 12—are not equally likely. For instance, we can get a sum of 2 in only one way: roll two "aces," for 1 + 1. Similarly, we have only one way to get a 12: roll two 6's, for 6 + 6.

However, there are six ways of getting a 7: (1, 6), (2, 5), (3, 4), (4, 3), (5, 2), and (6, 1). In other words, a sum of 7 happens more often than a sum of 2 or a sum of 12. The outcomes 2, 3, 4, 5, 6, 7, 8, 9, 10, 11, 12 are not equally likely.

Usually it is not practical to list all the outcomes in a sample space—or even just those in the event you're interested in. For example, if you are dealing out 5 playing cards from a 52-card deck, the sample space has 2,598,960 possibilities. You really do not want to list all of those outcomes. Mathematicians have developed rules for counting the number of outcomes in complex situations such as these. These rules do not play an important role in introductory statistics, and we do not include them in this book.

> **! Caution**
>
> **"Equally Likely Outcomes" Assumption**
> Just wishing it were true doesn't make it true. The assumption of equally likely outcomes is not always true. And if this assumption is wrong, your theoretical probabilities will not be correct.

## Combining Events with "AND" and "OR"

As you saw in Chapter 4, we are often interested in studying more than one variable at a time. Real people, after all, often have several attributes we want to study, and we frequently want to know the relationship among these variables. The words AND and OR can be used to combine events into new, more complex events. The real people in Figure 5.3a on the next page, for example, have two attributes we decided to examine. They are either wearing a hat, or not. Also, they are either wearing glasses, or not. In the photo, the people who are wearing hats AND glasses are raising their hands.

▲ **FIGURE 5.3 (a)** Raise your hand if you are wearing glasses AND a hat. **(b)** The people wearing both glasses AND a hat (Maria and David) appear in the intersection of the two circles in this Venn diagram.

---

**❗ Caution**

**Venn Diagrams**
The areas of the regions in Venn diagrams have no numerical meaning. A large area does not contain more outcomes than a small area.

---

Another way to visualize this situation is with a **Venn diagram**, as shown in Figure 5.3b. The rectangle represents the sample space, which consists of all possible outcomes if we were to select a person at random. The ovals represent events—for example, the event that someone is wearing glasses. The people who "belong" to *both* events are in the intersection of the two ovals.

The word **AND** creates a new event out of two other events. The probability that a randomly selected person in this photo is wearing a hat is 3/6, because three of the six people are wearing a hat. The probability that a randomly selected person wears glasses is 4/6. The probability that a person is wearing a hat AND glasses is 2/6, because only two people are in both groups. We could write this, mathematically, as

$$P(\text{wears glasses AND wears a hat}) = 2/6$$

**KEY POINT**   The word AND creates a new event out of two events A and B. The new event consists of *only* those outcomes that are in *both* event A and event B.

In most situations, you will not have a photo to rely on. A more typical situation is given in Table 5.3, which records frequencies of two attributes for the people in a random sample of a recent U.S. census (www.census.gov). The two attributes are highest educational level and current marital status. ("Single" means never married. All other categories refer to the respondent's current status. Thus a person who was divorced but then remarried is categorized as "Married.")

▶ **TABLE 5.3** Education and marital status for 665 randomly selected U.S. residents. "Less HS" means did not graduate from high school.

| Education Level | Single | Married | Divorced | Widow/Widower | Total |
|---|---|---|---|---|---|
| **Less HS** | 17 | 70 | 10 | 28 | 125 |
| **High school** | 68 | 240 | 59 | 30 | 397 |
| **College or higher** | 27 | 98 | 15 | 3 | 143 |
| **Total** | 112 | 408 | 84 | 61 | 665 |

## EXAMPLE 4 Education AND Marital Status

Suppose we select a person at random from the collection of 665 people categorized in Table 5.3.

QUESTIONS

**QUESTIONS**

a. What is the probability that the person is married?

b. What is the probability that the person has a college education or higher?

c. What is the probability that the person is married AND has a college education or higher?

**SOLUTIONS**   The sample space has a total of 665 equally likely outcomes.

a. In 408 of those outcomes, the person is married. So the probability that a randomly selected person is married is 408/665, or 61.4% (approximately).

b. In 143 of those outcomes, the person has a college education or higher. So the probability that the selected person has a college education or higher is 143/665, or 21.5%.

c. There are 665 possible outcomes. In 98 of them, the people are both married AND have a college degree or higher. So the probability that the selected person is married AND has a college degree is 98/665, or 14.7%.

**TRY THIS!**   Exercise 5.21

> **! Caution**
>
> **AND**
> P(A AND B) will always be less than (or equal to) P(A) and also less than (or equal to) P(B). If this isn't the case, you've made a mistake!

## Using "OR" to Combine Events

The people in Figure 5.4a were asked to raise their hands if they were wearing glasses OR wearing a hat. Note that people who are wearing both also have their hands raised. If we were to select one of these people at random, the probability that this person is wearing glasses OR wearing a hat would be 5/6, because we would count people who wear glasses, people who wear hats, and people who wear both glasses AND hats.

**(a)**

**(b)**

▲ **FIGURE 5.4 (a)** Raise your hand if you are wearing a hat OR glasses. This photograph illustrates the inclusive OR. **(b)** In this Venn diagram, note the orange region for "raise your hand if you are wearing a hat OR glasses."

In a Venn diagram, OR events are represented by shading all relevant events. Here Mike, Rena, Maria, Alan, and David appear in the orange area because each is wearing glasses OR wearing a hat.

The last example illustrates a special meaning of the word OR. This word is used slightly differently in mathematics and probability than you may use it in English. In statistics and probability, we use the **inclusive OR**. For example, the people in the photo shown in Figure 5.4a were asked to raise their hands if they had a hat OR glasses. This means that the people who raise their hands have a hat only, or have glasses only, or have both hats AND glasses.

**KEY POINT**   The word OR creates a new event out of the events A and B. The new event consists of all outcomes that are only in A, that are only in B, or that are in both.

## EXAMPLE 5 OR with Marital Status

Again, select someone at random from Table 5.3.

**QUESTION** What is the probability the person is single OR married?

**SOLUTION** The event of interest occurs if we select a person who is married, a person who is single, or a person who is both married AND single. There are 665 possible equally likely outcomes. Of these, 112 are single and 408 are married (and none are both!). Thus there are 112 + 408 = 520 people who are single OR married.

**CONCLUSION** The probability that the selected person is single OR married is 520/665, or 78.2%.

**TRY THIS!** Exercise 5.23

## EXAMPLE 6 Education OR Marital Status

Select someone at random from Table 5.3 on page 204 (which is shown again in Table 5.4 below).

**QUESTION** What is the probability that the person is married OR has a college degree?

**SOLUTION** Table 5.3 gives us 665 possible outcomes. The event of interest occurs if we select someone who is married, or someone who has a college degree, or someone who both is married AND has a college degree. There are 408 married people and 143 people with a college degree.

But wait a minute: There are not 408 + 143 different people who are married OR have a college degree—some of these people got counted twice! The people who are both married AND have a college degree were counted once when we looked at the married people, and they were counted a second time when we looked at the college graduates. We can see from Table 5.4 that 98 people *both* are married AND have a college degree. So we counted 98 people too many. Thus, we have 408 + 143 − 98 = 453 different outcomes in which the person is married or has a college degree. Table 5.4 is the same as Table 5.3 except for the added ovals, which are meant to be interpreted as in a Venn diagram.

▶ **TABLE 5.4** Here we reprint Table 5.3, with ovals for Married and for College added.

| Education Level | Single | Married | Divorced | Widow/Widower | Total |
|---|---|---|---|---|---|
| Less HS | 17 | 70 | 10 | 28 | 125 |
| High school | 68 | 240 | 59 | 30 | 397 |
| College or higher | 27 | 98 | 15 | 3 | 143 |
| Total | 112 | 408 | 84 | 61 | 665 |

The numbers in bold type represent the people who are married OR have a college degree. This Venn-like treatment emphasizes that one group (of 98 people) is in both categories and reminds us not to count them twice.

Another way to say this is that we count the 453 distinct outcomes by adding the numbers in the ovals in the table, but not adding any of them more than once:

$$70 + 240 + 98 + 27 + 15 + 3 = 453$$

**CONCLUSION** The probability that the randomly selected person is married OR has a college degree is 453/665, or 68.1%.

**TRY THIS!** Exercise 5.25

## Mutually Exclusive Events

Did you notice that the first example of an OR (single OR married) was much easier than the second (married OR has a college degree)? In the second example, we had to be careful not to count some of the people twice. In the first example, this was not a problem. Why?

The answer is that in the first example, we were counting people who were married OR single, and no person can be in both categories. It is impossible to be simultaneously married AND single. When two events have no outcomes in common—that is, when it is impossible for both events to happen at once—they are called **mutually exclusive events**. The events "person is married" and "person is single" are mutually exclusive.

The Venn diagram in Figure 5.5 shows two mutually exclusive events. There is no intersection between these events; it is impossible for both event A AND event B to happen at once. This means that the probability that both events occur at the same time is 0.

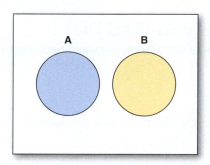

◀ **FIGURE 5.5** In a Venn diagram, two mutually exclusive events have no overlap.

## EXAMPLE 7 Mutually Exclusive Events: Education and Marital Status

Imagine selecting a person at random from those represented in Table 5.4.

**QUESTION** Name two mutually exclusive events, and name two events that are not mutually exclusive. Remember that marital status means a person's current marital status.

**SOLUTION** For mutually exclusive events, we can choose any two columns or any two rows. It is impossible for someone to be both divorced AND married (at the same time). It is impossible for someone to have less than a high school education AND to have a high school education. The events "person has a HS education" and "person has less than a HS education" are mutually exclusive, because no one can be in both categories at once. The probability that a randomly selected person has a HS education AND has less than a HS education is 0. We could have chosen other pairs of events as well.

To find two events that are not mutually exclusive, find events that have outcomes in common. There are 30 people who have a HS education AND are widows/widowers. Therefore, these events are *not* mutually exclusive. The events "person has HS education" AND "person is a widow/widower" are not mutually exclusive.

**TRY THIS!** Exercise 5.29

## Summary of Probability Rules

**Rule 4:** The probability that event A happens OR event B happens is

> (the probability that A happens) plus (the probability that B happens) minus (the probability that both A AND B happen).

If A and B are mutually exclusive events (for example, A is the event that the selected person is single, and B is the event that the person is married), then P(A AND B) = 0. Then the rule becomes simpler:

**Rule 4a:** If A and B are mutually exclusive events, the probability that event A happens OR event B happens is the sum of the probability that A happens and the probability that B happens.

Rule 4 in symbols:

> Always: P(A OR B) = P(A) + P(B) − P(A AND B)

Rule 4a in symbols:

> Only if A and B are mutually exclusive: P(A OR B) = P(A) + P(B)

## EXAMPLE 8 Rolling a Six-Sided Die

Roll a fair, six-sided die.

**QUESTIONS**

a. Find the probability that the die shows an even number OR a number greater than 4 on top.

b. Find the probability that the die shows an even number OR the number 5 on top.

**SOLUTIONS**

a. We could do this in two ways. First, we note that six equally likely outcomes are possible. The even numbers are (2, 4, 6) and the numbers greater than 4 are (5, 6). Thus the event "even number OR number greater than 4" has four different ways of happening: roll a 2, 4, 5, or 6. We conclude that the probability is 4/6.

> The second approach is to use Rule 4. The probability of getting an even number is 3/6. The probability of getting a number greater than 4 is 2/6. The probability of getting both an even number AND a number greater than 4 is 1/6 (because the only way for this to happen is to roll a 6). So

P(even OR greater than 4) = P(even) + P(greater than 4) − P(even AND greater than 4)

$$= 3/6 + 2/6 - 1/6$$
$$= 4/6$$

b. P(even OR roll 5) = P(even) + P(roll 5) − P(even AND roll 5)

It is impossible for the die to be both even AND a 5, because 5 is an odd number. So the events "get a 5" and "get an even number" are mutually exclusive. Therefore, we get

P(even number OR a 5) = 3/6 + 1/6 − 0 = 4/6

**CONCLUSIONS**

a. The probability of rolling an even number OR a number greater than 4 is 4/6 (or 2/3).

b. The probability of rolling an even number OR a 5 is 4/6 (or 2/3).

**TRY THIS!** Exercise 5.33

# Associations in Categorical Variables

Judging on the basis of our sample in Table 5.3, is there an association between marital status and having a college education? If so, we would expect the proportion of married people to be different for those who had a college education and those who did not have a college education. (Perhaps we would find different proportions of marital status for each category of education.)

In other words, if there is an association, we would expect the probability that a randomly selected college-educated person is married to be different from the probability that a person with less than a college education is married.

## Conditional Probabilities

Language is important here. The probability that "a college-educated person is married" is different from the probability that "a person is college-educated AND is married." In the AND case, we're looking at everyone in the sample and wondering how many have both a college degree AND are married. But when we ask for the probability that a college-educated person is married, we're taking it as given that the person is college-educated. We're *not* saying "choose someone from the whole collection." We're saying, "Just focus on the people with the college degrees. What proportion of those people are married?"

Probabilities such as these, where we focus on just one group of objects and imagine taking a random sample from that group alone, are called **conditional probabilities**.

For example, in Table 5.5 (which repeats Table 5.3), we've highlighted in red the people with college degrees. In this row, there are 143 people. If we select someone at random from among those 143 people, the probability that the person will be married is 98/143 (or about 0.685). We call this a conditional probability because we're finding the probability of being married *conditioned* on having a college education (that is, we are assuming we're selecting only from college-educated people).

| Education Level | Single | Married | Divorced | Widow/Widower | Total |
|---|---|---|---|---|---|
| Less HS | 17 | 70 | 10 | 28 | 125 |
| HS | 68 | 240 | 59 | 30 | 397 |
| College or higher | 27 | 98 | 15 | 3 | 143 |
| Total | 112 | 408 | 84 | 61 | 665 |

◄ **TABLE 5.5** What's the probability that a person with a college degree or higher is married? To find this, focus on the row shown in red and imagine selecting a person from this row.

### "Given That" vs. "AND"

Often, conditional probabilities are worded with the phrase *given that*, as in "Find the probability that a randomly selected person is married given that the person has a college degree." But you might also see it phrased as in the last paragraph: "Find the probability that a randomly selected person with a college degree is married." The latter phrasing is more subtle, because it implies that we're supposed to assume the selected person has a college degree: We must assume we are *given that* the person has a college degree.

Figure 5.6a on the next page shows a Venn diagram representing all of the data. The green overlap region represents the event of being married AND having completed college. By way of contrast, Figure 5.6b on the next page shows only those with college educations; it emphasizes that if we wish to find the probability of being married, given that the person has a college degree, we need to focus on only those with college degrees.

▶ **FIGURE 5.6** (a) The probability of being married AND having a college degree; (b) the probability that a person with a college degree is married.

**(a)**

**(b)**

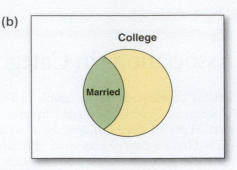

The mathematical notation for a conditional probability might seem a little unusual. We write

P(person is married | person has college degree) = 98/143

The vertical bar inside the probability notation is *not* a division sign. You should think of pronouncing it as "given that." This sentence reads, "The probability that the person is married, given that we know this person has a college degree, is 98/143." Some statisticians like to think of the vertical bar as meaning "within," and would translate this as "The probability that we randomly select a married person from within those who have a college degree." Either way you think about it is fine; use whichever makes the most sense to you.

**KEY POINT**      In the study of conditional probabilities, P(A | B) means to find the probability that event A occurs, but to restrict your consideration to those outcomes of A that occur within event B. It means "the probability of A occurring, given that event B has occurred."

## EXAMPLE 9 Teens and the Internet

Consider the following statements, which are based on a Pew Foundation report.

a. The probability that a randomly selected teenager (12–17 years old) from the United States will go online at some point during the month is about 93%. Event A: the selected person is a teenager. Event B: the selected person goes online during the month.

b. The probability that a randomly selected adult age 65 or older will go online during the month is 38%. Event A: the selected person is age 65 or older. Event B: the selected person will go online during the month.

c. The probability that a randomly selected resident of the United States is a teenager who will go on line this month is about 7%. Event A: the selected person is a teenager. Event B: the person will go online this month.

We wish to use this information to find the probability that a randomly selected resident of the United States will be 12–17 years old and will go online this month (Pew Foundation, July 2010).

**QUESTION**   For each of these three statements, determine whether the events in the question are used in a conditional probability or an AND probability. Explain. Write the statement using probability notation.

**SOLUTIONS**

a. This statement is asking about a conditional probability. It says that among all teenagers, 93% go online. We are "given that" the group we're sampling from are all teenagers.

b. This statement is also a conditional probability.

c. This statement, on the other hand, is asking us to assume nothing and, instead, once the person is selected from the entire United States, to determine whether that person has these two characteristics.

Using probability notation, these statements are

a. P(person goes online | person is teenager)

b. P(person goes online | person is adult aged 65 or older)

c. P(person goes online AND person is teenager)

   Exercise 5.45

## Finding Conditional Probabilities

If you are given a table like Table 5.5, you can find conditional probabilities as we did above: by isolating the group from which you are sampling. However, a formula exists that is useful for times when you do not have such complete information.

The formula for calculating conditional probabilities is

$$P(A \mid B) = \frac{P(A \text{ AND } B)}{P(B)}$$

## EXAMPLE 10 Education and Marital Status

Suppose a person is randomly selected from those represented in Table 5.3 on page 204.

**QUESTION**   Find the probability that a person with less than a high school degree (and no higher degrees) is married. Use the table, but then confirm your calculation with the formula.

**SOLUTION**   We are asked to find P(married | less than high school degree)—in other words, the probability a person with less than a HS degree is married. We are told to imagine taking a random sample from only those who have less than a high school degree. There are 125 such people, of whom 70 are married.

P(married | less HS) = 70/125 = 0.560

The formula confirms this:

$$P(\text{married} \mid \text{less HS}) = \frac{P(\text{married AND less HS})}{P(\text{less HS})} = \frac{\frac{70}{665}}{\frac{125}{665}} = \frac{70}{125} = 0.560$$

Interestingly, the probability that a college graduate is married (0.685) is greater than the probability that someone with less than a high school education is married (0.560).

   Exercise 5.47

With a little algebra, we can discover that this formula can serve as another way of finding AND probabilities:

$$P(A \text{ AND } B) = P(A)P(B \mid A)$$

We'll make use of this formula later.

**Summary of Probability Rules**

**Rule 5a:** $P(A|B) = \dfrac{P(A \text{ AND } B)}{P(B)}$

**Rule 5b:** $P(A \text{ AND } B) = P(B)\,P(A|B)$ and also $P(A \text{ AND } B) = P(A)\,P(B|A)$

Both forms of Rule 5b are true, because it doesn't matter which event is called A and which is called B.

## Flipping the Condition

A common mistake with conditional probabilities is thinking that $P(A \mid B)$ is the same as $P(B \mid A)$.

$$P(B|A) \neq P(A|B)$$

A second common mistake is to confuse conditional probabilities with fractions and think that $P(B|A) = 1/P(A|B)$.

$$P(B|A) \neq \dfrac{1}{P(A|B)}$$

For example, using the data in Table 5.3, we earlier computed that $P(\text{married}|\text{college}) = 98/143 = 0.685$. What if we wanted to know the probability that a randomly selected married person is college-educated?

$$P(\text{college}|\text{married}) = ?$$

From Table 5.3, we can see that if we know the person is married, there are 408 possible outcomes. Of these 408 married people, 98 are college-educated, so

$$P(\text{college}|\text{married}) = 98/408, \text{ or about } 0.240$$

Clearly, $P(\text{college}|\text{married})$, which is 0.240, does not equal $P(\text{married}|\text{college})$, which is 0.685.

Also, it is *not* true that $P(\text{married}|\text{college}) = 1/P(\text{college}|\text{married}) = 408/98 = 4.16$, a number bigger than 1! It is impossible for a probability to be bigger than 1, so obviously,

$$P(A|B) \text{ does not equal } 1/P(B|A).$$

**KEY POINT**     $P(B|A) \neq P(A|B)$

## Independent and Dependent Events

We saw that the probability that a person is married differs, depending on whether we know that person is a college grad or that she or he has less than a high school education. If we randomly select from different educational levels, we get a different probability of marriage. Another way of putting this is to say that marital status and education level are **associated**. We know they are associated because the conditional probabilities change depending on which educational level we condition on.

We call variables or events that are *not* associated **independent events**. Independent variables, and independent events, play a very important role in statistics. Let's first talk about independent events.

Two events are independent if knowledge that one event has happened tells you nothing about whether or not the other event has happened. In mathematical notation,

A and B are independent events means $P(A|B) = P(A)$.

In other words, if the event "a person is married" is independent from "a person has a college degree," then the probability that a married person has a college degree, P(college | married), is the same as the probability that any person in the sample has a college degree. We already found that

$$P(\text{college} \mid \text{married}) = 0.240 \text{ and } P(\text{college}) = 143/665 = 0.215$$

These probabilities are close—so close, in fact, that we might be willing to conclude they are "close enough." (You'll learn in Chapter 10 about how to make decisions like these.) We might conclude that the events "a randomly selected person is married" and "a randomly selected person is college-educated" are independent. However, for now we will assume that *probabilities have to be exactly the same for us to conclude independence.* We conclude that completing college and being married are *not* independent.

It doesn't matter which event you call A and which B, so events are also independent if $P(B \mid A) = P(B)$.

**KEY POINT**    To say that events A and B are independent means that $P(A \mid B) = P(A)$. In words: Knowledge that event B occurred does not change the probability of event A occurring.

## EXAMPLE 11 Dealing a Diamond

Figure 5.7 shows a standard deck of playing cards. When playing card games, players nearly always try to avoid showing their cards to the other players. The reason for this is that knowing the other player's cards can sometimes give you an advantage. Suppose you are wondering whether your opponent has a diamond. If you find out that one of the cards he holds is red, does this provide useful information?

**QUESTION**   Suppose a deck of cards is shuffled and one card is dealt face-down on the table. Are the events "the card is a diamond" and "the card is red" independent?

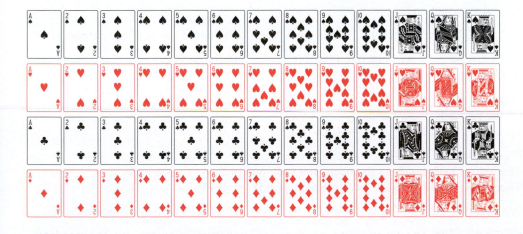

◀ **FIGURE 5.7** Fifty-two playing cards in a standard deck.

**SOLUTION**   To answer this, we must apply the definition of independent events and find

$$P(\text{card is a diamond})$$

and compare it to

$$P(\text{card is a diamond} \mid \text{card is red})$$

If these probabilities are different, then the events are not independent; they are associated events.

First we find P(card is a diamond).

Out of a total of 52 cards, 13 are diamonds. Therefore,

$$P(\text{card is a diamond}) = 13/52, \text{ or } 1/4$$

Now suppose we know the card is red. What's the probability that a red card is a diamond? That is, find

$$P(\text{card is a diamond} \mid \text{card is red})$$

The number of equally likely possible outcomes is reduced from 52 to 26, because there are 26 red cards. You are now limited to the 26 cards in the middle two rows of the picture. There are still 13 diamonds. Therefore, the probability that the card is a diamond, given that it is red, is $13/26 = 1/2$.

$$P(\text{card is diamond}) = 1/4$$

$$P(\text{card is diamond} \mid \text{card is red}) = 1/2$$

These probabilities are *not* equal.

**CONCLUSION** The events "select a diamond" and "select a red card" are associated, because P(select a diamond | color is red) is not the same as P(select a diamond). This means that if you learn, somehow, that the opponent's card is red, then you have gained some information that will be useful in deciding whether he has a diamond.

Note that we could also have compared P(card is red) to P(card is red | card is a diamond), and we would have reached the same conclusion.

**TRY THIS!** Exercise 5.55

## EXAMPLE **12** Dealing an Ace

A playing card is dealt face down. This time, you are interested in knowing whether your opponent holds an ace. You have discovered that his card is a diamond. Is this helpful information?

**QUESTION** Are the events "card is a diamond" and "card is an ace" independent?

**SOLUTION** Now we must find P(card is an ace) and compare it to P(card is an ace | card is a diamond).

P(card is an ace):
Of the 52 cards, 4 are aces.
Therefore, P(card is an ace) $= 4/52 = 1/13$.

P(card is an ace | card is a diamond):
There are 13 diamonds in the deck, and only one of these 13 is an ace.
Therefore, P(card is an ace | card is a diamond) $= 1/13$.

We find that P(card is an ace) $= 1/13 =$ P(card is an ace | card is a diamond).

**CONCLUSION** The events "card is a diamond" and "card is an ace" are independent. This means the information that your opponent's card is a diamond will not help you determine whether it is an ace.

Note that we could also compare P(card is a diamond) to P(card is a diamond | card is an ace), and we would have reached the same conclusion.

**TRY THIS!** Exercise 5.57

## Intuition about Independence

Sometimes you can use your intuition to decide whether two events are independent. For example, flip a coin twice. You should know that P(second flip is heads) = 1/2. But what if you know that the first flip was also a head? Then you need to find

P(second flip is heads | first flip was heads)

Intuitively, we know that the coin always has a 50% chance of coming up heads. The coin doesn't know what happened earlier. Thus

P(second is heads | first is heads) = 1/2 = P(second is heads)

The two events "second flip comes up heads" and "first flip comes up heads" are independent.

Although you can sometimes feel very confident in your intuition, you should check your intuition with data whenever possible.

## EXAMPLE 13 Education and Widows

Suppose we select a person at random from the sample of people asked about education and marital status in Table 5.3. Is the event "person selected has HS education" independent from the event "person selected is widowed"? Intuitively, we would think so. After all, why should a person's educational level affect whether his or her spouse dies first?

**QUESTION** Check, using the data in Table 5.3 on page 204, whether these two events are independent.

**SOLUTION** To check independence, we need to check whether P(A | B) = P(A). It doesn't matter which event we call A and which we call B, so let's check to see whether

P(person selected has HS education | person is widowed) = P(person has HS education)

From the table, we see that there are only 61 widows, so

$$P(HS \mid widowed) = 30/61 = 0.492$$
$$P(HS) = 397/665 = 0.597$$

The two probabilities are not equal.

**CONCLUSION** The events are associated. If you know the person is widowed, then the person is less likely to have a high school education than he or she would be if you knew nothing about his or her marital status. Our intuition was wrong, at least as far as these data are concerned. It is possible, of course, that these data are not representative of the population as a whole, or that our conclusion of association is incorrect because of chance variation in the people sampled.

**TRY THIS!** Exercise 5.59

## Sequences of Independent and Associated Events

A common challenge in probability is to find probabilities for sequences of events. By sequence, we mean events that take place in a certain order. For example, a married couple plans to have two children. What's the probability that the first will be a boy and the second a girl? When dealing with sequences, it is helpful to first determine whether the events are independent or associated.

If the two events are associated, then our knowledge of conditional probabilities is useful, and we should use Probability Rule 5b:

$$P(A \text{ AND } B) = P(A)P(B \mid A)$$

If the events are independent, then we know $P(B|A) = P(B)$, and this rule simplifies to $P(A \text{ AND } B) = P(A) P(B)$. This formula is often called the **multiplication rule**.

---

**Summary of Probability Rules**

**Rule 5c:** Multiplication Rule. If A and B are independent events, then

$$P(A \text{ AND } B) = P(A) P(B).$$

---

## Independent Events

When two events are independent, the multiplication rule speeds up probability calculations for events joined by AND.

For example, suppose that 51% of all babies born in the United States are boys. Then P(first child is boy) = 51%. What is the probability that a family planning to have two children will have two boys? In other words, how do we find the sequence probability

P(first child is boy AND second child is boy)?

Researchers have good reason to suspect that the genders of children in a family are independent (if you do not include identical twins). Because of this, we can apply the multiplication rule:

P(first child is boy AND second child is boy) = P(first is boy) P(second is boy)

$$= 0.51 \times 0.51 = 0.2601$$

The same logic could be applied to finding the probability that the first child is a boy and the second is a girl:

P(first child is boy AND second child is girl) = P(first is boy) P(second is girl)

$$= 0.51 \times 0.49 = 0.2499$$

## EXAMPLE  14  Three Coin Flips

Toss a fair coin three times. A fair coin is one in which each side is equally likely to land up when the coin is tossed.

**QUESTION**  What is the probability that all three tosses are tails? What is the probability that the first toss is heads AND the next two are tails?

**SOLUTION**  Using mathematical notation, we are trying to find P(first toss is tails AND second is tails AND third is tails). We know these events are independent (this is theoretical knowledge; we "know" this because the coin cannot change itself on the basis of its past). This means the probability is

P(first is tails) $\times$ P(second is tails) $\times$ P(third is tails) = $1/2 \times 1/2 \times 1/2 = 1/8$

Also, P(first is heads AND second is tails AND third is tails) is

P(heads) $\times$ P(tails) $\times$ P(tails) = $1/2 \times 1/2 \times 1/2 = 1/8$

**CONCLUSION**  The probability of getting three tails is the same as that of getting first heads and then two tails: 1/8.

**TRY THIS!**  Exercise 5.61

## EXAMPLE 15 Ten Coin Flips

Suppose I toss a coin 10 times and record whether each toss lands heads or tails. Assume that each side of the coin is equally likely to land up when the coin is tossed.

**QUESTION** Which sequence is the more likely outcome?

Sequence A: HTHTHTHTHT

Sequence B: HHTTTHTHHH

**SOLUTION** Because these are independent events, the probability that sequence A happens is

$$P(H)P(T)P(H)P(T)P(H)P(T)P(H)P(T)P(H)P(T) = \frac{1}{2} \times \frac{1}{2} \times \frac{1}{2} \times \frac{1}{2} \times \frac{1}{2} \times \frac{1}{2}$$

$$\times \frac{1}{2} \times \frac{1}{2} \times \frac{1}{2} \times \frac{1}{2}$$

$$= \left(\frac{1}{2}\right)^{10} = 0.0009766$$

The probability that sequence B happens is

$$P(H)P(H)P(T)P(T)P(T)P(H)P(T)P(H)P(H)P(H) = \frac{1}{2} \times \frac{1}{2} \times \frac{1}{2} \times \frac{1}{2} \times \frac{1}{2} \times \frac{1}{2}$$

$$\times \frac{1}{2} \times \frac{1}{2} \times \frac{1}{2} \times \frac{1}{2}$$

$$= \left(\frac{1}{2}\right)^{10} = 0.0009766$$

**CONCLUSION** Even though sequence A looks improbable, because it alternates between heads and tails, both outcomes have the same probability!

**TRY THIS!** Exercise 5.63

Another common probability question asks about the likelihood of "at least one" of a sequence happening a certain way.

## EXAMPLE 16 Harris Poll and "at Least One"

A Harris Poll (MetLife Survey of the American Teacher 2009) found that 67% of teachers are very satisfied with their careers. Suppose we select three teachers randomly and with replacement from the population of all teachers in the United States.

**QUESTIONS**

a. What is the probability that all three are very satisfied with their careers?

b. What is the probability that none of the three is satisfied?

c. What is the probability that at least one teacher is satisfied?

**SOLUTIONS**

a. We are asked to find P(first is satisfied AND second is satisfied AND third is satisfied). Because the teachers were selected at random from the population, these are independent events. (One teacher's answer won't affect the probability that the next one will answer one way or the other.) Because these are independent events, this is just P(first is satisfied) × P(second is satisfied) × P(third is satisfied) = 0.67 × 0.67 × 0.67 = 0.3008.

b. The probability that none is satisfied is trickier to determine. This event occurs if the first teacher is not satisfied AND the second is not AND the third is not. So we need to find P(first not satisfied AND second not satisfied AND third not satisfied) = P(firstnot) × P(second not) × P(third not) = (1 − 0.67) (1 − 0.67) (1 − 0.67) = 0.0359.

c. The probability that at least one teacher is satisfied is the probability that one is satisfied or two are satisfied or all three are satisfied. The calculation is easier if you realize that "at least one is satisfied" is the complement of "none is satisfied" because it includes all categories except "none."

$$1 - 0.0359 = 0.9641$$

**CONCLUSION**  The probability that all three randomly selected teachers are satisfied with their careers is 0.3008. The probability that none is satisfied is 0.0359. The probability that at least one is satisfied is 0.9641.

**TRY THIS!**  Exercise 5.65

---

### Watch Out for Incorrect Assumptions of Independence

Do not use the multiplication rule if events are not independent. For example, suppose we wanted to find the probability that a randomly selected person is female AND has long hair (say, more than 6 inches long).

About half of the population is female, so P(selected person is female) = 0.50. Suppose that about 35% of everyone in the population has long hair; then P(selected person has long hair) = 0.35. If we use the multiplication rule, we would find that P(selected person has long hair AND is female) = 0.35 × 0.5 = 0.175.

This relatively low probability of 17.5% makes it sound somewhat unusual to find a female with long hair. The reason is that we assumed that having long hair and being female are independent. This is a bad assumption: A woman is more likely than a man to have long hair. Thus "has long hair" and "is female" are associated, not independent, events. Therefore, once we know that the chosen person is female, there is a greater chance that the person has long hair.

### Associated Events with "AND"

If events are not independent, then we rely on Probability Rule 5b: P(A AND B) = P(A) P(B | A). Of course, this assumes that we know the value of P(B | A).

For example, the famous cycling race the Tour de France has in recent years been plagued with accusations against cyclists for taking steroids to boost their performance. Testing for these steroids is difficult, partly because these substances occur naturally in the human body, and partly because different individuals have different levels than other individuals. Even within an individual, the level of steroid varies during the day. Also, testing for the presence of steroids is expensive and time-consuming. For this reason, racers are chosen randomly for drug tests.

Let's imagine that 2% of the racers have taken an illegal steroid. Also, let's assume that if they took the drug, there is a 99% chance that the test will return a "positive" reading. (A "positive" reading here is a negative for the athlete, who will be disqualified.)

Keep in mind that these tests are not perfect. Even if the cyclist did not take a steroid, there is the probability that the test will return a positive. Let's suppose that, given that an athlete did *not* take a steroid, there is still a 3% chance that the test will return a positive (and the athlete will be unjustly disqualified).

What is the probability that a randomly chosen cyclist will be steroid-free and will still test positive for steroids? In other words, we need to find, for a randomly chosen cyclist,

P(cyclist does not take steroids AND tests positive)

To summarize, we have these pieces of information for a randomly selected athlete at this event:

$$P(\text{took steroid}) = 0.02$$

$$P(\text{did not take steroid}) = 0.98$$

$$P(\text{tests positive} \mid \text{took steroid}) = 0.99$$

$$P(\text{tests positive} \mid \text{did not take steroid}) = 0.03$$

Note that this is a sequence of events. First, the athlete either takes or doesn't take a steroid. Then a test is given, and the results are recorded.

According to Rule 5b,

$$P(A \text{ AND } B) = P(A) \, P(B \mid A)$$

$$
\begin{aligned}
P(\text{did not take steroid AND tests positive}) &= P(\text{did not take steroid}) \\
&\quad P(\text{tests positive} \mid \text{did not take steroid}) \\
&= 0.98 \times 0.03 \\
&= 0.0294
\end{aligned}
$$

We see that roughly 2.9% of all cyclists tested will be both steroid-free AND yet test positive for steroids.

Many people find it useful to represent problems in which events occur in sequence with a tree diagram. We can represent this sequence of all possible outcomes in the tree diagram shown in Figure 5.8.

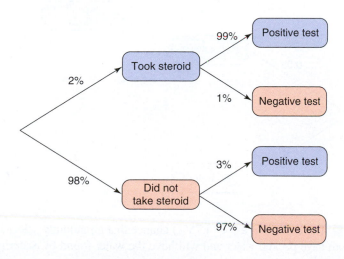

◄ **FIGURE 5.8** Tree diagram showing probabilities for the sequence of events in which an athlete either takes or does not take a drug, and then is tested for the presence of the drug.

The tree diagram shows all possible outcomes after selecting a cyclist (who either took steroids or did not) and then testing (and the test will be either positive or negative). Because the events "took steroid" and "tests positive" are associated, the probabilities of testing positive are different for the "took steroid" branch and the "did not take steroid" branch.

We want to find P(did not take steroid AND positive result), so we simply multiply along that branch of the tree that begins with "did not take steroid" and ends with "positive result":

$$
\begin{aligned}
P(\text{did not take steroid AND positive result}) &= P(\text{did not take steroid}) \times \\
&\quad P(\text{positive result} \mid \text{did not take steroid}) \\
&= 0.98 \times 0.03 \\
&= 0.0294
\end{aligned}
$$

## EXAMPLE 17 Airport Screeners

At many airports you are not allowed to take water through the security checkpoint. This means that Security screeners must check for people who accidentally pack water in their bags. Suppose that 5% of people accidentally pack a bottle of water in their bags. Also suppose that if there is a bottle of water in a bag, the Security screeners will catch it 95% of the time.

**QUESTION** What is the probability that a randomly chosen person with a backpack has a bottle of water in the backpack and Security finds it?

**SOLUTION** We are asked to find P(packed water AND Security finds the water). This is a sequence of events. First the person packs the water into the backpack. Later, Security finds (or does not find) the water.

We are given

$$P(\text{packs water}) = 0.05$$
$$P(\text{Security finds water} \mid \text{water packed}) = 0.95$$

Therefore,

$$P(\text{packs water and Security finds it}) = P(\text{packs water}) \times$$
$$P(\text{Security finds water} \mid \text{water packed})$$
$$= 0.05 \times 0.95$$
$$= 0.0475$$

The tree diagram in Figure 5.9 helps us show how to find this probability.

▶ **FIGURE 5.9** Tree diagram showing probabilities for the sequence of events in which a traveler either packs or does not pack water and then Security either finds or does not find the water.

**CONCLUSION** There is about a 5% (4.75%) chance that a randomly selected traveler will both have packed water and will have the water found by Security.

**TRY THIS!** Exercise 5.67

SECTION 5.4

# Finding Empirical Probabilities with Simulations

Empirical probabilities are useful when you're not sure what assumptions to make to find a theoretical probability or when the theoretical probability is too complex. Statisticians rely quite heavily on computer-based simulations, and learning to design simulations is an important skill in examining probabilities.

As we noted near the beginning of this chapter, a simulation is an experiment that, we hope, closely matches a real-life situation. We will simulate real-life probability problems to find empirical probabilities. As you will soon see, under the right conditions, these empirical probabilities can be good estimates of the theoretical probabilities.

The essential component of a simulation is some random act that matches real life, in the sense that our simulated outcome happens with the same relative frequency as the real-life outcome we are trying to simulate. Most computers and statistical calculators have very good random number generators that can be used to create these random outcomes. Another useful tool, if technology isn't available, is a random number table such as that used in Example 1.

## Coin Flip Simulation

Suppose we wish to know the probability that we will get three heads if we flip a coin three times. This is a probability we can figure out theoretically if we assume each side of the coin is equally likely (the solution is 1/8), so we can compare our empirical probability to the theoretical. As you will see from the Law of Large Numbers at the end of this section, if you do a large enough number of repetitions, the empirical probability can get fairly close to the theoretical result.

We will use the random number table (Appendix A). Each flip of a coin has two possible outcomes—heads or tails—and each outcome is equally likely. The trick, then, is to assign digits to represent "heads" and "tails" in such a way that these simulated outcomes are equally likely.

There are many ways to do this. Here are two:

1. Assign 0 to represent "tails" and 1 to represent "heads." The probability of a 1 is 0.50, so the probability of getting a simulated "heads" is 0.50.

2. Assign even numbers (0, 2, 4, 6, 8) to represent "tails" and odd numbers (1, 3, 5, 7, 9) to represent "heads." The probability of getting an odd number is 0.50.

For this example, we will use method 2, but it really doesn't matter which method you use.

Because we are flipping a coin three times, we read off three digits from the table: one for each flip. (An excerpt of the table is shown in Table 5.6.) Every time we see an even number, we record T, and when we see an odd number, we record H. (If we were using method 1, we would ignore the digits 2–9, because no outcome was assigned to these digits.)

| Line | | | | | | |
|------|-------|-------|-------|-------|-------|-------|
| 01 | 21033 | 32522 | 19305 | 90633 | 80873 | 19167 |
| 02 | 17516 | 69328 | 88389 | 19770 | 33197 | 27336 |
| 03 | 26427 | 40650 | 70251 | 84413 | 30896 | 21490 |
| 04 | 45506 | 44716 | 02498 | 15327 | 79149 | 28409 |
| 05 | 55185 | 74834 | 81172 | 89281 | 48134 | 71185 |

◀ **TABLE 5.6** The first five lines of the random number table, Appendix A.

After our three flips, we record the number of heads. We have just finished one trial.

We now repeat the above process many times—ideally, at least 100 times. At the end, we count how many times we saw three "heads" in our three flips. This number, divided by the number of trials, is our empirical probability.

Let's see what this looks like for 10 trials. Remember that odd values get translated to heads and even values to tails. Table 5.7 shows the results of our 10 trials using the first lines of the table:

◀ **TABLE 5.7** Simulating heads and tails with random numbers. Trials resulting in three heads are shown in red.

| Trial | Random Numbers | Translation | Number of Heads |
|-------|----------------|-------------|-----------------|
| 1 | 2 1 0 | THT | 1 |
| **2** | **3 3 3** | **HHH** | **3** |
| 3 | 2 5 2 | THT | 1 |
| 4 | 2 1 9 | THH | 2 |
| 5 | 3 0 5 | HTH | 2 |
| 6 | 9 0 6 | HTT | 1 |
| 7 | 3 3 8 | HHT | 2 |
| 8 | 0 8 7 | TTH | 1 |
| **9** | **3 1 9** | **HHH** | **3** |
| 10 | 1 6 7 | HTH | 2 |

We saw three heads in 2 of the 10 trials (shown in color in Table 5.7), so our empirical probability is 2/10, or 0.20. These two trials are shown in red in Table 5.7. This is quite a bit different from the theoretical probability of 0.125. As you shall soon see, this is not surprising, because with only 10 trials, there's quite a bit of variability in the empirical probabilities we might get. That is why it is best to do a large number of trials.

## Summary of Steps of a Simulation

A simulation consists of a number of trials. Each trial represents an action that has a random outcome. To design a simulation, follow these steps:

**Step 1:** Identify the actions that make up a single trial. This trial will be repeated many times. *For example, a trial consists of three flips of a coin.*

**Step 2:** Decide how to simulate a single trial. Explain how you will represent each possible outcome. *For example, each flip is represented by an odd (heads) or even (tails) digit.*

**Step 3:** Describe the response variable. The response variable is the outcome we will record. *For example, the response variable is the number of heads we see in three flips.*

**Step 4:** Describe the event of interest. The event of interest is the event that we want to compute a probability for. *The event of interest is "get three heads."*

**Step 5:** Carry out the simulation. Repeat very many times. *Best to do at least 100 trials if possible.*

**Step 6:** Calculate the proportion of trials in which the event of interest occurred. This is your empirical probability. *For example, we saw three heads in three flips 2 out of 10 times, for an empirical probability of 2/10 = 0.20.*

Step 2 (deciding how to simulate a single trial) is crucial. Because selecting an odd number is just as likely as getting a head when you flip a coin, our simulation is a good match to actually flipping a coin. However, if we had instead decided that "Digits 0, 1, 2 will represent heads, and 3, 4, 5, 6, 7, 8, 9 will represent tails," then we would have a very poor match to reality. One possibility that would have worked equally well is to let the digits 0 to 4 represent heads and 5 to 9 represent tails. This works because we are just as likely to get a number from 0 to 4 as a number from 5 to 9.

## You Try It

The SOCR website lets you repeat this simulation many times.

- Go to http://socr.ucla.edu/
- Click on the **Experiments** tab and select **Binomial Coin Experiment** from the dropdown menu that will appear.
- In the middle you'll see a slider next to "$n = 10$." This represents the number of coins to be tossed. Slide this to the left so that "$n = 3$."
- Click on the red **Play** button.

You will see the results of the three coin tosses. On the right, a red rectangle will show you how many heads occurred. Below that, a display keeps track of the results. In the "X" column you see the number of heads that might have come up: 0, 1, 2, or 3. In the column marked "Data," you see the percentage of times that those outcomes occurred. At this point, in the row corresponding to three heads, you will see either a 0 (if you did not get three heads) or a 1 (if you did). The middle column, labeled "Distribution," shows the theoretical probabilities for these outcomes.

- Click on the red **Play** button again. The displays update.
- Click on the **Fast Forward** button to repeat this 10 more times. You can change the selector from **Stop 10** to **Stop 100** if you want to repeat 100 times, or you can change it to **Stop 1000**, although this takes a little bit of time.

What proportion of the time did you get three heads? How close is this to the theoretical probability of 0.125?

## EXAMPLE 18 Dice Simulation

Use a simulation to find the approximate probability that a fair, six-sided die will land with the 6 showing on top. Do ten trials, using these random numbers:

```
44687   75032   83408   10239
80016   58250   91419   56315
```

**QUESTION** What is your empirical probability of getting a six? Compare this to the theoretical probability. Show all steps.

**SOLUTION**

**Step 1:** The trial consists of rolling a balanced die one time. (The die is balanced so that all six sides are equally likely to land face up.)

**Step 2:** Each number in the random number table will represent the roll of a die. The number in the table will represent the number that comes up on the die. We'll ignore the digits 0, 7, 8, and 9, because these are impossible outcomes when rolling a six-sided die.

**Step 3:** The response variable is the number that comes up on the die.

**Step 4:** The outcome of interest is whether the response variable is a 6.

Table 5.8 shows the results of each trial, using the random number table provided. Note that we simply skip the digits 0, 7, 8, and 9.

**Step 5:** A 6 occurred on 1 of the 10 trials.

**CONCLUSION** Our empirical probability of getting a 6 on the roll of a balanced die is 1/10, or 0.10. In contrast, the theoretical probability is 1/6, or about 0.167.

**TRY THIS!** Exercise 5.69

| Trial | Random Numbers | Response Variable |
|-------|----------------|-------------------|
| 1 | 4 | 4 |
| 2 | 4 | 4 |
| 3 | 6 | 6 |
| 4 | 5 | 5 |
| 5 | 3 | 3 |
| 6 | 2 | 2 |
| 7 | 3 | 3 |
| 8 | 4 | 4 |
| 9 | 1 | 1 |
| 10 | 2 | 2 |

▲ **TABLE 5.8** Simulations for rolling a six-sided die.

## You Try It

You can toss a die as many times as you like, without tiring your arm, by going to http://socr.ucla.edu/. Click on the **Experiments** tab and select **Dice Experiment** from the dropdown menu that will appear.

- Click on the **Play** button. A single die will be rolled. The graphic on the right shows you what the outcome was. Below the graphic is a table that shows all of the possible outcomes (in the column headed by "Y"); the theoretical probability of each outcome (labeled "Distribution"—these will all be the same, because this die is fair); and the empirical probability for each outcome in the column labeled "Data." At this point, if you rolled a 6, the Data will report a 1 because you got a 6 on 100% of your rolls. If you didn't roll a 6, you'll see a 0.

- Click again, or click on the **Fast Forward** button to repeat many times.

- Do about 100 simulations. How close is the empirical probability to the theoretical probability?

## The Law of Large Numbers

The **Law of Large Numbers** is a famous mathematical theorem that tells us that if our simulation is designed correctly, then the more trials we do, the closer we can expect our empirical probability to come to the true probability. The Law of Large Numbers shows that as we approach infinitely many trials, the true probability and the empirical probability approach the same value.

 **KEY POINT**

The Law of Large Numbers states that if an experiment with a random outcome is repeated a large number of times, the empirical probability of an event is likely to be close to the true probability. The larger the number of repetitions, the closer together these probabilities are likely to be.

The Law of Large Numbers is the reason why simulations are useful: Given enough trials, and assuming that our simulations are a good match to real life, we can get a good approximation of the true probability.

Table 5.9 shows the results of a very simple simulation. We used a computer to simulate flipping a coin, and we were interested in observing the frequency at which the "coin" comes up heads. We show the results at the end of each trial so that you can see how our approximation gets better as we perform more trials. For example, on the first trial we got a head, so our empirical probability is $1/1 = 1.00$. On the second and third trials we got heads, so the empirical probability of heads is still 1.00. On the fourth trial we got tails, and up to that point we've had 3 heads in 4 trials. Therefore, our empirical probability is $3/4 = 0.75$.

We can continue this way, but it is easier to show you the results by making a graph. Figure 5.10 shows a plot of the empirical probabilities against the number of trials.

Note that with a small number of trials (say, less than 75 or so), our empirical probability was relatively far away from the theoretical value of 0.50. Also, when the number of trials is small, the empirical probabilities can change a lot with each additional "coin flip." But eventually, things settle down, and the empirical probabilities get close to what we know to be the theoretical probability. If you were to simulate a coin flip just 20 times, you might not expect your empirical probability to be too close to the theoretical probability. But after 1000 flips, you'd expect it to be very close.

### How Many Trials Should I Do in a Simulation?

In our two examples of simulations we did 10 trials, which is not very many. As you can tell from Figure 5.10, if we had stopped at 5 flips, we would have had an empirical probability that was pretty far from the true value.

| Trial | Outcome | Empirical Probability of Heads |
|-------|---------|-------------------------------|
| 1 | H | $1/1 = 1.00$ |
| 2 | H | $2/2 = 1.00$ |
| 3 | H | $3/3 = 1.00$ |
| 4 | T | $3/4 = 0.75$ |
| 5 | H | $4/5 = 0.80$ |
| 6 | H | $5/6 = 0.83$ |
| 7 | T | $5/7 = 0.71$ |
| 8 | H | $6/8 = 0.75$ |
| 9 | T | $6/9 = 0.67$ |
| 10 | H | $7/10 = 0.70$ |

▲ **TABLE 5.9** Simulations of heads and tails with cumulative empirical probabilities

The number of trials you need to be confident that your empirical probability is close to the theoretical value can be fairly large. In general, the more rare the event, the more trials you'll need. Fortunately, simulations are very quick when done with computers, and a variety of software tools are available to help you do fast simulations. We recommend that you do at least 100 trials in most cases.

◄ **FIGURE 5.10** The Law of Large Numbers predicts that after many flips, the proportion of heads we get from flipping a real coin will get close to the true probability of getting heads. Because these empirical probabilities are "settling down" to about 0.50, this supports the theoretical probability of 0.50.

## You Try It

Toss a coin 10 times. What is your empirical probability of heads? How far from 0.50 is this? Repeat this a few times to get a sense of how far away the empirical probability can be from 0.50. What's the furthest you got?

What would happen if you were to toss a coin 20 times? Would you tend to get an empirical probability that is closer to 0.50? It's a bit time-consuming to do this with a real coin, so use a computer:

- Go to http://socr.ucla.edu/

- Click on the **Experiments** tab and select **Binomial Coin Experiment** from the dropdown menu that will appear.

- Click on the **Run** button to "toss" 10 "coins." The window below displays the number of heads and the proportion of heads.

- Move the slider beside "$n = 10$" so that "$n = 20$." The letter $n$ represents the number of tosses.

- Now click on the **Play** button a few times to get a sense of how far from 0.50 your empirical probabilities fall. What's the furthest you get? Typically, how far would you say your empirical probabilities are from 0.50? How does this compare to when you tossed the coin 10 times?

## What If My Simulation Doesn't Give the Theoretical Value I Expect?

You have no guarantee that your empirical probability will give you exactly the theoretical value. As you noticed if you did the above activity, even though it's not unusual to get 50% heads in 10 tosses, more often you get some other proportion.

When you don't get the "right" value, two explanations are possible:

1. Your theoretical value is incorrect.

2. Your empirical probability is just varying—that's what empirical probabilities do. You can make it vary less and get closer to the theoretical value by doing more trials.

How do we choose between these two alternatives? Well, that's what statistics is all about! This is one of the central questions of statistics: Are the data consistent with our expectations, or do they suggest that our expectations are wrong? We will return to this question in almost every chapter in the rest of this book.

## Some Subtleties with the Law

The Law of Large Numbers (LLN) is one law that cannot be broken. Nevertheless, many people think the law has been broken when it really hasn't, because interpreting the LLN takes some care.

The Law of Large Numbers tell us, for example, that with many flips of a coin, our empirical probability of heads will be close to 0.50. It tells us nothing about the *number* of heads we will get after some number of tosses and nothing about the order in which the heads and tails appear.

### Streaks: Tails Are Never "Due"

Many people mistakenly believe the LLN means that if they get a large number of heads in a row, then the next flip is more likely to come up tails. For example, if you just flipped five heads in a row, you might think that the sixth is more likely to be a tail than a head so that the empirical probability will work out to be 0.50. Some people might incorrectly say, "Tails are due."

This misinterpretation of the LLN has put many a gambler into debt. This is a misinterpretation for two reasons. First, the Law of Large Numbers is patient. It says that the empirical probability will equal the true probability after infinitely many trials. That's a lot of trials. Thus a streak of 10 or 20 or even 100 heads, though extremely rare, does not contradict the LLN.

### How Common Are Streaks?

Streaks are much more common than most people believe. At the beginning of the chapter, we asked you to toss a coin 20 times (or to simulate 20 tosses) and see how long your longest streak was. We now show results of a simulation to find the (empirical) probability that the longest streak is of length 6 or longer in 20 tosses of a coin.

In our 1000 trials, we saw 221 streaks that were of length 6 or greater. The longest streak we saw was of length 11. The results are shown in Figure 5.11. Our empirical probability of getting a longest streak of length 6 or more was 0.221.

There is another reason why you should not believe that heads are "due" after a long streak of tails. If this were the case, the coin would somehow have to know whether to come up heads or tails. How is the coin supposed to keep track of its past?

▶ **FIGURE 5.11** Longest streaks in 20 flips of a coin. In 1000 trials, 221 trials had streaks of length 6 or greater.

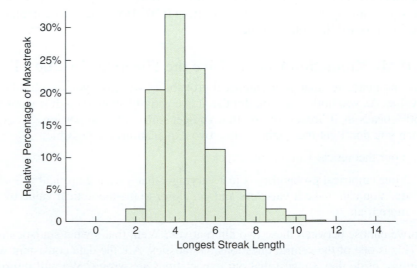

## CASE STUDY REVISITED

Sally Clark was convicted of murdering two children on the basis of the testimony of physician/expert Dr. Meadow. Dr. Meadow testified that the probability of two children in the same family dying of SIDS was extremely low and that, therefore, murder was a more plausible explanation. In Dr. Meadow's testimony, he assumed that the event "One baby in a family dies of SIDS" AND "a second baby in the family dies of SIDS" were independent. For this reason, he applied the multiplication rule to find the probability that two babies die of SIDS in the same family. The probability of one baby dying was 1/8543, so the probability of two independent deaths was ( 1/8543 ) $\times$ (1/8543), or about 1 in 73 million.

However, as noted in a press release by the Royal Statistical Society, these events may not be independent. "This approach (multiplying probabilities) . . . would only be valid if SIDS cases arose independently within families, an assumption that would need to be justified empirically. Not only was no such empirical justification provided in the case, but there are very strong *a priori* reasons for supposing that the assumption will be false. There may well be unknown genetic or environmental factors that predispose families to SIDS, so that a second case within the family becomes much more likely."

Dr. Meadow made several other errors in statistical reasoning, which are beyond the scope of this chapter but quite interesting nonetheless. There's a nice discussion that's easy to read at http://www.richardwebster.net/cotdeaths.html. For a summary and a list of supporting references, including a video by a statistician explaining the statistical errors, see http://en.wikipedia.org/wiki/Sally_Clark.

Sally Clark was released from prison but died in March 2007 of alcohol poisoning at the age of 43. Her family believes her early death was caused in part by the stress inflicted by her trial and imprisonment.

# EXPLORING STATISTICS
## CLASS ACTIVITY

## Let's Make A Deal: Stay or Switch?

| GOALS | MATERIALS |
|---|---|
| Simulate an experiment to estimate complex (and counterintuitive) probabilities. | Three playing cards—two of one color, the third of the other color—for each group of three students. |

### ACTIVITY

| Curtain A | Curtain B | Curtain C |
|---|---|---|

▲ FIGURE A Three curtains, and the contestant picked curtain A.

| Curtain A |  | Curtain C |
|---|---|---|

▲ FIGURE B The host, who knows there is a goat behind curtain B, reveals curtain B. Should the contestant stick with curtain A or switch to curtain C?

| Curtain A |  |  |
|---|---|---|

▲ FIGURE C The contestant switched and won.

In the popular game show "Let's Make a Deal," contestants are shown three curtains. Behind one curtain is a high-value prize, for example a car, and behind the other two are less desirable prizes, for example, goats. The contestant picks one of the three curtains. Then the host (who knows what is behind each curtain) raises another curtain to reveal a goat. The host then asks the contestant whether he or she wants to stay or switch to the other curtain before revealing what is behind the contestant's chosen curtain.

There are three possibilities: The contestant should stay; The contestant should switch; It doesn't matter whether the contestant stays or switches.

Form groups of three and choose roles: host, contestant, and recorder. The host lays the cards face down, with the two cards of the same color representing goats and the other card representing the car. The contestant chooses a card. The host then turns over a "goat" card (but not the one just selected). The contestant now must choose a strategy: stay or switch. After that decision is made, the host reveals whether the contestant won a goat or a car. Repeat this several times, giving the contestant several opportunities to try both strategies.

For each trial, the recorder will put a tally mark in one of four categories: Stayed and Won; Stayed and Lost; Switched and Won; Switched and Lost.

After a few games, change roles so that everyone gets to try all roles. Your instructor will compile results.

### BEFORE THE ACTIVITY

Which strategy do you think will be better: staying or switching? Or does it make no difference whatsoever? Why?

### AFTER THE ACTIVITY

Judging on the basis of your class data, what is the empirical probability of winning if the contestant stays? What is it if the contestant switches? Do these probabilities convince you to change your strategy? Explain.

# CHAPTER REVIEW

## Key Terms

Random, *197*
Probability, *198*
Theoretical probabilities, *198*
Empirical probabilities, *199*
Simulation, *199*

Complement, *200*
Sample space, *201*
Event, *201*
Venn diagram, *204*
AND, *204*

Inclusive OR, *205*
Mutually exclusive events, *207*
Conditional probabilities, *209*
Associated, *212*
Independent events, *212*

Multiplication rule, *216*
Law of Large Numbers, *224*

## Learning Objectives

After reading this chapter and doing the assigned homework problems, you should

- Understand that humans can't reliably create random numbers or sequences.

- Understand that a probability is a long-term relative frequency.

- Know the difference between empirical and theoretical probabilities—and know how to calculate them.

- Be able to determine whether two events are independent or associated and understand the implications of making incorrect assumptions about independent events.

- Understand that the Law of Large Numbers allows us to use empirical probabilities to estimate and test theoretical probabilities.

- Know how to design a simulation to estimate empirical probabilities.

## Summary

Random samples or random experiments must be generated with the use of outside mechanisms such as computer algorithms or by relying on random number tables. Human intuition cannot be relied on to produce reliable "randomness."

Probability is based on the concept of long-run relative frequencies: If an action is repeated infinitely many times, how often does a particular event occur? To find theoretical probabilities, we calculate these relative frequencies on the basis of assumptions about the situation and rely on mathematical rules. In finding empirical probabilities, we actually carry out the action many times or, alternatively, rely on a simulation (using a computer or random number table) to quickly carry out the action many times. The empirical probability is the proportion of times a particular event was observed to occur. The Law of Large Numbers tells us that the empirical probability becomes closer to the true probability as the number of repetitions is increased.

### Theoretical Probability Rules

**Rule 1:** A probability is always a number from 0 to 1 (or 0% to 100%) inclusive (which means 0 and 1 are allowed). It may be expressed as a fraction, a decimal, or a percent.

$$0 \le P(A) \le 1$$

**Rule 2:** For any event A,

$$P(A \text{ does } not \text{ occur}) = 1 - P(A \text{ does occur})$$

$A^c$ is the complement of A:

$$P(A^c) = 1 - P(A)$$

**Rule 3:** For equally likely outcomes:

$$P(A) = \frac{\text{Number of outcomes in A}}{\text{Number of all possible outcomes}}$$

**Rule 4:** Always: $P(A \text{ OR } B) = P(A) + P(B) - P(A \text{ AND } B)$
**Rule 4a:** Only if A and B are mutually exclusive:
$$P(A \text{ OR } B) = P(A) + P(B)$$
**Rule 5a:** Conditional probabilities

Probability of A given that B occurred: $P(A|B) = \dfrac{P(A \text{ AND } B)}{P(B)}$

**Rule 5b:** Always: $P(A \text{ AND } B) = P(B) P(A|B)$
**Rule 5c:** Multiplication Rule. If A and B are independent events, then

$$P(A \text{ AND } B) = P(A) P(B)$$

This applies for any (finite) number of events. For example, $P(A \text{ AND } B \text{ AND } C \text{ AND } D) = P(A) P(B) P(C) P(D)$ if A, B, C, D are independent of each other.

## Sources

MetLife Survey of the American Teacher, 2009, http://www.harrisinteractive .com

*Starr, Norton (1997),* "Nonrandom Risk: The 1970 Draft Lottery,"
*Journal of Statistics Education,* volume 5, number 2. Vietnam-era draft data can be found, along with supporting references, at www.amstat.org/ publications/jse/datasets/draft.txt.

Pew Foundation, *Social Media and Young Adults* report, http://pewinternet .org (accessed July 2010).

## SECTION EXERCISES

### SECTION 5.1

TRY **5.1 Simulation (Example 1)** If we flip a coin 10 times, how often do we get 6 or more heads? A first step to answering this question would be to simulate 10 flips. Use the random number table in Appendix A to simulate flipping a coin 10 times. Let odd digits (1, 3, 5, 7, 9) represent heads, and let even digits (0, 2, 4, 6, 8) represent tails. Begin with the first digit in the third row.

a. Write the sequence of 10 random digits.

b. Write the sequence of 10 "heads" and "tails." Write H for heads and T for tails.

c. Did you get 6 or more heads? How many heads did you get?

**5.2 Simulation** Suppose you are carrying out a randomized, controlled experiment to test whether guided meditation lowers blood pressure (treatment) compared to reading in a quiet room (control). Use the random number table in Appendix A to assign the first 12 subjects to treatment or control. You want each subject to have a 50% chance of being assigned to the treatment group. Let the digits 0, 1, 2, 3, 4 represent assignment to the meditation group, and let the digits 5, 6, 7, 8, 9 represent assignment to the control group. Begin with the first digit on the fourth line of the table.

a. Write the 12 numbers, and below them write M for meditation and C for control.

b. How many subjects were assigned to the meditation group?

c. A research assistant wants to use the digits 0–5 to represent assignment to the meditation group and the digits 6–9 for assignment to the control group. Explain why this will not work.

**5.3** A Monopoly player claims that the probability of getting a 4 when rolling a six-sided die is 1/6 because the die is equally likely to land on any of the six sides. Is this an example of a theoretical probability or an empirical probability? Explain.

**5.4** A person was trying to figure out the probability of getting two heads when flipping two coins. He flipped two coins 10 times, and in 2 of these 10 times both coins landed heads. On the basis of this outcome, he claims that the probability of two heads is 2/10, or 20%. Is this an example of an empirical probability or a theoretical probability? Explain.

**5.5** A friend flips a coin 10 times and says that the probability of getting a head is 60% because he got six heads. Is the friend referring to an empirical probability or a theoretical probability? Explain.

**5.6** A magician claims that he has a fair coin—"fair" because both sides, heads and tails, are equally likely to land face up when the coin is flipped. He tells you that if you flip the coin three times, the probability of getting three tails is 1/8. Is this an empirical probability or a theoretical probability? Explain.

### SECTION 5.2

**5.7 Criminal Court Judges** Criminal cases are assigned to judges randomly. The list of the criminal judges for Memphis, Tennessee (taken from the Tennessee Court System website, www.tsc.state.tn.us), is given in the table. Assume that only Carolyn Blackett and Paula Skahan are women and the rest are men. If you were a criminal defense attorney in Memphis, you might be interested in whether the judge assigned to your case was a man or a woman.

James C. Beasley

Carolyn W. Blackett

Lee V. Coffee

Chris Craft

John T. Fowlkes

W. Otis Higgs

James M. Lammey

Paula L. Skahan

W. Mark Ward

Suppose the names are put into a pot, and a clerk pulls a name out at random.

a. List the equally likely outcomes that could occur; last names are enough.

b. Suppose the event of interest, event A, is that a judge is a woman. List the outcomes that make up event A.

c. What is the probability that one case will be assigned to a female judge?

d. List the outcomes that are in the complement of event A.

**5.8 Random Assignment of Professors** A study looked at the way students attending the Air Force Academy were randomly assigned to different professors for Calculus I, with equal numbers of students assigned to each professor. Some professors were experienced, and some were relatively inexperienced. Suppose the names of the professors are Peters, Parker, Diaz, Nguyen, and Black. Suppose Diaz and Black are inexperienced and the others are experienced. Carrell and West reported that the students who had the experienced teachers for Calculus I did better in Calculus II. (Source: Scott E. Carrell and James E. West, *Does Professor Quality Matter? Evidence from Random Assignment of Students to Professors*, 2010)

a. List the equally likely outcomes that could occur for assignment of one student to a professor.

b. Suppose the event of interest, event A, is that a teacher is experienced. List the outcomes that make up event A.

c. What is the probability that one student will be assigned to an experienced teacher?

d. List the outcomes in the complement of event A. Describe this complement in words.

e. What is the probability that one student will be assigned to an inexperienced teacher?

**5.9** Which of the following numbers could *not* be probabilities, and why?

$$0.5, 1.3, 0.001, 75\%, 1/4, 150\%, -0.5$$

**5.10** Which of the following numbers could *not* be probabilities, and why?

$$0, 2/5, 7/4, 0.002, -10\%, 0.998, 45\%$$

TRY **5.11 Playing Cards (Example 2)** There are four suits: clubs (♣), diamonds (♦), hearts (♥), and spades (♠), and the following cards appear in each suit: Ace, 2, 3, 4, 5, 6, 7, 8, 9, 10, Jack, Queen, King. The Jack, Queen, and King are called face cards because they have a drawing of a face on them. Diamonds and hearts are red, and clubs and spades are black.

If you draw 1 card randomly from a standard 52-card playing deck, what is the probability that it will be:

a. A heart?

b. A red card?

c. An ace?

d. A face card (jack, queen, or king)?

e. A three?

**5.12 Playing Cards** Refer to Exercise 5.11 for information about cards. If you draw 1 card randomly from a standard 52-card playing deck, what is the probability that it will be:

a. A black card?

b. A diamond?

c. A face card (jack, queen, or king)?

d. A nine?

e. A king or queen?

### 5.13 Guessing on Tests

a. On a true/false quiz in which you are guessing, what is the probability of guessing correctly on one question?

b. What is the probability that a guess on one true/false question will be incorrect?

**5.14 Guessing on Tests** On a multiple-choice test with a total of four possible options for each question

a. What is the probability of guessing correctly on one question? (Assume that there are three incorrect options and one correct option.)

b. What is the probability that a guess on one question will be incorrect?

TRY **5.15 Four Children (Example 3)** The sample space given here shows all possible sequences for a family with 4 children, where B stands for boy and G stands for girl.

| | | | | |
|------|------|------|------|------|
| GGGG | GGGB | GGBB | GBBB | BBBB |
| | GGBG | GBGB | BGBB | |
| | GBGG | GBBG | BBGB | |
| | BGGG | BGGB | BBBG | |
| | | BGBG | | |
| | | BBGG | | |

Assume that all of the 16 outcomes are equally likely. Find the probability of having the following numbers of girls out of 4 children: (a) exactly 0 girls, (b) exactly 1 girl, (c) exactly 2 girls, (d) exactly 3 girls, (e) exactly 4 girls.

(*Hint:* The probability of having 3 girls and a boy is 4/16, or 25%, because the second column shows that there are 4 ways to have 3 girls and 1 boy.)

**5.16 Three Coins** The sample shows the possible sequences for flipping three fair coins or flipping one coin three times, where H stands for heads and T stands for tails.

| | | | |
|-----|-----|-----|-----|
| HHH | HHT | HTT | TTT |
| | HTH | THT | |
| | THH | TTH | |

Assume that all of the 8 outcomes are equally likely. Find the probability of having exactly the following numbers of heads out of the 3 coins: (a) exactly 0 heads, (b) exactly 1 head, (c) exactly 2 heads, (d) exactly 3 heads. What do the four probabilities add up to and why?

**5.17 Birthdays** What is the probability that a baby will be born on a Friday OR a Saturday OR a Sunday if all the days of the week are equally likely as birthdays?

**5.18 Playing Cards** If *one* card is selected from a well-shuffled deck of 52 cards, what is the probability that the card will be a club OR a diamond OR a heart? What is the probability of the complement of this event? (Refer to Exercise 5.11 for information about cards.)

**5.19 GSS: Political Party** The General Social Survey (GSS) is a survey done nearly every year at the University of Chicago. One survey, summarized in the table, asked each respondent to report her or his political party affiliation and whether she or he was liberal, moderate, or conservative. (Dem stands for Democrat, and Rep stands for Republican.)

| | Dem | Rep | Other | Total |
|--------------|-----|-----|-------|-------|
| Liberal | 306 | 26 | 198 | 530 |
| Moderate | 279 | 134 | 322 | 735 |
| Conservative | 104 | 309 | 180 | 593 |
| Total | 689 | 469 | 700 | 1858 |

a. If one person is chosen randomly from the group, what is the probability that the person is liberal?

b. If one person is chosen randomly from the group, what is the probability that the person is a Democrat?

**5.20 GSS: Political Party** Refer to the table given in Exercise 5.19.

a. If one person is chosen randomly from the group of 1858 people, what is the probability that the person is conservative?

b. If one person is chosen randomly from the group of 1858 people, what is the probability that the person is a Republican?

TRY **5.21 GSS: AND (Example 4)** Refer to the table given in Exercise 5.19. Suppose we select a person at random from this collection of 1858 people. What is the probability the person is conservative AND a Democrat? In other words, find P(person is conservative AND person is a Democrat).

**5.22 GSS: AND** Refer to the table given in Exercise 5.19. Suppose we select a person at random from this collection of 1858 people. What is the probability that the person is liberal AND a Republican? In other words, find P(person is liberal AND person is a Republican).

TRY **5.23 GSS: OR (Example 5)** Select someone at random from the 1858 people in the table given in Exercise 5.19. What is the probability that the person is a Democrat OR a Republican?

**5.24 GSS: OR** Select someone at random from the 2010 people in the table given in Exercise 5.19. What is the probability that the person is liberal OR conservative?

TRY **5.25 GSS: OR (Example 6)** Assume one person is chosen randomly from the 1858 people in the table given in Exercise 5.19. What is the probability that the person is liberal OR a Democrat? *See page 238 for guidance.*

**5.26 GSS: OR** Assume that one person is chosen randomly from the table given in Exercise 5.19. What is the probability that the person is conservative OR a Republican?

**5.27 Mutually Exclusive** Suppose a person is selected at random. Label each pair of events as *mutually exclusive* or *not mutually exclusive*.

a. The person has brown eyes; the person has blue eyes.

b. The person is 50 years old; the person is a U.S. senator.

**5.28 Mutually Exclusive** Suppose a person is selected at random. Label each pair of events as *mutually exclusive* or *not mutually exclusive*.

a. The person is a parent; the person is a toddler.

b. The person is a woman; the person is a CEO (chief executive officer).

TRY **5.29 GSS: Mutually Exclusive (Example 7)** Referring to the table given in Exercise 5.19, name a pair of mutually exclusive events that could result from selecting an individual at random from this sample.

**5.30 GSS: Mutually Exclusive** Referring to the table given in Exercise 5.19, name a pair of events that are not mutually exclusive that could result from selecting an individual at random from this sample.

**5.31 "OR"** In 2009, the Humane Society of the United States reported that 39% of households owned one or more dogs and 33% owned one or more cats. From this information, is it possible to find the percentage of households that owned a cat OR a dog? Why or why not?

**5.32 "OR"** Suppose you discovered that on your college campus, 6% of the female students were married and 4% of the female students had at least one child.

a. From this information, is it possible to determine the percentage of female students who were married OR had a child?

b. If your answer to part a is no, what additional information would you need to answer this question?

TRY **5.33 Fair Die (Example 8)** Roll a fair six-sided die.

a. What is the probability that the die shows on an odd number OR a number less than 3 on top?

b. What is the probability that the die shows on an odd number OR a number less than 2 on top?

**5.34 Roll a Die** Roll a fair six-sided die.

a. What is the probability the die shows on an odd number OR a number greater than 5 on top?

b. What is the probability the die shows on an odd number OR a number greater than 4 on top?

**5.35 Grades** Assume that the only grades possible in a history course are A, B, C, or lower than C. The probability that a randomly selected student will get an A in a certain history course is 0.18, the probability that a student will get a B in the course is 0.25, and the probability that a student will get a C in the course is 0.37.

a. What is the probability that a student will get an A OR a B?

b. What is the probability that a student will get an A OR a B OR a C?

c. What is the probability that a student will get a grade lower than a C?

**5.36 Changing Multiple-Choice Answers** One of the authors did a survey to determine the effect of students changing answers while taking a multiple-choice test on which there is only one correct answer for each question. Some students erase their initial choice and replace it with another. It turned out that 61% of the changes were from incorrect answers to correct and that 26% were from correct to incorrect. What percent of changes were from incorrect to incorrect?

**5.37 "AND" and "OR"** Consider these categories of people, assuming that we are talking about all the people in the United States:

Category 1: People who are currently married

Category 2: People who have children

Category 3: People who are currently married OR have children

Category 4: People who are currently married AND have children

a. Which of the four categories has the most people?

b. Which category has the fewest people?

**5.38 "AND" and "OR"** Assume that we are talking about all students at your college.

a. Which group is larger: students who are currently taking English AND math, or students who are currently taking English?

b. Which group is larger: students who are taking English OR math, or students who are taking English?

**5.39 Coin Flips** Let H stand for heads and let T stand for tails in an experiment where a fair coin is flipped twice. Assume that the four outcomes listed are equally likely outcomes:

$$HH, HT, TH, TT$$

What are the probabilities of getting:

a. 0 heads?

b. Exactly 1 head?

c. Exactly 2 heads?

d. At least 1 head?

e. Not more than 2 heads?

**5.40 Cubes** A hat contains a number of cubes: 15 red, 10 white, 5 blue, and 20 black. One cube is chosen at random. What is the probability that it is:

a. A red cube?

b. Not a red cube?

c. A cube that is white OR black?

d. A cube that is neither white nor black?

e. What do the answers to part a and part b add up to and why?

**5.41 Marital Status** The table gives the number of U.S. citizens (in thousands) who were 15 years of age and older and their marital status in 2009, according to the U.S Census Bureau. Each person is put in a single category on the basis of current status. Thus if a person was married, then became widowed, then remarried, and then divorced, he or she would be listed simply as divorced.

If one person 15 years of age or older is selected randomly from the people in the United States, what is the probability that:

a. The person is widowed?

b. The person is divorced?

c. The person is divorced OR widowed?

| | |
|---|---|
| Married, spouse present | 121,689 |
| Married, spouse absent | 3,348 |
| Widowed | 14,254 |
| Divorced | 23,266 |
| Separated | 5,410 |
| Never married | 72,065 |

**5.42 Marital Status** Refer to table in Exercise 5.41. If one person 15 years of age or older is selected randomly from the people in the United States, what is the probability that:

a. The person is not married (that is, the person is widowed, divorced, or never married)?

b. The person is not classified as widowed?

**\*5.43 Multiple-Choice Exam** An exam consists of 12 multiple-choice questions. Each of the 12 answers is either right or wrong. Suppose the probability that a student makes fewer than 3 mistakes on the exam is 0.48 and that the probability that a student makes from 3 to 8 (inclusive) mistakes is 0.30. Find the probability that a student makes:

a. More than 8 mistakes

b. 3 or more mistakes

c. At most 8 mistakes

d. Which two of these three events are complementary and why?

**\*5.44 Driving Exam** A driving exam consists of 30 multiple-choice questions. Each of the answers is either right or wrong. Suppose that the probability of making fewer than 7 mistakes is 0.23 and the probability of making from 7 to 15 mistakes is 0.41. Find the probability of making:

a. 16 or more mistakes

b. 7 or more mistakes

c. At most 15 mistakes

d. Which two of these three events are complementary? Explain.

## SECTION 5.3

TRY **5.45 Political Party, Again (Example 9)** The table summarizes results from a GSS survey that asked about political party affiliation and self-described political orientation. (Dem means Democrat, and Rep means Republican.)

| | Dem | Rep | Other | Total |
|---|---|---|---|---|
| Liberal | 306 | 26 | 198 | 530 |
| Moderate | 279 | 134 | 322 | 735 |
| Conservative | 104 | 309 | 180 | 593 |
| Total | 689 | 469 | 700 | 1858 |

A person is selected randomly from the sample summarized in the table. We want to find the probability that a conservative person is a Democrat. Which of the following statements best describes the problem? (Choose one.)

i. P(conservative | Democrat) "conservative given Democrat"

ii. P(Democrat | conservative) "Democrat given conservative"

iii. P(Democrat AND conservative)

**5.46 Political Party** Use the table in Exercise 5.45. A person is selected randomly from the sample summarized in the table. We want to determine the probability that the person is a moderate Republican. Which of the following statements best describes the problem?

i. P(moderate | Republican)

ii. P(Republican | moderate)

iii. P(moderate AND Republican)

TRY **5.47 Political Party, Again (Example 10)** Refer to the table for Exercise 5.45.

a. Find the probability that a randomly chosen respondent is a Democrat given that he or she is liberal. In other words, what percentage of the liberals are Democrats?

b. Find the probability that a randomly chosen respondent is a Democrat given he or she is conservative? In other words, what percentage of the conservatives are Democrats?

c. Which respondents are more likely to be Democrats: the liberal or the conservative respondents?

**5.48 Party, Again** Refer to the table for Exercise 5.45.

a. Find the probability that a randomly chosen respondent is conservative given that she or he is a Republican.

b. Find the probability that a randomly chosen respondent is a Republican given that he or she is conservative.

c. Find the probability that a randomly chosen respondent is both conservative AND a Republican.

**5.49 Left-handed** The percentage of left-handed people in the United States is estimated to be 10%. Men are about twice as likely to be left-handed as women. Are gender and handedness independent or associated? Explain.

**5.50 Smoking** According to the group International Network of Women Against Tobacco, 17% of adult women in the United States and 80% of adult women in Greenland smoked in 2007. Are smoking and whether the woman is from the United States or Greenland independent or associated?

**5.51 Independent?** Using your general knowledge, label the following pairs of variables as independent or associated. Explain your reasoning.

a. Hair color and age for adults 20 to 30 years old

b. Hair color and eye color for adults

**5.52 Independent?** A 2009 Gallup poll of over 14,000 people showed that 68% of weekly church attendees felt negatively about the economy. It also showed that 81% of those who seldom or never attended church felt negatively about the economy. According to the poll, are attendance at church and opinion about the economy independent? Explain.

**5.53 Same-Sex Marriage** The 2008 GSS found that 43% of women in the United States agreed that same-sex marriage should be allowed, but only 34% of men felt that way. Suppose that these figures are accurate (or nearly accurate) probabilities, and state whether gender and opinion about same-sex marriage are independent. Explain.

**5.54 Shopping** A survey published in *Vitality Magazine* suggested that 67% of women in the United States thought shopping was relaxing and enjoyable, but only 37% of men felt that way. Assuming these values are accurate, state whether or not gender and opinion about shopping are independent. Explain.

**TRY 5.55 Party: Independent? (Example 11)** Refer to the table in Exercise 5.45. If a person is selected at random, is the event that he or she is a Democrat independent of the event that he or she is liberal?

**\* 5.56 Party** A person is selected randomly from the sample summarized in the table in Exercise 5.45. The probability of being a Democrat given that the person is conservative is 17.5%, the probability of being conservative given that the person is a Democrat is 15.1%, and the probability of being both conservative and a Democrat is 5.6%. Why is the last probability the smallest?

**TRY 5.57 Hand Folding (Example 12)** When people fold their **g** hands together with interlocking fingers, most people are more comfortable with one of two ways. In one way, the right thumb ends up on top and in the other way, the left thumb is on top. The table shows the data from one group of people. M means man, and W means woman; Right means the right thumb is on top, and Left means the left thumb is on top. Judging on the basis of this data set, are the events "right thumb on top" and male independent or associated? Data were collected in a class taught by one of the authors but were simplified for clarity. The conclusion remains the same as that derived from the original data. *See page 238 for guidance.*

| | M | W |
|---|---|---|
| Right | 18 | 42 |
| Left | 12 | 28 |

**5.58 Dice** When two dice are rolled, is the event "the first die shows a 1 on top" independent of the event "the second die shows a 1 on top"?

**TRY 5.59 Happiness and Traditional Views (Example 13)** On the 2008 GSS, people were asked about their happiness and were also asked whether they agreed with the following statement: "In a marriage the husband should work, and the wife should take care of the home." The table summarizes the data collected.

| | Agree | Don't Know | Disagree |
|---|---|---|---|
| Happy | 345 | 261 | 656 |
| Unhappy | 34 | 22 | 44 |

a. Add the marginal totals and the grand total to the table. Show the complete table with marginal totals.

b. Determine whether being happy is independent of agreeing with the statement.

**5.60 Happiness** Using the table in Exercise 5.59, determine whether being unhappy is independent of disagreeing with the statement.

**TRY 5.61 Coin (Example 14)** Imagine flipping three fair coins.

a. What is the theoretical probability that all three will come up heads?

b. What is the theoretical probability the first toss is tails AND the next two are heads?

**5.62 Die** Imagine rolling a fair six-sided die three times.

a. What is the theoretical probability that all three rolls of the die show a 1 on top?

b. What is the theoretical probability that the first roll of the die shows a 6 AND the next two rolls show a 1 on top?

**TRY 5.63 Die Sequences (Example 15)** Roll a fair six-sided die five times and record the number of spots on top. Which sequence is more likely? Explain.

Sequence A: 66666

Sequence B: 16643

**5.64 Babies** Assume that babies born are equally likely to be boys (B) or girls (G). Assume a woman has 6 children, none of whom are twins. Which sequence is more likely? Explain.

Sequence A: GGGGGG

Sequence B: GGGBBB

**TRY 5.65 Seat Belt Use (Example 16)** According to the National Highway Traffic Safety Administration, the rate of seat belt use in the United States for 2009 was 84%. Suppose that you looked at two people, selected randomly and independently from the population, to see whether each had his or her seat belt fastened in 2009.

a. What is the probability that neither person had her or his belt fastened?

b. What is the probability that at least one person had her or his belt fastened? (*Hint:* "At least one had the belt fastened" is the complement of "neither one had the belt fastened.")

**5.66 Gender of Newborns** When a baby is born, the probability that it will be a boy is close to 51%.

a. What is the probability that a newborn will be a girl?

For the rest of the questions, assume that a woman gives birth to two babies who are not identical twins, and therefore the genders are independent.

b. What is the probability that they are both girls?

c. What is the probability of there being at least one boy?

**TRY \* 5.67 Cervical Cancer (Example 17)** According to a study published in *Scientific American*, about 8 women in 100,000 have cervical cancer (C), so P(C) = 0.00008. Suppose the chance that a Pap smear will detect cervical cancer when it is present is 0.84. Therefore,

$$P(\text{test pos} \mid C) = 0.84$$

What is the probability that a randomly chosen woman who has this test will both have cervical cancer AND test positive for it?

**\* 5.68 Cervical Cancer** About 8 women in 100,000 have cervical cancer (C), so P(C) = 0.00008 and P(no C) = 0.99992. The chance that a Pap smear will incorrectly indicate that a woman without cervical cancer has cervical cancer is 0.03. Therefore,

$$P(\text{test pos} \mid \text{no C}) = 0.03$$

What is the probability that a randomly chosen women who has this test will both be free of cervical cancer and test positive for cervical cancer (a false positive)?

## SECTION 5.4

**TRY 5.69 Simulating Coin Flips (Example 18)**

a. Simulate flipping a coin 20 times. Use the line of random numbers below to obtain and report the resulting list of heads and tails. Use odd numbers (1, 3, 5, 7, 9) for heads and even numbers for tails (0, 2, 4, 6, 8).

> 14709   93220   89547   95320

b. Judging on the basis of these 20 trials, what is the empirical probability of getting heads?

**★ 5.70 Simulation**

a. Explain how you could use digits from a random number table to simulate rolling a fair eight-sided die with outcomes 1, 2, 3, 4, 5, 6, 7, and 8 equally likely. Assume that you want to know the probability of getting a 1.

b. Carry out your simulation, beginning with line 5 of the random number table in Appendix A. Perform 20 repetitions of your trial. Using your results, report the empirical probability of getting a 1, and compare it with the theoretical probability of getting a 1.

**5.71 Law of Large Numbers** Refer to Histograms A, B, and C, which show the relative frequencies from experiments in which a fair six-sided die was rolled. One histogram shows the results from 20 rolls, one the results for 100 rolls, and another the results for 10,000 rolls. Which histogram do you think was for 10,000 rolls and why?

**Histogram A**

**Histogram B**

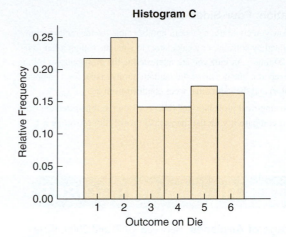

**Histogram C**

**5.72 Law of Large Numbers** The table shows the results of rolling a fair six-sided die.

| Outcome on Die | 20 Trials | 100 Trials | 1000 Trials |
|---|---|---|---|
| 1 | 8 | 20 | 167 |
| 2 | 4 | 23 | 167 |
| 3 | 5 | 13 | 161 |
| 4 | 1 | 13 | 166 |
| 5 | 2 | 16 | 172 |
| 6 | 0 | 15 | 167 |

Using the table, find the empirical probability of rolling a 1 for 20, 100, and 1000 trials. Report the theoretical probability of rolling a 1 with a fair six-sided die. Compare the empirical probabilities to the theoretical probability, and explain what they show.

**5.73 Coin Flips** Imagine flipping a fair coin many times. Explain what should happen to the proportion of heads as the number of coin flips increases.

**5.74 Coin Flips, Again** Refer to the figure.

a. After a large number of flips, the overall proportion of heads "settles down" to nearly what value?

b. Approximately how many coin flips does it take before the proportion of heads settles down?

c. What do we call the law that causes this settling down of the proportion?

d. From the graph, determine whether the first flip was heads or tails.

**Proportion of Heads When Flipping a Fair Coin**

**\* 5.75 Simulation: Four-Sided Die**

a. Explain how you could use a random number table (or the random numbers generated by software or a calculator) to simulate rolling a fair *four-sided* die 20 times. Assume you are interested in the probability of rolling a 1. Then report a line or two of the random number table (or a computer or calculator) and the values that were obtained from it.

b. Report the empirical probability of rolling a 1 on the four-sided die from part a, and compare it with the theoretical probability of rolling a 1.

**\* 5.76 Simulation: Six-Sided Die**

a. Explain how you could use a random number table to simulate rolling a fair six-sided die 20 times. Assume you wish to find the probability of rolling a 1. Then report a line or two of the random number table (or a computer or calculator) and the values that were obtained from it.

b. Report the empirical probability of rolling a 1 from part a, and compare it with the theoretical probability of rolling a 1.

# CHAPTER REVIEW EXERCISES

**5.77 Sabotage of Airplanes** Between 1950 and 2009, there were 1300 fatal commercial airplane accidents worldwide for which a cause could be determined. Of those, 117 were due to sabotage (www.planecrashinfo.com). Using these data, find the empirical probability that sabotage will be the cause of a fatal airplane accident.

**5.78 Pilot Error** Of the 1300 fatal commercial airplane accidents between 1950 and 2009 for which a cause could be determined, 377 were due to simple pilot error (without contributing weather problems or mechanical problems). Using these data, find the empirical probability that simple pilot error will be the cause of a fatal airplane accident.

**5.79 Independent Variables** Use your general knowledge to label the following pairs of variables as independent or associated. Explain.

a. For a sample of adults, gender and shoe size

b. For a sample of football teams, win/loss record for the coin toss at the beginning of the game and number of cheerleaders for the team

**5.80 Independent Variables** Use your general knowledge to label the following pairs of variables as independent or associated. Explain.

a. The outcome on flips of two separate, fair coins

b. Breed of dog and weight of dog for dogs at a dog show

**5.81 UFOs** When two people meet, they are sometimes surprised that they have similar beliefs. A survey of 1003 random adults conducted by the Scripps Survey Research Center at Ohio University in 2008 found that 62 percent of men and 50 percent of women believe in intelligent life on other planets. Well, actually, they said it is either "very likely" or "somewhat likely" that intelligent life exists on other planets (www.reporternews.com).

a. If a man and a woman meet, what is the probability that they both believe in intelligent life on other planets?

b. If a man and a woman meet, what is the probability that neither believes in intelligent life on other planets?

c. What is the probability that the man and woman agree about life on other planets?

d. If a man and a woman meet, what is the probability that they have opposite beliefs on this issue?

**5.82 Seat Belt Use** In 2009, the National Highway Traffic Safety Administration said that 84% of drivers buckled their seat belts. Assume that this percentage is still accurate.

a. If four drivers are randomly selected, what is the probability that they are all wearing their seat belts?

b. If four drivers are randomly selected, what is the probability that at least one is not wearing his or her seat belt?

**5.83 Internet Access** A 2009 Harris poll said that 80% of adults in the United States have Internet access. Assume that this is still correct.

a. If two people are randomly selected, what is the probability that they both have Internet access?

b. If the two people chosen were a married couple living in the same residence, explain why they would not be considered independent with regard to Internet access.

**5.84 SAT Scores** The probability of a randomly selected person having a grade of 500 or above on the quantitative portion of the SAT is 0.50.

a. If two students are chosen randomly and independently, what is the probability that they both score 500 or above?

b. If two students are selected from the same high school calculus class, do you think the probability of their both scoring 500 or above is different from your answer to part a? Explain.

**\* 5.85 Birthdays** Suppose all the days of the week are equally likely as birthdays. Alicia and David are two randomly selected, unrelated people.

a. What is the probability that they were both born on Monday?

b. What is the probability that Alicia OR David was born on Monday? *Hint:* The answer is not 2/7. Refer to Guided Exercise 5.25 if you need help.

**\* 5.86 Pass Rate of Written Driver's Exam** In California, about 92% of teens who take the written driver's exam fail the first time they take it (www.teendrivingcourse.com). Suppose that Sam and Maria are randomly selected teenagers taking the test for the first time.

a. What is the probability that they both pass the test?

b. What is the probability that Sam OR Maria passes the test?

**5.87 Opinion about Nurses** A Gallup Poll from December of 2009 estimated that 83% of all people thought nurses had high or very high ethical standards, putting nurses at the top of the professions with regard to this issue. If this rate is still correct and a new poll of 5000 people were obtained, how many out of those 5000 would you expect to think nurses have high or very high ethical standards?

**5.88 Climate Change** A Gallup poll from December of 2009 asked whether U.S. residents who are aware of climate change thought their government was doing enough to reduce emissions of cars and factories. Gallup estimated that 52% of all U.S residents felt that the government was not doing enough. If another poll were taken and there were 500 participants, how many would you expect to say that the United States was not doing enough, assuming the percentage remained the same?

* **5.89 Independent** Imagine rolling a red die and a blue die. From this trial, name a pair of independent events.

* **5.90 Mutually Exclusive** Imagine rolling a red die and a blue die. From this trial, name a pair of mutually exclusive events.

* **5.91** Construct a two-way table with 60 women and 80 men in which both groups show equal percentages of right-handedness.

* **5.92** Construct a two-way table with 60 women and 80 men in which there is a higher percentage of right-handed women.

* **5.93 Simulating Guessing on a Multiple-Choice Test**
Suppose a student takes a 10-question multiple-choice quiz, and for each question on the quiz there are five possible options. Only one option is correct. Now suppose the student, who did not study, guesses at random for each question. A passing grade is 3 (or more) correct. We wish to design a simulation to find the probability that a student who is guessing can pass the exam.

a. In this simulation, a single trial consists of a "student" guessing on 10 multiple-choice "questions." We will simulate this by selecting 10 single digits from the random number table. Using the random number table below, use 0 and 1 to represent correct answers, and use 2 through 9 to represent incorrect answers. Explain why this is a fair approach for the exam questions with five possible answers. (This completes the first two steps of the simulation given in Section 5.4.)

b. Write the sequence of numbers from the first trial. Also write how this translates to correct and incorrect answers.

c. Step 3 in a simulation is to identify the response variable. The outcome here is the number of correct responses the student gets. What was the outcome for the first trial?

d. Step 4 is to describe the event of interest. In this simulation, this is whether or not the student got 3 or more questions correct. Did we achieve that in the first trial?

e. Perform a second simulation of the student taking this 10-question quiz by guessing randomly. Use the second line of the table given. What score did your student get? Did the event of interest occur this time?

f. Repeat the trial twice more, using lines 3 and 4 below. On the basis of these four trials, what is the empirical probability of passing the exam by guessing?

```
11373   96871
52022   59093
14709   93220
31867   85872
```

* **5.94 Simulating Guessing on a True/False Test** Perform a simulation of a student guessing on a true/false quiz with 10 questions. You will use the same four lines that are given for the preceding question. Write out each of the six steps outlined in Section 5.4. What numbers will you use to represent correct answers and what numbers for incorrect answers? Explain why your choice is logical. Do four repetitions, each trial consisting of 10 questions. Find the empirical probability of getting more than 5 correct out of 10.

* **5.95 Law of Large Numbers** A famous study by Amos Tversky and Nobel laureate Daniel Kahneman asked people to consider two hospitals. Hospital A is small and has 15 babies born per day. Hospital B has 45 babies born each day. Over one year, each hospital recorded the number of days that it had more than 60% girls born. Assuming that 50% of all babies are girls, which hospital had the most such days? Or do you think both will have about the same number of days with more than 60% girls born? Answer, and explain. (Source: Amos Tversky. 2004. *Preference, Belief, and Similarity: Selected Writings*, ed. Eldar Shafir. Cambridge, Mass.: MIT Press, p. 205)

* **5.96 Law of Large Numbers** A certain professional basketball player typically makes 80% of his basket attempts, which is considered to be good. Suppose you go to several games at which this player plays. Sometimes the player attempts only a few baskets, say 10. Other times, he attempts about 60. On which of those nights is the player most likely to have a "bad" night, in which he makes much fewer than 80% of his baskets?

* **5.97 Red Light/Green Light** A busy street has three traffic lights in a row. These lights are not synchronized, so they run independently of each other. At any given moment, the probability that the light is green is 60%. Assuming there is no traffic, follow the steps below to design a simulation to estimate the probability that you will get three green lights.

a. Identify the actions that make up a single trial on the busy street.

b. Using the random number table in Appendix A, describe how to simulate a single trial. Which digits will represent a green light and which a non-green light?

c. The response variable is the number of "green lights" in one trial. Describe the event of interest—that is, the event for which we wish to estimate a probability.

d. Carry out 20 repetitions of your trial, beginning with the first digit on line 11 of the random number table. For each trial, list the random digits, the outcomes they represent, and the value of the response variable.

e. What is the empirical probability that you get three green lights?

* **5.98 Soda** A soda-bottling plant has a flaw in that 20% of the bottles it fills do not have enough soda in them. The sodas are sold in six-packs. Follow these steps to carry out a simulation to find the probability that three or more bottles in a six-pack have not enough soda.

a. Identify the actions that make up a single trial for a six-pack.

b. Using the random number table in Appendix A, describe how to simulate a single trial. Which digits will represent a "good" soda bottle and which will represent a "bad" (not full) soda bottle?

c. What is the response variable?

d. Describe the event of interest—that is, the event for which we wish to estimate a probability.

e. Carry out 10 repetitions of your trial, beginning with the first digit on line 15 of the random number table. For each trial, list the random digits, the outcomes they represent, and the value of the response variable.

f. What is the experimental probability that you get three or more "bad" bottles in a six-pack?

# gUIDED EXERCISES

g **5.25 GSS: OR** Referring to the table, assume one person is chosen randomly from the 1858 people in the table. Answer the question below the table by following the numbered steps.

|  | Dem | Rep | Other | Total |
|---|---|---|---|---|
| Liberal | 306 | 26 | 198 | 530 |
| Moderate | 279 | 134 | 322 | 735 |
| Conservative | 104 | 309 | 180 | 593 |
| Total | 689 | 469 | 700 | 1858 |

**QUESTION** What is the probability that the person is liberal OR a Democrat?

**Step 1 ▶** What is the probability that the person is liberal?

**Step 2 ▶** What is the probability that the person is a Democrat?

**Step 3 ▶** If the events liberal and a Democrat were mutually exclusive, you could just add the probabilities from step 1 and step 2 to find the probability that a person is liberal OR a Democrat. Are they mutually exclusive? Why or why not?

**Step 4 ▶** What is the probability that the person is liberal AND a Democrat? Use numbers from the table; do not assume independence.

**Step 5 ▶** Why should you subtract the probability that the person is liberal AND a Democrat from the sum as shown in the given formula?

$$P(\text{liberal OR a Democrat}) = P(\text{liberal}) + P(\text{a Democrat})$$
$$- P(\text{liberal AND a Democrat})$$

**Step 6 ▶** Do the calculation using the formula given in step 5 to find the probability that a person is liberal OR a Democrat. It is more accurate to use the numbers from the table (fractions) than to use your percentages, which are rounded off.

**Step 7 ▶** Finally, report the answer in a sentence.

g **5.57 Hand Folding** When people fold their hands together with interlocking fingers, most people are more comfortable doing it in one of two ways. In one way, the right thumb ends up on top, and in the other way, the left thumb is on top. The table shows the data from one group of people.

|  | M | W |
|---|---|---|
| Right | 18 | 42 |
| Left | 12 | 28 |

M means man, W means woman, Right means the right thumb is on top, and Left means the left thumb is on top.

**QUESTION** Say a person is selected from this group at random. Are the events "right thumb on top" and "male" independent or associated?

To answer, we need to determine whether the probability of having the right thumb on top given that you are a man is equal to the probability of having the right thumb on top (for the entire group). If so, the variables are independent.

**Step 1 ▶** Figure out the marginal totals and put them into the table.

**Step 2 ▶** Find the overall probability that the person's right thumb is on top.

**Step 3 ▶** Find the probability that the right thumb is on top given that the person is a man. (What percentage of men have the right thumb on top?)

**Step 4 ▶** Finally, are the variables independent? Why or why not?

# TechTips

## For All Technology

**EXAMPLE: GENERATING RANDOM INTEGERS** ▶ Generate four random integers from 1 to 6, for simulating the results of rolling a six-sided die.

### TI-83/84

**Seed First before the Random Integers**

If you do not seed the calculator, everyone might get the same series of "random" numbers.

1. Enter the last four digits of your Social Security number or cell phone number and press **STO →** .
2. Then press **MATH**, choose **PRB**, and **ENTER ENTER** (to choose 1:rand).

You only need to seed the calculator once, unless you **Reset** the calculator. (If you want the same sequence later on, you can seed again with the same number.)

**Random Integers**

1. Press **MATH**, choose **PRB**, and press **5** (to choose 5:randInt).
2. See Figure 5A. Enter **1 , 6 , 4)** and press **ENTER**. (The comma button is above the 7 button.)

▲ **FIGURE 5A** TI-83/84

The first two digits (1 and 6 in Figure 5A) determine the smallest and largest integers, and the third digit (4 in Figure 5A) determines the number of random integers generated. To get four more random integers, press **ENTER** again.

### MINITAB

**Random Integers**

1. **Calc > Random Data > Integer**
2. See Figure 5B. Enter:

   **Number of rows to generate**, **4**

   **Store in column(s)**, **c1**

   **Minimum value**, **1**

   **Maximum value**, **6**
3. Click **OK**.

▲ **FIGURE 5B** Minitab

### EXCEL

**Random Integers**

1. Click **fx**, select a category **All**, and **RANDBETWEEN**.
2. See Figure 5C.

   Enter: **Bottom**, **1**; **Top**, **6**.

   Click **OK**.

   You will get one random integer in the active cell in the spreadsheet.
3. To get more random integers, put the cursor at the lower right corner of the cell containing the first integer until you see a black cross (+), and drag downward until you have as many as you need.

▲ **FIGURE 5C** Excel

**239**

## Random Integers

1. **Data** > **Simulate Data** > **Discrete Uniform**.
2. See Figure 5D.

   Enter **Rows**, **4**; **Columns**, **1**; **minimum**, **1**; **maximum**, **6** and leave **Split across columns** and **Use preset generator**.
3. Click **Simulate**.

   You will get four random integers (from 1 to 6) in the first empty column.

▲ **FIGURE 5D** StatCrunch

# Modeling Random Events: The Normal and Binomial Models

## THEME

Probability distributions describe random outcomes in much the same way as the distributions we discussed in Chapter 2 describe collections of data. A probability distribution tells us the possible outcomes of a random trial and the probability that each of those outcomes will occur.

Random events can seem chaotic and unpredictable. If you flip a coin 10 times, there's no way you can say with absolute certainty how many heads will appear. There's no way your local weather forecaster can tell you with certainty whether it will rain tomorrow. Still, if we watch enough of these random events, patterns begin to emerge. We begin to learn how often different outcomes occur and to gain an understanding of what is common and what is unusual.

Science fiction writer Isaac Asimov is often quoted as having said, "The most exciting phrase to hear in science, the one that heralds new discoveries, is not Eureka! (I found it!) but rather 'hmm . . . that's funny . . . .'" When an outcome strikes us "funny" or unusual, it is because something unlikely occurred. This is exciting because it means a discovery has been made; the world is not the way we thought it was!

However, to know whether something unusual has happened, we first have to know how often it usually occurs. In other words, we need to know the probability. In the case study, you'll see a "homemade" example, in which one of the authors found that five ice cream cones ordered at a local McDonald's weighed more than the advertised amount. Was this a common occurrence, or did McDonald's perhaps deliberately understate the weight of the cones so that no one would ever be disappointed?

In this chapter we'll introduce a new tool to help us characterize probabilities for random events: the probability distribution. We'll examine the Normal probability distribution and the binomial probability distribution, two very useful tools for answering the question "Is this unusual?"

## CASE STUDY

### You Sometimes Get More Than You Pay For

A McDonald's restaurant near one of the authors sells ice cream cones, which, according to the "fact sheet" provided, weigh 3.18 ounces (converted from grams) and contain 150 calories. Do the ice cream cones really weigh exactly 3.18 ounces? To get 3.18 ounces for every cone would require a very fine-tuned machine or an employee with a *very* good sense of timing. Thus, we expect some natural variation in the weight of these cones. In fact, one of the authors bought five ice cream cones on different days. She found that each of the five cones weighed more than 3.18 ounces. Such an outcome, five out of five over the advertised weight, might occur just by chance. But how often? If such an outcome (five out of five) rarely happens, it's pretty surprising that it happened to us. In that case, perhaps McDonald's actually puts in more than 3.18 ounces. By the end of this chapter, after learning about the binomial and Normal probability distributions, you will be able to measure how surprising this outcome is—or is not.

# Probability Distributions Are Models of Random Experiments

A **probability model** is a description of how a statistician thinks data are produced. We use the word *model* to remind us that our description does not really explain how the data came into existence, but we hope that it describes the actual process fairly closely. We can tell whether a model is good by noting whether the probabilities that it predicts are matched by real-life outcomes. Thus, if a model says that the probability of getting heads when we flip a coin is 0.54, but in fact we get heads 50% of the time, we suspect that the model is not a good match.

A **probability distribution**, sometimes called a **probability distribution function (pdf),** is a tool that helps us by keeping track of the outcomes of a random experiment and the probabilities associated with those outcomes. For example, suppose the playlist on your mp3 player has 10 songs: 6 are Rock, 2 are Country, 1 is Hip-hop, and 1 is Opera. Put your player on shuffle. What is the probability that the first song chosen is Rock?

The way the question is worded ("What is the probability . . . is Rock"?) means that we care about only two things: Is the song classified as Rock or is it not? We could write the probabilities as shown in Table 6.1.

Table 6.1 is a very simple probability distribution. It has two important features: It tells us all the possible outcomes of our random experiment (Rock or Not Rock), and it tells us the probability of each of these outcomes. All probability distributions have these two features, though they are not always listed so clearly.

| Outcome | Probability |
|---------|-------------|
| Rock | 6/10 |
| Not Rock | 4/10 |

▲ **TABLE 6.1** Probability Distribution of Songs

**KEY POINT**    A probability distribution tells us (1) all the possible outcomes of a random experiment, and (2) the probability of each outcome.

↻ **Looking Back**

**Distributions of a Sample**
Probability distributions are similar to distributions of a sample, which were introduced in Chapter 2. A distribution of a sample tells us the values in the sample and their frequency. A probability distribution tells us the possible values of the random experiment and their probability.

In Chapter 1, we classified variables as either numerical or categorical. It is now useful to break the numerical variables down into two more categories. **Discrete outcomes** (or discrete variables) are numerical values that you can list or count. An example is the number of phone numbers stored on the phones of your classmates. **Continuous outcomes** (or continuous variables) cannot be listed or counted because they occur over a range. The length of time your next phone call will last is a continuous variable. Refer to Figure 6.1 for a visual comparison of these terms.

This distinction is important because if we can list the outcomes, as we can for a discrete variable, then we have a nice way of displaying the probability distribution. However, if we are working with a continuous variable, then we can't list the outcomes, and we have to be a bit more clever in describing the probability distribution function. For this reason, we treat discrete values separately from continuous variables.

### EXAMPLE 1 Discrete or Continuous?

Consider these variables:

a. The weight of a submarine sandwich you're served at a deli.

b. The elapsed time from when you leave your house to when you arrived in class this morning.

c. The number of people in the next passing car.

d. The blood alcohol level of a driver pulled over by the police in a random sobriety check. (Blood alcohol level is measured as the percent of the blood that is alcohol.)

e. The number of eggs laid by a randomly selected salmon as observed in a fishery.

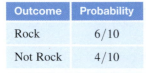

▲ **FIGURE 6.1** Visual representation of discrete and continuous changes in elevation. Note that for the staircase (discrete outcomes), you can count the stairs.

**QUESTION** Identify each of these numerical variables as continuous or discrete.

**SOLUTION** The continuous variables are variables a, b, and d. Continuous variables can take on any value in a spectrum. For example, the sandwich might weight 6 ounces or 6.1 ounces or 6.0013 ounces. The amount of time it takes you to get to class could be 5000 seconds or 5000.4 seconds or 5000.456. Blood-alcohol content can be any value between 0 and 1, including 0.13, 0.13333, 0.1357, and so on.

Variables c and e, on the other hand, are discrete quantities—that is, numbers that can be counted.

**TRY THIS!** Exercise 6.1

## Discrete Probability Distributions Can Be Tables or Graphs

A statistics class at UCLA was approximately 40% male and 60% female. Let's arbitrarily code the males as 0 and the females as 1. If we select a person at random, what is the probability that the person is female?

Creating a probability distribution for this situation is as easy as listing both outcomes (0 and 1) and their probabilities (0.40 and 0.60). The easiest way to do this is in a table, as shown as Table 6.2. However, we could also do this as a graph (Figure 6.2).

| Female | Probability |
|--------|-------------|
| 0 | 0.40 |
| 1 | 0.60 |

▲ **TABLE 6.2** Probability Distribution of Gender in a Class

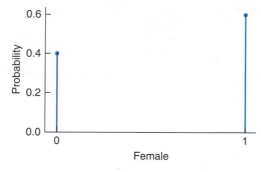

◀ **FIGURE 6.2** Probability distribution for selecting a person at random from a particular statistics class and recording whether the person is male (0) or female (1).

Amazon.com invites customers to rate books on a scale of 1 to 5 stars. At a visit to the site on July 5, 2010, *The Story of Edgar Sawtelle: A Novel*, by David Wroblewski, had received 1505 customer reviews. Suppose we randomly select a reviewer and base our decision whether to buy the book on how many stars that person gave to it. Table 6.3 and Figure 6.3 illustrate two different ways of representing the probability distribution for this random event. For example, we see that the most likely outcome is that the reviewer gave the book 1 star. The probability that the reviewer gave the book 1 star is 0.290, or about 29%.

| Number of Stars | Probability |
|-----------------|-------------|
| 1 | 0.290 |
| 2 | 0.146 |
| 3 | 0.163 |
| 4 | 0.195 |
| 5 | 0.206 |

▲ **TABLE 6.3** Distribution of Number of Stars

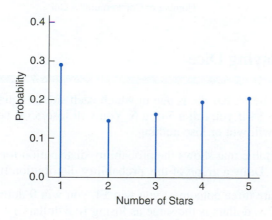

◀ **FIGURE 6.3** Probability distribution for the number of stars given, as a rating, to a particular book on amazon.com by a randomly selected customer.

Note that the probabilities in Table 6.3 add to 1. This is true of all probability distributions: When you add up the probabilities for all possible outcomes, they add to 1. They have to, because there are no other possibilities.

## Discrete Distributions Can Also Be Equations

What if we have too many outcomes to list in a table? For example, suppose a married couple decides to keep having children until they have a girl. How many children will they have, assuming that boys and girls are equally likely and that the gender of one birth doesn't depend on any of the previous births? It could very well turn out that their first child is a girl and they therefore have only one child. Or that the first is a boy but the second is a girl. Or, just possibly, they might never have a girl. The value of this experiment could be any number 1, 2, 3, . . . up to infinity. (Okay, in reality, it's impossible to have that many children. But we can imagine!)

We can't list all these values and probabilities in a table, and we can only hint at what the graph might look like. But we *can* write them in a formula:

The probability of having $x$ children before the first girl is $(1/2)^x$.

For example, the probability that they have 1 child is $(1/2)^1 = 1/2$.

The probability that they have 4 children is $(1/2)^4 = 1/16$.

The probability that they have 10 children is (thankfully) small: $(1/2)^{10} = 0.00098$.

In this book, we will give the probabilities in a table or graph whenever convenient. In fact, even if it's not especially convenient, we will often provide a graph to encourage you to visualize what the probability distribution looks like. Figure 6.4 is part of the graph of the probability distribution for the number of children a couple can have before having a girl. We see that the probability that the first child is a girl is 0.50 (50%). The probability that the couple has two children is half this: 0.25. The probabilities continue to decrease, and each probability is half the one before it.

▶ **FIGURE 6.4** Probability distribution of the number of children born until the first girl.

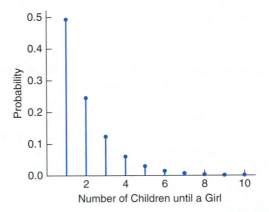

## EXAMPLE 2 Playing Dice

Roll a fair six-sided die. A fair die is one in which each side is equally likely to end up on top. You will win $4 if you roll a 5 or a 6. You will lose $5 if you roll a 1. For any other outcome, you will win or lose nothing.

**QUESTION** Give a table that shows the probability distribution for the amount of money you will win. Draw a graph of this probability distribution function.

**SOLUTION** There are three outcomes: you win $4, you win 0 dollars, you win −$5. (Winning negative five dollars is the same as losing five dollars.)

You win $4 if you roll a 5 or 6, so the probability is $2/6 = 1/3$.

You win $0 if you roll a 2, 3, or 4, so the probability is $3/6 = 1/2$

You "win" $-$5$ if you roll a 1, so the probability is $1/6$.

We can put the pdf in a table (Table 6.4), or we can represent the pdf as a graph (Figure 6.5).

◀ **FIGURE 6.5** Probability density function of the dice game.

| Winnings | Probability |
|----------|-------------|
| −5 | 1/6 |
| 0 | 1/2 |
| 4 | 1/3 |

▲ **TABLE 6.4** Probability density function of the dice game (Example 2).

( TRY THIS! )  Exercise 6.5

# Continuous Probabilities Are Represented as Areas under Curves

Finding probabilities for continuous outcomes is more complicated, because we cannot simply list all the possible values we might see. What we can list is the *range of values* we might see.

For example, suppose you want to know the probability that you will wait in line for between 3 and 4 minutes when you go to the coffee shop. You can't list all possible outcomes that could result from your visit: 1.0 minute, 1.00032 minute, 2.00000321 minute. It would take (literally) an eternity. But you can specify a range. Suppose this particular coffee shop has done extensive research and knows that everyone gets helped in under 5 minutes. So all customers get helped within the range of 0 to 5 minutes.

If we want to find probabilities concerning a continuous-valued experiment, we also need to give a range for the outcomes. For example, the manager wants to know the probability that a customer will wait less than 2 minutes; this gives a range of 0 to 2 minutes.

The probabilities for a continuous-valued random experiment are represented as areas under curves. The curve is called a **probability density curve**. The total area under the curve is 1, because this represents the probability that the outcome will be somewhere on the *x*-axis. To find the probability of waiting between 0 and 2 minutes, we find the area under the density curve and between 0 and 2 (Figure 6.6).

◀ **FIGURE 6.6** Probability distribution of times waiting in line at a particular coffee shop.

The $y$-axis in a continuous-valued pdf is labeled "Density." How the density is determined isn't important. What *is* important to know is that the units of density are scaled so that the area under the entire curve is 1.

You might wonder where this curve came from. How did we know the distribution was exactly this shape? In practice, it is very difficult to know which curve is correct for a real-life situation. Statisticians call these curves "probability models" because they are meant to mimic a real-life probability, but we don't know—and can never know for sure—that the curve is correct. On the other hand, we can compare our probability predictions to the actual frequencies that we see. If they are close, then our probability model is good. For example, if this probability model predicts that 45% of customers get coffee within 2 minutes, then we can compare this prediction to an actual sample of customers.

## Finding Probabilities for Continuous-Valued Outcomes

Calculating the area under a curve is not easy. If you have a formula for the probability density, then you can sometimes apply techniques from calculus to find the area. However, for many commonly used probability densities, basic calculus is not helpful, and computer-based approximations are required.

In this book, you will always find areas for continuous-valued outcomes by using a table or by using technology. In Section 6.2 we introduce a table that can be used to find areas for one type of probability density that is very common in practice: the Normal curve.

### EXAMPLE 3 Waiting for the Bus

The bus that runs near one of the authors' home arrives every 12 minutes. If the author arrives at the bus stop at a randomly chosen time, then the probability distribution for the number of minutes he must wait for the bus is shown in Figure 6.7.

▶ **FIGURE 6.7** Probability distribution function showing the number of minutes the author must wait for the bus if he arrives at a randomly determined time.

**QUESTION** Find the probability that the author will have to wait between 4 and 10 minutes for the bus. (*Hint:* Remember that the area of a rectangle is the product of the lengths of the two sides.)

**SOLUTION** The distribution shown in Figure 6.7 is called a uniform distribution. Finding areas under this curve is easy because the curve is just a rectangular shape. The area we need to find is shown in Figure 6.8. The area of a rectangle is width times height. The width is $(10 - 4) = 6$, and the height is 0.08333.

The probability that the author must wait between 4 and 10 minutes is $6 \times 0.08333 = 0.4998$, or about 0.500. Visually, we see that about half of the area in Figure 6.8 is shaded.

◀ **FIGURE 6.8** The shaded area represents the probability that the author will wait between 4 and 10 minutes if he arrives at the bus stop at a randomly determined time.

There is approximately a 50% chance that the author must wait between 4 and 10 minutes.

**TRY THIS!** Exercise 6.9

# The Normal Model

The **Normal model** is the most widely used probability model for continuous numerical variables. One reason is that many numerical variables in which researchers have historically been interested have distributions for which the Normal model provides a very close fit. Also, an important mathematical theorem (the Central Limit Theorem, to be introduced in Chapter 7) links the Normal model to several key statistical ideas, which provides good motivation for learning this model.

We begin by showing you what the Normal model looks like. Then we discuss how to find probabilities by finding areas underneath the Normal curve. We will illustrate these concepts with examples and also discuss why the Normal model is appropriate for these situations.

## Visualizing the Normal Distribution

Figure 6.9 on the next page shows several histograms of measurements taken from a sample of about 1400 adult men in the United States. All of these graphs have similar shapes: They are unimodal and symmetric. We have superimposed smooth curves over the histograms that capture this shape. You could easily imagine that if we continued to collect more and more data, the histogram would eventually fill in the curve and match the shape almost exactly.

The curve drawn on these histograms is called the **Normal curve**, or the **Normal distribution**. It is also sometimes called the Gaussian distribution, after Karl Friedrich Gauss (1777–1855), the mathematician who first derived the formula. Statisticians and scientists recognized that this curve provided a model that pretty closely described a good number of continuous-valued data distributions. Today, even though we have many other distributions to model real-life data, the Normal curve is still one of the most frequently used probability distribution functions in science.

## Center and Spread

In Chapters 2 and 3 we discussed the center and spread of a distribution of *data*. These concepts are also useful for studying distributions of *probability*. The mean of a probability distribution sits at the balancing point of the probability distribution. The

**Looking Back**

**Unimodal and Symmetric Distributions**
Symmetric distributions have histograms whose right and left sides are roughly mirror images of each other. Unimodal distributions have histograms with one mound.

**Details**

**The Bell Curve**
The Normal or Gaussian curve is also called the bell curve.

**(a)**

**(b)**

**(c)**

▲ **FIGURE 6.9** Measurements for a sample of men: **(a)** cholesterol levels, **(b)** diastolic blood pressure readings, and **(c)** height. For all three, a very similar shape appears: nearly symmetric and mound-shaped.

standard deviation of a probability distribution measures the spread of the distribution by telling us how far away, typically, the values are from the mean. The conceptual understanding you developed for the mean and standard deviation of a sample still apply to probability distributions.

The notation we use is slightly different, so that we can distinguish means and standard deviations of probability distributions from means and standard deviations of data. The **mean of a probability distribution** is represented by the Greek character $\mu$ (mu, pronounced "mew"), and the **standard deviation of a probability distribution** is represented by the character $\sigma$ (sigma). These Greek characters are used to avoid confusion of these concepts with their counterparts for samples of data, $\bar{x}$ and $s$.

### The Mean and Standard Deviation of a Normal Distribution

The exact shape of the Normal distribution is determined by the values of the mean and the standard deviation. Because the Normal distribution is symmetric, the mean is in the exact center of the distribution. The standard deviation determines whether the Normal curve is wide and low (large standard deviation) or narrow and tall (small standard deviation). Figure 6.10 shows two Normal curves that have the same mean but different standard deviations.

 **Looking Back**

**Mean and Standard Deviation**
In Chapter 3, you learned that the symbol for the mean of a *sample* of data is $\bar{x}$ and that the symbol for the standard deviation of a sample of data is $s$.

▶ **FIGURE 6.10** Two Normal curves with the same mean but different standard deviations.

A Normal curve with a mean of 69 inches and a standard deviation of 3 inches provides a very good match to the distribution of heights of all adult men in the United States. Surprisingly, a Normal curve with the same standard deviation of 3 inches, but a smaller mean of about 64 inches, describes the distribution of adult women's heights. Figure 6.11 shows what these Normal curves look like.

◀ **FIGURE 6.11** Two Normal curves. The blue curve represents the distribution of women's heights and has a mean of 64 inches and a standard deviation of 3 inches. The red curve represents the distribution of men's heights and has the same standard deviation, but the mean is 69 inches.

The only way to distinguish among different Normal distributions is by their means and standard deviations. We can take advantage of this fact to write a short-hand notation to represent a particular Normal distribution. The notation $N(\mu, \sigma)$ represents a Normal distribution that is centered at the value of $\mu$ (the mean of the distribution) and whose spread is measured by the value of $\sigma$ (the standard deviation of the distribution). For example, in Figure 6.11, the distribution of women's heights is $N(64, 3)$, and the distribution of men's heights is $N(69, 3)$.

**KEY POINT** ▶ The Normal distribution is symmetric and unimodal ("bell-shaped"). The notation $N(\mu, \sigma)$ tells us the mean and standard deviation of the Normal distribution.

## Finding Normal Probabilities

The Normal model $N(64, 3)$ gives a good approximation of the distribution of adult women's heights in the United States (where height is measured in inches). Suppose we were to select an adult woman from the United States at random and record her height. What is the probability that she is taller than a specified height?

Because height is a continuous numerical variable, we can answer this question by finding the appropriate area under the Normal curve. For example, Figure 6.12 shows a Normal curve that models the distribution of heights of women in the population—the same curve as in Figure 6.11. (We leave the numerical scale off the vertical axis from now on since it is not needed for doing calculations.) The area of the shaded region gives us the probability of selecting a woman taller than 62 inches. The entire area under the curve is 1.

In fact, Figure 6.12 represents both the probability of selecting a woman taller than 62 inches and also the probability of selecting a woman *62 inches tall or taller*. Because the areas for both regions (the one that is strictly greater than 62, and the other that includes 62) are the same, the probabilities are also the same. This is a convenient feature of continuous variables: We don't have to be too picky about our language when working with probabilities. This is in marked contrast with discrete variables, as you will soon see.

◀ **FIGURE 6.12** The area of the shaded region represents the probability of finding a woman taller than 62 inches from a $N(64, 3)$ distribution.

What if we instead wanted to know the probability that the chosen woman would be between 62 inches and 67 inches tall? That area would look like Figure 6.13.

▶ **FIGURE 6.13** The shaded area represents the probability that a randomly selected woman is between 62 and 67 inches tall. The probability distribution shown is the Normal distribution with mean 64 inches and standard deviation 3 inches.

Figure 6.14 shows the area corresponding to the probability that this randomly selected woman is less than 62 inches tall.

▶ **FIGURE 6.14** The shaded area represents the probability that the randomly selected woman is less than 62 inches tall.

**KEY POINT** ▶ When you are finding probabilities with Normal models, the first and most helpful step is to sketch the curve, label it appropriately, and shade in the region of interest.

We recommend that you always begin problems that concern Normal models by drawing a picture. One advantage to drawing a picture is sometimes that's all you need. For example, according to the McDonald's "Fact Sheet," each serving of ice cream in a cone weighs 3.18 ounces. Now, we know that in real life it is not possible to serve exactly 3.18 ounces. Probably the employees operating the machines (or the machines themselves) actually dispense a little more or a little less than 3.18 ounces. Suppose that the amount of ice cream dispensed follows a Normal distribution with a mean of 3.18 ounces. What is the probability that a hungry customer will actually get less than 3.18 ounces?

Figure 6.15a shows the situation. From it we easily see that the area to the left of 3.18 is exactly half of the total area. Thus we know that the probability of getting less than 3.18 ounces is 0.50. (We also know this because the Normal curve is symmetric, so the mean—the balancing point—must sit right in the middle. Therefore, the probability of getting a value less than the mean is 0.50.)

What if the true mean is actually larger than 3.18 ounces? How will that affect the probability of getting a cone that weighs less than 3.18 ounces? Imagine "sliding" the Normal curve to the right, which corresponds to increasing the mean. Does the area to the left of 3.18 go up or down as the curve slides to the right? Figure 6.15b shows that the area below 3.18 is now smaller than 50%. The larger the mean amount of ice cream dispensed, the less likely it is that a customer will complain about getting too little.

**(a)**

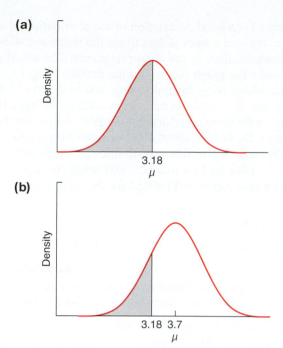

**(b)**

◄ **FIGURE 6.15 (a)** A $N$(3.18, 0.6) curve, showing that the probability of getting a cone that weighs less than 3.18 ounces is 50%. **(b)** A Normal curve with the same standard deviation (0.6) but a larger mean (3.7). The area below 3.18 is now much smaller.

## Finding Probability with Technology

Finding the area of a Normal distribution is best done with technology. Most calculators and many software packages will show you how to do this. We illustrate one such package, available free on the Internet, that you can use. We will also show you the "old-fashioned" way, which is useful when you do not have a computer handy. The old-fashioned way is also worth learning because it helps solidify your conceptual understanding of the Normal model.

Figure 6.16a is a screenshot from the SOCR (http://socr.stat.ucla.edu) calculator. It shows the probability that a randomly selected woman is between 62 and 67 inches

**(a)**

**Distribution Properties**
Normal (64.0, 3.0) Distribution
Mean: 64.000000
Median: 64.000000
Variance: 9.000000
Standard Deviation: 3.000000
Max Density: .132981

**Probabilities**

Left: .252493
Between (Red-Shaded): .588852
Right: .158655

**(b)**

**Distribution Properties**
Normal (64.0, 3.0) Distribution
Mean: 64.000000
Median: 64.000000
Variance: 9.000000
Standard Deviation: 3.000000
Max Density: .132981

**Probabilities**

Left: .252493
Between (Red-Shaded): .588852
Right: .158655

▲ **FIGURE 6.16 (a)** Screenshot showing a calculation of the area under a $N$(64, 3) curve between 62 and 67. **(b)** An enlarged view of the window below the graph, which shows the shaded and unshaded areas. The shaded area is given as .588852. The area to its left is given as .252493. (We will usually insert a 0 before the decimal point when we report or work with these values.)

tall if the $N(64, 3)$ model is a good description of the distribution of women's heights. To use the calculator, the user moves sliders to set the mean and standard deviation and then uses the mouse to shade in the appropriate area. The actual probability is given in a window below the graph (shown on the previous page).

This particular calculator prints the probability next to the word "Between": .588852. Thus, the probability that the woman is *between* 62 and 67 inches tall is about 59%. Note that we can also, with no extra effort, get the probability that this person will be shorter than 62 inches (.252493, or about 25%) or taller than 67 inches (about 16%.)

To use the SOCR calculator to find probabilities for the Normal model, go to http://socr.stat.ucla.edu, click on Distributions, and select Normal.

Figure 6.17 shows output from a TI-83/84 for the same calculation.

```
normalcdf(62,67,
64,3)
        .5888522734
```

▲ **FIGURE 6.17** TI-83/84 output showing that the probability that the woman is between 62 and 67 inches tall is about 59%.

## EXAMPLE 4  Baby Seals

Some research has shown that the mean length of a newborn Pacific harbor seal is 29.5 inches ($\mu = 29.5$ inches) and that the standard deviation is $\sigma = 1.2$ inches. Suppose that the lengths of these seal pups follow the Normal model.

**QUESTION**  Using the output in Figure 6.18, what is the probability that a randomly selected harbor seal pup is within 1 standard deviation of the mean length of 29.5 inches?

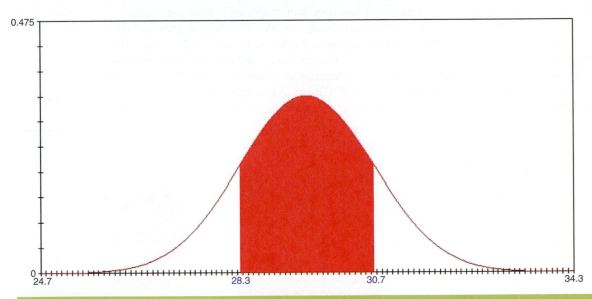

Distribution Properties
Normal (29.5, 1.2) Distribution
Mean: 29.500000
Median: 29.500000
Variance: 1.440000
Standard Deviation: 1.200000
Max Density: .332452

Probabilities

Left: .158655
Between (Red-Shaded): .682689
Right: .158655

▲ **FIGURE 6.18** The shaded region represents the area under the curve between 28.3 inches and 30.7 inches. In other words, it represents the probability that the length of a randomly selected seal pup is within 1 standard deviation of the mean.

**SOLUTION**  The phrase "within one standard deviation of the mean length" is one you will see often. (We used it in Chapter 3 when introducing the Empirical Rule.) It means that the pup's length will be somewhere between

mean minus 1 standard deviation

and

mean plus 1 standard deviation.

Because 1 standard deviation is 1.2 inches, this means the length must be between

$29.5 - 1.2 = 28.3$ inches

and

$29.5 + 1.2 = 30.7$ inches

From the results, we see that the probability that a randomly selected seal pup is within 1 standard deviation of mean length is about 68% (from Figure 6.18, it is 68.2689%).

**TRY THIS!**  Exercise 6.15

## Without Technology: The Standard Normal

Example 4 illustrates a principle that's very useful for finding probabilities from the Normal distribution without technology. This principle is the recognition that we don't need to refer to values in our distribution in the units in which they were measured. We can also refer to them in standard units. In other words, we can ask for the probability that the man's height is between 66 inches and 72 inches (measured units), but another way to say the same thing is to ask for the probability that his height is within 1 standard deviation of the mean, or between −1 and +1 standard units.

You can still use the Normal model if you change the units to standard units, but you must also convert the mean and standard deviation to standard units. This is easy, because the mean is 0 standard deviations away from itself, and any point 1 standard deviation away from the mean is 1 standard unit. Thus, if the Normal model was a good model, then when you convert to standard units, the $N(0, 1)$ model is appropriate.

This model—the Normal model with mean 0 and standard deviation 1—has a special name: the **standard Normal model**.

 **Looking Back**

**Standard Units**
Standard units (Chapter 3) tell us how many standard deviations from the mean an observation lies. Standard units are also called $z$-scores.

 **KEY POINT**   $N(0, 1)$ is the standard Normal model: a Normal model with a mean of 0 ($\mu = 0$) and a standard deviation of 1 ($\sigma = 1$).

The standard Normal model is an important concept, because it allows us to find probabilities for any Normal model. All we need to do is first convert to standard units. We can then look up the areas in a published table that lists useful areas for the $N(0, 1)$ model. One such table is available in Appendix A.

Table 6.5 shows an excerpt from this table. The values within the table represent areas (probabilities). The numbers along the left margin, when joined to the numbers

| z | .00 | .01 | .02 | .03 | .04 | .05 | .06 | .07 | .08 | .09 |
|-----|------|------|------|------|------|------|------|------|------|------|
| 0.9 | .8159 | .8186 | .8212 | .8238 | .8264 | .8289 | .8315 | .8340 | .8365 | .8389 |
| 1.0 | **.8413** | .8438 | .8461 | .8485 | .8508 | .8531 | .8554 | .8577 | .8599 | .8621 |
| 1.1 | .8643 | .8665 | .8686 | .8708 | .8729 | .8749 | .8770 | .8790 | .8810 | .8830 |
| 1.2 | .8849 | .8869 | .8888 | .8907 | .8925 | .8944 | .8962 | .8980 | .8997 | .9015 |

▲ **TABLE 6.5** Excerpt from the Normal table in Appendix A, which shows areas to the left of $z$ in a standard Normal distribution. For example, the area to the left of $z = 1.00$ is 0.8413, and the area to the left of $z = 1.01$ is 0.8438.

across the top, represent *z*-scores. For instance, the boldface value in this table represents the area under the curve *and to the left* of 1.00 standard unit. This represents the probability that a randomly selected person has a height *less than* 1 standard unit. Figure 6.19 shows what this area looks like.

▶ **FIGURE 6.19** The area of the shaded region represents the probability that a randomly selected person (or thing) has a value less than 1.00 standard deviation above the mean, which is about 84%.

For example, imagine we want to find the probability that a randomly selected woman is shorter than 62 inches. This person is selected from a population of women whose heights follow a $N(64, 3)$ distribution. Our strategy is

1. Convert 62 inches to standard units. Call this number *z*.

2. Look up the area below *z* in the table for the $N(0, 1)$ distribution.

# EXAMPLE 5 Small Pups

Small newborn seal pups have a lower chance of survival. Suppose that the length of a newborn seal pup follows a Normal distribution with a mean length of 29.5 inches and a standard deviation of 1.2 inches.

**QUESTION** What is the probability that a newborn pup selected at random is shorter than 28.0 inches?

**SOLUTION** Begin by converting the length 28.0 inches to standard units.

$$z = \frac{28 - 29.5}{1.2} = \frac{-1.5}{1.2} = -1.25$$

Next sketch the area that represents the probability we wish to find (see Figure 6.20). We want to find the area under the Normal curve and to the left of 28 inches, or, in standard units, to the left of $-1.25$.

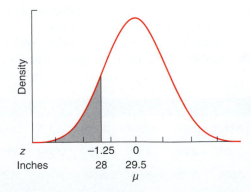

▶ **FIGURE 6.20** A standard Normal distribution, showing the shaded area that represents the probability of selecting a seal pup shorter than $-1.25$ standard deviations below the mean.

We can look this up in the standard Normal table in Appendix A. Table 6.6 on the next page shows the part we are interested in. We see that the area to the left of a *z*-score of $-1.25$ is 10.56%.

| z | .00 | .01 | .02 | .03 | .04 | .05 | .06 | .07 | .08 | .09 |
|------|------|------|------|------|------|------|------|------|------|------|
| **−1.3** | .0968 | .0951 | .0934 | .0918 | .0901 | .0885 | .0869 | .0853 | .0838 | .0823 |
| **−1.2** | .1151 | .1131 | .1112 | .1093 | .1075 | **.1056** | .1038 | .1020 | .1003 | .0985 |

▲ **TABLE 6.6** Part of the standard Normal table. The value printed in boldface type is the area under the standard Normal density curve to the left of −1.25.

The probability that a newborn seal pup will be shorter than 28 inches is about 11% (rounding up from 10.56%).

**TRY THIS!** Exercise 6.23

## EXAMPLE 6 A Range of Seal Pup Lengths

Again, suppose that the $N(29.5, 1.2)$ model is a good description of the distribution of seal pups' lengths.

**QUESTION** What is the probability that this randomly selected seal pup will be between 27 inches and 31 inches long?

**SOLUTION** This question is slightly tricky. The table gives us only the area *below* a given value. How do we find the area *between* two values?

We proceed in two steps. First we find the area less than 31 inches. Second, we "chop off" the area below 27 inches. The area that remains is the region between 27 and 31 inches. This process is illustrated in Figure 6.21.

To find the area less than 31 inches, we convert 31 inches to standard units:

$$z = \frac{(31 - 29.5)}{1.2} = 1.25$$

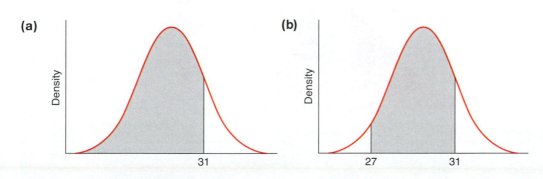

**(a)**      **(b)**

◄ **FIGURE 6.21** Steps for finding the area between 27 and 31 inches under a $N(29.5, 1.2)$ distribution. **(a)** The area below 31. **(b)** We chop off the area below 27, and the remaining area (shaded) is what we are looking for.

Using the Standard Normal Table in Appendix A, we find that this probability is 0.8944. Next, we find the area below 27 inches:

$$z = \frac{(27 - 29.5)}{1.2} = -2.08$$

Again, using the Standard Normal Table, we find this area to be 0.0188. Finally, we subtract (or "chop off") the smaller area from the big one:

$$
\begin{array}{r}
0.8944 \\
-0.0188 \\
\hline
0.8756
\end{array}
$$

▲ **FIGURE 6.22** TI-83/84 output for finding the probability that a newborn seal pup will be between 27 and 31 inches long.

> **Details**
>
> **Inverse Normal**
> Mathematicians sometimes call finding measurements from percentiles from Normal distributions finding "inverse Normal" values.

**CONCLUSION** The probability that a newborn seal pup will be between 27 inches and 31 inches long is about 88%. Figure 6.22 confirms that answer.

**TRY THIS!** Exercise 6.27

## Finding Measurements from Percentiles for the Normal Distribution

So far we have discussed how to find the probability that you will randomly select an object with a value that is within a certain range. Thus, in Example 5 we found that if newborn seal pups' lengths follow a $N(29.5, 1.2)$ distribution, then the probability that we will randomly select a pup shorter than 28 inches is (roughly) 11%. We found this by finding the area under the Normal curve that is to the left of 28 inches.

Sometimes, though, we wish to turn this around. We are given a probability, but we want to find the value that corresponds to that probability. For instance, we might want to find the length of a seal pup such that the probability that we'll see any pups shorter than that is 11%. Such a number is called a **percentile**. The 11th percentile for seal pup lengths is 28 inches, the length that has 11% of the area under the Normal curve to its left.

Finding measurements from percentiles is simple with the right technology. The screenshot in Figure 6.23 shows how to use SOCR to find the height of a woman in the 25th percentile, assuming that women's heights follow a $N(64, 3)$ distribution. Simply move the cursor until the shaded region represents the lower 25% of the curve. (This area is given after the words "Between (Red-Shaded)" below the graph.) We see that the value that has 25.2% (as close as we could get to 0.2500) below it is 61.99. So we say that 61.99 (or 62) is the 25th percentile for this distribution.

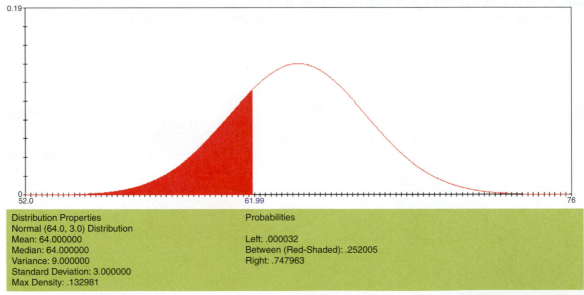

Distribution Properties
Normal (64.0, 3.0) Distribution
Mean: 64.000000
Median: 64.000000
Variance: 9.000000
Standard Deviation: 3.000000
Max Density: .132981

Probabilities

Left: .000032
Between (Red-Shaded): .252005
Right: .747963

▲ **FIGURE 6.23** Technology shows that 61.99 inches is the 25th percentile, because it has 25% of the area below it (to the left of it). Thus, if a woman is about 62 inches tall, there's a 25 percent chance that someone will be shorter than she.

Things are a little trickier if you don't have technology. You first need to use the standard Normal curve, $N(0, 1)$, to find the $z$-score from the percentile. Then you must convert the $z$-score to the proper units. So without technology, finding a measurement from a percentile is a two-step process:

**Step 1.** Find the $z$-score from the percentile.

**Step 2.** Convert the $z$-score to the proper units.

Example 8 illustrates these steps.

## EXAMPLE 7 Inverse Normal or Normal?

Suppose that the amount of money people keep in their online PayPal account follows a Normal model. Consider these two situations:

a. A PayPal customer wonders how much money he would have to put into the account to be in the 90th percentile.

b. A PayPal employee wonders what the probability is that a randomly selected customer will have less than $150 in his account.

**QUESTION** For each situation, identify whether the question asks for a measurement or a Normal probability.

**SOLUTIONS**

a. This situation gives a percentile (the 90th) and asks for the measurement (in dollars) that has 90% of the other values below it. This is an inverse Normal question.

b. This situation gives a measurement ($150) and asks for a probability.

**TRY THIS!** Exercise 6.37

## EXAMPLE 8 Finding Measurements from Percentiles by Hand

Assume that women's heights follow a Normal distribution with mean 64 inches and standard deviation 3 inches: $N(64, 3)$. Earlier, we used technology to find that the 25th percentile was approximately 62 inches.

**QUESTION** Using the Normal table in Appendix A, confirm that the 25th percentile height is 62 inches.

**SOLUTION** The question asks us to find a measurement (the height) that has 25% of all women's heights below it. Use the Normal table in Appendix A. This gives probabilities and percentiles for a Normal distribution with mean 0 and standard deviation 1: a $N(0, 1)$ or standard Normal distribution.

Step 1: Find the $z$-score from the percentile. To do this, you must first find the probability within the table. For the 25th percentile, use a probability of 0.25. Usually, you will not find exactly the value you are looking for, so settle for the value that is as close as you can get. This value, 0.2514, is underlined in Table 6.7, which is an excerpt from the Normal table in Appendix A.

| z | .00 | .01 | .02 | .03 | .04 | .05 | .06 | .07 | .08 | .09 |
|---|-----|-----|-----|-----|-----|-----|-----|-----|-----|-----|
| −0.7 | .2420 | .2389 | .2358 | .2327 | .2296 | .2266 | .2236 | .2206 | .2177 | .2148 |
| −0.6 | .2743 | .2709 | .2676 | .2643 | .2611 | .2578 | .2546 | .2514 | .2483 | .2451 |

▲ **TABLE 6.7** Part of the standard Normal table.

You can now see that the $z$-score corresponding to a probability of 0.2514 is −0.67. This relation between the $z$-score and the probability is shown in Figure 6.24 on the next page.

▶ **FIGURE 6.24** The percentile for 0.25 is −0.67 standard units, because 25% of the area under the standard Normal curve is below −0.67.

Step 2: Convert the $z$-score to the proper units.

A $z$-score of −0.67 tells us that this value is 0.67 standard deviations below the mean. We need to convert this to a height in inches.

One standard deviation is 3 inches, so 0.67 standard deviations is

$$0.67 \times 3 = 2.0 \text{ inches}$$

The height is 2.0 inches below the mean. The mean is 64.0 inches, so the 25th percentile is

$$64.0 - 2.0 = 62.0$$

**CONCLUSION** The woman's height at the 25th percentile is 62 inches, assuming that women's heights follow a $N(64, 3)$ distribution.

**TRY THIS!** Exercise 6.43

Figure 6.25 shows some percentiles and the corresponding $z$-scores to help you visualize percentiles.

▲ **FIGURE 6.25** $z$-Scores and percentiles—the 10th, 50th, and 90th percentiles. A percentile corresponds to the percentage in the area to the left under the curve.

## The Normal Model and the Empirical Rule

In Chapter 3 we mentioned the Empirical Rule, which was not so much a rule as a guideline for helping you understand how data are distributed. The Empirical Rule is meant to be applied to any symmetric, unimodal distribution. However, the Empirical Rule is based on the Normal model. For any arbitrary unimodal, symmetric distribution, the Empirical Rule is approximate. And sometimes very approximate. But if that distribution is the Normal model, the rule is exact.

In Example 4, we found that the area between $-1$ and $+1$ in a standard Normal distribution was 68%. This is exactly what the Empirical Rule predicts. We can also find the probability that a randomly selected observation will be between $-2$ and $2$ standard deviations of the mean, if that observation comes from the Normal model. Figure 6.26 shows a sketch of the $N(0, 1)$ model, with the region between $-2$ and $2$ (in standard units) shaded. From the table in the appendix, the area below $+2$ is 0.9772. Also from the table, the area below $-2$ is 0.0228. The difference is the shaded area and is $0.9772 - 0.0228 = 0.9544$, or about 95%, just as the Empirical Rule predicts.

These facts from the Empirical Rule help us interpret the standard deviation in the context of the Normal distribution. For example, because the heights of women are Normally distributed and have a standard deviation of about 3 inches, we know that a majority of women (in fact, 68%) have heights within 3 inches of the mean: between 61 inches and 67 inches. Because nearly all women are within 3 standard deviations of the mean, we know not to expect too many women to be taller than $64 + 3 \times 3 = 73$ inches tall (73 inches is 6 feet and 1 inch). Such women are very rare.

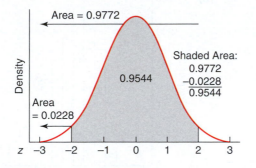

> **Looking Back**
>
> **The Empirical Rule**
> The Empirical Rule says that if a distribution of a sample of data is unimodal and roughly symmetric, then about 68% of the observations are within 1 standard deviation of the mean, about 95% are within 2 standard deviations of the mean, and nearly all are within 3 standard deviations of the mean.

◄ **FIGURE 6.26** The area between $z$-scores of $-2.00$ and $+2.00$ on the standard Normal curve.

## SNAPSHOT  THE NORMAL MODEL

| | |
|---|---|
| **WHAT IS IT?** ▶ | A model of a distribution for some numerical variables. |
| **WHAT DOES IT DO?** ▶ | Provides us with a model of the distributions of probabilities for many real-life numerical variables. |
| **HOW DOES IT DO IT?** ▶ | The probabilities are represented by the area underneath the bell-shaped curve. |
| **HOW IS IT USED?** ▶ | If the Normal model is appropriate, it can be used for finding probabilities or for finding measurements associated with particular percentiles. |

## Appropriateness of the Normal Model

The Normal model does not fit all distributions of numerical variables. For example, if we are randomly selecting people who submitted tax returns to the federal government, we cannot use the Normal model to find the probability that someone's return is higher than the mean value. The reason is that incomes are right-skewed, so the Normal model will not fit.

How do we know whether the Normal model is appropriate? Unfortunately, there is no checklist. However, the Normal model is a good first-choice model if you suspect that the distribution is symmetric and has one mode. Once you collected data, you can check to see whether the Normal model closely matches the data.

In short, the Normal model is appropriate if it produces results that match what we see in real life. If the data we collect match the Normal model fairly closely, then the model is appropriate. Figure 6.27a shows a histogram of the actual heights from a sample of more than 1400 women from the National Health and Nutrition Examination Survey (NHANES, www.cdc.gov/nchs/nhanes), with the Normal model superimposed over the histogram. Note that the model, though not perfect, is a pretty good description of the shape of the distribution. Compare this to Figure 6.27b, which shows the distribution of weights for the same women. The Normal model is not a very good fit for these data. The Normal model has the peak at the wrong place; specifically, the Normal model is symmetric, whereas the actual distribution is right-skewed.

Statisticians have several ways of checking whether the Normal model is a good fit to the population, but the easiest thing for you to do is to make a histogram of your data and see whether it looks unimodal and symmetric. If so, the Normal model is likely to be a good model.

▲ **FIGURE 6.27** **(a)** A histogram of data from a large sample of adult women in the United States drawn at random from the National Health and Nutrition Examination Survey. The red curve is the Normal curve, which fits the shape of the histogram very well, indicating that the Normal model would be appropriate for these data. **(b)** A histogram of weights for the same women. Here the Normal model is a bad fit to the data.

## SECTION 6.3

# The Binomial Model

The Normal model applies to many real-life, continuous-valued numerical variables. The **binomial probability model** is useful in many situations with discrete-valued numerical variables (typically counts, whole numbers). As with the Normal model, we will explain what the model looks like and how to calculate probabilities with it. We will also provide examples and discuss why the binomial model is appropriate to the situations.

The classic application of the binomial model is counting heads when flipping a coin. Let's stay I flip a coin 10 times. What is the probability I get 1 head? 2 heads? 10 heads? This is a situation that the binomial model fits quite well, and the probabilities of these outcomes are given by the binomial model. If we randomly select 10 people, what's the probability that exactly 5 will be Republicans? If we randomly select 100 students, what's the probability that 10 or fewer will be on the Dean's List? These are examples of situations where the binomial model applies.

How do you recognize a binomial model? The first sign that your random experiment is a candidate for the binomial model is that the outcome you are interested in is a count. If that's the case, then all four of the following characteristics must be present:

1. *A fixed number of trials.* We represent this number with the letter $n$. For example, if we flip a coin 10 times, then $n = 10$.

2. *Only two outcomes are possible at each trial.* We will call these two outcomes "success" and "failure." For example, we might consider the outcome of heads to be a success. Or we might be selecting people at random and counting the number of males; in this case, of the two outcomes "male" and "female," "male" would be considered a success.

3. *The probability of success is the same at each trial.* We represent this probability with the letter $p$. For example, the probability of getting heads after a coin flip is $p = 0.50$ and does not change from flip to flip.

4. *The trials are independent.* The outcome of one trial does not affect the outcome of any other trial.

If all four of these characteristics are present, the binomial model applies and you can easily find the probabilities by looking at the binomial probability distribution.

**KEY POINT**

The binomial model provides probabilities for random experiments in which you are counting the number of successes that occur. Four characteristics must be present:

1. Fixed number of trials: $n$
2. The only two outcomes are success and failure.
3. The probability of success, $p$, is the same at each trial.
4. The trials are independent.

## EXAMPLE 9 Extrasensory Perception (Mind Reading)

Zener cards are special cards used to test whether people can read minds (telepathy). Each card in a Zener deck has one of five special designs: a star, a circle, a plus-sign, a square, and three wavy lines (Figure 6.28). In an experiment, one person, the "sender," selects a card at random, looks at it, and thinks about the symbol on the card. Another person, the "receiver," cannot see the card (and in some studies cannot even see the sender), and guesses which of the symbols was chosen. A researcher records whether the guess was correct. The card is then placed back in the deck, the deck is shuffled, and another card is drawn. Suppose this happens 10 times (10 guesses are made). The receiver gets 3 guesses correct, and the researcher wants to know the probability of this happening if the receiver is simply guessing.

◄ **FIGURE 6.28** Zener cards (ESP cards) show one of five shapes. A deck has equal numbers of each shape.

QUESTION    Explain why this is a binomial experiment.

SOLUTION    First, we note that we are counting something: the number of successful guesses. We need to check that the experiment meets the four characteristics of a binomial model. (1) The experiment consists of a fixed number of trials: $n = 10$. (2) The outcome of each trial is success or failure: The receiver either gets the right answer or does not. (3) The probability of a success at a trial is $p = 1/5 = 0.20$, because there are 5 cards and so the receiver has a 1-in-5 chance of getting it correct, if we assume the receiver is guessing. (4) As long as the cards are put back in the deck and reshuffled (thoroughly), the probability of a success is the same for each trial, and each trial is independent.

All four characteristics are satisfied, so the experiment fits the binomial model.

TRY THIS!    Exercise 6.53

## EXAMPLE **10** Why Are They Not Binomial?

The following four experiments are almost, but not quite, binomial experiments.

a. Record the number of different eye colors in a group of 50 randomly selected people.

b. A married couple decides to have children until a girl is born, but to stop at five children if they do not have any girls. How many children will the couple have?

c. Suppose the probability that a flight will arrive on time (within 15 minutes of the scheduled arrival time) at O'Hare Airport in Chicago is 85%. How many flights arrive on time out of 300 flights scheduled to land on a day in January?

d. A student guesses on every question of a test that has 10 multiple-choice questions and 10 true-false questions. Record the number of questions the student gets right.

QUESTION    For each situation, explain which of the four characteristics is not met.

SOLUTIONS

a. This is not a binomial experiment because there are more than two eye colors, so more than two outcomes may occur at each trial. However, if we reduced the eye colors to two categories by, say, recording whether the eye color was brown or not brown, then this would be a binomial experiment.

b. This is not a binomial experiment because the number of trials is not fixed before the children are born. The number of "trials" depends on when (or whether) the first girl is born. The number of trials varies depending on what happens—the word *until* tells you that.

c. This is not binomial because the flights are not independent. If the weather is bad, the chance of arriving on time for all flights is lower. Therefore, if one flight arrives late, then another flight is more likely to arrive late.

d. This is not a binomial experiment because the probability of success on each trial is not constant; the probability of success is lower on multiple-choice questions than on true/false questions. Therefore, criterion 3 is not met. However, if the test were subdivided into two sections (multiple-choice and true/false), then each separate section could be called a binomial experiment if we assumed the student was guessing.

TRY THIS!    Exercise 6.55

## Visualizing the Binomial Distribution

All binomial models have the four characteristics listed above, but the list gives us flexibility in *n* and *p*. For example, if we had flipped the coin 6 times instead of 10, it would still be a binomial experiment. Also, if the probability of a success were 0.6 instead of 0.5, we would still have a binomial experiment. For different values of *n* and *p*, we have different binomial experiments, and the binomial distribution looks different in each case.

Figure 6.29 shows that the binomial distribution for $n = 3$ and $p = 0.5$ is symmetric. We can read from the graph that the probability of getting exactly 2 successes (2 heads in 3 flips of a coin) is almost 0.40, and the probability of getting 0 successes is the same as the probability of getting all successes.

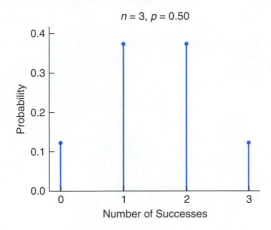

◀ **FIGURE 6.29** Binomial distribution with $n = 3$, $p = 0.50$.

If *n* is bigger but *p* remains fixed at 0.50, the distribution is still symmetric because the chance of a success is the same as the chance of a failure, as shown in Figure 6.30.

◀ **FIGURE 6.30** Binomial distribution with $n = 15$, $p = 0.50$.

If the probability of success is not 50%, the distribution might not be symmetric. Figure 6.31 on the next page shows the distribution for $p = 0.3$, which means we're less likely to get a large number of successes than a smaller number, so the probability "spikes" are taller for smaller numbers of successes. The plot is now right-skewed.

However, even if the distribution is not symmetric, if we increase the number of trials, it becomes symmetric. The shape of the distribution depends on both *n* and *p*. If

▶ **FIGURE 6.31** Binomial distribution with $n = 15$, $p = 0.30$.

we keep $p = 0.3$ but increase $n$ to 100, we get a more symmetric shape, as shown in Figure 6.32 shown below.

▶ **FIGURE 6.32** Binomial distribution with $n = 100$, $p = 0.30$. Note that we show $x$ only for values between 15 and 45. The shape is symmetric, even though $p$ is not 0.50.

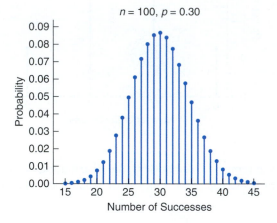

Binomial distributions have the interesting property that if the number of trials is large enough, the distributions are symmetric.

**KEY POINT**    The shape of a binomial distribution depends on both $n$ and $p$. Binomial distributions are symmetric when $p = 0.5$, but they are also symmetric when $n$ is large, even if $p$ is close to 0 or 1.

## Finding Binomial Probabilities

Because the binomial distribution depends only on the values of $n$ and $p$, once you have identified an experiment as binomial and have identified the values of $n$ and $p$, you can find any probability you wish. We use the notation $b(n, p, x)$ to represent the **binomial probability** of getting $x$ successes in a binomial experiment with $n$ trials and probability of success $p$. For example, imagine tossing a coin 10 times. If each side is equally likely, what is the probability of getting 4 heads? We represent this with $b(10, 0.50, 4)$.

The easiest and most accurate way to find binomial probabilities is to use technology. Statistical calculators and software have the binomial distribution built in and can easily calculate probabilities for you.

## EXAMPLE **11**  Stolen Bicycles

According to the website nationalbikeregistry.com, at the campus of UC Berkeley, only 3% of stolen bicycles are returned to owners.

**QUESTIONS** Accepting for the moment that the four characteristics of a binomial experiment are satisfied, write the notation for the probability that exactly 5 bicycles will be returned if 235 are stolen in the course of a year. What is this probability?

**SOLUTION** The number of trials is 235, the probability of success is 0.03, and the number of successes is 5. Therefore, we can write the binomial probability as $b(235, 0.03, 5)$.

The command and output for finding this probability on a TI-83/84 are shown in Figure 6.33. The command binompdf stands for "binomial probability distribution function." The results show us that the probability that exactly 5 bicycles will be returned, assuming that this is in fact a binomial experiment, is about 12.6%.

**TRY THIS!** Exercise 6.59

▲ **FIGURE 6.33** TI-83/84 output for a binomial trial.

Although we applied the binomial model to the stolen bicycles example, we had to make the assumption that the trials are independent. A "trial" in this context consists of a bicycle being stolen, and a "success" occurs if the bike is returned. If one person stole several bikes on a single day from a single location, this assumption of independent trials would be wrong, because if police found one of the bikes, they would have a good chance of finding the others that the thief took. Sometimes we have no choice but to make certain assumptions to complete a problem, but we need to be sure to check our assumptions when we can, and we must be prepared to change our conclusion if the assumptions were wrong.

Another approach to finding binomial probabilities is to use a table. Published tables are available that list binomial probabilities for a variety of combinations of values for $n$ and $p$. One such table is provided in Appendix A. This table lists binomial probabilities for values of $n$ between 2 and 15 and for several different values of $p$.

## EXAMPLE **12**  Recidivism in Texas

The three-year recidivism rate of parolees in Texas is 30% (www.lbb.state.tx.us). In other words, 30% of released prisoners return to prison within three years of their release. Suppose a prison in Texas released 15 prisoners.

**QUESTION** Assuming that whether one prisoner returns to prison is independent of whether any of the others returns, use Table 6.8 on the next page, which shows binomial probabilities for $n = 15$ and for various values of $p$, to find the probability that exactly 8 out of 15 will end up back in prison within three years.

**SOLUTION** Substituting the numbers, you can see that we are looking for $b(15, 0.30, 8)$. Referring to Table 6.8, on the next page, you can see—by looking in the table for $n = 15$, the row for $x = 8$, and the column for $p = 0.30$—that the probability that exactly 8 parolees will be back in prison within three years is 0.035, or about a 3.5% chance.

| x | 0.1 | 0.2 | 0.25 | 0.3 | 0.4 | 0.5 | 0.6 | 0.7 | 0.75 | 0.8 | 0.9 |
|---|-----|-----|------|-----|-----|-----|-----|-----|------|-----|-----|
| 6 | .002 | .043 | .092 | .147 | .207 | .153 | .061 | .012 | .003 | .001 | .000 |
| 7 | .000 | .014 | .039 | .081 | .177 | .196 | .118 | .035 | .013 | .003 | .000 |
| 8 | .000 | .003 | .013 | .035 | .118 | .196 | .177 | .081 | .039 | .014 | .000 |
| 9 | .000 | .001 | .003 | .012 | .061 | .153 | .207 | .147 | .092 | .043 | .002 |
| 10 | .000 | .000 | .001 | .003 | .024 | .092 | .186 | .206 | .165 | .103 | .010 |
| 11 | .000 | .000 | .000 | .001 | .007 | .042 | .127 | .219 | .225 | .188 | .043 |
| 12 | .000 | .000 | .000 | .000 | .002 | .014 | .063 | .170 | .225 | .250 | .129 |
| 13 | .000 | .000 | .000 | .000 | .000 | .003 | .022 | .092 | .156 | .231 | .267 |
| 14 | .000 | .000 | .000 | .000 | .000 | .000 | .005 | .031 | .067 | .132 | .343 |
| 15 | .000 | .000 | .000 | .000 | .000 | .000 | .000 | .005 | .013 | .035 | .206 |

▲ **TABLE 6.8** Binomial probabilities with a sample of 15 and *x*-values of 6 or higher.

```
binompdf(15,.3,8
)
          .0347700143
```

▲ **FIGURE 6.34** TI-83/84 output for *b*(15, 0.3, 8).

Using a TI-83/84 as shown in Figure 6.34, we can see another way to get the same answer.

**TRY THIS!** Exercise 6.61

## Finding (Slightly) More Complex Probabilities

### EXAMPLE 13 ESP with 10 Trials

For a test of psychic abilities, researchers have asked the sender to draw 10 cards at random from a large deck of Zener cards (see Example 9). Assume that the cards are replaced in the deck after each use and the deck is shuffled. Recall that this deck contains equal numbers of 5 unique shapes. The receiver guesses which card the sender has drawn.

**QUESTIONS**

a. What is the probability of getting *exactly 5* correct answers (out of 10 trials) if the receiver is simply guessing (and has no psychic ability)?

b. What is the probability that the receiver will get *5 or more* of the cards correct out of 10 trials?

c. What is the probability of getting *fewer than 5* correct in 10 trials with the ESP cards?

**SOLUTIONS**

a. In Example 9 we identified this as a binomial experiment. With that done, we must now identify *n* and *p*. The number of trials is 10, so $n = 10$. If the receiver is guessing, then the probability of a correct answer is $p = 1/5 = 0.20$. Therefore, we wish to find $b(10, 0.2, 5)$.

Figure 6.35a on the next page gives the TI-83/84 output, where you can see that the probability of getting 5 right out of 10 is only about 0.0264. Figure 6.35b shows Minitab output for the same question.

**(a)**

```
binompdf(10,.2,5
)
      .0264241152
■
```

**(b)**

| Probability Density Function |
|---|
| Binomial with n = 10 and p = 0.2 |
| |
| x         P |
| 5    0.0264241 |

◀ **FIGURE 6.35** Technology output for *b*(10, 0.2, 5). **(a)** Output from a TI-83/84. **(b)** Output from Minitab.

Figure 6.36a shows a graph of the pdf. The probability *b*(10, 0.2, 5) is so small that it is hard to read off the graph. The graph shows that it is unusual to get exactly 5 correct when the receiver is guessing.

b. The phrase *5 or more* means we need the probability that the receiver gets 5 correct **or** 6 correct **or** 7 **or** 8 **or** 9 **or** 10. The outcomes 5 correct, 6 correct, and so on are mutually exclusive, because if you get exactly 5 correct, you cannot possibly also get exactly 6 correct. Therefore, we can find the probability of 5 or more correct by adding the individual probabilities together:

$$b(10, 0.2, 5) + b(10, 0.2, 6) + b(10, 0.2, 7) + b(10, 0.2, 8) + b(10, 0.2, 9) + b(10, 0.2, 10)$$
$$= 0.026 \quad + \quad 0.006 \quad + \quad 0.001 \quad + \quad 0.000 \quad + \quad 0.000 \quad + \quad 0.000$$
$$= 0.033$$

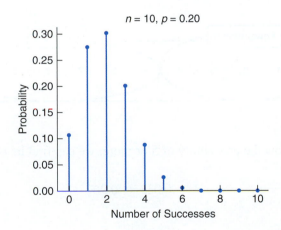

◀ **FIGURE 6.36(a)** The theoretical numbers of successes for 10 trials with the Zener deck, assuming guessing.

These probabilities are circled in Figure 6.36b.

◀ **FIGURE 6.36(b)** Summing the probabilities represented by the circled bars gives us the probability of getting 5 or more correct.

c. The phrase *fewer than 5 correct* means 4, 3, 2, 1, or 0 correct. These probabilities are circled in Figure 6.36c.

▶ **FIGURE 6.36(c)** Summing the probabilities represented by the circled bars gives the probability of getting fewer than 5 correct.

The event that we get fewer than 5 correct is the complement of the event that we get 5 or more correct, as you can see in Figure 6.37, which shows all possible numbers of successes with 10 trials. *Fewer than 5* is the same event as *4 or fewer* and is shown in the left oval in the figure.

▶ **FIGURE 6.37** The possible numbers of successes out of 10 trials with binomial data. Note that *fewer than 5* is the complement of *5 or more*.

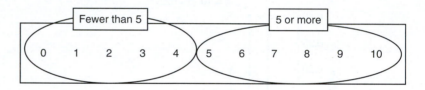

Because we know the probability of 5 or more, we can find its complement by subtracting from 1:

$$1 - 0.033 = 0.967$$

**CONCLUSIONS**

a. The probability of exactly 5 correct is 0.026.

b. The probability of 5 or more correct is 0.033.

c. The probability of fewer than 5 (that is, of 4 or fewer) is 0.967.

**TRY THIS!** Exercise 6.69

▲ **FIGURE 6.38** TI-83/84 output for $n = 10$, $p = 0.2$, and 4 or fewer successes.

Most technology also offers you the option of finding binomial probabilities for *x or fewer*. In general, probabilities of *x or fewer* are called **cumulative probabilities**. Figure 6.38 shows the cumulative binomial probabilities provided by the TI-83/84; notice the "c" in binom**c**df.

The probability of getting 5 or more correct is different from the probability of getting more than 5 correct. This very small change in the wording gives very different results. Figure 6.39 shows that *5 or more* includes the outcomes 5, 6, 7, 8, 9, and 10, whereas *more than 5* does not include the outcome of 5.

When finding Normal probabilities, we did not have to worry about such subtleties of language, because for a continuous numerical variable, the probability of getting 5 or more is the same as the probability of getting more than 5.

◀ **FIGURE 6.39** Interpretation of words for discrete counts with *n* of 10. **(a)** Results for *5 or more*; **(b)** results for *more than 5*. Note that they are different.

## Finding Binomial Probabilities by Hand

When the number of trials is small, we can find probabilities by relying on the probability rules in Chapter 5 and listing all possible outcomes. In fact, there is a general pattern we can exploit to find a formula that will give us the probabilities for any value of *n* or *p*. Listing all of the outcomes will help us see why this formula works.

Suppose we are testing someone using the Zener cards that have five different shapes. After each card is guessed, it is returned to the deck, and the deck is shuffled before the next card is drawn. We want to count how many correct guesses a potential psychic makes in four trials. We can have five different outcomes for this binomial experiment: 0 correct answers, 1 correct, 2 correct, 3 correct, and all 4 correct.

The first step is to list all possible outcomes after four attempts. At each trial, the guesser is either right or wrong. Using R for right and W for wrong, we list all possible sequences of right or wrong results in four trials.

| 4 Right | 3 Right | 2 Right | 1 Right | 0 Right |
|---------|---------|---------|---------|---------|
| RRRR | RRRW | RRWW | RWWW | WWWW |
| | RRWR | RWRW | WRWW | |
| | RWRR | RWWR | WWRW | |
| | WRRR | WWRR | WWWR | |
| | | WRWR | | |
| | | WRRW | | |

There are 16 possible outcomes, although these are not all equally likely.

Now we find the probabilities of getting 4 right, 3 right, 2 right, 1 right, and 0 right. Because there are five shapes, and all five are equally likely to be chosen, if the receiver is simply guessing and has no psychic ability, then the probability of a right answer at each trial is $1/5 = 0.2$, and the probability of guessing wrong is 0.8.

Four right means "right AND right AND right AND right." Successive trials are independent (because we replace the card and reshuffle every time), so we can multiply the probabilities using the multiplication rule.

$$P(RRRR) = 0.2 \times 0.2 \times 0.2 \times 0.2 = 0.0016$$

This probability is just $b(4, 0.2, 4) = 1(0.0016) = 1(0.2)^4$. (We multiply by 1 because there is only one way that we can get all 4 right to happen, which is also why there is only one outcome listed in the "4 right" column.)

 **Looking Back**

**And**

The multiplication rule was Probability Rule 5c in Chapter 5 and applies only to independent events: $P(A \text{ AND } B) = P(A) \, P(B)$.

The probability of getting 3 right and 1 wrong includes all four options in the second group. The probability for each of these options is obtained by calculating the probability of 3 right and 1 wrong, and all of these probabilities will be the same. To get the total probability, therefore, we multiply by 4.

$$P(RRRW) = 0.2 \times 0.2 \times 0.2 \times 0.8 = 0.0064$$

$$P(3 \text{ right and 1 wrong, in any order}) = 4(0.0064) = 0.0256$$

$$b(4, 0.2, 3) = 5(0.2)^3(0.8)^1 = 0.0256$$

The probability of getting 2 right and 2 wrong includes all six options in the third group. The probability for each of these options is obtained by calculating the probability of 2 right and 2 wrong; then, to get the total, we multiply by 6.

$$P(RRWW) = 0.2 \times 0.2 \times 0.8 \times 0.8 = 0.0256$$

$$P(2 \text{ right and 2 wrong, in any order}) = 6(0.0256) = 0.1536$$

$$b(4, 0.2, 2) = 6(0.2)^2(0.8)^2 = 0.1536$$

The probability of getting 1 right and 3 wrong includes all four options in the fourth group. We obtain the probability for each of the options by calculating the probability of 1 right and 3 wrong; then we multiply by 4 to get the total.

$$P(RWWW) = 0.2 \times 0.8 \times 0.8 \times 0.8 = 0.1024$$

$$P(1 \text{ right and 3 wrong, in any order}) = b(4, 0.2, 1) = 4(0.1024) = 0.4096$$

$$b(4, 0.2, 1) = 4(0.2)^1(0.8)^3 = 0.4096$$

Finally, the probability of getting all four wrong is

$$P(WWWW) = b(4, 0.2, 0) = 0.8 \times 0.8 \times 0.8 \times 0.8 = 0.4096$$

Because there is only one way for this to happen, we multiply 0.4096 by 1. Thus

$$b(4, 0.2, 0) = 1(0.8)^4 = 0.4096$$

Table 6.9 summarizes the results.

If you add the probabilities, you will see that they add to 1, as they should, because this list includes all possible outcomes that can happen.

Figure 6.40 shows a graph of the probability distribution. Note that the graph is right-skewed because the probability of success is less than 0.50. Also note that the probability of getting 4 out of 4 right with the Zener cards is very small.

You might compare these probabilities with those in the binomial table. You will find that they agree, if you round off the numbers we found "by hand" to three decimal places.

| Number Right | Probability |
|---|---|
| 4 right | 0.0016 |
| 3 right | 0.0256 |
| 2 right | 0.1536 |
| 1 right | 0.4096 |
| 0 right | 0.4096 |

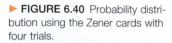 ▲ **TABLE 6.9** A summary of the probabilities of all possible numbers of successes in four trials with the Zener cards.

▶ **FIGURE 6.40** Probability distribution using the Zener cards with four trials.

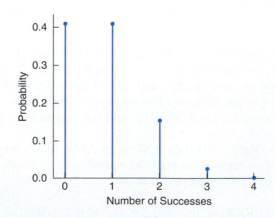

## The Formula

This approach of listing all possible outcomes is tedious when $n = 4$, and the tedium increases for larger values of $n$. (For $n = 5$ we have to list 32 possibilities, and for

$n = 6$ there are 64.) However, mathematicians have derived a formula that finds probabilities for the binomial distribution. The binomial table in this book and the results of your calculator are based on this formula:

$P(x$ successes in $n$ trials of a binomial experiment$) =$
(number of different ways of getting $x$ successes in $n$ trials) $p^x(1 - p)^{(n-x)}$

For example, for our alleged psychic who gets four guesses with the Zener cards, the probability of 3 successes ($x = 3$) in four trials ($n = 4$) is

(number of ways of getting 3 successes in 4 trials) $(0.2)^3(0.8)^1$

$= 4 (0.2)^3(0.8)^1$

$= 0.0256$

## The Shape of the Binomial Distribution: Center and Spread

Unlike the Normal distribution, the mean and standard deviation of the binomial distribution can be easily calculated. Their interpretation is the same as with all distributions: The mean tells us where the distribution balances, and the standard deviation tells us how far values are, typically, from the mean.

For example, in Figure 6.41, the binomial distribution is symmetric, so the mean sits right in the center at 7.5 successes.

Figure 6.41: A binomial distribution graph with n = 15, p = 0.5, Probability on the y-axis and Number of Successes on the x-axis.

◄ **FIGURE 6.41** A binomial distribution with $n = 15$, $p = 0.50$. The mean is at 7.5 successes.

If the distribution is right-skewed, the mean will be just to the right of the peak, closer to the right tail, as shown in Figure 6.42. This is a binomial distribution with $n = 15$, $p = 0.3$, and the mean sits at the balancing point: 4.5 successes.

Figure 6.42: A binomial distribution graph with n = 15, p = 0.3, Probability on the y-axis and Number of Successes on the x-axis.

◄ **FIGURE 6.42** A binomial distribution with $n = 15$, $p = 0.3$. The mean is at the balancing point of 4.5 successes.

The mean, $\mu$, of a binomial probability distribution can be found with a simple formula:

$$\mu = np$$

In words, the mean number of successes in a binomial experiment is the number of trials times the probability of a success.

This formula should make intuitive sense. If you toss a coin 100 times, you would expect about half of the tosses to be heads. Thus the "typical" outcome is $(1/2) \times 100 = 50$, or $np = 100 \times 0.5$.

The standard deviation, $\sigma$, of a binomial probability distribution, which measures the spread, is less intuitive:

$$\sigma = \sqrt{np(1 - p)}$$

For example, Figure 6.42 shows a binomial distribution with $n = 15$, $p = 0.30$. The mean is $15 \times 0.3 = 4.5$. The standard deviation is $\sqrt{15 \times 0.3 \times 0.7} = \sqrt{3.15} = 1.775$.

**KEY POINT**

For a binomial experiment, the mean is

$$\mu = np$$

For a binomial experiment, the standard deviation is

$$\sigma = \sqrt{np(1 - p)}$$

## Interpreting the Mean and Standard Deviation

The mean of any probability distribution, including the binomial, is sometimes called the **expected value**. This name gives some intuitive understanding. If you were to actually carry out a binomial experiment, you would *expect* about $\mu$ successes. If I flip a coin 100 times, I *expect* about 50 heads. If, in an ESP study, 10 trials are made and the probability of success at each trial is 0.20, we *expect* about 2 successes due to chance ($10 \times 0.2 = 2$).

Will I get exactly 50 heads? Will the ESP receiver get exactly 2 cards correct? Sometimes, yes. Usually, no. Although we expect $\mu$ successes, we usually get $\mu$ give or take some amount. That give-or-take amount is what is measured by $\sigma$.

In 100 tosses of a fair coin, we expect $\mu = 50$ heads, give-or-take $\sigma = \sqrt{100 \times 0.5 \times 0.5} = 5$ heads. We expect 50, but we will not be surprised if we get between 45 and 55 heads. In the Zener card experiment with 10 trials, we expect the receiver to guess about 2 cards correctly, but in practice we expect him or her to get 2 give or take 1.3, because

$$\sqrt{10 \times 0.2 \times 0.8} = \sqrt{1.6} = 1.26$$

## SNAPSHOT    THE BINOMIAL DISTRIBUTION

**WHAT IS IT?** ► A distribution for some discrete variables.

**WHAT DOES IT DO?** ► Gives probabilities for the number of successes observed in a fixed number of trials in a binomial experiment.

**HOW DOES IT DO IT?** ► If the conditions of a binomial experiment are met, once you identify $n$ (number of trials), $p$ (the probability of success), and $x$ (the number of successes), it gives the probability.

**HOW IS IT USED?** ► The probabilities are generally provided in the form of a table or a formula, but if you need to calculate them, use a calculator or technology.

## EXAMPLE 14 Basketball Free-Throw Shots

LeBron James has a free-throw success percentage of 74%. Assume the free-throw shots are independent; that is, success or failure on one shot does not affect the chance of success on another shot.

**QUESTION** If he has 600 free throws in an upcoming season, how many would you expect him to sink, give or take how many?

**SOLUTION** This is a binomial experiment. (Why?) The number of trials is $n = 600$, and the probability of success at each trial is $p = 0.74$. You would expect LeBron James to sink 74% of 600, or 444, free throws:

$$\mu = np = 600\,(0.74) = 444$$

The give-or-take amount is measured by the standard deviation:

$$\sigma = \sqrt{600 \times 0.74 \times 0.26} = 10.7443$$

You should expect him to hit about 444 free throws, give or take about 10.7.

$$444 + 10.7 = 454.7$$
$$444 - 10.7 = 433.3$$

**CONCLUSION** You expect James to sink between 433 and 455 out of 600 free-throw shots.

**TRY THIS!** Exercise 6.73

## Surveys: An Application of the Binomial Model

Perhaps the most common application of the binomial model is in survey sampling. Imagine a large population of people, say the 100 million or so registered voters in the United States. Some percentage of them, call it $p$, have a certain characteristic. If we choose 10 people at random (with replacement), we can ask what the probability is that all of them, or three of them, or six of them, have that characteristic.

## EXAMPLE 15 News Survey

The Pew Research Center says that 23% of people are "news integrators": people who get their news both from traditional media (television, radio, newspapers, magazines) and from the Internet (www.people-press.org). Suppose we take a random sample of 100 people.

**QUESTIONS** If, as is claimed, 23% of the population are news integrators, then how many people in our sample should we expect to be integrators? Give or take how many? Would you be surprised if 34 people in the sample turned out to be integrators?

**SOLUTION** Assuming that all four of the characeristics of a binomial distribution are satisfied, then we would expect 23% of our sample of 100 people to be integrators— that is, 23 people. The standard deviation is $\sqrt{100 \times 0.23 \times 0.77} = 4.2$. Thus we should expect 23 people, give or take about 4.2 people, to be integrators. This means that we shouldn't be surprised if we got as many as $23 + 4.2 = 27.2$, or about 27, people. However, 34 people is quite a bit more than 1 standard deviation above what we expect. In fact, it is almost 3 standard deviations away, so it would be a surprisingly large number of people.

**TRY THIS!** Exercise 6.75

The binomial model works well for surveys when people are selected *with* replacement (which means that, once they are selected, they have a chance of being selected again). If we are counting the people who have a certain characteristic, then the four characteristics of the binomial model are usually satisfied:

1. A fixed number of people are surveyed. In Example 15, $n = 100$.
2. The outcome is success (integrator) or failure (not an integrator).
3. The probability of a success is the same at each trial: $p = 0.23$.
4. The trials are independent. (This means that if one person is found to be an integrator, no one else in the sample will change their response.)

With surveys, we usually don't report the number of people who have an interesting characteristic; instead, we report the percentage of our sample. We would not report that 23 people in our sample of 100 were integrators; we would report that 23% of our sample were integrators. But the binomial distribution still applies, because we are simply converting our counts to percentages.

In reality, surveys don't select people with replacement. The most basic surveys sample people without replacement, which means that, strictly speaking, the probability of a success is different after each trial. Imagine that the first person selected is an integrator. Now there are fewer integrators left in the population, so it is no longer true that 23% of the population are integrators; the percentage is slightly less. However, as you can imagine, this is not a problem if the population is very large. In fact, if the population size is very large relative to the sample size (at least 10 times bigger), then this difference is so slight that characteristic 3 is essentially met.

Taking a random sample, either with or without replacement, of a large and diverse population such as all U.S. voters is quite complicated. In practice, the surveys that we read about in the papers or hear about on the news use a modified approach that is slightly different from what we've discussed here. Random selection is still at the heart of these modified methods, though, and the binomial distribution often provides a good approximation for probabilities, even under these more complex schemes.

## CASE STUDY REVISITED

McDonald's claims that its ice cream cones weigh 3.18 ounces. However, one of the authors bought five cones and found that all five weighed more than that. Is this surprising?

Suppose we assume that the amount of ice cream dispensed follows a Normal distribution centered on 3.18 ounces. In other words, typically, a cone weighs 3.18 ounces, but sometimes a cone weighs a little more, and sometimes a little less. If this is the case, then, because the Normal distribution is symmetric and centered on its mean, the probability that a cone weighs more than 3.18 ounces is 0.50.

If the probability that a cone weighs more than 3.18 ounces is 0.50, what is the probability that five cones all weigh more? We can think of this as a binomial experiment, if we assume the trials are independent. (This seems like a reasonable assumption.) We can then calculate $b(5, 0.5, 5)$ to find that the probability is 0.031.

If the typical McDonald's ice cream cone really weighs 3.18 ounces, then our outcome is fairly surprising: It happens only about 3% of the time. This means it is fairly rare and raises the possibility that this particular McDonald's might deliberately dispense more than 3.18 ounces.

# EXPLORING STATISTICS
## CLASS ACTIVITY

## ESP with Coin Flipping

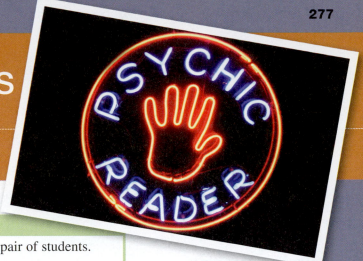

| GOALS | MATERIALS |
|---|---|
| Apply the binomial model to a real situation. | One coin for each pair of students. |

**ACTIVITY**

Choose a partner. One of you will play the role of the "sender" and the other will be the "receiver." The sender will flip a coin so that the receiver can't see the outcome (heads or tails). The sender will then look at the coin and concentrate, trying to mentally send a thought about whether the coin landed heads or tails. The receiver writes down (quickly) the outcome he or she believes was sent. (Just write the first thing that comes to mind.) The sender should make a tally of the number of right answers the receiver achieved in 10 trials. Now switch roles and try it again. Each of you should be prepared to report how many you got correct in your 10 trials.

**BEFORE THE ACTIVITY**

1. With 10 trials, how many would you expect to guess correctly if there is no ESP? If there are 20 people trying this, do you expect all of them to have the same results?

2. Find the standard deviation—the "give-or-take value"—of the number of correct guesses in 10 trials. Assuming that the receiver does not have ESP, what's the smallest number of correct guesses you might expect? What's the largest? How does the standard deviation help you determine this?

**AFTER THE ACTIVITY**

1. Make a histogram of the class results. Where is the distribution centered? Is this what you expected?

2. Are all of the results within the range of results predicted above?

3. Are any of the results unusually good? Does that show that the person with unusually good results has ESP? Why or why not? Explain.

## CHAPTER REVIEW

### Key Terms

Probability model, *244*
Probability distribution, *244*
Probability distribution function (pdf), *244*
Discrete outcomes, *244*
Continuous outcomes, *244*
Probability density curve, *247*

Normal model: Notation, $N(\mu, \sigma)$, *249*
Normal curve, *249*
Normal distribution, *249*
Mean of a probability distribution, $\mu$, *250*

Standard deviation of a probability distribution, $\sigma$, *250*
Standard Normal model Notation, $N(0, 1)$, *255*
Percentile, *258*
Binomial probability model Notation, $b(n, p, x)$, *262*

*n* is the number of trials
*p* is the probability of success on one trial
*x* is the number of successes
Cumulative probabilities, *270*
Expected value, *274*

### Learning Objectives

After reading this chapter and doing the assigned homework problems, you should

- Be able to distinguish between discrete and continuous-valued variables.

- Know when a Normal model is appropriate and be able to apply the model to find probabilities.

- Know when the binomial model is appropriate and be able to apply the model to find probabilities.

### Summary

Probability models try to capture the essential features of real-world experiments and phenomena that we want to study. In this chapter, we focused on two very useful models: the Normal model and the binomial model.

The Normal model is an example of a model of probabilities for continuous numerical variables. The Normal probability model is also called the Normal distribution, and the Gaussian curve. It can be a useful model when a histogram of data collected for a variable is unimodal and symmetric. Probabilities are found by finding the area under the appropriate region of the Normal curve. These areas are best calculated using technology. If technology is not available, you can also convert measures to standard units and then use the table of areas for the standard Normal distribution, provided in Appendix A.

The binomial model is an example of a model of probabilities for discrete numerical outcomes. The binomial model applies to binomial experiments, which occur when we are interested in counting the number of times some event happens. These four characteristics must be met for the binomial model to be applied:

1. There must be a fixed number of trials, *n*.
2. Each trial has exactly two possible outcomes.
3. Each of the trials must have the same probability of "success." This probability is represented by the letter *p*.
4. The trials must be independent of one another.

You can find binomial probabilities with technology or sometimes with a table, such as the one in the Appendix A.

It is important to distinguish between continuous and discrete numerical variables, because if the variable has discrete numerical outcomes, then the probability of getting, say, *5 or more* (5 OR 6 OR 7 OR . . .) is different from the probability of getting *more than* 5 (6 OR 7 OR . . .). This is not the case for a continuous numerical variable.

### Formulas

For converting to standardized scores: $z = \dfrac{x - \mu}{\sigma}$

*x* is a measurement.
$\mu$ is the mean of the probability distribution.
$\sigma$ is the standard deviation of the probability distribution.

For binomial models: $\mu = np$

$\mu$ is the binomial mean.
*n* is the number of trials.
*p* is the probability of success of one trial.

For binomial models: $\sigma = \sqrt{np(1 - p)}$

$\sigma$ is the binomial standard deviation.
*n* is the number of trials.
*p* is the probability of success of one trial.

### Sources

Men's cholesterol levels, blood pressures, and heights throughout this chapter: NHANES (www.cdc.gov/nchs/nhanes).

Women's heights and weights throughout this chapter: NHANES (www.cdc.gov/nchs/nhanes).

# SECTION EXERCISES

## SECTION 6.1

**6.1–6.4 Directions** Determine whether each of the following variables would best be modeled as continuous or discrete.

**TRY** **6.1 (Example 1)**

  a. The number of A's given to students in statistics one semester at your school

  b. The distance between two cars on the freeway

**6.2** a. The weight of a newborn puppy

  b. The number of newborn puppies in a litter

**6.3** a. The number of music CDs owned by a student

  b. The length of a newborn baby

**6.4** a. The height of a skyscraper in New York City

  b. The number of people who have climbed to the top of the skyscraper

**TRY** **6.5 Loaded Die (Example 2)** A magician has shaved an edge off one side of a six-sided die, and as a result, the die is no longer "fair." The figure shows a graph of the probability density function (pdf). Show the pdf in table format by listing all six possible outcomes and their probabilities.

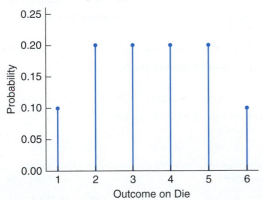

**6.6 Fair Die** Toss a fair six-sided die. The probability density function (pdf) in table form is given. Make a graph of the pdf for the die.

| Number of Spots | 1 | 2 | 3 | 4 | 5 | 6 |
|---|---|---|---|---|---|---|
| Probability | 1/6 | 1/6 | 1/6 | 1/6 | 1/6 | 1/6 |

**\*6.7 Fair Die, Again** Roll a fair six-sided die. You win $3 if you roll a 1, you lose $4 if you roll a 5 or a 6, and for any other outcome you win or lose nothing.

  a. Complete the table that shows the probability distribution.

  b. Complete the probability distribution graph.

| Winnings | Probability |
|---|---|
| $3 | 1/6 |
| $0 | — |
| −$4 | — |

**6.8 Fair Coin** Flip a fair coin. You win $10 if it comes up heads and lose $15 if it comes up tails.

  a. Create a table that shows the probability distribution of *winning*. For the loss of $15 use −$15.

  b. Create a graph that shows the probability density function.

  c. If someone offered to play this game with you, would it be sensible to play? Explain.

**TRY** **6.9 Snow Depth (Example 3)** Eric wants to go skiing tomorrow, but only if there are 3 inches or more of new snow. According to the weather report, any amount of new snow between 1 inch and 6 inches is equally likely. The probability density curve for tomorrow's new snow depth is shown. Find the probability that the new snow depth will be three inches or more tomorrow. Copy the graph, shade the appropriate area, and calculate its numerical value to find the probability. The total area is 1.

**6.10 Snow Depth** Refer to Exercise 6.9. What is the probability that the amount of new snow will be between 2 and 4 inches? Copy the graph from Exercise 6.9, shade the appropriate area, and report the numerical value of the probability.

## SECTION 6.2

**6.11 Applying the Empirical Rule with *z*-Scores** The Empirical Rule applies rough approximations to probabilities for any unimodal, symmetric distribution. But for the Normal distribution we can be more precise, as the figure shows. Use the figure and the fact that the Normal curve is symmetric to answer the questions. Do not use a Normal table or technology.

According to the Empirical Rule:

a. Roughly what percentage of z-scores are between −2 and 2?

   i. almost all    ii. 95%    iii. 68%    iv. 50%

b. Roughly what percentage of z-scores are between −3 and 3?

   i. almost all    ii. 95%    iii. 68%    iv. 50%

c. Roughly what percentage of z-scores are between −1 and 1.

   i. almost all    ii. 95%    iii. 68%    iv. 50%

d. Roughly what percentage of z-scores are more than 0?

   i. almost all    ii. 95%    iii. 68%    iv. 50%

e. Roughly what percentage of z-scores are between 1 and 2?

   i. almost all    ii. 13.5%    iii. 50%    iv. 2%

**6.12 IQs** Wechsler IQs are approximately Normally distributed with a mean of 100 and standard deviation of 15. Use the probabilities shown in the figure in Exercise 6.9 to answer the following questions. Do *not* use the Normal table or technology. You may want to label the figure with Empirical Rule probabilities to help you think about this question.

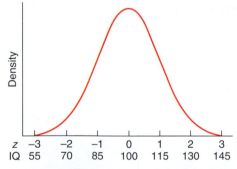

a. Roughly what percentage of people have IQs more than 100?

   i. almost all    ii. 95%    iii. 68%    iv. 50%

b. Roughly what percentage of people have IQs between 100 and 115?

   i. 34%    ii. 17%    iii. 2.5%    iv. 50%

c. Roughly what percentage of people have IQs below 55?

   i. almost all    ii. 50%    iii. 34%    iv. about 0%

d. Roughly what percentage of people have IQs between 70 and 130?

   i. almost all    ii. 95%    iii. 68%    iv. 50%

e. Roughly what percentage of people have IQs above 130?

   i. 34%    ii. 17%    iii. 2.5%    iv. 50%

f. Roughly what percentage people have IQs above 145?

   i. almost all    ii. 50%    iii. 34%    iv. about 0%

**6.13 SAT Scores** Quantitative SAT scores are approximately Normally distributed with a mean of 500 and a standard deviation of 100. On the horizontal axis of the graph, indicate the SAT scores that correspond with the provided z-scores. (See the labeling in Exercise 6.12.) Answer the questions using *only* your knowledge of the Empirical Rule and symmetry.

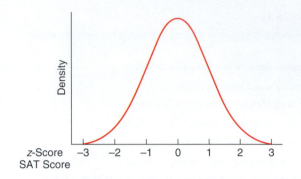

a. Roughly what percentage of students earn quantitative SAT scores more than 500?

   i. almost all    ii. 75%    iii. 50%    iv. 25%    v. about 0%

b. Roughly what percentage of students earn quantitative SAT scores between 400 and 600?

   i. almost all    ii. 95%    iii. 68%    iv. 34%    v. about 0%

c. Roughly what percentage of students earn quantitative SAT scores more than 800?

   i. almost all    ii. 95%    iii. 68%    iv. 34%    v. about 0%

d. Roughly what percentage of students earn quantitative SAT scores less than 200?

   i. almost all    ii. 95%    iii. 68%    iv. 34%    v. about 0%

e. Roughly what percentage of students earn quantitative SAT scores between 300 and 700?

   i. almost all    ii. 95%    iii. 68%    iv. 34%    v. 2.5%

f. Roughly what percentage of students earn quantitative SAT scores between 700 and 800?

   i. almost all    ii. 95%    iii. 68%    iv. 34%    v. 2.5%

**6.14 Women's Heights** Assume that college women's heights are approximately Normally distributed with a mean of 65 inches and a standard deviation of 2.5 inches. On the horizontal axis of the graph, indicate the heights that correspond to the z-scores provided. (See the labeling in Exercise 6.12.) Use only the Empirical Rule to choose your answers. Sixty inches is 5 feet and 72 inches is 6 feet.

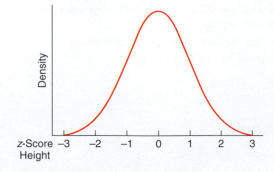

a. Roughly what percentage of women's heights are more than 72.5 inches?

   i. almost all    ii. 75%    iii. 50%    iv. 25%    v. about 0%

b. Roughly what percentage of women's heights are between 60 and 70 inches?

   i. almost all    ii. 95%    iii. 68%    iv. 34%    v. about 0%

c. Roughly what percentage of women's heights are between 65 and 67.5 inches?

   i. almost all    ii. 95%    iii. 68%    iv. 34%    v. about 0%

d. Roughly what percentage of women's heights are between 62.5 and 67.5 inches?

   i. almost all    ii. 95%    iii. 68%    iv. 34%    v. about 0%

e. Roughly what percentage of women's heights are less than 57.5 inches?

   i. almost all    ii. 95%    iii. 68%    iv. 34%    v. about 0%

f. Roughly what percentage of women's heights are between 65 and 70 inches?

   i. almost all    ii. 95%    iii. 47.5%    iv. 34%    v. 2.5%

**TRY 6.15 IQs (Example 4)** Wechsler IQs have a population mean of 100 and a population standard deviation of 15 and are approximately Normally distributed. Use one of the StatCrunch outputs to find the probability that a randomly selected person will have an IQ of 95 or above. State which output is correct for this question, figure (A) or figure (B).

(A)

(B)

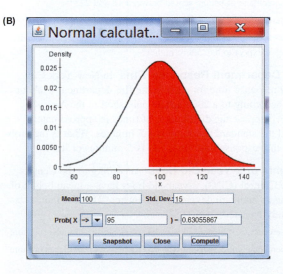

**6.16 SAT Scores** Quantitative SAT scores have a population mean of 500 and a population standard deviation of 100 and are approximately Normally distributed. Find the probability that a randomly selected person will have a quantitative SAT score of 650 or less by looking at the StatCrunch output.

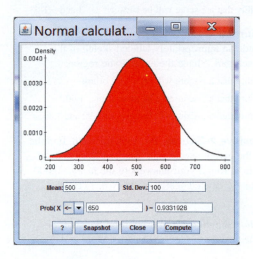

**6.17 Standard Normal** Use the table or technology to answer each question. Include an appropriately labeled sketch of the Normal curve for each part. Shade the appropriate region.

a. Find the area in a standard Normal curve to the left of 1.02 by using the Normal table. (See the excerpt provided.) Note the shaded curve.

b. Find the area in a standard Normal curve to the right of 1.02. Remember that the total area under the curve is 1.

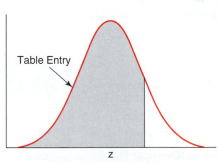

**Format of the Normal Table:** The area given is the area to the *left* of (less than) the given z-score.

| z | .00 | .01 | .02 | .03 | .04 |
|---|-----|-----|-----|-----|-----|
| **0.9** | .8159 | .8186 | .8212 | .8238 | .8264 |
| **1.0** | .8413 | .8438 | **.8461** | .8485 | .8508 |
| **1.1** | .8643 | .8665 | .8686 | .8708 | .8729 |

**6.18 Standard Normal** Use a table or technology to answer each question. Include an appropriately labeled sketch of the Normal curve for each part. Shade the appropriate region.

a. Find the area to the left of a z-score of −0.50.

b. Find the area to the right of a z-score of −0.50.

**6.19 Standard Normal** Use a table or technology to answer each question. Include an appropriately labeled sketch of the Normal curve for each part. Shade the appropriate region.

a. Find the probability that a *z*-score will be 1.76 or less.

b. Find the probability that a *z*-score will be 1.76 or more.

c. Find the probability that a *z*-score will be between −1.3 and −1.03.

**6.20 Standard Normal** Use a table or technology to answer each question. Include an appropriately labeled sketch of the Normal curve for each part. Shade the appropriate region.

a. Find the probability that a *z*-score will be −1.00 or less.

b. Find the probability that a *z*-score will be more than −1.00.

c. Find the probability that a *z*-score will be between 0.90 and 1.80.

**6.21 Extreme Positive *z*-Scores** For each question, find the area to the right of the given *z*-score in a standard Normal distribution. In this question, round your answers to the nearest 0.000. Include an appropriately labeled sketch of the *N*(0, 1) curve

a. *z* = 4.00

b. *z* = 10.00 (*Hint*: Should this tail proportion be larger or smaller than the answer to part a? Draw a picture and think about it.)

c. *z* = 50.00

d. If you had the *exact* probability for these tail proportions, which would be the biggest and which would be the smallest?

e. Which is equal to the area in part b: the area below (to the left of) *z* = −10.00 or the area above (to the right of) *z* = −10.00?

**6.22 Extreme Negative *z*-Scores** For each question, find the area to the right of the given *z*-score in a standard Normal distribution. In *this* question, round your answers to the nearest 0.000. Include an appropriately labeled sketch of the *N*(0, 1) curve.

a. *z* = −4.00

b. *z* = −8.00

c. *z* = −30.00

d. If you had the *exact* probability for these right proportions, which would be the biggest and which would be the smallest?

e. Which is equal to the area in part b: the area below (to the left of) *z* = 8.00 or the area above (to the right of) *z* = 8.00?

**TRY**
**g**
**6.23 Females' SAT Scores (Example 5)** According to data from the College Board, the mean quantitative SAT score for female college-bound high school seniors in 2009 was 500. Assume that SAT scores are approximately Normally distributed with a population standard deviation of 100. What percentage of female college-bound students had scores above 675? Please include a well-labeled Normal curve as part of your answer. *See page 287 for guidance.*

**6.24 Males' SAT Scores** According to data from the College Board, the mean quantitative SAT score for male college-bound high school seniors in 2009 was 530. Assume that SAT scores are approximately Normally distributed with a population standard deviation of 100. If a male college-bound high school senior is selected at random, what is the probability that he will score higher than 675?

**6.25 Stanford–Binet IQs** Stanford–Binet IQ scores for children are approximately Normally distributed and have *μ* = 100 and *σ* = 15. What is the probability that a randomly selected child will have an IQ below 115?

**6.26 Stanford–Binet IQs** Stanford–Binet IQs for children are approximately Normally distributed and have *μ* = 100 and *σ* = 15. What is the probability that a randomly selected child will have an IQ of 115 or higher?

**TRY** **6.27 Gestation Periods (Example 6)** The human gestation period (length of pregnancy) is 270 days with a standard deviation of 9 days and is approximately Normally distributed (assuming spontaneous labor). What is the probability that a randomly selected woman's gestation period will be between 261 and 279 days?

**6.28 Birth Weights** According to the National Vital Statistics, full-term babies' birth weights are approximately Normally distributed with a mean of 7.5 pounds and a standard deviation of 1.1 pounds. Some physicians believe that a birth weight of 5.5 pounds or less is dangerous.

a. For a randomly selected full-term pregnancy, what is the probability that the baby's birth weight is 5.5 pounds or less?

b. For a randomly selected full-term pregnancy, what is the probability that the baby's birth weight is between 6.4 and 8.6 pounds?

c. If there were 600 full-term babies born at a hospital in 2010, how many of them would you expect to weigh 5.5 pounds or less?

**6.29 Total Cholesterol** According to the Centers for Disease Control, the mean total cholesterol for men between the ages of 20 and 29 is 180 micrograms per deciliter with a standard deviation of 36.2. A healthy total cholesterol level is less than 200, 200–240 is borderline, and above 240 is dangerous. Assume the distribution is approximately Normal.

a. For a randomly selected man from this group, what is the probability that his total cholesterol level is 200 or more?

b. For a randomly selected man from this group, what is the probability that his total cholesterol level is 240 or more?

★ c. If two randomly selected men are chosen from this group, what is the probability that both will have a total cholesterol level of 200 or more? Assume independence.

d. If 750 randomly selected men are chosen from this group, how many (the count, not the percentage) would you expect to have a total cholesterol level of 200 or more?

**6.30 ACT Scores** In 2009, ACT reading scores had a mean of 21.4 (http://www.act.org). Suppose the population standard deviation is 5 and the distribution of ACT reading scores is approximately Normal.

a. What percentage of people scored between 18 and 22?

b. What percentage of people scored 25.8 or higher?

c. If scores for 880 randomly selected people were obtained, how many would you expect to be 25.8 or higher?

**6.31 Fire Department Response Time** In New York City the average response time for calls to the fire department is about 5 minutes, according to a 2008 article published in the *New York Daily News*. Suppose the distribution of times is approximately Normal and the standard deviation is 1.5 minutes. What is the probability that the response time for a call is 5 minutes or less?

**6.32 Heights of 3-Year-Old Boys** According to the Centers for Disease Control (CDC), 3-year-old boys have a mean height of 38 inches and a standard deviation of 2 inches. The distribution is approximately Normal.

a. What percentage of 3-year-old boys have a height of 39 inches or more?

b. What percentage of 3-year-old boys have a height of 38 inches or more?

**6.33 Heights of 10-Year-Old Girls** According to the CDC, 10-year-old girls have a mean height of 54.5 inches and a standard deviation of 2.5 inches. The distribution of heights is approximately Normal.

a. Find the percentage of 10-year-old girls with a height of 54.5 inches or more.

b. Find the percentage of 10-year-old girls with a height of 52.5 inches or more.

**6.34 Boston Weather** Boston's mean minimum daily temperature in February is 23°F (http://www.city-data.com). Suppose the standard deviation of the minimum temperature is 8°F and the distribution of minimum temperatures in February is approximately Normal. What percentage of days in February have minimum temperatures below freezing (32°F)?

**6.35 New York City Weather** New York City's mean minimum daily temperature in February is 27°F (http://www.ny.com). Suppose the standard deviation of the minimum temperature is 6°F and that the distribution of minimum temperatures in February is approximately Normal. What percentage of days in February has minimum temperatures below freezing (32°F)?

* **6.36 Women's Heights** Assume for this question that college women's heights are approximately Normally distributed with a mean of 64.6 inches and a standard deviation of 2.6 inches. Draw a well-labeled Normal curve for each part.

a. Find the percentage of women who should have heights of 63.5 inches or less.

b. In a sample of 123 women, according to the probability obtained in part a, how many should have heights of 63.5 inches or less?

c. The table shows the frequencies of heights for a sample of women, collected by statistician Brian Joiner in his statistics class. Count the women who have heights of 63.5 inches or less by looking at the table. They are in the oval.

d. Are the answers to parts b and c the same or different? Explain.

| Height | Inches | Frequency |
|--------|--------|-----------|
|        | 59     | 2         |
| 5′     | 60     | 5         |
|        | 61     | 7         |
| 5′2″   | 62     | 10        |
|        | 63     | 16        |
| 5′4″   | 64     | 23        |
|        | 65     | 19        |
| 5′6″   | 66     | 15        |
|        | 67     | 9         |
| 5′8″   | 68     | 6         |
|        | 69     | 6         |
| 5′10″  | 70     | 3         |
|        | 71     | 1         |
| 6′     | 72     | 1         |

**TRY 6.37 Probability or Measurement (Inverse)? (Example 7)** The Normal model $N(270, 9)$ describes the length of pregnancies, in days, for women who go into spontaneous labor. Which of these statements is asking for a measurement (is an inverse Normal question) and which is asking for a probability?

a. If 60% of women have a pregnancy length below this number of days, what is the number of days?

b. If we select, at random, a woman who goes into spontaneous labor, what is the probability that her pregnancy will be 279 days or fewer?

**6.38 Probability or Measurement (Inverse)?** The Normal model $N(100, 15)$ describes Wechsler IQs. Which of these statements is asking for a measurement (is an inverse Normal question) and which is asking for a probability?

a. What percentage of people have IQs above 130?

b. To get into Mensa, a high-IQ group, you must have an IQ in the upper 2%. What IQ is this?

**6.39 Inverse Normal, Standard** In a standard Normal distribution, if the area to the left of a z-score is about 0.6666, what is the approximate z-score?

First locate inside the table, the number closest to 0.6666. Then find the z-score by adding 0.4 and 0.03; refer to the table. Draw a sketch of the Normal curve, showing the area and z-score.

| z   | .00   | .01   | .02   | .03   | .04   | .05   |
|-----|-------|-------|-------|-------|-------|-------|
| 0.4 | .6554 | .6591 | .6628 | **.6664** | .6700 | .6736 |
| 0.5 | .6915 | .6950 | .6985 | .7019 | .7054 | .7088 |
| 0.6 | .7257 | .7291 | .7324 | .7357 | .7389 | .7422 |

**6.40 Inverse Normal, Standard** In a standard Normal distribution, if the area to the left of a z-score is about 0.1000, what is the approximate z-score?

**6.41 Inverse Normal, Standard** Assume a standard Normal distribution. Draw a separate, well-labeled Normal curve for each part.

a. Find the z-score that gives a left area of 0.7123.

b. Find the z-score that gives a left area of 0.1587.

**6.42 Inverse Normal, Standard** Assume a standard Normal distribution. Draw a separate, well-labeled Normal curve for each part.

a. Find an approximate z-score that gives a left area of 0.7000.

b. Find an approximate z-score that gives a left area of 0.9500.

**TRY 6.43 Females' SAT Scores (Example 8)** According to **g** the College Board, the mean quantitative SAT score for female college-bound high school seniors in 2009 was 500. SAT scores are approximately Normally distributed with a population standard deviation of 100. A scholarship committee wants to give awards to college-bound women who score at the 96th percentile or above on the SAT. What score does an applicant need? Include a well-labeled Normal curve as part of your answer. *See page 288 for guidance.*

**6.44 Males' SAT Scores** According to the College Board, the mean quantitative SAT score for male college-bound high school seniors in 2009 was 530. SAT scores are approximately Normally distributed with a population standard deviation of 100. What is the SAT score at the 96th percentile for male college-bound seniors?

**6.45 Males' Body Temperatures** A study showed that males' body temperatures are approximately Normally distributed with a mean of 98.1°F and a population standard deviation of 0.70°F. What body temperature does a male have if he is at the 90th percentile? Draw a well-labeled sketch to support your answer.

**6.46 Females' Body Temperatures** A study showed that females'body temperatures are approximately Normally distributed with a mean of 98.4°F and a population standard deviation of 0.70°F. Find the female body temperature at the 90th percentile. Draw a well-labeled sketch to support your answer.

**6.47 Women's Heights** Suppose college women's heights are approximately Normally distributed with a mean of 65 inches and a population standard deviation of 2.5 inches. What height is at the 20th percentile? Include an appropriately labeled sketch of the Normal curve to support your answer.

**6.48 Men's Heights** Suppose college men's heights are approximately Normally distributed with a mean of 70.0 inches and a population standard deviation of 3 inches. What height is at the 20th percentile? Include an appropriately labeled Normal curve to support your answer.

**6.49 Wechsler IQs** Wechsler IQs have a mean of 100 and a standard deviation of 15 and are approximately Normally distributed.

a. The Wechsler IQ at the 25th percentile is 90. What is the Wechsler IQ at the 75th percentile?

b. The interquartile range is Q3 minus Q1. Find the interquartile range for Wechsler IQs.

c. Is the interquartile range larger or smaller than the standard deviation, 15 IQ points?

**6.50 Gestation Period** The gestation period for humans is approximately Normally distributed with a mean of 270 days and population standard deviation of 9, assuming spontaneous labor. The gestation period at the 25th percentile is 264 days.

a. Find the gestation period at the 75th percentile.

b. Find the interquartile range of the gestation period.

c. Using the distribution of gestation periods, is the standard deviation larger or smaller than the interquartile range?

**6.51 Child and Adult Heights for Males** According to the National Center for Health Statistics, 3-year-old boys have a mean height of 38 inches and a standard deviation of 2 inches. Assume the distribution is approximately Normal.

a. Find the percentile measure for a height of 40 inches for a 3-year-old boy.

b. If this 3-year-old boy grows up to be a man with a height at the same percentile, what will his height be? Use a population mean of 70 inches and a population standard deviation of 3 inches.

**6.52 Child and Adult Heights for Females** According to the CDC, 10-year-old girls have a mean height of 54.5 inches and a standard deviation of 2.5 inches, and the distribution of heights is approximately Normal.

a. Find the percentile measure for a height of 57 inches for a 10-year-old girl.

b. If this 10-year-old girl grows up to be a woman with a height at the same percentile, what will her height be? Use a population mean of 65 inches and a population standard deviation of 2.5 inches.

## SECTION 6.3

TRY **6.53 Gender of Children (Example 9)** A married couple plans to have four children, and they are wondering how many boys they should expect to have. Assume none of the children will be twins or other multiple births. Also assume the probability that a child will be a boy is 0.50. Explain why this is a binomial experiment. Check all four required conditions.

**6.54 Coin Flip** A coin will be flipped three times, and the number of heads recorded. Explain why this is a binomial experiment. Check all four required conditions.

TRY **6.55 Coin Flips (Example 10)** A teacher wants to find out whether coin flips of pennies have a 50% chance of coming up heads. In the last 5 minutes of class, he has all the students flip pennies until the end of class and then report their results to him. Which condition or conditions for use of the binomial model is or are not met?

**6.56 Twins** In Exercise 6.53 you are told to assume that none of the children will be twins or other multiple births. Why? Which of the conditions required for a binomial experiment would be violated if there were twins?

**6.57 Divorce** Suppose that the probability that a randomly selected person who has recently married for the first time will get a divorce within 5 years is 0.2. Suppose we follow 12 married couples (24 people) for 5 years and record the number of people divorced. Why is the binomial model inappropriate for finding the probability that at least 7 of these 24 people will be divorced within 5 years? List all binomial conditions that are not met.

**6.58 Divorce** Suppose that the probability that a randomly selected person who has recently married for the first time will be divorced within 5 years is 0.2, and that the probability that a randomly selected person who has recently married for the second time will be divorced within 5 years is 0.30. Take a random sample of 10 people married for the first time and 10 people married for the second time. The sample is chosen such that no one in the sample is married to anyone else in the sample. Why is the binomial model inappropriate for finding the probability that exactly 4 of the 20 people in the sample will be divorced within 5 years? List all of the binomial conditions that are not met.

TRY **6.59 Identifying n, p, and x (Example 11)** For each situation, identify the sample size $n$, the probability of success $p$, and the number of successes $x$. Give the answer in the form $b(n, p, x)$. Do *not* go on to find the probability. Assume the four conditions for a binomial experiment are satisfied.

a. In the 2008 presidential election, 54% of the voters voted for President Obama. What is the probability that 65 out of 100 independently chosen voters voted for President Obama?

b. The manufacturer of LoJack Stolen Vehicle Recovery System claims that the probability that a stolen vehicle using LoJack will be recovered is 90%. What is the probability that exactly 9 out of 10 independently stolen vehicles with LoJack will be recovered?

c. A student is taking a 10-question multiple-choice test. Each question has four options: a, b, c, and d. One of these four options is correct and three of them are incorrect. What is the probability that the student correctly answers exactly 6 of the 10 questions on the test by guessing?

**6.60 Identifying n, p, and x** For each situation, identify the sample size $n$, the probability of success $p$, and the number of successes $x$.

Give the answer in the form $b(n, p, x)$. Do *not* go on to find the probability. Assume the four conditions for a binomial experiment are satisfied.

a. According to an article published on the *Daily Business Review* website, the rate of response to a jury summons is about 40% in Dade County, Florida. What is the probability that 40 out of 50 randomly selected people will respond?

b. Data from the National Comprehensive Auto Theft Research System (CARS) showed that 75% of motor vehicles stolen in Australia from 2005 to 2006 were recovered. What is the probability that in a sample of 10 stolen cars, 6 out of 10 will be recovered, assuming the same rate of recovery?

c. A student is taking a 10-question multiple-choice test. Each question has five options: a, b, c, d, and e. For each question, one of these five choices is correct and four of them are incorrect. What is the probability that the student correctly answers exactly 6 of the 10 questions on the test by guessing?

**TRY 6.61 Stolen Bicycles (Example 12)** According to the *Sidney Morning Herald*, 40% of bicycles stolen in Holland are recovered. (In contrast, only 2% of bikes stolen in New York City are recovered.) Find the probability that, in a sample of 6 randomly selected cases of bicycles stolen in Holland, exactly 2 out of 6 bikes are recovered.

**6.62 Florida Recidivism Rate** The three-year recidivism rate of parolees in Florida is about 30%; that is, 30% of parolees end up back in prison within three years (http://www.floridaperforms.com). Assume that whether one parolee returns to prison is independent of whether any of the others returns.

a. Find the probability that exactly 6 out of 20 parolees will end up back in prison within three years,

b. Find the probability that 6 or fewer out of 20 parolees will end up back in prison within three years.

**6.63 MBA Job Search** Historically, about 90% of MBAs (people who have earned a master's degree in business administration) from UCLA find a job within three months of graduation, according to an article published in *Bloomberg Businessweek* magazine. Assume for the sake of simplicity that whether a graduate finds a job within three months is independent of whether any of the other graduates find jobs.

a. Find the probability that at most 7 (this means 7 or fewer) out of 10 will find a job in their chosen field.

b. Find the probability that at least 8 (this means 8 or more) out of 10 will find a job in their chosen field.

c. Find the probability that anywhere from 7 to 9 out of 10 will find a job in their chosen field. The 7 to 9 is inclusive—that is, it includes the values for 7 and 9.

**6.64 Guessing on Exams** Suppose you are taking an exam with 10 questions and you are required to get 7 or more right answers to pass.

a. With a 10-question true/false test, what is the probability of getting at least 7 answers correct by guessing?

b. With a 10-question multiple-choice test where there are three possible choices for each question, what is the probability of getting at least 7 answers correct by guessing? Only one of the choices is correct for each question.

c. With a 10-question multiple-choice test where there are five possible choices for each question, only one of which is correct, what is the probability of getting at least 7 answers correct by guessing?

d. Which test (of those described in parts a, b, and c in this exercise) would be easiest to pass by guessing, which would be hardest, and why?

**6.65 Newborn Gender** Assume that half of all children born are male and half are female. In all of the following cases, we will assume that there are no twins (or triplets or more) and that the conditions of the binomial model are satisfied.

a. If a woman plans to have two children, what is the probability that both will be girls?

b. If a woman plans to have three children, what is the probability that all will be girls?

c. If a woman plans to have three children, what is the probability that she will have at least one boy? ("At least one boy" is the complement of "all girls.")

d. Does this mean that the more children a woman has, the more likely she will be to have at least one boy? Explain.

**6.66 2010 Employment Rates** According to a 2010 Gallup poll, almost 20% of people were underemployed. That means they either were unemployed or had a job working fewer hours than they would like. (Homemakers and other people not seeking employment are not included in the data.)

a. For one randomly selected person, what is the probability that the person is fully employed (not underemployed)?

For parts b and c, assume that you randomly select five people and inquire about their employment, but you do not include those not seeking employment.

b. What is the probability that all five people are fully employed?

c. What is the probability that at least one of the five is underemployed?

**6.67 Late Flights** For commercial flights in 2009, approximately 80% arrived on time (within 15 minutes of scheduled arrival time), according to the Bureau of Transportation Statistics.

a. Assuming that this success rate still holds, if you randomly select three flights and assume they are independent, what is the probability that all three will arrive on time?

b. What is the probability that at least one of the three flights will be late?

c. If all three flights are on the same day in December and all three are flights to Boston, explain why the binomial model is not appropriate for finding the probability that at least one flight will be late.

**6.68 Crime Clearance** In 2008, the overall U.S. crime clearance rate for violent crimes was 45%, according to the FBI. The term *clearance* means that an arrest is made, a crime is charged, and a case is referred to a court. Consider 10 violent crimes that are unrelated to each other, using the 45% clearance rate.

a. What is the probability that they are all cleared?

b. What is the probability that none of them is cleared?

c. What is the probability that at least one is cleared?

d. How many of the 10 would you expect to be cleared, on average?

**TRY 6.69 DWI Convictions (Example 13)** In New Mexico, about 70% of drivers who are arrested for driving while intoxicated (DWI) are convicted (http://www.drunkdrivingduilawblog.com).

a. If 15 independently selected drivers were arrested for DWI, how many of them would you expect to be convicted?

b. What is the probability that exactly 11 out of 15 independent selected drivers are convicted?

c. What is the probability that 11 or fewer are convicted?

\* **6.70 Internet Access** A 2010 Nielsen poll indicated that about 60% of U.S. households had access to a high-speed Internet connection.

a. Suppose 100 households were randomly selected from the United States. How many of the households would you expect to have access to a high-speed Internet connection?

b. If 10 households are selected randomly, what is the probability that exactly 6 have high-speed access?

c. If 10 households are selected randomly, what is the probability that 6 or fewer have high-speed access?

\* **6.71 Cell Phone Only** According to a Pew poll done in 2010, 50% of adults aged 25–29 had only a cell phone (no landline). Assume that two randomly selected adults (aged 25–29) are asked whether they have only a cell phone.

a. If a person has only a cell phone, we will record Y (for Yes). If not, we will record N. List all possible sequences of Y and N for this experiment.

b. For each sequence, find by hand the probability that it will occur, assuming each outcome is independent.

c. What is the probability that neither of the two randomly selected adults has only a cell phone?

d. What is the probability that exactly one person out of two has only a cell phone?

e. What is the probability that two out of two (both) have only a cell phone?

\* **6.72 Texting While Driving** According to a Pew poll in 2010, 1 in 4 teens of driving age have reported having sent or received a text message while driving. Assume that we randomly sample two teens of driving age.

a. If a teen has texted while driving, record Y; if not, record N. List all possible sequences of Y and N for this experiment.

b. For each sequence, find by hand the probability that it will occur, assuming each outcome is independent.

c. What is the probability that neither of the two randomly selected teens has texted?

d. What is the probability that exactly one out of the two teens has texted?

e. What is the probability that both have texted?

TRY **6.73 Coin Flip (Example 14)** A fair coin is flipped 50 times.

a. What is the expected number of heads?

b. Find the standard deviation for the number of heads.

c. How many heads should you expect, give or take how many? Give the range of the number of heads based on these numbers.

**6.74 Illinois Drivers** According to a 2010 *Chicago Sun Times* article, 20% of driver's license applicants fail the written test in Illinois. If 200 people take the driver's license exam, how many people should we expect to pass? Give or take how many?

TRY **6.75 Women without Children (Example 15)** According to a Pew Research poll in 2010, 20% of women in the United States have ended their childbearing years without having children. (In the 1970s, this number was 10%.)

a. If we randomly select 300 women, how many would we expect to have had no children? Give or take how many?

b. Give the range of likely values from 1 standard deviation above the mean to 1 standard deviation below the mean.

c. If you found that 62 out of 300 randomly sampled women ended their childbearing years without having children, would you be surprised? Why or why not?

**6.76 Interracial Marriage** Pew Research reported that 15% of marriages in the United States in 2008 were between people of different races. Suppose 100 random marriages were surveyed.

a. How many marriages would you expect to be between people of different races, give or take how many?

b. Give the range of likely values from 1 standard deviation above the mean to 1 standard deviation below the mean.

c. If you found that only 5 out of 100 randomly selected marriages were between people of different races, would you be surprised? Explain.

## CHAPTER REVIEW EXERCISES

**6.77 Birth Length** A study of U.S. births published on the website *Medscape from WebMD* reported that the average birth length of babies was 20.5 inches and the standard deviation was about 0.90 inch. Assume the distribution is approximately Normal. Find the percentage of babies with birth lengths of 22 inches or less.

**6.78 Birth Length** A study of U.S. births published on the website *Medscape from WebMD* reported that the average birth length of babies was 20.5 inches and the standard deviation was about 0.90 inch. Assume the distribution is approximately Normal. Find the percentage of babies who have lengths of 19 inches or less at birth.

**6.79 Males' Body Temperature** A study of human body temperatures using healthy men showed a mean of 98.1°F and a standard deviation of 0.70°F. Assume the temperatures are approximately Normally distributed.

a. Find the percentage of healthy men with temperatures below 98.6°F (that temperature was considered typical for many decades).

b. What temperature does a healthy man have if his temperature is at the 76th percentile?

**6.80 Females' Body Temperature** A study of human body temperatures using healthy women showed a mean of 98.4°F and a standard deviation of about 0.70°F. Assume the temperatures are approximately Normally distributed.

a. Find the percentage of healthy women with temperatures below 98.6°F (this temperature was considered typical for many decades).

b. What temperature does a healthy woman have if her temperature is at the 76th percentile?

**6.81 Baby's Gender** A woman with five sons and no daughters becomes pregnant. What is the likelihood that her next child will be a girl? Write a sentence or two explaining your answer.

**6.82 Coin Flip** You flip a coin and get heads four times in a row. What is the likelihood that the next flip will show heads? Write a sentence or two explaining your answer.

**6.83 Crime** According to a 2009 Gallup poll, 30% of U.S. households were the victims of a crime in the last year.

a. If there were 2000 independent households surveyed, how many would you expect to have been the victims of a crime, give or take how many?

b. Judging on the basis of your answer to part a, what range of households would you expect to have been the victims of a crime?

**6.84 Cesarean Births** According to the CDC, the rate of Cesarean births in the United States in 2007 was 32%.

a. If 200 independent births were surveyed in 2007, how many of those would you expect to be Cesarean, give or take how many?

b. Would it be unusual for a hospital with 200 births to have 110 Cesareans? Explain.

**6.85 Leukemia Survival Rate** The five-year survival rate for chronic myelogenous leukemia (CML) for those on the drug Gleevec is about 90%. Before Gleevec was introduced, the five-year survival rate was about 40%.

a. Suppose four patients have just been diagnosed with CML and will be using Gleevec. Assume that one patient's survival is independent of the survival of the others.

  i. What is the probability that the patients will all be alive in five years?

  ii. What is the probability that at least one will die within five years?

b. Suppose four patients had been diagnosed with CML 25 years ago, before Gleevec was available. Assume that one patient's survival was independent of the survival of the others.

  i. What is the probability that they all would have stayed alive for five years?

  ii. What is the probability that at least one would have been dead within five years?

c. Why is the answer to Question ii in part a smaller than the answer to Question ii in part b?

**6.86 Physician-assisted Suicide** A Field poll in 2005 showed that 70% of people polled in California answered yes to the question "Should incurably ill patients have the right to receive life-ending medication?" Assume that 70% is the correct population proportion agreeing with the statement.

a. If 300 randomly selected people in California were asked this question, how many of them would you expect to answer yes? Give or take how many?

b. Would you be surprised if 150 of the 300 answered yes to this question? Explain.

# gUIDED EXERCISES

**g 6.23 Females' SAT Scores** According to data from the College Board, the mean quantitative SAT score for female college-bound high school seniors in 2009 was 500. SAT scores are approximately Normally distributed with a population standard deviation of 100.

**QUESTION** What percentage of the female college-bound high school seniors had scores above 675? Answer this question by following the numbered steps.

**Step 1 ▶ Find the z-score**
To find the z-score for 675, subtract the mean and divide by the standard deviation. Report the z-score.

**Step 2 ▶ Explain the location of 500**
Refer to the Normal curve. Explain why the SAT score of 500 is right below the z-score of 0. The tick marks on the axis mark the location of z-scores that are integers from −3 to 3.

**Step 3 ▶ Label with SAT scores**
Carefully sketch a copy of the curve. Pencil in the SAT scores of 200, 300, 400, 600, and 700 in the correct places.

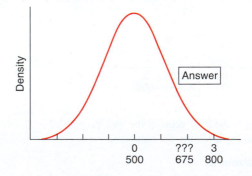

**Step 4 ▶ Add the line, z-score, and shading**
Draw a vertical line through the curve at the location of 675. Just above the 675 (indicated on the graph with "???") put in the corresponding z-score. We want to find what percentage of students had scores *above* 675. Therefore, shade the area to the *right* of this boundary, because numbers to the right are larger.

### Step 5 ▶ Use the table for the left area

Use the following excerpt from the Normal table to find and report the area to the left of the z-score that was obtained from an SAT score of 675. This is the area of the unshaded region.

### Step 6 ▶ Answer

Because you want the area to the right of the z-score, you will have to subtract the area you obtained in step 5 from 1. This is the area of the shaded region. Put it where the box labeled "Answer" is. Check to see that the number makes sense. For example if the shading is less than half the area, the answer should not be more than 0.5000.

### Step 7 ▶ Sentence

Finally, write a sentence telling what you found.

| z | .00 | .01 | .02 | .03 | .04 | .05 | .06 | .07 | .08 | .09 |
|---|-----|-----|-----|-----|-----|-----|-----|-----|-----|-----|
| 1.6 | .9452 | .9463 | .9474 | .9484 | .9495 | *.9505 | .9515 | .9525 | .9535 | .9545 |
| 1.7 | .9554 | .9564 | .9573 | .9582 | .9591 | **.9599** | .9608 | .9616 | .9625 | .9633 |
| 1.8 | .9641 | .9649 | .9656 | .9664 | .9671 | .9678 | .9686 | .9693 | .9699 | .9706 |

**g 6.43 Females' SAT scores**   According to the College Board, the mean quantitative SAT score for female college-bound high school seniors in 2009 was 500. SAT scores are approximately Normally distributed with a population standard deviation of 100. A scholarship committee wants to give awards to college-bound women who score at the 96th percentile or above on the SAT. What score does an applicant need? Include a well-labeled Normal curve as part of your answer.

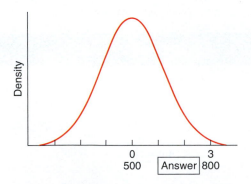

QUESTION   What is the SAT score at the 96th percentile? Answer this question by following the numbered steps.

### Step 1 ▶ Think about it

Will the SAT test score be above the mean or below it? Explain.

### Step 2 ▶ Label z-scores

Label the curve with integer z-scores. The tick marks represent the position of integer z-scores from −3 to 3.

### Step 3 ▶ Use the table

The 96th percentile has 96% of the area to the *left* because it is higher than 96% of the scores. The table below gives the areas to the *left* of z-scores. Therefore, we look for 0.9600 in the interior part of the table.

Use the excerpt of the Normal table given above for Exercise 6.23 to locate the area *closest* to 0.9600.

Then report the z-score for that area.

### Step 4 ▶ Add the z-score, line, and shading to the sketch

Add that z-score to the sketch and draw a vertical line above it through the curve. Shade the left side because the area to the left is what is given.

### Step 5 ▶ Find the SAT score

Find the SAT score that corresponds to the z-score. The score should be z standard deviations above the mean, so

$$x = \mu + z\sigma$$

### Step 6 ▶ Add the SAT score to the sketch

Add the SAT score on the sketch where it says "Answer."

### Step 7 ▶ Write a sentence

Finally, write a sentence stating what you found.

# TechTips

## For All Technology

All technologies will use the two examples that follow.

**EXAMPLE A: NORMAL** ▶ Wechsler IQs have a mean of 100 and standard deviation of 15 and are Normally distributed.

   a. Find the probability that a randomly chosen person will have an IQ between 85 and 115.

   b. Find the probability that a randomly chosen person will have an IQ that is 115 or less.

   c. Find the Wechsler IQ at the 75th percentile.

*Note:* If you want to use technology to find areas from standard units ($z$-scores), use a mean of 0 and a standard deviation of 1.

**EXAMPLE B: BINOMIAL** ▶ Imagine that you are flipping a fair coin (one that comes up heads 50% of the time in the long run).

   a. Find the probability of getting 28 or fewer heads in 50 flips of a fair coin.

   b. Find the probability of getting exactly 28 heads in 50 flips of a fair coin.

---

## TI-83/84

### NORMAL

#### a. Between Two Values

1. Press **2ND DISTR** (located below the four arrows on the keypad).
2. Select **2:normalcdf** and press **ENTER**.
3. Enter: (left boundary, right boundary, $\mu$, $\sigma$), in this example **85 , 115 , 100 , 15**) and press **ENTER**. (The comma button is above the 7 button.)

Your screen should look like Figure 6A, which shows that the probability that a randomly selected person will have a Wechsler IQ between 85 and 115 is equal to 0.6827.

**▲ FIGURE 6A** TI-83/84 normal<u>c</u>df (c stands for "cumulative")

#### b. Some Value or Less

1. Press **2ND DISTR**.
2. Select **2:normalcdf** and press **ENTER**.
3. Enter: (left boundary, right boundary, $\mu$, $\sigma$) and press **ENTER**.

The probability that a person's IQ is 115 *or less* has an *indeterminate* left boundary, for which you may use negative 1000000 or any extreme value that is clearly out of the range of data. Figure 6B shows the probability that a randomly selected person will have an IQ of 115 or less.

```
normalcdf(-10000
00,115,100,15)
            .8413447404
```

**▲ FIGURE 6B** TI-83/84 Normalcdf with indeterminate left boundary

*Caution:* The negative button (−) is to the left of the **ENTER** button and is not the same as the minus button that is above the plus button.

   (If you have an indeterminate right boundary, then to find the probability that the person's IQ is 85 or more, for example, use a right boundary (such as 1000000) that is clearly above all the data.)

#### c. Inverse Normal

If you want a measurement (such as an IQ) from a proportion or percentile:

1. Press **2ND DISTR**.
2. Select **3:invNorm** and press **ENTER**.
3. Enter: (left proportion, $\mu$, $\sigma$) and press **ENTER**. Be sure to include the commas between the numbers.

   Figure 6C shows the Wechsler IQ at the 75th percentile, which is 110. Note that the 75th percentile is entered as **.75**.

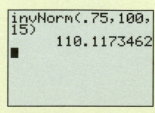

**▲ FIGURE 6C** TI-83/84 Inverse Normal

## Binomial

### a. Cumulative (or Fewer)

1. Press **2ND DISTR.**
2. Select **B:binomcdf** (you will have to scroll down to see it) and press **ENTER**. (On a TI-83, it is **A:binomcdf.**)
3. See Figure 6D. Enter: $(n, p, x)$ and press **ENTER**. Be sure to include the commas between the numbers.

   The answer will be the probability for *x or fewer*. Figure 6D shows the probability of 28 *or fewer* heads out of 50 flips of a fair coin. (You could find the probability of 29 *or more* heads by subtracting your answer from 1.)

▲ **FIGURE 6D** TI-83/4
binom**c**df (cumulative)

### b. Individual (Exact)

1. Press **2ND DISTR.**
2. Select **A:binompdf** and press **ENTER**. (On a TI-83, it is **0:binompdf.**)
3. Enter: $(n, p, x)$ and press **ENTER**. Be sure to include the commas between the numbers.

   Figure 6E shows the probability of *exactly* 28 heads out of 50 flips of a fair coin.

▲ **FIGURE 6E** TI-83/84
binom**p**df (individual)

---

## MINITAB

### Normal

#### a. Between Two Values

1. Enter the upper boundary, **115**, in the top cell of an empty column; here we use column C1, row 1. Enter the lower boundary, **85**, in the cell below; here column C1, row 2.
2. **Calc > Probability Distributions > Normal**.
3. See Figure 6F: Choose **Cumulative probability**. Enter: **Mean, 100; Standard deviation, 15; Input column, C1; Optional storage, C2.**
4. Click **OK**.
5. Subtract the lower probability from the larger shown in column C2.

   $0.8413 - 0.1587 = 0.6836$ is the probability that a Wechsler IQ is between 85 and 115.

#### b. Some Value or Less

1. The probability of an IQ of 115 *or less*, 0.8413, is shown in column C2, row1. (In other words, do as in part a above, except do *not* enter the lower boundary, 85.)

#### c. Inverse Normal

If you want a measurement (such as an IQ or height) from a proportion or percentile:

1. Enter the decimal form of the left proportion (**.75** for the 75th percentile) into a cell in an empty column in the spreadsheet; here we used column C1, row 1.
2. **Calc > Probability Distributions > Normal**.
3. See Figure 6G: Choose **Inverse cumulative probability**. Enter: **Mean, 100; Standard deviation, 15; Input column, c1; and Optional storage, c2** (or an empty column).
4. Click **OK**.

You will get **110**, which is the Wechsler IQ at the 75th percentile.

▲ **FIGURE 6F** Minitab Normal

▲ **FIGURE 6G** Minitab Inverse Normal

## Binomial

### a. Cumulative (or Fewer)

1. Enter the upper bound for the number of successes in an empty column; here we used column C1, row 1. Enter **28** to get the probability of 28 or fewer heads.
2. **Calc > Probability Distributions > Binomial**.
3. See Figure 6H. Choose **Cumulative probability**.

▲ **FIGURE 6H** Minitab Binomial

Enter: **Number of trials**, **50**; **Event probability**, **.5**; **Input column**, **c1**; **Optional storage**, **c2** (or an empty column).

4. Click **OK**.

Your answer will be 0.8389 for the probability of 28 *or fewer* heads.

### b. Individual (Exact)

1. Enter the number of successes at the top of column 1, **28** for 28 heads.
2. **Calc > Probability Distributions > Binomial**.
3. Choose **Probability** (at the top of Figure 6H) instead of **Cumulative Probability** and enter: **Number of trials**, **50**; **Event probability**, **.5**; **Input column**, **c1**; **Optional storage**, **c2** (or an empty column).
4. Click **OK**.

Your answer will be 0.0788 for the probability of *exactly* 28 heads.

## EXCEL

### Normal

Unlike the TI-83/84, Excel makes it easier to find the probability that a random person has an IQ of 115 or less) than to find than the probability that a random person has an IQ between 85 and 115). This is why, for Excel, part b appears before part a.

### b. Some Value or Less

1. Click *fx* (and **select a category All**).
2. Choose **NORMDIST**.
3. See Figure 6I. Enter: **X, 115; Mean, 100; Standard_dev, 15; Cumulative, true** (for 115 *or less*). The answer is shown as 0.8413. Click **OK** to make it show up in the active cell on the spreadsheet.

▲ **FIGURE 6I** Excel Normal

### a. Between Two Values

If you want the probability of an IQ between 85 and 115:

1. First, follow the instructions above for part b. *Do not change the active cell in the spreadsheet.*
2. You will see **=NORMDIST(115,100,15,TRUE)** in the *fx* box. Click in this box, to the right of **TRUE)** and put in a minus sign.
3. Now repeat the steps for part b starting by clicking *fx*, except enter 85 instead of 115 for X. The answer, **0.682689**, will be shown in the active cell.

    (Alternatively, just repeat steps 1−3 for part b using 85 instead of 115. Subtract the smaller probability value from the larger (0.8413 − 0.1587 = 0.6826).)

### c. Inverse Normal

If you want a measurement (such as an IQ or height) from a proportion or percentile:

1. Click *fx*.
2. Choose **NORMINV** and click **OK**.
3. See Figure 6J. Enter: **Probability**, **.75** (for the 75th percentile); **Mean, 100**: **Standard_dev**, **15**. You may read the answer off the screen or click **OK** to see it in the active cell in the spreadsheet.

The IQ at the 75th percentile is 110.

▲ FIGURE 6J Excel Inverse Normal

▲ FIGURE 6K Excel Binomial

## Binomial

### a. Cumulative (or Fewer)

1. Click $f_x$.
2. Choose **BINOMDIST** and click **OK**.
3. See Figure 6K. Enter: **Number_s, 28**; **Trials, 50**; **Probability_s, .5**; and **Cumulative, true** (for the probability of 28 *or fewer*).

    The answer (**0.8389**) shows up in the dialogue box and in the active cell when you click **OK**.

### b. Individual (Exact)

1. Click $f_x$.
2. Choose **BINOMDIST** and click **OK**.
3. Use the numbers in Figure 6K, but enter **False** in the Cumulative box. This will give you the probability of getting *exactly* 28 heads in 50 tosses of a fair coin.

    The answer (**0.0788**) shows up in the dialogue box and in the active cell when you click **OK**.

## STATCRUNCH

### Normal

Unlike the TI-83/84, StatCrunch makes it easier to find the probability for 115 or less than to find the probability between 85 and 115. This is why part b is done before part a.

### b. Some Value or Less

1. **Stat > Calculators > Normal**
2. See Figure 6L. To find the probability of having a Wechsler IQ of 115 or less, Enter: **Mean, 100**: **Std Dev, 15**. Make sure that the arrow to the right of **Prob(X** points left (for less than). Enter the **115** in the box above **Snapshot**.
3. Click **Compute** to see the answer, **0.8413**.

### a. Between Two Values

To find the probability of having a Wechsler IQ between 85 and 115, use steps 1, 2, and 3 again, but use **85** instead of **115** in the box above **Snapshot**. When you find that probability, subtract it from the probability found in Figure 6L.

$$0.8413 - 0.1587 = 0.6826$$

### c. Inverse Normal

If you want a measurement (such as an IQ or height) from a proportion or percentile:

1. **Stat > Calculators > Normal**
2. See Figure 6M. To find the Wechsler IQ at the 75th percentile, enter: Mean, **100**; **Std. Dev., 15**. Make sure that the arrow to the right of **Prob(X** points to the left, and enter **0.75** in the box above **Compute**.

▲ FIGURE 6L StatCrunch Normal

▲ FIGURE 6M StatCrunch Inverse Normal

3. Click **Compute** and the answer (**110**) is shown above **Snapshot**.

**Binomial**

### a. Cumulative (or Fewer)

1. **Stat > Calculators > Binomial**
2. See Figure 6N. To find the probability of 28 or fewer heads in 50 tosses of a fair coin, enter: **n, 50;** and **p, 0.5.** The arrow after **Prob(X** should point left (for *less than*). Enter **28** in the box above **Snapshot**.
3. Click **Compute** to see the answer (**0.8389**).

### b. Individual (Exact)

1. **Stat > Calculators > Binomial**
2. To find the probability of exactly 28 heads in 50 tosses of a fair coin, use a screen similar to Figure 6N, but to the right of **Prob(X** choose the equals sign. You will get **0.0788**.

▲ **FIGURE 6N** StartCrunch Binomial

# 7 Survey Sampling and Inference

**If survey subjects are chosen randomly, then we can use their answers to infer how the entire population would answer. We can also quantify how far off our estimate is likely to be.**

Somewhere in your town or city, possibly at this very moment, people are participating in a survey. Perhaps they are filling out a customer satisfaction card at a restaurant. Maybe their television is automatically transmitting information about which show is being watched so that marketers can estimate how many people are viewing their ads. They may even be text messaging in response to a television survey. Most of you will receive at least one phone call from a survey company that will ask whether you are satisfied with local government services or plan to vote for one candidate over another. The information gathered by these surveys is used to piece together, bit by bit, a picture of the larger world.

You've reached a pivotal point in the book. In this chapter, the data summary techniques you learned in Chapters 2 and 3, the probability you learned about in Chapter 5, and the Normal distribution, which you studied in Chapter 6, are all combined to allow us to generalize what we learn about a small sample to a larger group. Politicians rely on surveys of 1000 voters not because they care how those 1000 individuals will vote. Surveys are important to politicians only if they help them learn about *all* potential voters. In this and later chapters, we study ways to understand and measure just how reliable this projection from sample to the larger world is.

Whenever we draw a conclusion about a large group based on observations of some parts of that group, we are making an inference. Inferential reasoning lies at the foundation of science but is far from foolproof. As the following case study illustrates, when we make an inference we can never be absolutely certain of our conclusions. But applying the methods introduced in this chapter ensures that if we collect data carefully, we can at least measure how certain or uncertain we are.

## CASE STUDY

### Spring Break Fever: Just What the Doctors Ordered?

In 2006, the American Medical Association (AMA) issued a press release ("Sex and intoxication among women more common on spring break according to AMA poll") in which it concluded, among other things, that "eighty-three percent of the [female, college-attending] respondents agreed spring break trips involve more or heavier drinking than occurs on college campuses and 74 percent said spring break trips result in increased sexual activity." This survey made big news, particularly since the authors of the study claimed these percentages reflected the opinions not only of the 644 women who responded to the survey but of all women who participated in spring break.

The AMA's website claimed the results were based on "a nationwide random sample of 644 women who . . . currently attend college. . . . The survey has a margin of error of $+/-4$ percentage points at the 95 percent level of confidence." It all sounds very scientific, doesn't it?

However, some survey specialists were suspicious. After Cliff Zukin, a specialist who was president of the American Association for Public Opinion Research, corresponded with the AMA, it changed its website posting to say the results were based not on a random sample, but instead on "a nationwide sample of 644 women . . . who are part of an *online survey panel* . . . [emphasis added]." "Margin of error" is no longer mentioned.

Disagreements over how to interpret these results show just how difficult inference is. In this chapter you'll see why the method used to collect data is so important to inference, and how we use probability, under the correct conditions, to calculate a margin of error to quantify our uncertainty. At the end of the chapter, you'll see why the AMA changed its report.

# Learning about the World through Surveys

Surveys are probably the most often encountered application of statistics. Most news shows, newspapers, and magazines report on surveys or polls several times a week— and during a major election, several times a day. We can learn quite a bit through a survey if the survey is done correctly.

## Survey Terminology

A **population** is a group of objects or people we wish to study. Usually, this group is large—say, the group of all U.S. citizens, or all U.S. citizens between the ages of 13 and 18, or all senior citizens. However, it might be smaller, such as all phone calls made on your cell phone in January. We wish to know the value of a **parameter**, a numerical value that characterizes some aspect of this *population*. For example, political pollsters want to know the percentage of people who say they will vote in the next election. Drunk-driving opponents want to know the percentage of all teenagers with driver's licenses who have drunk alcohol while driving. Designers of passenger airplanes want to know the mean length of passengers' legs so that they can put the seats as close together as possible without having discomfort.

In this book we focus on two frequently used parameters: the mean of a population and the population proportion. This chapter deals with population proportions.

If the population is relatively small, we can find the exact value of the parameter by conducting a census. A **census** is a survey in which every member of the population is measured. For example, if you wish to know the percentage of people in your classroom who are left-handed, you can perform a census. The classroom is the population, and the parameter is the percentage of left-handers. We sometimes try to take a census with a large population (such as the U.S. Census), but such undertakings are too expensive for nongovernmental organizations and are filled with complications caused by trying to track down and count people who may not want to be found. (For example, the U.S. Census tends to undercount poor, urban-dwelling residents, as well as undocumented immigrants.)

In fact, most populations we find interesting are too large for a census. For this reason, we instead observe a smaller sample. A **sample** is a collection of people or objects taken from the population of interest.

Once a sample is collected, we measure the characteristic we're interested in. A **statistic** is a numerical characteristic of a sample of data. We use statistics to estimate parameters. For instance, we might be interested in knowing what proportion of all registered voters will vote in the next national election. The proportion of all registered voters who will vote in the next election is our *parameter*. Our method to estimate this parameter is to survey a small sample. The proportion of the sample who say they will vote in the next election is a *statistic*.

Statistics are sometimes called **estimators**, and the numbers that result are called **estimates**. For example, our *estimator* is the proportion of people in a sample who say they will vote in the next election. When we conduct this survey, we find, perhaps, that 0.75 of the sample say they will vote. This number, 0.75, is our *estimate*.

> **KEY POINT**    A statistic is a number that is based on data and used to estimate the value of a characteristic of the population. Thus it is sometimes called an estimator.

**Statistical inference** is the art and science of drawing conclusions about a population on the basis of observing only a small subset of that population. Statistical inference always involves uncertainty, so an important component of this science is measuring our uncertainty.

## EXAMPLE 1 Pew Poll: Age and the Internet

In January 2009, the Pew Research Center surveyed 1650 adult Internet users to estimate the proportion of Internet users between 18 and 32 years old (Gen Y "millenials"). Common wisdom holds that this group of young adults is among the heavier users of the Internet. The survey found that 30% of the respondents in the sample were between ages 18 and 32.

**QUESTIONS**  Identify the population and sample. What is the parameter of interest? What is the statistic?

**CONCLUSION**  The *population* we wish to study is all American adults who are Internet users. The *sample*, which is taken from the population, consists of the 1650 Internet users surveyed. The *parameter* of interest is the percentage of all Internet users who are in Gen Y (18 to 32 years old). The *statistic*, which is the percentage of the sample who are in Gen Y, is 30%.

**TRY THIS!**  Exercise 7.1

An important difference between statistics and parameters is that statistics are knowable. Any time we collect data, we can find the value of a statistic. In Example 1, we know that 30% of the people surveyed are between 18 and 32 years old. In contrast, a parameter is typically unknown. We do not know for certain the percentage of *all* Internet users between 18 and 32. The only way to find out would be to ask all adult Internet users, and we have neither the time nor the money to do this. Table 7.1 compares the known and the unknown in this situation.

| Unknown | Known |
|---|---|
| *Population*<br>All adult American Internet users | *Sample*<br>A small number of adult American Internet users |
| *Parameter*<br>Percentage of all Internet users who are in Gen Y<br>Mean age of all Americans | *Statistic*<br>Percentage of sample who are in Gen Y<br>Mean age of sample |

◀ **TABLE 7.1** Some examples of unknown quantities we might wish to estimate, and their knowable counterparts.

>  **Looking Back**
>
> **The Sample Proportion: $\hat{p}$**
> Recall that $\hat{p}$, pronounced "p-hat," is the sample proportion, which is the number of successes divided by the number of trials.

Statisticians have developed notation for keeping track of parameters and statistics. In general, Greek characters are used to represent population parameters. For example, $\mu$ (mu, pronounced "mew," like the beginning of *music*) represents the mean of a population. Also, $\sigma$ (sigma) represents the standard deviation of a population. Statistics (estimates based on a sample) are represented by English letters: $\bar{x}$

(pronounced "*x*-bar") is the mean of a sample, and *s* is the standard deviation of a sample, for instance.

One frequently encountered exception is the use of the letter *p* to represent the proportion of a population and $\hat{p}$ (pronounced "*p*-hat") to indicate the proportion of a sample. Table 7.2 summarizes this notation. You've seen most of these symbols before, but this table organizes them in a new way that is important for statistical inference.

▶ **TABLE 7.2** Notation for some commonly used statistics and parameters.

| Statistics (based on data) | | Parameters (typically unknown) | |
|---|---|---|---|
| Sample mean | $\bar{x}$ (x-bar) | Population mean | $\mu$ (mu) |
| Sample standard deviation | $s$ | Population standard deviation | $\sigma$ (sigma) |
| Sample variance | $s^2$ | Population variance | $\sigma^2$ |
| Sample proportion | $\hat{p}$ (p-hat) | Population proportion | $p$ |

## What Could Possibly Go Wrong? The Problem of Bias

Unfortunately, it is far easier to conduct a bad survey than to conduct a good survey. One of the many ways in which we can reach a wrong conclusion is to use a survey method that is biased.

A method is **biased** if it has a tendency to produce an untrue value. Bias can enter a survey in three ways. The first is through **sampling bias**, which results from taking a sample that is not representative of the population. A second way is **measurement bias**, which comes from asking questions that do not produce a true answer. For example, if we ask people their income, they are likely to inflate the value. In this case, we will get a positive (or "upwards") bias: Our estimate will tend to be too high. Measurement bias occurs when measurements tend to record values larger (or smaller) than the true value.

The third way occurs because some statistics are naturally biased. For example, if you use the statistic $10\bar{x}$ to estimate the mean, you'll typically get estimates that are ten times too big. Therefore, even when no measuring or sampling bias is present, you must also take care to use an estimator that is not biased.

> **! Caution**
>
> **Bias**
> Statistical bias is different from the everyday use of the term *bias*. You might perhaps say a friend is biased if she has a strong opinion that affects her judgment. In statistics, bias is a way of measuring the performance of a method over many different applications.

### Measurement Bias

In February 2010, the *Albany Times Union* newspaper reported on two recent surveys to determine the opinions of New York State residents on taxing soda. The Quinnipiac University Polling Institute asked, "There is a proposal for an 'obesity tax' or a 'fat tax' on non-diet sugary soft drinks. Do you support or oppose such a measure?" Forty percent of respondents said they supported the tax. Another firm, Kiley and Company, asked, "Please tell me whether you feel the state should take that step in order to help balance the budget, should seriously consider it, should consider it only as a last resort, or should definitely not consider taking that step: 'Imposing a new 18 percent tax on sodas and other soft drinks containing sugar, which would also reduce childhood obesity.'" Fifty-eight percent supported the tax when asked this question. One or both of these surveys have measurement bias.

A famous example occurred in 1993, when, on the basis of the results of a Roper Organization poll, many U.S. newspapers published headlines similar to this one from the *New York Times*: "1 in 5 in New Survey Express Some Doubt About the Holocaust" (April 20, 1993). Almost a year later, the *New York Times* reported that this alarmingly high percentage of alleged Holocaust doubters could be due to measurement error. The actual question respondents were asked contained a double negative: "Does it seem possible, or does it seem impossible to you, that the Nazi extermination of the Jews never happened?" When Gallup repeated the poll but did not use a double negative, only 9% expressed doubts (*New York Times* 1994).

## Sampling Bias

Writing good survey questions to reduce measurement bias is an art and a science. This book, however, is more concerned with sampling bias, which occurs when the estimation method uses a sample that is not representative of the population. (By "not representative" we mean that the sample is fundamentally different from the population.)

Have you ever heard of Alfred Landon? Unless you're a political science student, you probably haven't. In 1936, Landon was the Republican candidate for U.S. president, running against Franklin Delano Roosevelt. The *Literary Digest*, a popular news magazine, conducted a survey with over 10 million respondents and predicted that Landon would easily win the election with 57% of the vote. The fact that you probably haven't heard of Landon suggests that he didn't win, and in fact, he lost big, setting a record at the time for the lowest number of electoral votes received by a major-party candidate. What went wrong? The *Literary Digest* had a biased sample. The journal relied largely on polling its own readers, and its readers were more well-to-do than the general public and more likely to vote for a Republican. The reputation of the *Literary Digest* was so damaged that two years later it disappeared and was absorbed into *Time* magazine.

The U.S. presidential elections of 2004 and 2008 both had candidates who claimed to have captured the youth vote, and both times, candidates claimed the polls were biased. The reason given was that the surveys used to estimate candidate support relied on landline phones, and many young voters don't own land lines, relying instead on their cell phones. Reminiscent of the 1936 *Literary Digest* poll, these surveys were potentially biased because their sample systematically excluded an important part of the population: those who do not use land lines (Cornish, 2007).

In fact, the Pew Foundation conducted a study after the 2010 congressional elections. This study found that polls that excluded cell phones had a sampling bias in favor of Republican candidates.

Today, the most commonly encountered biased-sample surveys are probably Internet polls. These can be found on many news organization websites. ("Would You Want an Asteroid Named after You?" www.cnn.com, October 3, 2007; "Who Would Wall Street Like to See in the White House in 2008: A Democrat or a Republican?" www.foxnews.com, October 7, 2007). Internet polls suffer from what is sometimes called **voluntary-response bias**. People tend to respond to such surveys only if they have strong feelings about the results; otherwise, why bother? This implies that the sample of respondents is not necessarily representative of the population. Even if the population in this case is, for example, all readers of the Foxnews.com website, the survey may not accurately reflect their views, because the voluntary nature of the survey means the sample will probably be biased. This bias might be even worse if we took the population to be all U.S. residents. Readers of Internet websites may very well not be representative of all U.S. residents, and readers of particular websites such as Fox or CNN might be even less so.

To warn readers of this fact, most Internet polls have a disclaimer: "This is not a scientific poll." What does this mean? It means we should not trust the information reported to tell us anything about anyone other than the people who responded to the poll. (And remember, we can't even trust the counts on an Internet poll, because sometimes nothing prevents people from voting many times.)

A more subtle form of bias happens when those being surveyed fail to answer a question or respond to a survey. Such **nonresponse bias** occurs when people selected to participate in the survey refuse to do so. Refusal can happen for mundane reasons (they're eating dinner when the pollster calls) or for reasons that might have a more serious effect on the results (because people are embarrassed or offended by the question). Those who choose not to respond might have different views about the survey's topics than those who do. For example, questions about income might be refused by those who are embarrassed by what they feel is too low, or too high, an income. If a large percentage declines to participate, then a biased survey could result.

When reading about a survey, it is important to know

1. what percentage of people who were asked to participate actually did so
2. whether the researchers chose people to participate in the survey or people themselves chose to participate.

If a large percentage chosen to participate refused to answer questions, nonresponse bias could result. If people themselves chose whether to participate, there will be voluntary-response bias. If a survey suffers from these problems, its conclusions are suspect.

## Simple Random Sampling Saves the Day

How do we collect a sample that has as little bias as possible and is representative of the population? Only one way works: to take a random sample.

As we explained in Chapter 5, statisticians have a precise definition of *random*. A random sample does not mean that we stand on a street corner and stop whomever we like to ask them to participate in our survey. (Statisticians call this a **convenience sample**, for obvious reasons.) A random sample must be taken in such a way that every person in our population is equally likely to be chosen.

A true random sample is difficult to achieve. (That's a big understatement!) Pollsters have invented many clever ways of pulling this off, often with great success. One basic method that's easy to understand but somewhat difficult to put into practice is **simple random sampling (SRS)**.

In SRS, we draw subjects from the population at random and without replacement. **Without replacement** means that once a subject is selected for a sample, that subject cannot be selected again. This is like dealing cards from a deck. Once a card is dealt for a hand, no one else can get the same card. A result of this method is that every sample of the same fixed size is equally likely to be chosen. As a result, we can produce unbiased estimations of the population parameters of interest and can measure the precision of our estimator.

It can't be emphasized enough that if our sample is not random, there's really nothing we can learn about the population. We can't measure a survey's precision, and we can't know how large or small the bias might be. An unscientific survey is a useless survey for the purposes of learning about a population.

In theory, we can take an SRS by assigning a number to each and every member of the population. We then use a random number table or other random number generator to select our sample, ignoring numbers that appear twice.

> **Details**
>
> Simple random sampling is not the only valid method for statistical inference. Statisticians collect representative samples using other methods, as well (for example, sampling with replacement). What these methods all have in common is that they take samples randomly.

## EXAMPLE 2 Taking a Simple Random Sample

Alberto, Justin, Michael, Audrey, Brandy, and Nicole are in a class.

**QUESTION** Select an SRS of three names from these six names.

**SOLUTION** First assign each person a number, as shown:

| | |
|---|---|
| Alberto | 1 |
| Justin | 2 |
| Michael | 3 |
| Audrey | 4 |
| Brandy | 5 |
| Nicole | 6 |

Next, select three of these numbers without replacement. Figure 7.1 shows how this is done in StatCrunch, and almost all statistical technologies let you do this quite easily.

◄ **FIGURE 7.1** StatCrunch will randomly select, without replacement, three numbers from the six shown in the var1 column.

Using technology, we got these three numbers: 1, 2, and 6. These correspond to Alberto, Justin, and Nicole.

If technology is not available, a random number table, such as the one provided in Appendix A, can be used. Here are two lines from such a table:

7 7 5 9 8    2 9 5 1 1    9 8 1 4 9    6 3 9 9 1

3 1 9 4 2    0 4 6 8 4    6 9 3 6 9    5 0 8 1 4

You can start at any row or column you please. Here, we choose to start at the upper left (shown in bold face). Next, read off digits from left to right, skipping digits that are not in our population. Because no one has the number 7, skip this number, twice. The first person selected is number 5: Brandy. Then skip 9 and 8 and select number 2: Justin. Skip 9 and 5 (because you already selected Brandy) and select number 1: Alberto.

**CONCLUSION** Using technology, our sample consisted of Alberto, Justin, and Nicole. Using the random number table, we got a different sample: Brandy, Justin, and Alberto.

**TRY THIS!** Exercise 7.11

## Sampling in Practice

In practice, simple random samples are difficult to collect and often inefficient. In most situations, we can't make a list of all people in the United States and assign each of them a number. To get around this, we use alternative techniques. We'll discuss examples of some other sampling methods in Chapter 12.

In addition, random sampling does not cure all ills. Nonresponse bias can still be a problem, and the possibility always exists that methods of taking the random sample are flawed (as they are if only landline telephones are used when many in the population use cell phones).

## EXAMPLE 3 Survey on Sexual Harassment

A newspaper at a large college wants to determine whether sexual harassment is a problem on campus. The paper takes a simple random sample of 1000 students and asks each person whether he or she has been a victim of sexual harassment on campus. About 35% of those surveyed refuse to answer. Of those who do answer, 2% say they have been victims of sexual harassment.

**QUESTION**  Give a reason why we should be cautious about using the 2% value as an estimate for the population percentage of those who have been victims of sexual harassment.

**CONCLUSION**  There is a large percentage of students who did not respond. Those who did not respond might be different from those who did, and if their answers had been included, the results could have been quite different. When those surveyed refuse to respond, it can create a biased sample.

**TRY THIS!**  Exercise 7.17

There are always some people who refuse to participate in a survey, but a good researcher will do everything possible to keep the percentage of nonresponders as small as possible, to reduce this source of bias.

National Public Radio reported that a relatively large number of people have dropped their land lines and use only cell phones. Cell phone users are less likely to participate in polls, since federal law forbids the use of automated dialing for cell phone numbers. Because research shows that people who use cell phones rather than land lines tend to be different from the general population, bias can occur. In the 2008 Democratic primaries, the Barack Obama campaign feared Obama was polling second to Hillary Clinton because Obama's supporters were younger and more likely not to have land lines. Therefore, Obama supporters were less likely to be polled, which led to a bias. Another problem is that cell phone users are less likely to participate in a poll, in part because they must pay for the phone call, but also because they are more likely to be doing something else—driving a car, attending a meeting—than people using land lines at home.

## SECTION 7.2

# Measuring the Quality of a Survey

A frequent complaint about surveys is that a survey based on 1000 people can't possibly tell us what the entire country is thinking. This complaint raises interesting questions: How do we judge whether our estimators are working? What separates a good estimation method from a bad?

It's difficult, if not impossible, to judge whether any particular survey is good or bad. Sometimes we can find obvious sources of bias, but often we don't know whether a survey has failed unless we later learn the true parameter value. (This sometimes occurs in elections, when we learn that a survey must have had bias because it severely missed predicting the actual outcome.) Instead, statisticians evaluate the *method* used to estimate a parameter, not the outcome of a particular survey.

**KEY POINT**  Statisticians evaluate the method used for a survey, not the outcome of a single survey.

Before we talk about how to judge surveys, imagine the following scenario: We are not taking just one survey of 1000 randomly selected people. We are sending out an army of pollsters. Each pollster surveys a random sample of 1000 people, and they all use the same method for collecting the sample. Each pollster asks the same question and produces an estimate of the proportion of people in the population who would answer yes to the question. When the pollsters return home, we get to see not just a single estimate (as happens in real life) but a great many estimates. Because each estimate is based on a separate random collection of people, each one will differ slightly. We expect some of these estimates to be closer to the mark than others just because of random variation. What we really want to know is how the group did as a whole. For this reason, we talk about evaluating *estimation methods*, not estimates.

An estimation method is a lot like a golfer. To be a good golfer, we need to get the golf ball in the cup. A good golfer is both *accurate* (tends to hit the ball near the cup) and *precise* (even when she misses, she doesn't miss by very much.)

It is possible to be precise and yet be inaccurate, as shown in Figure 7.2b. Also, it is possible to aim in the right direction (be accurate) but be imprecise, as shown in Figure 7.2c. (Naturally, some of us are bad at both, as shown in Figure 7.2d.) But the best golfers can both aim in the right direction and manage to be very consistent, which Figure 7.2a shows us.

◀ **FIGURE 7.2 a.** Shots from a golfer with good aim and precision; the balls are tightly clustered and centered around the cup. **b.** Shots from a golfer with good precision but poor aim; the balls are close together but centered to the right of the cup. **c.** Shots from a golfer with good aim—the balls are centered around the cup—but bad precision. **d.** The worst-case scenario: bad precision *and* bad aim.

> **! Caution**
>
> **Estimator and Estimates**
> We often use the word **estimator** to mean the same thing as "estimation method." An **estimate**, on the other hand, is a number produced by our estimation method.

Think of the cup as the population parameter, and think of each golf ball as an estimate, a value of $\hat{p}$, that results from a different survey. We want an estimation method that aims in the right direction. Such a method will, on average, get the correct value of the population parameter. We also need a precise method so that if we repeated the survey, we would arrive at nearly the same estimate.

The aim of our method, which the *accuracy,* is measured in terms of the *bias.* The *precision* is measured by a number called the *standard error.* Discussion of simulation studies in the next sections will help clarify how accuracy and precision are measured. These simulation studies show how bias and standard error are used to quantify the uncertainty in our inference.

# Using Simulations to Understand the Behavior of Estimators

The three simulations that follow will help measure how well the sample proportion works as an estimator of the population proportion.

In the first simulation, imagine doing a survey of 4 people in a very small population with only 8 people. You'll see that the estimator of the population proportion is accurate (no bias) but, because of the small sample size, not terribly precise.

In the second simulation, the first simulation is repeated, using a larger population and sample. The estimator is still unbiased, and you will see a perhaps surprising change in precision. Finally, the third simulation will reveal that using a much larger sample size makes the result even more precise.

To learn how our estimation method behaves, we're going to create a very unusual, unrealistic situation: We're going to create a world in which we know the truth. In this world, there are two types of people: those who like dogs and those who like cats. No one likes both. Exactly 25% of the population are Cat People, and 75% are Dog People. We're going to take a random sample of people from this world and see what proportion of our sample are Cat People. Then we'll do it again. And again. From this repetition, we'll see some interesting patterns emerge.

## Simulation 1: Statistics Vary from Sample to Sample

To get started, let's create a very small world. This world has 8 people named 1, 2, 3, 4, 5, 6, 7, and 8. People 1 and 2 are Cat People.

▶ **FIGURE 7.3** The entire population of our simulated world; 25% are Cat People.

From this population, we use the random number table to generate four random numbers between 1 and 8. When a person's number is chosen, he or she steps out of the population and into our sample.

Before we tell who was selected, think for a moment about what you expect to happen. What proportion of our sample will be Cat People? Is it possible for 0% of the sample to be Cat People? For 100%?

Below is our random sample. Note that we sampled without replacement, as in a real survey. We don't want the same person to be in our sample twice.

| 6 | 8 | 4 | 5 |
|---|---|---|---|
| D | D | D | D |

None of those selected are Cat People, as Figure 7.4 indicates. The proportion of Cat People in our sample is 0%. We call this the *sample proportion* because it comes from the sample, not the population.

▶ **FIGURE 7.4** The first sample, which has 0% Cat People.

Let's take another random sample. It is possible that we will again get 0%, but it is also possible that we will get a different percentage.

| 7 | 2 | 6 | 3 |
|---|---|---|---|
| D | C | D | D |

This time, our sample proportion is 25%.
One more time:

| 2 | 8 | 6 | 5 |
|---|---|---|---|
| C | D | D | D |

Again, our sample proportion is 25%.

Table 7.3 shows what has happened so far. Even though we have done only three repetitions, we can make some interesting observations.

| Repetition | Population Parameter | Sample Statistics |
|---|---|---|
| 1 | $p = 25\%$ Cat People | $\hat{p} = 0\%$ Cat People |
| 2 | $p = 25\%$ Cat People | $\hat{p} = 25\%$ Cat People |
| 3 | $p = 25\%$ Cat People | $\hat{p} = 25\%$ Cat People |

◀ **TABLE 7.3** The results of three repetitions of our simulation.

First, notice that the population proportion, $p$, never changes. It can't, because in our made-up world, the population always has the same 8 people and the same 2 are Cat People. However, the sample proportion, $\hat{p}$, can be different in each sample. In fact, $\hat{p}$ is random, because it depends on a random sample.

**KEY POINT**  No matter how many different samples we take, the value of $p$ (the population proportion) is always the same, but the value of $\hat{p}$ changes from sample to sample.

This simulation is, in fact, a random experiment and $\hat{p}$ is our outcome. Because it is random, $\hat{p}$ has a probability distribution. The probability distribution of $\hat{p}$ has a special name: **sampling distribution**. This term reminds us that $\hat{p}$ is not just any random outcome; it is a statistic we use to estimate a population parameter.

Because our world has only 8 people in it and we are taking samples of 4 people, we can write down all of the possible outcomes. There are only 70. By doing this, we can see exactly how often $\hat{p}$ will be 0%, how often 25%, and how often 50%. (Notice that it can never be more than 50%.) These probabilities are listed in Table 7.4, which presents the sampling distribution for $\hat{p}$. Figure 7.5 visually represents this sampling distribution.

| Value of | Probability of Seeing That Value |
|---|---|
| 0% | 0.21429 |
| 25% | 0.57143 |
| 50% | 0.21429 |

▲ **TABLE 7.4** The sampling distribution for $\hat{p}$, based on our random sample.

◀ **FIGURE 7.5** Graphical representation of Table 7.4, the sampling distribution for $\hat{p}$ when $p$ is 0.25.

*[Figure 7.5: A plot of Probability (y-axis, 0.0 to 0.6) versus Sample Proportion (x-axis, 0% to 50%). Points at 0% (≈0.21), 25% (≈0.57), and 50% (≈0.21).]*

From Table 7.4 and Figure 7.5, we learn several things:

1. Our estimator, $\hat{p}$, is not always the same as our parameter, $p$. Sometimes $\hat{p}$ turns out to be 0%, sometimes it is 50%, and sometimes it hits the target value of 25%.

2. The mean of this distribution is 25%—the same value as $p$.

3. Even though $\hat{p}$ is not always the "true" value, $p$, we are never more than 25 percentage points away from the true value.

Why are these observations important? Let's consider each one separately.

The first observation reminds us that statistics based on random samples are random. Thus, we cannot know ahead of time, with certainty, exactly what estimates our survey will produce.

The second observation tells us that our estimator has no bias—that, *on average*, it is the same as the parameter value. **Bias** is measured as the distance between the mean value of the estimator (the center of the sampling distribution) and the population parameter. In this case, the center of the sampling distribution and the population parameter are both 0.25, so the distance is 0. In other words, there is no bias.

The third observation is about precision. We know that our estimator is, on average, the same as the parameter, but the sampling distribution tells us how far away, typically, the estimator might stray from average. **Precision** is reflected in the spread of the sampling distribution and is measured by using the standard deviation of the sampling distribution. In this simulation, the standard deviation is 0.16366, or roughly 16%. The standard deviation of a sampling distribution has a special name: the **standard error (SE)**.

The standard error measures how much our estimator typically varies from sample to sample. Thus, in the above example, if we survey 4 people, we usually get 25% Cat People, but this typically varies by plus or minus 16.4% (16.4 percentage points). Looking at the graph in Figure 7.5, we might think that the variability is typically plus or minus 25 percentage points, but we must remember that the standard deviation measures how spread out observations are from the average value. Many observations are identical to the average value, so the typical, or "standard," deviation from average is only 16.4 percentage points.

 **KEY POINT** Bias is measured using the center of the sampling distribution: It is the distance between the center and the population value.

Precision is measured using the standard deviation of the sampling distribution, which is called the standard error. When the standard error is small, we say the estimator is precise.

## SNAPSHOT SAMPLING DISTRIBUTION

**WHAT IS IT?** ▶ A special name for the probability distribution of a statistic.

**WHAT DOES IT DO?** ▶ Gives us probabilities for a statistic.

**WHAT IS IT USED FOR?** ▶ It tells us how often we can expect to see particular values of our estimator, and it also gives us important characteristics of the estimator, such as bias and precision.

**HOW IS IT USED?** ▶ It is used for making inferences about a population.

## Simulation 2: The Size of the Population Does Not Affect Precision

The first simulation was very simple, because our made-up world had only 8 people. Let's consider a more realistic example with a world that has 1000 people.

In our first simulation, the bias was 0, which is good; this means we have an accurate estimator. However, the precision was fairly poor (we had a large standard error). How can we improve precision? To understand, we need a slightly more realistic simulation.

This time, we'll use the same world but make it somewhat bigger. Let's assume we have 1000 people and 25% are Cat People ($p = 0.25$). (In other words, there are 250 Cat People.) We take a random sample of 10 people and find the sample proportion, $\hat{p}$, of Cat People.

Because we've already seen how this is done, we're going to skip a few steps and show the results. This time the different outcomes are too numerous to list, so instead we just do a simulation:

1. Take a random sample, without replacement, of 10 people.

2. Calculate $\hat{p}$: the proportion of Cat People in our sample.

3. Repeat steps 1 and 2 a total of 10,000 times. Each time, calculate $\hat{p}$ and record its value.

Here are our predictions:

1. We predict that $\hat{p}$ will not be the same value every time because it is based on a random sample, so the value of $\hat{p}$ will vary randomly.

2. We predict that the mean outcome, the typical value for $\hat{p}$, will be 25%—the same as the population parameter—because our estimator is unbiased.

3. Precision: This one is left to you. Do you think the result will be more precise or less precise than in the last simulation? In the last simulation, only 4 people were sampled and the variation, as measured by the standard error, was about 16%. This time more people (10) are being sampled, but the population is much larger (1000). Will the standard error be larger (less precise) or smaller (more precise) than 16%?

Figure 7.6 shows an estimate of the sampling distribution. Figure 7.6 is not the actual sampling distribution because the histogram is based on a simulation, not a list of all possible outcomes. The relative frequencies in the figure are estimates of the probabilities, not the true probabilities. Still, with 10,000 replications, it is a very good approximation of the actual sampling distribution.

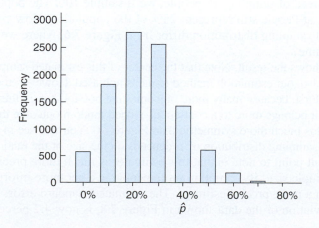

◄ **FIGURE 7.6** Simulation results for $\hat{p}$. This histogram is a simulation of the sampling distribution. The true value of $p$ is 25%. Each sample is based on 10 people, and we repeated the simulation 10,000 times.

The center of the estimated distribution is at 0.2501, which indicates that essentially no bias exists, because the population parameter is 0.25.

The outcomes ranged from $\hat{p} = 0$ to $\hat{p} = 0.8$. We can estimate the standard error by finding the standard deviation of our simulated $\hat{p}$. This turns out to be about 13.56%.

The value of the standard error tells us that if we were to take another sample of 10 people, we would expect to get about 25% Cat People, give or take 13.6 percentage points.

From Figure 7.6 we learn important information:

1. The bias of $\hat{p}$ is still 0, even though we used a larger population and a larger sample.

2. The variation of $\hat{p}$ is less; this estimator is more precise, even though the population is larger. In general, as long as the population is large relative to the sample size, the precision has *nothing* to do with the size of the *population*, but only with the size of the *sample*.

Many people are surprised to learn that precision is not affected by population size. How can the level of precision for a survey in a town of 10,000 people be the same as for one in a country of 210 *million* people?

Figure 7.7 provides an analogy. The bowls of soup represent two populations: a big one (a country, perhaps) and a small one (a city). Our goal is to taste each soup (take a sample from the population) to judge whether we like it. If both bowls are well stirred, the size of the bowl doesn't matter—using the same-size spoon, we can get the same amount of taste from either bowl.

► **FIGURE 7.7** The bowls of soup represent two populations and the sample size is represented by the spoons. The precision of an estimate depends only on the size of the sample, not the size of the population.

**KEY POINT**  The precision of an estimator does not depend on the size of the population; it depends only on the sample size. An estimator based on a sample size of 10 is just as precise in a population of 1000 people as in a population of a million.

## Simulation 3: Large Samples Produce More Precise Estimators

Let's do one more simulation. This time the only change we'll make is to increase the sample size. Instead of sampling 10 people, we'll sample 100. The population size is still 1000, and Cat People still represent 25% of the population. How will the precision of the simulated sampling distribution differ from Figure 7.6, where we sampled only 10 people at a time?

Figure 7.8 shows the result. Note that the center of this estimated sampling distribution is still at 25%. Also, our estimation method remains unbiased. However, the shape looks pretty different. First, because many more outcomes are possible for $\hat{p}$, this histogram looks as though it belongs more to a continuous-valued random outcome than to a discrete value. Second, it is much more symmetric than Figure 7.6. You will see in Section 7.3 that the shape of the sampling distribution of $\hat{p}$ depends on the size of the random sample.

An important point to note is that this estimator is much more precise because it uses a larger sample size. By sampling more people, we get more information, so we can end up with a more precise estimate. The estimated standard error, which is simply the standard deviation of the data shown in Figure 7.8, is now 4.2 percentage points.

Table 7.5 shows a summary of the three simulations.

| Simulation | Population Size | Sample Size | | | Mean | Standard Error | |
|:---:|:---:|:---:|:---:|:---:|:---:|:---:|:---:|
| 1 | 8 | 4 | | | 25% | 16.4% | |
| 2 | 1000 | 10 | Increasing Sample Size | | 25% | 13.5% | Increasing Precision |
| 3 | 1000 | 100 | | | 25% | 4.2% | |

◄ **TABLE 7.5** Increasing sample size results in increasing precision (measured as decreasing standard error).

Here is what we learned from Figure 7.8, which is based on sample sizes of 100 "people":

1. The estimator $\hat{p}$ is unbiased for all sample sizes (as long as we take random samples).

2. The precision improves as the sample size gets larger.

3. The shape of the sampling distribution is more symmetric for larger sample sizes.

**KEY POINT**   Surveys based on larger sample sizes have smaller standard error (SE) and therefore better precision. Increasing the sample size improves precision.

## Finding the Bias and the Standard Error

We've shown how to estimate bias and precision by running a simulation. But we can also do this mathematically, without running a simulation. Bias and standard error are easy to find for a sample proportion under certain conditions.

The bias of $\hat{p}$ is 0, and the standard error is

$$ SE = \sqrt{\frac{p(1-p)}{n}} $$

if the following two conditions are met:

**Condition 1.** The sample must be randomly selected from the population of interest, either with or without replacement. The population parameter to be estimated is the proportion of people (or objects) with some characteristic. This proportion is denoted as $p$.

**Condition 2.** If the sampling is without replacement, the population needs to be much larger than the sample size; at least 10 times bigger is a good rule of thumb.

## EXAMPLE 4 Pet World

Suppose that in Pet World, the population is 1000 people and 25% of the population are Cat People. Cat People love cats but hate dogs. We are planning a survey in which we take a random sample of 100 people, without replacement. We calculate the proportion of people in our sample who are Cat People.

**QUESTION** What value should we expect for our sample proportion? What's the standard error? How do we interpret these values?

**SOLUTION** The sample proportion is unbiased, so we expect it to be the same as the population proportion: 25%.

$$\text{The standard error is } \sqrt{\frac{p(1-p)}{n}} = \sqrt{\frac{0.25 \times 0.75}{100}} = \sqrt{\frac{0.1875}{100}}$$

$$= \sqrt{0.001875} = 0.04330, \text{ or about } 4.3\%.$$

This formula is appropriate because the population size is big with respect to the sample size. The population size is 1000, and the sample size is 100; $100 \times 10 = 1000$, so the population is 10 times larger than the sample size.

**CONCLUSION** We interpret the values to mean that if we were to take a survey of 100 people from Pet World, we would expect about 25% of them to be Cat People, give or take about 4.3%. The "give or take" means that if you were to draw a sample of 100 and I were to draw a sample of 100, our sample proportions would typically differ from the expected 25% by about 4.3 percentage points.

**TRY THIS!** Exercise 7.29

## Real Life: We Get Only One Chance

In simulations, we could repeat the survey many times to understand what might happen. In real life, we get just one chance. We take a sample, calculate $\hat{p}$, and then have to live with it.

It is important to realize that bias and precision are both measures of what happens if we could repeat our survey many times. Bias indicates the typical outcome of surveys repeated again and again. If the bias is 0, we will usually get the right value. If the bias is 0.10, then our estimate will characteristically be 10 percentage points too high. Precision measures how much our estimator will vary from the typical value if we do the survey again. To put it slightly differently, if someone else does the survey, precision helps determine how different her or his estimate could be from ours.

How small must the standard error be for a "good" survey? The answer varies, but the basic rule is that the precision should be small enough to be useful. A typical election poll has a sample of roughly 1000 registered voters and a standard error of about 1.5 percentage points. If the candidates are many percentage points apart, this is good precision. However, if they are neck and neck, this might not be good enough. In Section 7.4, we will discuss how to make decisions about whether the standard error is small enough.

In real life, we don't know the true value of the population proportion, $p$. This means we can't calculate the standard error. However, we can come pretty close by using the sample proportion. If $p$ is unknown, then

$$SE_{est} = \sqrt{\frac{\hat{p}(1 - \hat{p})}{n}},$$ where $SE_{est}$ is the estimated standard error,

is a useful approximation to the true standard error.

## SECTION 7.3

# The Central Limit Theorem for Sample Proportions

Remember that a probability tells us how often an event happens if we repeat an experiment an infinite number of times. For instance, the sampling distribution of $\hat{p}$ gives the probabilities of where our sample proportions will fall; that is, it tells us how often we would see particular values of $\hat{p}$ if we could repeat our survey infinitely many times. In the simulation, we repeated our fake survey 10,000 times. Ten thousand is a lot, but it's a far cry from infinity.

In the three simulations in Section 7.2, we saw that the shape of the sampling distribution (or our estimated version, based on simulations) changed as the sample size increased (compare Figures 7.5, 7.6, and Figure 7.8). If we used an even larger sample size than 100 (the sample size for the last simulation), what shape would the sampling distribution have? As it turns out, we don't need a simulation to tell us. For this statistic, and for some others, a mathematical theorem called the **Central Limit Theorem (CLT)** gives us a very good approximation of the sampling distribution without our needing to do simulations.

The Central Limit Theorem is helpful because sampling distributions are important. They are important because they, along with the bias and standard error, enable us to measure the quality of our estimation methods. Sampling distributions give us the probability that an estimate falls a specified distance from the population value. For example, we don't want to know simply that 18% of our customers are likely to buy new cell phones in the next year. We also want to know the probability that the true percentage might be higher than some particular value, say, 25%.

## Meet the Central Limit Theorem for Sample Proportions

The Central Limit Theorem has several versions. The one that applies to estimating proportions in a population tells us that if some basic conditions are met, then the sampling distribution of the sample proportion is close to the Normal distribution.

More precisely, when estimating a population proportion, $p$, we must have the same conditions that were used in finding bias and precision, but with one new condition:

Condition 1. *Random and Independent.* The sample is collected randomly from the population and observations are independent at each other. The sample can be collected either with or without replacement.

Condition 2. *Large Sample.* The sample size, $n$, is large enough that the sample expects at least 10 successes (yes's) and 10 failures (no's).

Condition 3. *Big Population.* If the sample is collected without replacement, then the population size must be much (at least 10 times) bigger than the sample size.

The sampling distribution for $\hat{p}$ is then approximately Normal, with mean $p$ (the population proportion) and standard deviation the same as the standard error:

$$SE = \sqrt{\frac{p(1 - p)}{n}}$$

**Looking Back**

**Normal Notation**
Recall that the notation N(mean, standard deviation) designates a particular Normal distribution.

> **KEY POINT**
>
> The Central Limit Theorem for Sample Proportions tells us that if we take a random sample from a population, and if the sample size is large and the population size much larger than the sample size, then the sampling distribution of $\hat{p}$ is approximately
>
> $$N\left(p, \sqrt{\frac{p(1-p)}{n}}\right)$$
>
> If you don't know the value of $p$, then you can substitute the value of $\hat{p}$ to calculate the standard error.

Figure 7.9 illustrates the CLT for proportions. Figure 7.9a is based on simulations in which the sample size was just 10 people, which is too small for the CLT to apply. In this case, the simulated sampling distribution does not look Normal; it is right-skewed and has large gaps between values. Figure 7.9b is based on simulations of samples of 100 observations. Because the true population proportion is $p = 0.25$, a sample size of 100 is large enough for the CLT to apply, and our simulated sampling distribution looks very close to the Normal model. Figure 7.9b is actually a repeat of Figure 7.8 with the Normal curve superimposed. Now that the graphs are shown on the same scale, we can see that the sample size of 100 gives better precision than the sample size of 10—the distribution is narrower.

▶ **FIGURE 7.9 a.** Revision of Figure 7.6, a histogram of 10,000 sample proportions, each based on $n = 10$ with a population percentage $p$ equal to 25% **b.** Revision of Figure 7.8, a histogram of 10,000 sample proportions, each based on $n = 100$ percentage and with the population percentage $p$ equal to 25%.

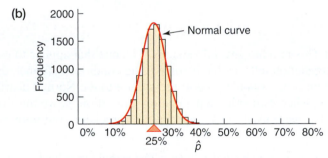

The Normal curve shown in Figure 7.9b has a mean of 0.25 because $p = 0.25$, and a standard deviation (also called the standard error) of 0.0433 because

$$\sqrt{\frac{0.25 \times 0.75}{100}} = 0.0433$$

Before illustrating how to use the CLT, we show how to check conditions to see whether the CLT applies.

## Checking Conditions for the Central Limit Theorem

The first condition states that the sample must be collected randomly. There is no way to check this just by looking at the data; you have to trust the researcher's report on how the data were collected, or, if you are the researcher, take care to use sound random sampling methods.

The second condition dictates that the sample size must be large enough. This we *can* check by looking at the data. The CLT says that the sample size needs to be sufficiently large to get at least 10 successes and 10 failures in our sample. If the probability of a success is $p$, then we would expect about $np$ successes and $n(1 - p)$ failures. One problem, though, is that we usually don't know the value of $p$. In this case, we instead check that

$$n\hat{p} \geq 10 \text{ and } n(1 - \hat{p}) \geq 10$$

For example, if our sample has 100 people and we are estimating the proportion of females in the population, and if our sample has 49% females, then we need to verify that both $100(0.49) \geq 10$ and $100(0.51) \geq 10$.

The third condition applies only to random samples done without replacement. In this case, the population must be at least 10 times bigger than the sample. In symbols, if $N$ is the number of people in the population and $n$ is the number in the sample, then

$$N \geq 10n$$

If this condition is not met, and the sample was collected without replacement, then the actual standard error will be a little smaller than what our formula says it should be.

In most real-life applications, the population size is much larger than the sample size. Over 300 million people live in the United States, so the typical survey of 1000 to 3000 easily meets this condition.

You can see how these conditions are used in the examples that follow.

## Using the Central Limit Theorem

The following examples use the CLT to find the probability that the sample proportion will be near (or far from) the population value.

### EXAMPLE 5  Pet World Revisited

Let's return to Pet World. The population is 1000 people, and the proportion of Cat People is 25%. We'll take a random sample of 100 people.

**QUESTION**  What is the approximate probability that the proportion in our sample will be bigger than 29%? Begin by checking conditions for the CLT.

**SOLUTION**  First we check conditions to see whether the Central Limit Theorem can be applied. The sample size is large enough because $np = 100(0.25) = 25$ is greater than 10 and $n(1 - p) = 100(0.75) = 75$, which is also greater than 10. Also, the population size is 10 times larger than the sample size, because $1000 = 10(100)$. Thus $N = 10(n)$; the population is just large enough. We are told that the sample was collected randomly.

According to the CLT, the sampling distribution will be approximately Normal. The mean is the same as the population proportion: $p = 0.25$. The standard deviation is the same as the standard error:

$$SE = \sqrt{\frac{p(1 - p)}{n}} = \sqrt{\frac{0.25 \times 0.75}{100}} = \sqrt{\frac{0.1875}{100}} = \sqrt{0.001875} = 0.0433$$

We can use technology to find the probability of getting a value larger than 0.29 in a $N(0.25, 0.0433)$ distribution. Or we can standardize.

In standard units, 0.29 is

$$z = \frac{0.29 - 0.25}{0.0433} = 0.924 \text{ standard unit}$$

In a $N(0,1)$ distribution, the probability of getting a number bigger than 0.924 is, from Table A in the appendix, about 0.18, or 18%. Figure 7.10 on the next page shows the results using technology.

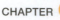

▶ **FIGURE 7.10** Output from SOCR (http://socr.stat.ucla.edu): There is about an 18% chance that $\hat{p}$ will be more than 4 percentage points above 25%.

| Distribution Properties | Probabilities |
|---|---|
| Normal (0.25, 0.0433) Distribution | |
| Mean: .250000 | Left: .822202 |
| Median: .250000 | Between (Red-Shaded): .177767 |
| Variance: .001875 | Right: .000032 |
| Standard Deviation: .043300 | |
| Max Density: 9.213448 | |

**CONCLUSION** With a sample size of 100, there is about an 18% chance that $\hat{p}$ will be more than 4 percentage points above 25%.

**TRY THIS!** Exercise 7.33

## SNAPSHOT  THE SAMPLE PROPORTION: $\hat{p}$ (*p*-HAT)

**WHAT IS IT?** ▶ The proportion of people or objects in a sample that have a particular characteristic in which we are interested.

**WHAT IT IS USED FOR?** ▶ To estimate the proportion of people or objects in a population that have that characteristic.

**WHY DO WE USE IT?** ▶ If the sample is drawn at random from the population, then the sample proportion is unbiased and has standard error $\sqrt{\dfrac{p(1-p)}{n}}$.

**HOW IS IT USED?** ▶ If, in addition to everything above, the sample size is fairly large, then we can use the Normal distribution to find probabilities concerning the sample proportion.

## EXAMPLE 6 Presidential Election Survey

In a hotly contested U.S. election, two candidates for president, a Democrat and a Republican, are running neck and neck; each candidate has 50% of the vote. Suppose a random sample of 1000 voters are asked whether they will vote for the Republican candidate.

**QUESTIONS** What percentage of the sample should be expected to express support for the Republican? What is the standard error for this sample proportion? Does the Central Limit Theorem apply? If so, what is the approximate probability that the sample proportion will fall within two standard errors of the population value of $p = 0.50$?

**SOLUTION** Because we have collected a random sample, the sample proportion has no bias (assuming there are no problems collecting the sample). Therefore, we expect that 50% of our sample supports the Republican candidate.

Because the sample size, $n = 1000$, is small relative to the population (which is over 100 million), we can calculate the standard error with

$$SE = \sqrt{\frac{(0.50)(0.50)}{1000}} = 0.0158$$

We can interpret this to mean that we expect our sample proportion to be 50%, give or take 1.58 percentage points.

Because the sample size is fairly large (the expected numbers for successes and failures are both equal to $np = 1000 \times 0.50 = 500$, which is much larger than 10), the CLT tells us we can use the Normal distribution—in particular, N(0.50, 0.0158).

We are asked to find the probability that the sample proportion will fall within two standard errors of 0.50. In other words, that it will fall somewhere between

$$0.50 - 2SE$$

and

$$0.50 + 2SE$$

Because this is a Normal distribution, we know the result will be very close to 95% (according to the Empirical Rule). But let's calculate the result anyway.

$$0.50 - 2SE = 0.50 - 2(0.0158) = 0.50 - 0.0316 = 0.4684$$

$$0.50 + 2SE = 0.50 + 0.0316 = 0.5316$$

That is, we want to find the area between 0.4684 and 0.5316 in a N(0.5, 0.0158) distribution. Figure 7.11 shows the result using technology, which tells us this probability is 0.9545.

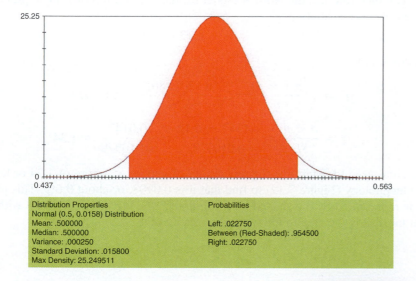

Distribution Properties
Normal (0.5, 0.0158) Distribution
Mean: .500000
Median: .500000
Variance: .000250
Standard Deviation: .015800
Max Density: 25.249511

Probabilities
Left: .022750
Between (Red-Shaded): .954500
Right: .022750

> **Looking Back**
>
> **Empirical Rule**
> Recall that the Empirical Rule says that roughly 68% of observations should be within one standard deviation of the mean, about 95% within two standard deviations of the mean, and nearly all within three standard deviations of the mean. In this context, the standard error is the standard deviation for the sampling distribution.

◀ **FIGURE 7.11** The probability that a sample proportion based on a random sample of 1000 people taken from a population in which p = 0.50 has about a 95% chance of falling within two standard errors of 0.50.

**CONCLUSION** If each candidate truly has 50% of the vote, then we'd expect our sample proportion to be about 0.50 (or 50%). There is about a 95% chance that the sample proportion falls within two standard errors of 50%.

**TRY THIS!** Exercise 7.35

The conclusion from Example 6 is useful because it implies that, in general, we can predict where $\hat{p}$ will fall, relative to $p$. It indicates that $\hat{p}$ is very likely to fall within two standard errors of the true value, as long as the sample size is large enough. If, in addition, we have a small standard error, we know that $\hat{p}$ is quite likely to fall close to $p$.

## EXAMPLE 7 Morse and the Proportion of E's

Samuel Morse (1791–1872), the inventor of Morse code, claimed that the letter used most frequently in the English language was E and that the proportion of E's was 0.12. We took a simple random sample with replacement from a modern-day book. Our sample consisted of 876 letters, and we found 118 E's, so $\hat{p} = 0.1347$.

**QUESTION** Find the probability that, if we were to take another random sample of 876 letters, the sample proportion would be greater than or equal to 0.1347. Assume that the true proportion of E's in the population is, as Morse claimed, 0.12. As a first step, check that the Central Limit Theorem can be applied in this case.

**SOLUTION** To check whether we can apply the Central Limit Theorem, we need to make sure the sample size is large enough. Because $p = 0.12$, we check:

$$np = 876(0.12) = 105.12, \text{ which is larger than 10,}$$

and

$$n(1 - p) = 876(0.88) = 770.88, \text{ which is also larger than 10}$$

The book contains far more than 8760 letters, so the population size is much larger than the sample size.

We can therefore use the Normal model for the distribution of sample proportions. The mean of this distribution is

$$p = 0.12$$

The standard error is

**▲ FIGURE 7.12** TI-83/84 output

$$SE = \sqrt{\frac{p(1 - p)}{n}} = \sqrt{\frac{0.12(0.880)}{876}} = 0.010979$$

$$z = \frac{\hat{p} - p}{SE} = \frac{0.1347 - 0.12}{0.010979} = \frac{0.0147}{0.010979} = 1.34$$

We therefore need to determine the probability of getting a $z$-score of 1.34 or larger. We use technology (Figure 7.12) to find that it is 0.0903, or about 9.0%. This probability is represented by the shaded area in Figure 7.13. We can also find it with the Normal table; it is the area to the right of a $z$-score of 1.34.

**▶ FIGURE 7.13** The shaded area represents the probability of finding a sample proportion of 0.1347 from a population with a proportion of 0.12.

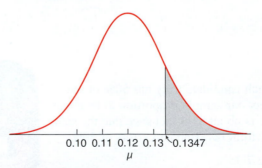

0.10  0.11  0.12  0.13  0.1347
$\mu$

**CONCLUSION** If the sample is 876 letters, the probability of getting a sample proportion of 0.1347 or larger, when the true proportion of E's in the population is 0.12, is about 9%.

**TRY THIS!** Exercise 7.37

# Estimating the Population Proportion with Confidence Intervals

Up to this point, all of our examples have described situations in which we know the value of the population parameter, $p$. In Example 7, for instance, $p$ represented the proportion of E's found in the English language, and we could assume that $p = 0.12$. In the real world, knowing $p$ is unusual. In fact, the whole point of carrying out a survey is that we *don't* know the value of $p$.

An example of a real survey illustrates this situation. The Pew Research Center took a random sample of 2928 adults in the United States in September 2008. In this sample, 53% of the 2928 people believed that reducing the spread of acquired immune deficiency disease (AIDS) and other infectious diseases was an important policy goal for the U.S. government. However, this percentage just tells us about our sample. What percentage of the population—that is, what percentage of all adults in the United States—believes this? How much larger or smaller than 53% might it be? Can we conclude that a majority (more than 50%) of Americans share this belief?

We don't know $p$, the population parameter. We do know $\hat{p}$ for this sample; it's equal to 53%. Here's what else we know from the preceding sections:

1. Our estimator is unbiased, so while our estimate of 53% may not be exactly equal to the population parameter, it's probably just a little higher or a little lower.

2. The standard error can be estimated as

$$\sqrt{\frac{\hat{p}(1 - \hat{p})}{n}} = \sqrt{\frac{(0.53)(0.47)}{2928}} = 0.0092, \text{ or about } 0.9\%.$$

   This means that the population parameter must be close by, because a standard error of 0.9 percentage point signifies a very precise estimator.

3. Because the sample size is large, we also realize that the probability distribution of $\hat{p}$ is pretty close to being Normally distributed and is centered around the true population parameter value. Thus, there's about a 68% chance that $\hat{p}$ is closer than 1 standard error away from the population proportion, and a 95% chance that it is closer than 2 standard errors away. (See Example 6.) Also, there is almost a 100% chance (99.7%, actually) that the sample proportion is closer than 3 standard errors from the population proportion. Thus we can feel very confident that the proportion of the population who support this policy goal is within 3 standard errors of 0.53. Three standard errors is $3(0.9\%) = 2.7\%$, so we can be almost certain that the value of the population parameter is within 2.7 percentage points of 53%.

In other words, we can be highly confident that the population parameter is between these two numbers:

$$53\% - 2.7\% \quad \text{to} \quad 53\% + 2.7\%, \text{ or}$$
$$50.3\% \quad\quad \text{to} \quad\quad 55.7\%$$

We have just calculated a **confidence interval**. Confidence intervals are often reported as the estimate plus or minus some amount:

$$53\% \text{ plus or minus } 2.7\%, \text{ or } 53\% \pm 2.7\%.$$

The "some amount," in this case the 2.7 percentage points, is called the **margin of error**. The margin of error tells how far from the population value our estimate can be.

A confidence interval provides two pieces of information: (1) a range of plausible values for our population parameter (50.3% to 55.7 %), and (2) a **confidence level**, which expresses (no surprise here) our level of confidence in this interval. Our high confidence level of 99.7% assures us we can be very confident that a majority of Americans consider fighting AIDS and other infectious diseases an important priority, because the smallest plausible level of support in the population is 50.3%, which is (just) bigger than a majority.

An analogy can help explain confidence intervals. Imagine a city park. In this park sit a mother and her daughter, a toddler. The mother sits in the same place every day, on a bench along a walkway, while her daughter wanders here and there. Most of the time, the child stays very close to her mother, as you would expect. In fact, our studies have revealed that 68% of the days we've looked, she is within 1 yard of her mother. Sometimes she strays a little bit farther, but on 95% of the days she is still within 2 yards of her mother. Only rarely does she move much farther; she is almost always within 3 yards of her mother.

One day the unimaginable happens, and the mother and the park bench become invisible. Fortunately, the child remains visible. The problem is to figure out where the mother is sitting.

Where is the mother? On 68% of the days, the child is within 1 yard of the mother, so at these times the mother must be within 1 yard of the child. If we think the mother is within 1 yard of the child on most days—that is, 68% of the days we observe—we will be right. But this also means we will be wrong on 32% of our visits. We could be more confident of being correct if we instead guessed that the mother is within 2 yards of the child. Then we would be wrong on only 5% of the days.

In this analogy, the mother is the population proportion. Like the mother, the population proportion never moves from its spot and never changes values. And just as we cannot see the invisible mother, we don't know where the parameter sits. The toddler is like our sample proportion, $\hat{p}$; we *do* know its value, and we know that it hangs out near the population proportion and moves around from sample to sample. Thus, even though we can't know exactly what the true population proportion is, we can infer that it is near the sample proportion.

## Setting the Confidence Level

The confidence level tells us how often the estimation method is successful. Our method is to take a random sample and calculate a confidence interval to estimate the population proportion. If the method has a 100% confidence level, that method always works. If the method has a 10% confidence level, it works in 10% of surveys. We say the method works if the interval captures the true value of the population parameter. In this case, the interval works if the true population proportion is inside the interval.

Think of the confidence level as the capture rate; it tells us how often a confidence interval based on a random sample will capture the population proportion. Keep in mind that the population proportion, like the mother on the park bench, does not move—it is always the same. However, the confidence interval does change with every random sample collected. Thus, the confidence level measures the success rate of the *method*, not of any one particular interval.

**KEY POINT** The confidence level measures the capture rate for our method of finding confidence intervals.

Figure 7.14 demonstrates what we mean by a 95% confidence level. Let's suppose that in the United States, 51% of all voters favor stricter laws with respect to buying and selling guns. We simulate taking a random sample of 1000 people. We calculate the percentage of the sample who favor stricter laws, and then we find the confidence interval that gives us a 95% confidence level. We do this again and keep repeating. Figure 7.14 shows 100 simulations.

Each blue point and each orange point represent a sample percentage. Note that the points are centered around the population percentage of 51%. The horizontal lines

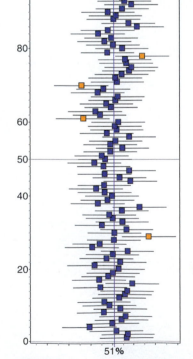

▲ **FIGURE 7.14** Results from 100 simulations in which we draw a random sample and then find and display a confidence interval with a 95% confidence level. The orange squares indicate "bad" intervals.

represent the confidence interval: the sample percentage plus or minus the margin of error. The margin of error was chosen so that the confidence level is 95%. Notice that most of the lines cross the vertical line at 51%. These are successful confidence intervals that capture the population value of 51% (in blue). However, a few sample percentages miss the mark; these are indicated by orange points. In 100 trials, our method failed 4 times and was successful 96 times. In other words, it worked in 96% of the trials. When we use a 95% confidence level, our method works in 95% of all surveys we conduct.

We can change the confidence level by changing the margin of error. The greater the margin of error, the higher our confidence level. For example, we can be 100% confident that the true percentage of Americans who favor stricter gun laws is between 0% and 100%. We're 100% confident in this interval because it can never be wrong. Of course, it will also never be useful. We really don't need to spend money on a survey to learn that the answer lies between 0% and 100%, do we?

It would be more helpful—more precise—to have a smaller margin of error than "plus or minus 50 percentage points." However, if the margin of error is too small, then we are more likely to be wrong. Think of the margin of error as a tennis racket. The bigger the racket, the more confident you are of hitting the ball. Choosing an interval that ranges from 0% to 100% is like using a racket that fills the entire court— you will definitely hit the ball, but not because you are a good tennis player. If the racket is too small, you are less confident of hitting the ball, so you don't want it too small. Somewhere between too big and too small is just right.

## Selecting a Margin of Error

We select a margin of error that will produce the desired confidence level. For instance, how can we choose a margin of error with a confidence level of 95%? We already know that if we take a large enough random sample and find the sample proportion, then the CLT tells us that 95% of the time, the sample proportion is within two standard errors of the population proportion. This is what we learned from Example 6. It stands to reason, then, that if we choose a margin of error that is two standard errors, then we'll cover the population proportion in 95% of our samples.

This means that

$$\hat{p} \pm 2SE$$

is a confidence interval with a 95% confidence level. More succinctly, we call this a 95% confidence interval.

Using the same logic, we understand that the interval

$$\hat{p} \pm 1SE$$

is a 68% confidence interval and that

$$\hat{p} \pm 3SE$$

is a 99.7% confidence interval.

Figure 7.15 shows four different margins of error for a sample in which $\hat{p} = 50\%$.

◀ **FIGURE 7.15** Four confidence intervals with confidence levels ranging from 99.99% (plus or minus 4 standard errors—top) to 68% (plus or minus 1 standard error). Notice how the interval gets wider with increasing confidence level.

This figure illustrates one reason why a 95% confidence interval is so desirable. If we increase the margin of error from 2 standard errors to 3, we gain only a small amount of confidence; the level goes from 95% to 99.7%. However, if we decrease from 2 standard errors to 1, we lose a lot of confidence; the level falls from 95% to 68%. Thus, the choice of 2 standard errors is very economical.

The margin of error has this structure:

$$\text{Margin of error} = z^*SE$$

where $z^*$ is a number that tells how many standard errors to include in the margin of error. If $z^* = 1$, the confidence level is 68%. If $z^* = 2$, the confidence level is 95%. Table 7.6 summarizes the margin of error for four commonly used confidence levels.

▶ **TABLE 7.6** We can set the confidence level to the value we wish by choosing the appropriate margin of error.

| Confidence Level | Margin of Error Is . . . |
|---|---|
| 99% | 2.58 standard errors |
| 95% | 1.96 (about 2) standard errors |
| 90% | 1.645 standard errors |
| 80% | 1.28 standard errors |

## Reality Check: Finding a Confidence Interval When $p$ Is Not Known

As we have seen, a confidence interval for a population proportion has this structure:

$$\hat{p} \pm m$$

where $m$ is the margin of error. Substituting for the margin of error, we can also write

$$\hat{p} \pm z^*SE$$

Finding the standard error requires us to know the value of $p$:

$$SE = \sqrt{\frac{p(1-p)}{n}}$$

However, in real life, we don't know $p$. So instead, we substitute our sample proportion and use this estimated standard error:

$$SE_{\text{est}} = \sqrt{\frac{\hat{p}(1-\hat{p})}{n}}$$

The result is a confidence interval with a confidence level close to, but not exactly equal to, the correct level. This tends to be close enough for most practical purposes.

In real life, then, Formula 7.1 is the method we use to find approximate confidence intervals for a population proportion.

**Formula 7.1:** $\hat{p} \pm m$ where $m = z^*SE_{\text{est}}$ and $SE_{\text{est}} = \sqrt{\frac{\hat{p}(1-\hat{p})}{n}}$

where:

$m$ is the margin of error
$\hat{p}$ is the sample proportion of successes, or the proportion of people in the sample with the characteristic we are interested in
$n$ is the sample size
$z^*$ is a multiplier that is chosen to achieve the desired confidence level (Table 7.6)
$SE_{\text{est}}$ is the estimated standard error

# EXAMPLE 8 Harris Poll on Teaching

Do a majority of Americans believe the profession of teaching is prestigious? A 2007 Harris Poll took a random sample of 1011 Americans and found that 49% of the sample believed that the profession of teaching "has very great prestige."

**QUESTION** Estimate the standard error. Find an approximate 95% confidence interval for the percentage of all Americans who believe teaching has very great prestige. Is it plausible to conclude that more than half of all Americans share this opinion?

**SOLUTION** We first make sure the conditions of the Central Limit Theorem apply. We are told that the Harris Poll took a random sample. We don't know whether the pollsters sampled with or without replacement, but because the population is very large, easily 10 times larger than the sample size, we don't need to worry about the replacement issue.

Next, we need to check that the sample size is large enough for the CLT. We do not know $p$, the proportion of all Americans who believe that teaching has great prestige. We know only $\hat{p}$ (equal to 0.49), which the Harris Poll found on the basis of its sample. This means that our sample has at least 10 successes (people who believe teaching has great prestige) because $1011(0.49) = 495$, which is much larger than 10. Also, we know we have at least 10 failures (people who don't believe), because $1011(0.51)$ is even bigger than 495.

The next step is to estimate the standard error.

$$SE_{est} = \sqrt{\frac{(0.49)(0.51)}{1011}} = 0.0157$$

The problem asks for a 95% confidence level, which can be achieved by using a margin of error roughly equal to 2 standard errors.

$$\hat{p} \pm 2SE_{est} = 0.49 \pm 2(0.0157) = 0.49 \pm 0.0314,$$
$$\text{or } 49\% \pm 3.1\%$$

Expressing this as an interval, we get

$$49\% - 3.1\% = 45.9\%$$
$$49\% + 3.1\% = 52.1\%$$

The 95% confidence interval is 45.9% to 52.1%.

**CONCLUSION** The confidence interval tells us which values are plausible for the population percentage. We have to conclude that it is indeed plausible that more than half (50%) of Americans think teaching has great prestige, since the confidence interval includes values greater than 50%. It is not plausible that a large majority believes this way, though, since the largest plausible value is 52.1%.

**TRY THIS!** Exercise 7.51

## Interpreting Confidence Intervals

A confidence interval for a sample proportion gives a set of values that are plausible for the population proportion. If a value is not in the confidence interval, we conclude that it is implausible. It's not impossible that the population value is outside the interval, but it would be pretty surprising.

Suppose a candidate for political office conducts a poll and finds that a 95% confidence interval for the proportion of voters who will vote for him is 42% to 48%. He would be wise to conclude that he does *not* have 50% of the population voting for him. The reason is that the value 50% is not in the confidence interval, so it is implausible to believe that the population value is 50%.

There are many common misinterpretations of confidence intervals you must avoid. The most common mistake that students (and, indeed, many others) make is trying to turn confidence intervals into some sort of probability problem. For example, if asked to interpret a 95% confidence interval of 45.9% to 53.1%, many people would mistakenly say, "This means there is a 95% chance that the population proportion is between 45.9% and 53.1%."

What's wrong with this statement? Remember that probabilities are long-run frequencies. This sentence claims that if we were to repeat this survey many times, then in 95% of the surveys the true population percentages would be a number between 45.9% and 53.1%. This claim is wrong, because the true population percentage doesn't change. It is either *always* between 45.9% and 53.1% or it is *never* between these two values. It can't be between these two numbers 95% of the time and somewhere else the rest of the time. In our story about the invisible mother, the mother, who represented the population proportion, *always* sat at the same place. Similarly, the population proportion (or percentage) is always the same value.

Another analogy will help make this clear. Suppose there is a skateboard factory. Say 95% of the skateboards produced by this factory are perfect, but 5% have no wheels. Once you buy a skateboard from this factory, you can't say that there is a 95% chance that it is a good board. Either it has wheels or it does not have wheels. It is not true that the board has wheels 95% of the time and, mysteriously, no wheels the other 5% of the time. A confidence interval is like one of these skateboards. Either it contains the true parameter (has wheels) or it does not. The "95% confidence" refers to the "factory" that "manufactures" confidence intervals: 95% of its products are good, 5% are bad.

Our confidence is in the process, not in the product.

**KEY POINT**
Our confidence is in the process that produces confidence intervals, not in any particular interval. It is incorrect to say that a particular confidence interval has a 95% (or any other percent) chance of including the true population parameter. Instead, we say that the *process* that produces intervals captures the true population parameter with a 95% probability.

## EXAMPLE **9** Underwater Mortgages

A mortgage is "underwater" if the amount owed is greater than the value of the property that is mortgaged. For a 2010 Rasmussen poll, investigators interviewed a random sample of 710 homeowners in the United States and found that 35% had mortgages that were underwater. Rasmussen reports that "The margin of sampling error is plus or minus 4 percentage points with a 95% level of confidence." [Source: http://www.rasmussenreports.com, viewed August 6, 2010]

**QUESTION** State the confidence interval in interval form. How would you interpret this confidence interval? What does "95%" mean?

**CONCLUSION** The margin of error, we are told, is 4 percentage points. In interval form, then, the 95% confidence interval is

$$35\% - 4\% \quad \text{to} \quad 35\% + 4\%, \text{ or}$$
$$31\% \quad \text{to} \quad 39\%$$

We interpret this to mean that we are 95% confident that the true proportion of all homeowners in the United States whose homes are worth less than the mortgage is between 31% and 39%. The 95% indicates that if we were to conduct not just this survey, but many, then 95% of them would result in confidence intervals that include the true population proportion.

**TRY THIS!** Exercise 7.59

Example 10 demonstrates the use of confidence intervals to make decisions about population proportions.

## EXAMPLE 10 Morse and E's

Recall from Example 7 that Morse believed the proportion of E's in the English language was 0.12 and that our sample showed 118 E's out of 876 randomly chosen letters from a modern-day book.

**QUESTION** Find a 95% confidence interval for the proportion of E's in the book. Is the proportion of E's in the book consistent with Morse's 0.12?

**SOLUTION** The best approach is to use technology. Figure 7.16 shows TI-83/84 output that gives a 95% confidence interval as

$$(0.112, 0.157) \quad \text{or} \quad (11.2\%, 15.7\%)$$

If you do not have access to statistical technology, then the first step is to find the sample proportion of E's: 118/876, or 0.1347.
  The estimated standard error is

$$SE_{est} = \sqrt{\frac{\hat{p}(1 - \hat{p})}{n}} = \sqrt{\frac{0.1347(0.8653)}{876}} = \sqrt{0.00012205} = 0.0115349$$

Because we want a 95% confidence level, our margin of error is plus or minus 1.96 standard errors:

$$\text{Margin of error} = 1.96 SE_{est} = 1.96(0.0115349) = 0.022608$$

The interval boundaries are

$$\hat{p} \pm 1.96 SE_{est} = 0.1347 \pm 0.0226$$

Upper end of interval: $0.1347 + 0.0226 = 0.1573$
Lower end of interval: $0.1347 - 0.0226 = 0.1121$

  This confirms the result we got through technology: A 95% confidence interval is (0.1121 to 0.1573). Note that this interval *does* include the value 0.12.

**CONCLUSION** We are 95% confident the proportion of E's in the modern book is between 0.112 and 0.157. This interval captures 0.12. This shows it is plausible that the population proportion of E's in the book is 0.12, as Morse suggested.

**TRY THIS!** Exercise 7.63

Tech

▲ **FIGURE 7.16** TI-83/84 output for a confidence interval for the proportion of E's.

## SECTION 7.5

# Margin of Error and Sample Size for Proportions

Quite often, researchers choose the size of the margin of error they wish to report for a survey and then collect a large enough sample to achieve it. For example, most of the surveys and polls reported on the news use a 95% confidence interval and a margin of error of 3 percentage points.
  To find the approximate sample size needed to get a margin of error of size $m$, we could use this formula:

$$n = \left(\frac{z^*}{m}\right)^2 \left(\frac{1}{4}\right)$$

In this formula, $z^*$ represents the multiplier we use in our confidence interval, chosen from Table 7.6. For instance, for a 95% confidence interval, $z^* = 1.96$, or about 2. For a 90% confidence interval, $z^* = 1.65$.

If we are using a 95% confidence level, this formula simplifies greatly. For a 95% confidence interval, $z^* = 1.96$. Rounding this up to 2.0, we get $(z^*)^2 = 2^2 = 4$, and 4 times $(1/4) = 1$, so the formula becomes

$$\textbf{Formula 7.2: } n = \frac{1}{m^2}$$

The resulting value for $n$ is the approximate sample size required to get a margin of error of $m$, assuming the confidence level is 95%.

### EXAMPLE 11 Finding the Sample Size

Many professional polls use a 95% confidence interval and report a margin of error of 0.03 (3 percentage points).

QUESTION Find the sample size needed in a survey of a large population if the margin of error for a 95% confidence interval must be 0.03 (3 percentage points).

SOLUTION Use Formula 7.2, with $m = 0.03$.

$$n = \frac{1}{0.03^2} = \frac{1}{0.0009} = 1111.11$$

Sample at least 1112 people.

CONCLUSION We must randomly sample 1112 people. Next time you see a poll published in a paper or on a website, check the sample size. You can bet it's around 1000 people.

For example, the Pew Research Center published a survey in January 2010 showing that 59% of Americans could correctly identify the country holding the most American debt (China). Pew used a sample size of $n = 1003$ to achieve a margin of error of 3 percentage points.

TRY THIS! Exercise 7.65

## CASE STUDY REVISITED

What was wrong with the American Medical Association's spring break survey? The AMA poll was actually based on an "online survey panel," which consists of a group of people who agree to take part in several different online surveys in exchange for a small payment. Marketing companies recruit people to join panels so that the marketers can investigate trends within various slices of the public. Such a sample may or may not be representative of the population we're interested in—we have no way of knowing. And because the sample is not chosen randomly, we also have no way of knowing how our estimate will behave from sample to sample.

For such a survey, it is impossible to find a confidence interval for the true proportion of women who "agree that spring break trips involve more or heavier drinking than occurs on college campuses" because (1) our estimate might be biased, and (2) the true percentage might lie much further from our estimate than two standard errors. For this reason, the AMA ended up removing the margin of error from its website and no longer claimed that the figures were a valid inference for all college-aged women.

# EXPLORING STATISTICS
## CLASS ACTIVITY

## Simple Random Sampling Prevents Bias

| GOALS | MATERIALS |
|---|---|
| In this activity, you'll see how the sampling method affects our estimation of a population mean. | • A list of the first four amendments to the U.S. Constitution with each word numbered<br><br>• A random number table or other method of obtaining random numbers |

**ACTIVITY**

James Madison (1751–1836), who became the fourth president of the United States, wrote the first ten amendments to the U.S. Constitution, which are known as the Bill of Rights. They went into effect on December 15, 1791.

Your teacher will give you a page where the first four amendments are printed, with each word numbered. Your goal is to estimate the mean length of the words that appear in these four amendments.

You will use two estimation procedures, compare them, and decide which one works better. One method is an informal method. The other is based on random sampling. Your instructor will give you detailed instructions about each method.

The first ten words of the Bill of Rights are shown in Figure A, with each word numbered.

| 001 | 002 | 003 | 004 | 005 | 006 | 007 | 008 | 009 | 010 |
|---|---|---|---|---|---|---|---|---|---|
| Congress | shall | make | no | law | respecting | an | establishment | of | religion, |

 **FIGURE A**  The First Ten Words of the First Amendment

**BEFORE THE ACTIVITY**

1. What is the mean length of the ten words in the excerpt shown in Figure A?

2. Select any two words you wish from this excerpt. How different is the average length of the two words you selected from the mean length of all ten words?

**AFTER THE ACTIVITY**

Your instructor will give you data from the entire class. The data will consist of a list of estimates of mean word length based on the "informal" method, and a list based on the simple random sampling method. Make a dotplot of the "informal" estimates and a second dotplot of the estimates obtained using simple random sampling.

1. Compare the distributions from the two methods. (Comment on the shape, center, and spread of the distributions.)

2. Judging on the basis of your comparison of the dotplots, why is simple random sampling preferred over the informal method for collecting data?

This exercise is based on an activity from an INSPIRE workshop, which was based on an activity from Workshop Statistics, © 2004, Dr. Allan Rossman and Dr. Beth Chance, California Polytechnic State University.

# CHAPTER REVIEW

## Key Terms

Population, *296*
Parameter, *296*
Census, *296*
Sample, *296*
Statistic, *296*
Estimator, *296*

Estimate, *296*
Statistical inference, *297*
Biased, *298*
Sampling bias, *298*
Measurement bias, *298*
Voluntary-response bias, *299*

Nonresponse bias, *299*
Convenience sample, *300*
Simple random sample (SRS), *300*
With and without replacement, *300*
Sampling distribution, *305*
Bias (accuracy), *306*

Precision, *306*
Standard error (SE), *306*
Central Limit Theorem (CLT), *311*
Confidence interval, *317*
Margin of error, *317*
Confidence level, *317*

## Learning Objectives

After reading this chapter and doing the assigned homework problems, you should

- Be able to estimate a population proportion from a sample proportion and quantify how far off the estimate is likely to be.

- Understand that random sampling reduces bias.

- Understand when the Central Limit Theorem for sample proportions applies and know how to use it to find approximate probabilities for sample proportions.

- Understand how to find, interpret, and use confidence intervals for a single population proportion.

## Summary

The Central Limit Theorem (CLT) for sample proportions tells us that if we take a random sample from a population, and if the sample size is large and the population size much larger than the sample size, then the sampling distribution of $\hat{p}$ is approximately

$$N\left(p, \sqrt{\frac{p(1-p)}{n}}\right)$$

This result is used to infer the true value of a population proportion on the basis of the proportion in a random sample. The primary means for doing this is with a confidence interval:

**Formula 7.1:** $\hat{p} \pm m$ where $m = z^* \times SE_{\text{est}}$

and $SE_{\text{est}} = \sqrt{\dfrac{\hat{p}(1-\hat{p})}{n}}$

where:

$\hat{p}$ is the sample proportion of successes, the proportion of people in the sample with the characteristic we are interested in
$m$ is the margin of error
$n$ is the sample size
$z^*$ is a multiplier that is chosen to achieve the desired confidence level

An important first step is to make sure that the sample size is large enough for the CLT to work. This means that we need the sample size times the sample proportion to be at least 10 and that we need the sample size times (1 minus the sample proportion) to be at least 10.

When designing a survey, we usually choose the margin of error in advance and then find the sample size that will give us this margin of error. A conservative formula for the required sample size is

**Formula 7.2:** $n = \dfrac{1}{m^2}$ for 95% confidence

A 95% confidence interval might or might not have the correct population value within it. However, we are confident that it does, because the method works for 95% of all samples.

## Sources

Cornish, A. Do polls miss views of the young and mobile? 2007. National Public Radio. October 1. http://www.npr.org (accessed March 29, 2010).

Crowley, C. 2010. Soda tax or flat tax? Questions can influence poll results, Cornell expert says. *Albany Times Union,* February 5. http://www.timesunion.com (accessed February 6, 2010).

Gallup. Shrunken majority now favors stricter gun laws. 2007. http://www.galluppoll.com October 11 (accessed March 29, 2010).

Jones, S., and S. Fox. 2009. Generations online in 2009. Pew Internet and American Life Project. January 28. http://pewresearch.org/pubs/1093/generations-online (accessed March 19, 2010).

Mystery Pollster. http://www.mysterypollster.com/ (accessed October 2, 2007).

*New York Times.* 1993. 1 in 5 in new survey express some doubt about the Holocaust. April 20.

*New York Times.* 1994. Pollster finds error on Holocaust doubts. May 20.

Pew Research Center for the People & the Press. 2010. The growing gap between landline and oval frame Election Polls. November 12, 2010, (accessed Dec. 20, 2010).

Pew Research Center for the People & the Press. 2008. Declining public support for global engagement. September 24. http://people-press.org (accessed March 29, 2010).

# SECTION EXERCISES

## SECTION 7.1

TRY **7.1 Parameter vs. Statistic (Example 1)** Explain the difference between a parameter and a statistic.

**7.2 Sample vs. Census** Explain the difference between a sample and a census. Every 10 years, the U.S. Census Bureau takes a census. What does that mean?

**7.3 $\bar{x}$ vs. $\mu$** Two symbols are used for the mean: $\mu$ and $\bar{x}$.
   a. Which represents a parameter and which a statistic?
   b. In determining the mean age of all students at your school, you survey 30 students and find the mean of their ages. Is this mean $\bar{x}$ or $\mu$?

**7.4 $\bar{x}$ vs. $\mu$** The mean GPA of all 5000 students at Uneeda College is 2.78. A sample of 50 GPAs from this school has a mean of 2.93. Which number is $\mu$ and which is $\bar{x}$?

* **7.5 Ages of Presidents** Suppose you knew the age at inauguration of *all* the past U.S. presidents. Could you use those data to make inferences about ages of past presidents? Why or why not?

* **7.6 Heights of Basketball Team** Suppose you find *all* the heights of the members of the men's basketball team at your school. Could you use those data to make inferences about heights of all men at your school? Why or why not?

**7.7 Sample vs. Census** You are receiving a large shipment of batteries and want to test their lifetimes. Explain why you would want to test a sample of batteries rather than the entire population.

**7.8 Sampling Strawberries** You are receiving a truckload of baskets of strawberries for your supermarket chain, and you want to check their quality. If you checked a few of the easily accessible baskets, would this necessarily be a good representative sample of the quality of all the berries? Explain.

**7.9 Sampling with and without Replacement** Explain the difference between sampling with replacement and sampling without replacement. Suppose you have the names of 10 students, each written on a 3 by 5 notecard, and want to select two names. Describe both procedures.

**7.10 Simple Random Sampling** Is simple random sampling usually done with or without replacement?

TRY **7.11 Finding a Random Sample (Example 2)** You need to select a simple random sample of four from eight friends who will participate in a survey. Assume the friends are numbered 1, 2, 3, 4, 5, 6, 7, and 8.

Select four friends, using the two lines of numbers in the next column from a random number table.

Read off each digit, skipping any digit not assigned to one of the friends. The sampling is without replacement, meaning that you cannot select the same person twice. Write down the numbers chosen. The first person is number 7.

```
0 7 0 3 3      7 5 2 5 0      3 4 5 4 6
7 5 2 9 8      3 3 8 9 3      6 4 4 8 7
```

Which four friends are chosen?

**7.12 Finding a Random Sample** You need to select a simple random sample of two from six friends who will participate in a survey. Assume the friends are numbered 1, 2, 3, 4, 5, and 6.

Use technology to select your random sample. Indicate what numbers you obtained and how you interpreted them.

If technology is not available, use the line from a random number table that corresponds to the day of the month on which you were born. For example, if you were born on the fifth day of any month, you would use line 05. Show the digits in the line and explain how you interpreted them.

**7.13 Criticize the Sampling** Marco is interested in whether Proposition P will be passed in the next election. He goes to the university library and takes a poll of 100 students. Since 58% favor Proposition P, Marco believes it will pass. Explain what is wrong with his approach.

**7.14 Criticize the Sampling** Maria opposes capital punishment and wants to find out if a majority of voters in her state support it. She goes to a church picnic and asks everyone there for their opinion. Because most of them oppose capital punishment, she concludes that a vote in her state would go against it. Explain what is wrong with Maria's approach.

**7.15 Bias** In 2003, shortly after the U.S. invasion of Iraq, www.TimeEurope.com posed a question to European readers of *Time* magazine on the Internet: "Who really poses the greatest danger to world peace? Iraq, North Korea, or the United States?" The site received 706,842 responses: 6.7% said North Korea, 6.3% Iraq, and 86.9% the United States. Identify the population, and explain why the results might not reflect true opinions in the population.

**7.16 Polling** When men and women live together, women are more likely to answer the telephone. When polling agencies call homes, they often ask to speak to the adult with the most recent birthday. Give at least one reason why polling agencies take this approach.

TRY **7.17 Questionnaire Response (Example 3)** A teacher at a community college sent out questionnaires to evaluate how well the administrators were doing their jobs. All teachers received questionnaires, but only 10% returned them. Most of the returned questionnaires contained negative comments about the administrators. Explain how an administrator could dismiss the negative findings of the report.

**7.18 Survey on Social Security** A phone survey asked whether Social Security should be continued or abandoned immediately. Only land lines (not cell phones) were called. Do you think this would introduce bias? Explain.

**7.19 Random Sampling?** If you walked around your school campus and asked people you met how many keys they were carrying, would you be obtaining a random sample? Explain.

**7.20 Biased Sample?** You want to find the mean weight of the students at your college. You calculate the mean weight of a sample of members of the football team. Is this method biased? If so, would the mean of the sample be larger or smaller than the true population mean for the whole school? Explain.

**7.21 Views on Capital Punishment** In carrying out a study on views of capital punishment, a student asked a question two ways:

1. With persuasion: "My brother has been accused of murder and he is innocent. If he is found guilty, he might suffer capital punishment. Now do you support or oppose capital punishment?"

2. Without persuasion: "Do you support or oppose capital punishment?"

Here is a breakdown of her actual data.

**Men**

| | With persuasion | No persuasion |
|---|---|---|
| For capital punishment | 6 | 13 |
| Against capital punishment | 9 | 2 |

**Women**

| | With persuasion | No persuasion |
|---|---|---|
| For capital punishment | 2 | 5 |
| Against capital punishment | 8 | 5 |

a. What percentage of those persuaded against it support capital punishment?

b. Find the percentage of those not persuaded who support capital punishment.

c. Compare the percentages in parts a and b. Is this what you expected? Explain.

**7.22 Views on Capital Punishment** Use the data given in Exercise 7.21.

Make the two given tables into one table by combining men for capital punishment into one group, men opposing it into another, women for it into one group, and women opposing it into another. Show your two-way table.

The student who collected the data could have made the results misleading by trying persuasion more often on one gender than on the other, but she did not do this. She used persuasion on 10 of 20 women (50%) and on 15 of 30 men (50%).

a. What percentage of the women support capital punishment? What percentage of the men support it?

b. On the basis of these results, if you were on trial for murder and did not want to suffer capital punishment, would you want men or women on your jury?

## SECTION 7.2

**7.23 Targets: Bias or Lack of Precision?**

a. If a rifleman's gunsight is adjusted incorrectly, he might shoot bullets consistently close to 2 feet left of the bull's-eye target. Draw a sketch of the target with the bullet holes. Does this show lack of precision or bias?

b. Draw a second sketch of the target if the shots are both unbiased and precise (have little variation).

The rifleman's aim is not perfect, so your sketches should show more than one bullethole.

**7.24 Targets: Bias or Lack of Precision?, Again**

a. If a rifleman's gunsight is adjusted correctly but he has shaky arms, the bullets might be scattered widely around the bull's-eye target. Draw a sketch of the target with the bullet holes. Does this show variation (lack of precision) or bias?

b. Draw a second sketch of the target if the shots are unbiased and have precision (little variation).

The rifleman's aim is not perfect, so your sketches should show more than one bullethole.

**\* 7.25 Bias?** Suppose that, when taking a random sample of 4 from 123 women, you get a mean height of only 60 inches (5 feet). The procedure may have been biased. What else could have caused this small mean?

**\* 7.26 Bias?** Four women selected from a photo of 123 were found to have a sample mean height of 71 inches (5 feet 11 inches). The population mean for all 123 women was 64.6 inches. Is this evidence that the sampling procedure was biased? Explain.

**7.27 Proportion of Odd Digits** A large collection of one-digit random numbers should have about 50% odd and 50% even digits because five of the ten digits are odd (1, 3, 5, 7, and 9) and five are even (0, 2, 4, 6, and 8).

a. Find the proportion of odd-numbered digits in the following lines from a random number table. Count carefully.

5 5 1 8 5     7 4 8 3 4     8 1 1 7 2
8 9 2 8 1     4 8 1 3 4     7 1 1 8 5

b. Does the proportion found in part (a) represent $\hat{p}$ (the sample proportion) or $p$ (the population proportion)?

c. Find the error in this estimate, the difference between $\hat{p}$ and $p$ (or $\hat{p} - p$).

**7.28 Proportion of Odd Digits** 1, 3, 5, 7, and 9 are odd and 0, 2, 4, 6, and 8 are even. Consider a 30-digit line from a random number table.

a. How many of the 30 digits would you expect to be odd, on average?

b. If you actually counted, would you get exactly the number you predicted in part (a)? Explain.

**TRY 7.29 M&Ms (Example 4)** According to the Mars company, packages of milk chocolate M&Ms contain 20% orange candies. Suppose we examine 100 random candies.

a. What value should we expect for our sample percentage of orange candies?

b. What is the standard error?

c. Use your answers to fill in the blanks:
We expect ____% orange candies, give or take _____%.

**7.30 Random Letters** Samuel Morse suggested in the nineteenth century that the letter "t" made up 9% of the English language. Assume this is still correct. A random sample of 1000 letters is taken from a randomly selected, large book and the t's are counted.

a. What value should we expect for our sample percentage of t's?

b. Calculate the standard error.

c. Use your answers to fill in the blanks:
We expect ____% t's, give or take _____%.

**\* 7.31 What Is the Proportion of Seniors?** A population of college students is taking an advanced math class. In the class are three juniors and two seniors. Using numbers 1, 2, and 3 to represent juniors and 4 and 5 to represent seniors, sample without replacement. Draw a sample of two people four times (once in each of parts a, b, c, and d), and then fill in the following table.

a. Use the first line (reprinted here) from the random number table to select your sample of two. (The selections are underlined.)

0 **2** 7 7 9     7 2 6 **4** 5     3 2 6 9 9     8 6 0 0 9

Report the percentage of seniors in the sample. (Count the number of 4s and 5s and divide by the sample size 2.)

b. Use the next line to select your sample of two.

**3 1** 8 6 7     8 5 8 7 2     9 1 4 3 0     4 5 5 5 4

Report the percentage of seniors in the sample.

c. Use the next line to select your sample of two.

0 7 0 **3** 3     7 5 2 5 0     3 4 5 4 6     7 5 2 9 8

Report the percentage of seniors in the sample.

d. Use the last line to select your sample of two.

0 9 0 8 4     9 8 9 4 8     0 9 5 4 1     8 0 6 2 3

Report the percentage of seniors in the sample.

e. Fill in the rest of the table below, showing the results of the four samples:

| Repetition | $p$ (Population Proportion of Seniors) | $\hat{p}$ (Sample Proportion of Seniors) | Error: $\hat{p} - p$ |
|---|---|---|---|
| 1 (from part a) | $2/5 = 0.4$ | $1/2 = 0.5$ | $0.5 - 0.4 = 0.1$ |
| 2 (from part b) | | | |
| 3 | | | |
| 4 | | | |

**7.32 Simulation** From a very large population (essentially infinite), of which half are men and half are women, you take a random sample, with replacement. Use the following random number table and assume each single digit represents selection of one person; the odd numbers represent women and the even numbers (0, 2, 4, 6, 8) men.

a. Start on the left side of the top line (with 118) and count 10 people. What percentage of the sample will be men?

1 1 8 4 8     8 0 8 0 9     2 5 8 1 8     3 8 8 5 7
2 3 8 1 1     8 0 9 0 2     8 5 7 5 7     3 3 9 6 3
9 3 0 7 6     3 9 9 5 0     2 9 6 5 8     0 7 5 3 0

b. Start in the middle of the second line (with 857) and count 20 people. What percentage of the sample will be men?

c. If you were to repeat parts a and b many times, which sample would typically come closer to 50%—the sample of 10 or the sample of 20? Why?

## SECTION 7.3

**\* 7.33 M & Ms (Example 5)** Return to Exercise 7.29 and find the approximate probability that the random sample of 100 will contain 24% or more orange candies.

**\* 7.34 Random Letters** Return to Exercise 7.30 and find the approximate probability that the random sample of 1000 letters will contain 8.1% or fewer t's.

**TRY 7.35 Jury Selection (Example 6)** Juries should have the same racial distribution as the surrounding communities. According to the U.S. Census, about 18% of residents in Minneapolis, Minnesota, are African American. Suppose a local court randomly selects 100 adult citizens of Minneapolis to participate in the jury pool.

Use the Central Limit Theorem (and the Empirical Rule) to find the approximate probability that the proportion of available African American jurors is more than two standard errors from the population value of 0.18.

The conditions for using the Central Limit Theorem are satisfied because the sample is random; the population is more than 10 times 1000; $n$ times $p$ is 18, and $n$ times (1 minus $p$) is 82, and both are more than 10.

**\* 7.36 Mercury in Freshwater Fish** According to an article from HuffingtonPost.com, some experts believe that 20% of all freshwater fish in the United States have such high levels of mercury that they are dangerous to eat. Suppose a fish market has 250 fish we consider randomly sampled from the population of edible freshwater fish.

Use the Central Limit Theorem (and the Empirical Rule) to find the approximate probability that the market will have a proportion of fish with dangerously high levels of mercury that is more than two standard errors above 0.20.

You can use the Central Limit Theorem because the fish were randomly sampled; the population is more than 10 times 250; and $n$ times $p$ is 50, and $n$ times (1 minus $p$) is 200, and both are more than 10.

**TRY \* g 7.37 The Oregon Bar Exam (Example 7)** According to the Oregon Bar Association, approximately 65% of the people who take the bar exam to practice law in Oregon pass the exam. Find the approximate probability that at least 67% of 200 randomly sampled people taking the Oregon bar exam will pass. (In other words, find the probability that at least 134 out of 200 will pass.) *See page 335 for guidance.*

**\* 7.38 Feeding Vegans** A survey of eating habits showed that approximately 4% of people in Portland, Oregon, are vegans. Vegans do not eat meat, poultry, fish, seafood, eggs, or milk. A restaurant in Portland expects 300 people on opening night, and the chef is deciding on the menu. Treat the patrons as a simple random sample from Portland and the surrounding area, which has a population of about 600,000.

If 14 vegan meals are available, what is the approximate probability that there will not be enough vegan meals—that is, the probability that 15 or more vegans will come to the restaurant? Assume the vegans are independent and there are no families of vegans.

**\* 7.39 Renting Chairs for a Meeting** You have sent out 4000 invitations to hear a speaker, and you must rent chairs for the people who come. In the past, usually about 8% of the people invited have come to hear the speaker.

a. On average, what proportion of those invited should we expect to attend?

b. Suppose you assume that 8.5% of those invited will attend, and so you rent 340 chairs (because 0.085 times 4000 is 340). What is the approximate probability that more than 8.5% of those invited will show up and you will not have enough chairs? Refer to the TI-83/84 output given. Recall that this gives the Normal cumulative probability in the following format:

Normalcdf (left boundary, right boundary, mean, standard deviation)

Draw a well-labeled sketch of the Normal curve, and shade the appropriate region to represent the probability.

c. What is the approximate probability that more than 9% of the 4000 invited will show up? How many chairs would you have to rent if exactly 9% of those invited attended?

d. Why is your answer to part c smaller than your answer to part b? Draw a Normal probability distribution with a center at a proportion of 0.08 and vertical lines at about 0.085 and 0.090 so that you can visualize the size of the right-tail area.

**7.40 High School Diplomas** A U.S. government survey in 2007 said that 87% of young Americans earn a high school diploma.

a. Assume you took a simple random sample of 2000 young Americans. What proportion would you expect to earn a high school diploma? How many young Americans is this? What proportion would you expect not to earn a high school diploma? How many young Americans is this? Are both numbers greater than or equal to 10, as required by the Central Limit Theorem?

b. If you took a simple random sample of 2000 young Americans, what is the approximate probability that 88% or more of the sample will have earned their high school diploma? Refer to Example A in the TechTips for one approach to this question.

\* **7.41 Passing a Test by Guessing** A true/false test has 40 questions. A passing grade is 60% or more correct answers.

a. What is the probability that a person will guess correctly on one true/false question?

b. What is the probability that a person will guess incorrectly on one question?

c. Find the approximate probability that a person who is just guessing will pass the test.

d. If a similar test were given with multiple-choice questions with four choices for each question, would the approximate probability of passing the test by guessing be higher or lower than the approximate probability of passing the true/false test? Why?

\* **7.42 Gender: Randomly Chosen?** A large community college district has 1000 teachers, of whom 50% are men and 50% are women. In this district, administrators are promoted from among the teachers. There are currently 50 administrators, and 70% of these administrators are men.

a. If administrators are selected randomly from the faculty, what is the approximate probability that the percentage of male administrators will be 70% or more?

b. If administrators are selected randomly from the faculty, what is the approximate probability that the percentage of female administrators will be 30% or less?

c. How are part a and part b related?

d. Do your answers suggest that it is reasonable to believe the claim that the administrators have been selected randomly from the teachers? Answer yes or no, and explain your answer.

# SECTION 7.4

**7.43 Wanderlust** In a Pew Research Center Poll (Oct. 3–11, 2008), 1040 of 2260 respondents said they would prefer to live somewhere else. For example, many who lived in cities would rather not live there. Using the output given, find a 95% confidence interval for the proportion who would rather live elsewhere. Copy the statement that follows, filling in the blanks:

I am 95% confident that the population proportion of people who would like to live somewhere else is between _____ and _____.

**7.44 Drug to Prevent Heart Attacks** In June of 2009, a Pew Poll asked 1005 people: Which over-the-counter drug do doctors recommend to help prevent heart attacks? Is it aspirin, cortisone, or antacids? (Options were rotated.)

Of 1005, 914 correctly chose aspirin. Report a 95% confidence interval for the population proportion who knew the correct answer. Use the output provided to find the interval. Copy the statement below, inserting the left and right endpoints of the interval.

I am 95% confident that the population proportion of people who know aspirin is used to prevent heart attacks is between _____ and _____.

**7.45 Economic Stimulus** A Gallup Poll of 1027 adults, conducted Jan. 30–Feb. 3, 2009, found that 770 of the respondents wanted Congress to pass a stimulus bill to rescue the economy.

a. What percentage of respondents wanted a stimulus bill?

b. Report a 95% confidence interval for the population percentage in favor of a stimulus bill. Refer to the Minitab output, and round the boundaries sensibly.

| Sample | X | N | Sample p | 95% CI |
|---|---|---|---|---|
| 1 | 770 | 1027 | 0.749757 | (0.723265, 0.776248) |

**7.46 Best President** A Gallup Poll of 1018 adults in February of 2009 asked, "If you had to choose, which of the following presidents would you regard as the greatest: Washington, Franklin Roosevelt, Lincoln, Kennedy, or Reagan?" The order was rotated to avoid biasing the response. Of the 1018 people surveyed, 224 chose Lincoln. Refer to the Minitab output.

| Sample | X | N | Sample p | 95% CI |
|---|---|---|---|---|
| 1 | 224 | 1018 | 0.220039 | (0.194591, 0.245488) |

a. What percentage of respondents chose Lincoln?

b. Report a 95% confidence interval for the population percentage who chose Lincoln. Refer to the figure.

**7.47 Voting** A random sample of likely voters showed that 55% planned to vote for Candidate X, with a margin of error of 2 percentage points and with 95% confidence.

a. Use a carefully worded sentence to report the 95% confidence interval for the percentage of voters who plan to vote for Candidate X.

b. Is there evidence that Candidate X could lose?

c. Suppose the survey was taken on the streets of New York City and the candidate was running for U.S. president. Explain how that would affect your conclusion.

**7.48 Voting** A random sample of likely voters showed that 49% planned to support Measure X. The margin of error is 3 percentage points with a 95% confidence level.

a. Using a carefully worded sentence, report the 95% confidence interval for the percentage of voters who plan to support Measure X.

b. Is there evidence that Measure X will fail?

c. Suppose the survey was taken on the streets of Miami and the measure was a Florida statewide measure. Explain how that would affect your conclusion.

**7.49 Estimating the Proportion of Odd Digits (with Given Standard Error)**

a. Find the proportion of odd-numbered digits in the following lines from a random number table. Count carefully; circle them!

```
5 5 1 8 5    7 4 8 3 4    8 1 1 7 2
8 9 2 8 1    4 8 1 3 4    7 1 1 8 5
```

b. The estimated standard error for the sample proportion of odd digits is 0.090. Find the margin of error for a 95% confidence interval for the proportion of odd digits in the entire random number table by multiplying the estimated standard error by 1.96.

c. Find a 95% confidence interval for the population proportion of odd-numbered digits in the entire random number table by adding the margin of error to the sample proportion to find the upper boundary and subtracting it to find the lower boundary.

d. The estimated standard error is 0.090. Find an 80% confidence interval for the population proportion of odd-numbered digits. For calculations by hand, multiply the standard error by 1.28 to find the margin of error because $z^*$ for 80% confidence is 1.28.

e. Find a 99% confidence interval for the population proportion of odd-numbered digits. For calculations by hand, $z^*$ for 99% is 2.58.

f. State which interval is widest and which is narrowest. Explain why.

**7.50 Proportion of Odd Digits (with Given Standard Error)**

a. Find the proportion of odd-numbered digits in the following lines from a random number table and then find a 95% confidence interval for the population proportion of odd-numbered digits in the entire random number table. Count carefully!

```
8 7 9 6 4    4 3 7 5 1    8 0 9 7 1
5 0 6 1 3    5 1 4 4 1    3 0 5 0 5
```

b. The estimated standard error is 0.090. Find an 80% confidence interval for the population proportion of odd-numbered digits in the full random number table. For calculations by hand, $z^*$ for 80% is 1.28.

c. Find a 99% confidence interval for the population proportion of odd-numbered digits. For calculations by hand, $z^*$ for 99% is 2.58.

d. State which interval is wider and which is narrower. Explain why.

**TRY 7.51 High School Diplomas (Example 8)** In a simple random sample of 1500 young Americans, 87% had earned a high school diploma.

a. What is the standard error for this estimate of the percentage of all young Americans who earned a high school diploma?

b. Find the margin of error, using a 95% confidence level, for estimating the percentage of all young Americans who earned a high school diploma.

c. Report the 95% confidence interval for the percentage of all young Americans who earned a high school diploma.

d. Suppose that in the past, 80% of all young Americans earned high school diplomas. Does the confidence interval you found in part c support or refute the claim that the percentage of young Americans who earn high school diplomas has increased? Explain.

**7.52 Diabetes** In a simple random sample of 1200 Americans age 20 and over, the proportion with diabetes was found to be 0.115 (or 11.5%).

a. What is the standard error for the estimate of the proportion of all Americans age 20 and over with diabetes?

b. Find the margin of error, using a 95% confidence level, for estimating this proportion.

c. Report the 95% confidence interval for the proportion of all Americans age 20 and over with diabetes.

d. According to the Centers for Disease Control, nationally, 10.7% of all Americans age 20 or over have diabetes. Does the confidence interval you found in part c support or refute this claim? Explain.

**7.53 Dreaming in Color** In a 2003 study on dreaming, an investigator attempted to replicate an experiment done by Middleton in 1942 (Schwitzgebel, *Perceptual and Motor Skills* 2003). In the study from 2003, 92 of 113 people said they dream in color.

a. What percentage of people in 2003 said they dreamed in color?

b. Find a 95% confidence interval for the percentage who dreamed in color in 2003. The margin of error is 0.0717.

c. In the study from 1942, 29.24% of people reported dreaming in color. Does the interval for the data from 2003 capture 29.24%? Is 29.24% plausible as the 2003 population percentage? Explain.

**7.54 Seat Belt Use** In a 2008 survey, the National Highway Traffic Safety Administration (NHTSA) reported that 83% of people used seat belts. The margin of error is 3 percentage points.

a. Assuming that the confidence level is 95% and the survey was random, find a 95% confidence interval for the percentage of people who used seat belts in 2008.

b. The NHTSA said the percentage of people using seat belts in 2000 was 71%. If 71% were suggested as the percentage for 2008, would you reject that as implausible? Why or why not? Does this suggest a change in seat belt use between 2000 and 2008? Explain.

**TRY 7.55 Understanding the Meaning of Confidence Levels: 90% (Example 9)** Each student in a class of 40 was randomly assigned one line of a random number table. Each student then counted the odd-numbered digits in a 30-digit line. (Remember that 0, 2, 4, 6, and 8 are even.)

a. *On average*, in the list of 30 digits, how many odd-numbered digits would each student find?

b. If each student found a 90% confidence interval for the percentage of odd-numbered digits in the entire random number table, how many intervals (out of 40) would you expect *not* to capture the population percentage of 50%?

**7.56 Understanding the Meaning of Confidence Levels: 80%** Each student in a class of 30 was assigned one random line of a random number table. Each student then counted the even-numbered digits in a 30-digit line.

a. *On average*, in the list of 30 digits, how many even-numbered digits would each student find?

b. If each student found an 80% confidence interval for the percentage of even-numbered digits, how many intervals (out of 30) would you expect *not* to capture 50%? Explain how you arrived at your answer.

**7.57 Percentage of Female U.S. Senators in 2008** In the fall of 2008, 16 of 100 U.S. senators were female. For the year 2008, find a 95% confidence interval for the percentage of U.S. senators who were female or explain why you should *not* find a confidence interval for the percentage of U.S. senators who were female in 2008.

**7.58 Past Presidential Vote** In the 1960 presidential election, 34,226,731 people voted for Kennedy; 34,108,157 for Nixon; and 197,029 for third-party candidates (www.uselectionatlas.org).

a. What percentage of voters chose Kennedy?

b. Would it be appropriate to find a confidence interval for the proportion of voters choosing Kennedy? Why or why not?

**7.59 Newspapers Closing** A Pew Poll asked for people's opinion on whether closing local newspapers would hurt civic life; 430 of 1001 respondents said it would hurt civic life a lot.

a. Find the proportion of the respondents who said that closing local papers would hurt civic life a lot.

b. Find a 95% confidence interval for the population proportion who believed closing newspapers would hurt civic life a lot. Assume the Pew Poll used a simple random sample (SRS). (In fact, it used random sampling, but a more complex method than SRS.)

c. Find an 80% confidence interval for the population proportion who believed closing newspapers would hurt civic life a lot.

d. Which interval is wider, and why?

**7.60 Human Cloning** In a Gallup Poll, 441 of 507 adults said it was "morally wrong" to clone humans.

a. What proportion of the respondents believed it morally wrong to clone humans?

b. Find a 95% confidence interval for the population proportion who believed it is morally wrong to clone humans. Assume that Gallup used a simple random sample.

c. Find an 80% confidence interval (using a $z^*$ of 1.28 if you are calculating by hand).

d. Which interval is wider, and why?

**7.61 Desire to Lose Weight** In a Gallup Poll from November of 2008, 595 of 1009 adults surveyed said they wanted to lose weight.

a. What percentage of the sample wanted to lose weight?

b. Find a 95% confidence interval for the proportion of people in the population who wanted to lose weight.

c. Would a 99% confidence interval be wider or narrower?

**7.62 Support for Nuclear Energy** Some people fear the use of nuclear energy because of potential accidents. However, a Gallup Poll in 2009 indicated that 74% of men supported the use of nuclear energy to produce electricity. The sample size was 506.

a. In this poll, how many men supported the use of nuclear energy? Round your answer to the nearest whole number.

b. Report a 95% confidence interval for the proportion of all men in the United States who support the use of nuclear energy.

c. Would a 90% interval be wider or narrower?

**TRY** **7.63 Do People Think Astrology Is Scientific? (Example 10)** In the 2008 General Social Survey, people were asked their opinions on astrology—whether it was very scientific, somewhat scientific, or not at all scientific. Of 1438 who responded, 74 said astrology was very scientific.

a. Find the proportion of people in the survey who believe astrology is very scientific.

b. Find a 95% confidence interval for the population proportion with this belief.

c. Suppose a TV news anchor said that 5% of people in the general population think astrology is very scientific. Would you say that is plausible? Explain your answer.

**7.64 Do People Think the Sun Goes around the Earth?** In the 2008 General Social Survey, people were asked whether they thought the sun went around the earth or vice versa. Of 1381 people, 310 thought the sun went around the earth.

a. What proportion of people in the survey believed the sun went around the earth?

b. Find a 95% confidence interval for the proportion of all people with this belief.

c. Suppose a scientist said that 30% of people in the general population believe the sun goes around the earth. Using the confidence interval, would you say that was plausible? Explain your answer.

## SECTION 7.5

**TRY** **7.65 Voters' Poll (Example 11)** A polling agency wants to determine the size of a random sample needed to estimate the percentage of voters that Candidate X will receive in the next city council election in a mid-sized city. The estimate should have a margin of error of no more than 1.5 percentage points at a 95% level of confidence.

a. How large a sample should the agency take? Because a 95% interval is desired, use the shortcut formula: $n = \dfrac{1}{m^2}$. Recall that for this formula, the margin of error must be in decimal form.

b. Would a 1% margin of error require a larger or a smaller sample than that in part a?

★ c. If the degree of confidence were lowered to 90%, with the margin of error no more than 1.5%, would that require a larger or a smaller sample size than the result for part a?

**7.66 Poll of Voters** A polling agency is deciding how many voters to poll.

a. The agency wants to estimate the percentage of voters in favor of extending tax cuts, and it wants to provide a margin of error of no more than 1.8 percentage points. Using 95% confidence, how many respondents must the agency poll?

b. If the margin of error is to be no more than 1.7%, with 95% confidence, should the sample be larger or smaller than that determined in part a? Explain your reasoning.

★ c. If the degree of confidence were lowered to 90%, with the margin of error no more than 1.8%, would that require a larger or a smaller sample size than the result for part a?

**\* 7.67 Ratio of Sample Sizes** Find the sample size required for a margin of error of 3 percentage points, and then find one for a margin of error of 1.5 percentage points; for both, use 95% confidence. Find the ratio of the larger sample size to the smaller sample size. To reduce the margin of error to half, what do you need to multiply the sample size by?

**\* 7.68 Ratio of Sample Sizes** Find the sample size required for a margin of error of 4 percentage points, and then find one for a margin of error of 1 percentage point; for both, use 95% confidence. Find the ratio of the larger sample size to the smaller sample size. To reduce the margin of error to one-quarter of what it was, what do you need to multiply the sample size by?

## CHAPTER REVIEW EXERCISES

**7.69 Heart Attacks and Exposure to Traffic** In 2004, *The New England Journal of Medicine* published a study of Germans who had suffered heart attacks and survived. Within the first 24 hours of the heart attack, the 691 subjects were interviewed about what they were doing before the heart attack. Of the 691 subjects, 515 reported exposure to traffic within an hour prior to the heart attack. Most of these people had been traveling in cars or trucks, although some had used public transportation and some had been riding motorcycles.

a. What percentage of people had been exposed to traffic?

b. Find a 95% confidence interval for the percentage of all heart attack victims exposed to traffic.

c. Can you conclude that traffic definitely causes heart attacks? Why or why not?

**7.70 Bariatric Surgery and Diabetes** A study published in *JAMA* in 2004 examined past results of other studies on bariatric surgery. Bariatric surgery is done to reduce the size of the stomach in various ways. It is typically used only on obese patients, and one form of surgery had a 1% mortality rate (death rate) caused by the surgery. However, in the studies reporting on the effect of this surgery on diabetes, 1417 of 1846 diabetic patients recovered completely from diabetes after surgery. These patients no longer needed diabetes medication such as insulin.

a. Find the percentage of patients who recovered from diabetes.

b. Find a 95% confidence interval for the percentage of recovery from diabetes.

c. If you had a morbidly obese relative with diabetes, what would you tell him or her about bariatric surgery? Explain.

**7.71–7.72 Polls: Sample Size and Margin of Error** A Gallup Poll on July 7–10, 2005, included the following statements (our numbering) in regard to the survey methods at the end of an article about banning smoking in some public places and banning it generally:

*Survey Methods* Results are based on telephone interviews with 1006 national adults, aged 18 and older, conducted July 7–10, 2005.

1. For results based on the sample of 1006 adults, one can say with 95% confidence that the maximum margin of sampling error is ±3 percentage points.

2. For results based on only 489 adults in the Form A, the maximum margin of sampling error is ±5 percentage points.

**\* 7.71** Verify Statement 1 by calculation. Find the margin of error from the sample size. Round your margin of error to the nearest whole percentage.

**\* 7.72** Verify Statement 2 by calculation. Find the margin of error from the sample size. Round your margin of error to the nearest whole percentage.

**7.73 Opinion on Banning Smoking** A Gallup Poll of 489 adults done in July of 2005 found that 78 of respondents answered yes to the following question.

*Should smoking in this country be made totally illegal, or not?*

a. What percentage said yes?

b. Find and interpret a 95% confidence interval for the percentage of American adults in the country who would say yes.

c. Is it plausible that a majority believe smoking should be made totally illegal?

d. Suppose the sample size were 978 (twice as large) and the number were 156 which is also twice as large; would the percentage that said yes be the same or different from the percentage reported in part a? Would the interval be wider or narrower? If you don't know, try it!

**7.74 to 7.76 Opinion on Banning Smoking** A Gallup Poll (July, 2005), asked 1006 American adults to choose from the following 4 options for regulating smoking:

1. Set-aside areas,

2. Totally ban,

3. No restrictions,

4. No opinion.

The numbers replying "Totally Ban" in the various locations were:

| | |
|---|---|
| Hotels and Motels | 342 responded "Totally Ban" |
| Workplaces | 412 responded "Totally Ban" |
| Restaurants | 543 responded "Totally Ban" |
| Bars | 292 Responded "Totally Ban" |

**7.74 Banning Smoking in Workplaces**

a. What percentage said totally ban in the workplace?

b. Find and interpret a 95% confidence interval for the percentage who believe smoking should be banned in the workplace.

c. Is it plausible that a majority believe that smoking should be banned in the workplace?

d. Suppose the sample size was 503 instead of 1006 and the number that said totally ban was 206. Is this percentage the same or different from the percentage reported in part a? Would the interval be wider or narrower?

**7.75 Banning Smoking in Bars** Answer questions a, b, and c from 7.74 but use the data from the bars.

d. If the sample size were 2012 instead of 1006 and the number that said yes was 584 (and so the percentage is the same) would the interval be wider or narrower?

**7.76 Banning Smoking in Restaurants** Answer questions a, b, and c from 7.74 but use the data from the restaurants.

   d. If you wanted a narrower interval how would you change this survey?

**7.77 Terri Schiavo** In 2005, Terri Schiavo was said to have been in a "persistent vegetative state" for many years. Her husband wanted her feeding tube removed, and her parents wanted it left in place. In a Gallup Poll taken on March 18–20, 2005, 56% of adults polled answered yes to the following question. "Should Terri Schiavo's feeding tube have been removed?" The margin of error was given as 4 percentage points, assuming a 95% confidence level. Assuming the sampling was random, does this suggest that a majority of adults believe the feeding tube should have been removed? Show your work and explain your answer.

**7.78 Wording Confidence Interval Statements** From Exercise 7.77, a 95% confidence interval for the percentage who answered yes to whether Terri Schiavo's feeding tube should have been removed was (52%, 60%).

   Which of the following is the best interpretation? Explain why the other wordings are incorrect.

   a. We are 95% confident that the boundaries for the interval are 52% and 60%.

   b. There is a 95% probability that the population percentage is between 52% and 60%.

   c. There is a 95% chance that the boundaries 52% and 60% capture the population percentage.

   d. Ninety-five percent of all sample percentages based on samples of the same size will be between 52% and 60%.

   e. We are 95% confident that the population percentage is between 52% and 60%.

# gUIDED EXERCISES

**g * 7.37 The Oregon Bar Exam** According to the Oregon Bar Association, approximately 65% of the people who take the bar exam to practice law in Oregon pass the exam.

**QUESTION** Find the approximate probability that at least 67% of 200 randomly sampled people who take the Oregon bar exam will pass it.

**Step 1 ▶ Population proportion**
The sample proportion is 0.67. What is the population proportion?

**Step 2 ▶ Check assumptions**
Because we are asked for an approximate probability, we might be able to use the Central Limit Theorem. In order to use the Central Limit Theorem for a proportion, we must check the assumptions.

  a. Randomly sampled: Yes

  b. Sample size: If a simple random sample of 200 independent people take the bar exam, how many of them would you expect to pass, on the average? Calculate $n$ times $p$. Also calculate how many you would expect to fail, $n$ times $(1 - p)$. State whether these are both more than 10.

  c. Assume the population size is at least 10 times the sample size, which would be at least 2000.

**Step 3 ▶ Calculate**
Part of the standardization follows. Finish it, showing all the numbers.
  First find the standard error:

$$SE = \sqrt{\frac{p(1 - p)}{n}} = \sqrt{\frac{0.65(0.35)}{?}} = ?$$

Then standardize:

$$z = \frac{\hat{p} - p}{SE} = \frac{0.67 - 0.65}{?} = ? \text{ standard units}$$

Find the approximate probability that at least 0.67 pass by finding the area to the right of the $z$-value of 0.59 in the Normal curve. Show a well-labeled curve, starting with what is given in Figure A.

**Step 4 ▶ Explain**
Explain why the tail area shaded in Figure A represents the correct probability.

**Step 5 ▶ Answer the question**
Find the area of the shaded region in Figure A.
  For another approach to this type of question, see Example A in the TechTips.

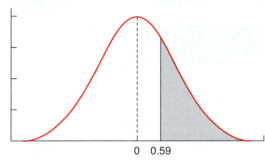

0   0.59

▲ **FIGURE A**

# TechTips

For generating random numbers, see the TechTips for Chapter 5.

## EXAMPLE A: ONE-PROPORTION PROBABILITY USING NORMAL TECHNOLOGY ►

A U.S. government survey in 2007 said that 87% of young Americans earn a high school diploma. If you took a simple random sample of 2000 young Americans, what is the approximate probability that 88% or more of the sample will have earned their high school diploma? To use the "Normal" steps from Chapter 6, we just need to evaluate the mean and standard error (the standard deviation). The population mean is 0.87. The standard error is

$$SE = \sqrt{\frac{p(1-p)}{n}} = \sqrt{\frac{0.87(1-0.87)}{2000}} = \sqrt{\frac{0.1131}{2000}} = 0.00752$$

Now you can use the "Normal" TechTips steps (for TI-83/84, Minitab, Excel, or StatCrunch) from Chapter 6 starting on page 289.

Figure 7A shows TI-83/84 output, and Figure 7B shows StatCrunch output.

Thus the probability is about 9%.

▲ **FIGURE 7A** TI-83/84 Normal Output

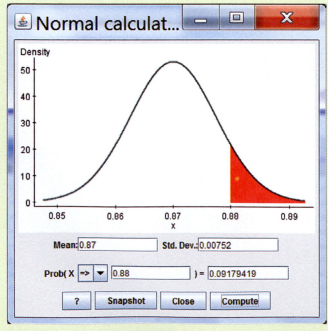

▲ **FIGURE 7B** StatCrunch Normal Output

## EXAMPLE B: FINDING A CONFIDENCE INTERVAL FOR A PROPORTION ►

Find a 95% confidence interval for a population proportion of heads when obtaining 22 heads in a sample of 50 tosses of a coin.

### Confidence Interval for One Proportion

1. Press **STAT**, choose **Tests, A: 1-PropZInt**.
2. Enter: **x, 22**; **n, 50**; **C-level**, **.95**.
3. Press **ENTER** when **Calculate** is highlighted.

Figure 7C shows the 95% confidence interval of (0.302, 0.578).

▲ **FIGURE 7C** TI-83/84 Output for a One-Proportion Z-interval

## MINITAB

### Confidence Interval for One Proportion

1. **Stat > Basic Statistics > 1 Proportion**.
2. Choose **Summarized data**.
   Enter: **Number of events**, **22**; **Number of trials**, **50**.
3. Click **Options** and Check **Use test and interval based on normal distribution**.
4. In the **1 Proportion – Options** box, if you want a confidence level different from 95%, change it.
5. Click **OK**.

The relevant part of the output is

```
         95% CI
(0.302411, 0.577589)
```

## EXCEL

### Confidence Interval for One Proportion

1. Click **Add-Ins**, **XLSTAT**, **Parametric Test**, **Tests for one proportion**.
2. See Figure 7D. Enter: **Frequency, 22**; **Sample size**, **50**; **Test Proportion**, **0.5.** Leave the other checked options as given and click **OK**.

   (If you wanted an interval other than 95%, you would click Options and change the significance level. For a 99% interval you would use 1. For a 90% interval you would use 10.)

The relevant part of the output is

95% confidence interval on the proportion (Wald):

$$(0.302, 0.578)$$

**▲ FIGURE 7D** XLSTAT Input for One Proportion

**Confidence Interval for One Proportion**
1. **Stat > Proportions > One sample > with Summary**
2. Enter: **number of successes, 22; number of observations, 50**.
3. Click **Next** and select the **Confidence Interval** option.
   Leave the default 0.95 for a 95% interval or change the **Level**.
   For **Method**, leave the default **Standard-Wald**.
4. Click **Calculate**.

The relevant part of the output is shown. "L. Limit" is the lower limit of the interval and "U. Limit" is the upper limit of the interval. Looking Back

| L. Limit | U. Limit |
|---|---|
| 0.30241108 | 0.5775889 |

# 8 Hypothesis Testing for Population Proportions

## THEME

In many scientific and business contexts, decisions must be made about the values of population parameters, even though our estimates of these parameters are uncertain. Hypothesis testing provides a method for making these decisions, while controlling for the probability of making certain types of mistakes.

In science, business, and everyday life, we often have to make decisions on the basis of incomplete information. For example, a chewing gum company might need to decide what percent of customers will buy a new flavor. The company can't test the new flavor on everyone in the country—it has to base its decision on a small sample. An educational psychologist wonders whether kids who receive music training become more creative than other children. The test she uses to measure creativity does in fact show an increase, but might this increase be explained by chance alone? A sample of people who watched violent TV when they were children turn out to exhibit more violent behavior than a comparison sample made up of people who did not watch violent TV when young. Could this difference be due to chance, or is something else going on?

Hypothesis testing is a type of statistical inference. In Chapter 7, we used confidence intervals to estimate parameters and provide a margin of error for our estimates. In this chapter we make decisions on the basis of the information provided by our sample. If we knew everything about the population, we would definitely know what decision to make. But seeing only a sample from the population makes this decision harder, and mistakes are inevitable. Just as we measured our uncertainty in Chapter 7, our next task is to measure our mistake rate when testing hypotheses. In this chapter, we continue to work with population proportions. In the next chapter, we'll see how to find confidence intervals and perform hypothesis tests for means.

## CASE STUDY

### Violence on TV

Does watching violence on TV as a child lead to violent behavior as an adult? Psychologists wishing to shed some light on this question examined 329 children and divided them into two groups: those who watched a high level of violent content on television, and those who did not. Fifteen years later, when the children were in their early twenties, the researchers interviewed them again. This time they asked about acts of aggression or violence (Husemann et al. 2003).

The researchers examined many different types of violence, and one form of violence recorded was whether the now-adults had pushed, grabbed, or shoved their spouse or partner. They found that about 38% of those who reported watching violent television had pushed, grabbed, or shoved their spouse or partner, while only 22% of those who had not watched a lot of TV violence reported engaging in the same behaviors. The difference between these two groups is 16 percentage points. Is this difference due to some real behavioral differences between these two groups of people? Or is the difference due merely to chance? After all, the researchers examined only 329 adults, which is a far cry from the entire population of all adults who watched television as children. Is it possible that even if there were no real differences between these two groups of people, we might see a difference of 16 percentage points or even more? How likely is this? Should we attribute this difference to random variation, or should we conclude that something deeper is going on?

# The Main Ingredients of Hypothesis Testing

A well-known story goes something like this: Four students missed the midterm for their statistics class. They went to the professor together and said, "Please let us make up the exam. We carpool together, and on our way to the exam, we got a flat tire. That's why we missed the exam." The professor didn't believe them, but instead of arguing he said, "Sure, you can make up the exam. Be in my office tomorrow at 8." The next day, they met in his office. He sent each student to a separate room and gave them an exam. The exam consisted of only one question: "Which tire?"

We don't know the outcome of this story, but let's imagine that all four students answer, "left rear tire." The professor is surprised. He had assumed that the students were lying. "Maybe," he thinks, "they just got lucky. After all, if they just guessed, they could still all choose the same tire." But then he does a quick calculation and figures out that the probability that all four students will guess the same tire is only 1.6%. Reluctantly, he concedes that the students were not lying, and now he must give all of them an A on the exam.

The statistics professor has just performed a hypothesis test. **Hypothesis testing** is a formal procedure that enables us to choose between two hypotheses when we are uncertain about our measurements. We call hypothesis testing a formal procedure because it is based on particular terminology and a rather well-specified set of steps. However, we hope to show you that this "formal" procedure has a generous helping of common sense supporting it.

This chapter is divided into five sections. In the first section, you will learn the main ingredients and basic terminology of hypothesis testing in the context of a simple example. Next we will develop the formal procedure in the context of two real situations, and in the third section, we will further explore some details. In the fourth section, you'll learn how to alter the basic steps of this formal procedure to accommodate a more complex situation. And in the final section, we will discuss some of the subtleties behind hypothesis testing.

But before reading further, get a penny.

Football games and tennis matches begin with a coin toss to determine which team or player gets to start by playing offense. Coin tosses, in which the coin is flipped high into the air, are used because flipping a coin is believed to be "fair." The coin is equally likely to land heads or tails, so both sides have an equal chance of winning.

But what if we spin the penny (on a hard, flat surface) rather than flipping it in the air? We claim that, because the heads side of a penny bulges out, the lack of symmetry will cause the spinning coin to land on one side more often than on the other. In other words, we believe a spun penny is not fair. Some people—and they may include you—will find this claim to be outrageous and will insist it is false. So let's collect data.

## 1. Start with: A Pair of Hypotheses

We now have a research hypothesis: A spun penny is not fair. This is a good first step toward stating what we will call an alternative hypothesis. It is only a first step, however, because it is not very precise. To be more precise, we need to explain what we mean by "fair."

> **Details**
>
> **Definition of Hypothesis**
> Merriam Webster's online dictionary defines a hypothesis as "a tentative claim made in order to draw out and test its logical or empirical consequences."

When a coin is flipped or spun, it has some probability of landing heads. Let's call that probability $p$. For a fair coin, $p = 0.5$. We are claiming that for a coin that is spun, $p$ is not equal to 0.5.

We now have a pair of formal hypotheses. The *alternative hypothesis* can be written as

$$p \neq 0.50$$

The commonsense belief, which most people would probably agree with at first glance, is that a spun coin *is* fair. This is the *null hypothesis*, which we can write as

$$p = 0.50$$

In a formal hypothesis test, hypotheses are *always* statements about population parameters. In this case, our population consists of infinitely many spins of our coin, and $p$ represents the probability of getting heads.

**KEY POINT** ▸ Hypotheses are *always* statements about population parameters; they are *never* statements about sample statistics.

Hypotheses come in pairs:

The **null hypothesis**, which we write $H_0$ (and pronounce "H-naught" or simply "the null hypothesis"), is the conservative, status-quo, business-as-usual statement about a population parameter. In the context of researching new ideas, the null hypothesis often represents "no change," "no effect," or "no difference."

The **alternative hypothesis**, $H_a$ (pronounced "H-A"), is the research hypothesis. It is usually a statement about the value of a parameter that we hope to demonstrate is true.

> **! Caution**
>
> **Pronunciation**
> Do *not* say "Ho" or "Ha," as Santa Claus might.

The most important step of a formal hypothesis test is choosing the hypotheses. In fact, there are really only two steps of a formal hypothesis test that a computer cannot do, and this is one of those steps. (The other step is checking to make sure that the conditions necessary for the probability calculations to be valid are satisfied.)

Hypothesis tests are like criminal trials. In a criminal trial, two hypotheses are placed before the jury: The defendant is not guilty, or he is guilty. These hypotheses are not given equal weight, however. The jury is told to assume the defendant is not guilty until the evidence overwhelmingly suggests this is not so. (Defendants charged with a crime in the United States must be found guilty "beyond all reasonable doubt.")

Hypothesis tests follow the same principles. The statistician plays the role of the prosecuting attorney, who hopes to show that the defendant is guilty. The hypothesis that the statistician or researcher hopes to establish plays the role of the alternative hypothesis. The null hypothesis is chosen to be a neutral, noncontroversial statement. Just as in a jury trial, where we ask the jury to believe that the defendant is not guilty unless the evidence against this belief is overwhelming, we will believe that the null hypothesis is true in the beginning. But once we examine the evidence, we may reject this belief if the evidence is overwhelmingly against it.

**KEY POINT** ▸ The null hypothesis always gets the benefit of the doubt and is *assumed* to be true throughout the hypothesis-testing procedure. If we decide at the last step that the observed outcome is extremely unusual under this assumption, then and only then do we reject the null hypothesis.

Our hypotheses about spinning a coin are competing predictions about what we will see when we collect data. If the null hypothesis is right, then when we spin a coin a number of times, about half of the outcomes should be heads. If the null hypothesis is wrong, we will see either a larger or a smaller proportion. Now stop reading this book and spin a penny 20 times. Record the number of heads and calculate the sample proportion of heads, $\hat{p}$. This will only take a minute. We'll wait.

Are your results consistent with the claim that $p = 0.5$?

## EXAMPLE 1 Internet Advertising

An Internet retail business is trying to decide whether to pay a search engine company to upgrade its advertising. In the past, 15% of customers who visited the company's webpage by clicking on the advertisement bought something (this is called a "click-through"). If the business decides to purchase premium advertising, then the search engine company will make that company's ad more prominent. The search engine company offers to do an experiment: For one day, customers will see the retail business's ad in the more prominent position. The retail business can then decide whether the advertising improves the percentage of click-throughs. At the end of the experiment, the business observed that 17% of the customers now bought something. A marketing executive wrote the following hypotheses to do a hypothesis test.

$$H_0: \hat{p} = 0.15$$
$$H_a: \hat{p} = 0.17$$

where $\hat{p}$ represents the proportion of the sample that bought something from the website.

**QUESTION** What is wrong with these hypotheses? Rewrite them so that they are correct.

**SOLUTION** First, these hypotheses are written about the sample proportion, $\hat{p}$. We *know* that 17% of the sample bought something, so there is no need to make a hypothesis about it. What we don't know is what proportion of the entire population of people who click on the advertisement will purchase something. The hypotheses should be written in terms of $p$, the proportion of the population that will purchase something.

A second problem is with the alternative hypothesis. The research question that the company wants to answer is not whether 17% of customers will purchase something. It wants to know whether the percentage of customers who do so has increased over what has happened in the past.

The correct hypotheses are

$$H_0: p = 0.15$$
$$H_a: p > 0.15$$

where $p$ represents the proportion of all customers who click on the advertisement and purchase a product.

**TRY THIS!** Exercise 8.5

## 2. Add In: Making Mistakes

Mistakes are an inevitable part of the hypothesis-testing process. The trick is not to make them too often.

One mistake we might make is to reject the null hypothesis when it is true. For example, even a fair coin sometimes turns up heads in 20 out of 20 flips. If that happened, we might conclude that the coin was unfair when it really was fair. We can't prevent this mistake from happening, but we can try to make it happen infrequently.

The **significance level** is the name for a special probability: It is the probability of rejecting the null hypothesis when, in fact, the null hypothesis is true. It is the probability of making a very particular type of mistake. In our experiment, the significance level is the probability that we conclude that spinning a coin is *not* fair when, in fact, it really *is* fair. In a criminal justice setting, the significance level is the probability that we conclude that the suspect is guilty when he is actually innocent. The significance level is used so often that it has its own symbol, the Greek lowercase alpha: $\alpha$.

Naturally, we want a procedure with a small significance level, because we don't want to make mistakes too often. How small? Most researchers and statisticians use a significance level of 0.05. In some situations it makes sense to allow the significance level to be bigger, and some situations require a smaller significance level. But $\alpha = 0.05$ is a good place to start.

**KEY POINT**  The significance level, $\alpha$ (Greek lowercase alpha), represents the probability of rejecting the null hypothesis when the null hypothesis is true. For many applications, $\alpha = 0.05$ is considered acceptably small, but 0.01 and 0.10 are also sometimes used.

## EXAMPLE 2 Significance Level for Internet Advertising

In Example 1, an Internet retail business gave a pair of hypotheses about $p$, the proportion of customers who click on an advertisement and then purchase a product from the company. Recall that in the past, the proportion of customers who bought the product was 0.15, and the company hopes this proportion has increased.

$$H_0: p = 0.15$$
$$H_a: p > 0.15$$

**QUESTION** Describe the significance level in context.

**SOLUTION** The significance level is the probability of rejecting $H_0$ when in fact it is true. In this context, this is the probability of the company concluding that a larger proportion of customers will buy its product than in the past when, in fact, the proportion has not changed.

**TRY THIS!** Exercise 8.11

## 3. Mix with: The Test Statistic

A **test statistic** compares our observed outcome to the null hypothesis. For example, in our coin-spinning example, we observed 7/20, or 0.35, heads. The null hypothesis tells us that we should expect half to be heads: 0.5. The test statistic compares the observed proportion to the expected proportion.

A useful test statistic in this situation is the **one-proportion z-test** statistic.

**Formula 8.1: The one-proportion z-test statistic**

$$z = \frac{\hat{p} - p_0}{SE}$$

$$\text{where } SE = \sqrt{\frac{p_0(1 - p_0)}{n}}$$

**Looking Back**

**Standard Error**
The standard error of the sample proportion, given in Chapter 7, is

$$SE = \sqrt{\frac{p(1-p)}{n}}$$

Formula 8.1 uses the symbol $p_0$ to remind us to use the value that the null hypothesis claims to be correct.

The symbol $p_0$ represents the value of $p$ that the null hypothesis claims is true. For example, for the coin-spinning example, $p_0$ is 0.5. Most of the test statistics you will see in this book have the same structure as Formula 8.1:

$$z = \frac{\text{observed value} - \text{null value}}{SE}$$

Because we observed 0.35 heads, the value for the test statistic for our experiment is

$$z_{\text{observed}} = \frac{0.35 - 0.5}{\sqrt{\frac{0.5(1-0.5)}{20}}} = -1.34$$

Note that the observed value of our test statistic is negative, because the proportion we observed was less than the proportion we expected.

## Why Is the *z*-Statistic Useful?

This test statistic is useful because it compares our observed sample proportion, $\hat{p}$, to the null hypothesis value, $p_0$. If the observed sample proportion is close to what the null hypothesis says we should see, then the test statistic will be close to 0. Thus, the farther the value of the test statistic is from 0 (that is, the larger its absolute value), the more suspicious we become of the null hypothesis.

 **KEY POINT**   If the null hypothesis is true, then the *z*-statistic will be close to 0. Therefore, the farther the *z*-statistic is from 0, the more the null hypothesis is discredited.

 **Looking Back**

**Sampling Distribution**
A sampling distribution is the probability distribution for a statistic.

The *z*-statistic is also useful because, as you will soon see, statisticians know a good approximation to the sampling distribution when the null hypothesis is true. This is an important requirement for a test statistic, because we need to know the sampling distribution to verify that we're getting the correct significance level.

## EXAMPLE 3 Test Statistic for Spinning Heads

The authors observed 7 heads when they spun their coin 20 times, and they calculated a test statistic of $z = -1.34$. A friend of the author observed 12 heads out of 20 spins.

**QUESTION** What value of the test statistic should the friend report?

**SOLUTION** The observed sample proportion is $12/20 = 0.60$. Therefore, the friend's observed test statistic is

$$\frac{0.60 - 0.50}{\sqrt{\frac{0.50 \times 0.50}{20}}} = 0.89$$

**CONCLUSION** The observed value of the test statistic is 0.89. Note that the friend has a positive-valued test statistic, because he observed more than the expected number of heads.

**TRY THIS!** Exercise 8.13

## 4. The Final Main Ingredient: Surprise!

No, the final main ingredient in hypothesis testing is not *a* surprise. The main ingredient is surprise itself.

Surprise happens when something unexpected occurs. The null hypothesis tells us what to expect when we look at our data. If we see something unexpected—that is, if we are surprised—then we should doubt the null hypothesis and, if we are really surprised, reject it altogether.

Figure 8.1 shows all possible outcomes of our coin-spinning experiment. According to the null hypothesis, getting between about 6 and 15 heads is not that unusual; these outcomes are quite common when one is flipping a fair coin. But getting less than 6 heads (5, 4, 3, 2, 1, or 0) or more than 15 (16, 17, 18, 19, or 20) is unusual. If this happened to us, we would be surprised and would probably reject the null hypothesis.

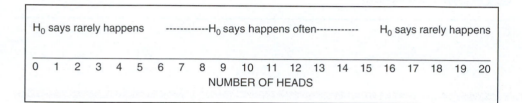

| $H_0$ says rarely happens | ----------$H_0$ says happens often------------ | $H_0$ says rarely happens |

0  1  2  3  4  5  6  7  8  9  10  11  12  13  14  15  16  17  18  19  20
NUMBER OF HEADS

◀ **FIGURE 8.1** Possible outcomes from spinning a coin 20 times and counting the number of heads. If the probability of heads is 0.50, then when spinning or tossing a coin 20 times, we would get between 6 and 14 heads quite often. The numbers 5 or fewer are rare, as are 15 or more.

Because we are statisticians, we have a way of measuring our surprise. The **p-value** is a number that measures our surprise by reporting the probability that if the null hypothesis is true, a test statistic will have a value as extreme as or more extreme than the value we actually observe. Small p-values (closer to 0) mean we have received a surprise. Large p-values (closer to 1) mean no surprise: The outcome happens fairly often. For instance, we can calculate that the probability of getting 4 or fewer heads when spinning a fair coin is 0.004. This tells us that getting 4 or fewer heads is highly unusual (and happens only 4 out of 1000 times). We would therefore be surprised if it were to happen.

> **Details**
>
> **"Often" and "Rarely"**
> If you are wondering exactly what we mean by "often" and "rarely," don't worry. We intend these to be informal terms.

**KEY POINT**  The p-value is a probability. Assuming the null hypothesis is true, the p-value is the probability that if the experiment were repeated, you would get a test statistic as extreme as or more extreme than the one you actually got. A small p-value suggests that a surprising outcome has occurred and discredits the null hypothesis.

For example, in our test to see whether a spun coin is fair, we spun a coin 20 times and saw 7 heads. The p-value associated with this outcome can be found to be about 0.18. This tells us that if spinning a coin really is fair, then the outcome we saw is not all that unusual. It happens about 1 time in 5. We would probably think that this does not discredit the hypothesis that spinning a coin is fair.

## EXAMPLE 4 Click-Throughs

The Internet business from Examples 1 and 2 wants to find out whether premium advertising will lead to a greater proportion of click-throughs—people who click on the advertisement and then actually purchase a product. Company investigators collect data and find that for one sample of customers, 18% click on the ad and then purchase a product. They perform a hypothesis test and find a p-value of 0.030. The null and alternative hypotheses are

$$H_0: p = 0.15$$

$$H_a: p > 0.15$$

where $p$ represents the proportion of all customers who click on the advertisement and purchase a product.

**QUESTION** Explain the meaning of this p-value, 0.030, in context.

**SOLUTION** The p-value of 0.030 means that if premium advertising really does nothing, then the probability of getting a proportion of purchases as high as or higher than 18% is 0.030. This tells us that if premium advertising does nothing, we have seen a fairly unusual outcome—an outcome that occurs only 3 out of 100 times.

**TRY THIS!** Exercise 8.15

---

## SECTION 8.2

# Characterizing p-Values

Understanding how to interpret the p-value is crucial to understanding hypothesis testing. The computer might compute the p-value for you, but you need to understand how the computer did this calculation if you are to successfully perform a hypothesis test.

### The p-Value Is All about Extremes

The meaning of the phrase "as extreme as or more extreme than" depends on the alternative hypothesis. There are three basic pairs of hypotheses, as Table 8.1 shows.

▶ **TABLE 8.1** These three pairs of hypotheses can be used in a hypothesis test.

| Two-tailed | One-tailed (Left) | One-tailed (Right) |
|---|---|---|
| $H_0: \quad p = p_0$ | $H_0: \quad p = p_0$ | $H_0: \quad p = p_0$ |
| $H_a: \quad p \neq p_0$ | $H_a: \quad p < p_0$ | $H_a: \quad p > p_0$ |

If the alternative hypothesis is

$H_a: p \neq p_0$ (the true value of $p$ is either bigger or smaller than what the null hypothesis claims)

then "as extreme as or more extreme than" means "even farther away from 0 than the value you observed." This corresponds to finding the probability in both tails of the N(0,1) distribution. This is called a **two-tailed hypothesis**.

If the alternative hypothesis is

$H_a: p < p_0$ (the true value is less than the value claimed by the null hypothesis)

then "as extreme as or more extreme than" means "less than or equal to the observed value." This corresponds to finding the probability in the left tail of N(0, 1). This is an example of a **one-tailed hypothesis**. This alternative is also sometimes called a **left-tailed** hypothesis, because the p-value area is in the left tail.

Finally, if the alternative hypothesis is

$H_a: p > p_0$ (the true value is greater than the value claimed by the null hypothesis) then "as extreme as or more extreme than" means "greater than or equal to the observed value." This corresponds to finding the probability in the right tail of N(0, 1). This is another one-tailed hypothesis. This alternative hypothesis is called a **right-tailed hypothesis**.

To illustrate, suppose the observed value of the test statistic is $z = 1.56$. Figure 8.2 shows the p-value (represented by the shaded region) for each of these three possible alternative hypotheses.

**(a)**

$H_a : p \neq p_0$

**(b)**

$H_a : p < p_0$

**(c)**

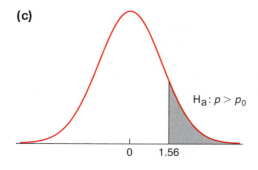

$H_a : p > p_0$

◄ **FIGURE 8.2** The shaded areas represent p-values for three different alternative hypotheses when the observed value of the test statistic is $z = 1.56$. **(a)** The p-value for a two-tailed alternative hypothesis (0.119). **(b)** The p-value for a left-tailed hypothesis (0.941). **(c)** The p-value for a right-tailed hypothesis (0.059).

Our coin-spinning experiment is an example of a two-tailed hypothesis. Figure 8.3 shows the area corresponding to the p-value for getting an outcome as extreme as or more extreme than a sample proportion of 0.35 for this two-tailed alternative hypothesis. (The units in Figure 8.3 are not "proportion of heads" but, rather, have been converted to standard units, using the test statistic above.)

As an example of a one-tailed p-value, consider Figure 8.4, which illustrates the p-value for one possible outcome in Example 4. In that example, an Internet retailer was wondering whether the new advertising campaign would result in a greater proportion of customers "clicking through" and purchasing its product. "As extreme as or more extreme than" the company planners' observed outcome means "as big as or bigger than" what the null hypothesis tells them to expect. (In Figure 8.4, the p-value shown is about 0.030. You can see that about 3% of the area under the curve is shaded.)

▲ **FIGURE 8.3** The shaded area represents the p-value: the probability of getting a proportion as extreme as or more extreme than 0.35 in 20 flips of a coin. Note that both the left and right tails are shaded.

◄ **FIGURE 8.4** The shaded area represents the p-value for a z-statistic of 1.84. Because the alternative hypothesis is one-tailed ("the proportion of customers who buy a product is *larger* than it used to be"), the p-value is the area in the right tail.

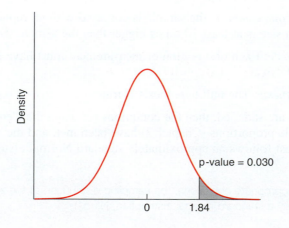

p-value = 0.030

## EXAMPLE 5 Identify the p-Values

Figure 8.5 shows sketches of three probability distributions. The shaded areas represent probabilities.

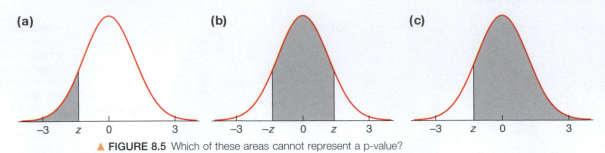

▲ **FIGURE 8.5** Which of these areas cannot represent a p-value?

**QUESTION** Which of these three sketches cannot represent a p-value? Explain your reasoning.

**SOLUTION** Figure 8.5b cannot represent a p-value because it is not a tail probability. Figure 8.5b represents the probability of getting an outcome between two values. Figure 8.5a represents the probability of getting an outcome as small or smaller than the observed outcome, and Figure 8.5c represents the probability of getting an outcome as big or bigger than the expected outcome. Both Figure 8.5a and Figure 8.5c, therefore, could represent p-values.

**TRY THIS!** Exercise 8.17

## Conditions Required for Calculating an Approximate p-Value

Note that in Figures 8.4 and 8.5 we used a Normal distribution to illustrate finding the p-value. This is no coincidence; under certain conditions, the standard Normal distribution, N(0, 1), provides a good model for the sampling distribution of the $z$-statistic.

These conditions are listed here and will be referred to throughout the remainder of the chapter.

1. *Random sample*: The sample is collected randomly from the population.

2. *Large enough sample size*: The sample size, $n$, is large enough that the sample has at least 10 expected successes and 10 expected failures; in other words, $np_0 \geq 10$ and $n(1 - p_0) \geq 10$.

3. *Without replacement*: If the sample is collected without replacement, then the population size is at least 10 times bigger than the sample size.

4. *Independence*: Each observation or measurement must have no influence on any others.

5. *Null hypothesis*: The null hypothesis is true.

If these conditions are satisfied, then the conditions for using the Central Limit Theorem for sample proportions (Chapter 7) have been met, and the $z$-statistic for the one-proportion $z$-test follows an approximately standard Normal distribution.

 **KEY POINT** Under the appropriate conditions, the sampling distribution of the $z$-statistic is approximately a standard Normal distribution, N(0, 1).

## Calculating the p-Value

To find p-values, because the sampling distribution is the standard Normal, use the same techniques described in Section 6.2. As always, we recommend that you use technology to find the area under the standard Normal curve, but if you do not have technology, you can use Table 2 in Appendix A.

Figure 8.6 illustrates how the p-value depends on the observed outcome of our coin-spinning study. Each graph represents the p-value for a different outcome, with the coin spun 20 times in each case. The null hypothesis in all cases is $p = 0.5$, and the alternative is the two-tailed hypothesis that the probability of heads is *not* 0.5. Note that the closer the number of heads is to 10, the closer the z-value is to 0 and the larger the p-value is. Also note that the p-value for an outcome of 11 heads is the same as for 9 heads, and the p-value for an outcome of 12 heads is the same as for 8 heads. This happens because the alternative hypothesis is two-tailed and the Normal distribution is symmetric.

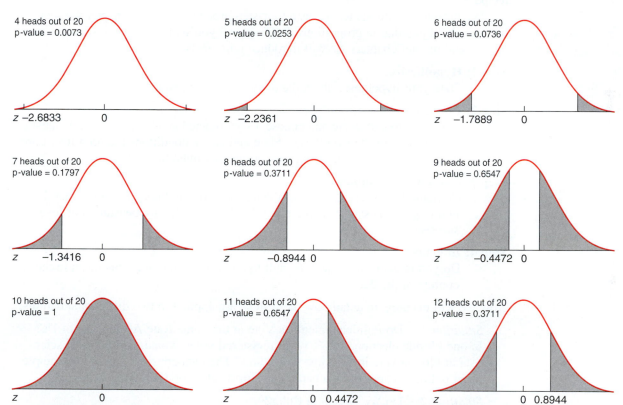

▲ **FIGURE 8.6** Each graph shows the p-value (shaded) for a different number of heads out of 20 spins of a coin, with the assumption that the coin is fair and using a two-tailed alternative hypothesis. Note that the closer the number of heads is to 10 (out of 20), the closer the z-value is to 0 and the larger the p-value is. Also, as the number of heads gets farther from 10, the z-value gets farther from 0 and the p-value gets smaller.

## EXAMPLE 6 p-Values for Coin Spinning

Two different students each did the coin-spinning experiment with a two-tailed alternative. Their test statistics follow.

$$\text{Study 1: } z = 1.98$$

$$\text{Study 2: } z = -2.02$$

QUESTION  Which of these test statistics has the smaller p-value and why?

> **! Caution**
>
> **So Many p's**
> $p$ is the population proportion.
> $p_0$ is the value of the population proportion according to the null hypothesis.
> $\hat{p}$ is the sample proportion.
> The p-value is the probability that if the null hypothesis is true, our test statistic will be as extreme as or more extreme than the value we actually observed.

**SOLUTION** If the null hypothesis is correct, then the test statistic should be close to 0. Values farther from 0 are more surprising and so have smaller p-values. Because $-2.02$ is farther from 0 than is 1.98, the area under the Normal curve in the tails is smaller for $-2.02$ than it is for 1.98. Thus $-2.02$ has the smaller p-value.

**TRY THIS!** Exercise 8.19

# SECTION 8.3

# Hypothesis Testing in Four Steps

Now that you know the main ingredients of hypothesis testing, it's time to learn the recipe.

We present hypothesis testing as a four-step procedure. In this chapter we focus on tests about population proportions. In Chapter 9 you will see tests about population means, and in later chapters we will introduce other tests.

**Step 1: Hypothesize.**
State your hypotheses about the population parameter.

**Step 2: Prepare.**
Get ready to test: State and choose a significance level. Choose a test statistic appropriate for the hypotheses. State and check conditions required for future computations, and state any assumptions that must be made.

**Step 3: Compute to compare.**
Compute the observed value of the test statistic, and compare it to what the null hypothesis said you would get. Find the p-value to measure your level of surprise.

**Step 4: Interpret.**
Do you reject or not reject the null hypothesis? What does this mean in the context of the data?

Now we will examine in detail how these steps are applied to two different situations.

*Situation 1:* Do Political Scientists Vote at the Same Rate As the Rest of Us? In one Florida election, 46.6% of all registered voters voted. Did political scientists in Florida vote in the same proportion? The data consist of a random sample of 54 political scientists in a particular Florida election (Schwitzgebel 2010).

*Situation 2:* Do We Dream in Color? Historically, before the age of television, color movies, and video games, 29% of the American population reported dreaming in color. A psychologist suspects that the present-day proportion might be higher, now that we are surrounded with color imagery. The data consist of a random sample of 113 people, taken in the year 2001 (Schwitzgebel 2003).

## Step 1: Hypothesize

Hypotheses, you will recall, consist of an alternative hypothesis that is based on the researchers' claim, and a null hypothesis that represents the status-quo position. The null hypothesis is assumed to be true until the very last of the four steps of our procedure.

Let's see how to create a pair of hypotheses for these two situations.

### Situation 1: Voting

Most people would probably think that political scientists are more likely to vote than the general public. However, some political scientists, a cynical lot, tend to think this group might be less likely to vote than the general public. The researcher therefore wonders whether it can be shown that political scientists do *not* vote in the same proportion as the general public. This is the alternative hypothesis. The neutral, status-quo hypothesis is that they *do* vote in the same proportion as the general public. (Recall that in the election considered, 46.6% of the registered voters voted.) So the researcher wonders whether political scientists are just as likely to vote as the general population or do not vote in the same proportion.

The null and alternative hypotheses can now be stated both in words and in symbols.

$H_0$: Political scientists vote in the same proportion as the public: 0.466.

$H_a$: Political scientists do not vote in the same proportion as the public.

$$H_0: p = 0.466$$
$$H_a: p \neq 0.466$$

This alternative hypothesis is a two-tailed hypothesis because it includes both the possibility that the population proportion might be bigger than 0.466 and the possibility that it might be smaller than 0.466.

## Situation 2: Dreaming

The researcher hopes to establish that a greater proportion of people today report that they dream in color than the proportion that reported dreaming in color historically. In the past, before color TV and movies, this proportion was 0.29. The researcher has interviewed a random sample of 113 people.

We state the null and alternative hypotheses in words and symbols.

$H_0$: The proportion of those who report dreaming in color in the population is the same as it has historically been: 0.29.

$H_a$: The proportion of those who report dreaming in color has increased.

$$H_0: p = 0.29$$
$$H_a: p > 0.29$$

This alternative hypothesis is a one-tailed hypothesis, because the hypothesis considers only the possibility that the true population proportion might be greater than—to the right of—the hypothesized value of 0.29.

>  **Details**
>
> **$H_0$ and the Equals Sign**
> In this book, the null hypothesis always has an equals sign, no matter which alternative hypothesis is used.

## EXAMPLE 7 Age Discrimination

According to the U.S. Census, about 30% of the population in Silicon Valley, a region in northern California, is between the ages of 40 and 65. A fired employee of a computer company who is in his fifties feels he is the victim of age discrimination. One piece of information he has is that only 2% of the 2100 employees at his former company are between the ages of 40 and 65 (www.workforce.com 2008). Here's what he wants to know: If the company hires at random, is this information evidence that it has a preference for younger employees?

**QUESTION** State the null and alternative hypotheses in words and symbols. Use the symbol $p$ to represent the proportion of all employees at this firm who are between the ages of 40 and 65.

**SOLUTION**

$H_0$: The proportion of employees at the company who are in this age group is 0.30, the same as that for the general population.

$H_a$: The proportion of employees at the company who are in this age group is less than that for the general population.

$$H_0: p = 0.30$$
$$H_a: p < 0.30$$

**TRY THIS!**    Exercise 8.23

## Step 2: Prepare

To prepare for the actual computations, we must choose a test statistic, and we must make sure we know what the sampling distribution of this test statistic is. But even before we do that, we need to set the significance level.

### Set the Significance Level

The significance level is the probability that we will make the mistake of rejecting the null hypothesis when, in fact, the null hypothesis is true. Most scientific journals require a relatively low probability of making this mistake: 0.05. In symbols,

$$\alpha = 0.05$$

### Select a Test Statistic

Choosing the correct test statistic is not a big deal at this point, because you have seen only one test statistic to choose from—the one-proportion $z$-test. This test statistic works well, because when the null hypothesis is true we tend to get small values (close to 0), and when the null hypothesis is not true we tend to get values far from 0. The test statistic is also useful because we know a good approximation to its sampling distribution and so can compute the significance level and the p-value.

In Section 8.4 you will study a new test statistic used when comparing two population proportions. Later in the book, you will see test statistics for comparing means, and then for comparing slopes and intercepts for regression lines. Thus it is important to understand which test statistic is appropriate for the context you are working with.

### Check the Sampling Distribution Conditions

We will need to know the sampling distribution of our test statistic. At this point, you should check that the conditions listed in Section 8.1 hold. These conditions are (1) random sample, (2) large enough sample size (3) sampling without replacement, and (4) independence. Condition (5), that the null hypothesis is true, is a condition we will assume to be met for the time being. If these conditions are not met, then other approaches are needed. (Some examples will be given in Chapter 13.)

> **KEY POINT**    If the null hypothesis is true, and if the four conditions for applying the Central Limit Theorem hold, then the test statistic $z$ follows a $N(0, 1)$ distribution. Extreme values are rare in a $N(0, 1)$ distribution, so if we see an extreme value, it is evidence that the null hypothesis is not true.

## EXAMPLE **8** Preparing for Situation 1: Voting

A researcher is wondering whether political scientists vote in the same proportion as the general public. The hypotheses for this test are

$$H_0: p = 0.466$$
$$H_a: p \neq 0.466$$

where $p$ is the proportion of all political scientists registered to vote in Florida. The data are from a random sample of 54 people.

QUESTION  Check that the conditions hold so that the sampling distribution of the $z$-statistic will approximately follow the standard Normal distribution.

SOLUTION  We are told that the data come from a random sample, and this satisfies the first condition. We must next check that the sample size of 54 is large enough to produce at least 10 successes and 10 failures.

If the null hypothesis is true, the probability of success is $p_0 = 0.466$. Because $n = 54$,

$$np_0 = 54 \times 0.466 = 25.16 > 10$$

$$n(1 - p_0) = 54 \times (1 - 0.466) = 28.84 > 10$$

The third condition is true if the population—all political scientists registered to vote in Florida—is more than 10 times bigger than the sample size; that is,

$$\text{Population size bigger than } 10 \times 54 = 540$$

We confess that we do not know the number of political scientists in Florida. We cannot check this condition, but we *assume* that it is true.

CONCLUSION  If the null hypothesis is true (as we assume when beginning these tests), then the conditions hold and the one-proportion $z$-test is appropriate to test the hypotheses.

TRY THIS!  Exercise 8.29

## Step 3: Compute to Compare

At last, we are at the step where we consider actual data. Our test statistic, usually calculated with technology, will compare the value of the statistic provided by the data ($\hat{p}$) with the value that the null hypothesis says we should see ($p_0$). We then find the p-value to determine whether this value is unusual, given our assumption that the null hypothesis is true.

### Find the Observed Value of the Test Statistic

Let's see how this works for the researcher studying political scientists' voting habits. In the sample of 54 political scientists, the researcher reported that 40 of them voted. In Example 8 we found that the one-proportion $z$-test is appropriate for these data. So let's find the observed value of the $z$-statistic and interpret it. For this test, the hypotheses are

$$H_0: p = 0.466$$

$$H_a: p \neq 0.466$$

The data: The researcher reports that in his sample of 54 political scientists, 40 of them voted. The sample proportion is therefore $\hat{p} = (40/54) = 0.7407$. We wish to compare this value to 0.466, the value that the null hypothesis leads us to expect. To do this, we find the $z$-statistic (Formula 8.1), which first requires that we find the standard error.

$$SE = \sqrt{\frac{p_0(1 - p_0)}{n}} = \sqrt{\frac{0.466 \times (1 - 0.466)}{54}} = 0.067884$$

Now we substitute the standard error into our $z$-statistic:

$$z_{observed} = \frac{\hat{p} - p_0}{SE} = \frac{0.740741 - 0.466}{0.067884} = 4.05$$

## Find the p-Value

How surprised are we by this outcome, if the null hypothesis is true? Recall from Section 8.1 that the p-value, one of the main ingredients of a hypothesis test, measures our surprise, because it is the probability of getting a value as extreme as or more extreme than what we observed—in other words, as extreme as or more extreme than 4.05.

### EXAMPLE 8 (continued)  Is the Vote Surprising?

In our example involving political scientists' voting behavior, our observed $z$-value was 4.05, and the alternative hypothesis was

$$H_a: p \neq 0.466$$

We have verified the conditions required for the sampling distribution to be well approximated by the N(0, 1) distribution.

**QUESTION**  Find the p-value.

**SOLUTION**  The p-value is the probability of getting a test statistic as extreme as or more extreme than 4.05. Because our alternative hypothesis is two-tailed, this translates to "the probability of getting a test statistic even farther from 0 than 4.05 is." Visually, it looks like Figure 8.7, where the shaded area corresponds to the p-value.

The p-value can be found using the methods discussed in Chapter 6 to find probabilities for the standard Normal distribution, or we can take advantage of the fact that many software packages provide the p-value automatically when asked to perform hypothesis tests. Figure 8.8 shows the p-value as calculated by the TI-83/84 one-proportion $z$-test.

**CONCLUSION**  If the null hypothesis is true, then the probability of getting a $z$-statistic as far as or farther from 0 than 4.05 is 0.0001. This is a rather small probability, which suggests that if the null hypothesis is true, the outcome is surprising.

**TRY THIS!**  Exercise 8.31

**▲ FIGURE 8.7** The p-value as a shaded area. This value is from a two-tailed test for which $z$ comes out to be 4.05. The area has been enlarged a bit so that it can be seen readily.

**▲ FIGURE 8.8** TI-83/84 output for the voting test for political scientists. The p-value is given in exponential notation, as 1 times 10 to the negative fourth power ($10^{-4}$), which is 0.0001. Note that the shaded area is so small that it is not noticeable.

z=4.0472    p=1E⁻⁴

### EXAMPLE 9  Dreaming in Color

In situation 2, the researchers are wondering whether a greater proportion of people now dream in color than did so before color television and movies became as prominent as they are today. The hypotheses they are checking are

$$H_0: p = 0.29$$
$$H_a: p > 0.29$$

The researchers took a random sample of 113 people. Of these 113 people, 92 reported dreaming in color.

**QUESTION**  Find the value of the sample proportion, $\hat{p}$. Find the observed value of the test statistic, and then find the p-value associated with this observed value. In the conclusion, interpret the p-value in context. Assume that the conditions that must be met in order for us to use the N(0, 1) distribution as the sampling distribution are satisfied.

**SOLUTION**

The sample proportion is $\hat{p} = \dfrac{92}{113} = 0.814159$.

We can then compute the observed value of the test statistic:

$$z_{observed} = \frac{0.814159 - 0.29}{\sqrt{\dfrac{0.29(1 - 0.29)}{113}}} = \frac{0.524159}{0.042686} = 12.28$$

The p-value is the area under the N(0, 1) curve to the right (above) 12.28. We know, without looking, that this area must be quite small, because it is extremely rare for a value from a N(0, 1) distribution to be greater than 3, so 12.28 is a very large number. Therefore, the p-value is quite small, and we report that it is less than 0.001.

**CONCLUSION**  If the proportion of people who now dream in color is the same as it was before color television, then the probability of getting a test statistic as large or larger than 12.28 is extremely small.

**TRY THIS!**  Exercise 8.33

> **Details**
>
> **Small p-Values**
> When small p-values, such as 0.0001, occur, many software packages round off and report the p-value as "p < 0.001" (or use some other small value). We will follow that practice in this book.

You should always draw a sketch before you compute the p-value, even if you use technology (as we strongly recommend) to find the probability.

## EXAMPLE 10 Improving Attendance in Calculus

A very large calculus class at a university has been troubled by a high attrition rate: Over the last 5 years, 15% of the students who started the class did not finish. An instructor thinks that one reason for this is that some students need help with algebra. She randomly selects 100 students and assigns them to meet weekly with teaching assistants to study algebra. At the end of the semester, she will see what proportion of these 100 students have dropped the course. If $p$ represents the proportion of all students who drop the course and had algebra tutoring, her hypotheses can be written as

$$H_0: p = 0.15$$

$$H_a: p < 0.15$$

After the study is finished, she finds that her test statistic is $z = -1.68$. Assume that all of the required conditions hold for the one-proportion $z$-test.

**QUESTION**  Three possible p-values are shown. One of them corresponds to the p-value for the alternative hypothesis $p > 0.15$, one to the alternative hypothesis $p < 0.15$, and one to the two-tailed alternative hypothesis $p \neq 0.15$. Without using any technology or referring to Normal tables, indicate which of the following numbers is the correct p-value. Explain.

   a. p-value = 0.0930
   b. p-value = 0.0465
   c. p-value = 0.9535

**SOLUTION**  Draw a sketch, as in Figure 8.9. This sketch shows that the p-value must be less than 50%, which eliminates answer c. Answer a is twice as big as answer b, so answer a must correspond to the two-tailed alternative. The correct answer is therefore answer b: p-value = 0.0465.

**TRY THIS!**  Exercise 8.35

> **Details**
>
> **Sign of z**
> $$z = \frac{\hat{p} - p}{SE}$$
> The sign of $z$ is determined by the difference $\hat{p} - p$. A negative sign for $z$ means that $\hat{p}$ was smaller than $p$.

p-value = 0.0465

$z$   −1.68    0

▲ **FIGURE 8.9** A representation of the p-value when the alternative hypothesis is left-tailed and the $z$-statistic is −1.68.

## Step 4: Interpret

The p-value measures our surprise (or lack of surprise) at the outcome of our test, but what should we do about this number? When is the outcome so unusual that we should reject the null hypothesis?

We apply a simple rule: Reject the null hypothesis if the p-value is smaller than (or equal to) the value chosen for the significance level, $\alpha$. If the p-value is larger than the significance level, do not reject the null hypothesis. For most applications, this means you reject the null hypothesis if the p-value is less than or equal to 0.05.

Following this rule ensures that our significance level is achieved. In other words, by following this rule and rejecting the null hypothesis when the p-value is less than or equal to 0.05, we know that there is only a 5% chance that we are mistakenly rejecting the null hypothesis (rejecting $H_0$ even though $H_0$ is true).

For the study of political scientists' voting pattern, our p-value was 0.0001, which is quite a bit smaller than 0.05. We therefore reject the null hypothesis and conclude that political scientists do not vote at the same rate as the general population.

 **KEY POINT** To achieve a significance level of $\alpha$, reject the null hypothesis if the p-value is less than (or equal to) $\alpha$. If the p-value is greater than $\alpha$, do not reject the null hypothesis.

You should always state your decision: reject the null hypothesis or do not reject the null hypothesis, and describe this decision in the context of the problem.

Now that we have illustrated all four steps, it's time to tackle an entire hypothesis test.

## EXAMPLE **11** Creative Kids and Music

Some arts educators reasoned that exposure to music instruction would help students think more creatively in a variety of educational situations than students who did not receive music instruction. To test this, elementary school children were randomly assigned, with their parents' consent, to participate either in a standard music instruction course (the treatment group) or in a recreation course (the control group). All students were given the Torrance Test of Creative Thinking, which measures several components of creativity, both before the music course began (in October) and after the course ended (in May). This test is fairly complex, and a student's score could change slightly just by chance. Of the 30 students in the treatment group, 19 increased their creativity scores by the end of the course. (The others' scores either were unchanged or decreased.) Is this evidence that the music program is effective? If the program had no effect, we would expect, due to chance, that about half of the children would exhibit an increase in creativity and that half would exhibit a decrease.

**QUESTION** Test the hypothesis that the probability that a child's creativity score will increase after participating in this music program is better than 50%. Assume the children are a random sample from a larger population of elementary school children. Use a 5% significance level.

**SOLUTION**

*Step 1: Hypothesize*
Let $p$ represent the probability that a child's creativity score will improve.

$$H_0: p = 0.50$$

$$H_a: p > 0.50$$

*Step 2: Prepare*

Use the one-proportion *z*-test. The children are assumed to be a random sample, and the measurements of their creativity scores can be assumed to be independent of each other. Thirty children is enough of a sample because $np_0 = 30(0.5) = 15$; and $n(1 - p_0) = 30(0.5) = 15$, and both are bigger than 10.

*Step 3: Compute to compare*

$$\hat{p} = \frac{19}{30} = 0.633333$$

We now compare the observed proportion to the hypothetical value of 0.50.

$$z_{observed} = \frac{0.633333 - 0.50}{\sqrt{\dfrac{(0.50)(1 - 0.50)}{30}}} = \frac{0.133333}{0.091287} = 1.46$$

The p-value is the area under a N(0, 1) curve and to the right of 1.46. This area, which is shown in Figure 8.10 and also in Figure 8.11, gives us

$$\text{p-value} = 0.072$$

*Step 4: Interpret*

Because the p-value is greater than 0.05, we cannot reject $H_0$. In other words, we do not have enough evidence to conclude that creativity scores increased. This is not the same as saying that the music program was ineffective; we are simply concluding that if it *is* effective, we do not have enough evidence to detect that effectiveness.

Most statistical software packages show you both *z* and the p-value at once. Some packages (see Figure 8.11) even show the Normal curve with the p-value shaded.

> TRY THIS!  Exercise 8.41

*p*-value = 0.072

▲ **FIGURE 8.10** The shaded area represents the p-value, the probability of getting a test statistic as large as or larger than 1.46 when the true proportion is 0.50.

▲ **FIGURE 8.11** TI-83/84 output for 19 out of 30, one-tailed, for which $H_a$ is $p > 0.50$. (Choose Draw to get the curve.)

**! Caution**

**Do Not Accept**
For reasons we'll explain in Section 8.5, it is incorrect to imply that we accept the null hypothesis. We do *not* say, "The music program was ineffective." Instead, we say, "We do not have enough evidence to show that the program is effective."

**SNAPSHOT  ONE-PROPORTION *z*-TEST**

**WHAT IS IT?** ▶ A procedure for choosing between two hypotheses about the true value of a single population proportion. The test statistic is

$$z = \frac{\hat{p} - p_0}{SE} \quad \text{where } SE = \sqrt{\frac{p_0(1 - p_0)}{n}}$$

**WHAT DOES IT DO?** ▶ Because estimates of population parameters are uncertain, a hypothesis test gives us a way of making a decision while knowing the probability that we will incorrectly reject the null hypothesis.

**HOW DOES IT DO IT?** ▶ The test statistic *z* compares the sample proportion to the hypothesized population proportion. Large values of the test statistic tend to discredit the null hypothesis.

**HOW IS IT USED?** ▶ When proposing hypotheses about a single population proportion. The data must be from an independent, random sample and the sample size must be sufficiently large. The N(0, 1) distribution is used to find the p-value for the observed test statistic.

# Comparing Proportions from Two Populations

You have now seen how to carry out a hypothesis test for a single population proportion. With very few changes, this procedure can be altered to accommodate a more interesting situation: comparing proportions from two populations.

Consider as an example embryonic stem cell research, medical research that shows great promise in the treatment of several major diseases. It is controversial, however, partly because it goes against many people's religious convictions. The Pew Forum on Religion & Public Life has, over time, conducted surveys to assess Americans' support for stem cell research. These surveys raise the question of whether Americans' support has changed since 2002, when 43% of Americans in the Pew sample expressed support for stem cell research. In a more recent survey conducted in August 2009, 58% of the sample expressed support (Pew Forum 2009). Can we conclude that support has changed in the population of all Americans? Or is this difference due to chance variation from the sampling procedure?

This problem involves two populations. One population consists of all Americans in 2002, and the second population consists of all Americans in 2009. Each population has a true proportion who support stem cell research, but in each case we cannot know this true value. Instead, we have a random sample taken from both populations, and we must estimate the two proportions from these two random samples.

Here are the changes we need to make to our "ingredients" in order to compare proportions from two populations.

## Changes to Ingredients: The Hypotheses

Because we now have two population proportions to consider, we need some new notation. We'll let $p_1$ represent the proportion of Americans who supported stem cell research in 2002, and we'll let $p_2$ represent the proportion who supported it in 2009.

We are not interested in the actual numerical values of $p_1$ and $p_2$, as we were when dealing with just one population proportion. We are interested only in their relation to each other. The conservative, status-quo, not-worth-a-headline hypothesis is that these proportions are the same. In other words, there has been no change in support. We write this as

$$H_0: p_1 = p_2$$

In words, the null hypothesis says that the proportion of Americans who support stem cell research was the same in 2009 as it was in 2002.

The alternative hypothesis is that the proportion of Americans who support stem cell research has changed. If so, the two proportions are no longer equal.

$$H_a: p_1 \neq p_2$$

One-tailed hypotheses are also possible. Our research question might instead have been "Has support for stem cell research increased?" If that had been our question, then we would have used

$$H_a: p_1 < p_2$$

And if we had wished to answer the question "Has support for stem cell research decreased?" we would have used this alternative:

$$H_a: p_1 > p_2$$

These options lead to three pairs of hypotheses, as shown in Table 8.2. You choose the pair that corresponds to the research question your study hopes to answer. Note that the null hypothesis is always $p_1 = p_2$ because the neutral position is always that the two proportions are the same.

| Hypothesis | Symbols | The Alternative in Words |
|---|---|---|
| Two-tailed | $H_0: p_1 = p_2$ <br> $H_a: p_1 \neq p_2$ | The proportions are different in the two populations. |
| One-tailed (left) | $H_0: p_1 = p_2$ <br> $H_a: p_1 < p_2$ | The proportion in population 1 is less than the proportion in population 2. |
| One-tailed (right) | $H_0: p_1 = p_2$ <br> $H_a: p_1 > p_2$ | The proportion in population 1 is greater than the proportion in population 2. |

◄ **TABLE 8.2** Possible hypotheses for a two-proportion hypothesis test.

## Changes to Ingredients: The Test Statistic

We are interested in how $p_1$ and $p_2$ differ, so our test statistic is based on the difference between our sample proportions from the two populations. The test statistic we will use has the same structure as the one-sample $z$-statistic:

$$z = \frac{\text{estimator} - \text{null value}}{SE}$$

However, the estimator for the **two-proportion $z$-test** is $\hat{p}_1 - \hat{p}_2$ because we are estimating the difference $p_1 - p_2$. Here $\hat{p}_1$ and $\hat{p}_2$ are just the sample proportions for the different samples. In our case, $\hat{p}_1$ is the sample proportion for the people surveyed in 2002 (reported as 0.43), and $\hat{p}_2$ is the sample proportion for the people surveyed in 2009 (reported as 0.58).

The null value is 0, because the null hypothesis claims these proportions are the same, so $p_1 - p_2 = 0$.

The standard error, SE, is more complicated than in the one-sample case, because the null hypothesis no longer tells us the value of the population proportion. All it tells us is that both populations have the same value. For this reason, when we estimate this single value, we pool the two samples together. Formula 8.2 shows you how to do this.

**Formula 8.2: The two-proportion $z$-test statistic**

$$z = \frac{\hat{p}_1 - \hat{p}_2 - 0}{SE}$$

where

$$SE = \sqrt{\hat{p}(1 - \hat{p})\left(\frac{1}{n_1} + \frac{1}{n_2}\right)}$$

$n_1$ = sample size in sample 1

$n_2$ = sample size in sample 2

$$\hat{p} = \frac{\text{number of successes in sample 1} + \text{number of successes in sample 2}}{n_1 + n_2}$$

$$\hat{p}_1 = \text{proportion of successes in sample 1} = \frac{\text{number of successes in sample 1}}{n_1}$$

$$\hat{p}_2 = \text{proportion of successes in sample 2} = \frac{\text{number of successes in sample 2}}{n_2}$$

Formula 8.2 is perhaps the most elaborate formula we have shown you so far. As usual, it is much more important to be able to use technology to perform this test than to apply the formula. Still, studying the formula does help us understand why the test statistic is useful.

## EXAMPLE 12 Pew Survey on Stem Cell Research

The researchers from the Pew study interviewed two random samples. Both samples, the one taken in 2002 and the one taken in 2009, had 1500 people. In 2002, 645 people expressed support for stem cell research. In 2009, 870 expressed support. These data are summarized in Table 8.3.

▶ **TABLE 8.3** Data for the Pew Study

|  | 2002 | 2009 | Total |
|---|---|---|---|
| Support Stem Cell Research | 645 | 870 | 1515 |
| Do Not Support Stem Cell Research | 855 | 630 | 1485 |
| Total | 1500 | 1500 | 3000 |

**Looking Back**

**Two-way Tables**
Two-way tables, such as the one used to summarize the data in Example 12, were first presented in Chapter 1.

**QUESTION** Find the observed value of the test statistic to test the hypotheses

$$H_0: p_1 = p_2$$
$$H_a: p_1 \neq p_2$$

where $p_1$ represents the proportion of Americans who supported stem cell research in 2002, and $p_2$ represents the proportion who supported this research in 2009.

**SOLUTION** We must bring all the pieces together and assemble them into the test statistic:

$$\hat{p}_1 = \frac{645}{1500} = 0.43 \text{ (a value we knew already from the original report)}$$

$$\hat{p}_2 = \frac{870}{1500} = 0.58 \text{ (another value we knew from the original report)}$$

$$\hat{p} = \frac{645 + 870}{1500 + 1500} = 0.505 \text{ (a pooled estimate of the sample proportion)}$$

$$SE = \sqrt{0.505(1 - 0.505)\left(\frac{1}{1500} + \frac{1}{1500}\right)} = 0.018257$$

**Details**

**The Sign of z**
In a two-proportion test, whether z is positive or negative depends on which population you call "1" and which you call "2." It's important to pay attention to which proportion is subtracted from which!

Now we assemble the pieces:

$$z_{observed} = \frac{0.43 - 0.58}{0.018257} = -8.22$$

**TRY THIS!** Exercise 8.49

## Changes to Ingredients: Checking Conditions

The conditions that we need to check for a two-sample test of proportions are similar to those for a one-sample test, but with some additional things to consider.

1. *Large Samples*: Both sample sizes must be large enough. Because we don't know the value of $p_1$ or $p_2$, we must use an estimate. The null hypothesis says that these two proportions are the same, so we use $\hat{p}$, the pooled sample proportion, to check this condition. Do not use $\hat{p}_1$ or $\hat{p}_2$. This means that we need

   a. $n_1\hat{p} \geq 10$ and $n_1(1 - \hat{p}) \geq 10$
   
   b. $n_2\hat{p} \geq 10$ and $n_2(1 - \hat{p}) \geq 10$

2. *Random Samples*: The samples are drawn randomly from the appropriate population. In practice, this condition is often impossible to check unless we were present when the data were collected. In such cases, we have to assume that it is true.

3. *Independent Samples*: The samples are independent of each other. This condition is violated if, for example, the same individuals are in both samples that we are comparing.

4. *Independent within Samples*: The observations within each sample must be independent of one another.

5. *Null Hypothesis*: The null hypothesis is true.

If the first four conditions hold, then, assuming the null hypothesis to be true, $z$ follows a $N(0, 1)$ distribution.

## EXAMPLE 13 Caffeine for Babies

Apnea of prematurity occurs when premature babies have shallow breathing or stop breathing for more than 20 seconds. One therapy for this condition is to give caffeine to the premature infants. Medical researchers conducted an international study in which one sample of premature infants was randomly assigned to receive caffeine therapy, and another sample received a placebo therapy. Researchers compared the rate of severely negative outcomes (death and severe disabilities) in the two groups to determine whether the caffeine therapy would lower the rate of such bad events.

The caffeine therapy group included 937 infants. Of these 937 infants, 377 suffered from death or disability. The placebo group had 932 infants, and of these, 431 suffered from death or disability (Schmidt et al. 2007). These data are summarized in Figure 8.12.

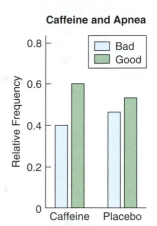

▲ FIGURE 8.12 Bar graph for caffeine treatment for apnea of prematurity.

QUESTION  Perform a four-step hypothesis test to test whether the caffeine therapy was effective (that is, whether it succeeded in lowering the death or disability proportion).

SOLUTION  As always, hypotheses are about populations, not samples. In this example, we have two samples, consisting of 937 and 932 infants, but our hypothesis is about *all* premature infants who might receive caffeine or placebo therapy.

Let $p_1$ represent the proportion of death or disability in all infants who could receive caffeine therapy, and let $p_2$ represent the proportion of death or disability in all infants who could receive the placebo therapy.

*Step 1: Hypothesize*
The null hypothesis is neutral; it states that the therapy changes nothing. If so, then the proportions of both populations are the same.

$$H_0: p_1 = p_2$$

The alternative hypothesis is that the therapy will help, so the proportion of bad events in the caffeine therapy group should be lower than that in the placebo group.

$$H_a: p_1 < p_2$$

*Step 2: Prepare*
We use a significance level of $\alpha = 0.05$.

We are comparing proportions from two populations, so we need to make sure that we can use the two-proportion $z$-test. Here are the conditions we check:

1. Are both samples large enough?
   We find that

$$\hat{p} = \frac{377 + 431}{937 + 932} = \frac{808}{1869} = 0.432317$$

First sample: $n_1 \times 0.432317 = 937 \times 0.432317 = 405$, which is greater than 10, and $937 \times (1 - 0.4323) = 532$, which is also greater than 10.

Second sample: $n_2 \times 0.432317 = 932 \times 0.4323 = 403$, which is greater than 10, and $932 \times (1 - 0.432317) = 529$, which is also greater than 10.

2. Are the samples are drawn randomly from the appropriate population?
   We are not told whether the samples are random, but for the purpose of this exercise we will assume that they are.

3. Are the samples independent of each other?
   The samples are independent of each other because the caffeine treatment is given to a different group of babies from those who received the placebo.

With these three conditions checked, we can proceed to step 3.

*Step 3: Compute to compare*
We must find the individual pieces of Formula 8.2. We defined $p_1$ to be the proportion of babies who could receive caffeine therapy in the population and who suffer from bad events. So let $\hat{p}_1$ be the proportion of babies in the caffeine therapy group who suffered these bad events. The sample size in this group was $n_1 = 937$, and 377 suffered bad events. Then

$$\hat{p}_1 = \frac{377}{937} = 0.4023$$

The sample proportion for the placebo group is

$$\hat{p}_2 = \frac{431}{932} = 0.4624$$

To find the standard error, we need to use $\hat{p}$, the proportion of bad events that happen in the sample if we ignore the fact that the babies belong to two different groups. Above we found that $\hat{p} = 0.4323$.

The standard error is then

$$SE = \sqrt{0.432317(1 - 0.432317)\left(\frac{1}{937} + \frac{1}{932}\right)} = 0.022918$$

Putting it all together yields

$$z_{observed} = \frac{0.402348 - 0.462446}{0.022918} = \frac{-0.060098}{0.022918} = -2.62$$

Now that we know the observed value, we must measure our surprise. The null hypothesis assumes the two population proportions are the same and that our sample proportions differed only by chance. The p-value will measure the probability of getting an outcome as extreme as or more extreme than $-2.62$, assuming that the population proportions are the same.

The p-value is calculated exactly the same way as with a one-proportion $z$-test. Our alternative hypothesis is a left-tailed hypothesis, so we need to find the probability of getting a value less than the observed value (Figure 8.13).

Figure 8.14 shows the $z$-value and the p-value from technology.

$$\text{p-value} = 0.004$$

*Step 4: Interpret*
The p-value is less than our significance level of 0.05, so we reject the null hypothesis. We conclude that the caffeine therapy does help: A lower proportion of babies will die or suffer disability with this therapy.

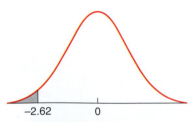

**▲ FIGURE 8.13** The shaded area represents the area to the left of a $z$-value of $-2.62$.

**▲ FIGURE 8.14** TI-83/84 output for the p-value of the test of caffeine on babies.

Note that because this is a controlled, randomized experiment (the researchers randomly assigned babies to treatment groups), we can make a cause-and-effect conclusion about the effectiveness of the treatment.

**TRY THIS!** Exercise 8.51

## SNAPSHOT    TWO-PROPORTION *z*-TEST

| | |
|---|---|
| **WHAT IS IT?** ▶ | A hypothesis test. |
| **WHAT DOES IT DO?** ▶ | Provides a procedure for comparing two population proportions. The null hypothesis is always that the proportions are the same, and this procedure gives us a way to reject or fail to reject that hypothesis. |
| **HOW DOES IT DO IT?** ▶ | The test statistic *z* compares the differences between the sample proportions and the value 0 (which is what the null hypothesis says this difference should be): |

$$z = \frac{\hat{p}_1 - \hat{p}_2 - 0}{SE} \text{ where } SE = \sqrt{\hat{p}(1 - \hat{p})\left(\frac{1}{n_1} + \frac{1}{n_2}\right)}$$

Values of *z* far from 0 tend to discredit the null hypothesis.

| | |
|---|---|
| **HOW IS IT USED?** ▶ | When comparing two proportions, each from a different population. The data must come from two independent, random samples, and each sample must be sufficiently large. Then $N(0, 1)$ can be used to compute the p-value for the observed test statistic. |

## SECTION 8.5

# Understanding Hypothesis Testing

Now that you've seen how to do a hypothesis test, you need to know about a few subtleties before you can become an expert.

## If Conditions Fail

If the conditions fail to be met for the hypothesis test, the *z*-statistic will not follow a Normal distribution when the null hypothesis is true. This means that we cannot find the p-value using the Normal curve (using Table 2 in Appendix A or technology). However, other approaches often exist.

## Sample Sizes Too Small

The Normal distribution is only an approximation to the true distribution of the *z*-statistic. If the sample sizes are large enough, then the approximation is very good. If the sample sizes are too small, then the approximation may not be good, but other tests can be used. Fisher's Exact Test is one such test that, under certain conditions, can find the exact p-value (not the approximate p-value found with the Normal curve) for two proportions. We will discuss Fisher's Exact Test in Chapter 10.

## Samples Not Independent

Suppose we survey a sample of consumers and determine the proportion that will buy a new car in the next year. This same sample then watches a car commercial, and by asking again whether they plan to buy a new car in the next year, we determine a second proportion. We do this because we want to know whether the commercial changed minds. In other words, has the proportion that will buy a new car increased after the sample watched the commercial? Two populations are being studied: those who have not seen the commercial and those who have. However, the two samples we collect consist of the same people, because we interview each person both before and after the commercial. Thus the samples are *not* independent. If you find yourself in this situation, all is not lost, but you should consult a statistician to consider your options.

## Samples Not Randomly Selected

If samples are not selected randomly, then it is not possible to make inferences about the populations the samples came from. That having been said, random samples are relatively rare in medical studies. For example, medical researchers cannot take a random sample of people with the medical condition they wish to study; they must rely on recruiting patients who come into hospitals. Psychologists at universities often study students, particularly students who are willing to submit to a study for a small amount of money or the chance to win a prize in a raffle. They cannot take a random sample of all people in the population they wish to study and fly them to the university to participate in their experiment.

These convenience samples are problematic for the statistical techniques in this book—and indeed for any statistical technique. Sometimes we get around this by assuming that the samples are random, or at the very least representative of the population. However, we have no guarantee that the conclusions we make with this assumption are valid or useful.

In many situations, though, researchers are not that interested in generalizing to the larger population. For example, in the study of apnea in premature infants, the researchers were interested in knowing whether caffeine therapy worked for this group of infants or if, instead, the differences between the caffeine therapy infants and the placebo infants could be explained by chance. They concluded that the differences were too large to be due to chance, so they felt confident concluding that at least for this group, the treatment was effective. Research with other groups is still needed to see whether the results obtained are **replicable**, but the study is at least an encouraging start because researchers know the therapy works with some infants. In Chapter 13 we study some tests that more thoroughly address this situation, in which samples are not necessarily drawn at random.

## Controlling Mistakes: The Role of Sample Size

In Section 8.1 we mentioned one type of mistake that we might make when carrying out a hypothesis test. This mistake occurs when we reject the null hypothesis even though it is true. The probability of making this mistake was called the significance level.

Another mistake we might make is to *fail* to reject the null hypothesis, even though it is not correct. We might make either of these two types of mistakes when doing a formal hypothesis test.

To better understand these errors, imagine sitting on a jury in a criminal trial. You are making the first type of mistake if you convict the defendant of being guilty when, in fact, he is innocent. You are making the second type if you declare the defendant not guilty when, in fact, he did commit the crime.

---

**📌 Details**

**Replicable Results**
The results of a study are said to have been replicated when researchers using new subjects come to the same conclusion.

---

**❗ Caution**

**Cause and Effect**
Remember that we can conclude that there is a cause-and-effect relationship between a treatment variable and a response variable only when we have a controlled experiment that uses random assignment, includes a placebo (or comparison) treatment, and is double-blind.

---

**📌 Details**

**Mistakes (Errors)**
We are not referring here to calculation mistakes. The mistakes we refer to occur because sometimes, just by chance, we get sample data that are very unusual and come to a mistaken conclusion about the population from which the data came.

But why be so negative? Rather than focusing on what can go wrong if we mistakenly fail to reject the null hypothesis, we instead focus on its complement: correctly rejecting the null hypothesis, or, in the context of a criminal trial, correctly convicting a guilty person. The probability of rejecting the null hypothesis when the null hypothesis is wrong—the probability of doing the right thing—is called the **power**.

We can ensure that the significance level is set at 5% just by rejecting $H_0$ whenever the p-value is less than or equal to 0.05, but the power is a little more difficult to control. This is because the null hypothesis can be wrong in many different ways, so it is harder to measure the probability of correctly rejecting the null.

The power depends on three factors:

1. Just how wrong the null hypothesis truly is. If the null hypothesis value, $p_0$, is very close to the true value of $p$, then the probability of correctly identifying that they are different is small, because the two values are very similar. So the power will be low and we are unlikely to make the right decision. On the other hand, if the true value of $p$ is very different from $p_0$, then it is easy to tell them apart, and the probability of making the correct decision is high. The basic idea here is that it is easier to confuse an alligator with a crocodile than to confuse it with a chicken.

2. The sample size. The larger the sample size, the bigger the power. More is always better (although more expensive) when it comes to sample size. With a larger sample, it's easier to tell the alligators from the crocodiles.

3. The significance level. The smaller the significance level, the smaller the power. (Remember: small power, bad; small significance level, good.) If you want a small probability of mistakenly rejecting the null hypothesis when it is in fact true, then you must live with a smaller probability that you will correctly reject it when it is false. We discuss this tradeoff after the next paragraph.

Calculating the power is tricky and somewhat complex, in part because it requires that we know the true value of the population proportion. We leave this calculation to a future statistics course. For now, be aware that if you do a hypothesis test and do not reject the null, then there is always the chance that you have made a mistake because your power is too low. You simply don't have enough evidence to tell the difference between the null hypothesis and the truth.

## The Tradeoff between Significance Level and Power

We are free to choose any value we wish for the significance level. Typically, we set this probability at 0.05, but sometimes we go as low as 0.01. But why don't we make it arbitrarily small? Say 0.0000001? That way, we'd almost never make this mistake.

We can't make the significance level as small as we would like because we have a price to pay. The price is that if we make the significance level smaller, then the power gets smaller too!

To see this, think about our criminal justice example. The null hypothesis is that the defendant is innocent. The alternative is that he is guilty. A mistake occurs when we convict an innocent man (mistakenly reject the null hypothesis). Another type of mistake occurs when we free a guilty man (fail to reject the null hypothesis even though it is false). We can make the significance level, the probability of convicting an innocent man, 0 by following a simple rule: Free every defendant. If everyone goes free, then it is impossible to convict an innocent person because you are convicting no one. But now the power—the probability of correctly convicting a guilty person—is 0%.

 **KEY POINT**    We cannot make the significance level arbitrarily small because doing so lowers the power—the probability that we will correctly reject the null hypothesis.

Of course, we could increase the power by changing course and convicting every defendant. Now the power is 100% because every guilty person will be convicted. But the significance level has gone to 100% as well, because every innocent person also gets convicted.

There is only one way out if you want to lower the significance level and keep the power high: Increase the sample size. This relationship between power and significance level is summarized in Table 8.4.

▶ **TABLE 8.4** Relationship between Power and Significance Level

|  | Reject $H_0$ | Fail to Reject $H_0$ |
|---|---|---|
| $H_0$ True | Bad (The probability of doing this is called the significance level.) | Good |
| $H_0$ False | Good (The probability of doing this is called the power.) | Bad |

## So What? Statistical Significance vs. Practical Significance

Researchers call a result "statistically significant" when they reject the null hypothesis. This means that the difference between their data-estimated value for a parameter and the null hypothesis value is so large that it cannot be explained by chance. However, just because a difference is statistically significant does not mean it is useful or meaningful. A *practically significant* result is both statistically significant and meaningful.

For example, suppose that the proportion of people who get a certain type of cancer is 1 in ten million. However, a statistical analysis finds that those who talk on their cell phones every day have a statistically significant greater risk of getting that cancer, and that the risk is doubled. It may be true that using your cell phone is therefore more dangerous than not using it, but would you really stop talking on the phone if your risk would change from 2 in 10 million to 1 in 10 million? That's a pretty big change of habit for a pretty small change in risk. Most people would conclude that the difference in risk is statistically significant, but not practically significant.

 **KEY POINT** — Statistically significant findings do not necessarily mean that the results are useful.

## Don't Change Hypotheses!

A researcher sets up a study to see whether caffeine affects our ability to concentrate. He has a large number of subjects, and he gives them a task to complete when they have not had any caffeine. The task takes some concentration to complete, and he records how long it takes them. Later, he asks them to complete the same task, only this time the subjects have had a dose of caffeine. Again he records the time, and he's interested in the proportion who take longer to complete the task with caffeine than without.

He isn't sure just what the effect of caffeine will be. It might help people concentrate, in which case only a small proportion of people will take longer. On the other hand, it might make people jittery so that a large proportion will take longer to complete the task. If caffeine has no effect, probably half will take the same amount of time or more, and half will take the same amount of time or less.

The researcher chooses a significance level of $\alpha = 0.05$ to test this pair of hypotheses:

$$H_0: p = 0.5$$

$$H_a: p \neq 0.5$$

The parameter $p$ represents the proportion of all people who would take longer to complete the task with caffeine than without. His alternative hypothesis is two-tailed because he does not know what the effect will be—that is, whether it will increase or decrease concentration.

He collects his data and gets a $z$-statistic of $-1.81$. This leads to a p-value of 0.07—and to a moral dilemma! (Figure 8.15 illustrates this p-value.) The researcher needs a p-value less than or equal to 0.05 if he is to publish this paper, because no one wants to hear about an insignificant result. Also, he needs to publish more papers if he is going to get that promotion he really wants.

However, it occurs to this researcher that if he had a different alternative hypothesis, his p-value would be different. Specifically, if he had used

$$H_a: p < 0.50$$

then the p-value would have been the area in just the lower tail. In that case, his p-value would be 0.035 and he would reject the null hypothesis.

What the researcher has just done is the sort of thing that small children (and politicians) do in a contest: They change the rules midway through so that they can win. As they say in the western movies, "You gotta dance with the gal/guy that brung ya." You can't choose your hypotheses to fit the data. By doing so, you are increasing the true significance level of your test.

▲ **FIGURE 8.15** The shaded areas represent the p-value of 0.070 for a test statistic of $z = -1.81$ in a two-tailed hypothesis test.

## Hypothesis-Testing Logic

Statisticians and scientists are rather touchy when it comes to talk about "proving" things. They often use softer words, such as "Our data demonstrate that . . ." or "Our data are consistent with the theory that . . . ." One reason is that in mathematical and scientific circles, the word *prove* has a very precise and very definite meaning. If something is proved, then it is absolutely, positively, and without any doubt true. However, in real life, and particularly in statistics and science (which we consider to be part of real life), you can never be completely certain. For this reason we avoid saying, for example, that we have proved that more people now believe they dream in color than thought so in the past.

On a similar note, it is improper (maybe even impolite!) to say that you have "accepted" the null hypothesis when your p-value is bigger than 0.05. Instead, we say "We have failed to reject $H_0$" or "We cannot reject $H_0$." The reason for this is that several factors might make it difficult to determine whether the null hypothesis is false. It could be that our statistical power was low, and we couldn't detect that the null hypothesis was false because we didn't have enough evidence.

In Example 11, we examined a study that asked whether music instruction improved children's creativity scores. We concluded that we could not reject the null hypothesis and that there was not enough evidence to show that the scores increased. Could we make a stronger statement? Specifically, could we conclude that the music instruction was *not* effective?

No. We might have had low power: That is, the probability of detecting a difference between the students' increase in scores for an effective music instruction program was too low. With a larger sample size, we might see that, in fact, the program is effective.

**KEY POINT** Don't say you "proved" something with statistics. Say you "demonstrated" it or "showed" it. Similarly, don't say you "accept the null hypothesis"; say, rather, that you "cannot reject the null hypothesis" or that you "failed to reject the null hypothesis" or that "there is insufficient evidence to reject the null hypothesis."

# CASE STUDY REVISITED

Does watching violence on TV as a child lead to violent behavior as an adult? To find out, researchers examined two independent groups of subjects. One group had watched a lot of violent television as children; the other group had not. The researchers recorded whether the subjects enacted a form of violence with their partners or spouses, including whether they had pushed or shoved a partner or spouse, which was defined as physical abuse. About 38% of those who had watched violent television as children said yes, compared to 22% of the watchers of nonviolent TV. Can we conclude, then, that these groups are truly different? Or might the observed difference be due simply to chance?

To find out, the researchers performed a hypothesis test. We can do a similar test with what we've learned in this chapter. The data are summarized in Table 8.5 and in Figure 8.16.

▶ TABLE 8.5 Two-way Table for the Relationship between TV Violence and Physical Abuse

|  | High TV Violence | Low TV Violence | Total |
|---|---|---|---|
| Yes, Physical Abuse | 25 | 57 | 82 |
| No, Physical Abuse | 41 | 206 | 247 |
| Total | 66 | 263 | 329 |

▶ FIGURE 8.16 Bar chart for the Relationship between TV Violence and Physical Abuse

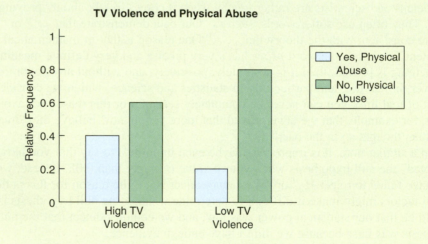

As it turns out, the differences between the two groups in the proportions of those who committed an act of violence is too large to be attributed to chance alone. Because this is an observational study, we cannot conclude that watching violent TV as a child causes adults to behave violently toward their spouses. Complicating factors could exist. For example, perhaps children raised by parents who behave violently toward each other are more likely to watch violent television *and* more likely to reenact their parents' behavior when they become adults. Still, we can rule out the possibility that these sample proportions differ only because of chance.

To come to this conclusion, we performed a two-proportion $z$-test with a significance level of $\alpha = 0.05$. The two populations consist of all the twenty-something adults who watched a lot of violent TV as children (population 1) and all the twenty-something adults who did not watch violent TV as children (population 2). The proportion of this first population that commit acts of violence directed at their partners we call $p_1$. The proportion of this second population that commit acts of violence against their partners we call $p_2$.

$H_0: p_1 = p_2$ Both groups have the same proportion of violent acts.

$H_a: p_1 > p_2$ The TV violence group has a greater proportion of violent acts.

A quick check shows that the sample sizes are large enough, and the other conditions can be assumed to hold as well. (One questionable condition is the one requiring random samples. However, we can assume the samples are random to see what the conclusion would be in this case. The fact that they are not really random means our results might not generalize to a larger group.)

The observed value of the $z$-statistic, calculated with technology, is $z = 2.72$ (see Figure 8.17). The p-value for this one-tailed hypothesis is 0.003. This small value tells us that the outcome was rather unusual if, in fact, the proportion of violent acts is the same in both populations. The fact that such an outcome is so rare leads us to reject the null hypothesis and conclude that those who watch violent TV as children do tend to commit more acts of violence toward their spouses as adults than do people who did not watch violent TV.

▲ **FIGURE 8.17** TI-83/84 output for a two-proportion $z$-test using a one-tailed alternative hypothesis, where the sample proportions are 25/66 and 57/263. Here $p_1$ is the proportion of those who watched violent TV who subsequently abused a partner, and $p_2$ is the proportion of those who did not watch violent TV who later abused a partner.

# EXPLORING STATISTICS
## CLASS ACTIVITY

### Identifying Flavors of Gum through Smell

| GOALS | MATERIALS |
|---|---|
| To use a hypothesis test to determine how well a person can distinguish between flavors based on smell alone. | • Gum (or candy) in two different flavors. We will call these flavor A and flavor B. Each student will need one piece of each flavor.<br>• A paper towel for each student. |

**ACTIVITY**

Pair up. One student will take the role of the sniffer, and the other will act as the researcher. Both students must know the two flavors that are being tested.

Researcher: Ask the sniffer to turn his or her back, then select a piece of gum at random and record the flavor of the gum on a sheet of paper. Place the gum on a sheet of paper towel.

Hold the gum about 2 inches below the sniffer's nose and ask him or her to identify which of the two flavors is in the hand. Sniffers may take as long as they like. Record whether the sniffer's response was correct next to the flavor that you wrote down for this trial.

Do 20 trials, then work together to determine the proportion of correct responses. If time permits, change roles.

**BEFORE THE ACTIVITY**

1. Let $p$ represent the probability that the sniffer makes a correct identification. If the sniffer is simply guessing and cannot tell the difference, what value would $p$ have?

2. If the sniffer is not guessing and really can tell the difference, would $p$ be bigger than, less than, or the same as the value you found in Question 1?

3. Suppose the sniffer cannot tell the difference between the two flavors. In 20 trials, about how many would you expect the sniffer to get right? What proportion is this? What is the greatest number of trials you'd expect the sniffer to get right if she or he were just guessing? What proportion is this?

4. How many trials would the sniffer have to get right before you would believe that he or she can tell the difference? All 20? 19? Explain.

5. Write a pair of hypotheses to test whether the sniffer is just guessing or can really tell the difference. Write the hypotheses in both words and symbols, using the parameter $p$ to represent the probability that the sniffer correctly identifies the scent.

**AFTER THE ACTIVITY**

1. Report the proportion of trials the sniffer got right.

2. Does this show that the sniffer could tell the difference in the scents? Explain.

3. If the sniffer is just guessing, what is the probability that he or she would have gotten as many right as, or more right than, the actual number recorded?

   a. Pretty likely: between 50% and 100%   b. Maybe: between 10% and 50%

   c. Fairly unlikely: between 5% and 10%   d. Very unlikely: between 0% and 5%

# CHAPTER REVIEW

## Key Terms

Hypothesis testing, *340*
Null hypothesis, $H_0$, *341*
Alternative hypothesis, $H_a$, *341*
Significance level, $\alpha$ (alpha), *343*

Test statistic, *343*
One-proportion *z*-test, *343*
*z*-statistic, *344*
p-value, *345*

Two-tailed hypothesis, *346*
One-tailed hypothesis, *346*
Left-tailed hypothesis, *346*
Right-tailed hypothesis, *347*

Two-proportion *z*-test, *359*
Replicable results, *364*
Power, *365*

## Learning Objectives

After reading this chapter and doing the assigned homework problems, you should

- Know how to test hypotheses concerning a population proportion and hypotheses concerning the comparison of two population proportions.

- Understand the meaning of p-value and how it is used.

- Understand the meaning of significance level and how it is used.

- Know the conditions required for calculating a p-value and significance level.

## Summary

Hypothesis tests are performed in the following four steps.

Step 1: Hypothesize.
Step 2: Prepare.
Step 3: Compute to compare.
Step 4: Interpret.

Step 1 is the most important, because it establishes the entire procedure. Hypotheses are *always* statements about parameters. The alternative hypothesis is the hypothesis that the researcher wishes to convince the public is true. The null hypothesis is the skeptical, neutral hypothesis. Each step of the hypothesis test is carried out assuming that the null hypothesis is true. For all tests in this book, the null hypothesis will always contain an equals (=) sign, whereas the alternative hypothesis can contain the symbol for "is greater than" (>), the symbol for "is less than" (<), or the symbol for "is not equal to" ($\neq$).

Step 2 requires that you decide what type of test you are doing. In this chapter, this means you are either testing the value of a proportion from a single population (one-proportion *z*-test) or comparing two proportions from different populations (two-proportion *z*-test). You must also check that the conditions necessary for using the standard Normal distribution as the sampling distribution are all met.

Step 3 is where the observed value of the statistics are compared to the null hypothesis. This step is most often handled by technology, which will compute a value of the test statistic and the p-value. These values are valid only if the conditions in step 2 are satisfied.

Step 4 requires you to compare the p-value, which measures our surprise at the outcome if the null hypothesis is true, to the significance level, which is the probability that we will mistakenly

reject the null hypothesis. If the p-value is less than (or equal to) the significance level, then you must reject the null hypothesis.

For a one-proportion *z*-test:

**Formula 8.1:** $z = \dfrac{\hat{p} - p_0}{SE}$

where

$$SE = \sqrt{\frac{p_0(1 - p_0)}{n}}$$

$p_0$ is the proposed population proportion
$\hat{p}$ (p-hat) is the sample proportion, $x/n$
$n$ is the sample size
For a two-proportion *z*-test:

**Formula 8.2:** $Z = \dfrac{\hat{p}_1 - \hat{p}_2 - 0}{SE}$

where

$$SE = \sqrt{\hat{p}(1 - \hat{p})\left(\frac{1}{n_1} + \frac{1}{n_2}\right)}$$

$$\hat{p} = \frac{\text{number of successes in both samples}}{n_1 + n_2}$$

$\hat{p}_1$ is the proportion of successes in the first sample, and $\hat{p}_2$ is the proportion in the second sample
Calculating the p-value depends on the which alternative hypothesis you are using. Figure 8.18 shows, from left to right, the p-values for a two-tailed test, a right-tailed test (one-tailed), and a left-tailed test (one-tailed).

▲ **FIGURE 8.18** Representations of possible p-values for three different alternative hypotheses. The area of the shaded regions represents the p-value.

## Sources

Husemann, L. R., J. Moise-Titus, C. Podolski, and L. D. Eron. 2003. Longitudinal relations between children's exposure to TV violence and their aggressive and violent behavior in young adulthood: 1977–1992. *Developmental Psychology* 39(2), 201–221.

Pew Forum. 2009. Declining majority of Americans favor embryonic stem cell research. The Pew Forum on Religion & Public Life Issues. http://people-press.org/report/528/] http://pewforum.org/docs/?DocID=317 (accessed March 13, 2009).

Schmidt, B. et al. for the Caffeine for Apnea of Prematurity Trial Group. 2007. Long-term effects of caffeine therapy for apnea of prematurity. *New England Journal of Medicine* 357, 1893–1902.

Schwitzgebel, E., and J. Rust. 2010. Do ethicists and political philosophers vote more often than other professors? *Review of Philosophy and Psychology,* 1, 189-199.

Schwitzgebel, E. 2003. Do people still report dreaming in black and white? An attempt to replicate a questionnaire from 1942. *Perceptual & Motor Skills* 96, 25–29).

www.workforce.com. 2008. California court to hear Google age discrimination case (accessed August 26, 2009).

# SECTION EXERCISES

## SECTION 8.1

**8.1** Choose one of the answers given. The null hypothesis is always a statement about a _____ (sample statistic or population parameter).

**8.2** Choose one of the answers in each case. In statistical inference, measurements are made on a _____ (sample or population), and generalizations are made to a _____ (sample or population).

**8.3** With a two-tailed test, if the test statistic (such as $z$) is far from 0, will the p-value be large (closer to 1) or small (closer to 0)?

**8.4** With a two-tailed test, if the test statistic is 0 or close to 0, will the p-value be large (closer to 1) or small (closer to 0)?

**TRY 8.5 Coin Flips (Example 1)** A coin is flipped 30 times and lands on heads 18 times. You want to test the hypothesis that the coin does not come up 50% heads in the long run.

Pick the correct null hypothesis for this test.

   i. $H_0: \hat{p} = 0.60$

   ii. $H_0: p = 0.50$

   iii. $H_0: p = 0.60$

   iv. $H_0: \hat{p} = 0.50$

**8.6 Die Rolls** You roll a six-sided die 30 times and land on an ace (a one) 6 times. You want to test the hypothesis that the die does not come up with an ace one-sixth of the time. Pick the correct null hypothesis.

  i. $H_0: \hat{p} = 1/5$    ii. $H_0: p = 1/5$    iii. $H_0: \hat{p} = 1/6$    iv. $H_0: p = 1/6$

**8.7** A friend claims he can predict the suit of a card drawn from a standard deck of 52 cards. There are four suits and equal numbers of cards in each suit. The parameter, $p$, is the probability of success, and the null hypothesis is that the friend is just guessing.

 a. Which is the correct null hypothesis?

   i. $p = 1/4$   ii. $p = 1/13$   iii. $p > 1/4$   iv. $p > 1/13$

 b. Which hypothesis best fits the friend's claim? (This is the alternative hypothesis.)

   i. $p = 1/4$   ii. $p = 1/13$   iii. $p > 1/4$   iv. $p > 1/13$

**8.8** A friend claims he can predict how a six-sided die will land. The parameter, $p$, is the long-run likelihood of success, and the null hypothesis is that the friend is guessing.

 a. Pick the correct null hypothesis.

   i. $p = 1/6$   ii. $p > 1/6$   iii. $p < 1/6$   iv. $p > 1/2$

 b. Which hypothesis best fits the friend's claim? (This is the alternative hypothesis.)

   i. $p = 1/6$   ii. $p > 1/6$   iii. $p < 1/6$   iv. $p > 1/2$

**8.9 Heart Attack Prevention** A new drug is being tested to see whether it can reduce the chance of heart attack in people who have previously had a heart attack. The rate of heart attack in the population of concern is 0.20. The null hypothesis is that $p$ (the population proportion using the new drug that have a heart attack) is 0.2.

Pick the correct alternative hypothesis.

   i. $p \neq 0.2$   ii. $p > 0.2$   iii. $p < 0.2$

**8.10 Stroke Survival Rate** The proportion of people who live after suffering a stroke is 0.85. Suppose there is a new treatment that is used to increase the survival rate. Use the parameter $p$ to represent the population proportion of people who survive after a stroke. For a hypothesis test of the treatment's effectiveness, researchers use a null hypothesis of $p = 0.85$. Pick the correct alternative hypothesis.

   i. $p \neq 0.85$   ii. $p > 0.85$   iii. $p < 0.85$

**TRY 8.11 ESP (Example 2)** We are testing someone who claims to have ESP by having that person predict whether a coin will come up heads or tails. The null hypothesis is that the person is guessing and does not have ESP, and the population proportion of success is 0.50. We test the claim with a hypothesis test, using a significance level of 0.05. Select an answer and fill in the blank.

The probability of concluding that the person has ESP when in fact she or he (does/does not)_____ have ESP is _____.

**\* 8.12 Multiple-Choice Test** A teacher is giving an exam with 20 multiple-choice questions, each with four possible answers. The teacher's null hypothesis is that the student is just guessing, and the population proportion of success is 0.25. Suppose we do a test with a significance level of 0.01. Write a sentence describing the significance level in the context of the hypothesis test.

**TRY 8.13 Votes for Independents (Example 3)** Judging on the basis of experience, a politician claims that 50% of voters in Pennsylvania have voted for an independent candidate in past elections. Suppose you surveyed 20 randomly selected people in Pennsylvania, and 12 of them reported having voted for an independent candidate. The null hypothesis is that the overall proportion of voters in Pennsylvania that have voted for an independent candidate is 50%. What value of the test statistic should you report?

**8.14 Votes for Independents** Refer to Exercise 8.13. Suppose 14 out of 20 voters in Pennsylvania report having voted for an independent candidate. The null hypothesis is that the population proportion is 0.50. What value of the test statistic should you report?

**TRY * 8.15 Texting While Driving (Example 4)** The mother of a teenager has heard a claim that 25% of teenagers who drive and use a cell phone reported texting while driving. She thinks that this rate is too high and wants to test the hypothesis that fewer than 25% of these drivers have texted while driving. Her alternative hypothesis is that the percentage of teenagers who have texted when driving is less than 25%.

$$H_0: p = 0.25$$

$$H_a: p < 0.25$$

She polls 40 randomly selected teenagers, and 5 of them report having texted while driving, a proportion of 0.125. The p-value is 0.034. Explain the meaning of the p-value in the context of this question.

**\* 8.16 True/False Test** A teacher giving a true/false test wants to make sure her students do better than they would if they were simply guessing, so she forms a hypothesis to test this. Her null hypothesis is that a student will get 50% of the questions on the exam correct. The alternative hypothesis is that the student is not guessing and should get more than 50% in the long run.

$$H_0: p = 0.50$$

$$H_a: p > 0.50$$

A student gets 30 out of 50 questions, or 60%, correct. The p-value is 0.079. Explain the meaning of the p-value in the context of this question.

## SECTION 8.2

**TRY 8.17 p-Values (Example 5)** For each graph, indicate whether the shaded area could represent a p-value. Explain why or why not. If yes, state whether the area could represent the p-value for a one-tailed or a two-tailed alternative hypothesis.

**(A)**

**(B)**

**8.18 p-Values** For each graph, state whether the shaded area could represent a p-value. Explain why or why not. If yes, state whether the area could represent the p-value for a one-tailed or a two-tailed alternative hypothesis.

**(A)**

**(B)**

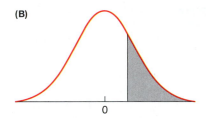

**TRY 8.19 p-Values (Example 6)** A researcher carried out a hypothesis test using a two-tailed alternative hypothesis. Which of the following z-scores is associated with the smallest p-value? Explain.

    i. $z = 0.50$   ii. $z = 1.00$   iii. $z = 2.00$   iv. $z = 3.00$

**8.20 Coin Flips** A test is conducted in which a coin is flipped 30 times to test whether the coin is unbiased. The null hypothesis is that the coin is fair. The alternative is that the coin is not fair. One of the accompanying figures represents the p-value after getting 16 heads out of 30 flips, and the other represents the p-value after getting 18 heads out of 30 flips. Which is which, and how do you know?

**(A)**

**(B)**

**8.21 ESP** Suppose a friend says he can predict whether a coin flip will result in heads or tails. You test him, and he gets 10 right out of 20. Do you think he can predict the coin flip (or has a way of cheating)? Or could this just be something that occurs by chance? Explain without doing any calculations.

**8.22 ESP Again** Suppose a friend says he can predict whether a coin flip will result in heads or tails. You test him, and he gets 20 right out of 20. Do you think he can predict the coin flip (or has a way of cheating)? Or could this just be something that is likely to occur by chance? Explain without performing any calculations.

## SECTION 8.3

TRY **8.23 Does Hand Washing Save Lives? (Example 7)** In the mid-1800s, Dr. Ignaz Semmelweiss decided to make the doctors wash their hands with a strong disinfectant between patients at a clinic with a death rate of 9.9%. Semmelweiss wanted to test the hypothesis that the death rate would go down after the new hand-washing procedure was used. What null and alternative hypotheses should he have used? Explain, using both words and symbols. Explain the meaning of any symbols you use.

**8.24 Healthcare Plan** Suppose you wanted to test the claim that more than half of U.S. voters support repealing the current U.S. health plan. Give the null and alternative hypotheses, and explain, using both words and symbols.

**8.25 Seatbelt Usage** The National Highway Traffic Safety Administration (NHTSA) claimed that the percentage of people who used their seatbelts in the year 2000 was 71%. The NHTSA also reported that a random sample of 3000 people showed that 2520 reported using their seatbelts in 2009.

Give the null and alternative hypotheses to test whether the percentage of people who used their seatbelts in 2009 *changed* from the 71% reported in the year 2000.

**8.26 Seatbelt Usage, Again** The NHTSA claims that the percentage of people who used their seatbelts in the year 2000 was 71%. The NHTSA recently reported that a random sample of 3000 people showed 2520 who reported using their seatbelts in 2009.

Give the null and alternative hypotheses to test whether the percentage of people who used their seatbelts in 2009 *increased* from the 71% reported in the year 2000.

**g 8.27 Gun Control** Historically, the percentage of U.S. residents who support stricter gun control laws has been 52%. A recent Gallup Poll of 1011 people showed 495 in favor of stricter gun control laws. Assume the poll was given to a random sample of people. Test the claim that the proportion of those favoring stricter gun control has changed. Perform a hypothesis test, using a significance level of 0.05. *See page 381 for guidance.* Choose one of the following conclusions:

   i. The percentage is not significantly different from 52%. (A significant difference is one for which the p-value is less than or equal to 0.050.)

   ii. The percentage is significantly different from 52%.

**8.28 Death Penalty** A Gallup Poll in May 2010 showed that 669 out of 1029 randomly chosen adults said that the death penalty was "morally acceptable." In 1996 the rate of support for the death penalty was estimated to be 78%.

   a. Using the data from 2010, test the hypothesis that the rate of support has changed since 1996. Use a value of 0.05 for the significance level, $\alpha$.

   b. Choose the appropriate conclusion.

      i. The rate of support is not significantly different from 78%.

      ii. The rate of support is significantly different from 78%.

TRY **8.29 Dreaming (Example 8)** A 2003 study of dreaming found that out of a random sample of 113 people, 92 reported dreaming in color. However, the rate of reported dreaming in color that was established in the 1940s was 0.29 (Schwitzgebel 2003). Check to see whether the conditions for using a one-proportion $z$-test are met assuming the researcher wanted test to see if the proportion dreaming in color had changed since the 1940s.

**8.30 Age Discrimination** About 30% of the population in Silicon Valley, a region in California, are between the ages of 40 and 65, according to the U.S. Census. However, only 2% of the 2100 employees at a laid-off man's former Silicon Valley company are between the ages of 40 and 65. Lawyers might argue that if the company hired people regardless of their age, the distribution of ages would be the same as if they had hired people at random from the surrounding population. Check whether the conditions for using the one-proportion $z$-test are met.

TRY **8.31 Coin Spinning (Example 8 continued)** Suppose you are testing the claim that a coin comes up tails more than 50% of the time when the coin is spun on a hard surface. Steps 1 and 2 of the hypothesis test are given. Suppose that you did this experiment and got 22 tails in 30 spins. Find the value of the test statistic $z$ and the corresponding p-value.

   $p$ is the proportion of tails.

   Step 1:  $H_0$: $p = 0.50$

            $H_a$: $p > 0.50$

   Step 2: Assume that the outcomes are random and the sample size is large enough because both $np$ and $n(1 - p)$ are 15.

**8.32 Coin Spinning** Repeat Exercise 8.31 assuming that you got 18 tails out of 30 spins.

TRY **8.33 Guessing on a True/False Test (Example 9)** A true/false test has 50 questions. Suppose a passing grade is 35 or more correct answers. Test the claim that a student knows more than half of the answers and is not just guessing. Assume the student gets 35 answers correct out of 50. Use a significance level of 0.05. Steps 1 and 2 of a hypothesis test procedure are given. Show steps 3 and 4, and be sure to write a clear conclusion.

   Step 1:  $H_0$: $p = 0.50$

            $H_a$: $p > 0.50$

   Step 2: Choose the one-proportion $z$-test. Sample size is large enough, because $np_0 = 50(0.5) = 25$ and $n(1 - p_0) = 50(.5) = 25$, and both are more than 10. Assume the sample is random.

**8.34 Guessing on a Multiple-Choice Test** A multiple-choice test has 50 questions with four possible options for each question. For each question, only one of the four options is correct. A passing grade is 35 or more correct answers.

   a. What is the probability that a person will guess correctly on one multiple-choice question?

   b. Test the hypothesis that a person who got 35 right out of 50 is not just guessing, using an alpha of 0.05. Steps 1 and 2 of the hypothesis testing procedure are given. Finish the question by doing steps 3 and 4.

      Step 1:  $H_0$: $p = 0.25$

                $H_a$: $p > 0.25$

Step 2: Choose the one-proportion $z$-test. $n$ times $p$ is 50 times 0.25, which is 12.5. This is more than 10, and 50 times 0.75 is also more than 10. Assume a random sample.

**TRY 8.35 Taste Test (Example 10)** A taste test was done in which a blindfolded subject was asked to identify the spread on English muffins. Twenty bites were taken, and they were randomly arranged such that exactly half of the bites contained butter and half contained margarine.

$$H_0: p = 0.50$$

$$H_a: p > 0.50$$

The subject got 13 out of 20 correct. The graphs show p-values (as shaded areas) corresponding to the alternative hypotheses $p \neq 0.50$, $p > 0.50$, and $p < 0.50$, *though not necessarily in that order.*

Indicate which graph matches the given alternative hypothesis, $p > 0.50$.

**(A)**

**(B)**

**(C)**

**8.36 Penny Spinning** Suppose we spin a penny 20 times to test whether it is biased. Let $p$ represent the probability of heads.

$$H_0: p = 0.50$$

$$H_a: p \neq 0.50$$

Suppose that 8 heads were obtained out of 20 spins. The graphs show p-values (as shaded areas) corresponding to the alternative hypotheses $p \neq 0.50$, $p > 0.50$, and $p < 0.50$. Indicate which graph matches the given alternative hypothesis, $p \neq 0.50$.

**(A)**

**(B)**

**(C)**

**8.37 Banning Sugary Foods in Schools** In a Rasmussen Poll of 1000 adults in July 2010, 520 of those polled said that schools should ban sugary snacks and soft drinks.

a. Do a majority of adults (more than 50%) support a ban on sugary snacks and soft drinks? Perform a hypothesis test using a significance level of 0.05.

b. Choose the best interpretation of the results you obtained in part a:

    i. The percentage of all adults who favor banning is significantly more than 50%.

    ii. The percentage of all adults who favor banning is not significantly more than 50%

**8.38 Year-Round School** In July of 2010, a Rasmussen Poll of 1000 adults showed that 630 opposed students attending school year-round without the traditional summer break.

a. Test the claim that more than half of adults oppose year-round school. Use a significance level of 0.05.

b. Choose the correct interpretation:

    i. The percentage opposing year-round school is significantly more than 50%.

    ii. The percentage opposing year-round school is not significantly more than 50%.

**8.39 Global Warming** Historically (from about 2001 to 2005), about 58% of Americans believed that the earth's temperature was rising ("global warming"). A March 2010 Gallup Poll wanted to determine whether this proportion had changed. The poll interviewed 1014 adult Americans, and 527 said they believed that global warming was real. (Assume these 1014 adults represented a simple random sample.)

a. What percentage in the sample believed global warming was real in 2010? Is this more or less than the historical 58%?

b. Test the hypothesis that the proportion of Americans who believe global warming is real has changed. Use a significance level of 0.05.

c. Choose the correct interpretation:

    i. In 2010, the percentage of Americans who believe global warming is real is not significantly different from 58%.

    ii. In 2010, the percentage of Americans who believe global warming is real has changed from the historical level of 58%.

**8.40 Plane Crashes** According to one source, 50% of plane crashes are due at least in part to pilot error (http://www.planecrashinfo.com). Suppose that in a random sample of 100 separate airplane accidents, 62 of them were due to pilot error (at least in part.)

a. Test the null hypothesis that the proportion of airplane accidents due to pilot error is not 0.50. Use a significance level of 0.05.

b. Choose the correct interpretation:

    i. The percentage of plane crashes due to pilot error is not significantly different from 50%.

    ii. The percentage of plane crashes due to pilot error is significantly different from 50%.

**TRY 8.41 Mercury in Freshwater Fish (Example 11)** Some experts believe that 20% of all freshwater fish in the United States have such high levels of mercury that they are dangerous to eat. Suppose a fish market has 250 fish tested, and 60 of them have dangerous levels of mercury. Test the hypothesis that this sample is *not* from a population with 20% dangerous fish. Use a significance level of 0.05.

Comment on your conclusion: Are you saying that the percentage of dangerous fish is definitely 20%? Explain.

**8.42 Newspapers Closing** A 2009 Pew Poll asked people whether losing their local newspapers would hurt civic life, and 430 out of 1001 said that it would hurt civic life a lot. Another source said that 44% of people would be hurt.

a. Find the percentage from the Pew Poll that said that closing local papers would hurt civic life a lot.

b. Test the hypothesis that fewer than 44% of people say it would hurt civic life a lot, using the Pew data and a significance level of 0.05.

Comment on your conclusion: Are you saying that the percentage is definitely 44%?

**8.43 Morse's Proportion of t's** Samuel Morse determined that the percentage of t's in the English language in the 1800's was 9%. A random sample of 600 letters from a current newspaper contained 48 t's. Test the hypothesis that the proportion of t's has changed in modern times, using the 0.05 level of significance. Treat the newspaper as a random sample of all letters used.

**8.44 Morse's Proportion of a's** Samuel Morse determined that the percentage of a's in the English language in the 1800s was 8%. A random sample of 600 letters from a current newspaper contained 60 a's. Test the hypothesis that the proportion of a's has changed in modern times, using the 0.05 level of significance. Treat the newspaper as a random sample of all letters used.

# SECTION 8.4

**g 8.45 Vioxx** In the fall of 2004, drug manufacturer Merck Pharmaceutical withdrew Vioxx, a drug that had been used for arthritis pain, from the market after a study revealed that its use was associated with an increase in the risk of heart attack. The experiment was placebo-controlled, randomized, and double-blind. Out of 1287 people taking Vioxx there were 45 heart attacks, and out of 1299 people taking the placebo, there were 25 heart attacks (*Los Angeles Times*, October 23, 2004). Perform a hypothesis test to test whether those who take Vioxx have a greater rate of heart attack than those who take a placebo. Use a level of significance of 0.05. *See page 381 for guidance.* Choose i or ii:

i. Those taking Vioxx did *not* have a significantly higher rate of heart attack than those taking the placebo.

ii. Those who took Vioxx had a significantly higher heart attack rate than those who took the placebo.

Can we conclude that Vioxx *causes* an increased risk of heart attack?

**8.46 Treating Traumatic Shock** Dopamine and norepinephrine are two drugs that are effective for treating traumatic shock, which results in dangerously low blood pressure and can cause death. A recent study compared the outcomes for patients who were given these drugs. After 28 days it was noted whether the patient was alive or dead. Of the 858 patients randomly assigned dopamine, 450 were alive. Of the 821 patients randomly assigned norepinephrine, 398 were alive. (Source: De Backer et al., Comparison of dopamine and norepinephrine in the treatment of shock, *New England Journal of Medicine*, vol. 362: 779–789, March 4, 2010)

a. Find and compare the sample percentages of patients alive for these groups.

b. Test the hypothesis that the rate of success (being alive) is different in the two groups, using a significance level of 0.05.

**8.47 Vaccine for Diarrhea** A vaccine to prevent severe rotavirus gastroenteritis (diarrhea) was given to African children within the first year of life as part of a drug study. The study reported that of the 3298 children randomly assigned the vaccine, 63 got the virus. Of the 1641 children randomly assigned the placebo, 80 got the virus. (Source: Madhi et al., Effect of human rotavirus vaccine on severe diarrhea in African infants, *New England Journal of Medicine*, vol. 362: 289–298, January 28, 2010)

a. Find the sample percentage of children who caught the virus in each group. Is the sample percent lower for the vaccine group, as investigators hoped?

b. Determine whether the vaccine is effective in reducing the chance of catching the virus, using a significance level of 0.05. Steps 1 and 2 of the hypothesis-testing procedure are given. Complete the question by doing steps 3 and 4.

Step 1: $H_0: p_v = p_p$ ($p_v$ is the proportion that got the virus among those who took the vaccine, and $p_p$ is the proportion that got the virus among those who took the placebo.)
$H_a: p_v < p_p$

Step 2: Although we don't have a random sample, we do have random assignment to groups.

$$\hat{p} = \frac{63 + 80}{3298 + 1641} = \frac{143}{4939} = 0.028953$$

$n_p \times \hat{p} = 3298 \times 0.028953 = 95.49$, which is more than 10

$n_p \times \hat{p} = 1641 \times 0.028953 = 47.51$, which is more than 10 (and the other two products are larger)

**★ 8.48 Radiation for Breast Cancer** Women with localized breast cancer are often given radiation treatment. A study tried to find out whether simplifying the treatment would lower the success rate. Each woman needing treatment was randomly assigned to receive whole-breast irradiation either at a standard dose of 50.0 Gy in 25 treatments over a period of 35 days (the control group) or at a dose of 42.5 Gy in 16 treatments over a period of 22 days (the new treatment group). The new treatment had smaller doses and also fewer treatments, so it was feared that the new treatment would be less effective than the standard treatment.

Among the 612 women randomly assigned to the standard (old) treatment, 41 had a recurrence within 10 years. Among the 622 women randomly assigned to the new treatment, 39 had a recurrence within 10 years. (Source: Whelan et al., Long-term results of hypofractionated radiation therapy for breast cancer, *New England Journal of Medicine*, vol. 362: 513–520, February 11, 2010)

a. Find the failure rate (rate of recurrence) for each treatment, and comment on the difference. Was this what was expected?

b. Determine whether the new treatment increases the rate of recurrence, using a significance level of 0.05. The first two steps of the hypothesis-testing procedure are given. You may refer to the TI-83/84 output to do steps 3 and 4 to complete the answer.

Step 1: $H_0: p_{New} = p_{Old}$ where $p_{New}$ is the probability of recurrence with the new treatment, and $p_{Old}$ is the probability of recurrence with the old treatment.

$$H_a: p_{New} > p_{Old}$$

Step 2: Although we don't have a random sample, we do have random assignment.

$$\hat{p} = \frac{41 + 39}{612 + 622} = \frac{80}{1234} = 0.064830$$

$n_{old}\hat{p} = 612 \times 0.064830 = 39.68$, which is more than 10.

$n_{new}\hat{p} = 622 \times 0.064830 = 40.32$, which is more than 10.
The other two products are larger.

```
2-PropZTest
 p1<p2
 z=.3061898787
 P=.620269896
 p̂1=.0669934641
 p̂2=.0627009646
↓p̂=.0648298217
```

**TRY** **8.49 Nicotine Gum (Example 12)** A study used nicotine gum to help people quit smoking. The study was placebo-controlled, randomized, and double-blind. Each participant was interviewed after 28 days, and success was defined as being abstinent from cigarettes for 28 days. The results showed that 174 out of 1649 people using the nicotine gum succeeded, and 66 out of 1648 using the placebo succeeded. Although the sample was not random, the assignment to groups was randomized. (Source Shiffman et al, Quitting by gradual smoking reduction using nicotine gum: A randomized controlled trial, *American Journal of Preventive Medicine*, vol. 36, issue 2, February, 2009)

  a. Find the proportion of people using nicotine gum that stopped smoking and the proportion of people using the placebo that stopped smoking, and compare them. Is this what the researchers had expected?

  b. Find the observed value of the test statistic, assuming that the conditions for a two-proportion $z$-test hold.

**8.50 Treating Prostate Cancer** A recently published article reported the results of an experiment on reducing the likelihood that men develop prostate cancer. The investigators randomly assigned 3305 men to receive the drug dutasteride (the generic name for the drug sold under the brand name Avodart) and assigned 3424 men to receive a placebo. Of those receiving the drug, 659 developed prostate cancer, and of those receiving the placebo, 858 developed prostate cancer. (Source: Andriole et al., Effect of dutasteride on the risk of prostate cancer, *New England Journal of Medicine*, vol. 362: 1192–1202, April 1, 2010)

  a. Find the percentage of men that developed prostate cancer in each group. Was the sample percentage for those taking the drug lower than the sample percentage for those who took the placebo?

  b. Find the observed value of the test statistic, assuming the conditions for a two-proportion $z$-test hold.

**TRY** **8.51 Smiling and Gender (Example 13)** In a 1997 study, people were observed for about 10 seconds in public places, such as malls and restaurants, to determine whether they smiled during the randomly chosen 10-second interval. The table shows the results for comparing males and females. (Source: M. S. Chappell, Frequency of public smiling over the life span, *Perceptual and Motor Skills*, vol. 45: 474, 1997)

|  | Male | Female |
|---|---|---|
| Smile | 3269 | 4471 |
| No Smile | 3806 | 4278 |

  a. Find and compare the sample percentages of women who were smiling and men who were smiling.

  b. Treat this as though it were a random sample, and test whether there are differences in the proportion of men and the proportion of women who smile. Use a significance level of 0.05.

  c. Explain why there is such a small p-value even though there is such a small difference in sample percentages.

**\* 8.52 Smiling and Age** Refer to the study discussed in Exercise 8.51. The accompanying table shows the results of the study for different age groups.

|  | Age Range | | | | |
|---|---|---|---|---|---|
|  | **0–10** | **11–20** | **21–40** | **41–60** | **61+** |
| Smile | 1131 | 1748 | 1608 | 937 | 522 |
| No Smile | 1187 | 2020 | 3038 | 2124 | 1509 |

  a. For each age group, find the percentage who were smiling.

  b. Treat this as a random sample of people, and merge the groups 0–10 and 10–20 into one group (0–20) and the groups 21–40, 41–60, and 61+ into another age group (21–65+). Then determine whether these two age groups have different proportions of people who smile in the general population, using a significance level of 0.05. Please comment on the results.

## SECTION 8.5

**8.53** If we reject the null hypothesis, can we claim to have *proved* that the null hypothesis is false? Why or why not?

**8.54** If we do not reject the null hypothesis, is it valid to say that we *accept* the null hypothesis? Why or why not?

**8.55** When a person stands trial for murder, the jury is instructed to assume that the defendant is innocent. Is this claim of innocence an example of a null hypothesis, or is it an example of an alternative hypothesis?

**8.56** When, in a criminal court, a defendant is found "not guilty," is the court saying with certainty that he or she is innocent? Explain.

**\* 8.57 Arthritis** A magazine advertisement claims that wearing a magnetized bracelet will reduce arthritis pain in those who suffer from arthritis. A medical researcher tests this claim with 233 arthritis suffers randomly assigned to wear either a magnetized bracelet or a placebo bracelet. The researcher records the proportion of each group who report relief from arthritis pain after 6 weeks. After analyzing the data, he fails to reject the null hypothesis. Which of the following are valid interpretations of his findings? There may be more than one correct answer.

  a. The magnetized bracelets are not effective at reducing arthritis pain.

  b. There's insufficient evidence that the magnetized bracelets are effective at reducing arthritis pain.

  c. The magnetized bracelets had exactly the same effect as the placebo at reducing arthritis pain.

  d. There were no statistically significant differences between the magnetized bracelets and the placebos in reducing arthritis pain.

**\* 8.58 No-carb Diet** A weight-loss diet claims that it causes weight loss by eliminating carbohydrates (breads and starches) from the diet. To test this claim, researchers randomly assign overweight subjects to two groups. Both groups eat the same amount of calories, but one group eats almost no carbs, and the other group includes carbs in their meals. After 2 months, the researchers test the claim that the no-carb diet is better than the usual diet. They record the proportion of each group that lost more than 5% of their initial weight. They then announce that they failed to reject the null hypothesis. Which of the following are valid interpretations of the researchers' findings?

a. There were no significant differences in effectiveness between the no-carb diet and the carb diet.

b. The no-carb diet and the carb diet were equally effective.

c. The researchers did not see enough evidence to conclude that the no-carb diet was more effective.

d. The no-carb diet was less effective than the carb diet.

**8.59** When comparing two sample proportions with a two-tailed alternative hypothesis, all other factors being equal, will you get a smaller p-value if the sample proportions are close together or if they are far apart? Explain.

**8.60** When comparing two sample proportions with a two-tailed alternative hypothesis, all other factors being equal, will you get a smaller p-value with a larger sample size or a smaller sample size? Explain.

* **8.61 Gun Control Laws** The Gallup organization frequently conducts polls in which they ask the following question:

"In general, do you feel that the laws covering the sale of firearms should be made more strict, less strict, or kept as they are now?"

In February 1999, 60% of those surveyed said "more strict," and on April 26, 1999, shortly after the Columbine High School shootings, 66% of those surveyed said "more strict."

a. Assume that both polls used samples of 560 people. Determine the number of people in the sample that said "more strict" in February 1999, before the school shootings, and the number that said "more strict" in late April 1999, after the school shootings.

b. Do a test to see whether the proportion that said "more strict" is statistically significantly different in the two different surveys, using a *significance level of 0.01*.

c. Repeat the problem, assuming that the sample sizes were both 1120.

d. Comment on the effect of different sample sizes on the p-value and on the conclusion.

**8.62 Weight Loss in Men** Many polls have asked people whether they are trying to lose weight. A Gallup Poll in November of 2008 showed that 22% of men said they were seriously trying to lose weight. In 2006, 24% of men (with the same average weight of 194 pounds as the men polled in 2008) said they were seriously trying to lose weight. Assume that both samples contained 500 men.

a. Determine how many men in the sample from 2008 and how many in the sample from 2006 said they were seriously trying to lose weight.

b. Determine whether the difference in proportions is significant at the 0.05 level.

c. Repeat the problem with the same proportions but a sample size of 5000 instead of 500.

d. Comment on the different p-values and conclusions with different sample sizes.

* **8.63 Effectiveness of Financial Incentives** A psychologist is interested in testing whether offering students a financial incentive improves their video-game-playing skills. She collects data and performs a hypothesis test to test whether the probability of getting to the highest level of a video game is greater with a financial incentive than without. Her null hypothesis is that the probability of getting to this level is the same with or without a financial incentive. The alternative is that this probability is greater. She gets a p-value from her hypothesis test of 0.003. Which of the following is the best interpretation of the p-value?

i. The p-value is the probability that financial incentives are *not* effective in this context.

ii. The p-value is the probability of getting exactly the result obtained, assuming that financial incentives are *not* effective in this context.

iii. The p-value is the probability of getting a result as extreme as or more extreme than the one obtained, assuming that financial incentives are *not* effective in this context.

iv. The p-value is the probability of getting exactly the result obtained, assuming that financial incentives *are* effective in this context.

v. The p-value is the probability of getting a result as extreme as or more extreme than the one obtained, assuming that financial incentives *are* effective in this context.

**8.64** Is it acceptable practice to look at your research results, note the direction of the difference, and then make the alternative hypothesis one-tailed in order to achieve a significant difference? Explain.

**8.65 Choosing a Test and Naming the Population** For each of the following, state whether a one-proportion z-test or a two-proportion z-test would be appropriate, and name the populations.

a. A polling agency takes a random sample to determine the proportion of people in Florida who support Proposition X.

b. A student asks men and women at her college whether they support capital punishment for some convicted murderers. She wants to determine whether the proportion of women who support capital punishment is less than the proportion of men who support it.

**8.66 Choosing a Test and Naming the Population** For each of the following, state whether a one-proportion z-test or a two-proportion z-test would be appropriate, and name the populations.

a. A student watches a random sample of men and women leaving a Milwaukee supermarket with carts to see whether they put the carts back in the designated area. She wants to compare the proportions of men and women who put the carts in the designated area.

b. The pass rate for the Oregon bar exam is 65%. A random sample of graduates from Oregon University School of Law is examined to see whether their pass rate is significantly higher than 65%.

**8.67 Choosing a Test and Giving the Hypotheses** Give the null and alternative hypotheses for each test, and state whether a one-proportion z-test or a two-proportion z-test would be appropriate.

a. You test a person to see whether he can tell tap water from bottled water. You give him 20 sips selected randomly (half from tap water and half from bottled water) and record the proportion he gets correct to test the hypothesis.

b. You test students at your college who stand on one foot with their eyes closed and determine who can stand for at least 10 seconds, comparing athletes and nonathletes.

**8.68 Choosing a Test and Naming the Population(s)** In each case, choose whether the appropriate test is a one-proportion $z$-test or a two-proportion $z$-test. Name the population(s).

a. A researcher takes a random sample of 4-year-olds to find out whether girls or boys are more likely to know the alphabet.

b. A pollster takes a random sample of all U.S. adult voters to see whether more than 50% approve of the performance of the current U.S. president.

c. A researcher wants to know whether a new heart medicine reduces the rate of heart attacks compared to an old medicine.

d. A pollster takes a poll in Wyoming about home schooling to find out whether the approval rate for men is equal to the approval rate for women.

e. A person is studied to see whether he or she can predict the results of coin flips better than chance alone.

## CHAPTER REVIEW EXERCISES

**8.69 Cola Taste Test** A student who claims he can tell cola A from cola B is blindly tested with 20 trials. At each trial, cola A or cola B is randomly chosen and presented to the student, who must correctly identify the cola. The experiment is designed so that the student will have exactly 10 sips from each cola. He gets 6 identifications right out of 20. Can he tell cola A from cola B at the 0.05 level of significance? Explain.

**8.70 Butter Taste Test** A man is tested to determine whether he can tell butter from margarine. He is blindfolded and given small bites of English muffin to identify. At each trial, an English muffin with either butter or margarine is randomly chosen. The experiment is designed so that he will have exactly 15 bites with butter and 15 with margarine. He gets 14 right out of 30. Can he tell butter from margarine at the 0.05 level? Explain.

**\* 8.71 Biased Coin?** A study is done to see whether a coin is biased. The alternative hypothesis used is two-tailed, and the obtained $z$-value is 2. Assuming that the sample size is sufficiently large and that the other conditions are also satisfied, use the Empirical Rule to approximate the p-value.

**\* 8.72 Biased Coin?** A study is done to see whether a coin is biased. The alternative hypothesis used is two-tailed, and the obtained $z$-value is 1. Assuming that the sample size is sufficiently large and that the other conditions are also satisfied, use the Empirical Rule to approximate the p-value.

**8.73 ESP** A researcher studying ESP tests 200 students. Each student is asked to predict the outcome of a large number of coin flips. For each student, a hypothesis test using a 5% significance level is performed. If the p-value for the student is less than or equal to 0.05, the researcher concludes the student has ESP. Out of 200 people who do *not* have ESP, about how many would you expect the researcher to declare *do* have ESP?

**8.74 Coin Flips** Suppose you tested 50 coins by flipping each of them many times. For each coin, you perform a significance test with a significance level of 0.05 to determine whether the coin is biased. Assuming that none of the coins is biased, about how many of the 50 coins would you expect to appear biased when this procedure is applied?

**8.75 1960 Presidential Election** In the 1960 presidential election 52% of male voters voted for Kennedy, and 48% voted for Nixon. Also, 49% of female voters voted for Kennedy and 51% voted for Nixon. Would it be appropriate to do a two-proportion $z$-test to determine whether the proportions of men

and women voting for Kennedy were significantly different (assuming we knew the number of men and women who voted)? Explain.

**8.76 Unemployment in 2009** In February 2009, the Bureau of Labor Statistics reported that the U.S. unemployment rate was 8.1% for men and 6.7% for women. Would it be appropriate to do a two-proportion $z$-test to determine whether the rates for men and women were significantly different (assuming we knew the total number of men and women)? Explain.

**8.77 Do Financial Incentives Help Smokers Quit?** A controlled, randomized study compared smoking cessation for a group in which participants received $350 if they quit and a group that did not receive money. The study participants had to submit to a biochemical test to prove they had quit. Of the 442 people randomly assigned not to receive money, 52 had quit smoking after 6 months. Of the 436 people randomly assigned to receive money, 91 had quit smoking after 6 months. (Source: Volpp et al., A randomized, controlled trial of financial incentives for smoking cessation, *New England Journal of Medicine*, vol. 361: 4, 331–333, 2009.) Determine whether the proportion of success was significantly greater for those who received money than for those who did not. Use a significance level of 0.05.

**8.78 Guns: Two Polls** In 2009 a Gallup Poll reported that 40% of people said they have a gun in their home. In the same year, a Pew Poll reported that 33% of people said they have a gun in the home. Assume that each poll used a sample size of 1000.

Do these polls disagree? Test the hypothesis that the two population proportions are different. Use a significance level of 0.05.

**8.79 Aspirin for Heart Disease in Women** The *New England Journal of Medicine* in March 2005 reported on a randomized, placebo-controlled study that was done to determine whether the use of aspirin could lower the risk of heart disease in women. It was already known that aspirin reduced heart disease in men. Researchers randomly assigned 39,876 healthy women 45 years of age or older to receive either placebo or 100 milligrams of aspirin on alternate days and then monitored them for 10 years for a first major cardiovascular event (such as a nonfatal myocardial infarction, a nonfatal stroke, or death from cardiovascular causes). The table shows the results. (Source: Ridke et al., A randomized trial of low-dose aspirin in the primary prevention of cardiovascular disease in women, *New England Journal of Medicine*, vol. 352: 1293–1304, March 31, 2005)

|  | Aspirin | Placebo |
|---|---|---|
| Heart Event | 477 | 522 |
| No Heart Event | 19,457 | 19,420 |
| Totals | 19,934 | 19,942 |

a. Find the percentage of those taking aspirin who had a "heart event," and compare it with the percentage of those taking a placebo who had a "heart event."

b. Use the two-proportion z-test to determine whether aspirin lowered the rate of women who suffered from cardiovascular events, using a significance level of 0.05.

**8.80 Support for Nuclear Energy** Some people fear the use of nuclear energy because of potential accidents. A Gallup Poll in 2009 showed that 74% of men and 47% of women surveyed supported the use of nuclear energy to produce electricity. The overall sample size was 1012.

a. Assume that exactly half the sample were men and half were women, and determine the numbers of men and women who said they supported the use of nuclear energy and the numbers of men and women who opposed it. Round your frequencies to whole numbers and report them in a table. Put the labels "Men" and "Women" across the top.

b. Now using the numbers from part a, determine whether women are less supportive of nuclear energy than men at the 0.05 level of significance.

* **8.81 Three-Strikes Law** California's controversial "three-strikes law" requires judges to sentence anyone convicted of three felony offenses to life in prison. Supporters say that this decreases crime both because it is a strong deterrent and because career criminals are removed from the streets. Opponents argue (among other things) that people serving life sentences have nothing to lose, so violence within the prison system increases. To test the opponents' claim, researchers examined data starting from the mid-1990s from the California Department of Corrections. "Three Strikes: Yes" means the person had committed three or more

felony offenses and was probably serving a life sentence. "Three Strikes: No" means the person had committed no more than two offenses. "Misconduct" includes serious offenses (such as assaulting an officer) and minor offenses (such as not standing for a count). "No Misconduct" means the offender had not committed any offenses in prison.

a. Compare the proportions of misconduct in these samples. Which proportion is higher, the proportion of misconduct for those who had three strikes or that for those who did not have three strikes? Explain.

b. Treat this as though it were a random sample and determine whether those with three strikes tend to have more offenses than those who do not. Use a 0.05 significance level.

|  | Three Strikes | |
|---|---|---|
|  | Yes | No |
| Misconduct | 163 | 974 |
| No Misconduct | 571 | 2214 |
| Totals | 734 | 3188 |

**8.82 Sleep Medicine for Shift Workers** Shift workers, who work during the night and must sleep during the day, often become sleepy when working and have trouble sleeping during the day. In a study done at Harvard Medical School, 209 shift workers were randomly divided into two groups; one group received a new sleep medicine (modafinil or Provigil), and the other group received a placebo. During the study, 54% of the workers taking the placebo and 29% of those taking the medicine reported accidents or near accidents commuting to and from work. Assume that 104 of the people were assigned the medicine and 105 were assigned the placebo. (Source: Czeisler et al., Modafinil for excessive sleepiness associated with shift-work sleep disorder, *New England Journal of Medicine*, vol. 353: 476–486, August 4, 2005)

a. State the null and alternative hypothesis. Is the alternative hypothesis one-tailed or two-tailed? Explain your choice.

b. Perform a statistical test to determine whether the difference in proportions is significant at the 0.05 level.

# gUIDED EXERCISES

**g 8.27 Gun Control** Historically, the percentage of U.S. residents who support stricter gun control laws has been 52%. A recent Gallup Poll of 1011 people showed 495 in favor of stricter gun control laws. Assume the poll was given to a random sample of people.

QUESTION  Test the claim that the proportion of those favoring stricter gun control has changed from 0.52. Perform a hypothesis test, using a significance level of 0.05, by following the steps.

**Step 1 ▶ Hypothesize**

H_0: The population proportion that supports gun control is 0.52, $p = $ _____.

H_a: $p$ _____.

**Step 2 ▶ Prepare**

Choose the one-proportion $z$-test.

Random sample: yes

Sample size: $np_0 = 1011(0.52) = $ about 526, which is more than 10, and $n(1 - p_0) = $ about ____, which is more than ____.

Population size is more than 10 times 1011.

**Step 3 ▶ Compute to compare**

$$\hat{p} = \underline{\quad}$$

$$SE = \sqrt{\frac{p_0(1 - p_0)}{n}} = \sqrt{\frac{0.52(\underline{\quad})}{1011}} = \underline{\quad}$$

$$z = \frac{\hat{p} - p_0}{SE} = \frac{0.4896 - \underline{\quad}}{SE} = \underline{\quad}$$

p-value = __

Please report your p-value with three decimal digits.

Check your answers with the accompanying figure. Don't worry if the last digits are a bit different (this can occur due to rounding).

z=⁻1.9339   p=.0531

TI-83/84 One-Proportion $z$-Test

**Step 4 ▶ Interpret**

Reject H_0 (if the p-value is 0.05 or less) or do not reject H_0 and choose one of the following conclusions:

i. The percentage is not significantly different from 52%. (A significant difference is one for which the p-value is less than or equal to 0.05.)

ii. The percentage is significantly different from 52%.

**g 8.45 Vioxx** In the fall of 2004, drug manufacturer Merck Pharmaceutical withdrew Vioxx, a drug that had been used for arthritis pain, from the market after a study revealed that its use was associated with an increase in the risk of heart attack. The experiment was placebo-controlled, randomized, and double-blind. Out of 1287 people taking Vioxx there were 45 heart attacks, and out of 1299 people taking the placebo there were 25 heart attacks (Source: *Los Angeles Times*, October 23, 2004).

QUESTION  Perform a hypothesis test to test whether those who take Vioxx have a greater rate of heart attack than those who take a placebo. Use a level of significance of 0.05. Follow the steps below to answer the question. In order to be able to use the typical four steps without changing the numbering, we have called the first step, step 0. Can we conclude that Vioxx *causes* an increased risk of heart attack?

**Step 0 ▶** Find the proportion of people in the sample taking Vioxx who had a heart attack and the proportion of people in the sample taking a placebo who had a heart attack. Compare these proportions.

**Step 1 ▶ Hypotheses**
Let $p_V$ be the population proportion of those taking Vioxx who had a heart attack, and let $p_p$ be the population proportion of those taking the placebo who had a heart attack.

$$H_0: p_V = p_p$$

$$H_a: \underline{\quad}$$

**Step 2 ▶ Prepare**
Choose the two-proportion $z$-test. Although we don't have a random sample, we have random assignment to groups. The pooled proportion of heart attacks is

$$\hat{p} = \frac{45 + 25}{1287 + 1299} = \frac{70}{2586} = 0.027069$$

$n_1 \times \hat{p} \geq 1287(0.027069) = 34.84$, which is more than 10
$n_1 \times (1 - \hat{p}) \geq$ __
$n_2 \times (1 - \hat{p}) \geq$ __
$n_2 \times \hat{p} \geq$ __

**Step 3 ▶ Compute to compare**
Refer to the accompanying figure.

$$z = \underline{\quad}$$

$$\text{p-value} = \underline{\quad}$$

```
2-PropZTest
 P1>P2
 z=2.462869689
 P=.0068915067
 ^p1=.034965035
 ^p2=.0192455735
↓^p=.0270688322
```

TI-83/84 Two-Proportion $z$-Test
1 is for Vioxx and 2 is for placebo.

**Step 4 ▶ Interpret**
Reject or do not reject the null hypothesis and choose i or ii:

i. Those taking Vioxx did *not* have a significantly higher rate of heart attack than those taking the placebo.

ii. Those who took Vioxx had a significantly higher heart attack rate than those who took the placebo.

**Causality**

Does the experiment satisfy the conditions required to conclude a cause-and-effect relationship exists between taking Vioxx and increased risk for heart attacks?

# TechTips

## General Instructions for All Technology

All technologies will use the examples that follow.

**EXAMPLE A** ▶ Do a one-proportion $z$-test to determine whether you can reject the hypothesis that a coin is a fair coin if 10 heads are obtained from 30 flips of the coin. Find $z$ and the p-value.

**EXAMPLE B** ▶ Do a two-proportion $z$-test: Find the observed value of the test statistic and the p-value that tests whether the proportion of people who support stem cell research changed from 2002 to 2007. In both years, the researchers sampled 1500 people. In 2002, 645 people expressed support. In 2007, 765 people expressed support.

### TI-83/84

#### One-Proportion z-Test

1. Press **STAT**, choose **Tests**, and choose **5: 1-PropZTest**.
2. See Figure 8A.

Enter: $p_0$, **.5**; **x**, **10**; **n**, **30**.
Leave the default $\neq p_0$.
Scroll down to **Calculate** and press **ENTER**.

▲ **FIGURE 8A** TI-83/84 Input for One-Proportion $z$-Test

You should get a screen like Figure 8B. If you choose **Draw** instead of **Calculate**, you can see the shading of the Normal curve.

```
1-PropZTest
 prop≠.5
 z=-1.825741858
 p=.0678890388
 p̂=.3333333333
 n=30
```

▲ **FIGURE 8B** TI-83/84 Output for One-Proportion $z$-Test

#### Two-Proportion z-Test

1. Press **STAT**, choose **Tests**, and choose **6: 2-PropZTest**.
2. See Figure 8C.

Enter: **x1**, **645**; **n1**, **1500**; **x2**, **765**; **n2**, **1500**.
Scroll down to **Calculate** (or **Draw**) and press **ENTER**.

```
2-PropZTest
 x1:645
 n1:1500
 x2:765
 n2:1500
 p1:≠p2  <p2  >p2
 Calculate
```

▲ **FIGURE 8C** TI-83/84 Input for Two-Proportion $z$-Test

You should get a screen like Figure 8D. The arrow down next to $\hat{p}$ means there is more information below what is given. To see that information, scroll down, using the down arrow on your keypad.

```
2-PropZTest
 p1≠p2
 z=-4.389689024
 p=1.1360415E-5
 p̂1=.43
 p̂2=.51
↓p̂=.47
```

▲ **FIGURE 8D** TI-83/84 Output for Two-Proportion $z$-Test

*Caution!* Beware of p-values that appear at first glance to be larger than 1. In Figure 8D, the p-value is 1.1 times 10 to the negative fifth power (−5), or 0.000011.

### MINITAB

#### One-Proportion z-Test

1. **Stat** > **Basic Statistics** > **1-Proportion**.
2. Refer to Figure 8E. Click **Summarized data**. Enter: **number of events**, **10**; **Number of trials**, **30**; check **Perform hypothesis test, hypothesized proportion**, **.5**

▲ **FIGURE 8E** Minitab Input for One-Proportion $z$-Test

3. Click **Options** and check the box that says **Use test and interval based on normal distribution**. (If you wanted to change the alternative hypothesis to one-tailed, you would do that also through **Options**.) Click **OK**.

You should get output that looks like Figure 8F. Note that you also get a 95% confidence interval (95% CI) for the proportion.

```
Test of p = 0.5 vs p not = 0.5

Sample   X   N   Sample p        95% CI           Z-Value   P-Value
1       10  30   0.333333  (0.164646, 0.502020)    -1.83     0.068

Using the normal approximation.
```

▲ **FIGURE 8F** Minitab Output for One-Proportion $z$-Test

**Two-Proportion z-Test**

1. **Stat > Basic Statistics > 2 Proportions**
2. See Figure 8G.

Click **Summarized data**.

| Enter | Events | Trials |
|-------|--------|--------|
| First | 645 | 1500 |
| Second | 765 | 1500 |

3. Click **OK**.

▲ **FIGURE 8G** Minitab Input for Two-Proportion z-test

Your output should look like Figure 8H. Note that it includes a 95% confidence interval for the difference between the proportions, as well as z and the p-value.

```
Test and CI for Two Proportions
Sample    X      N    Sample p
1       645   1500   0.430000
2       765   1500   0.510000


Difference = p (1) - p (2)
Estimate for difference:  -0.08
95% CI for difference:  (-0.115605, -0.0443955)
Test for difference = 0 (vs not = 0):  Z = -4.40   P-Value = 0.000
```

▲ **FIGURE 8H** Minitab Output for Two-Proportion z-Test and Interval

---

**EXCEL**

**One-Proportion z-Test**

1. Click **Add-ins**, **XLSTAT**, **Parametric Tests**, **Tests for one proportion**.
2. See Figure 8I.

Enter: **Frequency, 10**; **Sample size, 30**; **Test proportion, .5**.
(If you wanted a one-tailed hypothesis, you would click **Options**.)
Click **OK**.

▲ **FIGURE 8I** Input for One Proportion z-Test

When the output appears, you may need to change the column width to see the answers. Click **Home**, and in the **Cells** group click **Format** and **AutoFit Column Width**. The relevant part of the output is shown here.

| | |
|---|---|
| Difference | −0.167 |
| z (Observed value) | −1.826 |
| p-value (Two-tailed) | 0.068 |

**Two-Proportion z-Test**

1. Click **Add-ins**, **XLSTAT**, **Parametric Tests**, **Tests for two proportions**.
2. See Figure 8J.

Enter: **Frequency 1, 645**; **Sample size 1, 1500**; **Frequency 2, 765**; **Sample size 2, 1500**.
(If you wanted a one-tailed alternative, you would click **Options**.)
Click **OK**.

▲ **FIGURE 8J** XLSTAT Input for Two-Proportion z-Test

The relevant parts of the output are shown.

| | |
|---|---|
| Difference | −0.080 |
| z (Observed value) | −4.404 |
| p-value (Two-tailed) | <0.0001 |

## One-Proportion *z*-Test and Confidence Interval

1. **Stat** > **Proportions** > **One sample** > **with summary**
2. Enter: **number of successes, 10; number of observations, 30**.
   Click **Next** and select the **Hypothesis Test** or
   **Confidence interval** option.
   a. Leave Hypothesis test checked. Enter: **Null prop:=, 0.5**.
      Leave the **Alternative 2-tailed**, which is the default.
   b. For a Confidence interval, leave the **Level, 0.95** (the
      default). For **Method**, leave *Standard-Wald*, the default.
3. Click **Calculate**.
   Figure 8K shows the output for the hypothesis test.

▲ **FIGURE 8K** StatCrunch Output for On-Proportion *z*-Test

## Two-Proportion *z*-Test

1. **Stat** > **Proportions** > **Two samples** > **with summary**
2. Refer to Figure 8L.
   Enter:    **Sample 1: Number of successes, 645**;
                  **Number of observations, 1500**:
              **Sample 2: Number of successes, 765**;
                  **Number of observations, 1500**.
3. If you want to change the alternative hypothesis from the
   two-tailed default (or want a confidence interval), click **Next**.
   Otherwise, click **Calculate**.

▲ **FIGURE 8L** StatCrunch Input for Two-Proportion *z*-Test

Figure 8M shows the StatCrunch output.

▲ **FIGURE 8M** StatCrunch Output for Two-Proportion *z*-Test

# 9 Inferring Population Means

## THEME

The sample mean can be used to estimate the population mean. To understand how to form confidence intervals and carry out hypothesis tests, we need to understand the precision and accuracy of the sample mean as an estimate, and we need to know its probability distribution. To compare two means, we calculate the difference of the sample means, and we also need to know the accuracy, precision, and probability distribution for the difference of two samples means.

Brewing beer is a tricky business. Beer has only four main ingredients: malted barley, hops, yeast, and water. However, these four ingredients must be mixed in precise quantities at precise temperatures. Having some experience in trying to get this mix just right, in the late 1800s the Guinness brewery in Dublin, Ireland, began to hire the best and brightest science graduates to help them perfect the brewing process. One of these, hired in 1899 at the age of 23, was William Sealy Gosset (1876–1937), who had majored in chemistry and mathematics. One of Gosset's jobs was to measure the amount of yeast in a small sample of beer. On the basis of this sample, he was to estimate the mean amount of yeast in the beers produced. If this yeast amount was too high or too low, then something was wrong and the process would have to be fixed.

Naturally, uncertainty played a role. Suppose the average yeast count in his sample was too high or too low. Did this indicate that the mean yeast count in the entire factory was off? Or was his sample different just because of chance? The statistical science of the day knew how to answer this question if the number of samples was large, but Gosset worked in a context in which large samples were just too expensive and time-consuming to collect. A decision had to be based on a small sample. Gosset solved the problem, and his approach (which is called the *t*-test) is now one of the most widely used techniques in statistics.

Estimating one or more population means is still an important part of science and public policy. How do we compare the effects of different drugs on epilepsy? How do commuting times vary between cities? How do men and women differ in their abilities to sense their surroundings? To answer these questions, we need good estimates of the population means, and these estimates must be based on reliable data from small samples.

In Chapters 7 and 8 you learned two important techniques for statistical inference: the confidence interval and the hypothesis test. In those chapters, we applied statistical inference for population proportions. In this chapter, we use the same two techniques for making inferences about means of populations. We begin with inference for one population and conclude with inferring the difference between the means of two populations. In later chapters we develop these techniques further to see how they apply, for example, to comparing categorical variables, regression, and other situations.

## CASE STUDY

### Epilepsy Drugs and Children

Epilepsy is a neurological disorder that causes seizures, which can range from mild to severe. It has been estimated (wikipedia.org) that over 50 million people around the world suffer from epilepsy. It is usually treated with drugs, and four drugs that are commonly used for this purpose are carbamazepine, lamotrigine, phenytoin, and valproate. In 2009 researchers at the *New England Journal of Medicine* reported that pregnant mothers who take valproate might risk impairing the cognitive development of their children, compared to mothers who take one of the other drugs. As evidence, on the basis of a sample of pregnant women with epilepsy, they estimated the mean IQ of three-year-old children whose mothers took one of these four drugs during their pregnancies. They gave 95% confidence intervals (CI) for the mean IQ, as shown in Table 9.1 on the next page (Meador et al. 2009).

| Drug | 95% CI |
|------|--------|
| carbamazepine | (95, 101) |
| lamotrigine | (98, 104) |
| phenytoin | (94, 104) |
| valproate | (88, 97) |

▲ **TABLE 9.1** 95% confidence intervals for the average IQ of children whose mothers took various epilepsy drugs during their pregnancies.

Why did these four intervals lead the researchers to recommend that pregnant women not use valproate as a "first choice" drug for epilepsy? The researchers wrote that "Although the confidence intervals for carbamazepine and phenytoin overlap with the confidence interval for valproate, the confidence intervals for the differences between carbamazepine and valproate and between phenytoin and valproate do not include zero." What does this tell us?

In this chapter we discuss how confidence intervals can be used to estimate characteristics of a population—in this case, the population of all children of women with epilepsy who took one of these drugs during pregnancy. Confidence intervals can also be used to judge between hypotheses about the means and about differences between means. The population of pregnant women with epilepsy is large, and yet if conditions are right, we can make decisions and reach an understanding about the entire population on the basis of a small sample. At the end of this chapter, we will return to this study and see if we can better understand its conclusions.

# Sample Means of Random Samples

As you learned in Chapter 7, we estimate population parameters by collecting a random sample from that population. We use the collected data to calculate a statistic, and this statistic is used to estimate the parameter. Whether we are using the statistic $\hat{p}$ to estimate the parameter $p$ or are using $\bar{x}$ to estimate $\mu$, if we want to know how close our estimate is to the truth, we need to know how far away that statistic is, typically, from the parameter.

Just as we did in Chapter 7 with $\hat{p}$, we now examine three characteristics of the behavior of the sample mean: its accuracy, its precision, and its probability distribution. By understanding these characteristics, we'll be able to measure how well our estimate performs and thus make better decisions.

As a reminder, Table 9.2 shows the relation between some commonly used statistics and the parameters they estimate. (This table originally appeared as Table 7.2.)

▶ TABLE 9.2

| Statistic (based on data) | | Parameter (typically unknown) | |
|---------------------------|----|-------------------------------|----|
| Sample mean | $\bar{x}$ | Population mean | $\mu$ (mu) |
| Sample standard deviation | $s$ | Population standard deviation | $\sigma$ (sigma) |
| Sample variance | $s^2$ | Population variance | $\sigma^2$ |
| Sample proportion | $\hat{p}$ (p-hat) | Population proportion | $p$ |

 **Details**

**Mu-sings**
The mean of a population, represented by the Greek character $\mu$, is pronounced "mu" as in *music*.

This chapter uses much of the vocabulary introduced in Chapters 7 and 8, but we'll remind you of important terms as we proceed. To help you visualize how a sample mean based on randomly sampled data behaves, we'll make use of the by-now-familiar technique of simulation. Our simulation is slightly artificial, because to do a simulation, we need to know the population. However, after using a simulation to understand how the sample mean behaves in this artificial situation, we will discuss what we do in the real world when we do not know very much about the population.

# Accuracy and Precision of a Sample Mean

The reason why the sample mean is a useful estimator for the population mean is that the sample mean is accurate and, with a sufficiently large sample size, very precise. The accuracy of an estimator, you'll recall, is measured by the **bias**, and the precision is measured by the standard error. You will see in this simulation that

1. The sample mean is unbiased when estimating the population mean—-that is, on average, the sample mean is the same as the population mean.

2. The **precision** of the sample mean depends on the variability in the population, but the more observations we collect, the more precise the sample mean becomes.

For our simulation, we'll use the population that consists of the finishing times of everyone who ran the Cherry Blossom Ten Mile Run in 2009 (Kaplan 2009). (The finishing time is the amount of time it took to finish the race.) This race is held every spring in Washington, D.C. As in most such races, data are carefully collected on every participant. Rather than showing you the histogram of the finishing times for all 8600 runners, we're going to take advantage of the fact that the distribution closely follows a Normal model with a mean of 97 minutes and a standard deviation of 17 minutes: N(97, 17). Figure 9.1 shows the distribution of this population.

The population parameters are, in symbols,

$$\mu = 97 \text{ minutes}$$

$$\sigma = 17 \text{ minutes}$$

For our simulation, we will randomly sample 30 runners and calculate the average of their finishing times. We'll then repeat this many, many times. (The exact number of repetitions we performed isn't important for this discussion.) We are interested in two questions: (1) What is the typical value of the sample mean? If it is 97 minutes—the value of the population mean—then the sample mean is unbiased. (2) Typically, how far away is a sample mean from 97? In other words, how much spread is in the distribution of sample means? This spread helps us measure the precision of the sample mean as an estimator of the population mean.

**🔄 Looking Back**

**Bias and Precision**
Bias is the mean distance between the sample statistic and the parameter it is estimating. Precision is measured by the standard error, which tells us how far the statistic typically deviates from its center. Review Figure 7.2 to help visualize these two properties.

N(97, 17)

40   60   80   97 100 120   140   160
Time (minutes)

**◄ FIGURE 9.1** Finishing times in the Cherry Blossom Ten Mile Run follow a Normal model with a mean of 97 minutes and a standard deviation of 17 minutes.

For example, our very first sample of 30 runners had a mean finishing time of 95.1 minutes. We plotted this sample mean, as well as the means of the many other samples we took, in Figure 9.2 on the next page. The plot is on the same scale as Figure 9.1, the picture of the population distribution, so that you can see how much narrower this distribution of sample means is.

From this dotplot of sample means, we learn that the typical value of the sample means is the same as the population mean of 97 minutes. And we see that the sample mean is a relatively precise estimator: All of the sample means are within about 10 minutes (either above or below) the true mean value of 97 minutes.

**Sample Mean (minutes)**

▶ **FIGURE 9.2** Each dot represents a sample mean based on 30 runners who were randomly selected from the population whose distribution is shown in Figure 9.1. Note that the spread of this distribution is much smaller than the spread of the population, but the center looks to be at about the same place: 97 minutes.

**↻ Looking Back**

**Sampling Distribution**
In Chapter 7 we introduced the sampling distribution of sample proportions. The sampling distribution of sample means is the same concept: It is a distribution that gives us probabilities for sample means drawn from a population. The sampling distribution is the distribution of all possible sample means.

Figure 9.2 is a very approximate picture of the **sampling distribution** of the sample mean for samples of size 30. Recall that a sampling distribution is a probability distribution of a statistic; in this case, the statistic is the sample mean. You can think of the sampling distribution as the distribution of *all* possible sample means that would result from drawing repeated random samples of a certain size from the population.

When the mean of the sampling distribution is the same value as the population mean, we say that the statistic is an **unbiased estimator**. This appears to be the case here, because both the mean of the distribution of sample means in Figure 9.2 and the population mean are about 97 minutes.

The standard deviation of the sampling distribution is what we call the **standard error**. The standard error measures the precision of an estimator by telling us how much the statistic varies from sample to sample. For the sample mean, the standard error is smaller than the population standard deviation. We can see this because the spread for the sampling distribution is smaller than the spread of the population distribution. Soon you'll see how to calculate the standard error.

What happens to the center and spread of the sampling distribution if we increase the sample size? Let's start all over with the simulation. But this time, we take a random sample of 100 runners and calculate the mean. We then repeat this many hundreds of times. Figure 9.3 shows the results for this simulation and also for two new

**(a)**

**(b)**

**(c)**

▶ **FIGURE 9.3** **(a)** Histogram of a large number of sample means. Each sample mean is based on a sample of 100 randomly selected runners. **(b)** Sample means based on samples of 500 runners. **(c)** Sample means based on samples of 1000 runners. Each time, the sampling distribution gets narrower, reflecting a smaller standard error.

simulations where each sample mean is based on 500 runners and then on 1000 runners. The scale of the *x*-axis is the same as in Figure 9.1. Note that the spread of the distributions becomes quite small—so small, in fact, that we can't get a good look at the shape of the distributions.

## What Have We Demonstrated with These Simulations?

Because the sampling distributions are always centered at the population mean, we have demonstrated that the sample mean is an unbiased estimator of the population mean. We saw this for only one type of population distribution: the Normal distribution. But in fact, this is the case for any population distribution.

We have demonstrated that the standard deviation of the sampling distribution, which is called the **standard error** of the sample mean, gets smaller with larger sample size. This is true for any population distribution.

We can be more precise. If the symbol $\mu$ represents the mean of the population and if $\sigma$ represents the standard deviation of the population, then

1. The mean of the sampling distribution is also $\mu$ (which tells us that the sample mean is unbiased when estimating the population mean).

2. The standard error is $\dfrac{\sigma}{\sqrt{n}}$ (which tells us that the standard error depends on

the population distribution and is smaller for larger samples).

>  **Looking Back**
>
> **Sample Proportions from Random Samples**
> Compare the properties of the sample mean to those of the sample proportion, $\hat{p}$, given in Chapter 7. The sample proportion is also an unbiased estimator (for estimating the population proportion, $p$). It has standard error
> $$\sqrt{\dfrac{p(1-p)}{n}}.$$

 **KEY POINT**  For all populations, the sample mean is unbiased when estimating the population mean. The standard error of the sample mean is $\dfrac{\sigma}{\sqrt{n}}$, so the sample mean is more precise for larger sample sizes.

## EXAMPLE **1** iTunes Library Statistics

A student's iTunes library of mp3s has a very large number of songs. The mean length of the songs is 243 seconds, and the standard deviation is 93 seconds. The distribution of song lengths is right-skewed. Using his mp3 player, this student will create a playlist that consists of 25 randomly selected songs.

**QUESTIONS**

a. Is the mean value of 243 minutes an example of a parameter or a statistic? Explain.

b. What should the student expect the average song length to be for his playlist?

c. What is the standard error for the mean song length of 25 randomly selected songs?

**SOLUTIONS**

a. The mean of 243 is an example of a parameter, because it is the mean of the population that consists of all of the songs in the student's library.

b. The sample mean length can vary, but is typically the same as the population mean: 243 seconds.

c. The standard error is $\dfrac{\sigma}{\sqrt{n}} = \dfrac{93}{\sqrt{25}} = \dfrac{93}{5} = 18.6$ seconds.

**TRY THIS!**  Exercise 9.9

# The Central Limit Theorem for Sample Means

In the last simulation, all of the approximate sampling distributions (Figures 9.2 and 9.3) were Normal. This probably doesn't surprise you, because the population distribution was Normal.

What might surprise you is that the sampling distribution of the mean is always Normal (or at least approximately Normal), regardless of the shape of the population distribution. (If the sample size is small, however, the approximation can be pretty lousy.) This is the conclusion of the Central Limit Theorem, an important mathematical theorem that tells us that as long as the sample size is large, we can use the Normal distribution to perform statistical inference, regardless of the population the data are sampled from.

The **Central Limit Theorem (CLT)** assures us that no matter what the shape of the population distribution, if a sample is selected such that the following conditions are met, then the distribution of sample means follows an approximately Normal distribution. The mean of this distribution is the same as the population mean. The standard deviation (also called the standard error) of this distribution is the population standard deviation divided by the square root of the sample size. As a rule of thumb, sample sizes of 25 or more may be considered "large."

> Condition 1: *Random Sample and Independence*. Each observation is collected randomly from the population, and observations are independent of each other. The sample can be collected either with or without replacement.
>
> Condition 2: *Normal*. Either the population distribution is Normal or the sample size is large.
>
> Condition 3: *Big Population*. If the sample is collected without replacement, then the population must be at least 10 times larger than the sample size.

**KEY POINT**

The sampling distribution of $\bar{x}$ is approximately $N\left(\mu, \dfrac{\sigma}{\sqrt{n}}\right)$, where $\mu$ is the mean of the population and $\sigma$ is the standard deviation of the population. The larger the sample size, $n$, the better the approximation. If the population is Normal to begin with, then the sampling distribution is exactly a Normal distribution.

## Visualizing Distributions of Sample Means

The histogram in Figure 9.4 shows the distribution of in-state tuition and fees for all two-year colleges in the United States for the 2008–2009 academic year (*The Chronicle of Higher Education, Facts & Figures* 2008). Note that the distribution has an unusual feature: It has a mode near 0. This is because all two-year colleges in California charged $647 for California residents. The distribution is right-skewed.

### Looking Back

**CLT for Proportions**
In Chapter 7, you saw that the Central Limit Theorem applies to sample proportions. Here you'll see that it also applies to sample means.

▶ **FIGURE 9.4** Distribution of annual tuitions and fees at all two-year colleges in the United States for the 2008–2009 academic year.

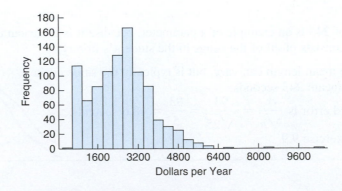

This histogram represents the distribution of a population, because it includes *all* two-year colleges. The mean of this population—the "typical" tuition of all two-year colleges—is $2535.

Using this distribution, we now show the results of a simulation that should be starting to feel familiar. First, we take a random sample of 30 colleges. The distribution of this sample is shown in Figure 9.5. We find the mean tuition of the 30 colleges in the sample and record this figure; for example, the sample mean for the sample shown in Figure 9.5 is about $2380.

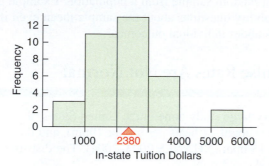

◀ **FIGURE 9.5** Distribution of a sample of 30 colleges taken from the population of all colleges. The mean of this sample, $2380, is indicated.

We repeat this activity (that is, we sample another 30 colleges from the population of all colleges and record the mean tuition of the sample) 200 times. When we are finished, we have 200 sample mean tuitions, each sample mean based on a sample of 30 colleges. Figure 9.6a shows this distribution. Figure 9.6b shows the distribution of averages when, instead of sampling 30 colleges, we double the number and sample 60 colleges. What differences do you see among the population distribution (Figures 9.4), the distribution of one of the samples (Figure 9.5), the sampling distribution when the sample size is 30 (Figure 9.6a), and the sampling distribution when the sample size is 60 (Figure 9.6b)?

▲ **FIGURE 9.6 (a)** Distribution of sample means, where each sample mean is based on a sample size of $n = 30$ college tuitions and is drawn from the population shown in Figure 9.4. This is (approximately) the sampling distribution of $\bar{x}$ when $n = 30$. A Normal curve is superimposed. **(b)** The approximate sampling distribution of $\bar{x}$ when $n = 60$.

Both of the sampling distributions in Figures 9.6a and 9.6b show us the values and relative frequencies for $\bar{x}$, but they are based on different sample sizes. We see that even though the *population* distribution has an unusual shape (Figure 9.4), the sampling distributions for $\bar{x}$ are fairly symmetric and unimodal. Although the Normal curve that is superimposed doesn't match the histogram very closely when $n = 30$, the match is pretty good for $n = 60$.

This is exactly what the CLT predicts. When the sample size is large enough, we can use the Normal distribution to find approximate probabilities for the values we might see for $\bar{x}$ when we take a random sample from the population.

The more observations in your sample, the better an approximation the Normal distribution provides. Generally, the CLT provides a useful approximation of the true

↻ **Looking Back**

**Distribution of a Sample vs. Sampling Distribution**
Recall that these are two different concepts. The *distribution of a sample*, from Chapter 3, is the distribution of one single sample of data (Figure 9.5). The *sampling distribution*, on the other hand, is the probability distribution of an estimator or statistic such as the sample mean (Figures 9.6a and 9.6b).

probabilities if the sample size is 25 or more. But this is just a rule of thumb. Be aware that you might need larger sample sizes in some situations. Unlike in Chapter 7, where we worked with sample proportions, we can't provide a hard-and-fast rule for sample size. For nearly all examples in this book, though not always in real life, 25 is large enough.

## Applying the Central Limit Theorem

The Central Limit Theorem helps us find probabilities for sample means when those means are based on a random sample from a population. Example 2 demonstrates how we can answer probability questions about the sample mean even if we can't answer probability questions about individual outcomes.

### EXAMPLE 2 Pulse Rates Are Not Normal

According to one very large study done in the United States, the mean resting pulse rate of adult women is about 74 beats per minute (bpm), and the standard deviation of this population is 13 bpm (NHANES 2003–2004). The distribution of resting pulse rates is known to be skewed right.

QUESTIONS

a. Suppose we take a random sample of 36 women from this population. What is the approximate probability that the average pulse rate of this sample will be below 71 or above 77 bpm? (In other words, what is the probability that it will be more than 3 bpm away from the population mean of 74 bpm?)

b. Can you find the probability that a single adult woman will have a resting pulse rate more than 3 bpm away from the mean value of 74?

SOLUTION

a. It doesn't matter that the population distribution is not Normal. Because the sample size of 36 women is relatively large, the distribution of sample means will be approximately (though not exactly) Normal.

The mean of this Normal distribution will be the same as the population mean: $\mu = 74$ bpm. The standard deviation of this distribution is the standard error:

$$SE = \frac{\sigma}{\sqrt{n}} = \frac{13}{\sqrt{36}} = \frac{13}{6} = 2.167$$

To use the Normal table to find probabilities requires that the values of 71 bpm and 77 bpm be converted to standard units:

$$z = \frac{\bar{x} - \mu}{SE} = \frac{71 - 74}{2.167} = \frac{-3}{2.167} = -1.38$$

Figure 9.7 shows that the area that corresponds to the probability that the sample mean pulse rate will be more than 1.38 standard errors away from the population mean pulse rate. This probability is calculated to be about 17%.

▶ **FIGURE 9.7** Area of the Normal curve outside of z-scores of −1.38 and 1.38.

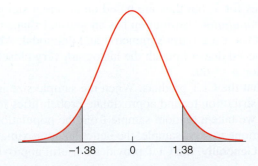

**CONCLUSIONS**

a. The approximate probability that the average pulse of 36 adult women will be more than 3 bpm away from 74 bpm is about 17%.

b. We cannot find the probability for a single woman because we do not know the probability distribution. We know only that it is "right-skewed," which is not enough information to find actual probabilities.

**TRY THIS!**  Exercise 9.11

> **! Caution**
>
> **CLT Not Universal**
> The CLT does not apply to all statistics you run across. It does not apply to the sample median, for example. No matter how large the sample size, you cannot use the Normal distribution to find a probability for the median value. It also does not apply to the sample standard deviation.

## Many Distributions

It's natural at this point to feel that you have seen a confusingly large number of types of distributions, but it's important that you keep them straight. The *population distribution* is the distribution of values from the population. Figure 9.4 (two-year college tuitions) is an example of a population distribution because it shows the distribution of *all* two-year colleges. Figure 9.1 (runners' times for all competitors in a race) is another example of a population distribution. For some populations, we don't know precisely what this distribution is. Sometimes we assume (or know) that it is Normal, sometimes we know it is skewed in one direction or the other, and sometimes we know almost nothing.

We then take a random sample of *n* observations. We can make a histogram of these data. This histogram gives us a picture of the *distribution of the sample*. If the sample size is large, and if the sample is random, then the sample will be representative of the population, and the distribution of the sample will look similar to (but not the same as!) the population distribution. Figure 9.5 is an example of the distribution of a sample of size *n* = 30 taken from the population of two-year college tuitions.

The *sampling distribution* is more abstract. If we take a random sample of data and find the sample mean (the center of the distribution of the sample), and then repeat this many, many times, we will get an idea what the sampling distribution looks like. Figures 9.6a and 9.6b are examples of approximate sampling distributions for the sample mean, based on samples from the two-year college tuition data. Note that these do not share the same shape as the population or the sample; they are both approximately Normal.

## EXAMPLE 3 Identify the Distribution

Figure 9.8 shows three distributions. One distribution is a population. The other two distributions are (approximate) sampling distributions. One sampling distribution is based on sample means of size 10, and the other is based on sample means of size 25.

◀ **FIGURE 9.8** Three distributions, all on the same scale. One is a population distribution, and the other two are sampling distributions for means sampled from the population. (Source: Rice Virtual Lab in Statistics, http://onlinestatbook.com/)

**QUESTION** Which graph (a, b, or c) is the population distribution? Which shows the sampling distribution for the mean with $n = 10$? Which with $n = 25$?

**SOLUTION** The Central Limit Theorem tells us that sampling distributions for means are approximately Normal. This implies that Figure 9.8b is not a sampling distribution, so it must be the population distribution from which the samples were taken. We know that the sample mean is more precise for larger samples, and because Figure 9.8a has the larger standard error (is wider), it must be the graph associated with $n = 10$. This means that Figure 9.8c is the sampling distribution of means with $n = 25$.

**TRY THIS!** Exercise 9.13

## SNAPSHOT    THE SAMPLE MEAN ($\overline{x}$)

|  |  |
|---|---|
| **WHAT IS IT?** ▶ | The arithmetic average of a sample of data. |
| **WHAT DOES IT DO?** ▶ | Estimates the mean value of a population, $\mu$. The mean is used as a measure of what is "typical" for a population. |
| **HOW DOES IT DO IT?** ▶ | If the sample was a random sample, then the sample mean is unbiased, and we can make the precision of the estimator as good as we want by taking a large enough sample size. |
| **HOW IS IT USED?** ▶ | If the sample size is large enough (or the population is Normal), we can use the Normal distribution to find the probability that the sample mean will take on a value in any given range. This lets us know how wrong our estimate could be. |

## The *t*-Distribution

The hypothesis tests and confidence intervals that we will use for estimating and testing the mean are based on a statistic called the ***t*-statistic:**

$$ t = \frac{\overline{x} - \mu}{\left(\dfrac{s}{\sqrt{n}}\right)} $$

The *t*-statistic is very similar to a *z*-score for the sample mean. In the numerator, we subtract the population mean from the sample mean. Then we divide not by the standard error but, rather, by an *estimate* of the standard error.

It would be nice if we could divide by the true standard error. But in real life, we almost never know the value of $\sigma$, the population standard deviation. So instead, we replace it with an estimate: the sample standard deviation, $s$. This gives us an estimate of the standard error:

$$ SE_{\text{EST}} = \frac{s}{\sqrt{n}} $$

Compare the *t*-statistic to the *z*-statistic, and you will see that we simply replaced $\sigma$ in the *z*-statistic with $s$.

$$ z = \frac{\overline{x} - \mu}{\left(\dfrac{\sigma}{\sqrt{n}}\right)} $$

The *t*-statistic does *not* follow the Normal distribution. One reason for this is that the denominator changes with every sample. For this reason, the *t*-statistic is

**Looking Back**

**Sample Standard Deviation**
In Chapter 3 we gave the formula for the sample standard deviation:

$$ s = \sqrt{\frac{\Sigma(x - \overline{x})^2}{n - 1}} $$

more variable than the *z*-statistic (whose denominator is always the same.) Instead, the *t*-statistic follows a distribution called—surprise!—the ***t*-distribution**. This was Gosset's great discovery at the Guinness brewery. When small sample sizes were used to make inferences about the mean, even if the population was Normal, the Normal distribution just didn't fit the results that well. Gosset discovered a new distribution, which he called the *t*-distribution, that turned out to be a better model than the Normal for the sampling distribution of $\bar{x}$ when $\sigma$ is not known.

The *t*-distribution shares many characteristics with the $N(0, 1)$ distribution. Both are symmetric, are unimodal, and might be described as "bell-shaped." However, the *t*-distribution has thicker tails. This means that in a *t*-distribution, it is more likely that we will see extreme values (values far from 0) than it is in a standard Normal distribution.

The *t*-distribution's shape depends on only one parameter, called the **degrees of freedom (df)**. The number of degrees of freedom is (usually) an integer: 1, 2, 3, and so on. If df is small, then the *t*-distribution has very thick tails. As the degrees of freedom get larger, the tails get thinner. Ultimately, when df is infinitely large, the *t*-distribution is exactly the same as the $N(0, 1)$ distribution.

Figure 9.9 shows *t*-distributions with 1, 10, and 40 degrees of freedom. In each case, the *t*-distribution is shown with a $N(0, 1)$ curve so that you can compare them. The *t*-distribution is the one whose tails are "higher" at the extremes. Note that by the time the degrees of freedom reaches 40 (Figure 9.9c), the *t*-distribution and the $N(0, 1)$ distribution are too close to tell apart (on this scale.)

**(a)**

**(b)**

**(c)**

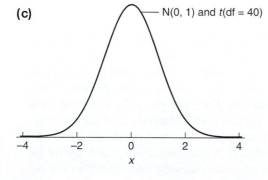

◀ **FIGURE 9.9 (a)** A *t*-distribution with 1 degree of freedom, along with a $N(0, 1)$ distribution. The *t*-distribution has much thicker tails. **(b)** The degrees of freedom are now equal to 10, and the tails are only slightly thicker in the *t*-distribution. **(c)** The degrees of freedom are now 40, and the two distributions are visually indistinguishable.

## SECTION 9.3

# Answering Questions about the Mean of a Population

Do you commute to work? How long does it take you to get there? Is this amount of time typical for others in your state? Which state has the greatest commuting times? This information is important not just to those of us who must fight traffic every day, but also to business leaders and politicians who make decisions about quality of living and the cost of doing business. The U.S. Census performs periodic surveys that determine, among other things, commuting times around the country. In 2007 (the last time the survey was done), the state of New York had the greatest mean commuting time, which was 31.5 minutes. Vermont was lowest at 21.2 minutes.

These means are estimates of the mean commuting time for all residents in these states who work outside of their homes. We can learn the true mean commuting time only by asking all residents, which is clearly too time-consuming to do very often. Instead, the U.S. Census takes a random sample of U.S. residents to estimate these values.

In this section we present two techniques for answering questions about the population mean. Confidence intervals are used for estimating values. Hypothesis tests are used for deciding whether the value is one thing or another. These are the same methods that were introduced in Chapter 7 (confidence intervals) and Chapter 8 (hypothesis tests) for population proportions, but here you'll see how they are modified to work with means.

## Estimation with Confidence Intervals

**Confidence intervals** are a technique for communicating an estimate of the mean along with a measure of our uncertainty in that estimate. The job of a confidence interval is to provide us with a range of values that, according to the data, are highly plausible values for the unknown population mean. For instance, the range of values for the mean commuting time for all Vermont commuters is 21.1 to 21.3 minutes.

Not all confidence intervals do an equally good job; the "job performance" of a confidence interval is therefore measured with something called the **confidence level**. The higher this level, the better the confidence interval performs. The confidence interval for mean Vermont commuting times is 90%, which means we can be extremely confident that this interval contains the true mean.

Sometimes, you will be in a situation in which you will know only the sample mean and sample standard deviation. In these situations, you can use a calculator to find the confidence interval. However, if you have access to the actual data, you are much better off using statistical software to do all the calculations for you. We will show you how to respond to both situations.

No matter which situation you are in, you will need to judge whether a confidence interval is appropriate for the situation, and you will need to interpret the confidence interval. Therefore, we will discuss these essential skills before demonstrating the calculations.

### When Are Confidence Intervals Useful?

A confidence interval is a useful answer to the following questions: "What's the typical value for a variable in this large group of objects or people? And how far away from the truth might this estimate of the typical value be?" You should provide a confidence interval whenever you are estimating the value of a population parameter on the basis of a random sample from that population. For example, judging on the basis of a random sample of 30 adults, what's the typical body temperature of all healthy adults? On the basis of a survey of a random sample of Vermont residents, what's the typical commuting time for all Vermont residents? A confidence interval is useful for answering questions such as these because it communicates the uncertainty in our estimate and provides a range of plausible values.

A confidence interval is not appropriate if there is no uncertainty in your estimate. This would be the case if your "sample" were actually the entire population. For example, it is not necessary to find a confidence interval for the mean score on your class's statistics exam. The population is your class, and all of the scores are known. Thus the population mean is known, and there is no need to estimate it.

### Checking Conditions

In order to measure the correct confidence level, the following conditions must hold:

Condition 1: *Random Sample and Independence.* The data must be collected randomly, and each observation must be independent of the others.

Condition 2: *Normal*. Either the population must be Normally distributed or the sample size must be fairly large (at least 25).

If these conditions do not hold, then we cannot measure the job performance of the interval; the confidence level may be incorrect. This means that we may advertise a 95% confidence level when, in fact, the true performance is much worse than this.

To check the first condition, you must know how the data were collected. This is not always possible, so rather than checking these conditions, you must simply assume that they hold. If they do not, your interval will not be valid.

The requirement for independence means that measurement of one object in the sample does not affect any other. Essentially, if we know the value of any one observation, this knowledge should tell us nothing about the values of other observations. This condition might be violated if, say, we randomly sampled several schools and gave all of the students math exams. The individual math scores would not be independent, because we would expect that students within the same school might have similar scores.

The second condition is due to the Central Limit Theorem. If the population distribution is Normal (or very close to it), then we have nothing to worry about. But if it is non-Normal, then we need a large enough sample size so that the sampling distribution of sample means is approximately Normal. For many applications, a sample size of 25 is large enough, but for extremely skewed distributions, you might need an even larger sample size.

## EXAMPLE 4 Is the Cost of College Rising?

Many cities and states are finding it more difficult to offer low-cost college educations. Did the mean cost of attending two-year colleges increase in the United States from the 2007–2008 academic year to 2009–2010? In 2007–2008, the mean cost of *all* two-year colleges was $2429. A random sample of 35 two-year colleges in the United States found that the average tuition charged in 2008–2009 was $2380, with a standard deviation of $1160. Figure 9.10 shows the distribution of this sample. On the basis of this, a 90% confidence interval for the mean cost of attending two-year colleges in 2008–2009 is $2048 to $2711.

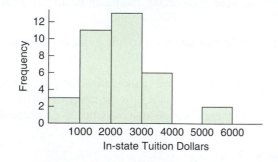

◀ **FIGURE 9.10** Distribution of in-state tuition for a random sample of 35 two-year colleges during the academic year 2008–2009.

**QUESTIONS**

a. Describe the population. Is the number $2380 an example of a parameter or a statistic?

b. Verify that the conditions for a valid confidence interval are met.

**SOLUTIONS**

a. The population consists of all two-year college tuitions (for in-state residents) in the academic year 2008–2009. (There are roughly 1000 two-year colleges in the United States.) The number $2380 is the mean of a sample of only 35 colleges. Because it is the mean of a sample, it is a statistic.

b. The first condition is that the data represent a random sample of independent observations. We are told the sample was collected randomly, so we assume this is true.

Independence also holds, because knowledge about any one school's tuition tells us nothing about other schools in the sample. The second condition requires that the population be roughly Normally distributed or the sample size be equal to or larger than 25. We do not know the distribution of the population, but because the sample size is large enough (bigger than 25), this condition is satisfied.

**TRY THIS!** Exercise 9.15

## Interpreting Confidence Intervals

To understand confidence intervals, you must know how to interpret a confidence interval and how to interpret a confidence level.

A confidence *interval* can be interpreted as a range of plausible values for the population parameter. In other words, in the case of population means, we can be confident that if we were to someday learn the true value of the population mean, it would be within the range of values given by our confidence interval. For example, the U.S. Census estimates that the mean commuting time for Vermont residents is 21.1 minutes to 21.3 minutes, with a 90% confidence level. We interpret this to mean that we can be fairly confident that the true mean commute for *all* Vermont residents is between 21.1 and 21.3 minutes. Yes, we could be wrong. The mean might be less than 21.1 minutes, or it might be more than 21.3 minutes. However, we would be rather surprised to find this was the case; we are highly confident that the mean is within this interval.

**KEY POINT** A confidence interval can be interpreted as a range of plausible values for the population parameter.

## EXAMPLE 5 Evidence for Changing College Costs

Based on a random sample of 35 two-year colleges, a 90% confidence interval for the mean in-state tuition at two-year colleges for the 2009–2010 academic year is $2048 to $2711. In the academic year 2007–2008, the mean of all two-year colleges' in-state tuitions was $2429.

**QUESTION** Does the confidence interval provide evidence that the mean tuition has increased?

**SOLUTION** No, it does not. Although we cannot know the population mean of all tuitions in 2009/2010, we are highly confident that it is in the range of $2048 to $2711. This range includes the value $2429, so there is not enough evidence to conclude that the mean tuition has changed.

**TRY THIS!** Exercise 9.17

## Measuring Performance with the Confidence Level

The confidence *level*, which in the case of the interval for the mean Vermont commuting time was 90%, tells us about the method used to find the interval. A value for the level of 90% tells us that the U.S. Census used a method that works in 90% of all samples. In other words, if we were to take many same-sized samples of Vermont residents, and for each sample calculate a 90% confidence interval, then 90% of those intervals would contain the population mean.

The confidence level does *not* tell us whether the interval (21.1 to 21.3) contains or does not contain the population mean. The "90%" just tells us that the method that produced this interval is a pretty good method.

Suppose you decided to purchase an mp3 player online. You have your choice of several manufacturers, and they are rated in terms of their performance level. One manufacturer has a 90% performance level, which means that 90% of the players they produce are good ones, and 10% are defective. Some other manufacturers have lower levels: 80%, 60%, and worse. From whom do you buy? You choose to buy from the manufacturer with the 90% level, because you can be very confident that the player they send you will be good. Of course, once the player arrives at your home, the confidence level isn't too useful. Your player either works or does not work; there's no 90% about it.

Confidence levels work the same way. We prefer confidence intervals that have 90% or higher confidence levels, because then we know that the process that produced these levels is a good process, and therefore, we are confident in any decisions or conclusions we reach. But the level doesn't tell us whether this one particular interval sitting in front of us is good or bad. In fact, we shall never know that, unless we someday gain access to the entire population.

**KEY POINT** The confidence level is a measure of how well the method used to produce the confidence interval performs. We can interpret the confidence level to mean that if we were to take many random samples of the same size from the same population, and for each random sample calculate a confidence interval, then the confidence level is the proportion of intervals that "work"—the proportion that contain the population parameter.

Figure 9.11 illustrates this interpretation of confidence levels. From the population of 107 movies that were playing during the week of April 3, 2009 (http://www.the-numbers.com/), we took a random sample (with replacement) of 30 movies and calculated the mean revenue per theater in this sample. Because the samples were random, each sample produced a different sample mean. For each sample we also calculated a 95% confidence interval. We repeated this process 100 times, and each time we made a plot of the confidence interval. Figure 9.11a shows the results from the first 10 samples of 30 theaters. All 10 samples produced "good" intervals—intervals that contain the true population mean of $2429. Figure 9.11b shows what happened after we collected 100 95% confidence intervals. With a 95% confidence interval, we'd expect about 95% of the intervals to be good and 5% to be bad. And in fact, four intervals (shown in red) were bad.

> **! Caution**
>
> **Confidence Levels Are Not Probabilities**
> A confidence level, such as 90%, is not a probability. Saying we are 90% confident the mean is between 21.1 minutes and 21.3 minutes does *not* mean that there is a 90% chance that the mean is between these two values. It either is, or isn't. There's no probability about it.

▲ **FIGURE 9.11** **(a)** Ten different 95% confidence intervals, each based on a separate random sample of 30 movie theaters. The population mean of $2500 is shown with a vertical bar. All ten intervals are good because they include this population mean. **(b)** One hundred confidence intervals, each based on a random sample of 30 theaters. Because we are using a 95% confidence level, we expect about 95% of the intervals to be good. In fact, 96 of the 100 turned out to be good, this time. The red intervals are "bad" intervals that do not contain the population mean.

## EXAMPLE 6 iPad Batteries

A consumer group wishes to test Apple's claim that the iPad has a 10-hour battery life. With a random sample of 5 iPads, and running them under identical conditions, the group finds a 95% confidence interval for the mean battery life of an iPad to be 9.5 hours to 12.5 hours. One of the following statements is a correct interpretation of the confidence level. The other is a correct interpretation of the confidence interval.

  (i) We are very confident that the mean battery life of all iPads is between 9.5 and 12.5 hours.

  (ii) In about 95% of all samples of 5 iPads, the resulting confidence interval will contain the mean battery life of all iPads.

QUESTION Which of these statements is a valid interpretation of a confidence interval? Which of these statements is a valid interpretation of a confidence level?

SOLUTION Statement (i) interprets a confidence *interval* (9.5, 12.5). Statement (ii) tells us the meaning of the 95% confidence *level*.

TRY THIS! Exercise 9.21

### Calculating the Confidence Interval

Confidence intervals for means have the same basic structure as they did for proportions:

$$\text{Estimator} \pm \text{margin of error}$$

As in Chapter 7, the margin of error has the structure

$$\text{Margin of error} = (\text{multiplier}) \times SE$$

The standard error ($SE$) is $SE = \dfrac{\sigma}{\sqrt{n}}$. Because we usually do not know the standard deviation of the population and hence the $SE$, we replace the $SE$ with its estimate. This leads to a formula similar in structure, but slightly different in details, from the one you learned for proportions.

---

**Formula 9.1: One-Sample *t*-Interval**

$$\bar{x} \pm m$$

where $\qquad m = t^{*}SE_{\text{EST}}$ and $SE_{\text{EST}} = \dfrac{s}{\sqrt{n}}$

The multiplier $t^{*}$ is a constant that is used to fine-tune the margin of error so that it has the level of confidence we want. This multiplier is found using a $t$-distribution with $n - 1$ degrees of freedom. $SE_{\text{EST}}$ is the estimated standard error.

To compute a confidence interval for the mean, you first need to choose the level of confidence. After that, you need either the original data or these four pieces of information:

  1. The sample average, $\bar{x}$, which you calculate from the data.

  2. The sample standard deviation, $s$, which you calculate from the data.

  3. The sample size, $n$, which you know from looking at the data.

  4. The multiplier, $t^{*}$, which you look up in a table (or use technology) and which is determined by your confidence level and the sample size $n$. The value of $t^{*}$ tells us how wide the margin of error is, in terms of standard errors. For example, if $t^{*}$ is 2, then our margin of error is two standard errors wide.

---

The first three steps are pretty straightforward, so let's spend a minute on finding $t^{*}$, the multiplier for the margin of error.

The multiplier is based on a *t*-distribution with $n - 1$ degrees of freedom. The correct values can be found in Table 4 in the Appendix, or you can use technology. Table 4 is organized such that each row represents possible values of *t** for each degree of freedom. The columns contain the values of *t** for a given confidence level. For example, for a 95% confidence level and a sample size of $n = 30$, we use $t^* = 2.045$. We find this in the table by looking in the row with df $= n - 1 = 30 - 1 = 29$ and using the column for a 95% confidence level. Refer to Table 9.3, which is from the table in the Appendix.

## EXAMPLE 7 Finding the Multiplier *t**

Suppose we collect a sample of 30 iPads and wish to calculate a 90% confidence interval for the mean battery life.

QUESTION   Using Table 9.3, which is from the table in Appendix A, find *t** for a 90% confidence interval when $n = 30$.

| DF | Confidence Level | | | |
|---|---|---|---|---|
|    | 90% | 95% | 98% | 99% |
| 28 | 1.701 | 2.048 | 2.467 | 2.763 |
| 29 | <u>1.699</u> | 2.045 | 2.462 | 2.756 |
| 30 | 1.697 | 2.042 | 2.457 | 2.750 |
| 34 | 1.691 | 2.032 | 2.441 | 2.728 |

◄ **TABLE 9.3** Critical Values of *t*.

SOLUTION   We find the number of degrees of freedom from the sample size,

$$\text{df} = n - 1 = 30 - 1 = 29$$

And so we find, from Table 9.3, $t^* = 1.699$ (shown underlined)

TRY THIS!   Exercise 9.23

It is best to use technology to find the multiplier, because most tables stop at 35 or 40 degrees of freedom. For a 95% confidence level, if you do not have access to technology and the sample size is bigger than 40, it is usually safe to use $t^* = 1.96$—the same multiplier that we used for confidence intervals for sample proportions (for 95% confidence). The precise value, if we used a computer, is 2.02, but this is only 0.06 unit away from 1.96, so the result is probably not going to be affected in a big way.

The wider the confidence interval, the more confident we will be that it covers the true parameter value. We can always increase our level of confidence by making the margin of error bigger. We do this by choosing larger values for *t**.

> ↻ **Looking Back**
>
> **Why Not 100%?**
> In Chapter 7, you learned that one reason why a 95% confidence level is popular is that increasing the confidence level only a small amount beyond 95% requires a much larger margin of error.

 ## EXAMPLE 8 College Tuition Costs

A random sample of 35 two-year colleges in 2008–2009 had a mean tuition (for in-state students) of $2380, with a standard deviation of $1160.

QUESTION   Find a 90% confidence interval and a 95% confidence interval for the mean in-state tuition of all two-year colleges in 2008–2009. Interpret the intervals. Assume the necessary conditions hold.

**SOLUTION**  We are given the desired confidence level, the standard deviation, and the sample mean, $\bar{x}$, so the next step is to calculate the estimated standard error.

$$\bar{x} = \$2380$$

$$SE_{EST} = \frac{1160}{\sqrt{35}} = 196.0758$$

$$\bar{x} \pm m = \bar{x} \pm t^*SE_{EST}$$

We find the appropriate values of $t^*$ (from Table 9.3 above):

$$t^* \text{ (for 90\%)} = 1.691$$
$$t^* \text{ (for 95\%)} = 2.032$$

For the 90% confidence interval,

$$\bar{x} \pm t^*SE_{EST} \text{ becomes } 2380 \pm (1.691 \times 196.0758) = 2380 \pm 331.5642$$

Lower limit: $2380 - 331.5642 = 2048.44$
Upper limit: $2380 + 331.5642 = 2711.56$

A 90% confidence interval for the mean tuition of all two-year colleges in the 2008–2009 academic year is ($2048, $2712). That is, we are 90% confident that the mean tuition (the typical tuition) of all two-year colleges in 2008–2009 was between $2048 and $2712.

For the 95% confidence interval,

$$\bar{x} \pm t^*SE_{EST} \text{ becomes } 2380 \pm (2.032 \times 196.0758) = 2380 \pm 398.4260$$

Lower Limit: $2380 - 398.4260 = 1981.570$
Upper Limit: $2380 + 398.4260 = 2778.4260$

**CONCLUSION**  The 90% confidence interval is ($2048, $2712). The 95% confidence interval is ($1982, $2778), which is wider. We are 90% confident that the mean tuition of all two-year colleges is between $2048 and $2712. We are 95% confident that the mean tuition could be as low as $1982 or as high as $2778.

**TRY THIS!**  Exercise 9.25

Tech

If you have access to the original data (and not just to the summary statistics, as we were given in Example 8), then it is always best to use a computer to find the confidence interval for you. Figure 9.12 shows StatCrunch output that shows us what we would see if we had access to the full data for Example 8 and asked the software

**▶ FIGURE 9.12** StatCrunch output showing the 95% confidence interval for the mean in-state tuition at two-year colleges in the academic year 2008–2009.

One sample T statistics with data

**Options**

**95% confidence interval results:**
μ : mean of Variable

| Variable | Sample Mean | Std. Err. | DF | L. Limit | U. Limit |
|---|---|---|---|---|---|
| Instate_08 | 2380.8857 | 196.01347 | 34 | 1982.5385 | 2779.233 |

to find a 95% confidence interval for the mean tuition of all two-year colleges. The output shows us the estimated mean ($2380.8857), the standard error ($196.01347), the degrees of freedom (34), the lower limit of the confidence interval ($1982.5385), and the upper limit ($2779.233). Note that these are not exactly the values we calculated in Example 8. The StatCrunch values are more accurate because there is less rounding.

## Reporting and Reading Confidence Intervals

There are two ways of reporting confidence intervals. Professional statisticians tend to report (lower boundary, upper boundary). This is what we've done so far in this chapter. Thus we reported the 95% confidence interval for the mean of two-year college tuitions in 2008–2009 as ($1982, $2778).

In the press, however, and in some scholarly publications, you'll also see confidence intervals reported as

$$\text{Estimate} \pm \text{margin of error}$$

For the two-year college tuitions, we calculated the margin of error to be $398.426 for 95% confidence. Thus we could also report the confidence interval as

$$\$2380 \pm \$398$$

This form is useful because it shows our estimate for the mean ($2380) as well as our uncertainty (the mean could plausibly be $398 lower or $398 more).

You're welcome to choose whichever you think best, although you should be familiar with both forms.

## Testing a Mean

In Chapter 8 we laid out the foundations of hypothesis testing. Here, you'll see that the same four steps can be used to test hypotheses about means of populations. These four steps are

**Step 1: Hypothesize.**
State your hypotheses about the population parameter.

**Step 2: Prepare.**
Get ready to test: Choose and state a significance level. Choose a test statistic appropriate for the hypotheses. State and check conditions required for the computations, and state any assumptions that must be made.

**Step 3: Compute to compare.**
Compute the observed value of the test statistic in order to compare the null hypothesis value to our observed value. Find the p-value to measure your level of surprise.

**Step 4: Interpret.**
Do you reject or not reject the null hypothesis? What does this mean?

As an example of testing a mean, consider this "study" one of the authors did. McDonald's advertises that its ice cream cones have a mean weight of 3.2 ounces ($\mu = 3.2$). A human server starts and stops the machine that dispenses the ice cream, so we might expect some variation in the amount. Some cones might weight slightly more, some cones slightly less. If we weighed all of the McDonald's ice cream cones at a particular store, would the mean be 3.2 ounces, as the company claims?

One of the authors collected a sample of five ice-cream cones and weighed them on a postage scale. The weights were (in ounces)

$$4.2, 3.6, 3.9, 3.4, \text{ and } 3.3$$

We summarize these data as

$$\bar{x} = 3.68 \text{ ounces}, s = 0.3701 \text{ ounce}$$

Do these observations support the claim that the mean weight is 3.2 ounces? Or is the mean a different value? We'll apply the four steps of the hypothesis test to make a decision.

**Step 1: Hypothesize**
Hypotheses come in pairs and are always statements about the population. In this case, the population consists of all ice cream cones that have been, will be, or could be dispensed from a particular McDonald's. In this chapter, our hypotheses are about the mean values of populations.

The null hypothesis is the status-quo position, which is the claim that McDonald's makes. An individual cone might weigh a little more than 3.2 ounces, or a little less, but after looking at a great many cones, we would find that McDonald's is right and the mean weight is 3.2 ounces.

We state the null hypothesis as

$$H_0: \mu = 3.2$$

Recall that the null hypothesis always contains an equals sign.

The alternative hypothesis, on the other hand, says that the mean weight is different from 3.2 ounces:

$$H_a: \mu \neq 3.2$$

This is an example of a two-tailed hypothesis. We will reject the null hypothesis if the average of our sample cones is very big (suggesting that the population mean is bigger than 3.2) or very small (suggesting the population mean is less than 3.2). It is also possible to have one-tailed hypotheses, as you will see later in this chapter.

 **KEY POINT** Hypotheses are always statements about population parameters. For the test you are about to learn, this parameter is always $\mu$, the mean of the population.

 **Details**

**What Value for $\alpha$?**
For most situations, using a significance level of 0.05 is a good choice and is recommended by many scientific journals. Values of 0.01 and 0.10 are also commonly used.

**Step 2: Prepare**
The first step is to set the significance level $\alpha$ (alpha), as we discussed in Chapter 8. The significance level is a performance measure that helps us evaluate the quality of our test procedure. It is the probability of making the mistake of rejecting the null hypothesis when, in fact, the null hypothesis is true. In this case, this is the probability that we will say McDonald's cones do not weigh an average of 3.2 ounces when, in fact, they really do.

The test statistic, called the one-sample $t$-test, is very similar in structure to the test for one proportion and is based—not surprisingly, given the name of the test—on the $t$-statistic introduced in Section 9.2. The idea is simple: Compare the observed value of the sample mean, $\bar{x}$, to the value claimed by the null hypothesis, $\mu_0$.

**Formula 9.2: Test Statistic for the One-Sample $t$-Test**

$$t = \frac{\bar{x} - \mu_0}{SE_{EST}}, \qquad \text{where} \qquad SE_{EST} = \frac{s}{\sqrt{n}}$$

If conditions hold, the test statistic follows a $t$-distribution with df $= n - 1$

**Looking Back**

**The $z$-Test**
Compare this to the $z$-test statistic for one proportion in Chapter 8, which has a very similar structure:

$$z = \frac{\hat{p} - p_0}{SE}$$

This test statistic works because it compares the value of the parameter that the null hypothesis says is true, $\mu_0$, to the estimate of that value that we actually observed in our data. If the estimate is close to the null hypothesis value, then the $t$-statistic is close to 0. But if the estimate is far from the null hypothesis value, then the $t$-statistic is far from 0. The farther $t$ is from 0, the worse things look for the null hypothesis.

Anyone can make a decision, but only a statistician can measure the probability that the decision is right or wrong. To do this, we need to know the sampling distribution of our test statistic.

The sampling distribution will follow the *t*-distribution under these conditions:

Condition 1: *Random Sample and Independence.* The data must be a random sample from a population, and observations must be independent of one another.

Condition 2: *Normal.* The population distribution must be Normal or the sample size must be large. For most situations, 25 is large enough.

Now let's apply this to our ice cream problem. The population for testing the mean ice cream cone weight is somewhat abstract, because a constant stream of ice cream cones is being produced by McDonald's. However, it seems logical that if some cones weigh slightly more than 3.2 ounces and some weigh slightly less, then this distribution of weights should be symmetric and not too different from a Normal distribution. Because our population distribution is Normal, the fact that we have a small sample size, $n = 5$, is not a problem here.

The ice cream cone weights are independent of each other because we were careful, when weighing, to recalibrate the scale, and each cone was obtained on a different day. The cones were not, strictly speaking, randomly sampled, although because the cones were collected on different days and at different times, we will assume that they behave as though they come from a random sample. (But if we're wrong, our conclusions could be *very* wrong!)

### Step 3: Compute to compare

The conditions of our data tell us that our test statistic should follow a *t*-distribution with $n - 1$ degrees of freedom. Therefore, we proceed to do the calculations necessary to compare our observed sample mean to the hypothesized value of the population mean, and to measure our surprise.

To find the observed value of our *t*-statistic, we need to find the sample mean and the standard deviation of our sample. These values are given above, but you can easily calculate them from the data.

$$SE_{EST} = \frac{0.3701}{\sqrt{5}} = 0.1655$$

$$t = \frac{3.68 - 3.2}{0.1655} = \frac{0.48}{0.1655} = 2.90$$

The observed sample mean was 2.90—almost 3 standard errors above the value expected by the null hypothesis.

 **KEY POINT**  The *t*-statistic measures how far away (how many standard errors) our observed mean, $\bar{x}$, lies from the hypothesized value, $\mu_0$. Values far from 0 tend to discredit the null hypothesis.

How unusual is such a value, according to the null hypothesis? The p-value tells us exactly that—the probability that we would get a *t*-statistic as extreme as or more extreme than what we observed, if in fact the mean is 3.2 ounces.

Because our alternative hypothesis says we should be on the lookout for *t*-statistic values that are much bigger or smaller than 0, we must find the probability in both tails of the *t*-distribution. The p-value is shown in the small shaded tails of Figure 9.13. Our sample size is $n = 5$, so our degrees of freedom are $n - 1 = 5 - 1 = 4$.

The p-value of 0.044 tells us that if the typical cone really weighs 3.2 ounces, our observations are somewhat unusual. We should be surprised.

### Step 4: Interpret

The last step is to compare the p-value to the significance level and decide whether to reject the null hypothesis. If we follow a rule that says we will reject whenever the

▲ **FIGURE 9.13** The tail areas above 2.90 and below −2.90 are shown as the small shaded areas on both sides. The p-value is 0.0441, the probability that if $\mu = 3.2$, a test statistic will be bigger than 2.90 or smaller than −2.90. The distribution shown is a *t*-distribution with 4 degrees of freedom.

## Looking Back

**p-Values**

In Chapter 8 you learned that the p-value is the probability that when the null hypothesis is true, we will get a test statistic as extreme as or more extreme than what we actually saw. (What is meant by "extreme" depends on the alternative hypothesis.) The p-value measures our surprise at the outcome.

p-value is less than or equal to the significance level, then we know that the probability of mistakenly rejecting the null hypothesis will be the value of $\alpha$.

Our p-value (0.044) is less than the significance level we chose (0.05), so we should reject the null hypothesis and conclude that at this particular McDonald's, cones do *not* weigh, on average, 3.2 ounces.

This result makes some sense from a public relations standpoint. If the mean were 3.2 ounces, about half of the customers would be getting cones that weighed too little. By setting the mean weight a little higher than what is advertised, McDonald's can give everyone more than they thought they were getting.

## One- and Two-Tailed Alternative Hypotheses

The alternative hypothesis in the ice cream cone test was two-tailed. As you learned in Chapter 8, alternative hypotheses can also be one-tailed. The exact form of the alternative hypothesis depends on the research question. In turn, the form of the alternative hypothesis tells us how to find the p-value.

You will always use one of the following three pairs of hypotheses for the one-sample *t*-test:

| Two-Tailed | One-Tailed (Left) | One-Tailed (Right) |
|---|---|---|
| $H_0: \mu = \mu_0$ | $H_0: \mu = \mu_0$ | $H_0: \mu = \mu_0$ |
| $H_a: \mu \neq \mu_0$ | $H_a: \mu < \mu_0$ | $H_a: \mu > \mu_0$ |

You choose the pair of hypotheses on the basis of your research question. For the ice cream cone example, we asked if the mean weight is *different* from the value advertised, and so used a two-tailed alternative hypothesis. Had we instead wanted to know whether the mean weight was less than 3.2 ounces, we would have used a one-tailed (left) hypothesis.

Your choice of alternative hypothesis determines how you calculate the p-value. Figure 9.14 shows how to find the p-value for each alternative hypothesis, all using the same *t*-statistic value of $t = 2.1$ and the same sample size of n = 30.

▶ **FIGURE 9.14** The distributions are *t*-distributions with $n - 1 = 29$ degrees of freedom. The shaded region in each graph represents a p-value when $t = 2.1$ for **(a)** a two-tailed alternative hypothesis, **(b)** a one-tailed (left) hypothesis and **(c)** a one-tailed (right) hypothesis.

**(a)**

**(b)**

**(c)**

## Looking Back

**What Does "as extreme as or more extreme than" Mean?**

See Chapter 8 for a detailed discussion of how the p-value depends on the alternative hypothesis.

Note that the p-value is always an "extreme" probability; it's always the probability of the tails (even if the tail is pretty big, as it is in Figure 9.14b).

##  EXAMPLE 9 Dieting

Americans who want to lose weight have many different diets among which to choose. In one study (Dansinger et al. 2005), researchers compared results from four different diets. In this example, though, we look at only a small part of these data to examine whether one of the more popular diets, the Weight Watchers diet, is effective. The researchers examined 40 subjects who were randomly assigned to this diet. Researchers recorded the change in weight after 12 months. The distribution of amount of weight lost in this sample is shown in Figure 9.15. Only 26 of the 40 subjects stayed with the diet for that long, so we have data on only these 26 people.

QUESTION  Test the hypothesis that people on the Weight Watchers diet tend to lose weight. Summary statistics are given below. (A negative weight change means the person lost weight.)

$$\bar{x} = -4.6 \, \text{kg}, \quad s = 5.4 \, \text{kg}$$
(4.6 kilograms is about 10 pounds.)

SOLUTION  From Figure 9.15, we see that although a small number of people actually gained weight, the typical experience was a loss in weight. After a year, the average change in weight of the 26 people who stayed on the diet was $-4.6$ kilograms (about 10 pounds), with a standard deviation of 5.4 kilograms.

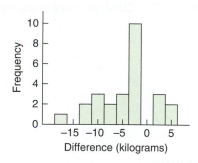

◀ **FIGURE 9.15** Change in weight for subjects after one year on a low-calorie diet (the Weight Watchers diet). Note that most values are negative, representing people who lost weight.

Our population of interest is the group of all overweight people who might go on the Weight Watchers diet and stick to the diet for one year. Could the mean weight change of this population be negative? If so, then, on average, we could say that people do lose weight on this diet.

*Step 1: Hypothesize*
Let $\mu$ represent the mean weight change of the population.

$$H_0: \mu = 0$$

$$H_a: \mu < 0$$

The null hypothesis says that the mean is 0, because the neutral position here is that no change occurs, on average. This is the same as saying that the diet is ineffective, or not different from no diet at all.

   The alternative is that the mean change is negative. This differs from the ice cream cone example, where we only wanted to know whether or not the mean weight was 3.2 ounces. Here we care about the direction of the weight change: Did it go down?

*Step 2: Prepare*

We will test using a 5% significance level.

We need to check the conditions to see whether the *t*-statistic will follow the *t*-distribution (at least approximately).

Condition 1: *Random Sample and Independence.* The subjects in this study were not selected randomly from the population of all dieters. But they were selected randomly from a larger group of dieters, because one-fourth of the subjects in this study were randomly assigned to Weight Watchers and the rest to other diets. We will assume that the researchers took care so that observations are independent.

Condition 2: *Normal.* The distribution of the sample does not look Normal, which makes us suspect that the population distribution is not Normal. But because the sample size is larger than 25, this condition is satisfied.

*Step 3: Compute to compare*

$$SE = \frac{s}{\sqrt{n}} = \frac{5.4}{\sqrt{26}} = 1.0590$$

$$t = \frac{\bar{x} - \mu}{SE} = \frac{-4.6}{1.0590} = -4.34$$

This tells us that our test statistic is 4.34 standard errors below what the null hypothesis expected.

Our p-value here is the area below the observed value, because the alternative cares only whether we get values that are smaller than expected. (Remember, this is a one-tailed hypothesis.) We find the p-value, using technology, to be 0.0001 (Figure 9.16). We use a *t*-distribution with $n - 1 = 25$ degrees of freedom.

► **FIGURE 9.16** The p-value (shaded) for the diet data. The area has been enlarged so that it can be seen.

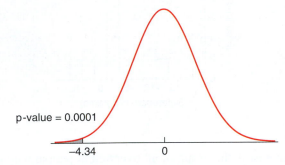

p-value = 0.0001

−4.34       0

*Step 4: Interpret*

The p-value is much smaller than 0.05, so we conclude that the mean weight change is in fact negative, meaning that people do tend to lose weight after one year on this diet.

**TRY THIS!** Exercise 9.29

Tech

If you have access to the original data, then you should use statistical software to perform the *t*-test. Figure 9.17 shows the results from using StatCrunch to carry out the hypothesis test. We had to tell the computer only the null and alternative hypotheses, and it did the rest. Note that, because of rounding, the values are not precisely the same as those we calculated "by hand" in the example.

○ ○ ○          One sample T statistics with data          ⚠

Options

**Hypothesis test results:**

Where: DIET="Weight Watchers" and weightloss <>0

μ : mean of Variable

$H_0 : \mu = 0$

$H_A : \mu < 0$

| Variable | Sample Mean | Std. Err. | DF | T-Stat | P-value |
|---|---|---|---|---|---|
| weightloss | -4.5923076 | 1.0575868 | 25 | -4.3422513 | 0.0001 |

◀ **FIGURE 9.17** Output from StatCrunch, showing the test that the mean weight loss under Weight Watchers was a negative value (Example 8). The small p-value leads us to conclude that, on average, the dieters really lost weight.

**SNAPSHOT   ONE SAMPLE *t*-TEST**

| | |
|---|---|
| **WHAT IS IT?** ▶ | $t = \dfrac{\bar{x} - \mu_0}{SE_{EST}}$,     where     $SE_{EST} = \dfrac{s}{\sqrt{n}}$ |
| **WHAT DOES IT DO?** ▶ | It tests hypotheses about a mean of a single population. |
| **HOW DOES IT DO IT?** ▶ | If the estimated mean differs from the hypothesized value, then the test statistic will be far from 0. Thus, values of the *t*-statistic that are unexpectedly far from 0 (in one direction or the other) discredit the null hypothesis. |
| **HOW IS IT USED?** ▶ | When proposing values about a single population mean. The observed value of the test statistic is compared to a *t*-distribution with $n - 1$ degrees of freedom to calculate the p-value. If the p-value is small, you should be surprised by the outcome and reject the null hypothesis. |

## SECTION 9.4

# Comparing Two Population Means

Does your ability to smell depend on whether you are sitting up or lying down? Do gas prices in Denver differ from those in San Francisco? These questions can be answered, in part, by comparing the means of two populations. Although we could construct separate confidence intervals to estimate each mean (for example, the mean smelling ability of people who are lying down compared to the mean smelling ability of people who are sitting), we can construct more precise estimates by focusing on the difference between the two means.

When comparing two populations, it is important to pay attention to whether the data sampled from the populations are two **independent samples** or are, in fact, one sample of related pairs (paired samples). With **paired (dependent) samples**, if you know the value that a subject has in one group, then you know something about the other group, too. In such a case, you have somewhat less information than you might have if the samples were independent. We begin with some examples to help you see which is which.

Usually, dependence occurs when the objects in your sample are all measured twice (as is common in "before and after" comparisons), or when the objects are related somehow (for example, if you are comparing twins, siblings, or spouses), or when the experimenters have deliberately matched subjects in the groups to have similar characteristics.

## EXAMPLE **10** Independent or Dependent Samples?

Here are four descriptions of research studies.

a. Subjects were tested for their sense of smell twice: once when lying down, once while sitting up. Researchers want to know whether the mean ability to detect smells differs depending on whether one is sitting up or lying down.

b. Men and women each had their sense of smell measured. Researchers want to know whether, typically, men and women differ in their ability to sense smells.

c. Researchers randomly assigned overweight people to one of two diets: Weight Watchers and Atkins. Researchers want to know whether the mean weight loss on Weight Watchers was different from that on Atkins.

d. The numbers of years of education for husbands and wives are compared to see whether the means are different.

**QUESTION**  For each study, state whether it involves two independent samples or paired (that is, dependent) samples.

**SOLUTIONS**

a. This study has two populations: One population consists of people lying down, the other of people sitting up. However, the samples for the population actually consist of just one group of people. Each person has her or his sense of smell measured twice: once sitting up and once lying down. It seems reasonable to expect that a person who has a very good (or very bad) sense of smell while sitting up might also have a better (or worse) than average sense of smell while lying down. Thus, these samples are *paired* (or *dependent*); knowledge about a value in one sample could give us some information about the other value (because the same people are in both samples).

b. The two populations consist of men in one and women in the other. As long as the people are not related, knowledge about a measurement of a man could not tell us anything about any of the women. These are *independent* samples.

c. The two populations are people on the Weight Watchers diet and people on the Atkins diet. We are told that the two samples consist of different people; subjects are randomly assigned to one diet but not to the other. These are *independent* samples.

d. The populations are matched. Each husband is coupled with one particular wife, so the samples are *paired* (or *dependent*).

**TRY THIS!**  Exercise 9.45

As you shall see, we analyze paired data differently from data that come from two independent samples. Paired data are turned into "difference" scores: We simply subtract one value in each pair from the other. We now have just a single variable, and we can analyze it using the one-sample techniques discussed in Section 9.3. The weight loss data in Example 9 was an example of this. Patients were measured before and after the diet, and these two paired values were subtracted to produce a weight loss measure.

## Estimating the Mean Difference with Confidence Intervals (Independent Samples)

The Weight Watchers diet is a very traditional, low-calorie diet. The Atkins diet, on the other hand, limits the amount of carbohydrates. Which diet is more effective? Researchers compared these two diets (as well as two others) by randomly assigning overweight subjects to the two diet groups. The boxplots in Figure 9.18 show summary statistics for the samples' weight losses after one year. The mean weight loss of the Weight Watchers dieters was 4.6 kilograms (about 10 pounds), and the mean loss of the Atkins dieters was 3.9 kilograms (Dansinger et al. 2005).

◀ **FIGURE 9.18** Weight change (kilograms) for people randomly assigned to the Atkins diet or the Weight Watchers diet. The medians are close, but because of the skew in the distributions, the sample means are slightly less close.

To guarantee a particular confidence level—for example, 95%—requires that certain conditions hold:

**Condition 1:** *Random Samples and Independence.* Both samples are randomly taken from their populations, and each observation is independent of any other.

**Condition 2:** *Independent Samples.* The two samples are independent of each other (not paired).

**Condition 3:** *Normal.* The populations are approximately Normal, or the sample size in each sample is at least 25. (In special cases, you might need even larger sample sizes.)

If these conditions hold, we can use the following procedure to find an interval with a 95% confidence level.

The formula for a confidence interval comparing two means, when the data are from independent samples, is the same structure as before:

$$(\text{Estimate}) \pm \text{margin of error}$$

which is

$$(\text{Estimate of difference}) \pm t^*(SE_{\text{estimate of difference}})$$

We estimate the difference with

$$(\text{Mean of first sample}) - (\text{mean of second sample})$$

The standard error of this estimator depends on the sample sizes of both samples and also on the standard deviations of both samples:

$$SE_{\text{EST}} = \sqrt{\frac{s_1^2}{n_1} + \frac{s_2^2}{n_2}}$$

We can put these together into a confidence interval:

### Formula 9.3: Two-Sample *t*-Interval

$$(\bar{x}_1 - \bar{x}_2) + t^* \sqrt{\frac{s_1^2}{n_1} + \frac{s_2^2}{n_2}}$$

The multiplier $t^*$ is based on an approximate *t*-distribution. If a computer is not available, you can conservatively calculate the degrees of freedom for the $t^*$ multiplier as the smaller of $n_1 - 1$ and $n_2 - 1$, but a computer provides a more accurate value.

Choosing the value of $t^*$ (the critical value of *t*) by hand to get your desired level of confidence is tricky. For reasons requiring some pretty advanced mathematics to explain, the sampling distribution is not a *t*-distribution, but only approximately a *t*-distribution. To make matters worse, to get the approximation to be good requires using a rather complex formula to find the degrees of freedom. If you must do these calculations by hand, we recommend taking a "fast and easy" (but also safe and conservative) approach instead. For $t^*$, use a *t*-distribution with degrees of freedom equal to the smaller of $n_1 - 1$ and $n_2 - 1$. That is, use the smaller of the two samples, and subtract 1. For a 95% confidence level, if both samples contain 40 or more observations, you can use 1.96 for the multiplier.

## EXAMPLE 11 Comparing Men's and Women's Senses of Smell

Researchers studying people's sense of smell devised a measure of smelling ability (Lundström et al. 2006). If you score high on this scale, you can detect smells better than others. Although it was not a goal of the study, we can use the data collected by these researchers to determine whether women and men differ in their ability to detect smells. The fact that the researchers felt it was important to record the gender of the study participants suggests that there may be some reason to think this sense might vary by gender.

For this example, we compare men and women whose sense of smell was measured while they were lying down. (The subjects' sense of smell was also measured when they were sitting up.) The summary statistics are

Men:   $\bar{x} = 10.0694$, $s = 3.3583$, $n = 18$
Women: $\bar{x} = 11.1250$, $s = 2.7295$, $n = 18$

Boxplots are shown in Figure 9.19.

▶ FIGURE 9.19 Distribution of smelling ability for men and women. There is slight skew, and there is one potential outlier (indicated by the asterisk).

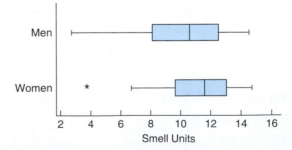

QUESTION  It looks as if women tend to have the more sensitive sense of smell. But could this difference be due to chance? Find a 95% confidence interval for the mean difference in smelling ability between men and women. Interpret the interval. Assume that the participants in this sample are random samples from the population of all adult men and women.

SOLUTION  These data consist of two independent samples: 18 men and 18 women. You might expect that the ability to smell would be Normally distributed across the population, with some people a little better than average and some people a little worse. The boxplots show some skew (and one potential outlier), but with a sample of size 18, it can be hard to tell whether a distribution is Normal on the basis of the box-plot (or histogram). Thus, even though we are not certain that the Normal condition is fulfilled, we will proceed by assuming that it is. Our assumption is based on some theoretical beliefs about how biological traits such as sense of smell are distributed.

Because the sample sizes of both groups are the same (18), our number of degrees of freedom for $t^*$ is conservatively estimated as the smaller of $18 - 1$ and $18 - 1$, which equals 17. For an approximate 95% confidence interval, we use Table 4 in Appendix A to find $t^* = 2.110$.

Let's call the group of men group 1. (It doesn't matter which we choose for group 1 and which for group 2.)

$$\text{Estimate of difference: } 10.0694 - 11.1250 = -1.0556$$

$$m = t^*\sqrt{\frac{s_1^2}{n_1} + \frac{s_2^2}{n_1}} = t^*\sqrt{\frac{3.3583^2}{18} + \frac{2.7295^2}{18}} = t^* 1.0200$$

$$m = 2.110 \times 1.0200 = 2.1522$$

Therefore, a 95% confidence interval is

$$-1.0556 \pm 2.1522, \text{ or about } (-3.2, 1.1)$$

Because the interval contains zero, we cannot rule out the possibility that the mean difference in the population is 0. This suggests that men and women may not differ in their ability to smell.

**TRY THIS!** Exercise 9.51

With access to the full data set, and not just to the summary statistics that were provided in Example 11, we can use statistical software to get more accurate calculations. Figure 9.20 shows StatCrunch output for the 95% confidence interval for the difference of the mean smelling ability of men and that of women. The confidence interval is (−3.1, 1.0), which is slightly different (and narrower) than what we found "by hand." The computer-produced interval is more accurate.

**Tech**

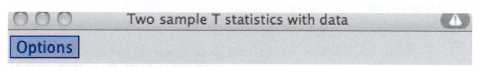

◀ **FIGURE 9.20** StatCrunch output for smelling ability confidence interval.

### Two sample T statistics with data

**Options**

**95% confidence interval results:**

$\mu_1$ : mean of Lying Down where Sex="man"
$\mu_2$ : mean of Lying Down where Sex="woman"
$\mu_1 - \mu_2$ : mean difference
(with pooled variances)

| Difference | Sample Mean | Std. Err. | DF | L. Limit | U. Limit |
|---|---|---|---|---|---|
| $\mu_1 - \mu_2$ | -1.0555556 | 1.0200313 | 34 | -3.1285086 | 1.0173975 |

## Testing Hypotheses about Mean Differences

Hypothesis tests to compare two means from independent samples follow the same structure we discussed in Chapter 8, although now we have more parameters to compare. We show this structure by revisiting the study to investigate whether men and women differ in their ability to detect smells. In Example 11 you found a confidence interval for the difference in the mean smelling ability of men and women. Here we approach the same data with a hypothesis test.

In Example 11 we used boxplots to investigate the shape of the distribution of smelling ability. Here, we examine histograms (Figure 9.21), which show a more detailed picture of the distributions. Both distributions have roughly the same amount of spread, and the histograms show only a little left skew.

▶ **FIGURE 9.21** Distributions of smelling ability for a sample of 18 men and 18 women.

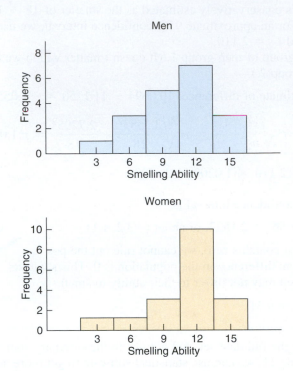

We call men "population 1" and women "population 2." Then the symbol $\mu_1$ represents the mean ability to smell for *all* men (while lying down), and $\mu_2$ represents the mean ability to smell for *all* women (while lying down.)

### Step 1: Hypothesize

$H_0: \mu_1 = \mu_2$ (men and women have the same means for the sense of smell)

$H_a: \mu_1 \neq \mu_2$ (men and women have different means for the sense of smell)

### Step 2: Prepare

The conditions for testing two means are not very different those for from testing one mean and are identical to those for finding confidence intervals of the difference of two means.

Condition 1: *Random Samples and Independent Observations.* Observations are taken at random from two populations, producing two samples. Observations within a sample are independent of one another, which means that knowledge of one value tells us nothing about other observed values in that sample.

Condition 2: *Independent Samples.* The samples are independent of each other. Knowledge about a value in one sample does not tell us anything about any value in the other sample.

Condition 3: *Normal.* Both populations are approximately Normal, or both sample sizes are 25 or more. (In extreme situations, larger sample sizes may be required.)

We might expect the population distributions to be Normally distributed, because measures of biological traits, such as sense of smell, are often Normally distributed.

The sample distributions show a little bit of left skew, but even Normal populations sometimes produce skewed samples when the sample sizes are small, as they are here. The skew is not so great that we would doubt that the populations are Normally distributed, so we assume that condition 2 holds. We must assume that condition 1 (random samples and independence) holds, because we don't know whether the people were randomly sampled. (In fact, they probably were not, because this is rather difficult to do for such studies.) We will assume it is true, understanding that if these people are not representative of the population, we might have substantial bias in our results. It is also safe to assume that condition 3, independent samples, holds, because these are two distinct groups of people. (If researchers had sampled married couples, for example, then this assumption would have been violated.)

Another step in our preparation is to choose a significance level. It is common to use $\alpha = 0.05$, and we will do so for this example.

### Step 3: Compute to compare

The test statistic used to test this hypothesis is based on the difference between the sample means. Basically, the test statistic measures how far away the observed difference in sample means is from the hypothesized difference in population means. Yes, you guessed it: The distance is measured in terms of standard errors.

$$t = \frac{(\text{difference in sample means} - \text{what null hypothesis says the difference is})}{SE_{\text{EST}}}$$

Using the test statistic is made easier by the fact that the null hypothesis almost always says that the difference is 0.

> **Details**
>
> **Null Hypotheses for Two Means**
> Mathematically, we can easily adjust our test statistic if the null hypothesis claims that the difference in means is some value other than 0. But in almost all scientific, business, and legal settings, the null hypothesis value will be 0.

---

**Formula 9.4: Two-Sample *t*-Test**

$$t = \frac{\bar{x}_1 - \bar{x}_2 - 0}{SE_{\text{EST}}}, \quad \text{where} \quad SE_{\text{EST}} = \sqrt{\frac{s_1^2}{n_1} + \frac{s_2^2}{n_2}}$$

If conditions hold, the test statistic follows an approximate *t*-distribution, where the degrees of freedom are conservatively estimated to be the smaller of $n_1 - 1$ and $n_2 - 1$.

---

To compare the sense of smell between men and women:

Difference in sample means $= \bar{x}_1 - \bar{x}_2 = 10.0694 - 11.125 = -1.0556$

$$SE_{\text{EST}} = \sqrt{\frac{s_1^2}{n_1} + \frac{s_2^2}{n_2}} = \sqrt{\frac{3.3583^2}{18} + \frac{2.7295^2}{18}} = 1.0200$$

$$t = \frac{-1.0556}{1.0200} = -1.03$$

This statistic tells us that the observed difference, $-1.0556$, is about one standard error below what the null hypothesis told us to expect.

We now measure how surprising this is if the null hypothesis is true. To do this, we need to know the sampling distribution of the test statistic *t*, because we measure surprise by finding the probability that if the null hypothesis is true, we would see a value as extreme as or more extreme than the value we observed. In other words, we need to find the p-value.

If the conditions listed in the Prepare step hold, then *t* follows, approximately, a *t*-distribution with minimum $(n_1 - 1, n_2 - 1)$ degrees of freedom. This approximation can be made even better by adjusting the degrees of freedom, but this adjustment is, for most cases, too complex for a "by hand" calculation. For this reason, we recommend using technology for two-sample hypothesis tests, because you will get more accurate p-values.

▲ **FIGURE 9.22** The shaded area represents the p-value for this test: the probability of getting a *t*-statistic more than 1.03 standard errors away from 0 when the null hypothesis is true.

Both sample sizes are 18, so $n_1 = 18$ and $n_2 = 18$. Thus we use $18 - 1 = 17$ for the degrees of freedom.

Our alternative hypothesis is two-tailed and says that the true difference might be much bigger than 0 or much smaller than 0. We therefore find the area under both tails of the *t*-distribution. Figure 9.22 shows this probability as the shaded area under the appropriate *t*-distribution.

The p-value (found with technology) is 0.3083.

**Step 4: Interpret**

Again, we compare the p-value to the significance level, $\alpha$. If the p-value is less than or equal to $\alpha$, we reject the null hypothesis. In this example, the p-value, 0.308, is bigger than 0.05. If men and women have the same mean smelling ability, then test statistics as extreme as ours occur fairly often. Our test statistic was not a surprise to the null hypothesis—it is pretty much what the null told us we would get. Therefore, we do not reject the null hypothesis.

The previous analysis was done using only the summary statistics provided. If you have the raw data, then you should use computer software to do the analysis. You will get more accurate values and save yourself lots of time. Figure 9.23 shows StatCrunch output for testing whether the mean sense of smell is different for men than for women.

▶ **FIGURE 9.23** StatCrunch output in a test of whether, on average, men's ability to smell is different from women's.

Tech

Two sample T statistics with data

**Options**

**Hypothesis test results:**

$\mu_1$ : mean of Lying Down where Sex="man"
$\mu_2$ : mean of Lying Down where Sex="woman"
$\mu_1 - \mu_2$ : mean difference
$H_0 : \mu_1 - \mu_2 = 0$
$H_A : \mu_1 - \mu_2 \neq 0$
(without pooled variances)

| Difference | Sample Mean | Std. Err. | DF | T-Stat | P-value |
|---|---|---|---|---|---|
| $\mu_1 - \mu_2$ | -1.0555556 | 1.0200313 | 32.636784 | -1.0348266 | 0.3084 |

> **! Caution**
>
> **Don't Accept!**
> Remember from Chapter 8 that we do not "accept" the null hypothesis. It is possible that the sample size is too small (the test has low power) to detect the real difference that exists. Instead, we say that there is not enough evidence for us to reject the null.

> **! Caution**
>
> **Don't Pool**
> When using software to do a two-sample *t*-test, make sure it does the unpooled version. You might have to tell the software explicitly. The unpooled version is more accurate in more situations than the pooled version.

## Into the Pool

Some software packages, and some textbooks too, provide for another version of this *t*-test called the "pooled two-sample *t*-test." We have presented the unpooled version (you can see this in the StatCrunch output above the table, where it says "without pooled variances"). The unpooled version is preferred over the other version because the pooled version works only in special circumstances (when the population standard deviations are equal). The unpooled version works reasonably well in all situations, as long as the listed conditions hold.

## Test of Two Means: Dependent Samples

With **paired samples**, we turn two samples into one. We do this by finding the difference in each pair.

Recall the study to evaluate smelling ability. Earlier, you saw there were no differences in mean smelling ability between men and women. Are there differences, however, that depend on position? Researchers carried out this study to determine whether people differ in their ability to smell depending on whether they are sitting up or lying down.

## SNAPSHOT TWO SAMPLE *t*-TEST (FROM INDEPENDENT SAMPLES)

**WHAT IS IT?** ▶ A procedure for deciding whether two means, estimated from independent samples, are different. The test statistic used is

$$t = \frac{\bar{x}_1 - \bar{x}_2 - 0}{SE_{EST}}, \text{ where } SE_{EST} = \sqrt{\frac{s_1^2}{n_1} + \frac{s_2^2}{n_2}}$$

**WHAT DOES IT DO?** ▶ Provides us with a decision on whether to reject the null hypothesis that the two means are the same and lets us do so knowing the probability that we are making a mistake.

**HOW DOES IT DO IT?** ▶ Compares the observed difference in sample means to 0, the value we expect if the population means are equal.

**HOW IS IT USED?** ▶ The observed value of the test statistic can be compared to a *t*-distribution.

To test this idea, they measured each subject's sense of smell twice: once while sitting and once while lying down. This is a test of two means, because we are interested in two populations:

Population 1: All people lying down; $\mu_1$ represents the mean ability to smell while lying down.
Population 2: All people sitting up; $\mu_2$ represents the mean ability to smell while sitting up.

However, even though we have two populations, we do not have two *independent* samples. Rather, we have one sample of people who were measured twice. Thus we can change the problem slightly so that, instead of measuring the ability to smell in each position, we measure the *difference* in ability when a person goes from sitting up to lying down.

The first few lines of the data are shown in Table 9.4a.

| Subject Number | Sex | Sitting | Lying |
|---|---|---|---|
| 1 | woman | 13.5 | 13.25 |
| 2 | woman | 13.5 | 13 |
| 3 | woman | 12.75 | 11.5 |
| 4 | man | 12.5 | 12.5 |

◀ **TABLE 9.4a** Smelling ability for the first four people sitting and lying.

We create a new variable, call it *difference*, and define it to be the difference between smelling ability sitting up and smelling ability lying down. We show this new variable in Table 9.4b.

| Subject Number | Sex | Sitting | Lying | Difference |
|---|---|---|---|---|
| 1 | woman | 13.5 | 13.25 | 0.25 |
| 2 | woman | 13.5 | 13 | 0.50 |
| 3 | woman | 12.75 | 11.5 | 1.25 |
| 4 | man | 12.5 | 12.5 | 0 |

◀ **TABLE 9.4b** Difference between smelling ability while sitting up and smelling ability while lying down.

Our hypotheses are now about just one mean: the mean of *difference*.

$$H_0: \mu_{\text{difference}} = 0 \quad (\text{or } \mu_{\text{sitting}} = \mu_{\text{lying}})$$

$$H_a: \mu_{\text{difference}} \neq 0 \quad (\text{or } \mu_{\text{sitting}} \neq \mu_{\text{lying}})$$

Our test statistic is the same as for the one-sample *t*-test:

$$t = \frac{\overline{x}_{\text{difference}} - 0}{SE_{\text{difference}}} \quad \text{where} \quad SE_{\text{difference}} = \frac{s_{\text{difference}}}{\sqrt{n}}$$

We find $\overline{x}$ by averaging the change variable: $\overline{x} = 0.8681$.

We find $s_{\text{difference}}$ by finding the standard deviation of the difference variable: $s = 2.3946$.

There were 36 participants altogether, so

$$SE_{\text{EST}} = \frac{2.3946}{\sqrt{36}} = 0.3991$$

and then

$$t = \frac{0.8681}{0.3991} = 2.18$$

▲ **FIGURE 9.24** A *t*-distribution with $n - 1 = 35$ degrees of freedom. The shaded area represents the p-value for the smell study (sitting vs. lying) and illustrates that if there is no difference in our ability to smell, then our outcome was very unusual and surprising.

To find the p-value, we use a *t*-distribution (assuming the conditions for a one-sample *t*-test hold) with $n - 1$ degrees of freedom, where $n$ is the number of data pairs. Figure 9.24 shows a *t*-distribution with 35 degrees of freedom. The shaded areas represent the (two-sided) p-value of 0.0365.

Because the p-value is less than 0.05, we reject the null hypothesis. There is evidence that our sense of smell is affected by the position of our body.

## Paired *t*-Test vs. Two-Sample *t*-Test

If you have paired data and (incorrectly) do the two-sample *t*-test, you will generally get a p-value that is too big. Figures 9.25 and 9.26 compare the results of doing a two-sample *t*-test on paired data (Figure 9.25) and doing a paired *t*-test on the same data (Figure 9.26.) Note that the test statistic is much larger when you (correctly) use the paired *t*-test to test the paired data; as a result, the p-value is much smaller.

▶ **FIGURE 9.25** TI-83/84 output for two-sample *t*-test.

▲ **FIGURE 9.26** TI-83/84 output for paired *t*-test.

The tests produce different values because when we convert the paired data to differences, the resulting differences have a smaller standard deviation than does either sample by itself. This smaller standard deviation leads to a smaller standard error. So even though the numerators of both *t*-statistics (the paired and the two-sample) are the same, the paired *t*-statistic is larger because its denominator is smaller. In Chapter 12 you'll learn more about why studies designed with paired data can be powerful.

## SNAPSHOT PAIRED t-TEST (DEPENDENT SAMPLES)

**WHAT IS IT?** ▶ A procedure for deciding whether two dependent (paired) samples have different means. Each pair is converted to a difference. The test statistic is the same as for the one-sample *t*-test, except that the null-hypothesis value is 0:

$$t = \frac{\bar{x}_{\text{difference}} - 0}{SE_{\text{difference}}}$$

**WHAT DOES IT DO?** ▶ Lets us make decisions about whether the means are different, while knowing the probability that we are making a mistake.

**HOW DOES IT DO IT?** ▶ The test statistic compares the observed average difference, $\bar{x}_{\text{difference}}$, with the average difference we would expect if the means were the same: 0. Large values discredit the null hypothesis.

**HOW IS IT USED?** ▶ If the required conditions hold, the value of the observed test statistic can be compared to a *t*-distribution with $n - 1$ degrees of freedom.

## SECTION 9.5

# Overview of Analyzing Means

We hope you've been noticing a lot of repetition. The hypothesis test for two means is very similar to the test for one mean, and the hypothesis test for paired data is really a special case of the one-sample *t*-test. Also, the hypothesis tests use almost the same calculations as the confidence intervals, and they impose the same conditions, arranged slightly differently.

All the test statistics (for one proportion, for one mean, for two means, and for two proportions) have this structure:

$$\text{Test statistic} = \frac{(\text{estimated value}) - (\text{null hypothesis value})}{SE}$$

All the confidence intervals have this form:

$$\text{Estimated value} \pm (\text{multiplier}) \, SE_{\text{EST}}$$

Not all confidence intervals used in statistics have this structure, but most that you will encounter do.

The method for computing a p-value is the same for all tests, although different distributions are used for different situations. The important point is to pay attention to the alternative hypothesis, which tells you whether you are finding a two-tailed or a one-tailed (and *which* tail) p-value.

## Confidence Intervals and Hypothesis Tests

In the preceding examples, we reached exactly the same conclusion about men and women and their sense of smell, whether we used a confidence interval or a hypothesis test. This is no coincidence. In fact, it has to be that way. If you have a *two-tailed* alternative hypothesis, then you actually have two choices for how to do the test. Both choices always reach the same conclusion.

| Confidence Level | Equivalent $\alpha$ (Two-Tailed) |
|---|---|
| 99% | 0.01 |
| 95% | 0.05 |
| 90% | 0.10 |

▲ **TABLE 9.5** Equivalences between confidence intervals and tests with two-tailed alternative hypotheses.

Choice 1: Perform the hypothesis test as described above with significance level $\alpha$.

Choice 2: Find a $(1 - \alpha) \times 100\%$ confidence interval (using methods given above). Reject the null hypothesis if the value does *not* appear in the interval.

> **KEY POINT** A 95% confidence interval is equivalent to a test with a two-tailed alternative with a significance level of 0.05. Table 9.5 shows some other equivalences. All are true only for *two-tailed* alternative hypotheses.

## EXAMPLE 12 Calcium Levels in the Elderly

The boxplots in Figure 9.27 show the results of a study to determine whether calcium levels differ substantially between senior men and senior women (all older than 65 years). Calcium is associated with strong bones, and people with low calcium levels are believed to be more susceptible to bone fractures. The researchers carried out a hypothesis test to see whether the mean calcium levels for men and women were the same. Figure 9.28 shows the results. Calcium levels (the variable *cammol*) are measured in millimoles per liter (mmol/L).

▶ **FIGURE 9.27** Boxplots of calcium levels (mmol/L) for males and females.

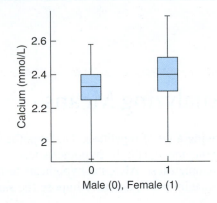

▶ **FIGURE 9.28** StatCrunch output for testing whether mean calcium levels in men and women differ. The difference estimated is mean of females minus mean of males.

**Two sample T statistics with data**

**Options**

**Hypothesis test results:**

$\mu_1$ : mean of cammol where female=1

$\mu_2$ : mean of cammol where female=0

$\mu_1 - \mu_2$ : mean difference

$H_0 : \mu_1 - \mu_2 = 0$

$H_A : \mu_1 - \mu_2 \neq 0$

(without pooled variances)

| Difference | Sample Mean | Std. Err. | DF | T–Stat | P-value |
|---|---|---|---|---|---|
| $\mu_1 - \mu_2$ | 0.07570534 | 0.019790715 | 168.45009 | 3.825296 | 0.0002 |

**QUESTIONS**

a. Assuming that all conditions necessary for carrying out *t*-tests and finding confidence intervals hold, what conclusion should the researchers make on the basis of this output? Use a significance level of 0.05.

b. Suppose the researchers calculate a confidence interval for the difference of the two means. Will this interval include the value 0? If not, will it include all negative values or all positive values? Explain.

**SOLUTIONS**

a. The p-value, 0.0002, is less than 0.05, so the researchers should reject the null hypothesis and conclude that men and women have different calcium levels.

b. Because we rejected the null hypothesis, we know that confidence interval cannot include the value 0. If it did, then 0 would be a plausible difference between the means, and our hypothesis test says 0 is not plausible. The estimated difference between the two means is, from the output, 0.0757. Because this value is positive, and because the interval cannot include 0, all values in the interval must be positive.

**TRY THIS!**  Exercise 9.65

## Hypothesis Test or Confidence Interval?

If you can use either a confidence interval or a hypothesis test, how do you choose? First of all, remember that these two techniques produce the same results only when you have a two-tailed alternative hypothesis, so you need only make the choice when you have a two-tailed alternative.

These two approaches answer slightly different questions. The confidence interval answers the questions "What's the estimated value? And how much uncertainty do you have in this estimate?" Hypothesis tests are designed to answer the question "Is the parameter's value one thing, or another?"

For many situations, the confidence interval provides much more information than the hypothesis test. It not only tells us whether or not we should reject the null hypothesis but also gives us a plausible range for the population value. The hypothesis test, on the other hand, simply tells us whether to reject or not (although it does give us the p-value, which helps us see just how unusual our result is if the null hypothesis is true).

For example, in our test of whether people have a different sense of smell depending on whether they are sitting up or lying down, we rejected the null hypothesis and concluded that the sense of smell was affected by body position. But that is all we can say with a two-tailed test: The means are different. If we want to ask, "How much does the sense of smell change when we sit up?" then a confidence interval for the difference in these two means helps. The 95% confidence interval (smelling ability while sitting up minus smelling ability while lying down) is (0.06, 1.68). This interval tells us that the ability to smell doesn't change very much with body position.

## CASE STUDY REVISITED

The researchers reported 95% confidence intervals for the mean IQ of three-year-old children whose mothers took one of four drugs for epilepsy, as shown in Table 9.6.

◄ **TABLE 9.6** Confidence intervals for IQs.

| Drug | 95% CI |
|------|--------|
| carbamazepine | (95, 101) |
| lamotrigine | (98, 104) |
| phenytoin | (94, 104) |
| valproate | (88, 97) |

The researchers could not observe all women on these drugs, so they based their observations on a random sample. If we think of these women as being sampled randomly, then the confidence intervals represent a range of plausible values for the mean IQ for the population of all three-year-olds whose mothers took these drugs.

It is helpful to display these confidence intervals graphically (as the researchers do in their paper), as shown in Figure 9.29.

◀ **FIGURE 9.29** Four confidence intervals for mean IQs of children from mothers taking different drugs for epilepsy.

From the figure, we see that the confidence interval for valproate does not overlap with that for lamotrigine. This suggests to us visually that it is *not* plausible that the mean IQ for children whose mothers took these drugs could be the same. The confidence interval for valproate has little overlap with the others, which makes us wonder how much different the mean IQs for the children of valproate users are from those for the children of the other mothers.

For this reason, we need to focus our attention on the differences between the means, not on the individual values for the means. If we do this, as the researchers did, we will find the confidence intervals shown in Table 9.7.

◀ **TABLE 9.7** Confidence intervals for differences in mean IQs.

| Difference | 95% CI |
|---|---|
| carbamazepine – valproate | (0.6, 12.0) |
| lamotrigine – valproate | (3.1, 14.6) |
| phenytoin – valproate | (0.2, 14.0) |

The first interval tells us that the difference between the mean IQ of those under carbamazepine and that of those under valproate could be as small as 0.6 IQ point or as large as 12.0 points. None of these intervals contains 0. This tells us that if we do a hypothesis test that the means are the same, we will have to reject the null hypothesis and conclude that the means are different.

# EXPLORING STATISTICS
## CLASS ACTIVITY

## Pulse Rates

| GOALS | MATERIALS |
| --- | --- |
| Learn to use a confidence interval (and/or a hypothesis test) to compare the means of two populations. | • A clock on the wall with a second hand or a watch with a second hand (for the instructor).<br>• A computer or calculator |

**ACTIVITY**

You will take your pulse rate before and after an activity to measure the effect of the activity on your pulse rate and/or to compare pulse rates between groups.

Try to find your pulse; it is usually easiest to find in the neck on one side or the other.

After everyone has found his or her pulse, your instructor will say, "Start counting," and you will count beats until the instructor says, "Stop counting." If your instructor uses a 30-second interval, double the count to get beats per minute.

Option A: Breathe in and out ten times, taking slow and deep breaths. Now measure your pulse rate again.

Option B: Stand up and sit down five times and then measure your pulse rate again.

Your instructor will collect this data and display the values (before and after each activity) for the class.

**BEFORE THE ACTIVITY**

1. Try finding your pulse (in your neck) to see how to do it.

2. Do you think that either activity (breathing slowly or standing up and sitting down) will change your heart rate? If so, by how much, and will it raise or lower it?

3. How would you measure the typical pulse rate of the class before and after the activity? How would you measure the change in pulse rate after each activity?

4. Do you think men and women have different mean heart rates before the activities?

5. Do you think the change in pulses will be different for men and women? (*Note:* If the class includes only one gender, your instructor may ask you to compare athletes to nonathletes or the taller half to the shorter half.)

**AFTER THE ACTIVITY**

1. State a pair of hypotheses (in words) for testing whether the breathing activity changes the mean pulse rate of the class. Do the same for the standing and sitting activity.

2. State a pair of hypotheses (in words) for whether men and women have the same mean resting pulse rate.

3. Calculate a 95% confidence interval for the change in pulse rates after the activity. What does this confidence interval tell us about the effect of the activity on the mean heart rate? Suppose you did a two-sided hypothesis test. On the basis of the confidence intervals, can you tell what the conclusions of the hypothesis test will be?

## CHAPTER 9

## CHAPTER REVIEW

## Key Terms

You may want to review the following terms, which were introduced in Chapters 7 and 8.

**Chapter 7**: statistic, estimator, bias, precision, sampling distribution, standard error, confidence interval, confidence level, margin of error
**Chapter 8**: null hypothesis, alternative hypothesis, significance level, test statistic, p-value, one-tailed hypothesis, two-tailed hypothesis

| | | | |
|---|---|---|---|
| Bias, *389* | Standard error, *390* | *t*-Distribution, *397* | Independent samples, *411* |
| Precision, *389* | Central Limit Theorem | Degrees of freedom (df), *397* | Paired (dependent) samples, *411* |
| Sampling distribution, *390* | (CLT), *392* | Confidence intervals, *398* | |
| Unbiased estimator, *390* | *t*-Statistic, *396* | Confidence level, *398* | |

## Learning Objectives

After reading this chapter and doing the assigned homework problems, you should

- Understand when the Central Limit Theorem for sample means applies and know how to use it to find approximate probabilities for sample means.

- Know how to test hypotheses concerning a population mean and concerning the comparison of two population means.

- Understand how to find, interpret, and use confidence intervals for a single population mean and for the difference of two population means.

- Understand the meaning of the p-value and of significance levels.

- Understand how to use a confidence interval to carry out a two-tailed hypothesis test for a population mean or for a difference of two population means.

## Summary

The sample mean gives us an unbiased estimator of the population mean, provided that the observations are sampled randomly from a population and are independent of each other. The precision of this estimator, measured by the standard error (the standard deviation of the sampling distribution), improves as the sample size increases. If the population distribution is Normal, then the sampling distribution is Normal also. Otherwise, according to the Central Limit Theorem, the sampling distribution is approximately Normal, although for small sample sizes the approximation can be very bad. If the population distribution is not Normal, we recommend that you use a sample size of 25 or more.

Although we did not go into the (fairly complex) mathematics, our ability to measure confidence levels (which tell us how well confidence intervals perform), p-values, and significance levels ($\alpha$) depends on the Central Limit Theorem (CLT). If the conditions for applying the CLT are not satisfied, then our reported values for these performance measures may be wrong.

Confidence intervals are used to provide estimates of parameters, along with a measure of our uncertainty in that estimate. The confidence intervals in this chapter differ only a little from those for proportions (Chapter 7). In both chapters, all are of the form

$$\text{Estimate} \pm \text{margin of error}$$

One thing that is different is that now you must decide whether your two samples are independent or paired before performing your analysis.

Another difference between confidence intervals for means and those for proportions is that the multiplier in the margin of error is based on the *t*-distribution, not on the Normal distribution.

By this point you have learned several different hypothesis tests, including the *z*-test for one-sample proportions and for two-sample proportions; and the t-test for one-sample mean, for two means from independent samples, and for two means from dependent samples. (You may also find pooled and unpooled versions of

the two-mean independent-samples *t*-test, but you should always use the unpooled version.) It is important to learn which test to choose for the data you wish to analyze.

Hypothesis tests follow the structure described in Chapter 8, and, just as for confidence intervals, you must decide whether you have independent samples or paired samples. Tests of two means based on independent samples are based on the difference between the means. The null hypothesis is (almost) always that the difference is 0. The alternative hypothesis depends on the research question.

To find the p-value for a test of two means, use the *t*-distribution. To use the *t*-distribution, you must know the degrees of freedom (df), and this depends on whether you are doing a test for one mean (df $= n - 1$); two means from independent samples (use your computer, or, if working by hand, use the df for the smaller of $n_1 - 1$, $n_2 - 1$); or two means from paired data (number of pairs $-1$).

### Formulas

Samples are selected randomly from each population and are independent. Population distributions are Normal, or if not, sample sizes need to be 25 or bigger for each sample.

**Formula 9.1: One-Sample Confidence Interval for Mean**

$$\bar{x} \pm m$$

where $m = t^* SE_{\text{EST}}$ and $SE_{\text{EST}} = \dfrac{s}{\sqrt{n}}$

The multiplier $t^*$ is a constant that is used to fine-tune the margin of error so that it has the level of confidence we want. It is chosen on the basis of a *t*-distribution with $n - 1$ degrees of freedom. $SE_{\text{EST}}$ is the estimated standard error.

**Paired:** $\bar{x}_{\text{difference}} \pm m$, where $m = t^* SE_{\text{EST}}$ and

$$SE_{\text{EST}} = \frac{s_{\text{difference}}}{\sqrt{n}}$$

(where $\bar{x}_{\text{difference}}$ is the average difference, $s_{\text{difference}}$ is the standard deviation of the differences, and $n$ is the number of data pairs)

**Formula 9.2: The One-Sample $t$-Test for Mean**

$$t = \frac{\bar{x} - \mu}{SE_{\text{EST}}}, \text{ where } SE_{\text{EST}} = \frac{s}{\sqrt{n}} \text{ and, if conditions hold,}$$

$t$ follows a $t$-distribution with df $= n - 1$

**Paired:** $t = \dfrac{\bar{x}_{\text{difference}} - 0}{SE_{\text{difference}}}$, where $SE_{\text{difference}} = \dfrac{s_{\text{difference}}}{\sqrt{n}}$

(where $\bar{x}_{\text{difference}}$ is the average difference, $s_{\text{difference}}$ is the standard deviation of the differences, and $n$ is the number of data pairs)

If conditions hold, $t$ follows a $t$-distribution with degrees of freedom df $= n - 1$ (where $n$ is the number of data pairs)

**Formula 9.3: Two-Sample Confidence Interval**

$$(\bar{x}_1 - \bar{x}_2) + t^*\sqrt{\frac{s_1^2}{n_1} + \frac{s_2^2}{n_2}}$$

If conditions hold, $t^*$ is based on a $t$-distribution. If no computer is available, the degrees of freedom are conservatively estimated as the smaller of $n_1 - 1$ and $n_2 - 1$.

**Formula 9.4: Two-Sample $t$-Test (Unpooled)**

$$t = \frac{\bar{x}_1 - \bar{x}_2}{SE_{\text{EST}}}, \quad \text{where } SE_{\text{EST}} = \sqrt{\frac{s_1^2}{n_1} + \frac{s_2^2}{n_2}}$$

If conditions hold, $t$ follows an approximate $t$-distribution. If no computer is available, the degrees of freedom, df, are conservatively estimated as the smaller of $n_1 - 1$ and $n_2 - 1$
(Do not use the pooled version.)

## Sources

*The Chronicle of Higher Education, Facts & Figures.* 2008. http://chronicle .com/premium/stats/tuition/2008 (accessed July 2009).

Dansinger, M., J. Gleason, J. Griffith, H. Selker, and E. Schaefer. 2005. Comparison of the Atkins, Ornish, Weight Watchers, and Zone diets for weight loss and heart disease risk reduction: A randomized trial. *Journal of the American Medical Association* 293(1), 43–53.

Kaplan, D. 2009. www.cherryblossom.org/results (accessed July 2009).

Lundström, J., Boyle, J., and Jones-Gotman, M. 2006. Sit up and smell the roses better: olfactory sensitivity to phenyl ethyl alcohol is dependent on body position", *Chemical Senses*, 31(3), March 2006: 249-252. doi: 10.1093/chemse/bjj025.

Meador, K. J., et al. 2009. Cognitive function at 3 years of age after fetal exposure to antileptic drugs, *New England Journal of Medicine* 360(16), 1597–1605.

National Health and Nutrition Examination Survey (NHANES). Centers for Disease Control and Prevention (CDC). National Center for Health Statistics (NCHS). National Health and Nutrition Examination Survey Data. Hyattsville, MD: U.S. Department of Health and Human Services, Centers for Disease Control and Prevention, 2003–2004.

# SECTION EXERCISES

## SECTION 9.1

**9.1 Library Books** An exhaustive cataloging of all hardbound books at a college library revealed that the mean age was 54.4 years, with a standard deviation of 15.5 years. Recently, the librarian commissioned a study to determine the condition of the books. A random sample of 500 books was assembled, and the mean age of this collection was found to be 50.2 years, with a standard deviation of 18.0 years.

  a. Which of these numerical values are parameters?

  b. Which of these numerical values are statistics?

**9.2 American Time Use** In 2010, the American Time Use Survey reported that "employed persons worked an average of 7.5 hours on the days they worked." Is the number 7.5 a statistic or a parameter? Explain. (Source: http://www.bls.gov/tus/, accessed January 16, 2011)

**9.3 Exam Scores** The distribution of the scores on a certain exam is N(70, 10), which means that the exam scores are Normally distributed with a mean of 70 and standard deviation of 10.

  a. Sketch the curve and label, on the $x$-axis, the position of the mean, the mean plus or minus one standard deviation, the mean plus or minus two standard deviations, and the mean plus or minus three standard deviations.

  b. Find the probability that a randomly selected score will be bigger than 80. Shade the region under the Normal curve whose area corresponds to this probability.

**9.4 Exam Scores** The distribution of the scores on a certain exam is N(70, 10), which means that the exam scores are Normally distributed with a mean of 70 and standard deviation of 10.

  a. Sketch the curve and label, on the $x$-axis, the position of the mean, the mean plus or minus one standard deviation, the mean plus or minus two standard deviations, and the mean plus or minus three standard deviations.

  b. Find the probability that a randomly selected score will be between 50 and 90. Shade the region under the Normal curve whose area corresponds to this probability.

**9.5 Bats** A biologist is interested in studying the effects that applying insecticide to a fruit farm has on the local bat population. She collects 23 bats from a grove of fruit trees with the insecticide and finds the mean weight of this sample to be 503.4 grams. Assuming the selected bats are a random sample, she concludes that because the sample mean is an unbiased estimator of the population mean, the mean weight of bats in the population is also 503.4 grams. Explain why this is an incorrect interpretation of what it means to have an unbiased estimator.

**9.6 Cellphone Calls** Answers.com claims that the mean length of all cell phone conversations in the United States is 3.25 minutes (3 minutes and 15 seconds). Assume that this is correct, and also assume that the standard deviation is 4.2 minutes. (Source: wiki.answers.com, accessed January 16, 2011)

* a. Describe the shape of the distribution of the length of cell phone conversations in this population. Do you expect it to be approximately Normally distributed, right-skewed, or left-skewed? Explain your reasoning.

b. Suppose that, using a phone company's records, we randomly sample 100 phone calls. We calculate the mean length from this sample and record the value. We repeat this thousands of times. What will be the (approximate) mean value of the distribution of these thousands of sample means?

c. Refer to part b. What will be the standard deviation of this distribution of thousands of sample means?

**9.7 Retirement Income** Several times during the year, the U.S. Census Bureau takes random samples from the population. One such survey is the American Community Survey. The most recent such survey, based on a large (several thousand) sample of randomly selected households, estimates the mean retirement income in the United States to be $21,201 per year. Suppose we were to make a histogram of all of the retirement incomes from this sample. Would the histogram be a display of the population distribution, the distribution of a sample, or the sampling distribution of means?

**9.8 Time Employed** A human resource manager for a large company takes a random sample of 50 employees from the company database. She calculates the mean time that they have been employed. She records this value and then repeats the process: She takes another random sample of 50 names and calculates the mean employment time. After she has done this 1000 times, she makes a histogram of the mean employment times. Is this histogram a display of the population distribution, the distribution of a sample, or the sampling distribution of means?

TRY **9.9 Women's Weights (Example 1)** Twenty-year-old women's weights have a population mean, $\mu$, of 128 pounds and a population standard deviation, $\sigma$, of 20 pounds (http://www.kidsgrowth.com). The distribution is right-skewed. Suppose we take a random sample of 100 twenty-year-old women's weights.

a. What value should we expect for the mean weight of this sample of 100 women? Why?

b. Of course, the actual sample mean will not be exactly the value you gave in part a. The amount it typically differs from this value is given by the standard error. What is the standard error for a sample mean taken from this population?

**9.10 Men's Weights** Twenty-year-old men's weights have a population mean, $\mu$, of 155 pounds and a population standard deviation, $\sigma$, of 22 pounds (http://www.kidsgrowth.com). The distribution is right-skewed. Suppose we take a random sample of 100 twenty-year-old men's weights.

a. What value should we expect for the mean weight of this sample of 100 men? Why?

b. Of course, the value that we actually see for the sample mean based on 100 men will not be exactly he same as your answer to part a. By how much, typically, will it vary from the value in part a? In other words, what is the standard error?

# SECTION 9.2

TRY **9.11 Women's Weights (Example 2)** Twenty-year-old women's weights have a population mean of 128 pounds and a population standard deviation of 20 pounds. The distribution is right-skewed. Suppose we take a random sample of 100 twenty-year-old women's weights. You will find the probability that the sample mean will be more than 4 pounds from the population mean. The Normal curve shown in figure (A) is the sampling distribution of means and therefore is not skewed.

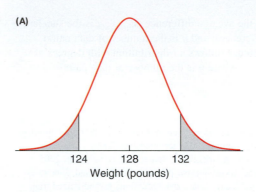

(A)

124    128    132
Weight (pounds)

a. The sample is random. Why is the sample size large enough for us to use the Central Limit Theorem for means? Explain.

b. What are the mean and standard deviation of the sampling distribution. (*Hint:* Refer to Exercise 9.9.)

c. Find the probability that the sample mean will be more than 132 pounds or less than 124 pounds, as shown in figure (A). Figure (B) shows TI-83/84 output for finding the probability that the mean will be *between* 124 and 132, but you want the probability that the mean will be less than 124 or more than 132.

(B)

```
normalcdf(124,13
2,128,2)
          .954499876
■
```

TI-83/84 Normal Curve Output

* **9.12 Men's Weights** Twenty-year-old men's weights have a population mean of 155 pounds and a population standard deviation of 22 pounds (http://www.kidsgrowth.com). The distribution is right-skewed. Suppose we take a random sample of 100 twenty-year-old men's weights.

a. Explain why the Central Limit Theorem is applicable.

b. Sketch the sampling distribution of means. Label the mean, the mean plus or minus one standard deviation, the mean plus or minus two standard deviations, and the mean plus or minus three standard deviations. (*Hint:* The standard deviation for the distribution of means—which is the standard error—was calculated in Exercise 9.10.)

c. Using your sketch of the sampling distribution, shade the area that corresponds to the probability that the sample mean will be between 152.8 pounds and 157.2 pounds.

d. What is this probability?

**TRY** **9.13 Length of Cell Phone Calls (Example 3)** One histogram shows the distribution of the length of cell phone calls for one of the authors in one month, and the other shows many sample means, in which each is the mean length for 10 randomly selected calls from the author's phone records for that month. Which is which? Explain.

**(A)**

**(B)**

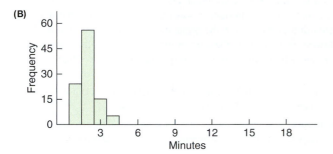

**9.14 Used Car Ages** One histogram shows the distribution of ages of all 638 used cars for sale in the *Ventura County Star* Sunday newspaper in 2008. The other three graphs show distributions of means from random samples taken from the same population of used cars. One histogram shows means based on samples of 2 cars, another shows means based on samples of 5 cars, and another shows means based on samples of 10 cars. Each graph based on means was done with many repetitions. Which distribution is which, and why?

**(A)**

**(B)**

**(C)**

**(D)**

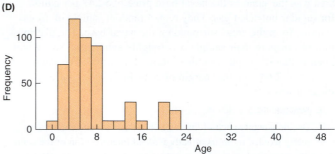

**TRY** **9.15 Used Car Ages (Example 4)** The mean age of all 638 used cars for sale in the *Ventura Country Star* one Saturday in 2008 was 7.9 years, with a standard deviation of 7.7 years. The distribution of ages is right-skewed. For a study to determine the reliability of classified ads, a reporter randomly selects 40 of these used cars and plans to visit each owner to inspect the cars. He finds that the mean age of the 40 cars he samples is 8.2 years and the standard deviation of those 40 cars is 6.0 years.

  a. Which of these four numerical values are parameters and which are statistics?

  b. $\mu = ?\ \sigma = ?\ s = ?\ \bar{x} = ?$

  c. Are the conditions for using the CLT fulfilled? What would be the shape of the approximate sampling distribution of a large number of means, each from a sample of 40 cars?

**9.16 Student Ages** The mean age of all 2550 students at a small college is 22.8 years with a standard deviation is 3.2 years, and the distribution is right-skewed. A random sample of 4 students' ages is obtained, and the mean is 23.2 with a standard deviation of 2.4 years.

  a. $\mu = ?\ \sigma = ?\ \bar{x} = ?\ s = ?$

  b. Is $\mu$ a parameter or a statistic?

  c. Are the conditions for using the CLT fulfilled? What would be the shape of the approximate sampling distribution of many means, each from a sample of 4 students? Would the shape be right-skewed, Normal, or left-skewed?

## SECTION 9.3

**TRY** **9.17 GPAs of Statistics Students (Example 5)** A random sample of 30 students taking statistics at a community college was asked their GPA. The sample mean was 3.25, and the margin of error for a 95% confidence interval was 0.18.

  a. Decide whether each of the following three statements a correctly *worded* or an incorrectly worded interpretation of the confidence interval, and fill in the blanks for the correct one(s).

    i. We are 95% confident that the *population* mean GPA of *statistics* students at the school is between ____ and ____.

ii. We are 95% confident that the recorded *sample* mean GPA of *statistics* students at the school is between ____ and ____.

iii. We are 95% confidence that the population mean GPA for *all* the students in the school is between ____ and ____.

b. What does the interval tell us about the population mean GPA for *statistics* students at the school? Can you reject 2.81 as the mean for statistics students? Explain.

**9.18 Gas Prices** Some students who lived on a naval base wanted to see whether the mean price of gas in the surrounding area was the same as the naval base price of $2.47 per gallon for regular unleaded gas. They took a random sample of 13 gas stations from the areas surrounding the naval base. The distribution of prices in their sample was roughly symmetric, so we will assume that the population distribution is Normal. The sample mean was 2.64, and the margin of error for a 95% confidence interval was 0.09.

a. Describe the population.

b. Decide whether each of the following three statements is a correctly worded or an incorrectly worded interpretation of the confidence interval, and fill in the blanks for any option(s) with correct wording.

i. There is a 95% probability that all the next random samples of 13 stations will have a sample mean between _____ and _____.

ii. We are 95% confident that the population mean is between ____ and ____.

iii. We are 95% confident that the boundaries for the interval are ____ and ____.

c. On the basis of this confidence interval, what should the students conclude about whether gas prices in surrounding areas are different from those at the naval base? Explain.

**9.19 Oranges** A statistics instructor randomly selected four bags of oranges, each bag labeled 10 pounds, and weighed the bags. They weighed 10.2, 10.5, 10.3, and 10.3 pounds. Assume that the distribution of weights is Normal. Find a 95% confidence interval for the mean weight of all bags of oranges. Use technology for your calculations.

a. Decide whether each of the three statements is a correctly worded interpretation of the confidence interval, and fill in the blanks for the correct option(s).

i. I am 95% confident the population mean is between ____ and ____.

ii. There is a 95% chance that all intervals will be between ___ and ___.

iii. I am 95% confident that the sample mean is between ___ and ___.

b. Does the interval capture 10 pounds? Is there enough evidence to reject the null hypothesis that the population mean weight is 10 pounds? Explain your answer.

**9.20 Carrots** The weights of four randomly chosen bags of horse carrots, each bag labeled 20 pounds, were 20.5, 19.8, 20.8, and 20.0 pounds. Assume that the distribution of weights is Normal. Find a 95% confidence interval for the mean weight of all bags of horse carrots. Use technology for your calculations.

a. Decide whether each of the three statements is a correctly interpretation of the confidence interval, and fill in the blanks for the correct option(s).

i. 95% of all sample means based on samples of the same size will be between ___ and ____.

ii. I am 95% confident that the population mean is between ____ and ____.

iii. We are 95% confident that the boundaries are ____ and ____ .

b. Can you reject a population mean of 20 pounds? Explain.

**TRY 9.21 Human Body Temperature (Example 6)** A random sample of 10 people's body temperatures was taken in degrees Fahrenheit, and the 95% confidence interval for the mean was (97.5, 98.8). Distributions of human body temperatures are approximately Normal. Which of the following statements is the correct interpretation of the confidence *level*, and which is the correct interpretation of the confidence *interval*?

a. In about 95% of all samples of 10 temperatures, the resulting confidence interval will contain the population mean body temperature.

b. We are very confident that the mean body temperature is between 97.5 and 98.8.

**9.22 Ice Cream Cones** A random sample of 5 ice cream cones from McDonald's were weighed, and the 95% confidence interval for the mean weight in ounces was (4.2, 5.4). Assume that the distribution of weights in the population is Normal. Which of the following statements is the correct interpretation of the confidence *level*, and which is the correct interpretation of the confidence *interval*?

a. We are very confident that the mean weight of all McDonald's cones is between 4.2 and 5.4 ounces.

b. In about 95% of all samples of 5 cones, the resulting confidence intervals will contain the mean weight of all McDonald's cones.

**TRY 9.23 *t*(Example 7)**

a. A researcher collects a sample of 30 measurements from a population and wishes to find a 90% confidence interval for the population mean. What value should he use for $t^*$? (Recall that $df = n - 1$ for a one-sample *t*-test or interval.)

b. If he instead decides to use a 95% confidence interval, will the interval be wider, be narrower, or stay the same? Why?

| df | C-level | | |
|----|---------|---------|---------|
|    | 90%     | 95%     | 99%     |
| 28 | 1.701   | 2.048   | 2.763   |
| 29 | 1.699   | 2.045   | 2.756   |
| 30 | 1.697   | 2.042   | 2.750   |
| 34 | 1.691   | 2.032   | 2.728   |

**9.24 *t***

a. A researcher collects a sample of 35 measurements from a population and wishes to find a 99% confidence interval for the population mean. What value should he use for $t^*$?

b. If he instead decides to use a 95% confidence interval, will the interval be wider or be narrower or stay the same? Why?

**TRY ★ 9.25 Dancers' Heights (Example 8)** A random sample of 20 independent female college-aged dancers showed a sample mean height of 63.3 inches and a sample standard deviation of 2.17 inches. Assume that this distribution of heights is Normal.

| | C-Level | | |
|---|---|---|---|
| **df** | **90%** | **95%** | **99%** |
| **14** | 1.761 | 2.145 | 2.977 |
| **15** | 1.753 | 2.131 | 2.947 |
| **16** | 1.746 | 2.120 | 2.921 |
| **17** | 1.740 | 2.110 | 2.898 |
| **18** | 1.734 | 2.101 | 2.878 |
| **19** | 1.729 | 2.093 | 2.861 |

a. Find a 95% confidence interval for the population mean height of dancers. Interpret the interval. Refer to the table.

b. Find a 99% confidence interval for the population mean height. Interpret the interval.

c. Which interval is wider and why?

**＊ 9.26 Women's Pulses** A random sample of 15 women's resting pulse rates from the National Health and Nutrition Examination Survey (NHANES) showed a mean of 73.5 beats per minute and standard deviation of 17.1. Assume that pulse rates are Normally distributed.

a. Are the mean and standard deviation that are provided statistics or parameters? Why?

b. Find a 95% confidence interval for the population mean pulse rate of women, and report it in a sentence. For $t^*$ refer to the table for Exercise 9.25.

c. Find a 90% confidence interval.

d. Which interval is wider and why?

**9.27 Confidence Interval Changes** State whether each of the following changes would make a confidence interval wider or narrower. (Assume that nothing else changes.)

a. Changing from a 90% confidence level to a 99% confidence level.

b. Changing from a sample size of 30 to a sample size of 200.

c. Changing from a standard deviation of 20 pounds to a standard deviation of 25 pounds.

**9.28 Confidence Interval Changes** State whether each of the following changes would make a confidence interval wider or narrower. (Assume that nothing else changes.)

a. Changing from a 95% level of confidence to a 90% level of confidence.

b. Changing from a sample size of 30 to a sample size of 20.

c. Changing from a standard deviation of 3 inches to a standard deviation of 2.5 inches.

**TRY** **9.29 Human Body Temperatures (Example 9)** A random **g** sample of 10 independent healthy people showed the following ✿ body temperatures (in degrees Fahrenheit):

98.5, 98.2  99.0, 96.3, 98.3, 98.7, 97.2, 99.1, 98.7, 97.2

Test the hypothesis that the population mean is not 98.6°F; use a significance level of 0.05. *See page 439 for guidance.*

✿ **9.30 Reaction Distance** Data on the disk and website show reaction distances in centimeters for the dominant hand for a random

sample of 40 independently chosen college students. Smaller distances indicate quicker reactions.

a. Make a graph of the distribution of the sample, and describe its shape.

b. Find, report, and interpret a 95% confidence interval for the population mean.

c. Suppose a professor said that the population mean should be 10 centimeters. Test the hypothesis that the population mean is not 10 cm, using the four-step procedure, with a significance level of 0.05.

**9.31 Male Height** In the United States, the population mean height for 3-year-old boys is 38 inches (http://www.kidsgrowth .com). Suppose a random sample of 15 non-U.S. 3-year-old boys showed a sample mean of 37.2 inches with a standard deviation of 3 inches. The boys were independently sampled. Assume that heights are Normally distributed in the population.

a. Determine whether the population mean for non-U.S. boys is significantly different from the U.S. population mean. Use a significance level of 0.05.

b. Now suppose the sample consists of 30 boys instead of 15, and repeat the test.

c. Explain why the $t$-values and p-values for parts a and b are different.

**9.32 Vegetarians' Weights** The mean weight of all 20-year-old women is 128 pounds (http://www.kidsgrowth.com). A random sample of 40 vegetarian women who are 20 years old showed a sample mean of 122 pounds with a standard deviation of 15 pounds. The women's measurements were independent of each other.

a. Determine whether the mean weight for 20-year old vegetarian women is significantly less than 128, using a significance level of 0.05.

b. Now suppose the sample consists of 100 vegetarian women who are 20 years old, and repeat the test.

c. Explain what causes the difference between the p-values for parts a and b.

✿ **9.33 GPAs** Thirty GPAs from a randomly selected sample of statistics students at Oxnard College are available on the disk and website. Assume that the population distribution is approximately Normal. The technician in charge of records claimed that the population mean GPA for the whole college is 2.81.

a. What is the sample mean? Is it higher or lower than the population mean of 2.81?

b. The chair of the mathematics department claims that statistics students typically have higher GPAs than the typical college student. Use the four-step procedure and the data provided to test this claim. Use a significance level of 0.05.

✿ **9.34 Dancers' Heights** A random sample of 20 independent female college-aged dancers was obtained and their heights (in inches) were measured. Assume the population distribution is Normal.

a. What is the sample mean? Is it above or below 64.5 inches?

b. Some people claim that the physical demands on dancers are such that dancers tend to be shorter than the typical person in the population. Use the four-step procedure to test the hypothesis that dancers have a smaller population mean height than 64.5 inches. Use a significance level of 0.05.

✿ **9.35 Pulse and Meditation** Eleven students took their pulses while in class. Then they were asked to meditate by sitting with their eyes closed and counting their breaths for 5 minutes. After this, they took their pulse rates again. (The pulse was counted for 15 seconds and multiplied by 4 to get the number of beats per minute.) The data recorded provide differences: pulse before meditation minus pulse after meditation. So a negative value means that the pulse rates increased after meditation. Treat these students as a random sample of college students and perform a hypothesis test to determine whether the mean pulse rate decreased after meditation. In other words, determine whether the mean difference is positive. Clearly state all four steps. Assume that the conditions for using a *t*-test are met.

$$12, 4, 4, 0, -8, -8, 0, 12, 8, 0, 8$$

**9.36 Amazon Textbook Prices** Amelia Suragairin, a statistics student, was interested in the prices of textbooks and went online to the Amazon.com U.S. and Amazon.com U.K. websites to compare prices. The prices in the U.K. were in pounds and were converted to dollars using the exchange rate at that time. The data provide the differences (U.S. minus U.K.) in price between the two countries (rounded to the nearest dollar). Negative signs indicate that the U.K. price was higher. Treat these samples as random samples from their populations and perform a hypothesis test to determine whether the mean difference is significantly different from 0, using a significance level of 0.05. Clearly state all four steps. Assume that the conditions for using a *t*-test are met.

$$-4, 0, 0, -22, -21, -14, -12, 14, 3, -7, 13, -6, -12,$$
$$-11, -7, 12, -2, 2, 1, -13$$

**9.37 Carrots, Three Tests** A random and independently chosen sample of four bags of horse carrots, each bag labeled 20 pounds, had weights of 20.5, 19.8, 20.8, and 20.0 pounds. Assume that the distribution of weights in the population is Normal. In each case, report the alternative hypothesis, the *t*-value, the p-value, and the conclusion for each of the three parts.

a. Test the hypothesis that the population mean weight is not 20 pounds.

b. Test the hypothesis that the population mean is less than 20 pounds.

c. Test the hypothesis that the population mean is more than 20 pounds.

**9.38 Soft Drink Serving Size** A consumer advocate wants to find out whether the soft drinks sold at a fast-food restaurant contain less than the advertised 16 ounces. A random sample of 10 independently chosen drinks produced a mean of 16.5 ounces with a standard deviation of 0.4 ounce. Assume that the distribution of serving sizes in the population is Normally distributed.

a. State the null hypothesis in both words and symbols, using $\mu$ as the population mean weight of all the drinks labeled 16 ounces at this fast-food restaurant.

b. State the alternative hypothesis in both words and symbols.

c. Carry out the appropriate test using a 5% significance level. Explain what conclusion the consumer advocate should reach.

d. Explain why no calculations are needed to show that the population mean is not significantly less than 16 ounces.

**9.39 UCLA Textbook Prices** A 95% confidence interval for the mean price of textbooks at UCLA in the spring quarter 2010, based on a random sample taken by statistician David Diez, was ($58.30, $86.14). Assuming that nothing else changed, what would have been the effect on the confidence interval if David had used a smaller sample size? Assume that the necessary conditions for finding a confidence interval hold, and pick the best answer.

a. The confidence interval would have been wider.

b. The confidence interval would have been narrower.

c. The confidence level would have been higher.

d. The confidence level would have been lower.

e. Answers a and c are both correct.

f. Answers b and d are both correct.

**\* 9.40 UCLA Textbook Prices** A 95% confidence interval for the mean price of textbooks at UCLA in the spring quarter 2010, based on a random sample taken by David Diez (a statistician) was ($58.30, $86.14). To obtain this confidence interval, he multiplied the standard error by 2.04, which was the value of $t^*$. What would have been the effect if he had multiplied the standard error by 1.70 instead? Choose the best answer.

a. The confidence interval would have been wider.

b. The confidence interval would have been narrower.

c. The confidence level would have been higher.

d. The confidence level would have been lower.

e. Answers a and c are both correct.

f. Answers b and d are both correct.

**9.41 Student Ages** Suppose that 200 statistics students each took a random sample (with replacement) of 50 students at their college and recorded the ages of the students in their sample. Then each student used his or her data to calculate a 95% confidence interval for the mean age of all students at the college. How many of the 200 intervals would you expect to capture the true population mean age, and how many would you expect not to capture the true population mean? Explain by showing your calculation.

**9.42 Ages of Married Couples** Suppose that 300 statistics students each collected a random sample of 40 married couples from their state and recorded the ages. From the ages of these couples, they calculated the difference in ages: husband's age minus wife's age. Then each found a 90% confidence interval for the population mean age difference. How many of the 300 intervals would you expect to capture the true population mean difference, and how many would you expect to miss the true population mean difference? Explain by showing your calculation.

**9.43 Presidents' Ages at Inauguration** A 95% confidence interval for the ages of the first six presidents at their inaugurations is (56.2, 59.5). Either interpret the interval or explain why it should not be interpreted.

**9.44 Chief Justices' Ages at Installation** A 95% confidence interval for the age of the first six chief justices at their installation is (45.3, 58.4). Either interpret the interval or explain why it should not be interpreted.

## SECTION 9.4

**9.45 Independent or Paired? (Example 10)** State whether each situation has independent or paired (dependent) samples.

a. A researcher wants to know whether men and women at University X have different mean GPAs. She gathers two random samples: one of GPAs from 50 men and the other from 50 women.

b. A researcher wants to know whether husbands and wives have different mean GPAs. He collects a sample of husbands and wives who have both earned a bachelor's degree and has each person report his or her overall GPA in college.

**9.46 Independent or Paired?** State whether each situation has independent or paired (dependent) samples.

a. A researcher wants to know whether hypnosis tends to reduce pain. Research subjects place their hands in ice water to see how long they can tolerate it before the pain becomes unbearable. Then the subjects are hypnotized and again put their hands in ice water to see whether they tend to leave them in the water longer after hypnosis.

b. A researcher wants to know whether male or female professors at his college tend to have longer recorded greetings on their telephone answering machines. He calls 40 random male and 40 random female professors over the weekend and times their messages.

**g 9.47 Televisions** The table shows the Minitab output for a two-sample *t*-test for the number of televisions owned in households of random samples of students at two different community colleges. Each individual was randomly chosen independently of the others; the students were not chosen as pairs or in groups. One of the schools is in a wealthy community (MC), and the other (OC) is in a less wealthy community. Test the hypothesis that the population means are not the same, using a significance level of 0.05. *See page 440 for guidance.*

```
Two-sample T for OCTV vs MCTV

       N     Mean    StDev   SE Mean
OCTV   30    3.70    1.49    0.27
MCTV   30    3.33    1.49    0.27

Difference = mu OCTV - mu MCTV
Estimate for difference: 0.367
95% CI for difference: (-0.404, 1.138)
T-Test of difference = 0 (vs not =): T-Value = 0.95  P-Value = 0.345
```

**9.48 Pulse Rates** Using data from NHANES, we looked at pulse rates of nearly 800 people to see whether men or women tended to have higher pulse rates. Refer to the Minitab output provided.

a. Report the sample means and state which group had the higher sample mean pulse rate?

b. Use the Minitab output to test the hypothesis that pulse rates for men and women are not equal, using a significance level of 0.05. The samples are large enough so that Normality is not an issue.

```
Two-sample T for Pulse

Sex        N   Mean   StDev   SE Mean
Female   384   76.3   12.8    0.65
Male     372   72.1   13.0    0.67

Difference = mu (Female) - mu (Male)
Estimate for difference: 4.248
95% CI for difference: (2.406, 6.090)
T-Test of difference = 0 (vs not =): T-Value = 4.53  P-Value = 0.000
```

**9.49 Triglycerides** Triglycerides are a form of fat found in the body. Using data from NHANES, we looked whether men have higher triglyceride levels than women.

a. Report the sample means and state which group had the higher sample mean triglyceride level. Refer to the Minitab output in figure (A).

b. Carry out a hypothesis test to determine whether men have a higher mean triglyceride level than women. Refer to the Minitab output provided

in figure A. Output for three different alternative hypotheses is provided—see figures (B), (C), and (D)—and you must choose and state the most appropriate output.

**(A)**

```
Two-Sample T-Test and CI: Triglycerides, Gender

Two-sample T for Triglycerides

Gender   N    Mean   StDev   SE Mean
Female   44   84.4   40.2    6.1
Male     48   139.5  85.3    12

Difference = mu (Female) - mu (Male)
Estimate for difference: -55.1
95% CI for difference: (-82.5, -27.7)
```

**(B)**

```
9.49 B: T-Test of difference = 0 (vs <): T-Value = -4.02  P-Value = 0.000
```

**(C)**

```
9.49 C: T-Test of difference = 0 (vs >): T-Value = -4.02  P-Value = 1.000
```

**(D)**

```
9.49 D: T-Test of difference = 0 (vs not =): T-Value = -4.02  P-Value = 0.000
```

**9.50 Systolic Blood Pressures** When you have your blood pressure taken, the larger number is the systolic blood pressure. Using data from NHANES, we looked at whether men and women have different systolic blood pressure levels.

a. Report the two sample means and state which group had the higher sample mean systolic blood pressure. Refer to the Minitab output in figure (A).

b. Refer to the Minitab output given in figure (A) to test the hypothesis that the mean systolic blood pressures for men and women are not equal, using a significance level of 0.05. Although the distribution blood pressures in the population are right-skewed, the sample size is large enough for us to use *t*-tests. Choose from figures (B), (C), and (D) for your p-value, and explain.

**(A)**

```
Two-sample T for BPSys

gender    N    Mean   StDev   SE Mean
Female   404   116.8  22.7    1.1
Male     410   118.7  18.0    0.89

Difference = mu (Female) - mu (Male)
Estimate for difference: -1.93
95% CI for difference: (-4.75, 0.89)
```

**(B)**

```
9.50 B: T-Test of difference = 0 (vs not =): T-Value = -1.34  P-Value = 0.180
```

**(C)**

```
9.50 C: T-Test of difference = 0 (vs >): T-Value = -1.34  P-Value = 0.910
```

**(D)**

```
9.50 D: T-Test of difference = 0 (vs <): T-Value = -1.34  P-Value = 0.090
```

TRY **9.51 Triglycerides, Again (Example 11)** Report and interpret the 95% confidence interval for the difference in mean triglycerides between men and women (refer to the Minitab output in Exercise 9.49). Does this support the hypothesis that men and women differ in mean triglyceride level? Explain.

**9.52 Blood Pressures, Again** Report and interpret the 95% confidence interval for the difference in mean systolic blood pressure for men and women (refer to the Minitab output in Exercise 9.50). Does this support the hypothesis that men and women differ in mean systolic blood pressure? Explain.

**9.53 Clothes Spending** A random sample of 14 college women and a random sample of 19 college men were separately asked to estimate how much they spent on clothing in the last month. The table shows the data.

Test the hypothesis that the population mean amounts spent on clothes are different for men and women. Use a significance level of 0.05. Assume that the distributions are Normal enough for us to use the *t*-test.

| Sex | $clothes | Sex | $clothes |
|-----|----------|-----|----------|
| m | 175 | f | 80 |
| f | 200 | m | 200 |
| m | 150 | m | 80 |
| f | 200 | m | 100 |
| f | 100 | m | 120 |
| f | 100 | m | 80 |
| f | 200 | m | 25 |
| m | 100 | f | 80 |
| m | 100 | m | 50 |
| f | 200 | m | 100 |
| m | 200 | m | 30 |
| m | 200 | f | 20 |
| m | 200 | f | 50 |
| f | 250 | m | 60 |
| f | 150 | f | 100 |
| m | 100 | f | 350 |
| m | 0 | | |

**9.54 College Athletes' Weights** A random sample of male college baseball players and a random sample of male college soccer players were obtained independently and weighed. The table shows the unstacked weights (in pounds). The distributions of both

data sets suggest that the population distributions are roughly Normal. Determine whether the difference in means is significant, using a significance level of 0.05.

| Baseball | Soccer | Baseball | Soccer |
|----------|--------|----------|--------|
| 190 | 165 | 186 | 156 |
| 200 | 190 | 210 | 168 |
| 187 | 185 | 198 | 173 |
| 182 | 187 | 180 | 158 |
| 192 | 183 | 182 | 150 |
| 205 | 189 | 193 | 172 |
| 185 | 170 | 200 | 180 |
| 177 | 182 | 195 | 184 |
| 207 | 172 | 182 | 174 |
| 225 | 180 | 193 | 190 |
| 230 | 167 | 190 | 156 |
| 195 | 190 | 186 | 163 |
| 169 | 185 | | |

* **9.55 Clothes Spending** In Exercise 9.53 you could not reject the null hypothesis that the mean amount spent by men and the mean amount spent by women for clothing are the same, using a two-tailed test with a significance level of 0.05.

a. If you found a 95% confidence interval for the difference between means, would it capture 0? Explain.

b. If you found a 99% confidence interval, would it capture 0? Explain.

c. Now go back to Exercise 9.53. Find a 95% confidence interval for the difference between means, and explain what it shows.

* **9.56 College Athletes' Weights** In Exercise 9.54, you could reject the null hypothesis that the mean weights of soccer and baseball players were equal using a two-tailed test with a significance level of 0.05.

a. If you found a 95% confidence interval for the difference between means, would it capture 0? Explain.

b. If you found a 90% interval, would it capture 0? Explain.

c. Now go back to Exercise 9.54. Find a 95% confidence interval for the difference between means, and explain what it shows.

**9.57 Criminology** In April 2005, the *Journal of Experimental Criminology* published the results of a 1998 study in which 121 rental properties that had already been the target of drug law enforcement were randomly divided into two groups. In the experimental group, the tenants received a letter from the police describing the enforcement tactics in place. For the control group, there was no letter. The table gives summary statistics for the number of crimes reported over a 30-month interval.

| | Experimental | Control |
|---|--------------|---------|
| Mean | 3.2 | 5.1 |
| SD | 4.1 | 6.3 |
| *n* | 79 | 42 |

Determine whether the letter from the police was effective in *reducing* the number of crimes at the 0.05 level. Although the distribution of number of crimes is not Normal, assume that the sample size is large enough for the Central Limit Theorem to apply. (Source: David P. Farrington and Brandon C. Welsh, Randomized experiments in criminology: What have we learned in the last two decades?" *Journal of Experimental Criminology* (2005), 1: 9–38)

**9.58 Criminology** In April 2005, the *Journal of Experimental Criminology* reported on a study done in Washington in which 171 children were randomly divided into two groups. The experimental group went to a special cognitive preschool, and the control group went to a traditional preschool. The summary statistics for the number of self-reported criminal offenses by the age of 15 are given in the table. Although the number of criminal offenses is not Normally distributed, sample sizes are large enough that the Central Limit Theorem applies. Test the hypothesis that the mean number of self-reported offenses is not the same for each group, using a 0.05 level of significance. (Source: David P. Farrington and Brandon C. Welsh, Randomized experiments in criminology: What have we learned in the last two decades?" *Journal of Experimental Criminology* (2005), 1: 9–38)

|  | Experimental | Control |
|---|---|---|
| Mean | 12.62 | 11.05 |
| SD | 14.27 | 13.66 |
| n | 90 | 81 |

**g  9.59 Females–Pulse Rates before and after a Fright** In a statistics class taught by one of the authors, students took their pulses before and after being frightened. The frightening event was having the teacher scream and run from one side of the room to the other. The pulse rates (beats per minute) of the women before and after the scream were obtained separately and are shown in the table. Treat this as though it were a random sample of female community college students. Test the hypothesis that the mean of college women's pulse rates is higher after a fright, using a significance level of 0.05. *See page 440 for guidance.*

| Women | | Women | |
|---|---|---|---|
| Pulse Before | Pulse After | Pulse Before | Pulse After |
| 64 | 68 | 84 | 88 |
| 100 | 112 | 80 | 80 |
| 80 | 84 | 68 | 92 |
| 60 | 68 | 60 | 76 |
| 92 | 104 | 68 | 72 |
| 80 | 92 | 68 | 80 |
| 68 | 72 | | |

**9.60 Males–Pulse Rates before and after a Fright** Follow the instructions for Exercise 9.59, but use the data for the men in the class. Test the hypothesis that the mean of college men's pulse rates is higher after a fright, using a significance level of 0.05.

| Men | |
|---|---|
| Pulse Before | Pulse After |
| 50 | 64 |
| 84 | 72 |
| 96 | 88 |
| 80 | 72 |
| 80 | 88 |
| 64 | 68 |
| 88 | 100 |
| 84 | 80 |
| 76 | 80 |

**9.61 Textbook Prices, UCSB vs. CSUN** The prices of a sample of books at University of California at Santa Barbara (UCSB) were obtained by statistics students Ricky Hernandez and Elizabeth Alamillo. Then the cost of books for the same subjects (at the same level) were obtained for California State University at Northridge (CSUN). Assume that the distribution of differences is Normal enough to proceed, and assume that the sampling was random.

  a. First find both sample means and compare them.

  b. Test the hypothesis that the population means are different, using a significance level of 0.05.

**9.62 Textbook Prices. OC vs. CSUN** The prices of a random sample of comparable textbooks from two schools were recorded. We are comparing the prices at OC (Oxnard Community College) and CSUN (California State University at Northridge). Assume that the population distribution of differences is approximately Normal. Each book was priced separately; there were no books "bundled" together.

  a. Compare the sample means.

  b. Determine whether the mean prices of all books are significantly different. Use a significance level of 0.05.

**9.63 Ages of Brides and Grooms** Data for the ages of grooms and their brides for a random sample of 31 couples in Ventura County, California, were obtained.

  a. Compare the sample means.

  b. Test the hypothesis that there is a significant difference in mean ages of brides and grooms, using a significance level of 0.05.

  c. If the test had been done to determine whether the mean for the grooms was significantly larger than the mean for the brides, how would that change the alternative hypothesis and the p-value?

**9.64 Navy Commissary Prices** Amber Sanchez, a statistics student, collected data on the prices of the same items at the Navy commissary on the naval base in Ventura County, California, and a nearby Kmart. The items were matched for content, manufacturer, and size and were priced separately.

  a. Report and compare the sample means.

  b. Assume that they are a random sample of items, and use a significance level of 0.05 to test the hypothesis that the Navy commissary has a lower mean price. Assume that the population distribution of differences is approximately Normal.

**TRY** **9.65 Self-Reported Heights of Men (Example 12)** A random sample of students at Oxnard College reported what they believed to be their heights in inches. Then the students measured each others' heights in centimeters, without shoes. The data shown are for the men. Assume that conditions for *t*-tests hold.

a. Convert heights in inches to centimeters by multiplying inches by 2.54. Find a 95% confidence interval for the mean difference as measured in centimeters. Does it capture 0? What does that show?

b. Perform a *t*-test to test the hypothesis that the means are not the same. Use a significance level of 0.05, and show all four steps.

| Men | | | Men | |
|-----|-----|-----|-----|-----|
| Centimeters | Inches | | Centimeters | Inches |
| 166 | 66 | | 178 | 70 |
| 172 | 68 | | 177 | 69 |
| 184 | 73 | | 181 | 71 |
| 166 | 67 | | 175 | 69 |
| 191 | 76 | | 171 | 67 |
| 173 | 68 | | 170 | 67 |
| 174 | 69 | | 184 | 72 |
| 191 | 76 | | | |

**9.66 Female Self-Reported Heights** Follow the instructions for Exercise 9.65, but use the data for the women. Assume that heights are Normally distributed in the population and the sample is random.

| Women | | Women | |
|-------|------|-------|------|
| Cm | Inch | Cm | Inch |
| 160 | 63 | 158 | 63 |
| 160 | 63 | 159 | 62 |
| 158 | 63 | 180 | 70 |
| 155 | 61 | 156 | 61 |
| 172 | 67 | 155 | 60 |
| 167 | 66 | 159 | 63 |
| 153 | 61 | 179 | 71 |
| 161 | 64 | 175 | 70 |
| 156 | 62 | 163 | 64 |
| 157 | 62 | 159 | 61 |
| 161 | 63 | 162 | 64 |
| 155 | 62 | 154 | 61 |
| 175 | 70 | 159 | 62 |

**9.67 Surfers** Surfers and statistics students Rex Robinson and Sandy Hudson collected data on the number of days on which surfers surfed in the last month for 30 longboard (L) users and 30 shortboard (S) users. Treat these data as though they were from two independent random samples. Test the hypothesis that the mean days surfed for all longboarders is larger than the mean days surfed for all shortboarders (because longboards can go out in many different surfing conditions). Use a level of significance of 0.05. The data can be found in Exercise 3.11 on page 114.

**9.68 Eating Out** Jacqueline Loya, a statistics student, asked students with jobs how many times they went out to eat in the last week. There were 25 students who had part-time jobs and 25 students who had full-time jobs. Carry out a hypothesis test to determine whether the mean number of meals out per week for students with full-time jobs is greater than that for those with part-time jobs. Use a significance level of 0.05. Assume that the conditions for a two-sample *t*-test hold. The data can be found in Exercise 3.12 on page 114.

## CHAPTER REVIEW EXERCISES

**9.69 Which *t*-Test?** State whether you would use a one-sample, two-sample, or paired *t*-test for each part, assuming the conditions required for *t*-tests hold.

a. The weights of a random sample of 25 people are measured before and after a diet. You want to test the hypothesis that weights tend to be lower at the end of the diet.

b. The number of felony convictions for each prisoner in a random sample of 10 prisoners at Prison A and a random sample of 10 prisoners at Prison B are obtained. You want to test the claim that the prisoners in Prison A tend to have more felony convictions than the prisoners in Prison B.

**9.70 Which *t*-Test?** State whether you would use a one-sample, two-sample, or paired *t*-test for each part, assuming the conditions required for *t*-test hold.

a. The weights of a random sample of 10 male weight-lifters at Gold's Gym are measured. You want to test the hypothesis that the weight-lifters tend to weigh more than men in the general population, for which the population mean is 180 pounds.

b. The ages of the people in a randomly selected group of community college students are obtained. You want to find out whether the male and female students have significantly different mean ages.

c. The ages at marriage for brides and grooms are obtained from a randomly selected group of marriage licenses. You want to determine whether the mean for grooms is significantly higher than the mean for their brides.

**9.71 Marathon Times** The results for a Forest City marathon in Ontario contain times for a sample of both men and women. Below is the Minitab output for a two-sample *t*-test.

a. Test the hypothesis that men (M) are quicker, on average, than women (F), at the 0.05 level of significance. The times are in seconds. (15,455 seconds is about 4 hours and 18 minutes.)

b. Interpret the 95% CI (confidence interval).

```
Two-Sample T-Test and CI: time in secs, gender

Two-sample T for time in secs

gender    N    Mean   StDev   SE Mean
F        158   15455   1949      155
M        298   13671   1973      114


Difference = mu (F) - mu (M)
Estimate for difference: 1784
95% CI for difference: (1405, 2163)
T-Test of difference = 0 (vs not =): T-Value = 9.26   P-Value = 0.000
```

**9.72 Risks** In the Chapter 3 case study, we considered a study of the perceived risks of using appliances for both men and women. We observed that the mean perceived risk was slightly higher for women in the sample.

a. Use the Minitab output to determine whether the mean risk of using appliances for women is significantly different from the mean for men. The women are indicated with a 0 and the men with a 1.

b. Interpret the 95% CI (confidence interval).

```
Two-Sample T-Test and CI: appl, gender

gender    N   Mean   StDev   SE Mean
0        353   20.6    20.8     1.1
1        214   18.3    20.0     1.4


Difference = mu (0) - mu (1)
Estimate for difference: 2.30
95% CI for difference: (-1.16, 5.76)
T-Test of difference = 0 (vs not =): T-Value = 1.30   P-Value = 0.193
```

**9.73 Heart Rate before and after Coffee** Elena Lucin, a statistics student, collected the data in the table showing heart rate (beats per minute) for a random sample of coffee drinkers before and 15 minutes after they drank coffee. Carry out a complete analysis, using the techniques you learned in this chapter. Use a 5% significance level to test whether coffee increases heart rates. The same amount of caffeinated coffee was served to each person, and you may assume that conditions for a *t*-test hold.

| Before | After | | Before | After |
|--------|-------|--|--------|-------|
| 90     | 92    | | 74     | 78    |
| 84     | 88    | | 72     | 82    |
| 102    | 102   | | 72     | 76    |
| 84     | 96    | | 92     | 96    |
| 74     | 96    | | 86     | 88    |
| 88     | 100   | | 90     | 92    |
| 80     | 84    | | 80     | 74    |
| 68     | 68    | |        |       |

**★ 9.74 Exam Grades** The final exam grades for a sample of daytime statistics students and evening statistics students at one college are reported. The classes had the same instructor, covered the same material, and had similar exams. Using graphical and numerical summaries, write a brief description about how grades differ for these two groups. Then carry out a hypothesis test to determine whether the mean grades are significantly different for evening and daytime students. Assume that conditions for a *t*-test hold. Select your significance level.

Daytime grades: 100, 100, 93, 76, 86, 72.5, 82, 63, 59.5, 53, 79.5, 67, 48, 42.5, 39

Evening grades: 100, 98, 95, 91.5, 104.5, 94, 86, 84.5, 73, 92.5, 86.5, 73.5, 87, 72.5, 82, 68.5, 64.5, 90.75, 66.5

**9.75 Hours of Television Viewing** The number of hours per week of television viewing for random samples of fifth grade boys and fifth grade girls were obtained. Each student logged his or her hours for one Monday-through-Friday period. Assume that the students were independent; for example, there were no pairs of siblings who watched the same shows.

Using graphical and numerical summaries, write a brief description of how the hours differed for the boys and girls. Then carry out a hypothesis test to determine whether the mean hours of television viewing are different for boys and girls. Evaluate whether the conditions for a *t*-test are met, and state any assumptions you must make in order to carry out a *t*-test.

**9.76 Reaction Distances** Reaction distances in centimeters for a random sample of 40 college students were obtained. Shorter distances indicate quicker reactions. Each student tried the experiment both with his or her dominant hand, and with his or her nondominant hand, catching the meter stick. The subjects all started with their dominant hand.

Examine summary statistics, and explain what we can learn from them. Then do an appropriate test to see whether the mean reaction distance is shorter for the dominant hand. Use a significance level of 0.05.

**9.77 Work Hours: Men and Women** Random samples of 50 men and 50 women reported how many hours they worked in the last week as part of the 2008 General Social Survey. (People who did not work were not included.)

a. Two different analyses were performed to test the claim that the population means are not the same. Choose the correct output, and carry out the hypothesis test. Both analyses were done using two-tailed alternative hypotheses.

b. Use the correct output to report and interpret an appropriate 95% confidence interval for the difference in means.

```
Two-sample T for 50 men vs 50 women

            N   Mean  StDev  SE Mean
50 men     50   49.8   16.8      2.4
50 women   50   37.8   12.2      1.7

Difference = mu (50 men) - mu (50 women)
Estimate for difference: 12.02
95% CI for difference: (6.18, 17.86)
T-Test of difference = 0 (vs not =): T-Value = 4.09  P-Value = 0.000
```

**Output A (above)**

```
Paired T for 50 men - 50 women

             N   Mean  StDev  SE Mean
50 men      50  49.80  16.81     2.38
50 women    50  37.78  12.24     1.73
Difference  50  12.02  19.91     2.82

95% CI for mean difference: (6.36, 17.68)
T-Test of mean difference = 0 (vs not = 0): T-Value = 4.27  P-Value = 0.000
```

**Output B**

**9.78 Number of Children** Random samples of 100 men and 100 women were collected from the 2008 General Social Survey, with each person reporting the number of children they had.

a. What do the sample means tell us about the differences between men and women in terms of the number of children they report having?

b. Two analyses were carried out to test the claim that the mean number of children for men is not the same as the mean number of children for women. Only one analysis is appropriate for these data. Carry out an appropriate hypothesis test, including all four steps. The sample sizes are large enough so that the shapes of the population distributions are not a concern.

c. Report and interpret the 95% confidence interval for the difference in means.

```
Two-sample T for 100 men vs 100 women

              N   Mean  StDev  SE Mean
100 men     100   1.49   1.76     0.18
100 women   100   1.85   1.57     0.16

Difference = mu (100 men) - mu (100 women)
Estimate for difference: -0.360
95% CI for difference: (-0.825, 0.105)
T-Test of difference = 0 (vs not =): T-Value = -1.53  P-Value = 0.128
```

**Output A (above)**

```
Paired T for 100 men - 100 women

               N    Mean  StDev  SE Mean
100 men      100   1.490  1.755    0.176
100 women    100   1.850  1.572    0.157
Difference   100  -0.360  2.259    0.226

95% CI for mean difference: (-0.808, 0.088)
T-Test of mean difference = 0 (vs not = 0): T-Value = -1.59  P-Value = 0.114
```

**Output B**

**9.79 Drug Testing** In 2005, the *Journal of Experimental Criminology* reported on a 1992 study in Pima County Arizona that looked at drug testing in criminal defendants. Out of 231 pretrial

defendants observed, 153 were randomly selected to have frequent pretrial drug testing (the experimental group), and 78 were merely released (the control group). The summary statistics are given in the table, and the numbers represent pretrial arrests per person. The mean of 0.04 means that on average, one pretrial defendant in 25 (4%) was arrested.

Test the hypothesis that drug testing caused the rate of arrests to go down, using a level of significance of 0.05. Assume that conditions for *t*-tests hold. Begin by choosing the appropriate output, the TI-83/84 output shown in figure (A) or that shown in figure (B), and explain your choice. In the output, 1 indicates the experimental group, and 2 indicates the control group.

|        | Experimental | Control |
|--------|:------------:|:-------:|
| Mean   | 0.04         | 0.12    |
| SD     | 0.19         | 0.32    |
| $n$    | 153          | 78      |

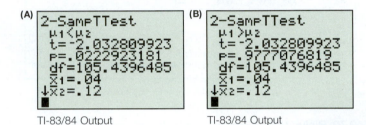

(A) TI-83/84 Output     (B) TI-83/84 Output

**★ 9.80 Drug Testing** Refer to the study description in Exercise 9.79. A similar study was performed in Maricopa County, Arizona; the data are shown in the table. Follow the instructions in Exercise 9.79 to test the hypothesis that drug testing caused the rate of arrests to go down, using a level of significance of 0.05. In the output, 1 refers to the experimental group, and 2 refers to the control group. Begin by choosing the correct TI-83/84 output—that is, the output shown in figure (A) or that shown in figure (B).

|        | Experimental | Control |
|--------|:------------:|:-------:|
| Mean   | 0.45         | 0.37    |
| SD     | 0.50         | 0.48    |
| $n$    | 425          | 465     |

(A) TI-83/84 Output     (B) TI-83/84 Output

**★ 9.81 Why Is $n - 1$ in the Sample Standard Deviation?** Why do we calculate $s$ by dividing by $n - 1$, rather than just $n$?

$$s^2 = \frac{\sum (x - \bar{x})^2}{n - 1}$$

(A)

TI-83/84 Output

The reason is that if we divide by $n - 1$, then $s^2$ is an unbiased estimator of $\sigma^2$, the population variance.

We want to show that $s^2$ is an unbiased estimator of $\sigma^2$, sigma squared. The mathematical proof that this is true is beyond the scope of an introductory statistics course, but we can use an example to demonstrate that it is.

First we will use a very small population that consists only of these three numbers: 1, 2, and 5.

You can determine that the population standard deviation, $\sigma$, for this population is 1.699673 (or about 1.70), as shown in the TI-83/84 output. So the population variance, sigma squared, $\sigma^2$, is therefore 2.888889 (or about 2.89).

Now take all possible samples, with replacement, of size 2 from the population, and find the sample variance, $s^2$, for each sample. This process is started for you in the table. Average these

| Sample | $s$ | $s^2$ |
|--------|------|------|
| 1, 1 | 0 | 0 |
| 1, 2 | 0.7071 | 0.5 |
| 1, 5 | 2.8284 | 8.0 |
| 2, 1 | | |
| 2, 2 | | |
| 2, 5 | | |
| 5, 1 | | |
| 5, 2 | | |
| 5, 5 | | |

sample variances ($s^2$), and you should get approximately 2.88889. If you do, then you have demonstrated that $s^2$ is an unbiased estimator of $\sigma^2$, sigma squared.

Show your work by filling in the accompanying table and show the average of $s^2$.

**9.82 Is $s$ an Unbiased Estimator of $\sigma$?** Use the data given in Exercise 9.81 to find out whether $s$ is a biased or an unbiased estimator of $\sigma$, sigma. To do this, add the standard deviations ($s$) for the nine samples, and divide by 9 to get the average. Report this average. Is it larger than, smaller than, or the same size as 1.69967? Is it a biased estimator of sigma or an unbiased estimator?

\* **9.83** Construct two sets of body temperatures (in degrees Fahrenheit, such as 96.2°F), one for men and one for women, such that the sample means are different but the hypothesis test shows the population means are not different. Each set should have three numbers in it.

\* **9.84** Construct heights for 3 or more sets of twins (6 or more people). Make the twins similar, but not exactly the same, in height. Put all of the shorter twins in set A and all of the taller twins in set B. Create the numbers such that a two-sample $t$-test will *not* show a significant difference in the mean heights of the shortest of each pair and the mean heights of the tallest of each pair but the paired $t$-test *does* show a significant difference. (*Hint:* Make one of the pairs really tall, one of the pairs really short, and one of the pairs in between.) Report all the numbers and the $t$- and p-values for the tests. Explain why the paired $t$-test shows a difference and the two-sample $t$-test does not show a difference. Remember that 5 feet is 60 inches and that 6 feet is 72 inches.

# gUIDED EXERCISES

g **9.29 Human Body Temperatures** A random sample of 10 independent healthy people showed body temperatures (in degrees Fahrenheit) as follows:

98.5, 98.2, 99.0, 96.3, 98.3, 98.7, 97.2, 99.1, 98.7, 97.2

The Minitab output of the results of a one-sample $t$-test is shown.

```
One-Sample T: Sample of 10
Test of mu = 98.6 vs not = 98.6

Variable   N    Mean    StDev   SE Mean      95% CI          T       P
sam10      10  98.1200  0.9187   0.2905  (97.4628, 98.7772)  -1.65  0.133
```

QUESTION  Test the hypothesis that the population mean is not 98.6°F, using a significance level of 0.05. Write out the steps given, filling in the blanks.

**Step 1 ▶ Hypothesize**

$H_0: \mu = 98.6$

$H_a:$ _____

**Step 2 ▶ Prepare**

A stemplot is shown that is not strongly skewed, suggesting that the distribution of the population is also approximately Normal. (A histogram would also be appropriate.) Comment on the data collection, and state the test to be used. State the significance level.

```
96 | 3
97 | 22
98 | 23577
99 | 01
```

### Step 3 ► Compute to compare

$$t = \underline{\hspace{2cm}}$$

p-value = \underline{\hspace{2cm}}

### Step 4 ► Interpret

Reject or do not reject $H_0$, and choose interpretation i, ii, or iii:

i. We cannot reject 98.6 as the population mean from these data at the 0.05 level.

ii. The population mean is definitely 98.6 on the basis of these data at the 0.05 level.

iii. We can reject the null hypothesis on the basis of these data, at the 0.05 level. The population mean is not 98.6.

**g** **9.47 Televisions** Minitab output is shown for a two-sample *t*-test for the number of televisions owned in households of random samples of students at two different community colleges. Each individual was randomly chosen independently of the others; the students were not chosen as pairs or in groups. One of the schools is in a wealthy community (MC), and the other (OC) is in a less wealthy community.

```
Two-Sample T-Test and CI: OCTV, MCTV

         N     Mean    StDev   SE Mean
OCTV    30     3.70    1.49     0.27
MCTV    30     3.33    1.49     0.27

Difference = mu OCTV - mu MCTV
Estimate for difference: 0.367
95% CI for difference: (-0.404, 1.138)
T-Test of difference = 0 (vs not =): T-Value = 0.95  P-Value = 0.345
```

QUESTION Complete the steps to test the hypothesis that the mean number of televisions per household is different in the two communities, using a significance level of 0.05.

### Step 1 ► Hypothesize

Let $\mu_{oc}$ be the population mean number of televisions owned by families of students in the less wealthy community (OC), and let $\mu_{mc}$ be the population mean number of televisions owned by families of students in the wealthier community (MC).

$$H_0: \mu_{oc} = \mu_{mc}$$
$$H_a: \underline{\hspace{2cm}}$$

### Step 2 ► Prepare

Choose an appropriate *t*-test. Because the sample sizes are 30, the Normality condition of the *t*-test is satisfied. State the other conditions, indicate whether they hold, and state the significance level that will be used.

### Step 3 ► Compute to compare

$$t = \underline{\hspace{2cm}}$$

p-value = \underline{\hspace{2cm}}

### Step 4 ► Interpret

Reject or do not reject the null hypothesis. Then choose the correct interpretation:

i. At the 5% significance level, we cannot reject the hypothesis that the mean number of televisions of all students in the wealthier community is the same as the mean number of televisions of all students in the less wealthy community.

ii. At the 5% significance level, we conclude that the mean number of televisions of all students in the wealthier community is different from the mean number of televisions of all students in the less wealthy community.

**Confidence Interval**
Report the confidence interval for the difference in means given by Minitab, and state whether it captures 0 and what that shows.

**g** **9.59 Female Pulse Rates before and after a Fright** In a statistics class taught by one of the authors, students took their pulses before and after being frightened. The frightening event was having the teacher scream and run from one side of the room to the other. The pulse rates (beats per minute) of the women before and after the scream were obtained separately and are shown in the table. Treat this as though it were a random sample of female community college students.

QUESTION Test the hypothesis that the mean of college women's pulse rates is higher after a fright, using a significance level of 0.05, by following the steps below.

| Women | | Women | |
|---|---|---|---|
| **Pulse Before** | **Pulse After** | **Pulse Before** | **Pulse After** |
| 64 | 68 | 84 | 88 |
| 100 | 112 | 80 | 80 |
| 80 | 84 | 68 | 92 |
| 60 | 68 | 60 | 76 |
| 92 | 104 | 68 | 72 |
| 80 | 92 | 68 | 80 |
| 68 | 72 | | |

### Step 1 ► Hypothesize

$\mu$ is the mean number of beats per minute.

$$H_0: \mu_{before} = \mu_{after}$$
$$H_a: \mu_{before} \underline{\hspace{2cm}} \mu_{after}$$

### Step 2 ► Prepare

Choose a test: Should it be a paired *t*-test or a two-sample *t*-test? Why? Assume that the sample was random and that the distribution of differences is sufficiently Normal. Mention the level of significance.

### Step 3 ► Compute to compare

$$t = \underline{\hspace{2cm}}$$

p-value = \underline{\hspace{2cm}}

### Step 4 ► Interpret

Reject or do not reject $H_0$. Then write a sentence that includes "significant" or "significantly" in it. Report the sample mean pulse rate before the scream and the sample mean pulse rate after the scream.

## CHECK YOUR TECH

## Pulse Rates after Exercise: Understanding *t*

Pulse rates were observed for 35 people before and after running in place for 2 minutes. The Minitab output for a paired *t*-test is shown. We will check that the test statistic value reported by Minitab is correct by using a formula. We will not focus on the four steps.

```
Paired T-Test and CI: Pulse Before Run, Pulse After Run

Paired T for Pulse Before Run - Pulse After Run

                N       Mean     StDev   SE Mean
Pul Bef Run    35    73.6000   11.4358    1.9330
Pul Aft Run    35    92.5143   18.9432    3.2020
Difference     35   -18.9143   15.0497    2.5439


95% CI for mean difference: (-24.0840, -13.7445)
T-Test of mean difference = 0 (vs not = 0): T-Value = -7.44   P-Value = 0.000
```

Paired *t*-test formula:

$$t = \frac{\bar{x}_{\text{diff}} - 0}{SE_{\text{diff}}}$$

where

$$SE_{\text{diff}} = \frac{s_{\text{diff}}}{\sqrt{n}}$$

$\bar{x}_{\text{diff}}$ is the average difference

$s_{\text{diff}}$ is the standard deviation of the differences

$n$ is the number of data pairs

You will verify that *t* is –7.44 by following the steps below, using the formula given.

**Step 1 ▶ Find the numerator of *t*, $\bar{x}_{\text{diff}}$**
Find the difference between the average pulse before the run and the average pulse after the run. This is the numerator. Retain four decimal digits.

**Step 2 ▶ Find the denominator of *t*, $SE_{\text{diff}}$**
To verify the standard error (*SE*) of the difference, take the standard deviation of the difference (StDev) and divide by the square root of 35, which is the sample size. (You may round to four decimal digits, which is what Minitab does.)

**Step 3 ▶ Find *t***
Divide the numerator (from step 1) by the denominator from step 2 and see whether you get –7.44 (or close to it), the *t* reported in the output.

**Step 4 ▶ Understanding What Influences *t***
In answering these questions, consider whether the variable is in the numerator or in the denominator.

a. If the difference between means were larger, if all else were the same, would that cause *t* to be larger or smaller? Why?

b. If the standard deviation ($s_{\text{diff}}$) were larger, if all else were the same, would that cause *t* to be larger or smaller? Why?

c. If the sample size were larger, if all else were the same, would that cause *t* to be larger or smaller? Why?

# TechTips

## General Instructions for All Technology

Because of the limitations of the algorithms, precision, and rounding involved in the various technologies, there can be slight differences in the outputs. These differences can be noticeable, especially for the calculated p-values involving *t*-distributions. It is suggested that the technology that was used be reported along with the p-value, especially for two-sample *t*-tests.

**EXAMPLE A (ONE-SAMPLE *t*-TEST AND CONFIDENCE INTERVAL):** ▶ McDonald's sells ice cream cones, and the company's fact sheet says that these cones weigh 3.2 ounces. A random sample of 5 cones was obtained, and the weights were 4.2, 3.6, 3.9, 3.4, and 3.3 ounces. Test the hypothesis that the population mean is 3.2 ounces. Report the *t*- and p-values. Also find a 95% confidence interval for the population mean.

**EXAMPLE B (TWO-SAMPLE *t*-TEST AND CONFIDENCE INTERVAL):** ▶ Below are the GPAs for random samples of men and women.

Men: 3.0, 2.8, 3.5

Women: 2.2, 3.9, 3.0

Perform a two sample *t*-test to determine whether you can reject the hypothesis that the population means are equal. Find the *t*- and p-values. Also find a 95% confidence interval for the difference in means.

**EXAMPLE C (PAIRED *t*-TEST):** ▶ Here are the pulse rates (in beats per minute) before and after exercise for three randomly selected people.

| Person | Before | After |
|--------|--------|-------|
| A | 60 | 75 |
| B | 72 | 80 |
| C | 80 | 92 |

Determine whether you can reject the hypothesis that the population mean change is 0 (in other words, that the two population means are equal). Find the *t*- and p-values.

## TI-83/84

### One-Sample *t*-Test

1. Press **STAT** and choose **EDIT**, and type the data into **L1** (list one).

2. Press **STAT**, choose **TESTS**, and choose **2: T-Test**.

3. Note this but don't do it: If you did not have the data in the list and wanted to enter summary statistics such as $\bar{x}$, *s*, and *n*, you would put the cursor over **Stats** and press **ENTER**, and put in the required numbers.

4. See Figure 9A. Because you have raw data, put the cursor over **Data** and press **ENTER**.

Enter: $\mu_0$, **3.2**; **List**, **L1**; **Freq: 1**; put the cursor over ≠ and press **ENTER**; scroll down to **Calculate** and press **ENTER**.

▲ **FIGURE 9A** TI-83/84 Input for One-Sample *t*-Test

Your output should look like Figure 9B.

▲ **FIGURE 9B** TI-83/84 Output for One-sample *t*-Test

### One-Sample *t*-Interval

1. Press **STAT**, choose **TESTS**, and choose **8:TInterval**.

2. Choose **Data** because you have raw data. (If you had summary statistics, you would choose **Stats**.) Choose the correct **List** (to select **L1**, press **2nd** and **1**) and **C-Level**, here 0.95. Leave **Freq:1**, which is the default. Scroll down to **Calculate** and press **ENTER**.

The 95% confidence interval reported for the mean weight of the cones (ounces) is (3.2204, 4.1396).

### Two-Sample *t*-Test

1. Press **STAT**, choose **EDIT**, and put your data (GPAs) in two separate lists (unstacked). We put the men's GPAs into **L1** and the women's GPAs into **L2**.

2. Press **STAT**, choose **TESTS**, and choose **4:2-Samp-TTest**.

3. For **Inpt**, choose **Data** because we put the data into the lists. (If you had summary statistics, you would choose **Stats** and put in the required numbers.)

4. In choosing your options, be sure the lists chosen are the ones containing the data; leave the **Freq**s as 1, choose ≠ as the alternative, and choose **Pooled No** (which is the default). Scroll down to **Calculate** and press **ENTER**.

You should get the output shown in Figure 9C. The arrow down on the left-hand side means that you can scroll down to see more of the output.

▲ **FIGURE 9C** TI-83/84 Output for Two-Sample *t*-Test

### Two-Sample *t*-Interval

1. After entering your data into two lists, press **STAT**, choose **TESTS**, and choose **0:2-SampTInt**.

2. Choose **Data** because you have raw data. (If you had summary statistics, you would choose **Stats**.) Make sure the lists chosen are the ones with your data. Leave the default for **Freq1:1**, **Freq2:1** and **Pooled No**. Be sure the **C-Level** is **.95**. Scroll down to **Calculate** and press **ENTER**.

The interval for the GPA example will be (−1.744, 1.8774) if the men's data corresponded to **L1** and the women's to **L2**.

### Paired *t*-Test

1. Enter the data given in Example 3 into **L1** and **L2** as shown in Figure 9D.

▲ **FIGURE 9D** Obtaining the List of Differences for the TI-83/84

2. See Figure 9D. Use your arrows to move the cursor to the top of **L3** so that you are in the label region. Then press **2ND L1 - 2ND L2**. For the minus sign, be sure to use the button above the plus button. Then press **ENTER**, and you should see all the differences in **L3**.

3. Press **STAT**, choose **TESTS**, and choose **2: T-Test**.

4. See Figure 9E. For **Inpt** choose **Data**. Be sure $\mu_0$ is **0** because we are testing to see whether the mean difference is 0. Also be sure to choose **L3**, if that is where the differences are. Scroll down to **Calculate** and press **ENTER**.

▲ **FIGURE 9E** TI-83/84 Input for Paired *t*-Test

Your output should look like Figure 9F.

▲ **FIGURE 9F** TI-83/84 Output for Paired *t*-Test

### One-Sample *t*-Test and Confidence Interval

1. Type the weight of the cones in **C1** (column 1).

2. **Stat > Basic Statistics > 1-Sample t**

3. See Figure 9G. Click in the empty white box below **Samples in columns:**. A list of columns containing data will appear to the left. Double click **C1** to choose it. Check **Perform hypothesis test**, and put in **3.2** as the **Hypothesized mean**. (If you wanted a one-tailed test or a confidence level other than 95%, you would use **Options**.)

4. Click **OK**.

▲ **FIGURE 9G** Minitab Input Screen for One-Sample *t*-Test

The output is shown in Figure 9H.

```
One-Sample T: C1
Test of mu = 3.2 vs not = 3.2


Variable   N    Mean   StDev   SE Mean      95% CI        T      P
C1         5    3.680  0.370   0.166    (3.220, 4.140)  2.90   0.044
```

▲ **FIGURE 9H** Minitab Output for One-Sample *t*-Test and Confidence Interval

Note that the Minitab output (Figure 9H) includes the 95% confidence interval for the mean weight (3.22, 4.14) as well as the *t*-test.

### Two-Sample *t*-Test and Confidence Interval

Use stacked data with all the GPAs in one column. The second column will contain the categorical variable that designates groups: male or female.

1. Upload the data from the disk or use the following procedure: Enter the GPAs in the first column. In the second column put the corresponding **m** or **f**. (Complete words or coding are also allowed for the second column but you must decide on a system for one data set and stick to it. For example using F one time and f the other times within one data set will create problems.) Use headers for the columns: **GPA** and **Gender.**

2. **Stat > Basic Statistics > 2-Sample t**

3. Refer to Figure 9I. Choose **Samples in one column,** because we have stacked data. Click in the small box to the right of **Samples** at the top to activate the box. Then double click **GPA** and then double click **Gender** to get it into the **Subscripts** box. If you wanted to do a one-tailed test or to use a confidence level other than 95%, you would click **Options.**

   (If you had unstacked data, you would choose **Samples in different columns**, and choose both columns of data.)

4. Click **OK.**

▲ **FIGURE 9I** Minitab Input Screen for Two-Sample *t*-Test and Confidence Interval

The output is shown in Figure 9J. Note that the confidence interval is included.

```
Two-Sample T-Test and CI: GPA, Gender
Two-sample T for GPA
Gender  N   Mean   StDev   SE Mean
f       3   3.033  0.850    0.49
m       3   3.100  0.361    0.21

Difference = mu (f) - mu (m)
Estimate for difference:  -0.067
95% CI for difference:  (-2.361, 2.228)
T-Test of difference = 0 (vs not =): T-Value = -0.13 P-Value = 0.912 DF = 2
```

▲ **FIGURE 9J** Minitab Output for Two-Sample *t*-Test and Confidence Interval

### Paired *t*-Test and Confidence Interval

1. Type the numbers (pulse rates) in two columns. Label the first column "**Before**" and the second column "**After**."

2. **Stat > Basic Statistics > Paired t**

3. Double click **Before** and then double click **After.** (If you wanted a one-tailed test or a confidence level other than 95%, you would click **Options.**)

4. Click **OK.**

Figure 9K shows the output.

```
Paired T-Test and CI: Before, After
Paired T for Before - After

            N    Mean   StDev   SE Mean
Before      3   70.67   10.07    5.81
After       3   82.33    8.74    5.04
Difference  3  -11.67    3.51    2.03

95% CI for mean difference: (-20.39, -2.94)
T-Test of mean difference = 0 (vs not = 0): T-Value = -5.75 P-Value = 0.029
```

▲ **FIGURE 9K** Minitab Output for Paired *t*-Test and Confidence Interval

We are saving the one-sample *t*-test for last, because we have to treat it strangely.

### Two-Sample *t*-Test

1. Type the GPAs in two columns side by side.
2. Click on **Data** and **Data Analysis,** and then scroll down to **t-Test: Two-Sample Assuming Unequal Variances** and double click it.
3. See Figure 9L. For the **Variable 1 Range** select one column of numbers (don't include any labels). Then click inside the box for **Variable 2 Range**, and select the other column of numbers.
   You may leave the hypothesized mean difference empty, because the default value is 0, and that is what you want.
4. Click **OK**.

▲ **FIGURE 9L** Excel Input for Two-Sample *t*-Test

To see all of the output you may have to click **Home**, **Format** (in the **Cells** group), and **AutoFit Column Width**.

Figure 9M shows the relevant part of the output.

```
t Stat               0.125
P(T<=t)one-tail      0.454214709
P(T<-t)two-tail      0.908429419
```

▲ **FIGURE 9M** Part of the Excel Output for Two-Sample *t*-Test

For a one-tailed alternative hypothesis Excel always reports one-half the p-value for the two-tailed hypothesis. This is the correct p-value only when the observed value of the test statistic is consistent with the direction of the alternative hypothesis. (In other words, if the alternative hypothesis is ">", then the observed test statistic is positive; if "<", then observed value is negative.) If this is not the case, to find the correct p-value, calculate 1 minus the reported one-tailed p-value.

### Paired *t*-Test

1. Type the data into two columns.
2. Click on **Data**, **Data Analysis**, and **t test: Paired Two Sample for Means**, and follow the same procedure as for the two-sample *t*-test.

### One-Sample *t*-Test

1. You need to use a trick to force Excel to do this test. Enter the weights of the cones in column A, and put zeros in column B so that the columns are equal in length, as shown in Figure 9N.

| A | B |
|---|---|
| 4.2 | 0 |
| 3.6 | 0 |
| 3.9 | 0 |
| 3.4 | 0 |
| 3.3 | 0 |

▲ **FIGURE 9N** Excel Data Entry for One-Sample *t*-Test

2. Click **Data**, **Data Analysis**, and **t-Test: Paired Two Sample for Means**.
3. See Figure 9O. After selecting the two groups of data, you need to put the hypothesized mean in the box labeled **Hypothesized Mean Difference**. For the way we set up this example, it is **3.2**. (If you had entered 3.2's in column B, you would put in 0 for the Hypothesized Mean Difference.)
4. Click **OK**.

▲ **FIGURE 9O** Excel Input for One-Sample *t*-Test

Figure 9P shows the relevant part of the Excel output.

```
Hypothesized Mean Difference     3.2
Df                               4
P(T<=t) one-tail                 0.022063708
P(T<=t) two-tail                 0.044127415
```

▲ **FIGURE 9P** Relevant Part of the Excel Output for One-sample *t*-Test

Again the p-value for the one-tailed alternative hypothesis is consistent with the alternative hypothesis that the mean weight was *more* than 3.2 ounces.

### One-Sample *t*-Test

1. Type the weights of the cones into the first column.
2. **Stat > T statistics > one sample > with data**

   (If you had summary statistics, then after **one sample**, you would choose **with summary**.)
3. Click on the column containing the data, **var1**, and click **Next**.
4. See Figure 9Q. Put in the **Null mean**, which is **3.2** for the ice cream cones. Leave the default not equal for the **Alternative**, and click **Calculate**.

▲ **FIGURE 9Q** Part of StatCrunch Input for One-Sample *t*-Test

You will get the output shown in Figure 9R.

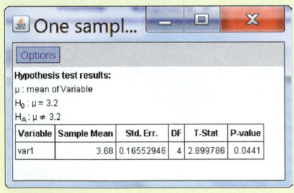

▲ **FIGURE 9R** StatCrunch Output for One-Sample *t*-Test

### One-Sample Confidence Interval

Go back and perform the same steps as for the one-sample *t*-test, but when you get to step 4, check **Confidence Interval**. You may change the confidence level from the default 0.95 if you want. See Figure 9Q.

### Two-Sample *t*-Test

Use stacked data with all the GPAs in one column. The second column will contain the categorical variable that designates groups: male or female.

1. Upload the data from the disk or follow these steps: Enter the GPAs in column 1 (**var1**). Put **m** or **f** in column 2 (**var2**) as appropriate. (Complete words or coding for column 2 are

also allowed but whatever system you use must be maintained within the data set.) Put labels at the top of the columns, change **var1** to **GPA** and **var2** to **Gender**.

2. **Stat > T statistics > Two sample > with data**
3. See Figure 9S. For **Sample 1** choose **GPA**. For the first **Where** put **Gender=m**. For **Sample 2** choose **GPA** and for the second **Where** put **Gender=f**. Click off **Pool Variances**. (If you had unstacked data you would choose the two lists for the two samples and not use the **Where** boxes.)

▲ **FIGURE 9S** StatCrunch Input for Two-Sample *t*-Test (Stacked Data)

4. Click **Next** if you want to change the alternative (the default is not equal) or want a confidence interval.
5. Click **Calculate**.

Figure 9T shows the output.

▲ **FIGURE 9T** StatCrunch Output for Two-Sample *t*-test

### Two-Sample Confidence Interval

1. Go back and do the preceding first three numbered steps.
2. When you get to step 4, click **Next**, and check **Confidence Interval**.
3. Click **Calculate**.

### Paired *t*-Test

1. Type the pulse rates Before in column 1 and the pulse rates After in column 2. The headings for the columns are not necessary.
2. **Stat > T statistics > Paired**
3. Select the two columns. Ignore the **Where** and **Group by** boxes. If you want to change your alternative hypothesis from the default two-tailed, click **Next**.
4. Click **Calculate.**

Figure 9U shows the output.

▲ **FIGURE 9U** StatCrunch Output for Paired *t*-Test

# 10 Associations between Categorical Variables

## Distributions of categorical variables are often summarized in two-way tables. We can make inferences about these distributions by calculating how many observations we would expect in each cell if the null hypothesis were true, and then comparing this to the actual counts.

As we have shown in the last few chapters, statistical tests are comparisons between two different views. In one view, if two samples have different means, then it might be because the populations are truly different. In the other view, randomness rules, and this difference is due merely to chance. A sample of people who had watched violent TV as children may behave more aggressively toward their partners when adults. But this might just be due to chance. Changing the placement of an advertisement on an Internet search results page might lead to a greater number of "follow-through" visits to the company that placed the ad. Or the difference could be due to chance—that is, if we were to repeat the study, we might see a very different outcome.

In this chapter we ask the same question as before: Are the results we see due to chance, or might something else be going on? In Chapter 8, we asked this question about one-sample and two-sample proportions. We now ask the same question with respect to categorical variables with multiple categories. For example, we might compare two categorical variables: the results of popping a batch of popcorn ("good" and "bad") for different amounts of oil ("no oil," "medium amount of oil," and "maximum oil"). Or we might compare a single categorical variable with a proposed model: Do the proportions of people hospitalized for swine flu for different age groups match the proportions of people in those age groups in the general population? If yes, this suggests that swine flu affects all age groups the same.

One of the nice surprises in this chapter is that only one test statistic is needed to analyze data collected under several different situations. This test statistic can be used to compare distributions of categorical variables, to compare population proportions, and to test whether two categorical variables are independent. You'll also be introduced in this chapter to a hypothesis test that requires no assumptions about the distribution of the population or the sample.

## CASE STUDY

# Popping Better Popcorn

You're planning a movie night and decide to cook up some popcorn. Many factors might affect how good the popcorn tastes: the brand, for example, or how much oil you use, or how long you let the popcorn pop before stopping. Some researchers (Kukuyeva, et al. 2008) investigated factors that might determine how to pop the perfect batch of popcorn. They decided that if more than half of the kernels in a bag were popped in the first 75 seconds, then the batch was a "success." If fewer than half, then it was a "failure." They popped 36 bags under three different treatments: no oil, "medium" amount of oil (1/2 tsp), and "maximum" oil (1 tsp). The bags were randomly assigned to an oil group. Each bag had 50 kernels. The outcome is

| Result | Oil Amount | | |
|---|---|---|---|
| | No Oil | Medium Oil | Maximum Oil |
| Failure | 23 | 22 | 33 |
| Success | 13 | 14 | 3 |

shown in the accompanying table. From the table it looks as though it is bad to use too much oil. But might this just be due to chance? In other words, if other investigators were to do this experiment the same way, would they also find so few successful bags with the maximum amount of oil?

This question—Is the outcome due to chance?—is one we've asked before. In this study the two variables are the amount of oil (with three values: none, medium, and maximum), and the result (success or failure). Both *Oil* and *Result* are categorical variables. In this chapter, we will see how to test the hypothesis that the amount of oil had an effect on the outcome.

# The Basic Ingredients for Testing with Categorical Variables

 **Details**

**Pronunciation**
*Chi* is pronounced "kie" (rhymes with "pie"), with a silent h.

Hypothesis tests that involve categorical variables follow the same four-step recipe you studied in Chapters 8 and 9. However, the basic ingredients are slightly different. Here we introduce these ingredients: data, expected counts, the chi-square statistic (our test statistic), and the chi-square distribution.

To provide a context for our discussion, we will test whether a six-sided die is "fair." In other words, is each side of a thrown die equally likely to end on top? We think so, because of the symmetry of a die. However, the manufacturing process must be very precise to achieve this. Surely, the dice we buy have a few slight defects here and there that would lead to them being biased.

We took a die from our favorite board game and decided to test whether it was fair. Our null hypothesis was that the die is balanced so that each side is equally likely to land face up. The alternative hypothesis was that one or more sides are favored. To test whether this die was fair, we rolled it 60 times. For each outcome, we counted the number of dots showing on top of the die.

Before we look at our data, let's take a moment to think about what the outcome would look like in a perfect world if the die really were fair. If we rolled a perfect die infinitely many times, the outcome should look like Figure 10.1—a uniform distribution, because each outcome is equally likely.

▶ **FIGURE 10.1** Probability distribution for an ideal, fair die. Each outcome is equally likely (with probability 1/6).

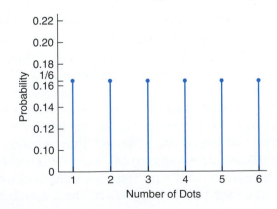

**Details**

**Raw Data**
The raw data look like this: "three dots," "two dots," "six dots," "two dots" . . . ." There are 60 entries.

How did our real die compare to the perfect, ideal die? Table 10.1 provides a summary of the data.

▶ **TABLE 10.1** Summary of outcomes with a six-sided die.

| Outcome: | One dot | Two dots | Three dots | Four dots | Five dots | Six dots |
|---|---|---|---|---|---|---|
| **Frequency:** | 12 | 8 | 10 | 11 | 9 | 10 |

Figure 10.2 shows the graph of the results with this real die, tossed 60 times.

◄ **FIGURE 10.2** Frequency counts for a real die, tossed 60 times. Comparing this to Figure 10.1, we see that it is much less uniform. Is this irregularity due to chance, or is the die unfair?

You can see more clearly from the graph than from the table that we did not get a perfectly uniform distribution. The question now is whether these deviations from the "perfect" model in Figure 10.1 occur because the die is not fair or are due merely to chance.

## 1. The Data

Recall that categorical variables are those whose values are categories, as opposed to numbers. The outcome of the roll of a die can be thought of as a category, particularly if we don't care about the numerical values on the die.

When two categorical variables are analyzed, as we are doing in this chapter, they are often displayed in a **two-way table**, a summary table that displays frequencies for the outcomes. Such tables can give the impression that the data are numerical, because you are seeing numbers in the table. But it's important to keep in mind that the numbers are *summaries* of variables whose values are *categories*.

For instance, the General Social Survey (GSS) asked respondents, "There are always some people whose ideas are considered bad or dangerous by other people. For instance, somebody who is against churches and religion. If such a person wanted to make a speech in your (city/town/community) against churches and religion, should he be allowed to speak, or not?" These respondents were also asked about their income and, depending on their response, were assigned to one of four annual income levels: $0–$20K, $20–$40K, $40–$70K, and $70K and more. The first four lines of the actual data would look something like Table 10.2. A value on the boundary would be assigned to the group with the larger incomes, so an income of 40 thousand would be in the 40–60 group.

| Observation ID | Response | Income Level |
|---|---|---|
| 1 | allowed | 20–40 |
| 2 | allowed | 20–40 |
| 3 | not allowed | 70 and over |
| 4 | allowed | 0–20 |
| . . . | . . . | . . . |

◄ **TABLE 10.2** The first four lines of the raw data (based on an actual data set) showing responses to two questions on the General Social Survey.

We have chosen to display another variable, *Observation ID*, which is useful simply for keeping track of the order in which the observations are stored.

A two-way table summarizes the association between the two variables *Response* and *Income Level* as in Table 10.3.

| Response | Income 0–20 | 20–40 | 40–70 | 70+ |
|---|---|---|---|---|
| Allowed | 2404 | 978 | 1015 | 971 |
| Not Allowed | 898 | 347 | 226 | 129 |

## 2. Expected Counts

The **expected counts** are the numbers of observations we would see in each cell of the summary table if the null hypotheses were true. In a few pages, we will set up some formal hypothesis tests, and you will see more precisely how to calculate the expected counts.

For example, Figure 10.2 shows the actual counts we observed when we rolled a six-sided die 60 times. If the die were truly fair, then because each outcome would have the same probability, we should see the same number of outcomes in each category.

When the die is rolled 60 times, the expected counts are therefore 10 in each outcome category. Because the probability for each category is $1/6$, we expect $1/6$ of the total outcomes to fall in each category: $(1/6) \times 60 = 10$. Table 10.4 compares these expected counts with our actual counts.

**Details**

**Expectations**
Expected counts are actually long-run averages. When we say that we "expect" 10 observations in a cell of a table, we mean that if the null hypothesis were true and we were to repeat this data collection many, many times, then, on average, we would see 10 observations in that cell.

▶ **TABLE 10.4** The expected (top) and observed (bottom) counts for rolling a fair die 60 times.

| Outcome | One Dot | Two Dots | Three Dots | Four Dots | Five Dots | Six Dots |
|---|---|---|---|---|---|---|
| Expected Counts | 10 | 10 | 10 | 10 | 10 | 10 |
| Observed Counts | 12 | 8 | 10 | 11 | 9 | 10 |

For a slightly more complex (but still quite typical) example, consider a study of the link between TV viewing and violent behavior. (We introduced these data in Chapter 8.) Researchers compared TV viewing habits for children, and then interviewed them about violent behavior many years later (Husemann et al. 2003). Table 10.5 summarizes the results.

▶ **TABLE 10.5** A two-way summary of TV violence and later abusive behavior.

| | High TV Violence | Low TV Violence | Total |
|---|---|---|---|
| Yes, Physical Abuse | 25 | 57 | 82 |
| No Physical Abuse | 41 | 206 | 247 |
| Total | 66 | 263 | 329 |

There are two categorical variables: *TV Violence* ("High" or "Low") and *Physical Abuse* ("Yes" or "No"). (Recall that the subjects were asked whether they had pushed, grabbed, or shoved their partner.) We wish to know whether these variables are associated. The null hypothesis says that they are not—that these variables are independent. In other words, any patterns you might see are due purely to chance.

What counts should we expect if these variables are truly not related to each other? That is, if the null hypothesis is true, what would we expect the table to look like? There are two ways of answering this question, and they both lead to the same answer. Let's look at them both.

## Starting with the *Physical Abuse* Variable

We notice that out of the entire sample, 82/329 (a proportion of 0.249240, or about 24.92%) said that they had physically abused their partners. If abuse is independent of TV watching—that is, if there is no relationship between these two variables—then we should expect to find the same percentages of violent abuse in those who watched high TV violence and in those with low TV violence.

So when we consider the 66 people who watched high TV violence as children, we expect 24.92% of them to have abused their partners. This translates to $(0.249240 \times 66) = 16.4498$ people. When we consider the 263 people who watched low TV violence, we expect again that 24.92%, or $0.249240 \times 263 = 65.5501$ people to have abused their partners.

We can use similar reasoning to find the other expected counts. We know that if a proportion of 0.249240 abused their partners, then $1 - 0.249240 = 0.750760$ did not. This proportion should be the same no matter which level of TV violence was watched. Thus, we expect the following:

In the "no physical abuse and high TV violence" group,

$$0.750760 \times 66 = 49.5502$$

In the "no physical abuse and low TV violence" group,

$$0.750760 \times 263 = 197.4499$$

(We are avoiding rounding in these intermediate steps so that our answers for the expected values will be as accurate as possible.)

We summarize these calculations in Table 10.6, which shows the expected counts in parentheses. We include the actual counts in the same table so that we can compare. Note that we rounded to two decimal digits for ease of presentation.

|  | High TV Violence | Low TV Violence | Total |
|---|---|---|---|
| **Yes, Physical Abuse** | 25 (16.45) | 57 (65.55) | 82 |
| **No Physical Abuse** | 41 (49.55) | 206 (197.45) | 247 |
| **Total** | 66 | 263 | 329 |

◀ **TABLE 10.6** TV Violence: A summary including expected values (in parentheses).

If we call the values 82 and 247 row totals (for obvious reasons, we hope!), 66 and 263 column totals, and 329 the grand total, we can generate a formula for automatically finding expected counts for each cell. This formula is rarely needed. First, you can and should always think through the calculations as we did here. Second, most software will do this automatically for you.

**Formula 10.1:**   $\text{Expected count for a cell} = \dfrac{(\text{row total}) \times (\text{column total})}{\text{grand total}}$

## Starting with the *TV Violence* Variable

The other way of finding the expected counts is to begin by considering the *TV Violence* variable, rather than beginning with the *Physical Abuse* variable. We see that 66/329 = 0.200608, or 20.06%, watched high TV violence. The rest, 263/329 = 0.799392, or 79.94%, watched low TV violence.

If these variables are not related, then when we look at the 82 people who committed physical abuse, we should expect about 20.06% of them to fall in the High TV Violence category and the rest to fall in the Low TV Violence category.

Also, when we look at the 247 who did not commit physical abuse, we should expect 20.06% of them to have viewed high TV violence.

Among the abusive, the expected number with high TV violence is $0.200608 \times 82 = 16.45$. Among the nonabusive, the expected number with high TV violence is $0.200608 \times 247 = 49.55$.

> **Details**
>
> **Fractions of People**
> Does it bother you that we have fractions of people in each category? It *is* a little strange, until you think about this in terms of an ideal model. These expected counts are like averages. We say the average family has 2.4 children, and we know very well that there is no single family with a 0.4 child. This number 2.4 is a description of the collection of all families. Our claim that we expect 16.45 people (in this group) who have seen high TV violence to be abusive is a similar idealization.

You see that you will get the same result no matter which variable you consider first.

## EXAMPLE 1 Gender and Opinion on Same-Sex Marriage

Do men and women feel differently about the issue of same-sex marriage? In 2008, the General Social Survey took a random sample of 1326 Americans, recording their gender and their level of agreement with the statement "Homosexuals should have the right to marry." The results are summarized in Table 10.7.

▶ **TABLE 10.7** Summary of gender, and opinion that homosexuals should have the right to marry.

| Opinion | Male | Female | Total |
|---|---|---|---|
| Strongly Agree | 80 | 143 | 223 |
| Agree | 124 | 173 | 297 |
| Neutral | 79 | 91 | 170 |
| Disagree | 100 | 113 | 213 |
| Strongly Disagree | 213 | 210 | 423 |
| Total | 596 | 730 | 1326 |

**QUESTION** Assuming that the two variables *Opinion* and *Gender* are *not* associated, find the expected number of males who would strongly agree and the expected number of females who would strongly agree.

**SOLUTION** We consider *Gender* first. The percentage of men in the sample is $(596/1326) \times 100\% = 44.9472\%$. The percentage of women is therefore $100\% - 44.9472\% = 55.0528\%$. If Gender is not associated with *Opinion*, then if we look at the 223 people who strongly agree, we should see that about 44.9% of those who strongly agree are male, and about 55.1% are female. In other words:

**CONCLUSION** Expected count of males who strongly agree $= 223 \times 0.449472 = 100.2323$, or about 100.23.
Expected count of females who strongly agree $= 223 \times 0.550528 = 122.7677$, or about 122.77.

**TRY THIS!** Exercise 10.9

## 3. The Chi-Square Statistic

Let's examine again our table of the relation of expected die tosses to our actual die tosses (Table 10.8).

▶ **TABLE 10.8** Sixty rolls of a six-sided die.

| Outcome: | One Dot | Two Dots | Three Dots | Four Dots | Five Dots | Six Dots |
|---|---|---|---|---|---|---|
| **Expected Counts (*E*)** | 10 | 10 | 10 | 10 | 10 | 10 |
| **Observed Counts (*O*)** | 12 | 8 | 10 | 11 | 9 | 10 |

We note that in the first category, we saw two more "aces" ("One Dot") than expected. On the other hand, we saw two fewer "Two Dots" and exactly the expected number of "Three Dots" and "Six Dots." Are these differences big or small? If they are small, then we can believe that the deviations between actual and expected counts are just due to chance. But if they are big, then maybe the die is not fair.

The **chi-square statistic** is a statistic that measures the amount that our expected counts differ from our observed counts. This statistic is shown in Formula 10.2.

$$\textbf{Formula 10.2:} \quad X^2 = \sum_{\text{cells}} \frac{(O - E)^2}{E}$$

where

$O$ is the observed count in each cell
$E$ is the expected count in each cell
$\Sigma$ means add the results from each cell

Why does this statistic work? The term $(O - E)$ is the difference between what we observe and what we expect under the null hypothesis. To measure the total amount of deviation between Observed and Expected, it is tempting to just add together the individual differences. But this doesn't work, because the expected counts and the observed counts always add to the same value; if we sum up the differences, they will always add to 0.

You can see that the differences between Observed and Expected add to 0 in Table 10.9, where we've added a row of differences (Observed Counts minus Expected Counts). You'll notice that $2 - 2 + 0 + 1 - 1 + 0 = 0$.

| Outcome: | One Dot | Two Dots | Three Dots | Four Dots | Five Dots | Six Dots |
|---|---|---|---|---|---|---|
| **Expected Counts** | 10 | 10 | 10 | 10 | 10 | 10 |
| **Observed Counts** | 12 | 8 | 10 | 11 | 9 | 10 |
| **Observed minus Expected** | 2 | −2 | 0 | 1 | −1 | 0 |

▲ **TABLE 10.9** Sixty rolls of a six-sided die, emphasizing the observed minus expected values.

One reason why the chi-square statistic uses squared differences is that by squaring the differences, we always get a positive value, because even negative numbers multiplied by themselves result in positive numbers:

$$2^2 + (-2)^2 + 0^2 + 1^2 + (-1)^2 + 0^2 = 4 + 4 + 0 + 1 + 1 + 0 = 10$$

Why divide by the expected count? The reason is that a difference between the expected and actual counts of, say, 2 is a small difference if we were expecting 1000 counts. But if we were expecting only 5 counts, then this difference of 2 is substantial. By dividing by the expected count, we're controlling for the size of the expected count. Basically, for each cell, we are finding what proportion of the expected count the squared difference is.

If we apply this formula to our test of whether the die is unbalanced, we get $X^2 = 1.0$. We must still decide whether this value discredits the null hypothesis that the die is fair. Keep reading.

## EXAMPLE 2 Viewing Violent TV as a Child and Abusiveness as an Adult

Table 10.10 on the next page shows summary statistics from a study that asked whether there was an association between watching violent TV as a child and aggressive behavior toward one's spouse later in life. The table shows both actual counts and expected counts (in parentheses).

▶ **TABLE 10.10** A two-way summary of the effect of viewing TV violence on later abusiveness (expected values are shown in parentheses).

|  | High TV Violence | Low TV Violence | Total |
|---|---|---|---|
| **Yes, Physical Abuse** | 25 (16.45) | 57 (65.55) | 82 |
| **No Physical Abuse** | 41 (49.55) | 206 (197.45) | 247 |
| **Total** | 66 | 263 | 329 |

**QUESTION** Find the chi-square statistic to measure the difference between the observed counts and expected counts for the study of the effect of violent TV on future behavior.

**SOLUTION** We use Formula 10.2 with the values for $O$ and $E$ taken from Table 10.10.

$$X^2 = \sum \frac{(\text{Observed} - \text{Expected})^2}{\text{Expected}}$$

$$= \frac{(25 - 16.45)^2}{16.45} + \frac{(57 - 65.55)^2}{65.55} + \frac{(41 - 49.55)^2}{49.55} + \frac{(206 - 197.45)^2}{197.45} = 7.4047$$

**CONCLUSION**

$$X^2 = 7.40$$

Later we will see whether this is an unusually large value for two independent variables.

**TRY THIS!** Exercise 10.15

---

As you might expect, for tables with many cells, these calculations can quickly become tiresome. Fortunately, technology comes to our rescue. Most statistical software will calculate the chi-square statistic for you, given data summarized in a two-way table or presented as raw data as in Table 10.2, and some software will even display the expected counts alongside the observed counts. Figure 10.3 shows the output from StatCrunch for these data.

What happens to the chi-square statisitic when the expected counts are exactly the same as the observed counts for every cell of a table? In our test to see whether a die was fair, we rolled the die 60 times. We expected 10 outcomes in each category. If we had gotten exactly 10 (for each cell), than our observations would have matched our expectations perfectly. In that case, the chi-square statistic would equal 0, because

$$(\text{Observed} - \text{Expected})^2 = (10 - 10)^2 = 0 \text{ in each cell}$$

Thus, when expectations and reality are exactly the same, the chi-square statistic is 0.

Even when the null hypothesis is true, our real-life observations will differ slightly from the expected counts just by chance. When this happens, the chi-square statistic will be a small value.

If reality is very different from what our null hypothesis claims, then our observed counts should differ substantially from the expected counts. When that happens, the chi-square statistic is a big value.

The trick, then, is to decide what values of the chi-square statistic are "big." Big values discredit the null hypothesis. To determine whether an observed value is big, we need to know its probability distribution when the null hypothesis is true.

◀ **FIGURE 10.3** StatCrunch output for TV violence and abusiveness. The expected values are below the observed values.

**Tech**

> **KEY POINT** If the data conform to the null hypothesis, then the value of the chi-square statistic will be small. For this reason, large values of the chi-square statistic make us suspicious of the null hypothesis.

## 4. The Sampling Distribution of the Chi-Square Statistic

We are wondering whether our die is fair or biased. If it's fair, then the deviations we see from our expected counts should be small, and the chi-square statistic should be small. We found that $X^2 = 1.0$ for the die. Is this small?

We are wondering whether television-viewing habits as a child are associated with violent behavior as an adult. If there is *no* association, then the observed counts in Table 10.10 should be close to the expected counts, and our chi-square statistic should be small. We found that $X^2 = 7.40$. Is this small?

To help us determine whether a particular value for a chi-square statistic is big or small, we need to know the probability distribution of $X^2$. Recall that the probability distribution of a statistic, such as $X^2$, is called a sampling distribution. Using the sampling distribution for $X^2$, we can find the probability that, just through chance, a chi-square statistic would have the value of 1.0 or bigger. If this probability is large, it means that 1.0 is not an unusual value. If the probability is small, it suggests that the value might be unusually large.

If the sample size is large enough, there is a probability distribution that gives a fairly good approximation to the sampling distribution. Not surprisingly, this approximate distribution is called the **chi-square distribution**. The chi-square distribution is often represented with the Greek lowercase letter chi ($\chi$) raised to the power of 2—that is, $\chi^2$.

Unlike the normal distribution and the *t*-distribution, the $\chi^2$ distribution allows for only positive values. It also differs from the other sampling distributions you've seen in that it is (usually) not symmetric and is instead right-skewed.

> **! Caution**
>
> **Symbols**
> The symbol $X^2$ is used to represent the chi-square *statistic*. The symbol $\chi^2$ is used to represent the chi-square *distribution*. Do not confuse the two. One is a statistic whose value is based on data; the other provides (approximate) probabilities for the values of that statistic.

> **Details**
>
> **Degrees of Freedom (df)**
> The degrees of freedom for the chi-square distribution are determined by the number of categories, not by the number of observations. For example, when a six-sided die is tossed, there are six categories.

Like the shape of the *t*-distribution, the shape of the chi-square distribution depends on a parameter called the **degrees of freedom**. The lower the degrees of freedom, the more skewed the shape of the chi-square distribution. Figure 10.4 shows the chi-square distribution for several different values of the degrees of freedom. Sometimes the degrees of freedom are indicated using this notation: $\chi^2_{df}$. For example, $\chi^2_6$ represents a chi-square distribution with 6 degrees of freedom, as in Figure 10.4b.

The degrees-of-freedom parameter is different for different tests, but in general it depends on the number of categories in the summary table.

▶ **FIGURE 10.4** Three chi-square distributions for **(a)** df = 1, **(b)** df = 6, and **(c)** df = 20. Note that the shape becomes more symmetric as the degrees of freedom (df) increase. No negative values are possible with the chi-square distribution; the smallest possible value is 0.

The chi-square distribution $\chi^2$ is only an approximation of the true sampling distribution of the statistic $X^2$. The approximation is usually quite good if all of the expected counts are 5 or higher.

> **KEY POINT**  The chi-square distribution provides a good approximation to the sampling distribution of the chi-square statistic only if the sample size is large. For many applications, the sample size is large enough if each expected count is 5 or higher.

At the start of this section, we asked whether the value we saw for the chi-square statistic when comparing an ideal die to our real die was big or small. Recall that the value was $X^2 = 1.0$. As you will see in Section 10.2, df = 5 for this problem. Figure 10.5 shows

▶ **FIGURE 10.5** A chi-square distribution with 5 degrees of freedom. The shaded area is the area above the value 1.0. This suggests that 1.0 is not a very large value for this distribution; most values are bigger than 1.0.

that 1.0 is not a very big value. Specifically, about 96% of the total area of the chi-square distribution is above 1.0.

We also asked whether the value $X^2 = 7.40$ was small for the problem of determining whether viewing TV violence was associated with abusiveness later in life. We'll answer this question in Section 10.3.

## Finding p-Values without Technology

Not all calculators will produce the analyses we're about to show you. Also, not all calculators will let you find probabilities with the chi-square distribution. If you find yourself without technology, you will have to rely on a published table of probabilities, such as Table 5 in Appendix A.

Tables of chi-square probabilities cannot be complete, because there are too many different chi-square distributions (one for each integer). The best we can do, then, is to approximate the p-value.

For example, Table 10.11 is from Table 5 in Appendix A and shows only a few "right- tailed" chi-square probabilities for a few different degree-of-freedom parameters (df = 4, df = 5, and df = 6).

| df | 0.10 | 0.05 | 0.025 | 0.01 | 0.005 |
|----|------|------|-------|------|-------|
| 4 | 7.78 | 9.49 | 11.14 | 13.28 | 14.86 |
| 5 | 9.24 | **11.07** | 12.83 | 15.09 | 16.75 |
| 6 | 10.64 | 12.59 | 14.45 | 16.81 | 18.55 |

◀ **TABLE 10.11** Selected values of chi-square probabilities.

For example, the table tells us that in a chi-square distribution with 5 degrees of freedom, the probability that we will get a value bigger than 11.07 is 0.05, or 5%. This distribution is illustrated in Figure 10.6.

◀ **FIGURE 10.6** The table of chi-square probabilities in Appendix A shows you which values have right-tailed probabilities of 0.10, 0.05, 0.025, 0.01, and 0.005. This figure illustrates that the value 11.07 has a right-tailed probability of 0.05. This means that the probability that the chi-square statistic will be bigger than 11.07 is 0.05.

How can we use Table 5 in Appendix A to find the p-value for our chi-square statistic of 1.0 with 5 degrees of freedom? (This was the chi-square statistic we saw after tossing a die 60 times and comparing the outcomes to the expected counts from a fair die.) We can't get the exact p-value, but we can see that the p-value is going to be fairly large. If $X^2 = 1.0$, we can see from Table 10.11 that this value is smaller than 9.24, the smallest value for the df = 5 row. Because the probabilities get larger as you move to the left, this means that whatever the p-value is, it must be bigger than 0.10. Thus, we can report simply that the p-value is bigger than 0.10, but using this table, we cannot be more precise.

Table 10.11 also tells us that if we plan to reject the null hypothesis whenever the p-value is less than 0.05, then if we have 5 degrees of freedom, we would reject the null hypothesis whenever the chi-square statistic is bigger than (or equal to) 11.07.

# The Chi-Square Test for Goodness of Fit

We now show how these ingredients can be combined to do a hypothesis test to determine whether a distribution of categorical variables is following any proposed distribution. These tests are called **goodness-of-fit tests**. Our experiment with the potentially unfair die was a goodness-of-fit test. We had a proposed distribution (each outcome is equally likely and happens with probability 1/6) and an experimental distribution that we got from rolling a real die.

The goodness-of-fit test is a hypothesis test, and it follows the same four-step procedure as the tests we presented in Chapters 8 and 9. It relies on the chi-square statistic and can be applied when you are comparing the distribution of counts of a categorical variable (*one* categorical variable) with a distribution of expected counts.

**Step 1: Hypothesize**
The hypotheses for a goodness-of-fit test are always the same:

> $H_0$: The population distribution of the variable is the same as the proposed distribution.
> $H_a$: The distributions are different.

For our die-testing example, we could state these hypotheses as

> $H_0$: The die is fair (each outcome is equally likely).
> $H_a$: The die is not fair.

**Step 2: Prepare**
We plan to use the chi-square goodness-of-fit test statistic to compare the observed number of outcomes in each category ($O$) with the expected number of outcomes ($E$).

Recall that an important feature of a hypothesis test is being able to measure how often our conclusions will be wrong. To do this, we need to know the probability distribution of our test statistic so that we can find the p-value and significance level for our test. Under the following conditions, the chi-square distribution with $k - 1$ degrees of freedom gives good approximations for the p-value and significance level. The letter $k$ is used to represent the number of categories for our variable. In the die example, $k = 6$, so we had $6 - 1 = 5$ degrees of freedom.

Conditions:

1. *Random Sample*. The sample was collected randomly.

2. *Independent Measurements*. Each measurement on an individual is independent of all other measurements.

3. *Large Sample*. The expected count is at least 5 in each cell.

The first condition shouldn't surprise you. In order for us to learn about the population, our sample must be representative of that population, so it must must be collected randomly.

The second condition is there to make sure that each observation provides a useful additional bit of information. This condition would be violated if, for example, you had randomly sampled married couples, instead of individuals, and recorded their marital status. Then, if you knew the marital status of one person in your sample, you would definitely know the marital status of another person. The "measurement" of assessing marital status would not be independent for these two people.

These first two conditions are difficult to check unless you were present when the data were collected or have access to very detailed notes about how they were collected. However, we often know enough to reasonably *assume* that these conditions hold.

Unlike the first two conditions, the third condition can be checked by looking at the data. This third condition is another way of saying that a large sample is required. It must be large enough for us to *expect* at least 5 observations in each cell (each category). Note that we don't need to actually get 5 or more observations in each cell. We just need to *expect* 5 or more.

## Goodness of Fit

We use the chi-square statistic (Formula 10.2) for the test statistic:

$$X^2 = \sum_{\text{cells}} \frac{(O - E)^2}{E}$$

If the expected counts are large enough, the chi-square distribution with degree of freedom = (number of categories − 1) is a good approximation for the distribution of $X^2$.

To complete the "Prepare" step, you need to calculate the expected counts and verify that all are greater than or equal to 5. This can be time-consuming, and we recommend that you use technology.

Another important part of the "Prepare" step is to choose a significance level. This is nearly always chosen to be 0.05, but you are allowed to choose any small probability, as long as you state it clearly.

### Step 3: Compute to compare

The chi-square statistic compares the observed counts to the expected counts. Calculating the value of this statistic is best done with technology, although for tables with only a few cells, it is not too tedious to do it by hand. Once we have the observed value of the chi-square statistic, we compute the p-value, which measures whether we should be surprised by how large the chi-square statistic is (if the null hypothesis is true.)

The p-value is the probability that, if the null hypothesis is true, a chi-square statistic will be as big as or bigger than the observed value. In other words, it is always the area under the $\chi^2$ probability curve for values as big or bigger than the one you observed. Figure 10.5 is an example of this. The shaded area represents the p-value for our die-rolling experiment. If the die really is fair (so that each outcome is equally likely), then the probability of getting a chi-square statistic as big as or bigger than 1.0 is about 96%.

### Step 4: Interpret

This step is the same, no matter which test we use. If the p-value is less than or equal to the significance level, $\alpha$(alpha), then we reject the null hypothesis. If the p-value is bigger than $\alpha$, then we do not reject.

 ## EXAMPLE 3 Ages of Swine Flu Victims

Table 10.12 shows the distribution of ages (by category) of all 619 reported swine flu victims in the United States between May 1, 2009, and June 11, 2009 (http://www.cdc .gov/mmwr/). The distribution is shown in Figure 10.7 on the next page. The age categories were created by medical experts.

| Age Group | Observed Frequency | Expected Frequency |
|-----------|--------------------|--------------------|
| Under 5 | 51 | 49.52 |
| 5–14 | 204 | 68.09 |
| 15–29 | 250 | 132.47 |
| 30–44 | 68 | 146.08 |
| 45–59 | 36 | 121.32 |
| 60 and over | 10 | 101.52 |

**Tech**

◀ **TABLE 10.12** The observed number of swine flu victims in each age category, and the number of victims that would be expected if the swine flu cases followed the same distribution as the general U.S. population.

◀ **FIGURE 10.7** Distribution of swine flu cases by age group, May 1–May 5, 2009, United States.

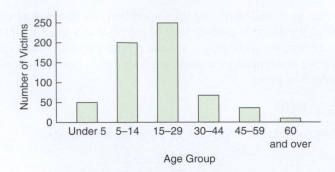

Does the swine flu affect people of some ages more than others? The Expected Frequency column shows the number of people we would expect in each age category if we simply took a random sample of 619 people, based on U.S. Census data. If these two distributions are the same, it suggests that the swine flu affects all ages equally. If they are different, then it might mean that some age groups are more (or less) susceptible to this strain of the flu.

**QUESTION** Test whether the observed distribution fits the distribution of ages in the United States (shown in Table 10.12), using a significance level of 0.05.

**SOLUTION** The *Age Group* variable has six categories. Overall, there were 619 cases of swine flu. The age group with the most cases was the group containing people 15–29 years old.

*Step 1: Hypothesize*
$H_0$: The distribution of ages of swine flu victims follows that of the U.S. population.
$H_a$: The distribution of ages of swine flu victims differs from that of the U.S. population.

*Step 2: Prepare*
Because we have a single random sample of independent observations (or we assume so) and are measuring a single categorical variable (*Age Group*), we choose a chi-square goodness-of-fit test to compare the distribution of ages to a theoretical distribution. In this case, the theoretical distribution is the age distribution of all people in the United States. The smallest expected count is 49.52, which is much more than the required minimum expected count of 5, so the chi-square distribution will provide a good approximation to the p-value.

*Step 3: Compute to compare*
Figure 10.8 shows both the data as entered in StatCrunch and the output from the goodness-of-fit test. StatCrunch, like some other software packages, requires that you give it not the raw data but a column of the observed counts and another column of expected counts.

We see that the chi-square statistic is 559.85. This is a very large value. The small p-value (less than 0.0001) tells us that if it were true that the age distribution of swine flu victims was the same as that of the general U.S. population, and if these victims were randomly selected from that population, then this is a very unlikely outcome. Put differently, if the null hypothesis is correct and the distribution of swine flu victims across age categories is the same as it is for the general U.S. population, then we should be very surprised at getting such a large chi-square statistic.

Note that the degree of freedom (labeled "DF" in Figure 10.8) is the number of categories, 6, minus 1.

*Step 4: Interpret*
Because of the small p-value (much less than our significance level of 0.05), we reject the null hypothesis. We find that the distribution of ages of swine flu victims differs

 **Looking Back**

**p-Value**
Recall that the p-value is the probability of getting a test statistic as extreme as or more extreme than what was observed, assuming that the null hypothesis is true. For the chi-square statistic, "as extreme as or more extreme than" means "as big as or bigger than."

| StatCrunch | Data | Stat | Graphics | Help | | |
|---|---|---|---|---|---|---|
| Row | age group (years) | | swine flu count | | expected count | |
| 1 | Under 5 | | 51 | | 49.52 | |
| 2 | 5–14 | | 204 | | 68.09 | |
| 3 | 15–29 | | 250 | | 132.466 | |
| 4 | 30–44 | | 68 | | 146.084 | |
| 5 | 45–59 | | 36 | | 121.324 | |
| 6 | 60 and over | | 10 | | 101.516 | |
| 7 | | | | | | |
| 8 | | | | | | |
| 9 | | | | | | |
| 10 | | | | | | |
| 11 | | | | | | |
| 12 | | | | | | |
| 13 | | | | | | |
| 14 | | | | | | |
| 15 | | | | | | |
| 16 | | | | | | |

Chi-Square test

Options

**Chi-Square goodness-of-fit results:**
Observed: swine flu count
Expected: expected count

| N | DF | Chi-Square | P-Value |
|---|---|---|---|
| 619 | 5 | 559.85474 | <0.0001 |

◄ **FIGURE 10.8** Data to calculate the goodness-of-fit test, and also the output from that test, using StatCrunch.

from the distribution of ages in the general U.S. population, which suggests that the effects of the swine flu depended on the victim's age.

**CONCLUSION** The distribution of ages of swine flu victims does not match that of the general U.S. population. In fact, evidence appeared in early 2009 that older people were more likely than others to be immune to the H1N1 virus that causes swine flu.

**TRY THIS!** Exercise 10.19

## SNAPSHOT  CHI-SQUARE TEST OF GOODNESS-OF-FIT

**WHAT IS IT?** ► A test of whether a distribution of a single categorical variable follows a proposed distribution.

**WHAT DOES IT DO?** ► Using the chi-square statistic, it compares the observed counts with the counts we would expect if the proposed distribution were the true distribution.

**HOW DOES IT DO IT?** ► If the sample size is large enough and basic conditions are met, the chi-square statistic follows a chi-square distribution with df = (number of categories − 1). The p-value is the probability of getting a value as large as or larger than the observed chi-square statistic value, using the chi-square distribution.

**HOW IS IT USED?** ► To compare a single categorical variable to a theoretical distribution.

*Note:* The chi-square statistic is designed so that it ranges from 0 to potentially very large positive values. If the null hypothesis is true, the chi-square statistic will be close to 0. If the alternative hypothesis is true, the chi-square statistic should be big. For these reasons, when finding the p-value, we always use the area to the right of the observed test statistic.

## SECTION 10.3

# Chi-Square Tests for Associations between Categorical Variables

There are two tests to determine whether two categorical variables are associated. Which test you used depends on how the data were collected.

We often summarize two categorical variables in a two-way table, in part because it helps us see whether associations exist between these variables. When we examine a two-way table (or any data summary, for that matter), one aspect that gets hidden is the method used to collect the data. We can collect data that might appear in a two-way table in either of two ways.

The first method is to collect two or more distinct, independent samples, one from each population. Each object sampled has a categorical value that we record. For example, we could collect a random sample of men and a distinct random sample of women. We could then ask them to what extent they agree with the statement that same-sex marriage should be allowed: Strongly Agree, Agree, Neutral, Disagree, or Strongly Disagree. We now have one categorical response variable, *Opinion*. We also have another categorical variable, *Gender*, that keeps track of which population the response belongs to. Hence we have two samples, one categorical response variable, and one categorical grouping variable.

The second method is to collect just one sample. For the objects in this sample, we record two categorical response variables. For example, we might collect a large sample of people and record their marital status (single, married, divorced, or widowed) and their educational level (high school, college, graduate school). From this one sample, we get two categorical variables: *Marital Status* and *Educational Level*.

In both data collection methods, we are interested in knowing whether the two categorical variables are related or unrelated. However, because the data collection methods are different, the ways in which we test the relationship between variables differ also. That's the bad news. The good news is that this difference is all behind the scenes. The result of careful calculation shows us that no matter which method we use to collect data, we can use the same chi-square statistic and the same chi-square distribution to test the relation between variables.

These two methods have different names. If we test the association, based on two independent samples, between the grouping variable and the categorical response variable (the first method), the test is called a test of **homogeneity**. If we base the test on one sample (the second method), the test is called a test of **independence**. Two different data collection methods, two different names—but, fortunately, the same test!

**Details**

**Homogeneity**
The word *homogeneity* is based on the word *homogeneous*, which means "of the same, or similar, kind or nature."

**KEY POINT** — There are two tests to determine whether two categorical variables are associated. For two or more samples and one categorical response variable, we use a test of homogeneity. For one sample and two categorical response variables, we use a test of independence.

## EXAMPLE 4 Independence or Homogeneity?

The Pew Foundation conducts a regular survey to measure consumers' attitudes toward certain products. In 2006, Pew randomly sampled 2000 people and asked whether they considered a TV set a luxury or a necessity. In 2009, while the U.S. was in the middle of one of the most severe recessions in its history, Pew asked another random sample of 1003 people the same question. The researchers want to know whether consumers' attitudes about TV sets have changed since 2006, and they plan to carry out a formal hypothesis test to see whether an association exists between the variables *Year* and *Attitude*.

**QUESTION** Will this be a test of independence or a test of homogeneity?

**SOLUTION** The Pew Foundation took two distinct samples of people: the first sample in 2006 and the second in 2009. The variable *Year* simply tells us which group the respondents belonged to. There is one response variable: *Attitude*. So this test has two samples and one response variable, and is a test of homogeneity.

**CONCLUSION** The proposed hypothesis test is a test of homogeneity

**TRY THIS!** Exercise 10.27

# EXAMPLE 5 Independence or Homogeneity?

The 2008 U.S. presidential election was historic for many reasons, but one was that a woman (Hillary Clinton) and an African American (Barack Obama) both had good chances of winning the Democratic nomination. Before the Democratic National Convention, the General Social Survey polled a random sample of 1286 people eligible to vote and asked whether they would (a) vote for a woman and (b) vote for an African American. One question of interest is whether those who would take the historic leap of voting for a woman for president would also be willing to vote for an African American. (You can imagine that the Obama campaign, expecting a close battle with the Clinton campaign, might have been interested in this question.)

**QUESTION** If we test whether a person's willingness to vote for a woman for president is associated with that person's willingness to vote for an African American for president, is this a test of homogeneity or of independence?

**CONCLUSION** There is one sample, consisting of 1286 people. These people are asked two questions, so there are two outcome variables: "Would you vote for a woman?" and "Would you vote for an African American? This is a test of independence.

**TRY THIS!** Exercise 10.29

## Tests of Independence and Homogeneity

Again, the tests follow the four-step procedure of all hypothesis tests. We'll give you an overview and then fill in the details with an example.

### Step 1: Hypothesize

The hypotheses are always the same:

$H_0$: There is *no* association between the two variables (the variables are independent).

$H_a$: There is an association between the two variables (the variables are not independent).

Although the hypotheses are always the same, you should phrase these hypotheses in the context of the problem.

### Step 2: Prepare

Whether you are testing independence or homogeneity, the test statistic you should use to compare counts is the chi-square statistic, shown in Formula 10.2 and repeated here.

$$X^2 = \sum_{\text{cells}} \frac{(O - E)^2}{E}$$

If the conditions are right, then this statistic follows, approximately, a chi-square distribution with

$$df = (\text{number of rows} - 1)(\text{number of columns} - 1)$$

Conditions:

1. *Random Samples.* All samples were collected randomly.

2. *Independent Samples and Observations.* All samples are independent of each other. Always, the observations within a sample must be independent of each other.

3. *Large Samples.* The expected count must be 5 or more in each cell.

Note that in a test of independence, there is always only one sample. But a test of homogeneity might have several independent samples.

### Step 3: Compute to compare

This step is best done with technology. The p-value is the probability, assuming the null hypothesis is true, of getting a value as large as or larger than the observed chi-square statistic. In other words, the p-value is the probability, if the variables really are not associated, that we would see an outcome as large or larger than the one observed. A small p-value therefore means a large test statistic, which casts doubt on the hypothesis that the variables are not associated.

### Step 4: Interpret

If the p-value is less than or equal to the stated significance level, we reject the null hypothesis and conclude that the variables are associated.

 ## EXAMPLE **6** Education and Marital Status

Does a person's educational level affect his or her decision about marrying? With observational data, we can't know for certain, but we can see whether the data are consistent with or contradict the idea that people's educational level affects their decisions about marrying. From the U.S. Census data, we took a random sample of 665 people and measured their marital status (single, married, divorced, or widow/widower) and their educational level (less than high school, high school degree, college degree or higher). You saw these data first in Chapter 5. Figure 10.9 shows output from StatCrunch.

QUESTION  Use the provided output to test whether marital status and educational level are associated. Is this a test of homogeneity or of independence?

SOLUTION  Because there is one sample with two response variables, this is a test of independence.

*Step 1: Hypothesize*

> $H_0$: Among all U.S. residents, marital status and educational level are independent.
> $H_a$: Among all U.S. residents, marital status and educational level are associated.

*Step 2: Prepare*

We can see from the output that the expected counts are 5 or more. (The smallest expected count is 11.47.) This means that the chi-square distribution will provide a good approximation for the p-value. The data are summarized in Figure 10.9. This table displays the actual counts for a random sample of 665 people, the expected counts, and the calculated values.

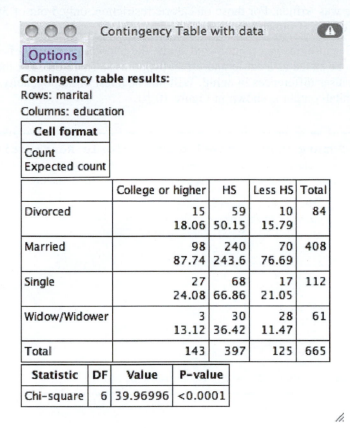

**Contingency Table with data**

**Options**

**Contingency table results:**
Rows: marital
Columns: education

| Cell format |
| --- |
| Count |
| Expected count |

| | College or higher | HS | Less HS | Total |
| --- | --- | --- | --- | --- |
| Divorced | 15 | 59 | 10 | 84 |
| | 18.06 | 50.15 | 15.79 | |
| Married | 98 | 240 | 70 | 408 |
| | 87.74 | 243.6 | 76.69 | |
| Single | 27 | 68 | 17 | 112 |
| | 24.08 | 66.86 | 21.05 | |
| Widow/Widower | 3 | 30 | 28 | 61 |
| | 13.12 | 36.42 | 11.47 | |
| Total | 143 | 397 | 125 | 665 |

| Statistic | DF | Value | P-value |
| --- | --- | --- | --- |
| Chi-square | 6 | 39.96996 | <0.0001 |

▲ **FIGURE 10.9** Summary table with expected counts, chi-square statistic, and p-value to test whether marital status and educational level are associated.

Tech

*Step 3: Compute to compare*
The chi-square statistic is $X^2 = 39.97$ (rounding to two decimal digits). The chi-square distribution has 6 degrees of freedom, and the p-value is reported as smaller than 0.0001.

*Step 4: Interpret*
Because the p-value is less than 0.05, we reject the null hypothesis.

**CONCLUSION** Marital status and educational level are associated.

**TRY THIS!** Exercise 10.33

# EXAMPLE **7** Hungry Monkeys

Research in the past has suggested that mice and rats that are fed less food live longer and healthier lives. Recently, a study of Rhesus monkeys was done that involved caloric restriction (less food). It is believed that monkeys have many similarities to humans, which is what makes this study so interesting.

Seventy-six Rhesus monkeys, all young adults, were randomly divided into two groups. Half of the monkeys (38) were assigned to caloric restriction. Their food was

► **FIGURE 10.10** Minitab output showing *Diet* (normal or caloric) and *Aging* (died from age-related causes or not) for 76 monkeys.

decreased about 10% per month for three months, and as a result they were fed about 30% less food than the other 38 monkeys for the duration of the experiment.

For those on the normal diet, 14 out of 38 had died of age-related causes by the time the article was written. For those on caloric restriction, only 5 out of 38 had died of age-related causes (Colman et al. 2009).

**QUESTION** Because this is a randomized study, the hypothesis that diet is associated with aging can be stated as a cause-and-effect hypothesis. Therefore, test the hypothesis that diet causes differences in aging. Will this be a test of homogeneity or of independence? Minitab output is shown in Figure 10.10.

```
Chi-Square Test: normal diet, caloric Restriction
Expected counts are printed below observed counts
Chi-Square contributions are printed below expected counts

            Normal       Caloric
             Diet      Restriction   Total
Died          14             5        19
             9.50          9.50
             2.132         2.132

Not           24            33        57
            28.50         28.50
             0.711         0.711

Total         38            38        76

Chi-Sq = 5.684, DF = 1, P-Value = 0.017
```

**SOLUTION** There are two samples (monkeys with caloric restriction and monkeys without) and one outcome variable: whether the monkey died of age-related causes. Therefore, this is a test of homogeneity.

*Step 1: Hypothesize*

> $H_0$: For these monkeys, the amount of calories in the diet is independent of aging.
> $H_a$: For these monkeys, the amount of calories in the diet causes differences in aging. Because we do not have a random sample, our results do not generalize beyond this group of monkeys. But because we have randomized assignment to treatment groups, we are able to conclude that any differences we see in the aging process are caused by the diet.

*Step 2: Prepare*
Because we wish to use the chi-square test, we must confirm that the expected counts are all 5 or more. This is the case here, as Figure 10.10 confirms.

*Step 3: Compute to compare*
From the Minitab output, Figure 10.10, you can see that the chi-square value is 5.68 and the p-value is 0.017.

*Step 4: Interpret*
Because the p-value is less than 0.05, we can reject the null hypothesis.

**CONCLUSION** The monkeys' deaths from age-related causes were caused by differences in the number of calories in the diet.

**TRY THIS!** Exercise 10.39

The article contains other information that suggests that monkeys on a restricted diet are generally healthier than monkeys on a normal diet. Figure 10.11 shows a photo of two monkeys. The one on the right had the restricted diet and shows fewer characteristics of old age.

◀ **FIGURE 10.11** The healthier monkey (on the right) was one of those on caloric restriction.

## Random Samples and Randomized Assignment

You have now seen randomization used in two different ways. Random sampling is the practice of selecting objects in our sample by choosing them at random from the population, as is done in many surveys. We can make generalizations about the population only if the sample is selected randomly, because this is the only way of ensuring that the sample is representative of the population. The General Social Survey is an example of studies based on random sampling. When we conclude, as we did in Example 6, that marital status and educational level are associated, we are stating a conclusion about the entire population—in this case, all adults in the United States. We are confident that these variables are associated in the population, because our sample was selected at random.

In Example 7, on the other hand, there was no random sample. However, the monkeys were randomly assigned to a treatment group (low-calorie diet) or the control group (normal diet). Because the monkeys were not selected randomly, we have no means of generalizing about the population as a whole, statistically speaking. (There might be a biological argument, or an assumption, that a diet that works on one group of monkeys would work on any other group, but as statisticians, we have no data to support this assumption.) However, the researchers performed this study because they were interested in a cause-and-effect relationship: Does changing the calories in a monkey's diet change the monkey's health and longevity?

Because researchers controlled which monkeys got which diet, this is a controlled experiment. And because they used randomized assignment, and because we rejected the null hypothesis that diet and health were independent, we can conclude that in fact the caloric restriction *did* affect the monkeys' health.

> 🔁 **Looking Back**
>
> **Data Collection**
> Controlled experiments are those in which experimenters determine how subjects are assigned to treatment groups. In contrast, observational studies are those in which subjects place themselves into treatment groups, by behavior or innate characteristics such as gender. Causal conclusions cannot be based on a single observational study.

| | |
|---|---|
| **WHAT IS IT?** ▶ | A test of whether two categorical variables are associated. |
| **WHAT DOES IT DO?** ▶ | Using the chi-square statistic, we compare the observed counts in each outcome category with the counts we would expect if the variables were *not* related. If observations are too different from expectations, then the assumption that there is no association between the categorical variables looks suspicious. |
| **HOW DOES IT DO IT?** ▶ | If the sample size is large enough and basic conditions are met, then the chi-square statistic follows, approximately, a chi-square distribution with df = (number of rows − 1) × (number of columns − 1). The p-value is the probability of getting a value as large as or larger than the observed chi-square statistic, using the chi-square distribution. |
| **HOW IS IT USED?** ▶ | To compare distributions of two categorical variables. |

## Relation to Tests of Proportions

In the special case in which both categorical variables have only two categories, the test of homogeneity is identical to a $z$-test of two proportions, using a two-tailed alternative hypothesis. The following analysis illustrates this.

In a landmark study of a potential AIDS vaccine published in 2009, researchers from the U.S. Army and the Thai Ministry of Health randomly assigned about 8200 volunteers to receive a vaccine against AIDS and another 8200 to receive a placebo. (We rounded the numbers slightly to make this discussion easier.) Both groups received counseling on AIDS prevention measures and were promised life-time treatment should they contract AIDS. Of those who received the vaccine, 51 had AIDS at the end of the study (three years later). Of those that received the placebo, 74 had AIDS (http://www.hivresearch.org/, accessed September 29, 2009). We will show two ways of testing whether an association existed between receiving the vaccine and getting AIDS. The data are summarized in Table 10.13.

▶ **TABLE 10.13** Effect of the vaccine on AIDS.

| | Vaccine | No Vaccine | Total |
|---|---|---|---|
| **AIDS** | 51 | 74 | 125 |
| **No AIDS** | 8149 | 8126 | 16275 |
| **Total** | 8200 | 8200 | 16400 |

If we use the approach of this chapter, we would recognize that this is a test of homogeneity, because there are two samples (*Vaccine* and *Placebo*) and one outcome variable (*AIDS*). Although we cannot generalize to a larger population (because the volunteers were not randomly selected), we can make a cause-and-effect conclusion about whether differences in AIDS rates are due to the vaccine, because this is a controlled, randomized study.

As a first step, we calculate the expected counts, under the assumption that the two variables are not associated.

Because the proportion of those who got AIDS was 125/16400 = 0.007622, if the risk of getting AIDS had nothing to do with the vaccine, then we should see about the same proportion of those getting AIDS in both groups. If the proportion of people who got AIDS in the *Vaccine* group was 0.007622, then we would expect 8200 × 0.007622 = 62.5 people to get AIDS in the *Vaccine* group.

Both groups are the same size, so we would expect the same number of AIDS victims in the *Placebo* group. This means that in both groups, we would expect $8200 - 62.5 = 8137.5$ not to get AIDS.

The results, with expected counts in parentheses to the right of the observed counts, are shown in Table 10.14.

|  | Vaccine | No Vaccine | Total |
|---|---|---|---|
| **AIDS** | 51 (62.5) | 74 (62.5) | 125 |
| **No AIDS** | 8149 (8137.5) | 8126 (8137.5) | 16275 |
| **Total** | 8200 | 8200 | 16400 |

◀ **TABLE 10.14** Expected counts, assuming no association between variables, are shown in parentheses.

Note that all expected counts are much greater than 5.

The chi-square statistic is not difficult to calculate:

$$X^2 = \sum \frac{(\text{Observed} - \text{Expected})^2}{\text{Expected}}$$

$$= \frac{(51 - 62.5)^2}{62.5} + \frac{(74 - 62.5)^2}{62.5} + \frac{(8149 - 8137.5)^2}{8137.5} + \frac{(8126 - 8137.5)^2}{8137.5}$$

$$= 4.26$$

The degree of freedom of the corresponding chi-square distribution is

$$(\text{Number of rows} - 1)(\text{number of columns} - 1) = (2 - 1)(2 - 1) = 1 \times 1 = 1$$

The p-value is illustrated in Figure 10.12. It is the area under a chi-square distribution with 1 degree of freedom and to the right of 4.26. The p-value turns out to be 0.039. We therefore reject the null hypothesis and conclude that there is an association between getting the vaccination and contracting AIDS. The difference in the numbers of AIDS victims was caused by the vaccine.

◀ **FIGURE 10.12** The area to the right of 4.26 represents the p-value to test whether there is an association between receiving the AIDS vaccine and contracting AIDS. The distribution is a chi-square distribution with 1 degree of freedom. The p-value is 0.039.

One thing that's disappointing about this conclusion is that the alternative hypothesis only states that the variables are associated. That's nice, but what we really want to know is *how* they are associated. Did the vaccine decrease the number of people who got AIDS? That's what the researchers wanted to know. They didn't want to know merely whether there was an association. They had a very specific direction in mind for this association.

One drawback with chi-square tests is that they reveal only whether two variables are associated, not how they are associated. Fortunately, when both categorical variables have only two categories, we can instead do a two-proportion z-test.

Define $p_1$ to be the probability that someone given the vaccine will get AIDS, and define $p_2$ to be the probability that someone given the placebo will get AIDS. Then

## Step 1: Hypothesize

$H_0: p_1 = p_2$

$H_a: p_1 < p_2$

In words, the null hypothesis says that the probability of getting AIDS is the same whether or not the individual gets the vaccine. The alternative hypothesis is one-tailed: If the individual gets the vaccine, the probability of contracting AIDS is less than if she or he got the placebo.

## Step 2: Prepare

We plan to use the $z$-test for two independent proportions. Under the null hypothesis, the probability of getting AIDS is the same for both groups, and we estimate this probability with

$$\hat{p} = \frac{(51 + 74)}{16400} = 0.007622$$

(This is the same proportion we calculated above.)

We must check that both samples are large enough—both expected values must be bigger than 10. This is straightforward, because both samples are the same size:

$$8200 \times 0.007622 = 62.5$$

This means that we can use the standard Normal distribution (with mean 0 and standard deviation 1) to approximate the p-value.

## Step 3: Calculate to compare

The observed value of the $z$-statistic is

$$z = \frac{\hat{p}_1 - \hat{p}_2}{\sqrt{\hat{p}\left(1 - \hat{p}\right)\left(\frac{1}{n_1} + \frac{1}{n_2}\right)}}$$

Note that $\hat{p}_1 = 51/8200$, $\hat{p}_2 = 74/8200$, and both $n_1$ and $n_2$ are equal to 8200. Substituting, we get

$$z = -2.065068, \text{ or about } -2.07$$

It is no coincidence that if you square this value, you get the same value as our chi-square statistic. Recall that $X^2 = 4.26$ and

$$z^2 = (-2.065068)^2 = 4.26$$

For the p-value, we find the area under a N(0, 1) curve below $-2.06$. This probability is 0.0194582, or about 0.019.

## Step 4: Interpret

Based on the p-value, we reject the null hypothesis and conclude that people who receive the vaccine are less likely to contract AIDS than those who do not. Because this was a randomized, controlled study, we can conclude that the vaccine caused a decrease in the probability of people getting AIDS.

Note that if we had doubled the p-value—that is, if we had instead used a two-tailed alternative hypothesis—we would have got $2 \times 0.019458 = 0.039$, almost exactly what we got for the chi-square test.

Although medical researchers were very excited about this study—not too many years before, it was thought to be impossible to develop an AIDS vaccine—they caution that this vaccine lowers risk by only about 30%. Many medical professionals consider a vaccine to be useful only if it lowers risk by at least 60%. There was also some controversy over this study: Some argued that a few of the subjects who were dropped from the initial analysis (because of preexisting medical conditions) should have been included. If these subjects had been included, then the vaccine would no longer have been judged effective, statistically speaking.

**KEY POINT** For a 2-by-2 two-way table of counts, a two-proportion *z*-test with a two-tailed alternative is equivalent to a test of homogeneity.

When should you use the two proportion *z*-test, and when should you use the test for homogeneity? If you need to use a one-tailed alternative hypothesis, then you should use the *z*-test. However, if you plan to use a two-tailed alternative hypothesis, then it doesn't matter which test you use.

## SECTION 10.4

# Hypothesis Tests When Sample Sizes Are Small

We demonstrate two approaches for dealing with data in which the expected values are less than 5. One approach is to combine categories so that each of the new, larger categories has an expected count of 5 or more. The other approach is to use Fisher's Exact Test. Unlike the chi-square approach, which gives us approximate p-values, Fisher's Exact Test gives the actual p-value.

## Combining Categories

The next example illustrates how to combine categories with small expected counts so that you can do a chi-square test.

 **EXAMPLE 8** Swine Flu Hospitalizations

In Example 3 we determined with a goodness-of-fit test that swine flu did not affect all age groups equally. Swine flu was found less frequently in older people, possibly because of immunity they gained during previous flu seasons. Now we will look at rates of hospitalization for swine flu in different age categories (http://www.cdc.gov/mmwr). Different hospitalization rates might suggest that the severity of the flu is different by age group. Perhaps older people are less likely to get sick, but when they do, they get a more severe form. Table 10.15 looks at the relation between *Age Category* and *Hospitalized* for all swine flu victims in the United States (before May 5, 2009). Presumably, hospitalization occurs when the victim has a more severe form of the illness.

| | | Age Category | | | | | | |
| | | Under 5 | 5–14 | 15–29 | 30–44 | 45–60 | over 60 | Totals |
|---|---|---|---|---|---|---|---|---|
| Hospitalized? | **Yes** | 7 | 9 | 9 | 9 | 1 | 0 | 35 |
| | **No** | 44 | 195 | 241 | 59 | 35 | 10 | 584 |
| | **Totals** | 51 | 204 | 250 | 68 | 36 | 10 | 619 |

◀ **TABLE 10.15** Summary of swine flu cases in the United States from May 1, 2009, to May 5, 2009.

**QUESTION** Test the hypothesis that hospitalization for swine flu is associated with age category, using a significance level of 0.05. Use the Minitab output provided in Figure 10.13.

**SOLUTION** This is a test of independence. There is one sample (swine flu victims) and two response variables (*Age Category* and *Hospitalized*—yes or no.)

*Step 1: Hypothesize*

$H_0$: Age category and whether a swine flu victim is hospitalized are independent.
$H_a$: Age category and whether a swine flu victim is hospitalized are associated.

*Step 2: Prepare*
Our first choice would be a chi-square test of independence. However, several of the expected values shown in Figure 10.13 are less than 5. Minitab even alerts us to this fact (as most, but not all, packages do). For emphasis, we highlighted this warning and the low expected values in red ink.

▶ **FIGURE 10.13** Minitab output for swine flu hospitalizations, showing low expected values. "Yes" means the people were hospitalized, and "No" means they were not hospitalized.

```
Chi-Square Test: Under 5, 5-14, 15-29, 30-44, 45-60, Over 60
Expected counts are printed below observed counts
Chi-Square contributions are printed below expected counts

           Under 5    5-14   15-29   30-44   45-60   Over 60   Total
  Yes           7       9       9       9       1         0      35
             2.88   11.53   14.14    3.84    2.04      0.57

  No           44     195     241      59      35        10     584
            48.12  192.47  235.86   64.16   33.96      9.43

Total          51     204     250      68      36        10     619

Chi-Sq = 17.280, DF = 5
WARNING: 1 cells with expected counts less than 1. Chi-Square approximation
probably invalid.
```

Our solution, which can be a little unsatisfying, is to combine categories so that each expected count is 5 or more. For example, you can see that if we combine the Under 5 and 5–14 categories, we get a new category, "Under 15," with expected count 2.88 + 11.53 = 14.41. Similarly, we can combine the upper three categories into a new category of "30 and older," with expected count 3.84 + 2.04 + 0.57 = 6.45.

The result is shown in Figure 10.14. We now have only three age categories, but the expected counts are large enough for us to use the chi-square distribution to reliably find a p-value.

▶ **FIGURE 10.14** Minitab output with groups combined so that none of the expected values is less than 5. "Yes" means the people were hospitalized, and "No" means they were not hospitalized.

```
Chi-Square Test: Under 15, 15-29, 30 and older
Expected counts are printed below observed counts
Chi-Square contributions are printed below expected counts

                              30 and
          Under 15   15-29    older    Total
  Yes           16       9       10       35
             14.42   14.14     6.45
             0.173   1.866    1.960

  No           239     241      104      584
            240.58  235.86   107.55
             0.010   0.112    0.117

Total         255     250      114      619

Chi-Sq = 4.239, DF = 2, P-Value = 0.120
```

*Step 3: Compute to compare*
Chi-square = 4.24, p-value = 0.120

*Step 4: Interpret*
Do not reject $H_0$.

CONCLUSION With these limited number of categories, there is not enough evidence to conclude that there is an association between age and hospitalization.

TRY THIS! Exercise 10.49

## Advantages and Disadvantages of Combining Categories

The biggest advantage of combining categories is that you are able to use a procedure, the chi-square test, that otherwise would give you very inaccurate p-values and might therefore lead to the wrong conclusion. Had we not combined cells, the p-value for the chi-square test shown in Figure 10.13 would have been 0.004. We would have rejected the null hypothesis and concluded that there was an association. But this conclusion would have been based on a potentially very inaccurate p-value.

The disadvantage is that our knowledge becomes more crude. Having an age category of "Under 15" is somewhat disappointing, because we know that children change a lot between birth and their early teenage years. The same can be said of those "30 and older." Thirty-year-olds should have very different immune systems than 90-year-olds. Unfortunately, we cannot say more without more data. (However, given that "more data" means more swine flu victims, it is not necessarily a bad thing that we can't tell more about this association.)

## Fisher's Exact Test

Another approach that works in some cases is Fisher's Exact Test. This test is called exact because in many situations we can find the exact p-value, not the approximate p-value found by the chi-square test.

**Fisher's Exact Test** relies on the fact that if the null hypothesis is true and there is no association between the variables, then a summary table of the data should look somewhat like a table in which the counts were thrown down randomly. More precisely, we imagine that the row and column totals are fixed, but the numbers within the table can be any value (as long as the totals stay the same). Mix those values up at random, and you can simulate the distribution for the null hypothesis.

To illustrate this approach, consider an example based on an actual study of the effectiveness of an experimental antivenom for scorpion stings in children (Boyer et al. 2009). The researchers randomly assigned 8 children suffering from scorpion stings to receive the antivenom, and 7 to receive a placebo. After several hours, they recorded which children showed substantial improvement. (The sting is not fatal, and the antivenom was not known to be a successful treatment.) The data are given in Table 10.16.

|  | Antivenom | Placebo | Total |
|---|---|---|---|
| **No Improvement** | 1 | 6 | 7 |
| **Improvement** | 7 | 1 | 8 |
| **Total** | 8 | 7 | 15 |

◀ **TABLE 10.16** Summary of data on children with scorpion bites.

The expected counts are too small for us to use the chi-square test. Besides that, the chi-square test can tell us only whether there is an association between the two variables. Here we want to know more—we want to know whether the antivenom group did better than the placebo group.

On the face of it, it looks like the antivenom group in fact did do better: 7 out of 8 children improved, as opposed to only 1 out of 7 who received the placebo. But could this difference be due to chance?

Note that it would have been possible to make an even stronger case for the antivenom if all of the children who received it had improved. To see what this would look like, in the cell where "Improvement and Antivenom" intersect, we replace the 7 with an 8. This means changing all of the other cells in order to keep the row and column totals the same. Table 10.17 shows this change.

▶ **TABLE 10.17** A hypothetical outcome even more extreme than the actual outcome and with the same marginal totals as in Table 10.16.

|  | Antivenom | Placebo | Total |
|---|---|---|---|
| **No Improvement** | 0 | 7 | 7 |
| **Improvement** | 8 | 0 | 8 |
| **Total** | 8 | 7 | 15 |

If, instead, we had seen only 6 improvements in the antivenom group, this would have meant weaker support for the antivenom, because a lower proportion of children receiving the treatment would have recovered.

Now imagine that these cases are assigned to the table at random. The only restriction is that you always need 8 cases in the Antivenom column, 7 cases in the Placebo column, 7 cases in the No Improvement row, and 8 cases in the Improvement row. Maybe you throw darts or (more practically) use a computer to randomly assign cases to categories. After doing this many times, you'll see what the distribution of counts looks like when the null hypothesis is true and there is no association between variables.

To calculate an approximate p-value (the probability of getting a result as extreme as or more extreme than what we observed when the variables are not associated), we use this generated distribution to calculate the approximate probability of getting the observed number of "Improved and Antivenom," or greater. In fact, we don't even need to use a computer simulation. A mathematical approach can provide an exact answer. (This is why it's called Fishers *Exact* Test!) These probabilities are found using a distribution called the hypergeometric function. Many software packages and on-line statistical calculators will find these probabilities for you. For example, to find the p-value using the SOCR website, we input the observed table, as in Figure 10.15.

▶ **FIGURE 10.15** Data input for SOCR's Fisher's Exact Test. Go to socr.stat.ucla.edu and click on Analyses.

The result is shown in Figure 10.16. The p-value is 0.009, which is smaller than 0.05, so we reject the null hypothesis and conclude that the antivenom is beneficial. (We can conclude that the antivenom *caused* this improvement because the researchers randomly assigned subjects to receive the treatment or a placebo.)

Fisher's Exact Test can be used for tables with more than two rows or columns, although then some technical considerations arise that are beyond an introductory course. You may have noticed that the two-tailed p-value is not simply twice the one-tailed p-value. This is one way in which Fisher's Exact Test differs from the *t*-test or *z*-test in earlier chapters.

The condition that the rows and columns remain fixed might strike you as somewhat odd. If the condition is not in fact met, the calculated p-value will be larger than the true p-value. This means that the test is conservative, which, for most purposes, is a good thing. (It also means that the test may not be very powerful—it might fail to detect real associations.) However, when the condition does hold, the p-value is exactly right. We do not need to rely on approximations, as we do with the chi-square test.

**Results of Fisher's Exact Test**

**Number of Rows = 2**
**Number of Columns = 2**

|         | Column 1 | Column 2 | Row Total |
|---------|----------|----------|-----------|
| Row 1   | 1.0      | 6.0      | 7         |
| Row 2   | 7.0      | 1.0      | 8         |
| Col Total | 8      | 7        | 15        |

P-Value Cutoff = .009

1-Tail P-Value = .009

2-Tail P-Value = .010

◄ **FIGURE 10.16** Output from SOCR website for Fisher's Exact Test. We used the one-tailed p-value.

## CASE STUDY REVISITED

To better understand the ideal conditions for popping corn, experimenters designed a randomized, controlled study to observe how well the popcorn popped under different conditions. Of interest here was how the amount of oil affected the outcome. A summary of results was shown in the Case Study at the beginning of the chapter, where we observed that it looked as though using the maximum amount of oil did not work well, because so few popcorn bags were successful by the criteria adopted for success.

Did the amount of oil affect the resulting quality of the popcorn? To test this using the methods of this chapter, we carry out a test of homogeneity, because we have three independent samples (no oil, medium oil, and maximum oil) and one categorical response variable (*Result*: success or failure). We have three independent samples, because bags were randomly assigned to one of these three groups. Each group was assigned 36 bags of popcorn, and each bag had 50 kernels.

The hypotheses follow.

$H_0$: The quality of popcorn and the amount of oil are independent.
$H_a$: The amount of oil affects the quality of the popcorn.

The results (and some of the raw data) are shown in Figure 10.17 in output generated by StatCrunch. The output shows that the expected counts are all greater than 5, so our sample sizes are large enough for the chi-square distribution (with 2 degrees of freedom) to produce a good approximation to the p-value.

▶ **FIGURE 10.17** StatCrunch output shows the results of the analysis in the foreground and the raw data in the background. The expected values in the table are given below the observed values.

**popcornexperiment.txt**

| StatCrunch | Data | Stat | Graphics | Help |

| Row | brand | time | oil | container | good | result | var7 | var8 |
|---|---|---|---|---|---|---|---|---|
| 1 | 1 | 0 | maximum | 1 | 8 | Failure | | |
| 2 | 0 | 1 | medium | 0 | 26 | Success | | |
| 3 | 0 | 0 | maximum | 1 | 5 | Failure | | |
| 4 | 1 | 0 | none | 1 | 8 | Failure | | |
| 5 | 1 | 0 | none | 0 | 7 | Failure | | |
| 6 | 1 | 1 | medium | 1 | 20 | Failure | | |
| 7 | 1 | | | | | | | |
| 8 | 0 | | | | | | | |
| 9 | 1 | | | | | | | |
| 10 | 0 | | | | | | | |
| 11 | 1 | | | | | | | |
| 12 | 0 | | | | | | | |
| 13 | 0 | | | | | | | |
| 14 | 1 | | | | | | | |
| 15 | 0 | | | | | | | |
| 16 | 0 | | | | | | | |

**Description:**
popcorn popped under different randomly
more than 25, then result was deemed a

**Source:**
Irina A. Kukuyeva and Jean Wang and Yul

**Tags:**
anova, chi-square, design

○ ○ ○    Contingency Table with data    Java Applet Window ⚠

Options

**Contingency table results:**
Rows: result
Columns: oil

| Cell format |
|---|
| Count |
| Expected count |

| | maximum | medium | none | Total |
|---|---|---|---|---|
| Failure | 33 | 22 | 23 | 78 |
| | 26 | 26 | 26 | |
| Success | 3 | 14 | 13 | 30 |
| | 10 | 10 | 10 | |
| Total | 36 | 36 | 36 | 108 |

| Statistic | DF | Value | P-value |
|---|---|---|---|
| Chi-square | 2 | 10.246154 | 0.006 |

From the output, we see that the test statistic has a value of 10.25, with a p-value of 0.006. This is quite small. Certainly it is less than 0.05, so at the 5% significance level we reject the null hypothesis. We conclude that quality is affected by the amount of oil used. (At least, it is if you believe that quality is measured by the number of kernels popped after 75 seconds.)

# EXPLORING STATISTICS
## CLASS ACTIVITY

## Skittles

| GOALS | MATERIALS |
|---|---|
| Apply a chi-square test to check whether two bags of Skittles candies contain the same proportions of colors. | • One small bag of Skittles for each student<br>• Computer or TI-83/84 |

**ACTIVITY**

Open a bag of Skittles and count how many of each color are in your bag. Fill in the numbers in the table below. Then find a partner and fill in the colors from his or her bag. (If someone does not have a partner, form a group of three and use three rows.)

|  | Purple | Red | Orange | Yellow | Green | Total |
|---|---|---|---|---|---|---|
| Yours | | | | | | |
| Partner's | | | | | | |

**BEFORE THE ACTIVITY**

1. Do you think that you and your partner(s) will get exactly the same number in each category?

2. Do you think you and your partner(s) will have significantly different distributions of colors? Why or why not?

**AFTER THE ACTIVITY**

1. Perform a hypothesis test to test whether the two bags have a significantly different distribution of colors, using a significance level of 0.05.

2. Throw away, save, or eat the Skittles.

## CHAPTER REVIEW

### Key Terms

Two-way table, *451*
Expected counts, *452*
Chi-square statistic, *455*

Chi-square distribution, *457*
Degrees of freedom, *458*

Goodness-of-fit tests, *460*
Homogeneity, *464*

Independence, *464*
Fisher's Exact Test, *475*

### Learning Objectives

After reading this chapter and doing the assigned homework problems, you should

- Understand when a goodness-of-fit test is needed and appropriate, and know how to perform the test and interpret results.

- Distinguish between tests of homogeneity and tests of independence.

- Understand when it is appropriate to use a chi-square statistic to test whether two categorical variables are associated; know how to perform this test and interpret the results.

- Know when and how to perform Fisher's Exact Test.

### Summary

We presented three types of tests for analyzing categorical variables. The goodness-of-fit test is used to compare a proposed distribution for one categorical variable with the observed sample distribution. Although the test of homogeneity is conceptually different from the test of independence, they are exactly the same in terms of the calculations required. Both of these tests attempt to determine whether two categorical variables are associated. The only difference is in the way the data for the study were collected. When researchers collect two or more independent samples and measure one categorical response variable, they are performing a test of homogeneity. When instead they collect one sample and measure two categorical response variables, they are performing a test of independence.

All three tests rely on the chi-square statistic. For each cell of a two-way summary table, we compare the observed count with the count we would expect if the null hypothesis were true. If the chi-square statistic is big, it means that these two counts don't agree, and it discredits the null hypothesis.

An approximate p-value is calculated by finding the area to the right of the observed chi-square statistic using a chi-square distribution. To do this, you need to know the degrees of freedom for the chi-square distribution, and this depends on which test you are using.

If the sample size is too small to use the chi-square test, you might consider combining categories. Fisher's Exact Test is another alternative you can try.

### Formulas

#### Expected Counts

**Formula 10.1:** Expected count for a cell $= \dfrac{(\text{row total}) \times (\text{column total})}{\text{grand total}}$

#### Chi-Square Statistic

**Formula 10.2:** $X^2 = \sum\limits_{\text{cells}} \dfrac{(O - E)^2}{E}$

### Goodness-of-Fit Test

#### Hypotheses

$H_0$: The true distribution is the same as the proposed distribution.
$H_a$: The true distribution is different from the proposed distribution.

#### Conditions

1. *Random Sample.* The sample was collected randomly

2. *Independent Measurements.* Each measurement on an individual is independent of all other measurements

3. *Large Sample.* The expected count is at least 5 in each category.

#### Sampling Distribution

If conditions hold, then the sampling distribution of this statistic follows a chi-square distribution with degrees of freedom = number of categories − 1.

### Test of Homogeneity and Independence

#### Hypotheses

$H_0$: The variables are independent.
$H_a$: The variables are associated.

#### Conditions (Homogeneity)

1. *Random Samples.* Two or more samples, all sampled randomly.

2. *Independent Samples and Observations.* Samples are independent of each other. The observations within each sample are independent.

3. *Large Samples.* At least 5 expected counts in each cell of the summary table.

**Conditions (Independence)**

1. *Random Sample.* One sample, selected randomly.

2. *Independent Observations.* Observations are independent of each other.

3. *Large Sample.* There are at least 5 expected counts in each cell of the summary table.

**Sampling Distribution**

If conditions hold, the sampling distribution follows a chi-square distribution with degrees of freedom = (number of rows − 1) × (number of columns − 1).

## Sources

Boyer, L. V., et al. 2009. Antivenom for critically ill children with neuro-toxicity from scorpion stings. *New England Journal of Medicine* 360, 2090–2098.

Colman, R. J., et al. 2009. Caloric restriction delays disease onset and mortality in Rhesus monkeys. *Science* 325, 201.

Husemann, L. R., J. Moise-Titus, C. Podolski, and L. D. Eron. 2003. Longitudinal relations between children's exposure to TV violence and their aggressive and violent behavior in young adulthood: 1977–1992. *Developmental Psychology* 39(2), 201–221.

I. A. Kukuyeva, J. Wang, and Y. Yaglovskaya. 2008. Popcorn popping yield: An investigation, JSM

# SECTION EXERCISES

## SECTION 10.1

**10.1 Tests**

a. In Chapter 8, you learned some tests of proportions. Are tests of proportions used for categorical or numerical data?

b. In this chapter, you are learning to use chi-square tests. Do these tests apply to categorical or numerical data?

**10.2** In Chapter 9, you learned some tests of means. Are tests of means used for numerical or categorical data?

**10.3 Crime and Gender** A statistics student conducted a study in Ventura County, California, that looked at criminals on probation who were under 15 years of age to see whether there was an association between the type of crime (violent or nonviolent) and gender. Violent crimes involve physical contact such as hitting or fighting; nonviolent crimes include vandalism, robbery, and verbal assault. The raw data are shown in Exercise 1.45 on page 31.

Create a two-way table to summarize these data. Are the two variables numerical or categorical? If you are doing this by hand, create a table with two rows and two columns. Label the columns Boy and Girl (across the top). Label the rows Violent and Nonviolent. Begin with a big table, making a tally mark in one of the four cells for each observation, and then summarize the tally marks as counts.

**10.4 Red Cars and Stop Signs** The table shows the raw data for the results of a student survey of 22 cars and whether they stopped completely at a stop sign or not. In the Color column, "Red" means the car was red and "No" means the car was not red. In the Stop column, "Stop" means the car stopped, and "No" means the car did not stop fully.

Create a two-by-two table to summarize these data. Use Red and No for the columns (across the top) and Stop and No for the rows.

(We gave you an orientation of the table so that your answers would be easy to compare.) Are the two variables categorical or numerical?

| Color | Stop | Color | Stop |
|---|---|---|---|
| Red | Stop | No | No |
| Red | Stop | Red | Stop |
| Red | No | Red | No |
| Red | No | Red | No |
| Red | No | No | Stop |
| No | Stop | No | Stop |
| No | Stop | No | Stop |
| No | Stop | No | Stop |
| No | No | Red | Stop |
| Red | Stop | Red | No |
| Red | No | Red | No |

**10.5** The table summarizes the outcomes of a study that students carried out to determine whether humanities students had a higher mean GPA than science students. Identify both of the variables, and state whether they are numerical or categorical. If numerical, state whether they are continuous or discrete.

| | Mean GPA |
|---|---|
| Science | 3.4 |
| Humanities | 3.5 |

**10.6 Finger Length** There is a theory that relative finger length depends on testosterone level. The table shows a summary of the outcomes of an observational study that one of the authors carried out to determine whether men or women were more likely to have a ring finger that appeared longer than their index finger. Identify both of the variables, and state whether they are numerical or categorical. If numerical, state whether they are continuous or discrete.

|  | Men | Women |
|---|---|---|
| Ring Finger Longer | 23 | 13 |
| Ring Finger Not Longer | 4 | 14 |

**10.7 Cell Phone Only** According to a 2010 Pew Poll, 50% adults aged 25–29 had access only to a cell phone (no landline). Assume that you randomly sample 50 adults aged 25–29 and ask whether they have access only to a cell phone. Suppose 30 say yes and 20 say no.

a. What is the observed proportion of people who say yes?

b. What is the observed frequency (count) of people who say yes?

c. What is the expected frequency of people who say yes if the Pew Poll is right?

**10.8 Coin Flips** You flip a coin 100 times and get 58 heads and 42 tails.

a. What is the observed proportion of heads?

b. What is the observed frequency (count) of heads?

c. What is the expected frequency of heads if the coin is not biased?

**TRY 10.9 Effects of Television Violence on Men (Example 1)**
A study done by Husemann et al. and published in *Developmental Psychology* in 2003 compared men who viewed high levels of television violence as children with those who did not in order to study the differences with regard to physical abuse of their partners as adults. The men categorized as physically abusive had hit, grabbed, or shoved their partners.

|  | High TV Violence | Low TV Violence |
|---|---|---|
| Yes, Physical Abuse | 13 | 27 |
| No Physical Abuse | 18 | 95 |

a. Find the row, column, and grand totals, and prepare a table showing these values as well as the counts given.

b. Find the percentage of men overall that were abusive.

c. Find the expected number of men exposed to high levels of television violence who should say yes, if the variables are independent. Multiply the proportion overall that were abusive times the number of men exposed to high levels of television violence. Do not round off to a whole number. Round to two decimal digits.

d. Find the other expected values by knowing that the expected values must add to the row and column totals. Report them in a table with the same orientation as the one given for the data.

**10.10 Effects of Television Violence on Women** Refer to Exercise 10.9. This data table compares women who viewed high levels of television violence as children with those who did not in order to study the differences with regard to physical abuse of their partners as adults. The women categorized as physically abusive had hit, grabbed, or shoved their partners.

|  | High TV Violence | Low TV Violence |
|---|---|---|
| Yes, Physical Abuse | 12 | 30 |
| No Physical Abuse | 23 | 111 |

a. Find the row, column, and grand totals and report the table showing these values.

b. Find the percentage of all women who were abusive (who answered yes).

c. Find the expected number of women exposed to high levels of television violence who should say yes, if the variables are independent. Do not round off to a whole number. Round to two decimal digits.

d. Find the other expected values. Report them in a table with the same orientation as the one for the data.

**10.11 Mummies with Heart Disease** According to the website MedicalNewsToday.com, coronary artery disease accounts for about 40% of deaths in the United States. Many people believe this is due to modern-day factors such as high-calorie fast food and lack of exercise. However, a study published in the *Journal of the American Medical Association* in November 2009 (www.medicalnewstoday.com) reported on 16 mummies from the Egyptian National Museum of Antiquities in Cairo. The mummies were examined, and 9 of them had hardening of the arteries, which seems to suggest that hardening of the arteries is not a new problem.

a. Calculate the expected number of mummies with artery disease (assuming the rate is the same as in the modern day). Then calculate the expected number of mummies without artery disease (the rest).

b. Calculate the observed value of the chi-square statistic for these mummies.

**10.12 Cell Phone Only** According to a Pew Poll done in 2010, 50% of adults aged 25–29 had access only to a cell phone (no landline). Assume that you randomly sample 50 adults aged 25–29, and ask whether they have access only to a cell phone. Suppose that 30 say yes and 20 say no. Calculate the observed value of the chi-square statistic for testing the hypothesis that 50% have a cell phone only.

**10.13 Violins** Stradivarius violins, made in the 1700s by a man of the same name, are worth millions of dollars. They are prized by music lovers for their uniquely rich, full sound. In September 2009, an audience of experts took part in a blind test of violins, one of which was a Stradivarius. There were four other violins (modern-day instruments) made of specially treated wood. When asked to pick the Stradivarius after listening to all five violins, 39 got it right and 113 got it wrong (*Time Magazine*, November 23, 2009).

a. If this group were just guessing, how many people (out of the 152) would be expected to guess correctly? And how many would be expected to guess incorrectly?

b. Calculate the observed value of the chi-square statistic by hand, showing your work.

**10.14 Coin Flips** You flip a coin 100 times and get 58 heads and 42 tails. Calculate the chi-square statistic by hand, showing your work.

TRY **10.15 Effects of Television Violence on Men (Example 2)** Refer to Exercise 10.9. The data table compares men who viewed high levels of television violence as children with those who did not, in order to study the differences with regard to physical abuse of their partners as adults. Report the observed value of the chi-square statistic.

|                     | High TV Violence | Low TV Violence |
|---------------------|:----------------:|:---------------:|
| **Yes, Physical Abuse** | 13           | 27              |
| **No Physical Abuse**   | 18           | 95              |

**10.16 Effects of Television Violence on Women** Refer to Exercise 10.10. The data table compares women who viewed high levels of television violence as children with those who did not, in order to study the differences with regard to physical abuse of their partners as adults. Calculate the observed value of the chi-square statistic.

|                     | High TV Violence | Low TV Violence |
|---------------------|:----------------:|:---------------:|
| **Yes, Physical Abuse** | 12           | 30              |
| **No Physical Abuse**   | 23           | 111             |

## SECTION 10.2

**10.17** Fill in the blank by choosing one of the options given: Chi-square goodness-of-fit tests are applicable if the data consist of _____ (one categorical variable, two categorical variables, one numerical variable, or two numerical variables).

**10.18** Fill in the blank by choosing one of the options given: Chi-square goodness-of-fit data are often summarized with _____ (one row or one column of observed counts (but not both), at least two rows and at least two columns of observed counts).

TRY **10.19 Are Humans Like Random Number Generators?**
g **(Example 3)** One of the authors collected data from a class to see whether humans made selections randomly, as a random number generator would. Each of 38 students had to pick an integer from one to five. The data are summarized in the table.

| Integer:      | One | Two | Three | Four | Five |
|---------------|:---:|:---:|:-----:|:----:|:----:|
| Times Chosen: | 3   | 5   | 14    | 11   | 5    |

A true random number generator would create roughly equal numbers of all five integers. Do a goodness-of-fit analysis to test the hypothesis that humans are not like random number generators. Use a significance level of 0.05, and assume these data were from a random sample of students. *See page 493 for guidance.*

**10.20 Is the Random Number Table Really Random?** We counted ones, twos, threes, fours, and fives from a few lines of a random number table, and we should expect to get equal numbers of each. (We ignored the sixes, sevens, eights, nines, and zeros.) There were 14 ones, 12 twos, 16 threes, 11 fours, and 8 fives, which is 61 numbers in the categories selected. First, find the expected values, which should all be the same.

TI-84 GOF Output

Then test the hypothesis that the random number table does not generate equal proportions of ones, twos, threes, fours, and fives, using a significance level of 0.05. Refer to the goodness-of-fit (GOF) output shown.

**10.21 Coin Spins** A penny was spun on a hard, flat surface 50 times, landing on 15 heads and 35 tails. Using a chi-square test for goodness of fit, test the hypothesis that the coin is biased, using a 0.05 level of significance. Refer to the goodness-of-fit (GOF) output shown.

TI-84 GOF Output

**10.22 Teens Texting** According to a 2010 Pew Poll, 1 in 4 teens of driving age have reported having sent or received a text message while driving. Suppose we randomly sample 100 teens, and 18 say they have texted while driving and 82 say they have not. Test the hypothesis that the percentage that have texted is not the 25% that Pew claims, using goodness of fit and a 0.05 level of significance. Refer to the output shown.

TI-84 GOF Output

**10.23 Is the Six-sided Die Fair?** The table shows the results of rolling a six-sided die 120 times.

| Outcome on Die | Frequency |
|---|---|
| 1 | 27 |
| 2 | 20 |
| 3 | 22 |
| 4 | 23 |
| 5 | 19 |
| 6 | 9 |

Test the hypothesis that the die is not fair. A fair die should produce equal numbers of each outcome.

Use the four-step procedure with a significance level of 0.05, and state your conclusion clearly. Refer to the output shown.

TI-84 GOF Output

**10.24 Is the Six-sided Die Fair?** Repeat the chi-square test (all four steps) from Exercise 10.23, but this time assume that you got exactly 20 outcomes in each of the six categories. Refer to the figure. Explain.

TI-84 GOF Output for "Draw"

*** 10.25 Violins** Professional musicians listened to five violins being played, without seeing the instruments. One violin was a Stradivarius, and the other four were modern-day violins. When asked to pick the Stradivarius (after listening to all five), 39 got it right and 113 got it wrong.

  a. Use the chi-square goodness-of-fit test to test the hypothesis that the experts are not simply guessing. Use a significance level of 0.05. Refer to the output shown.

  b. Perform a one-proportion z-test with the same data, using a one-tailed alternative that the experts should get more than 20% correct. Use a significance level of 0.05.

  c. Compare your p-values and conclusions.

TI-84 GOF Output

**10.26 Mummies with Heart Disease** Exercise 10.11 on artery disease in mummies showed that 9 out of 16 mummies showed heart disease (hardening of the arteries).

Test the hypothesis that the population proportion of mummies with hardening of the arteries is not the same as in the modern United States (that it is not 40%). Use a significance level of 0.05. Refer to the output shown.

TI-84 Output

## SECTION 10.3

TRY **10.27 Political Party and Right Direction? (Example 4)** A December 2009 poll conducted by Associated Press and Roper asked a random sample of Americans whether they thought the country was heading in the right direction. Each respondent answered "Right Direction" or "Wrong Direction" and was classified as Republican, Democrat, or Independent. If we were interested in testing whether party affiliation was associated with the answer to the question, would this be a test of homogeneity or of independence? Explain.

**10.28 Antibiotic or Placebo** A large number of surgery patients get infections after surgery, which can sometimes be quite serious. Researchers randomly assigned some surgery patients to receive a simple antibiotic ointment after surgery, others to receive a placebo, and others to receive just cleansing with soap. If we wanted to test the association between treatment and whether or not patients get a, infection after surgery, would this be a test of homogeneity or of independence? Explain. (Source: Hospitals could stop infections by tackling bacteria patients bring in, studies find. *New York Times*, January 6, 2010.)

TRY **10.29 Jet Lag Drug (Example 5)** A recent study was conducted to determine whether the drug Nuvigil was effective at helping east-bound jet passengers adjust to jet lag. Subjects were randomly assigned either to one of three different doses of Nuvigil (low, medium, high) or to a placebo, flown to France in a plane in which they could not drink alcohol or coffee or take sleeping pills, and then examined in a lab where their state of wakefulness was measured and classified into categories (low, normal, alert). If we test whether treatments for jet lag are associated with wakefulness, are we doing a test of independence or of homogeneity? Explain. (Source: A drug's second act: Battling jet lag. *New York Times*, January 6, 2010.)

**10.30 Most Important Problems** A November 2009 AP–Stanford University Poll took a random sample of 1005 Americans and asked what they considered to be the most important problem facing the country today. There were 24 different categories of responses. The economy was the most popular choice. Respondents were classified by the region of the country they lived in: Northeast, Midwest, South, or West. If we wanted to test whether there was an association between the response to the question and the region of the country, would this be a test of homogeneity or of independence? Explain.

**10.31 Country of Origin** The table shows the country of origin and the percentage of foreign-born people in the United States in 2000 and 2007 for the four countries of origin with the highest percentages (*2009 World Almanac and Book of Facts*). Give two reasons why you should not do a chi-square test with these data.

|      | Mexico | China | Philippines | India |
|------|--------|-------|-------------|-------|
| **2000** | 29.5 | 4.9 | 4.4 | 3.3 |
| **2007** | 30.8 | 5.1 | 4.5 | 3.9 |

**10.32 Unemployment Rates** U.S. Unemployment rates for all residents 20 years old and older are given in table 10.32 as a percentage. The table does not include people who are not actively seeking employment (*2009 World Almanac and Book of Facts*). Give two reasons why a chi-square test is not appropriate for this set of data.

|      | Female | Male |
|------|--------|------|
| **1990** | 4.9 | 5.0 |
| **2000** | 3.6 | 3.3 |
| **2009** | 9.5 | 10.5 |

TRY
g **10.33 Obesity and Marital Status (Example 6)** A study reported in the medical journal *Obesity* in 2009 analyzed data from the National Longitudinal Study of Adolescent Health. Obesity was defined as having a body mass index (BMI) of 30 or more. The research subjects were followed from adolescence to adulthood, and all the people in the sample were categorized in terms of whether they were obese and whether they were dating, cohabiting, or married. Test the hypothesis that relationship status and obesity are associated, using a significance level of 0.05. Can we conclude from these data that living with someone is making some people obese and that marrying is making people even more obese? Can we conclude that obesity affects relationship status? Explain. *See page 493 for guidance.*

|          | Dating | Cohabiting | Married |
|----------|--------|------------|---------|
| **Obese** | 81 | 103 | 147 |
| **Not Obese** | 359 | 326 | 277 |

(Source: N. S. The and P. Gordon-Larsen. 2009. Entry into romantic partnership is associated with obesity. *Obesity* 17(7), 1441–1447.)

**10.34 Opinions on Household Responsibilities** In the 2008 General Social Survey, respondents were asked whether they agreed or disagreed with the following statement: "Husbands should work and wives should take care of the home." Each respondent was classified by gender. The table shows the responses.

|          | Men | Women |
|----------|-----|-------|
| **Agree** | 183 | 192 |
| **Neither Agree nor Disagree** | 143 | 140 |
| **Disagree** | 288 | 412 |

```
              Men       Women      Total
Agree         183       192        375
              169.55    205.45

Neither       143       140        283
              127.95    155.05

Disagree      288       412        700
              316.49    383.51

Total         614       744        1358
Chi-Sq = 9.859, DF = 2, P-Value = 0.007
```

a. Write a sentence containing the percentages of men and women who agreed with this statement, and compare them.

b. Test the hypothesis that gender and opinion are associated, using the 0.05 level of significance. Assume the sample is random. Refer to the Minitab output, which shows expected values below the observed counts.

**10.35 Effects of Television Violence on Men** The data table compares men who viewed television violence with those who did not, in order to study the differences in physical abuse of the spouse. For the men in the table, test whether television violence and abusiveness are associated, using a significance level of 0.05. Refer to the Minitab output.

|          | High TV Violence | Low TV Violence |
|----------|------------------|-----------------|
| **Yes, Physical Abuse** | 13 | 27 |
| **No Physical Abuse** | 18 | 95 |

(Source: L. R. Husemann et al. 2003. Longitudinal relations between children's exposure to TV violence and their aggressive and violent behavior in young adulthood: 1977–1992. *Developmental Psychology* 39(2), 201–221.)

```
Chi-Square Test: High TV Violence, Low TV Violence
Expected counts are printed below observed counts
Chi-Square contributions are printed below expected counts

           High TV    Low TV
           Violence   Violence   Total
Yes, Ab       13         27        40
              8.10       31.90
              2.957      0.751

No            18         95        113
              22.90      90.10
              1.047      0.266

Total         31         122       153

Chi-Sq = 5.021, DF = 1, P-Value = 0.025
```

**10.36 Effects of Television Violence on Women** The data table compares women who viewed television violence with those who did not, in order to study the differences in physical abuse of the spouse (Husemann et al. 2003). Test whether television violence and abusiveness are associated, using a significance level of 0.05. Refer to the Minitab output.

|  | High TV Violence | Low TV Violence |
|---|---|---|
| **Yes, Physical Abuse** | 12 | 30 |
| **No Physical Abuse** | 23 | 111 |

```
Chi-Square Test: High TV Violence, Low TV Violence
Expected counts are printed below observed counts
Chi-Square contributions are printed below expected counts

         High TV    Low TV
         Violence   Violence   Total
Yes, ab     12         30        42
           8.35      33.65
          1.593       0.395

No          23        111       134
          26.65     107.35
           0.499      0.124
Total       35        141       176

Chi-Sq = 2.612, DF = 1, P-Value = 0.106
```

**10.37 Gender and Happiness of Marriage** The table shows the results of a cross tabulation of gender and whether a person is happy in his or her marriage, according to data obtained from the 2008 General Social Survey.

**Happiness of Marriage × Respondent's Gender Cross Tabulation**

| | | | Respondent's Gender | | |
|---|---|---|---|---|---|
| | | | **Male** | **Female** | **Total** |
| **Happiness of Marriage** | Very happy | Count | 297 | 299 | 596 |
| | | Expected Count | 289.1 | 306.9 | 596.0 |
| | Pretty happy | Count | 160 | 183 | 343 |
| | | Expected Count | 166.4 | 176.6 | 343.0 |
| | Not too happy | Count | 13 | 17 | 30 |
| | | Expected Count | 14.6 | 15.4 | 30.0 |
| **Total** | | Count | 470 | 499 | 969 |

a. If we carry out a test to determine whether these variables are associated, is this a test of independence, homogeneity, or goodness of fit?

b. Do a chi-square test with a significance level of 0.05 to determine whether gender and happiness of marriage are associated.

c. Does this suggest that women and men tend to have different levels of happiness or that their rates of happiness in marriage are about equal?

**10.38 Is Smiling Independent of Age?** Randomly chosen people were observed for about 10 seconds in several public places, such as malls and restaurants, to see whether they smiled during that time. The table shows the results for different age groups.

| | Age Group | | | | |
|---|---|---|---|---|---|
| | **0–10** | **11–20** | **21–40** | **41–60** | **61+** |
| **Smile** | 1131 | 1748 | 1608 | 937 | 522 |
| **No Smile** | 1187 | 2020 | 3038 | 2124 | 1509 |

(Source: M. S. Chapell. 1997. Frequency of public smiling over the life span. *Perceptual and Motor Skills* 85, 1326.)

a. Find the percentage of each age group that were observed smiling, and compare these percentages.

b. Treat this as a single random sample of people, and test whether smiling and age group are associated, using a significance level of 0.05. Comment on the results.

TRY **10.39 Preschool Attendance and High School Graduation Rates (Example 7)** The Perry preschool project was created in the early 1960s by David Weikart in Ypsilanti, Michigan. One hundred and twenty three African American children were randomly assigned to one of two groups: One group enrolled in the Perry preschool, and one did not enroll. Follow-up studies were done for decades to answer the research question of whether attendance at preschool had an effect on high school graduation. The table shows whether the students graduated from regular high school or not. Students who received GEDs were counted as not graduating from high school. This table includes 121 of the original 123. This is a test of homogeneity, because the students were randomized into two distinct samples.

| | Preschool | No Preschool |
|---|---|---|
| **HS Grad** | 37 | 29 |
| **No HS Grad** | 20 | 35 |

a. For those who attended preschool, the high school graduation rate was 37/57, or 64.9%. Find the high school graduation rate for those not attending preschool, and compare the two. Comment on what the rates show for *these* subjects.

b. Are attendance at preschool and high school graduation associated? Use a 0.05 level of significance.

(Source: L. J. Schweinhart et al. 2005. Lifetime effects: The High/Scope Perry Preschool Study through age 40. *Monographs of the High/Scope Educational Research Foundation,* 14. Ypsilanti, Michigan: High/Scope Press.)

**10.40 Preschool Attendance and High School Graduation Rates for Females** The Perry preschool project data presented in Exercise 10.39 (Schweinhart et al. 2005) can be divided to see whether the preschool attendance effect is different for males and females. The table shows a summary of the data for females, and the figure shows Minitab output that you may use.

```
Chi-Square Test: Preschool, No Preschool: Girls
Expected counts are printed below observed counts

                          No
         Preschool   Preschool   Total

Grad         21          8         29
           14.50      14.50

No Grad       4         17         21
           10.50      10.50

Total        25         25         50

Chi-Sq = 13.875, DF = 1, P-Value = 0.000
```

Minitab Output for the Girls Only

|            | Preschool | No Preschool |
|------------|-----------|--------------|
| HS Grad    | 21        | 8            |
| HS Grad No | 4         | 17           |

a. Find the graduation rate for those females who went to preschool, and compare it with the graduation rate for those who did not go to preschool.

b. Test the hypothesis that preschool and graduation rate are associated, using a significance level of 0.05.

**10.41 Preschool Attendance and High School Graduation Rates for Males** The Perry preschool project data presented in Exercise 10.39 can be divided to see whether there are different effects for males and females. The table shows a summary of the data for males (Schweinhart et al. 2005).

|            | Preschool | No Preschool |
|------------|-----------|--------------|
| HS Grad    | 16        | 21           |
| HS Grad No | 16        | 18           |

a. Find the graduation rate for males who went to preschool, and compare it with the graduation rate for those who did not go to preschool.

b. Test the hypothesis that preschool and graduation are associated, using a significance level of 0.05.

c. Exercise 10.40 showed an association between preschool and graduation for just the females in this study. Write a sentence or two giving your advice to parents with preschool-eligible children about whether attending preschool is good for their children's future academic success, based on this dataset.

**10.42 Drug for Atrial Fibrillation** Atrial fibrillation is a heart problem in which the atrium (part of the heart) beats very rapidly.

Researchers compared the drug valsartan with a placebo. People in the study had had at least one bout of atrial fibrillation in the previous six months. They were randomly assigned to receive valsartan or placebo, and the counts are shown below. A "bad" outcome is when atrial fibrillation recurred again within one year, and a "good" outcome is when it did not recur during the one-year period.

|      | Valsartan | Placebo |
|------|-----------|---------|
| Bad  | 371       | 375     |
| Good | 351       | 345     |

a. What percentage of those assigned valsartan had a bad outcome? What percentage of those assigned the placebo had a bad outcome? Compare these values.

b. Perform a hypothesis test to determine whether treatment is associated with the outcome. Use a significance level of 0.05.

c. Would you recommend valsartan for a friend with atrial fibrillation?

(Source: The GISSI-AF Investigators. 2009. Valsartan for prevention of recurrent atrial fibrillation. *New England Journal of Medicine* 360, 1606–1617.)

**10.43 Gastric Surgery and Diabetes** A study published in the *Journal of the American Medical Association* in 2008 reported on the effect of lap band surgery on diabetes. In this study, 55 obese patients with type 2 diabetes were randomly assigned to have lap band surgery or not. After two years, researchers recorded whether or not the subjects still had diabetes.

|                       | Lap Band Surgery | No Surgery |
|-----------------------|------------------|------------|
| Free from Diabetes    | 22               | 4          |
| Not Free from Diabetes| 7                | 22         |

a. What percentage of those who had the surgery were diabetes-free? What percentage of those who did not have the surgery were diabetes-free? Compare the two sample percentages.

b. Note the observed value of 4. Does this prevent us from using a chi-square test? Explain.

c. Test the hypothesis that lap bands and diabetes are associated, using a significance level of 0.05.

(Source: J. B. Dixon et al. 2008. *Adjustable gastric banding and conventional therapy for type 2 diabetes, Journal of the American Diabetes Association* 299(3), 316–323.)

**10.44 Progressive Rewards** A study was done on the effect of rewards on smokers' ability to refrain from smoking. All 60 smokers were adults who were *not* trying to quit smoking. Whether they were smoking or not was determined by a carbon monoxide meter. The smokers were studied for five days and tested three times each day. Each smoker was randomly assigned to one of three groups. The first group had "progressive rewards," in which the amount of money received per test gradually increased as they continued to refrain from smoking. The second group had "fixed rewards," in which they earned a fixed amount of money each time the test showed that they had refrained from smoking. The total dollar amounts for the first two groups were the same, assuming no smoking during the experiment. The control group did not receive any monetary reward.

The table shows whether the person smoked at all during the five days of the experiment.

|  | Progressive | Fixed | Control |
|---|---|---|---|
| **No Smoking** | 10 | 6 | 1 |
| **Some Smoking** | 10 | 14 | 19 |

a. Compare the percentage of success (no smoking) for each group.

b. Note the observed value of 1, which is less than 5. Does this mean we cannot use chi-square on this data set? Explain.

c. Determine whether the treatment (group) was associated with the outcome. Use the 0.05 level of significance.

d. Can we conclude that the type of encouragement changed smoking behaviors?

(Source: L. Shaffer and M. R. Merrens. 2001. *Research Stories for Introductory Psychology*, Allyn and Bacon, Boston. Original Source: J. M. Roll et al. 1996. An experimental comparison of three different schedules of reinforcement of drug abstinence using cigarette smoking as an exemplar. *Journal of Applied Behavioral Analysis* 29, 495–505.)

**10.45 Confederates and Compliance** A study was done to see whether participants would ignore a sign that said, "Elevator may stick between floors. Use the stairs." The people who used the stairs were said to be compliant, and those who used the elevator were noncompliant. The study was done in a university dormitory on the ground floor of a three-story building. There were three different situations, two of which involved confederates. A confederate (Conf) is a person who is secretly working with the experimenter. In the first situation, there was no other person using the stairs or elevator—that is, no confederate. In the second, there was a compliant confederate (one who used the stairs). In the third, there was a noncompliant confederate (one who used the elevator). A summary of the data is given in the table and TI-83/84 output is given.

|  | No Conf | Compliant Conf | Noncompliant Conf |
|---|---|---|---|
| **Participant Used Stairs** | 6 | 16 | 5 |
| **Participant Used Elevator** | 12 | 2 | 13 |

a. Find the percentage of participants who used the stairs in all three situations. What do these sample percentages say about the association between compliance and the existence of confederates?

b. From the figure, is the p-value 2.69? Explain. Report the actual p-value.

```
X²-Test
 X²=16.44444444
 P=2.6861747E-4
 df=2
```

c. Determine whether there is an association between the three situations and whether the participant used the stairs (was compliant) or not. Use a significance level of 0.05.

(Source: L. Shaffer and M. R. Merrens. 2001. *Research Stories for Introductory Psychology*. Allyn and Bacon, Boston. Original Source: M. S. Wogalter, et al. 1987. Effectiveness of warnings. *Human Factors* 29, 599–612.)

**\*10.46 Effects of Intensive Glucose Control in the ICU**
Adults admitted to a hospital intensive care unit (ICU) who were expected to be there three or more days were randomly assigned to undergo either intensive glucose (blood sugar) control or conventional glucose control. Intensive glucose control means an attempt was made to make sure the patients' glucose levels were within a narrow range of values (81–108 milligrams per deciliter). Conventional control means an attempt was made to keep the glucose level below 180 mg per dl. The subjects were observed for 90 days to see whether they died from any cause.

|  | Intensive Glucose Control | Conventional Glucose Control |
|---|---|---|
| **Died** | 829 | 751 |
| **Did Not Die** | 2181 | 2261 |

a. What percentage of those who had intensive glucose control died? And what percentage of those who had conventional glucose control died? Compare these percentages, and comment on what the comparison shows.

b. Do a hypothesis test to see whether the treatment is associated with whether the patient died. Use a significance level of 0.05.

c. Judging on the basis of these data, do you think that the glucose control method affects survival?

(Source: The NICE-SUGAR Study Investigators. 2009. *Intensive versus Conventional Glucose Control in Critically Ill Patients. New England Journal of Medicine* 360(13), 1283–1297.)

**10.47 Light at Night and Tumors** A study was done on female mice to see whether the amount of light affects the risk of developing tumors. Fifty mice were randomly assigned to a regimen of 12 hours of light and 12 hours of dark (LD). Fifty similar mice were assigned to 24 hours of light (LL). The study began when the mice were 2 months old. The mice were observed for about two years. Four of the LD mice developed tumors, and 15 of the LL mice developed tumors.

a. What percentage of the LD mice in the sample developed tumors? And what percentage of the LL mice developed tumors? Compare these percentages and comment.

b. Create a two-way table showing the observed values. Label the columns (across the top) with LD and LL.

c. Test the hypothesis that the amount of light is associated with tumors (at the 0.05 level of significance).

d. Some researchers are now starting to investigate whether this phenomenon occurs in humans. They are concerned about shift workers—workers (such as nurses) who may have night shifts. Why would this be a concern?

(Source: D. Baturin et al. 2001. The effect of light regimen and melatonin on the development of spontaneous mammary tumors. *Neuroendocrinology Letters* 22, 441–447.)

**10.48 Antidepressants for Bipolar Disorder** Bipolar disorder is one in which patients cycle between mania and depression. In this study, done by G. S. Sachs et al. in 2007, the participants were randomly assigned to receive a mood stabilizer plus antidepressant or a mood stabilizer plus placebo for 26 weeks. A durable recovery means that the patients had a reasonably happy mood without being manic.

|  | Antidepressant | Placebo |
|---|---|---|
| **Durable Recovery** | 42 | 51 |
| **Not Durable Recovery** | 137 | 136 |

a. What percentage of the patients on the antidepressant had a durable recovery? What percentage of the patients on the placebo had a durable recovery? Compare these percentages, and comment.

b. Test the hypothesis that the treatment is associated with the outcome, using a significance level of 0.05.

c. If you had a friend with bipolar disorder, would you suggest that she or he ask about taking an antidepressant in addition to the mood stabilizer already prescribed? Explain why or why not.

(Source: G. S. Sachs et al. 2007. Effectiveness of adjunctive antidepressant treatment for bipolar depression. *New England Journal of Medicine* 356(17), 1711–1722.)

## SECTION 10.4

TRY **10.49 Swine Flu Hospitalizations (Example 8)** The data on hospitalizations for swine flu, obtained in May 2009 from the Centers for Disease Control, is summarized in the accompanying table. The table looks at the relation between Age Group and Hospitalization, for all swine flu victims in the United States (www.cdc.gov).

| Hospitalized? | Age Group | | | | | | |
|---|---|---|---|---|---|---|---|
|  | Under 5 | 5–14 | 15–29 | 30–44 | 45–60 | over 60 | Totals |
| **Yes** | 7 | 9 | 9 | 9 | 1 | 0 | 35 |
| **No** | 44 | 195 | 241 | 59 | 35 | 10 | 584 |
| **Totals** | 51 | 204 | 250 | 68 | 36 | 10 | 619 |

Because some of the expected values are less than 5, we should combine some groups. For this question, combine all the people who are 30 years old or older in one group and all the people who are less than 30 years old in another group. Show your new table. Then test the new table to see whether there is an association between the variables, using a significance level of 0.05. Assume that this is a random sample from the population of all future and past swine flu victims.

**10.50 2008 General Social Survey: Happiness and Political Views** The table shows the counts (in black) and expected values (in red). All respondents reported their level of happiness. VH is Very Happy, FH is Fairly Happy, NTH is Not Too Happy, and NH is Not at All Happy. Each respondent also reported his or her political view: Liberal, Moderate, or Conservative. We have merged some categories to get these three. For example, we combined the categories Extremely Liberal and Liberal into our category Liberal.

|  | Liberal | Moderate | Conservative | Total |
|---|---|---|---|---|
| VH | 84 / 110.30 | 346 / 303.46 | 131 / 147.25 | 561 |
| FH | 98 / 68.03 | 127 / 187.16 | 121 / 90.82 | 346 |
| NTH | 11 / 15.53 | 58 / 42.73 | 10 / 20.74 | 79 |
| NH | 4 / 3.15 | 11 / 8.65 | 1 / 4.20 | 16 |

a. We would like to perform a chi-square test with the data in the table. Check whether the conditions hold. If you cannot use the table or a chi-square test, explain why you cannot.

b. Combine the Very Happy and Fairly Happy categories into one category (Happy). Also combine Not Very Happy and Not at All Happy into one category (Not Happy). Create a new two-way table. (The table you create should have two rows and three columns.)

c. What percentage of liberals are happy? What percentage of moderates are happy? What percentage of conservatives are happy? Compare them.

d. Do a chi-square test with a significance level of 0.05 to see whether the variables *Political View* and *Happiness* are associated.

**10.51 Gender and Political Party Affiliation** The data in the table on the next page are from the 2008 General Social Survey and concern gender and political party.

a. Find the expected values and report the smallest. Could we use the table as is, without combining categories?

b. Create a new table, using the data from the table shown, with fewer categories. Merge the Strong Democrat and the Not Strong Democrat categories into one group called Democrats. Merge the Strong Republican and the Not Strong Republican categories into one group called Republicans. Put all the Independent categories and the Other category into one group called Other.

c. What percentage of the women and what percentage of the men are Democrat? Compare them.

d. Do a chi-square test with a significance level of 0.05 to see whether the variables *Gender* and *Political Party* are associated using your new grouping.

**Political Party and Gender Counts**

| | | Respondent's Gender | | |
|---|---|---|---|---|
| | | **Male** | **Female** | **Total** |
| Political Party | Strong Democrat | 150 | 240 | 390 |
| | Not str Democrat | 149 | 182 | 331 |
| | Ind, near Dem | 128 | 134 | 262 |
| | Independent | 139 | 183 | 322 |
| | Ind, near Rep | 98 | 64 | 162 |
| | Not str Republican | 140 | 163 | 303 |
| | Strong Republican | 92 | 110 | 202 |
| | Other party | 30 | 8 | 38 |
| Total | | 926 | 1084 | 2010 |

**10.52 Children and Happiness** The data in the table come from the 2008 General Social Survey. The top row is the number of children reported for the respondents. The respondents also reported their level of happiness; Very H means Very Happy, and so on. The counts are shown in the table. Is happiness associated with having at least one child?

| | Number of Children | | | | | | | | |
|---|---|---|---|---|---|---|---|---|---|
| | **0** | **1** | **2** | **3** | **4** | **5** | **6** | **7** | **8 or more** |
| **Very H** | 129 | 99 | 151 | 112 | 51 | 24 | 6 | 7 | 8 |
| **Fairly H** | 188 | 110 | 189 | 98 | 53 | 21 | 9 | 5 | 5 |
| **Not Very H** | 22 | 13 | 20 | 12 | 7 | 6 | 0 | 2 | 0 |
| **Not at All H** | 6 | 3 | 3 | 1 | 2 | 2 | 0 | 1 | 0 |

a. Merge all the Number of Children categories into two groups: those who have 0 children and those who have at least 1 child. For the rows, merge the very happy and fairly happy into one group called Happy. Merge the not very happy and the not at all happy into one group called Unhappy. Report the new table, which should have two rows and two columns.

b. We wish to test whether happiness is associated with having children. Why was it necessary to merge categories?

c. With the merged data, determine whether there is an association between happiness and whether a person has at least one child. Use a significance level of 0.05.

\* **10.53 Scorpion Antivenom** A study was done on children, aged 6 months to 18 years, who had nonlethal scorpion stings. Each child was randomly assigned to receive an experimental antivenom or a placebo. "Good" results were no symptoms after four hours and no detectable plasma venom.

| | **Antivenom** | **Placebo** | **Total** |
|---|---|---|---|
| **No Improvement** | 1 | 6 | 7 |
| **Improvement** | 7 | 1 | 8 |
| **Total** | 8 | 7 | 15 |

The alternative hypothesis is that the antivenom leads to improvement. The p-value for a one-tailed Fisher's Exact Test with these data is 0.009.

a. Suppose the study had turned out differently, as in the following table.

| | **Antivenom** | **Placebo** |
|---|---|---|
| **Bad** | 0 | 7 |
| **Good** | 8 | 0 |

Would Fisher's Exact Test have led to a p-value larger or smaller than 0.009? Explain.

b. Suppose the study had turned out differently, as in the following table.

| | **Antivenom** | **Placebo** |
|---|---|---|
| **Bad** | 2 | 5 |
| **Good** | 6 | 2 |

Would Fisher's Exact Test have led to a p-value larger or smaller than 0.009? Explain.

c. Try the two tests, and report the p-values. Were you right? Search for a Fisher's Exact Test calculator on the Internet, and use it.

(Source: L. V. Boyer et al. 2009. Antivenom for critically ill children with neurotoxicity from scorpion stings. *New England Journal of Medicine* 360(20), 2090–2098.)

\* **10.54 Young Criminals and Violence** A statistics student performed a study on male and female criminals 15 years old and under who were on probation in Ventura County, California. The purpose was to see whether there was an association between type of crime (violent or nonviolent) and gender. Violent crimes involve physical contact such as hitting or fighting. Nonviolent crimes include vandalism, robbery, and verbal assault.

The raw data were shown in Exercise 1.45 on page 31. Use a two-tailed alternative.

a. Make a two-way table that summarizes the data; use Boy and Girl across the top.

b. Compare the percentages of girls and boys who expressed violent behavior.

c. Test the hypothesis that gender and type of crime are associated, using the chi-square statistic. Use a significance level of 0.05.

d. Do a Fisher's Exact Test for the data with the same significance level. Report the two-tailed p-value and your conclusion. (Search for a Fisher's Exact Test calculator on the Internet.)

e. Compare the p-values from the chi-square test and Fisher's Exact Test. Which is more accurate?

f. Suppose the study had turned out differently. Create a new two-way summary table with the same row and column totals that yields a more extreme outcome in the direction you found. Report the p-value for this test, and compare it with the p-value you got in part d.

# CHAPTER REVIEW EXERCISES

**\*10.55 Perry Preschool Arrests** The Perry preschool project discussed in Exercises 10.39–10.41 found that 8 of the 58 students who attended preschool had at least one felony arrest by age 40 and that 31 of the 65 students who did not attend preschool had at least one felony arrest (Schweinhart et al. 2005).

a. Compare the percentages descriptively. What does this comparison suggest?

b. Create a two-way table from the data and do a chi-square test on it, using a significance level of 0.05. Test the hypothesis that preschool attendance is associated with being arrested.

c. Do a two-proportion *z*-test. Your alternative hypothesis should be that preschool attendance lowers the chances of arrest.

d. What advantage does the two-proportion *z*-test have over the chi-square test?

**\*10.56 Parental Training and Criminal Behavior of Children** In Montreal, Canada, an experiment was done with parents of children who were thought to have a high risk of committing crimes when they became teenagers. Some of the families were randomly assigned to receive parental training, and the others were not. Out of 43 children whose parents were randomly assigned to the parental training group, 6 had been arrested by the age of 15. Out of 123 children whose parents were not in the parental training group, 37 had been arrested by age 15.

a. Find and compare the percentages of children arrested by age 15. Is this what researchers might have hoped?

b. Create a two-way table from the data, and test whether the treatment program is associated with arrests. Use a significance level of 0.05.

c. Do a two-proportion *z*-test, testing whether the parental training lowers the rate of bad results. Use a significance level of 0.05.

d. Explain the difference in the results of the chi-square test and the two-proportion *z*-test.

e. Can you conclude that the treatment causes the better result? Why or why not?

(Source: R. E. Tremblay et al. 1996. From childhood physical aggression to adolescent maladjustment: The Montreal prevention experiment. In R.D. Peters and R. J. McMahon. *Preventing childhood disorders, substance use and delinquency.* Thousand Oaks, California: Sage, pp. 268–298.)

**10.57 Diet Drug** A randomized, placebo-controlled study of the diet drug Meridia (sibutramine) was done on overweight or obese subjects and reported in the *New England Journal of Medicine*. The patients were all 55 years old or older with a high risk of cardiovascular events. Those who had a heart event experienced a heart attack, stroke, or death from heart-related factors. The table gives the counts.

|  | Drug | Placebo |
|---|---|---|
| **Heart Event** | 559 | 490 |
| **No Heart Event** | 4347 | 4408 |

a. Compare the rates of heart event for the drug group and the placebo group.

b. Test for association of the drug and a heart event, using a significance level of 0.05.

c. Can you conclude that the drug causes the difference in the rate of heart events? Why or why not?

(Source: W. P. James et al. 2010. Effect of sibutramine on cardiovascular outcomes in overweight and obese subjects. *New England Journal of Medicine* 363, 905–917.)

**10.58 Statin for Prevention of Blood Clots?** A group of 17,802 healthy people with normal cholesterol readings were randomly assigned to take the cholesterol-lowering drug Crestor (a statin) or a placebo. They were then observed for about two years to see whether they developed pulmonary embolism, or blood clots. Assume that exactly half were assigned the drug, and half were assigned the placebo.

When the results were reported in 2009, 34 of those taking Crestor and 60 of those in the placebo group had developed pulmonary embolism, or blood clots.

a. Create a two-way table showing the numbers of those who developed blood clots. Use Crestor and Placebo as labels for the columns.

b. Compare the percentages that had blood clots for these two treatments.

c. Do a test to find out whether treatment and outcome are associated, using a significance level of 0.05.

d. Can you conclude that using Crestor caused the difference? Why or why not?

(Source: R. J. Glynn et al. 2009. A randomized trial of rosuvastatin in the prevention of venous thromboembolism. *New England Journal of Medicine* 360(18), 1851–1861.)

* **10.59 Scared Straight** In the 1980s, the program Scared Straight was created. In this program young delinquents would go to a prison and meet prisoners who told them how difficult prison life is. The aim of the program was to scare the kids so that they would not commit crimes. A study was done in which half of the kids were randomly assigned to a Scared Straight program and half had no treatment. Then all the kids were observed for 12 months to see whether they were arrested.

Forty-three out of 53 of the Scared Straight kids were arrested, and 36 out of 53 of the group that did not see the prison were arrested.

(A)
```
2-PropZTest
 P1<P2
 z=1.560470435
 p=.9406756147
 p̂1=.8113207547
 p̂2=.679245283
↓p̂=.7452830189
```

(B)
```
2-PropZTest
 P1>P2
 z=1.560470435
 p=.0593243853
 p̂1=.8113207547
 p̂2=.679245283
↓p̂=.7452830189
```

a. Find the percentage of each group that was arrested, and comment on it.

b. Perform a chi-square test of homogeneity to determine whether the treatment and outcome (Arrest or No Arrest) are independent. Use a significance level of 0.05.

c. Do a one-tailed two-proportion $z$-test to see whether Scared Straight decreases arrest rates. Which output shown is the correct one, output A or output B? Number 1 refers to the Scared Straight group, and number 2 refers to the control group with no treatment.

d. Compare the results of parts b and c. Include a comparison of p-values. Is there a relationship between the observed values of the $z$-statistic and the chi-square statistic? If so, what is it?

(Source: R. V. Lewis. 1983. Scared Straight—California style: Evaluation of the San Quentin Squires program. *Criminal Justice and Behavior* 10, 209–226.)

* **10.60 Boot Camp** In 1997 the California Youth Authority reported the results of a randomized experiment with young offenders. Some of the offenders were randomly assigned to a "boot camp," and some were simply confined. After they were all released, they were observed for 12 months to see whether they had been arrested. Of the 313 assigned to boot camp, 190 were arrested; of the 243 assigned to confinement, 141 were arrested.

a. Find the percentage of each group that was arrested, and comment on it.

(A)

(B)

b. Perform a chi-square test of homogeneity to determine whether the variables *Treatment* and *Arrest* are associated. Use a significance level of 0.05.

c. Do a two-proportion $z$-test to see whether there was a significantly lower rate of arrest for the boot camp group, using a significance level of 0.05. Which of the two TI-83/84 outputs shown is correct?

d. Compare the results of parts b and c. Include a comparison of p-values. Also, is there a relationship between $z$ and chi-square? If so, what is it?

(Source: California Youth Authority. 1997. *LEAD: A boot camp and intensive parole program: The final impact evaluation.* Sacramento, California: Department of the Youth Authority.)

* **10.61 Robot Cockroaches** Cockroaches tend to rest in groups and prefer dark areas. In a study published in *Science Magazine* in November 2007, cockroaches were introduced to a brightly lit, enclosed area with two different available shelters, one darker than the other. Each time a group of cockroaches was put into the brightly lit area will be called a trial. When groups of 16 real cockroaches were put in a brightly lit area, in 22 out of 30 trials, all the cockroaches went under the same shelter. In the other 8 trials, some of the cockroaches went under one shelter and some under the other one.

Another group consisted of a mixture of real cockroaches and robot cockroaches (4 robots and 12 real cockroaches). The robots did not look like cockroaches but had the odor of male cockroaches, and they were programmed to prefer groups (and brighter shelters). There were 30 trials. In 28 of the trials, all the cockroaches and robots rested under the same shelter, and in 2 of the trials they split up.

|  | Cockroaches Only | Robots Also |
|---|---|---|
| **One Shelter Used** | 22 | 28 |
| **Both Shelters Used** | 8 | 2 |

Is the inclusion of robots associated with whether they all went under the same shelter or not? To answer these questions, assume the cockroaches are a random sample of all cockroaches.

a. Use a chi-square test for homogeneity with a significance level of 0.05 to see whether the presence of robots is associated with whether roaches went into one shelter or two.

b. Repeat the question using Fisher's Exact Test. (If your software will not perform the test for you, search for Fisher's Exact Test on the Internet to do the calculations.) Conduct a two-tailed hypothesis test so that the test is consistent with the test in part a.

c. Compare the p-values and conclusions from part a and part b. Which statistical test do you think is the better procedure in this case? Why?

(Source: J. Halloy et al. 2007. Social integration of robots into groups of cockroaches to control self-organized choices. *Science Magazine* 318(5853), 1155–1158.)

**\*10.62 Robot Cockroaches** Refer to the description in Exercise 10.61. There were 22 trials with only cockroaches (no robots) that went under one shelter. In 16 of these 22 trials, the group chose the darker shelter, and in 6 of the 22 the group chose the lighter shelter. There were 28 trials with a mixture of real cockroaches and robots that all went under one shelter. In 11 of these trials, the group chose the darker shelter, and in 17 the group chose the lighter shelter. The robot cockroaches were programmed to choose the lighter shelter (as well as preferring groups).

| | All under One Shelter | |
| --- | --- | --- |
| | **Cockroaches Only** | **Robots Also** |
| **Darker** | 16 | 11 |
| **Lighter** | 6 | 17 |

Is the introduction of robot cockroaches associated with the type of shelter when the group went under one shelter? Assume cockroaches were randomly sampled from some meaningful population of cockroaches.

a. Use the chi-square test to see whether the presence or absence of robots is associated with whether they went under the darker or the brighter shelter. Use a significance level of 0.05

b. Do Fisher's Exact Test with the data. If your software does not do Fisher's Exact Test, search the Internet for a Fisher's Exact Test calculator and use it.

c. Compare the p-values for parts a and b. Which do you think is the more accurate procedure? The p-values that result from the two methods in this question are closer than the p-values in the previous question. Why do you think that is?

# g UIDED EXERCISES

**g  10.19 Are Humans Like Random Number Generators?**
One of the authors collected data from a class to see whether humans made selections randomly like a random number generator would. Each of 38 students had to pick an integer from one to five. The data are summarized in the table.

| Integer: | One | Two | Three | Four | Five |
| --- | --- | --- | --- | --- | --- |
| **Times Chosen:** | 3 | 5 | 14 | 11 | 5 |

A true random number generator would create roughly equal numbers of all five integers.

QUESTION  Test the hypothesis that humans are not like random number generators by following the steps below. Use a significance level of 0.05, and assume the data were collected from a random sample of students.

**Step 1 ▶ Hypothesize**

$H_0$: Humans are like random number generators and produce numbers in equal quantities.

$H_a$: ?

TI-84 GOF Output

**Step 2 ▶ Prepare**
Choose chi-square goodness of fit (GOF). We have only one variable: *Integer Chosen*. Use a significance level of 0.05. We are assuming randomness. We must check to see that all the expected values are 5 or more. There were 38 students. Explain why all the expected values are 7.6.

**Step 3 ▶ Compute to compare**

$$X^2 = \frac{(3 - 7.6)^2}{7.6} + \frac{(5 - 7.6)^2}{7.6}$$
$$+ \underline{\quad} + \underline{\quad} + \underline{\quad} = \underline{\quad}$$

Complete the calculation of the chi-square statistic. Then check your calculated chi-square with the output. Obtain the p-value from the TI-84 output.

**Step 4 ▶ Interpret**
Reject or do not reject the null hypothesis, and pick an interpretation from those below.

i. Humans have not been shown to be different from random number generators.

ii. Humans have been shown to be different from random number generators.

**g  10.33 Obesity and Marital Status** A study reported in the medical journal *Obesity* in 2009 analyzed data from the National Longitudinal Study of Adolescent Health. Obesity was defined as having a body mass index (BMI) of 30 or more. The research subjects were followed from adolescence to adulthood, with all individuals in the sample categorized in terms of whether or not they were obese and whether they were dating, cohabiting, or married.

| | Dating | Cohabiting | Married |
| --- | --- | --- | --- |
| **Obese** | 81 | 103 | 147 |
| **Not Obese** | 359 | 326 | 277 |

QUESTION  Test the hypothesis that the variables *Relationship Status* and *Obesity* are associated, using a significance level of 0.05. Also consider whether the study shows causality. The steps will guide you through the process. Minitab output is provided.

**Step 1 ▶ Hypothesize**

$H_0$: Relationship status and obesity are independent.

$H_a$: ?

```
Chi-Square Test: Dating, Cohabiting, Married
Expected counts are printed below observed counts

         Dating  Cohabiting  Married  Total
Obese        81         103      147    331
          112.64      109.82   108.54

Not         359         326      277    962
          327.36      319.18   315.46

Total       440         429      424   1293

Chi-Sq = 30.829, DF = 2, P-Value = 0.000
```

**Step 2 ▶ Prepare**

We choose the chi-square test of independence because the data were from *one* random sample in which the people were classified two different ways. We do not have a random sample or a random assignment, so we will test to see whether these results could easily have occurred by chance. Find the smallest expected value and report it. Is it more than 5?

**Step 3 ▶ Compute to compare**

Refer to the output given.

$X^2 = $ _____

p-value = _____

**Step 4 ▶ Interpret**

Reject or do not reject the null hypothesis, and state what that means.

**Causality**

Can we conclude from these data that living with someone is making some people obese and that marrying is making even more people obese? Can we conclude that obesity affects your relationship status? Explain why or why not.

**Percentages**

Find and compare the percentages obese in the three relationship statuses.

# TechTips

## General Instructions for All Technology

### EXAMPLE A (CHI-SQUARE TEST FOR TWO-WAY TABLES): PERRY PRESCHOOL AND GRADUATION FROM HIGH SCHOOL ▶ In the 1960s an experiment was started in which a group of children were randomly assigned to attend preschool or not to attend preschool. They were studied for years, and whether they graduated from high school is shown in Table A.

We will show the chi-square test for two-way tables to see whether the factors are independent or not.

|            | Preschool | No Preschool |
|------------|-----------|--------------|
| **Grad HS**    | 37        | 29           |
| **No Grad HS** | 20        | 35           |

▲ **TABLE A** Two-way Table for Preschool and Graduation from High School

### Discussion of Data

Much of technology is set up so that you can use the table summary (such as Table A) and find the calculated results. However, it is also possible to start with a spreadsheet containing the raw data. Table B shows the beginning of the raw data, for which there would be 121 rows for the 121 children.

| Preschool | Graduate HS |
|-----------|-------------|
| Yes       | No          |
| Yes       | Yes         |
| No        | Yes         |
| No        | Yes         |
| Yes       | Yes         |
| No        | No          |

▲ **TABLE B** Some Raw Data

## EXAMPLE B CHI-SQUARE FOR GOODNESS OF FIT: ARE HUMANS LIKE RANDOM NUMBER GENERATORS? ▶ One of the authors collected data from a class to see whether humans made selections randomly, as a random number generator would. Each of 38 students had to pick one of the integers one, two, three, four, or five. Table C summarizes the collected (*Observed*) data. If the students picked randomly, we would expect each number (one through five) to be chosen an equal number of times (uniform distribution). Because there were 38 students and 5 choices, each *Expected* value is $38 \times (1/5)$, or 7.6, as listed in Table C.

| Integer: | One | Two | Three | Four | Five |
|---|---|---|---|---|---|
| Times *Observed* to Be Chosen: | 3 | 5 | 14 | 11 | 5 |
| Times *Expected* to Be Chosen: | 7.6 | 7.6 | 7.6 | 7.6 | 7.6 |

▲ **TABLE C** Goodness-of-Fit Data

### TI-83/84

#### Chi-Square for Two-way Tables

You will not put the data into the lists. You will use a matrix (table), and the data must be in the form of a summary such as Table A.

1. Press **2ND** and **MATRIX** (or **MATRX**).

2. Scroll over to **EDIT** and press **ENTER** when **1:** is highlighted.

3. See Figure 10A. Put in the dimensions. Because the table has two rows and two columns, press **2**, **ENTER**, **2**, **ENTER**. (The first number is the number of rows, and the second number is the number of columns.)

4. Enter each of the four numbers in the table, as shown in Figure 10A. Press **ENTER** after typing each number.

5. Press **STAT**, and scroll over to **TESTS**.

6. Scroll down (or up) to **C: $\chi^2$-Test** and press **ENTER**.

7. Leave the **Observed** as **A** and the **Expected** as **B**. Scroll down to **Calculate** and press **ENTER**.

You should get the output shown in Figure 10B.

▲ **FIGURE 10A** TI-83/84 Input for Two-way Table
▲ **FIGURE 10B** TI-83/84 Output for a Chi-Square Test for Two-way Tables

8. To see the expected values, click **2ND**, **MATRIX**, scroll over to **EDIT**, scroll down to **2: [B]**, and press **ENTER**. You may have to scroll to the right to see some of the numbers. They will be arranged in the same order as the table of observed values. Check these numbers for the required minimum value of 5.

## Goodness-of-Fit Test Using a TI-84
(The TI-83 does not provide the goodness-of-fit test.)

1. Put the observed counts in one list, such as **L1**, and put the expected values in another list, such as **L2**, using the same order.

2. Then choose **Stat**, **Tests**, and **D: X²GOF-Test**.

3. Make sure the **Observed** list is the one in which you have the observed counts and that the **Expected** list is where you have the expected values. You will have to put in the value for **df**, which is the number of categories minus 1. In Example B there were five numbers to choose from, so df is 4 because $5 - 1 = 4$.

4. Then choose **Calculate** (or **Draw**).

5. The chi-square value comes out to be 11.47368, and the p-value is 0.0217258

### MINITAB

#### Two-way Tables

For Minitab you may have your data as a table summary (as shown in Table A) or as raw data (as shown in Table B).

**TABLE SUMMARY**

1. Type a summary of your data into the columns.

| C1 | C2 |
|---|---|
| 37 | 29 |
| 20 | 35 |

You may put labels above the numbers if you want to.

2. **Stat > Tables > Chi-square Test > Two-Way Table in Worksheet**

3. Select both columns (by double clicking them) and click **OK**. Figure 10C shows the output.

```
Chi-Square Test: C1, C2
Expected counts are printed below observed counts
Chi-Square contributions are printed below expected counts

        C1     C2   Total
 1      37     29      66
     31.09  34.91
      1.123  1.000

 2      20     35      55
     25.91  29.09
      1.348  1.200

Total   57     64     121
Chi-Sq = 4.671, DF = 1, P-Value = 0.031
```

▲ **FIGURE 10C** Minitab Output for Chi-Square Test for Two-way Tables

**RAW DATA**

1. Make sure your raw data are in the columns. See Table B.

2. **Stat > Tables > Cross Tabulation and Chi-square**.

3. See Figure 10D: Click either C1 **For rows** and C2 **For columns** (or vice versa). Ignore **For layers**.

▲ **FIGURE 10D** Minitab Input for Cross Tabulation and Chi-Square

4. Click **Chi-Square**.

5. Select both **Chi-Square analysis** and **Expected cell counts**.

6. Click **OK** and click **OK**.

### Goodness of Fit

1. Type your observed counts in one column, such as **C1**.

2. If you have expected values that are not equal, type them in another column, such as **C2**. (If all the expected values are the same, as in our Example B, they don't have to be typed in.)

3. **Stat > Tables > Chi-Square Goodness-of-Fit Test (One Variable)**

4. See Figure 10E. Click in the box next to **Observed Counts**, and then double click the column with the observed counts.

5. If there are **Equal proportions**, leave it. If you have a column of expected counts, click **Proportions specified by historical counts** and then choose the column containing the expected values.

   Your results will show bar charts (as well as chi-square output) unless you click on **Graphs** and turn the graphs off.

▲ **FIGURE 10E** Minitab Goodness-of-Fit Options

6. Click **OK**.

   You should get a chi-square value of 11.4737and a p-value of 0.022.

### Two-way Tables

1. Type a summary of your data into two (or more) columns, as shown in columns A and B in Figure 10F.

|   | A | B |
|---|---|---|
| 1 | 37 | 29 |
| 2 | 20 | 35 |

▲ **FIGURE 10F** Excel Input for Two-way Table

2. To get the total of 66 (from 37 + 29), click in the box to the right of 29, and then click **fx**, double click **SUM**, and click **OK**. You can do the same thing for the other sums, each time starting from the cell in which you want to put the sum. Save the grand total for last. Your table with the totals should look like columns A, B, and C in Figure 10G. Alternatively, you could simply add to get the totals.

3. To get the expected values, you will be using the formula

$$\text{Expected value} = \frac{(\text{row marginal total}) \times (\text{column marginal total})}{(\text{grand total})}$$

To get the first expected value, 31.09091, click in the cell where you want the expected value to be placed (here, cell E1). (An empty column, such as column D in Figure 10G, improves the clarity.) Then type = and click on the **66** in the table, type **\*** (for multiplication), click on the **57**, type **/** (for division), click on the **121**, and press **Enter**. Figure 10G shows part of the process for getting the expected value that goes below the 31.09. For each of the expected values, you start from the cell you want filled, and you click on the row total, **\*** (for multiply), the column total, **/** (for divide), and the grand total and press **Enter**. Alternatively you could figure out the expected values by hand.

|   | A | B | C | D | E |
|---|---|---|---|---|---|
| 1 | 37 | 29 | 66 |   | 31.09091 |
| 2 | 20 | 35 | 55 |   | =A3*C1/C3 |
| 3 | 57 | 64 | 121 |   |   |

▲ **FIGURE 10G** Excel, Including Totals and One Expected Value

After you have all four expected values, be sure they are arranged in the same order as the original data:

4. Click **fx**.

5. Select a category: **Statistical** or **All**.

6. Choose **CHISQ.TEST**. For the **Actual_range**, highlight the table containing the observed counts, but do *not* include the row and column totals or the grand total. For the **Expected range**, highlight the table with the expected values.

   You will see the p-value (0.030671). Press **OK** and it will show up in the active cell in the worksheet.

   The previous steps for Excel will give you the p-value but not the value for chi-square. If you want the numerical value for chi-square, continue with the steps that follow.

7. Click in an empty cell.

8. Click **fx**.

9. Select a category: **Statistical** or **All**.

10. Choose **CHISQ.INV.RT** (for inverse, right tail).

11. For the **Probability**, click on the cell from Step 6 that shows the p-value of **0.030671**. For **Deg_freedom**, put in the degrees of freedom (df). For two-way tables,

$$df = (\text{number of rows} - 1)(\text{number of columns} - 1).$$

For Example A, df is 1. Click **OK**. You should get a chi-square of 4.67 for Example A.

### Excel Goodness-of-Fit Test

1. Type your observed counts in one column and your expected counts in another column, in the same order.

2. Follow steps 4, 5, and 6 above.

   You will get a p-value of 0.021726.

   If you want the numerical value of chi-square, follow the instructions in steps 7–11 above. However, in step 11, for goodness-of-fit,

$$df = \text{number of categories} - 1$$

   For Example B, the number of categories is 5, so df is 4. You will get a chi-square value of 11.47.

### Two-way Tables

#### TABLE SUMMARY

1. Enter your data summary as shown in Figure 10H. Note that you can have column labels (Preschool or No Preschool) and also row labels (GradHS or NoGrad).

▲ **FIGURE 10H** StatCrunch Input for Two-way Table

2. **Stat > Tables > Contingency > with summary**

3. See Figure 10I. Select the columns that contain the summary counts, and select the column that contains the **Row labels**, here **var1**.

▲ **FIGURE 10I** StatCrunch Two-way-Table Options

4. Click **Next** and click **Expected Count**.

5. Click **Calculate**.

   Figure 10J shows the well-labeled output.

▲ **FIGURE 10J** StatCrunch Output for Two-way Table

#### RAW DATA

1. Be sure you have raw data in the columns; see Table B.

2. **Stat > Tables > Contingency > with data**

3. Select both columns.

4. Click **Next** and check **Expected Count**.

5. Click **Calculate**.

#### GOODNESS-OF-FIT TEST

1. Enter the observed counts in one column and the expected counts in another column, in the same order.

2. **Stat > Goodness-of-fit > Chi-square test**

3. Select the column for **Observed**, and select the column for **Expected**.

4. Click **Calculate**.

   With the data on integer selection you should get a chi-square of 11.47 and a p-value of 0.0217.

# 11 Multiple Comparisons and Analysis of Variance

## THEME

**Comparing the means of more than two groups is a method for examining whether a numerical variable is associated with a categorical variable. Although it is tempting to compare groups pair-by-pair, this leads to problems. ANOVA is a procedure that compares means across groups, while avoiding these problems.**

In Chapter 9, you learned how to compare the means of two groups using confidence intervals and hypothesis tests. There is nothing magical about the number two: We often need to compare three, four, or even more groups with each other. Analysis of variance, or ANOVA, is a method for doing just that. The name *analysis of variance* is slightly misleading, because although we will indeed use the variance (recall that the variance is just the standard deviation squared), it is really the *means* we are interested in.

ANOVA is a method for testing whether there is an association between a categorical variable that identifies different groups, and a numerical variable. For example, do you want to know which diet (the categorical variable) is best for losing weight (the numerical variable)? One approach is to randomly assign people to one of several diets and then compare the mean weight lost by each dieting type. If there is an association between the categorical variable *Diet Type* and the numerical variable

*Weight Loss*, then it may be that the type of diet you choose determines the amount of weight you could lose. Is success in sports determined in part by the color of the team uniform? To examine this question, we could compare the success rates of several different teams after grouping the teams by the categorical variable *Uniform Color*.

We begin this chapter by introducing a famous difficulty in statistical inference: the problem of multiple comparisons. Comparing several groups requires some care. We then discuss a basic mindset for thinking about the variation within and between groups, and we show how this leads to a test statistic that can compare the means of several different groups without raising the problem of multiple comparisons. This test statistic works best under certain conditions, and we discuss these in detail. Finally, we consider post-hoc tests, which allow us to explore the data at a greater level of detail.

## CASE STUDY

## Seeing Red

Does the color of your team uniform affect your performance? That's what researchers in England claimed, after studying data from 68 soccer teams over 56 seasons. On average, teams wearing red won 53% of their home games. Teams wearing blue won, on average, 51%, and teams wearing yellow/orange won only 50%. The researchers compared the performance of teams wearing four different color groups (blue, orange or yellow, red, white) and found significant differences in performance, as measured by the percentage of home games won. (Teams wear their team colors only for home games.) Further investigation showed that teams wearing red did the best of all (Attrill et al. 2008).

You have seen two groups compared before, but how do we compare means across *four* different groups? This chapter covers techniques for doing so. Then we will return to this case study at the end of the chapter and see whether red is, statistically speaking, the "winningest" color.

# Multiple Comparisons

In earlier chapters, we compared two groups to each other, either with a *t*-test for comparing means or with a confidence interval for the difference of two means. We are now going to be examining situations that involve more than two groups. As it turns out, special care is needed, and in this section we will show you why. We begin by showing what the data you encounter might look like for these situations.

## Data for Multiple Comparisons

In this chapter, we analyze data that have a very specific structure. The response variable is always numerical. Examples include the number of kernels of corn that popped, the amount of weight a dieter lost, the temperature of the planet Earth in a given decade, and the height of children in different age groups. We're comparing these numerical values across different groups, and these groups are represented by a single categorical variable.

Table 11.1 shows the first six rows of data from the popcorn-popping experiment in the Case Study of Chapter 10. One factor that the experimenters examined was the amount of oil. They tried three different amounts of oil: none, 1/2 teaspoon, and 1 teaspoon. For our purposes, we call these groups None, Medium, and Maximum. Then they popped 36 bags using no oil, another 36 using a medium amount of oil, and another 36 using the maximum amount. Each bag contained exactly 50 kernels of popcorn. After 75 seconds, the bags were removed from heat, and the experimenters counted and recorded the number of kernels popped. Does the amount of oil that is used affect the number of kernels that get popped?

The data set has two variables: a categorical variable (*Amount of Oil*) and a numerical variable (*Number of Kernels Popped*).

We might summarize these data by showing the mean and standard deviation of the number of kernels popped in each group. Graphically, we might show boxplots to summarize the shape of the distributions. Table 11.2 shows the summary statistics. Figure 11.1 shows boxplots of these data.

| Amount of Oil | Number of Kernels Popped |
|---|---|
| Maximum | 8 |
| Medium | 26 |
| Maximum | 5 |
| None | 8 |
| None | 7 |
| Medium | 20 |

▲ **TABLE 11.1** The first six lines of the popcorn data set.

▶ **TABLE 11.2** The mean and standard deviation (in parentheses) for the number of kernels popped for each amount of oil.

| Group: | None | Medium | Maximum |
|---|---|---|---|
| Mean (SD) | 19.75 (11.76) | 19.75 (11.83) | 13.47 (9.30) |

▶ **FIGURE 11.1** Boxplots showing the five-number summary of the distribution of number of kernels popped. The group in which the maximum amount of oil was used had a lower median number of good kernels than the other two groups.

## The Problem of Multiple Comparisons

Figure 11.2 shows another perspective on the popcorn data. Each dot represents a bag of popcorn and shows us how many kernels were popped. The mean number of popped kernels in each group is indicated by a red dot. We want to know whether the amount of oil affects the number of kernels popped. As you can see, this varies from bag to bag. But are the population means different? Or are observed differences of sample means due to chance? That is, if we were to pop a very large number of bags under these three different conditions, might the mean number of kernels popped all be the same?

◀ **FIGURE 11.2** Each dot represents a bag with 50 popcorn kernels in it. The bags were heated for 75 seconds, and the graph shows the number of kernels that had popped after this time. Different amounts of oil were used. The large red dots show the mean number of kernels popped.

In Chapter 9, we would have answered these questions by doing a two-sample *t*-test. But as the name implies, two-sample *t*-tests are used for comparing only two groups. We need to compare three. We might try doing three separate two-sample *t*-tests. For example, we might test

**Hypothesis 1**
$H_0$: Mean with no oil = mean with medium amount of oil
$H_a$: Mean with no oil ≠ mean with medium amount of oil

**Hypothesis 2**
$H_0$: Mean with no oil = mean with maximum oil
$H_a$: Mean with no oil ≠ mean with maximum oil

**Hypothesis 3**
$H_0$: Mean with medium oil = mean with maximum oil
$H_a$: Mean with medium oil ≠ mean with maximum oil

To compare all three groups, we need three different hypothesis tests. This is called a **multiple comparison**, because we are comparing multiple pairs of means. But this causes a problem.

To understand the problem, recall that whenever we do a significance test, there is a small probability that we will make the mistake of rejecting the null hypothesis when it is true. This probability is called the significance level, and we usually set it to be 0.05, or 5%.

The basic idea with multiple comparisons is that even though the probability of something going wrong on one occasion is small, if we keep repeating the experiment, eventually something *will* go wrong. Essentially, by doing multiple tests, we are creating more opportunities to mistakenly reject the null hypothesis. The more tests we do, the greater the probability that we'll mistakenly reject the null hypothesis at least once.

The probability that we mistakenly reject the null hypothesis at least once after doing several hypothesis tests is called the **overall significance level**. Making "at least one" such mistake means that we make this error for one test, for two tests, or for all tests. When three hypothesis tests are performed, each with a significance level of 5%, the overall significance level is about 14%. In other words, the probability that we will conclude that at least one amount of oil is more effective than another, when the truth is that all amounts are equally effective, is 14%. This is not terribly high, but we were shooting for a 5% error rate, and 14% is quite a bit higher.

**Details**

**Six Comparisons from Four Groups**
The groups are A, B, C and D; the comparisons are AB, AC, AD, BC, BD, and CD.

When we have only three groups to compare, we need to make only three comparisons, and the overall significance level goes from 5% (for one comparison) to 14%. But if we have more groups, the overall significance level error goes up dramatically. If we have four groups, then we need to make six comparisons. The overall significance level is now about 26%. If we have five groups, then we need to make ten comparisons, and there is about a 40% chance that we will mistakenly reject at least one of the ten null hypotheses.

Figure 11.3 shows how the overall significance level increases as the number of groups increases. You can see that if you have ten groups, then the probability that you will mistakenly reject at least one null hypothesis is 90%. You are almost certain to make a mistake!

▶ **FIGURE 11.3** The overall significance level—the probability that at least once we'll reject the null hypothesis when in fact it is true—increases dramatically as the number of groups increases. If there are ten groups and all of the means are equal, then the probability that we will mistakenly conclude that at least one is different from the others is about 90%.

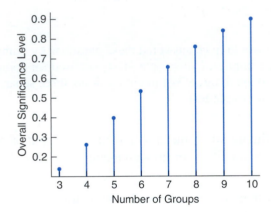

## One Solution: The Bonferroni Correction

Through the years, statisticians have worked out different ways of solving the problem of multiple comparisons. Some methods are fairly complex (both mathematically and computationally) and are best left to a computer. We'll talk about one of these in Section 9.4. But one method, the Bonferroni Correction, is fairly straightforward.

The **Bonferroni Correction** is done by using a smaller significance level for each individual test. If you want an overall significance level of $\alpha$, and you are doing $m$ comparisons, then you should do each individual comparison with significance level $\alpha/m$. If you do this, then the overall significance level will be $\alpha$ or smaller. (Usually, it will be smaller.)

### Finding the Number of Comparisons

The number of comparisons required depends on the number of groups that you are comparing. In the popcorn experiment, there were three groups and three comparisons. The fact that in this case the number of comparisons is the same as the number of groups is a bit of a coincidence, and it happens only when there are three groups. In general, the number of comparisons is the same as the number of different ways you can select two objects from a group of $k$ objects. Formula 11.1 shows how to find this number.

$$\text{Formula 11.1: Number of comparisons} = \frac{k(k-1)}{2}$$

where $k$ is the number of groups you are comparing.

## EXAMPLE 1 Health of Sea Lions

Biologists collected data on the health of sea lions living in four distinct regions on the northern Pacific coast. They took measurements that reflect the overall health of sea lion colonies living in these regions. They wish to test the hypothesis that the mean health is the same for each region.

**QUESTIONS**

a. If they choose to do *t*-tests comparing each region to every other region, how many comparisons must they make?

b. If they wish to have an overall significance level of $\alpha = 0.05$, applying the Bonferroni Correction, what significance level should they use for each individual *t*-test?

**SOLUTIONS**

a. We have $k = 4$ groups, so we need

$$\text{Number of comparisons} = \frac{k(k-1)}{2} = \frac{4(3)}{2} = \frac{12}{2} = 6$$

b. The Bonferroni Correction says to divide the overall significance by the number of comparisons, so we should use $\alpha = (0.05/6) = 0.0083$ for each individual test.

**TRY THIS!** Exercise 11.3

**KEY POINT** When comparing more than two groups with hypothesis tests, in order to keep the overall significance level at or below the given value of $\alpha$ (usually 0.05), the Bonferroni Correction divides $\alpha$ by the number of comparisons, not the number of groups.

## EXAMPLE 2 Popping Corn in Oil

Below are results from three *t*-tests that compare all possible pairs of groups to test which amount of oil works best for popping corn. Each group had 36 bags of popcorn, each with the same number of kernels. The researchers recorded the number of kernels popped after a fixed amount of time and then tested whether the mean amount differed for different amounts of oil.

Hypothesis 1
$H_0: \mu_{none} = \mu_{med}$
$H_a: \mu_{none} \neq \mu_{med}$
$t = 0$, df $= 66$, p-value $= 1.000$

Hypothesis 2
$H_0: \mu_{none} = \mu_{max}$
$H_a: \mu_{none} \neq \mu_{max}$
$t = 2.51$, df $= 66$, p-value $= 0.014$

Hypothesis 3
$H_0: \mu_{med} = \mu_{max}$
$H_a: \mu_{med} \neq \mu_{max}$
$t = 2.50$, df $= 66$, p-value $= 0.015$

**QUESTION** Apply the Bonferroni Correction to test whether the mean number of kernels popped for each group is different from those for the other groups. Use an overall significance level of $\alpha = 0.05$.

**SOLUTION** There are three groups, so we are making three comparisons. This means that to have an overall significance level of 0.05, we conduct each individual test with a significance level of $0.05/3 = 0.0167$. Thus we reject the null hypothesis if the p-value is less than or equal to 0.0167.

The p-value for Hypothesis 1 (1.000) is clearly bigger than 0.0167, so we do not reject the null hypothesis.

The p-value for Hypothesis 2 (0.014) is smaller than 0.0167, so we can reject the null hypothesis.

The p-value for Hypothesis 3 (0.015) is also smaller than 0.0167, so we can reject the null hypothesis.

**CONCLUSION** We find that the mean for no oil is significantly different from the mean for maximum oil and also that the mean for medium oil is significantly different from the mean for maximum oil. However, there is no significant difference between the means for medium oil and no oil.

In practice, you should remember to check whether the conditions necessary for using the *t*-test hold before carrying out these tests. For this example, we skipped these details to focus on the main idea: When doing multiple comparisons, we need to make adjustments to keep the overall significance level to a reasonable value.

**TRY THIS!** Exercise 11.5

One drawback of the Bonferroni Correction is that it is very conservative. This can be good, because it means that the true overall significance level is even lower than $\alpha$. The bad news is that it also means that the statistical power—the probability that we will correctly acknowledge that two groups have different means—is lower.

**Looking Back**

**Power**

Recall that, all other factors being equal, the **power**, which is the probability of correctly rejecting the null hypothesis when the null hypothesis is false, decreases if the significance level goes down and increases if the significance level goes up.

## Bonferroni Intervals

In Chapter 9 you learned that you can compare the means of two independent groups either by using a two-sample *t*-test or by constructing a two-sample confidence interval for the difference between means. Confidence intervals are useful because they give more information than a *t*-test, such as estimates about the size of the difference between the means.

When examining several groups, we can also construct confidence intervals for each pair of groups. However, we have the same problem that we have with the hypothesis test. If we construct a 95% confidence interval for each pair of groups, it is true that we're 95% confident that each interval includes the true difference. This means that if we were to repeat this procedure with new data, there would be only a 5% chance that any given interval would *not* include the true difference in means. Unfortunately, there would be more than a 5% chance that *at least one* of the confidence intervals we constructed would not include the true difference in means.

The Bonferroni approach can fix this. The idea is that we make each individual confidence interval slightly wider (use a larger margin of error) so that we achieve the desired overall confidence level.

To see how to do this, it helps to use a special notation and write the confidence level like this:

$$L = (1 - \alpha)100\%$$

For example, if $\alpha = 0.05$, then the confidence level is

$$L = (1 - 0.05)100\% = 0.95 \times 100\% = 95\%$$

The Bonferroni Correction says that if you want an overall confidence level of $L$, then you must make each individual confidence interval with this level:

$$(1 - \alpha/m)100\%$$

where $m$ is the number of intervals you are making.

For example, suppose we have $m = 3$ comparisons, and so three intervals are needed. If we want an overall confidence level of 95%, then $\alpha = 0.05$ (because $95\% = 100\% - 5\%$, so $\alpha = 0.05$). The Bonferroni Correction says to make each interval with confidence level

$$(1 - 0.05/3)100\% = (1 - 0.00167)100\% = 98.33\%$$

Below you can see the output for Bonferroni intervals for the popcorn example. Because we have three comparisons and are finding three intervals, we use $m = 3$. Each is a 98.33% confidence interval. Together, these are called *95%* **simultaneous confidence intervals**, because if we follow the Bonferroni approach, we can be 95% confident that *all three* of the intervals include the true difference in means. Note that we come to the same conclusion as we did with the *t*-tests.

|  | Lower Boundary | Upper Boundary | Conclusion |
|---|---|---|---|
| No oil to medium oil | −6.82 | 6.82 | Not different |
| No oil to maximum oil | 0.14 | 12.41 | *No oil* pops more, on average, than maximum oil |
| Medium oil to maximum oil | 0.12 | 12.44 | Medium oil pops more, on average, than maximum oil |

To interpret these intervals, we use the same approaches as in previous chapters. If the interval includes the value 0, then it is plausible that the difference in means is 0. In this case, we would say that there's not enough evidence to conclude that the means are different.

If, on the other hand, the interval does not include 0, we can conclude (with a 95% overall level of confidence) that the means are different, because their difference is not plausibly equal to 0.

The interval for the difference in the mean number of kernels popped with no oil and the mean number popped with medium amounts of oil captures 0, so there is no evidence that you will get different results using no oil and using a medium amount of oil. On the other hand, the other two intervals do not include 0 and include all positive values. So we can be confident that using no oil produces, on average, more popped kernels than using the maximum amount of oil, and using medium amounts of oil results in a greater number of popped kernels, on average, than using the maximum amount of oil.

**KEY POINT**  We can use the Bonferroni Correction to produce a collection of simultaneous confidence intervals so that we can be 95% confident that all of the intervals include the population parameters.

The ANOVA procedure (which we introduce next), like the Bonferroni approach, "solves" the problem of multiple comparisons. ANOVA has the advantage of providing a higher power than the Bonferroni approach.

**Looking Back**

**Confidence Level**
A confidence level of $L$ tells us that if we were to take many different samples, and for each compute a confidence interval with confidence level $L$, then $L\%$ of those intervals would include the population parameter, and $100\% - L$ would not.

Tech

# The Analysis of Variance

The analysis of variance, also known as ANOVA, is a procedure for comparing the means of several groups. In this section, we introduce the fundamental concepts that support ANOVA and introduce a test statistic that we will use to test whether the mean of a variable is different for different groups. In the next section, we will introduce the mechanics for carrying out the ANOVA procedure.

ANOVA is a more powerful approach than the Bonferroni Correction described in the last section. This means that if there really is a difference in the means, then ANOVA has a better chance of finding it. ANOVA is also quite a bit less tedious than doing a series of *t*-tests.

When we are comparing means across several different groups (one categorical variable and one numerical variable), we use a **one-way analysis of variance**. It is also possible to have two-way ANOVAs, which are used to compare means when there are two categorical variables and one numerical response variable. Even "higher-way" ANOVAs are possible. These are all usually covered in later statistics courses.

The null and alternative hypotheses are the same for every ANOVA.

$H_0$: There is no association between the categorical variable and the numerical variable.

$H_a$: There is an association; the population means of the categories differ.

Another way of stating this is to consider the different groups of the categorical variable. Suppose we have $k$ different groups. For example, for the popcorn experiment, $k = 3$: no oil, medium oil, maximum oil. Then the hypotheses are

$H_0$: $\mu_1 = \mu_2 = \cdots = \mu_k$

$H_a$: At least one of the means is not equal to another.

In the context of popping corn, the null hypothesis says that no matter what amount of oil you use, you'll get the same mean number of kernels popped.

The alternative hypothesis claims that the two variables are associated and that, therefore, at least one mean is different from another. Different amounts of oil are associated with differing numbers of kernels popped. In a controlled study, we can claim that if the alternative hypothesis is true, then if you use different amounts of oil, you can expect a different number of kernels to be popped for at least one of the amounts of oil.

Rejecting the null hypothesis for an ANOVA is usually unfulfilling. We learn that at least one of the means is different from another, but we do not learn which one. We do not learn whether there is more than one. We don't learn which is the biggest, which is the smallest, or really any information other than that the null hypothesis is not true. For this reason, an ANOVA test is often followed up with an exploration to look further at the relationship of the groups to one another. We'll give some approaches for this exploration, which is often called post-hoc analysis, in Section 11.4.

**KEY POINT** ANOVA tests whether a categorical variable is associated with a numerical variable. This is the same as testing whether the mean value of a numerical variable is different in different groups.

The ANOVA is based on comparing the amount of variation that exists between groups with the variation within each group. To help you understand what this means, let's think about our analysis visually.

**Details**

**Pronunciation**
*ANOVA* is pronounced with the emphasis on the second syllable, an-**OH**-va.

**Looking Back**

**Variance vs. Variation**
Both *standard deviation* and *variance* have specific mathematical definitions. Standard deviation and variance are methods for measuring *variation*, which is itself a more general term and has no mathematical definition.

# Visualizing It

As the name suggests, the analysis of variance determines whether the means from the groups are the same or different by comparing the variances of the groups. Basically, if at least one of the means is very far away from another, we want to reject the null hypothesis. Here *very far* means "very far, relative to the amount of variation within the data."

Figure 11.4a shows a plot of means from four groups. (The data are simulated.) Do the means look similar or different to you?

◄ **FIGURE 11.4a** The means of four different groups. The means are slightly different, but are they close enough together for the observed differences to be due to chance? Or are they so far apart that we conclude they're actually different?

Figure 11.4b shows the same four means, but now we also show the data points themselves. You can see that each group has lots of variability. We would say that these means are actually fairly close together, because the means are more similar to each other than are individual values within any of the groups. In this case, we would expect our ANOVA *not* to reject the null hypothesis. There is not enough evidence to convince us that the means are different from each other. The small differences we see could easily be explained by chance.

▲ **FIGURE 11.4b** With the actual data points displayed, we now see that the means are actually close together. The distances between means are much less than the distances between points within any of the groups.

In fact, the standard deviation between the four means is 1.38. The standard deviation within each of the groups is about 13.6. Thus the amount of variation between the means is much smaller than the amount of variation within each group.

Figure 11.4c on the next page shows another scenario. The means *are exactly the same as in Figures 11.4a and 11.4b,* but in this new data set, the variability is so small that the group means look very different from each other. The standard deviation between the group means is still 1.38. But now the standard deviation within each group is only 0.10. The amount of variation between the group means is much greater than the variation within each group, so the means are more different from each other. In this case, we would expect our ANOVA procedure to reject the null hypothesis.

▲ **FIGURE 11.4c** These data have the same means as in Figures 11.4a and 11.4b, but there is much less variation *within* each group. In this context, the means are very different from each other.

▶ **FIGURE 11.5** These boxplots show us mostly the same information as in Figure 11.4c. We see that the medians of the groups are very different with respect to the variation within each group. Because the distributions are symmetric, the medians (represented by the horizontal lines in the middle of the boxes) are very close to the means.

Although ANOVA focuses on means, boxplots are often useful as a first step for visualizing ANOVA. This is true even though, as you'll recall, boxplots display the median, not the mean, of a sample of data. Particularly if the distributions of the groups are roughly symmetric, then the mean and median are nearly equal, so in this case the boxplots help us see how different the means are. Figure 11.5 shows boxplots for the four groups in Figure 11.4c.

## Putting a Number on It

To perform a hypothesis, test we need a test statistic. This means that these informal ideas—variation between groups and variation within groups—need to be turned into a number. The *F*-statistic, the test statistic we will use for ANOVA, does just that.

$$F = \frac{\text{Variation between groups}}{\text{Variation within groups}}$$

The *F*-statistic compares the variation between groups to the variation within groups. If the variation between groups is big relative to the variation within groups (as in Figure 11.4c), then *F* will tend to be a large number. If the variation between groups is small relative to that within groups (as in Figure 11.4b), then *F* will be a small number. We will reject the null hypothesis if *F* gets to be too big.

Soon we will show you more precisely how the *variation between groups* and the *variation within groups* are measured in order to calculate the *F*-statistic.

**KEY POINT**  Large values of the *F*-statistic discredit the null hypothesis of no association, because a large value suggests that there is more variation between the groups than within each group.

## EXAMPLE 3 Large and Small *F*-Values

We calculated the *F*-statistics for Figures 11.4b and 11.4c. One of these had an *F*-value of 1.03, and the other had an *F*-value of 17344.81.

**QUESTION**  Which observed *F*-statistic belongs to which figure, and why?

**CONCLUSION**  The null hypothesis is that the means are the same. Large values of the *F*-statistic discredit the null hypothesis. Although the means in Figures 11.4b and 11.4c have the same numerical value, the variation between groups is much greater in Figure 11.4c when compared to the variation within each group. For this reason, the large $F$-statistic of 17344.81 must belong to Figure 11.4c.

**TRY THIS!** Exercise 11.9

Most statistical software packages produce something called an **ANOVA table**. This table includes lots of information (some of it redundant), including the value of the $F$-statistic.

 **EXAMPLE 4 Book Ratings**

The website Shelfari (www.shelfari.com) invites visitors to list books they've read and to rate the books. Books are also "tagged" with keywords that identify the type of book. Figure 11.6 shows ratings for three different tags. Figure 11.7 shows the output from a one-way ANOVA analysis using StatCrunch.

◀ **FIGURE 11.6** Ratings as posted by visitors to shelfari.com. The tags help identify the genre of the book.

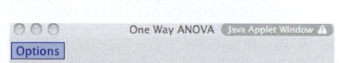

**Analysis of Variance results:**
Responses stored in rating.
Factors stored in tag.
**Factor means**

| tag | n | Mean | Std. Error |
|---|---|---|---|
| dinosaurs | 15 | 4.2066665 | 0.07284469 |
| pirates | 15 | 3.93 | 0.095593385 |
| vampires | 15 | 4.076667 | 0.07235275 |

**ANOVA table**

| Source | df | SS | MS | F–Stat | P-value |
|---|---|---|---|---|---|
| Treatments | 2 | 0.5747778 | 0.2873889 | 2.920713 | 0.0649 |
| Error | 42 | 4.1326666 | 0.09839682 | | |
| Total | 44 | 4.7074447 | | | |

◀ **FIGURE 11.7** StatCrunch output for one-way ANOVA of the shelfari data.

**QUESTION** State the null hypothesis and the alternative hypothesis. What is the value of the *F*-statistic?

**CONCLUSION** The null hypothesis is that the tag (Dinosaurs, Pirates, and Vampires) is independent of the rating of the book; in other words, the population means are all equal. The alternative hypothesis is that the mean ratings are not the same for all of the tags.

From the ANOVA table, we see that the *F*-statistic, given in the column labeled F-Stat, is 2.920713, or about 2.92. You will learn, in Section 11.3, how to determine whether this is a large enough value for *F* so that we should reject the null hypothesis and conclude that there is an association between the variables *Tag* and *Book Rating*.

**TRY THIS!** Exercise 11.11

## ANOVA in Context: A Tour of the ANOVA Table

Researchers in nutrition put 126 people on a diet (Dansinger et al. 2005). They weighed these people before the diet and again two months later. This allowed them to see how much weight each person lost. Obviously, the amount varied from person to person. Why did different people lose different amounts of weight? What is the explanation for this variation in weight loss?

### Explained variation + unexplained variation = total variation

You can probably think of lots of reasons why weight loss varied. Maybe gender matters. Maybe some people had more weight to lose. Maybe some people had different metabolisms, or some people exercised more than others. Maybe there are just unknown, immeasurable factors that influence how much weight a person will lose. One possible explanation—the one the researchers were most interested in—is the type of diet.

ANOVA measures the total amount of variation with a number called the **Total Sum of Squares**. If the variation between groups is big—that is, if the different diets have very different results—then most of this total variation will be due to different diets producing different outcomes. For this reason, another phrase used for "the variation between groups" is "the variation explained by the categorical variable."

Most statistical software packages produce an ANOVA table. This table breaks the total variation (measured by the Total Sum of Squares) into two pieces: the **explained variation** and the **unexplained variation**. The explained variation is just the **variation between groups**. The farther apart the means of the groups are, the bigger the explained variation will be. The explained variation is sometimes called the **treatment variation**, because it is the amount of variation that is explained by the treatment (in this case, the diet to which the subjects were assigned.)

The unexplained variation is just the **variation within groups**. This is sometimes called the **error variation** (as in Figure 11.8) or the **residual variation**. It is residual because it is the variation that is "left over" after we have explained some of the variation with the *Treatment*.

Figure 11.8 shows the total variation in weight loss (1378.5836) in the row labeled Total under the column labeled SS. You can see that if you add the variation due to treatments (10.115185) to the variation due to error (1368.4684), you get the total variation.

### The Mean Sum of Squares

The column labeled MS in the ANOVA table contains the **mean sum of squares**. This number is simply the SS value divided by the df value. We need to divide the SS by

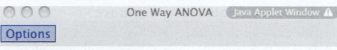

◄ **FIGURE 11.8** ANOVA output for weight loss of four groups of dieters.

One Way ANOVA    Java Applet Window ⚠

Options

**Analysis of Variance results:**
Responses stored in weightloss.
Factors stored in DIET.

**Factor means**

| DIET | n | Mean | Std. Error |
|---|---|---|---|
| Atkins | 31 | -4.680645 | 0.5290368 |
| Ornish | 29 | -5 | 0.5630013 |
| Weight Watchers | 33 | -4.2 | 0.66733867 |
| Zone | 33 | -4.6 | 0.60028404 |

**ANOVA table**

| Source | df | SS | MS | F-Stat | P-value |
|---|---|---|---|---|---|
| Treatments | 3 | 10.115185 | 3.3717282 | 0.30059212 | 0.8249 |
| Error | 122 | 1368.4684 | 11.216954 | | |
| Total | 125 | 1378.5836 | | | |

something in order to compare the "treatment" variation and the "error" variation. This is because the SS Treatment gets bigger as the number of groups gets bigger, and the SS Error gets bigger as the sample size gets bigger. For this reason, we need to adjust the SS due to treatments for the number of groups and adjust the SS due to error by the number of observations.

 **KEY POINT** Total variation as measured by the sum of squares can be broken into two pieces: explained variation and unexplained variation. If the explained variation is large, it indicates that the means of the groups are very different.

The df value, the degrees of freedom, does the adjustment. The df for SS Treatment is the number of groups minus 1. The df for SS Error is the sample size minus the number of groups. By dividing the SS by their degrees of freedom, we get a sort of average sum of squares. This is why it is called the mean sum of squares.

The $F$-statistic is calculated with the ratio of these two values:

$$F = \frac{\text{Variation between groups}}{\text{Variation within groups}} = \frac{MS_{\text{between}}}{MS_{\text{within}}} = \frac{3.3717282}{11.216954} = 0.300592$$

## In Other Words

There are many different phrases used in ANOVA to describe the same thing.

*Variation between groups* is also called *variation due to treatment, explained variation*, and *variation due to factors*.

*Variation within groups* is also called *variation due to error, unexplained variation*, and *residual variation*.

Which term is used depends on the software you are using, but many authors use all of these terms interchangeably. Different software label the columns and rows of an ANOVA table with slightly different labels.

### Relation of Total Sum of Squares to Variance

ANOVA is called the analysis of *variance*. We have discussed measuring variation and have shown that the ANOVA table measures variation with the sum of squares and with the mean sum of squares. What does this have to do with the variance and standard deviation that we used earlier to measure variation?

In fact, they are very closely related. The Total Sum of Squares is simply the numerator in the formula for finding the variance. It is the sum of the squared deviations from the mean.

Recall that the variance is the square of the standard deviation:

$$s^2 = \frac{\sum (x - \bar{x})^2}{n - 1} = \frac{\sum (\text{deviation})^2}{n - 1} = \frac{SS}{n - 1}$$

This means that we can get the variance of the weight loss of all dieters by dividing the total sum of squares (1378.5826) by the total sample size minus 1, which for our dieters is 125 (because $n$ is 126 and $126 - 1 = 125$). There is no need to do this calculation, but we mention it to emphasize that essentially, the ANOVA table uses the same measure of variation that you learned in Chapter 3.

 ## EXAMPLE 5 Fuel Economy

Does fuel economy on the highway (measured in the United States by the number of miles a car typically travels on one gallon of fuel) vary by the number of cylinders in a car's engine? Data from the U.S. Department of Energy are used to perform an ANOVA. The sum of squares due to treatments has been hidden in Figure 11.9.

QUESTION What is the value of the sum of squares due to treatments?

► **FIGURE 11.9** ANOVA table to test whether mean mileage differs for different types of cars. Produced by StatCrunch.

One Way ANOVA

Options

**Analysis of Variance results:**
Responses stored in hwyfeguideconventionalfuel.
Factors stored in cyl.

**Factor means**

| cyl | n | Mean | Std. Error |
|---|---|---|---|
| 4 | 364 | 29.76923 | 0.22043692 |
| 5 | 42 | 25.97619 | 0.6717185 |
| 6 | 400 | 24.315 | 0.13559277 |
| 8 | 243 | 20.074074 | 0.15366969 |

**ANOVA table**

| Source | df | SS | MS | F-Stat | P-value |
|---|---|---|---|---|---|
| Treatments | 3 | 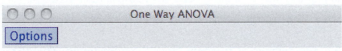 | 4771.3896 | 432.80005 | <0.0001 |
| Error | 1045 | 11520.568 | 11.024467 | | |
| Total | 1048 | 25834.738 | | | |

SOLUTION Total SS = SS due to treatments + SS due to error.

Therefore,

SS due to treatments = total SS − SS due to error

SS due to treatments = 25834.738 − 11520.568 = 14314.17

**CONCLUSION** The sum of squares due to treatments is 14314.17.

**TRY THIS!** Exercise 11.13

The ANOVA table actually has many redundant parts. Table 11.3 shows a generic ANOVA table and helps us see how the different parts are related.

| Source | df | SS | MS | F-Statistic | p-value |
|---|---|---|---|---|---|
| Treatment/Factor/Explained | $k - 1$ | xxxxx | $SS_{tmt}/(k - 1)$ | $MS_{tmt}/MS_{error}$ | xxxx |
| Error/Residual/Unexplained | $N - k$ | xxxxx | $SS_{error}/(N - k)$ | | |
| Total | $N - 1$ | $\sum(x - \bar{\bar{x}})^2$ | | | |

◄ **TABLE 11.3** Generic ANOVA Table

The $x$ with the double bar, $\bar{\bar{x}}$, is used to represent the average of all the observations pooled together; $N$ is the total number of observed values; and $k$ is the number of groups. We do not give the formulas for the SS because these require introducing a great deal of special notation, and the details are more than you need at this point.

## SECTION 11.3

# The ANOVA Test

You've seen that if the null hypothesis is true and the mean value is the same for all categories, then the $F$-statistic will be fairly small. Large values of the $F$-statistic discredit this null hypothesis, but how do we tell whether a particular $F$-statistic is large because the null hypothesis is wrong and the means are different, or is it large just because of chance?

## Finding the p-value

The p-value measures our surprise at the outcome and is used to determine whether the $F$-statistic is "large." Recall that the p-value is the probability of getting a test statistic as extreme as or more extreme than the observed value, assuming the null hypothesis is true. For ANOVA, the p-value is the probability that you get an $F$-statistic equal to or larger than the observed value, assuming all of the means are equal.

To find this probability, we need to know the sampling distribution of the $F$-statistic. This requires that certain conditions hold.

1. *Random Sample and Independent Measurements.* Observations can be thought of as a random sample from a population, and individual measurements within a group are independent of each other.

2. *Independent Groups.* The groups are independent of each other.

3. *Same Variance.* The population variances (or, if you prefer, the standard deviations) of the groups are the same.

4. *Normal Distribution or Large Sample.* The distribution of observations is Normal in each group's population or the sample size is large (at least 25 in each group).

Under these conditions, the distribution of the *F*-statistic follows something called the *F*-distribution (no surprise here). The *F*-distribution has two parameters, and both are called the degrees of freedom. These are found in the ANOVA table. The first df parameter is the one given for treatment/explained variation, and the second is the one given for the error/residual/unexplained variation.

The p-value is usually given in the ANOVA table, as you can see from Figures 11.8 and 11.9. Still, it's good to keep a visual in mind so that you understand where the p-value comes from. Figure 11.10 shows the *F*-distribution with 3 and 102 degrees of freedom, to test the hypothesis that the four dieting groups all have the same means. The shaded area represents the p-value, which is 0.8249.

▶ **FIGURE 11.10** An *F*-distribution with 3 and 102 degrees of freedom. The shaded area is the probability that an *F*-statistic will be larger than 0.3005. This value is about 82%. This is the p-value calculated for the ANOVA shown in Figure 11.8 to test whether the diets differed in mean weight loss.

Clearly, this p-value is so large that we would not reject the null hypothesis. This shows that even though these diets use very different strategies for weight loss, they seem to have the same results. Although the amount of weight lost varies from person to person, this variation doesn't seem to have anything to do with the type of diet (at least after two months of dieting).

## Checking Conditions

Of the four conditions required for finding the p-value, the first two (independent groups and independent observations within groups) are the most important. Unfortunately, these two are also the hardest to check. Independence of groups and independence of observations within groups are results of the data collection method, so if this is well documented, we can gain some idea whether these conditions hold. Sometimes we just have to assume that they do, understanding that if we are wrong, our results could be very wrong.

The first condition, random sample and independent measurements, is similar to the requirement for the one-sample *t*-test of Chapter 9. Having objects within a group that are associated with each other violates this condition. For example, if we randomly sample family members within each group, then for many purposes these measurements would be associated with each other, not independent.

The second condition, independent groups, would be violated if, say, the same objects were measured in each group. This is a fairly common data structure, particularly in medical research. For example, subjects might receive three different medications, one every six months. Researchers measure the subjects' reaction to the medications. The "groups" are the different medications, but because the same subjects appear in each group, the groups are not independent. A procedure called repeated measures ANOVA exists for analyzing data such as these, but it is beyond the scope of this book.

The third condition, that the variances or standard deviations of the groups be equal, is sometimes called the homoskedastic condition. **Homoskedasticity** is a fancy word for "having the same variance." One rule of thumb for checking this condition is that the largest standard deviation must be no more than twice the smallest.

**Details**

**Same-Variance**
One advantage of doing *t*-tests for all comparisons, and applying a correction such as the Bonferroni correction, is that we do not need to require equal variances for the groups.

For example, the standard deviations of the dieting groups are

| | |
|---|---|
| Atkins | 2.9 kg |
| Ornish | 3.0 kg |
| Weight Watchers | 3.8 kg |
| Zone | 3.4 kg |

The smallest SD is 2.9 kg, so as long as the largest SD is smaller than $2 \times 2.9 = 5.8$ kg, we may assume this condition holds. Another way to see this is to divide the largest SD by the smallest: $3.8/2.9 = 1.3$, which is smaller than 2.

In other books, you may come across other rules of thumb for checking this condition. The ANOVA procedure is fairly resistant to slight deviations from homoskedasticity, which is why some rules of thumb are slightly different.

The final condition is that the observations in each group be sampled from a population that follows a Normal distribution. This can be checked by examining a histogram for each group. However, unless the sample sizes are fairly large in each group, this can be hard to check because histograms from Normal populations don't always look Normal when the sample size is small. Another approach requires more work but gives better insight: Subtract each observation from the average of its group. Then make a histogram of all of these differences (called **residuals**). If these residuals look Normal, the condition holds. The *F*-test is more sensitive to departures from Normality than other procedures we've studied up to this point, so some caution is needed.

## What If Conditions Are Not Satisfied?

If either of the assumptions about independence is violated, then a one-way ANOVA is not the correct procedure to use. Other procedures exist for different situations, and these are usually covered in more advanced statistics courses.

If the same-variance condition is not satisfied, then transforming the data by replacing each value with either its square root or the logarithm will sometimes fix the problem. This transformation approach also sometimes works if the Normality condition fails. After replacing the data with their transformed values, you need to check the conditions again. If the conditions still fail, then you need a different procedure. In Chapter 13 we will discuss some procedures (called nonparametric procedures) that can sometimes be used in this situation, and we will demonstrate some transformations.

## EXAMPLE 6 Checking Conditions

Atmospheric scientists often use anomaly temperatures to study global climate change. These temperatures are the difference between the readings at a particular location and the average global temperature for the entire twentieth century (1901–2000). A negative anomaly temperature is lower than the reference point, and a positive value means it is above that reference point. These are also called "residual" temperatures. All temperatures are measured in degrees Celsius (°C). The boxplots in Figure 11.11 on the next page show global surface temperatures grouped by "double-decades" for the last century (that is, 1880–1899, 1900–1919, and so on). The data are global monthly averages—the average temperature on land around the world for a particular month. The temperatures are collected by a network of sensors at fixed locations around the planet. Figure 11.12 on the next page shows a histogram of the residuals from an ANOVA. It is tempting to do an ANOVA, because we have a numerical response variable (*Temperature Anomaly*) and a categorical predictor variable (*Double-Decade*). However, the conditions do not hold.

QUESTION Which conditions are not satisfied for using ANOVA to test whether the mean global surface temperature is different for the given time periods?

CONCLUSION Judging on the basis of the boxplots, the amount of variation in each group seems about the same, so we would expect the standard deviations to be about the

▶ **FIGURE 11.11** Mean global surface temperatures for each "double-decade." Temperatures were measured relative to a reference temperature, which was the mean surface temperature for the twentieth century. Temperatures are in degrees Celsius.

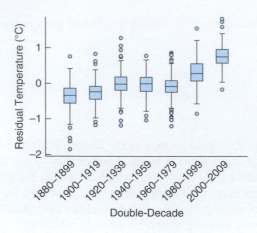

▶ **FIGURE 11.12** Residual temperatures. Each observation is the temperature minus the average temperature for its double-decade.

same. This means that the third condition—same variance—is met. The histogram of residuals looks roughly Normal, so the fourth condition is met. However, the first two conditions are not met. From month to month, temperatures are not independent. One unusually warm month is likely to be followed by another unusually warm month, so observations within groups might be dependent. Also, because the same objects were measured in each group, the groups are not independent. A method other than ANOVA must be used to decide whether differences in mean temperatures are due solely to chance variation.

**TRY THIS!**  Exercise 11.17

## Carrying Out an ANOVA Test

ANOVA is just another hypothesis test, like all of the others we've covered. Example 7 shows you how to do a complete ANOVA by applying the same four-step procedure we've done for other hypothesis tests.

> **Looking Back**
>
> **Four Steps**
> The four steps of a hypothesis test are (1) hypothesize, (2) prepare, (3) compute to compare, and (4) interpret.

## EXAMPLE **7** Testing Popped Popcorn

Does the amount of oil used affect the number of kernels popped when one is making popcorn? Recall that researchers randomly assigned bags with 50 unpopped kernels to be popped with no oil, a medium amount of oil (1/2 tsp), or the maximum amount of oil (1 tsp). Thirty-six bags were assigned to each group. After 75 seconds, the popped kernels were counted.

**QUESTION**  Using a significance level of 5%, carry out an ANOVA test, showing each of the four steps of a hypothesis test.

*Step 1: Hypothesize*

$$H_0: \mu_{none} = \mu_{medium} = \mu_{max}$$

$H_a$: The mean number of kernels popped differs by amount of oil used.

In words, the null hypothesis says that the mean number of kernels popped has nothing to do with the amount of oil used. The alternative hypothesis says that the number of kernels popped is associated with the amount of oil used.

*Step 2: Prepare*

In order for the F-statistic to follow the F-distribution, we need to check conditions. The first condition, independence between groups, holds because bags were randomly assigned to each group. You must assume that the second condition, independence within groups, holds. This condition would fail if, for example, the oven was still heating up while some bags in a group were tested. For the third condition, equal variances, refer to Figure 11.1 on page 500, which shows boxplots of the three groups. The interquartile ranges are about the same, so you can expect the standard deviations to be roughly equal as well. Using technology, you could check this by computing the standard deviation for each group, but many statistical packages make this difficult to do. The final condition, Normal distribution, is also difficult to check with most statistical software packages. We note here that the response variable consists of counts, which are not Normally distributed because the Normal distribution has continuous, not discrete, values. However, because the counts have a large range (from 0 to 50), these distributions might be roughly Normal. You might not have the tools to check, so you'll have to make an assumption that the distributions are close enough to Normal for you to continue.

*Step 3: Compute to compare*

We used StatCrunch to do the ANOVA, and the output is shown in Figure 11.13.

One Way ANOVA

Options

**Analysis of Variance results:**
Responses stored in good.
Factors stored in oil.
**Factor means**

| oil | n | Mean | Std. Error |
|---|---|---|---|
| maximum | 36 | 13.472222 | 1.5494424 |
| medium | 36 | 19.75 | 1.971775 |
| none | 36 | 19.75 | 1.9596627 |

**ANOVA table**

| Source | df | SS | MS | F-Stat | P-value |
|---|---|---|---|---|---|
| Treatments | 2 | 945.85187 | 472.92593 | 3.890878 | 0.0234 |
| Error | 105 | 12762.473 | 121.547356 | | |
| Total | 107 | 13708.324 | | | |

◀ **FIGURE 11.13** ANOVA output (StatCrunch) for the popcorn experiment.

The F-statistic compares the between-groups (Treatments) variation to the within-groups variation (Error). For these data, the between-groups variation is almost four times greater than the within-groups variation, as measured by the mean sum of squares.

$$F = 3.89$$

If the means are equal, then the probability of getting an F-statistic this large or larger is 0.0234.

$$\text{p-value} = 0.023$$

*Step 4: Interpret*
Because the p-value is small (smaller than our significance level of 0.05), we reject the null hypothesis and conclude that the mean amount of kernels popped differs by amount of oil used.

**CONCLUSION**   Because we reject the null hypothesis, we conclude that the amount of oil used affects the number of kernels that get popped. We can make a causal conclusion here—that the amount of oil caused the differences we observed—because the bags were randomly assigned to the groups.

**TRY THIS!**   Exercise 11.21

## SNAPSHOT   ANALYSIS OF VARIANCE (ANOVA)

**WHAT IS IT?** ▶ ANOVA is a procedure for testing whether the population means for several groups differ.

**WHAT DOES IT DO?** ▶ It compares the amount of variation between groups to the amount of variation within groups.

**HOW DOES IT DO IT?** ▶ If the population means of the groups are different, the variation between groups will be larger than the variation within groups.

**HOW IS IT USED?** ▶ The *F*-statistic is a ratio of the variation between groups (measured with the mean sum of squares) to the variation within groups (also measured with its mean sum of squares). If the *F*-statistic is large, reject the null hypothesis and conclude that at least one mean is different from another.

## SECTION 11.4

# Post-Hoc Procedures

As we noted earlier, rejecting the null hypothesis of an ANOVA can be anticlimactic. We learn that oil affects the results of popping popcorn, but we don't know which amount of oil works best. For this reason, after rejecting the null hypothesis of an ANOVA, we next do a **post-hoc analysis**. *Post-hoc* is Latin for "after this." The phrase reminds us that we do this analysis after we have looked at the data and determined (with a chance of making an error) that at least one of the means is too different from another to be convincingly explained by chance. If we do not reject the null hypothesis, then there is no need to do a post-hoc procedure, because we do not have enough evidence to conclude that the means differ.

The goal of a post-hoc analysis is to determine which groups have different means from the others, which groups have the highest means, which the lowest means, and so on. The spirit of post-hoc analysis is that we go into it knowing nothing other than that at least one of the means is different from another. As you saw in Section 11.1, we must be cautious when doing multiple comparisons of groups.

A popular approach for post-hoc analysis is based on the Bonferroni Correction. However, in this section, we will use an approach similar to Bonferroni but more powerful, called the Tukey Honestly Significant Difference (Tukey HSD) approach. "More powerful" means that with Tukey HSD, you are more likely to correctly conclude that two means are different than you are with the Bonferroni Correction. Also, most software packages automatically produce the Tukey HSD intervals by default, or at least they make these an easy option to choose.

Rather than doing hypothesis tests for our post-hoc analysis, we will instead find confidence intervals for the differences between all possible pairs of means. This is because we learn more by looking at confidence intervals. Not only do we learn that one mean is higher than another, but we also get an estimate of how much higher it might be.

**KEY POINT**    A post-hoc analysis is performed after rejecting the null hypothesis from an ANOVA and concluding that at least one of the group means is different from another. The goal is to determine *which* means are different by comparing all possible pairs of means. An appropriate correction must be made in order to keep the overall significance level at a specified value.

## Tukey Honestly Significant Difference Confidence Intervals

The **Tukey Honestly Significant Difference (HSD)** approach provides a set of simultaneous 95% confidence intervals similar to the Bonferroni simultaneous intervals in Section 9.1. Statisticians prefer the Tukey HSD to the Bonferroni Correction, because in many situations, the Tukey HSD intervals tend to be narrower—and therefore more precise—estimates. In a post-hoc analysis, the Tukey HSD intervals are used to determine which means are greater than the others.

Figure 11.14 on the next page shows an extended ANOVA output for examining the popcorn experiment. Two additional tables provide confidence intervals for the difference of means using the Tukey HSD approach. The first of these tables gives us confidence intervals for comparing these mean differences:

- (Medium oil) minus (maximum oil)

- (No oil) minus (maximum oil)

The next table tells us the mean difference for

- (No oil) minus (medium oil)

For example, the confidence interval for (none minus maximum) is (0.0999 to 12.4557). Because this interval includes only positive numbers, we can be confident that the mean number of kernels popped is higher with no oil than with the maximum amount of oil.

Interestingly, the mean number of kernels popped was the same for both the None and the Medium Oil groups, and for this reason, the confidence intervals are the same.

The fact that the confidence interval that compares medium amounts of oil to using no oil includes 0 tells us that no significant differences appear between using no oil and using medium amounts of oil.

From the confidence intervals, we conclude that using medium oil or no oil is better than using the maximum amount of oil. Also, it does not matter whether you use no oil or medium amounts of oil.

▶ **FIGURE 11.14** ANOVA Table for popcorn data with Tukey HSD intervals included (StatCrunch).

 **Tech**

○ ○ ○       One Way ANOVA    **Java Applet Window** ⚠

**Options**

**Analysis of Variance results:**
Responses stored in good.
Factors stored in oil.
**Factor means**

| oil | n | Mean | Std. Error |
|-----|-----|------|------------|
| maximum | 36 | 13.472222 | 1.5494424 |
| medium | 36 | 19.75 | 1.971775 |
| none | 36 | 19.75 | 1.9596627 |

**ANOVA table**

| Source | df | SS | MS | F–Stat | P–value |
|--------|-----|-----|-----|--------|---------|
| Treatments | 2 | 945.85187 | 472.92593 | 3.890878 | 0.0234 |
| Error | 105 | 12762.473 | 121.547356 | | |
| Total | 107 | 13708.324 | | | |

**Tukey 95% Simultaneous Confidence Intervals**

maximum subtracted from

| | Lower | Upper |
|-----|-------|-------|
| medium | 0.09988753 | 12.455668 |
| none | 0.09988753 | 12.455668 |

medium subtracted from

| | Lower | Upper |
|-----|-------|-------|
| none | -6.1778903 | 6.1778903 |

## EXAMPLE 8 Selenium in Ohio Animals

Selenium is an element that is necessary for good health. We ingest selenium through plants grown in high-selenium soil and also by eating animals containing selenium. Researchers were interested in knowing how much selenium was present in game animals (animals that are hunted and eaten by humans) in Ohio, a low-selenium area of the country. Of particular interest was whether the amount of selenium found in game animals depended on the type of animal. They sampled 42 deer (coded VEN for venison) and 30 squirrels (SQU) and compared these game meats to nongame meat: 53 samples of local beef (RRB for "regional-raised beef"), and 19 samples of nonregional beef (NRB). Selenium is measured in micrograms per 100 grams of meat (Holben 2002). Assume that the conditions for ANOVA hold. Figure 11.15 shows a boxplot of the recorded selenium levels by type of animal.

▶ **FIGURE 11.15** Selenium content of samples of game animals found in Ohio. Selenium content is measured in micrograms per 100 grams of meat.

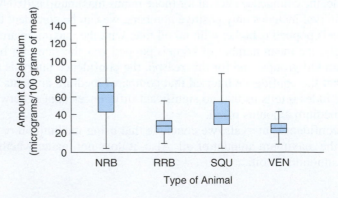

a. Figure 11.16 shows output from an ANOVA and post-hoc analysis. Carry out an ANOVA.

b. Carry out a post-hoc analysis to determine which groups are different from others. How are they different?

c. Is there evidence that selenium levels are low in Ohio-based animals?

a. The null hypothesis is that mean selenium levels do not vary between the different animal groups. The alternative hypothesis is that some animals have more (or less) selenium than others, on average. We see from the output that the variability between groups, as measured by the mean sum of squares for treatment, is much larger than the variability within groups, measured by the sum of squares for error. This results in a fairly large $F$-statistic. The p-value, the probability of getting such

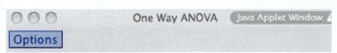

One Way ANOVA    Java Applet Window

**Options**

**Analysis of Variance results:**
Responses stored in selen.
Factors stored in type.

**Factor means**

| type | n | Mean | Std. Error |
|------|-----|-----------|-----------|
| NRB | 19 | 62.046314 | 7.14626 |
| RRB | 53 | 29.08302 | 1.4253525 |
| SQU | 30 | 43.245667 | 3.561792 |
| VEN | 42 | 25.875477 | 1.2394295 |

**ANOVA table**

| Source | df | SS | MS | F-Stat | P-value |
|--------|-----|-----------|-----------|-----------|---------|
| Treatments | 3 | 21261.828 | 7087.2764 | 27.001184 | <0.0001 |
| Error | 140 | 36747.227 | 262.4802 | | |
| Total | 143 | 58009.055 | | | |

**Tukey 95% Simultaneous Confidence Intervals**

NRB subtracted from

| | Lower | Upper |
|-----|-----------|-----------|
| RRB | -44.227478 | -21.699116 |
| SQU | -31.151836 | -6.4494624 |
| VEN | -47.817772 | -24.523907 |

RRB subtracted from

| | Lower | Upper |
|-----|-----------|-----------|
| SQU | 4.5379176 | 23.787378 |
| VEN | -11.910109 | 5.495023 |

SQU subtracted from

| | Lower | Upper |
|-----|-----------|-----------|
| VEN | -27.440182 | -7.3002 |

◄ **FIGURE 11.16** Output from ANOVA analysis for selenium data (StatCrunch).

a large or larger value if the means are truly equal, is extremely small. We therefore reject the null hypothesis and conclude that the mean selenium content varies between the different animal types.

b. The first table of Tukey HSD intervals shows us the mean selenium content of each animal type minus the mean selenium content for nonregional beef (NRB). Positive values mean that animals have more selenium, on average, than does NRB. Negative values mean that NRB has higher levels of selenium, on average. Because none of the intervals include 0 and all are negative, we are confident that the mean selenium level is *higher* in NRB than in all three types of meat. This is not surprising, given the boxplot.

The next table subtracts the mean selenium of regional-raised (RRB) from the remaining types. The values for squirrel are all positive, which means that RRB has a *lower* mean selenium content than squirrel. However, the interval comparing RRB to venison includes 0, so we conclude that the mean selenium levels have not been shown to be different for regionally raised beef and venison.

The last table compares squirrel to venison, and from this we are confident that squirrel has a higher mean selenium content than venison.

c. The highest level of selenium is from the NRB, the beef from a region outside of Ohio. This suggests that animals hunted or raised in Ohio contain less selenium than animals from at least one other location.

**TRY THIS!**  Exercise 11.31

## Visualizing Simultaneous Confidence Intervals

A useful trick for sorting out simultaneous confidence intervals like the Tukey HSD is first to write the names of the groups from the group with the lowest sample mean on the left to the group with the highest sample mean on the right. (If sample means are equal, you can put them in any convenient order.) For example, the animal types sorted from lowest to highest sample mean selenium content (refer to Figure 11.16) are

VEN   RRB   SQU   NRB

Next we draw lines underneath any groups whose confidence intervals include 0. (This tells us that their means are so close together that we can't tell the population means apart.)

<u>VEN   RRB</u>   SQU   NRB

This helps us see that venison and regional-raised beef are about the same in mean selenium content, that nonregional beef is higher than squirrel, and that squirrel is higher than venison and regional-raised beef.

We can apply this technique to the popcorn experiment too:

Max Oil   <u>Medium Oil   No Oil</u>

## Final Thoughts on Post-Hoc Analyses

Different software packages provide very different options for performing the post-hoc analysis for an ANOVA. Most do some version of what we've presented: They give you simultaneous confidence intervals to compare pairs of means. A few, such as Minitab, default to showing you confidence intervals for *individual* groups, not for the *differences* between groups. We recommend that you ignore these individual intervals, because they fail to take into account the necessary post-hoc adjustments and because, unless you take great care, you can easily reach the wrong conclusion. Instead, ask the software to find simultaneous confidence intervals for *differences* between all pairs of means.

The post-hoc analyses we've discussed are intended to show us all possible comparisons between groups. This is appropriate if, when you began the ANOVA, you did not know which groups would be bigger or smaller than the others. Sometimes, researchers have very definite hypotheses about the order of groups. For example, medical researchers testing a drug might be interested in studying whether low-dose, medium-dose, or high-dose patients recover more quickly from their illness. They suspect that the high-dose patients will recover quickest, followed by medium-dose patients, followed by low-dose patients. In more advanced books on ANOVA, you will learn that we can do more focused analyses that concentrate on comparing particular groups, not all possible pairs of groups.

## CASE STUDY REVISITED

The researchers were interested in understanding whether the color of a soccer team's uniform affects its performance. They cited a history of research that suggests that some colors are more successful than others. Some researchers explain this by citing biological or evolutionary factors (many animals display red as a sign of dominance, and humans associate red with the emotions of anger and aggression). The researchers examined 55 years of soccer results in the United Kingdom. Part of their analysis focused on how the percentage of games won varied between color groups. A boxplot of the data, Figure 11.17, shows that the median values differ and supports the hypothesis that teams with red jerseys are slightly more successful than other teams. The boxplot shows that the same-variance requirement is satisfied, but the number of outliers and a suggestion of skewness for certain groups suggests that the Normal condition might not be satisfied. Also, the data might not really be independent, because the teams play each other. (If one team wins a lot, the other teams must be winning less.) Because of these possible problems, the researchers decided to use a different approach, examples of which you'll see in Chapter 13. But it's instructive first to consider what we might learn from the ANOVA approach in this chapter. The output is shown in Figure 11.18 on the next page.

▶ **FIGURE 11.17** The percentages of home game wins for teams with jerseys of different colors: blue (B), red (R), white (W), and yellow or orange (YO). Teams wearing red shirts seem to have a slightly higher winning percentage than other teams, on average.

The p-value of 0.0361 means that, with a significance level of 5%, we reject the null hypothesis and conclude that the mean number of wins differs by color of the team jersey.

We visualize the post-hoc analyses by writing the groups in order of their mean percentage of wins and underlining groups that have (statistically speaking) the same mean.

YO B W R

▶ FIGURE 11.18 Output from ANOVA to test whether teams with uniforms of different colors have different winning percentages, on average.

One Way ANOVA    Java Applet Window

Options

**Analysis of Variance results:**
Responses stored in percentwin.
Factors stored in color.

**Factor means**

| color | n | Mean | Std. Error |
|---|---|---|---|
| B | 23 | 51.239132 | 0.42154145 |
| R | 16 | 53.025 | 1.066634 |
| W | 11 | 51.29091 | 0.71343654 |
| YO | 18 | 50.283333 | 0.37593022 |

**ANOVA table**

| Source | df | SS | MS | F–Stat | P-value |
|---|---|---|---|---|---|
| Treatments | 3 | 65.40642 | 21.80214 | 3.0189104 | 0.0361 |
| Error | 64 | 462.19888 | 7.2218575 | | |
| Total | 67 | 527.6053 | | | |

**Tukey 95% Simultaneous Confidence Intervals**

B subtracted from

| | Lower | Upper |
|---|---|---|
| R | -0.5218412 | 4.0935802 |
| W | -2.5468976 | 2.650455 |
| YO | -3.186618 | 1.275024 |

R subtracted from

| | Lower | Upper |
|---|---|---|
| W | -4.5105963 | 1.0424145 |
| YO | -5.1773243 | -0.30600926 |

W subtracted from

| | Lower | Upper |
|---|---|---|
| YO | -3.7205107 | 1.7053592 |

This visualization shows us that the only confidence interval that does not include 0 is the one that compares red teams with yellow/orange teams. From this, we are confident that teams with red uniforms win a greater percentage of home games, on average, than teams with yellow/orange uniforms.

Although not all of the conditions for an ANOVA were satisfied, the conclusions are supported by more careful analyses that the authors of this study performed in addition to the basic ANOVA.

# EXPLORING STATISTICS
## CLASS ACTIVITY

## Recovery Heart Rate

| GOALS | MATERIALS |
|---|---|
| The goal is to determine, using methods you have learned in this chapter, whether those who feel they are athletic have a better recovery heart rate. | A clock or watch with a second hand is needed. The instructor can call out the starting and ending times. |

**ACTIVITY**

To do this activity, you need to record your pulse in terms of beats per minute. To practice this, make sure you can find your pulse. Count how many beats you feel in 15 seconds, and multiply by 4.

Now stand up and sit down five times, and take your pulse immediately.

Sit and rest for one minute. Take your pulse.

Your recovery heart rate is the difference between the two pulse rates: the pulse after exercise (standing up and sitting down) minus the pulse rate after resting. Some physicians believe that the larger the difference, in general, the more fit you are.

Your instructor will collect your recovery pulse rate and also ask you to classify your general activity level.

To classify your activity level, think about how often you do an activity that raises your pulse rate for 20 minutes or more. If it is once a week or less, then you are in Group A. If it is twice a week to three times per week, you are in Group B. If it is four times a week or more, you are in Group C.

**BEFORE THE ACTIVITY**

1. Why can't we just use a *t*-test for the analysis of this data set?

2. What do you think will happen? Which group will have the highest mean recovery pulse rate?

**AFTER THE ACTIVITY**

Was your prediction about recovery rates correct?

## CHAPTER REVIEW

### Key Terms

Multiple comparisons, *501*
Overall significance
    level, *501*
Bonferroni Correction, *502*
Simultaneous confidence
    intervals, *505*

One-way analysis of variance
    (ANOVA), *506*
ANOVA table, *509*
Total Sum of Squares, *510*
Explained variation (variation
    between groups, treatment
    variation), *510*

Unexplained variation (variation
    within groups, error variation,
    residual variation), *510*
Mean Sum of Squares, *510*
Homoskedasticity, *514*
Residuals, *515*

Post-hoc analysis, *518*
Tukey Honestly Significant
    Difference (HSD), *519*

### Learning Objectives

After reading this chapter and doing the assigned homework problems, you should

- Understand the problem of multiple comparisons and the overall significance level.

- Know how to apply the Bonferroni Correction to multiple hypothesis tests and confidence intervals.

- Understand how the *F*-statistic allows us to compare means of multiple groups by comparing the variance within a group to the variance between groups.

- Be able to perform and interpret an ANOVA procedure to analyze potential associations between a categorical predictor variable and a numerical response variable.

### Summary

When comparing means from more than two groups, we need to be aware that with each comparison, there is a probability that we will mistakenly conclude the means are different when in fact they are not. In order to keep the probability of making such an error across all comparisons less than or equal to $\alpha$ (alpha), usually 0.05, we need to make an adjustment.

The Bonferroni Correction is one method that adjusts for multiple comparisons by using a smaller significance level for each individual comparison. To achieve an overall significance level of $\alpha$, each comparison is made with a significance level of $\alpha/$(number of comparisons).

ANOVA (one-way) is an overall test to determine whether the means of three or more groups differ. If you reject the null hypothesis done with ANOVA, you may go on to do post-hoc tests to estimate which means are significantly different from which others. If you cannot reject the null hypothesis that all the means are the same, you should not go on to do post-hoc tests.

A variety of different post-hoc procedures are available, and you will be limited, to some extent, by which procedures your software provides. Most software provides something similar to the Tukey HSD method, which finds confidence intervals for the difference in means for all possible pairs of groups.

### Sources

Attrill, M. J., K. A. Gresty, R. A. Hill, and R. A. Barton. 2008. Red shirt
    colour is associated with long-term team success in English football.
    *Journal of Sports Science* 26(6), 577–582.
Dansinger, M. L., J. A. Gleason, J. L. Griffith, H. P. Selker, and E. J.
    Schaefer. 2005. Comparison of the Atkins, Ornish, Weight Watchers,
    and Zone diets for weight loss and heart disease risk reduction.
    *Journal of the American Medical Association* 293(1), 43–53.

Holben, D. H. 2002. Selenium content of venison, squirrel, and beef purchased or produced in Ohio, a low-selenium region of the United States. *Journal of Food Science* 67, 431–433 (posted by jph422@sc on Statcrunch.com).

## SECTION EXERCISES

### SECTION 11.1

*For all t-tests in the section, do not assume equal variances.*

*In Exercises 11.1 and 11.2, for each situation, choose the appropriate test: one-sample t-test, two-sample t-test, ANOVA, or chi-square test.*

#### 11.1 Choosing a Test

a. You wish to test whether an association exists between a categorical variable (such as rank of a professor at a university, assuming four ranks) and a numerical variable (such as yearly salary).

b. You wish to test whether the means of a numerical variable are different for two possible values of a categorical variable (such as yearly salary for gender).

#### 11.2 Choosing a Test

a. You wish to test whether an association exists between two categorical variables.

b. You wish to test whether the sample mean of a numerical variable is not equal to a known population mean.

TRY **11.3 Bonferroni Correction (Example 1)** Suppose you have five groups of observations, and you do hypothesis tests (*t*-tests) to compare all possible pairs of means.

    a. How many pairwise comparisons can be done with five groups? List all comparisons with five groups labeled A, B, C, D, and E, starting with AB, AC, etc.

    b. Using the Bonferroni Correction, what significance level should you use for each hypothesis test if you want an overall significance level of 0.05?

**11.4 Bonferroni Correction** Suppose you have four groups of data, and you want to do hypothesis tests (*t*-tests) to compare all possible pairs of means.

    a. How many pairwise comparisons can be done with four groups called A, B, C, and D? Show all possible pairs, starting with AB.

    b. Using the Bonferroni Correction, which significance level should you use for each comparisons if you want an overall significance level of 0.05?

TRY **11.5 Hours of Study (Example 2)** Three independent random
g samples of community college students were obtained to find
out how many hours the students spent each week doing math homework outside of the classroom. The samples were made up of students enrolled in pre-algebra (PreAlg), elementary algebra (ElemAlg), and intermediate algebra (InterAlg).

    Use three two-sample *t*-tests, applying the appropriate Bonferroni Correction to achieve an overall significance level of 0.05, to compare all possible pairs of means. Assume the conditions for two-sample *t*-tests are met. *See page 535 for guidance.*

**11.6 Commuting Times** Mark Bates, a statistics professor at Oxnard College, recorded his commuting times using three different routes from home to work. The routes are named for the streets on which he travelled, and the times are in seconds.

    a. For the boxplots given, compare the medians, interquartile ranges, and shapes, and mention any potential outliers.

    b. Assuming that the overall significance level is 0.05, what is the Bonferroni corrected level of significance for each pair of comparisons?

    c. Carry out *t*-tests to compare the means of all pairs, and summarize your findings by reporting *t*-statistics, p-values, and conclusions. Assume that the conditions for using two-sample *t*-tests are met.

| Oxnard | Rose | Rice |
|--------|------|------|
| 732 | 869 | 694 |
| 842 | 648 | 629 |
| 736 | 1045 | 863 |
| 732 | 674 | 748 |
| 736 | 821 | 767 |
| 833 | 708 | 574 |
| 655 | 840 | 628 |
| 688 | 1029 | 637 |
| 727 | 735 | 620 |
| 721 | 745 | 752 |
| 695 | 794 | 608 |
| 707 | 652 | 983 |
| 843 | 552 | 765 |
| 852 | 732 | 666 |
| 789 | 578 | 727 |
|  | 661 | 729 |
|  | 657 | 605 |
|  | 869 | 717 |
|  |  | 679 |

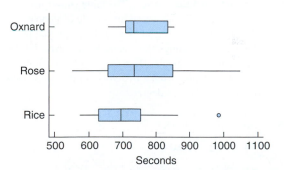

**11.7 Comedy Television Programming** The Nielsen organization did a poll to determine whether men and women in different age groups watched different amounts of comedy television. The data, minutes of comedy television viewed in one week, are given in the table on the next page. Assume that the samples are Normal enough for *t*-tests, and use an overall significance level of 0.05.

| Men 18–34 | Men 55+ | Wom 18–34 | Wom 55+ |
|---|---|---|---|
| 414 | 203 | 387 | 412 |
| 359 | 202 | 375 | 397 |
| 331 | 199 | 369 | 385 |
| 327 | 177 | 364 | 383 |
| 249 | 177 | 358 | 372 |
| 244 | 176 | 356 | 351 |
| 242 | 174 | 339 | 344 |
| 239 | 173 | 332 | 323 |
| 234 | 168 | 330 | 303 |
| 234 | 61 | 327 | 299 |

a. Compare the sample mean number of minutes of comedy television watched by each group.

b. Find the number of pairwise comparisons, and report the Bonferroni corrected significance level to be used to perform a two-sample $t$-test for all pairs to test whether the population mean times are different.

c. Test all six pairs to see whether there are any significant differences, reporting the $t$-statistic, the p-value, and the conclusion for each.

d. Suppose you work for a cable channel and wish to sell advertising during a comedy show. You are interested in what kinds of advertisements to show, a decision that depends on which types of people tend to watch the show more often. Write a short note reporting the target audience for advertisements, and explain.

**11.8 Work Hours and Education** The table shows the number of work hours "last week" from a random sample of people, as reported by the 2008 General Social Survey. Unemployed people were not included. The headings show educational level: LTHS stands for "less than high school," HS refers to a high school degree, JC refers to a degree from a two-year college, and Bach refers to a four-year bachelor's degree.

| LTHS | HS | JC | Bach |
|---|---|---|---|
| 50 | 49 | 45 | 60 |
| 40 | 41 | 35 | 36 |
| 61 | 40 | 40 | 40 |
| 40 | 65 | 55 | 12 |
| 60 | 40 | 50 | 40 |
| 6 | 60 | 45 | 40 |
| 35 | 38 | 89 | 32 |
| 20 | 41 | 37 | 35 |
| 62 | 23 | 30 | 46 |
| 80 | 43 | 40 | 30 |

a. Compare the sample mean numbers of work hours for these groups.

b. Find the number of pairwise comparisons. Report the Bonferroni corrected significance level required to carry out two-sample $t$-tests to test

whether each pair of means is different. The desired overall significance level is 0.05.

c. Assume that conditions for two-sample $t$-tests are met. Use an overall significance level of 0.05, and for each pair, test whether the means are different. Report the value of the $t$-statistic, the p-value, and the conclusion for each test. Are education levels and work hours associated?

# SECTION 11.2

**TRY 11.9 Comparing $F$-Values from Boxplots (Example 3)** Refer to the figure. Assume that all distributions are symmetric (therefore the sample mean and median are approximately equal) and that all the samples are the same size. Imagine carrying out two ANOVAs. The first compares the means based on samples A, B, and C (above the horizontal line), and the second is based on samples L, M, and N (below the horizontal line). One of the calculated values of the $F$ statistic is 9.38, and the other is 150.00. Which value is which? Explain.

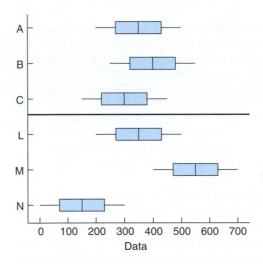

**11.10 Comparing $F$-Values from Boxplots** Refer to the figure. Assume that all data sets are symmetric and that all the samples are the same size. Imagine carrying out two ANOVAs. The first is compares the means based on samples A, B, and C (above the horizontal line), and the other is based on samples G, H, and K (below the horizontal line). One of the calculated values of the $F$ statistic is 9.38, and the other is 25.00. Which value is which? Explain.

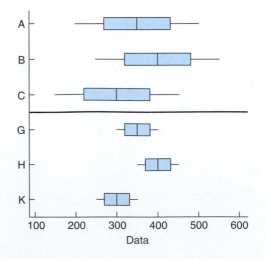

TRY **11.11 Marital Status and Cholesterol (Example 4)** Refer to the StatCrunch output from National Health and Nutrition Examination Survey (NHANES) data, which shows the effect of marital status on cholesterol. Assume the population distributions are close enough to Normal to justify using ANOVA.

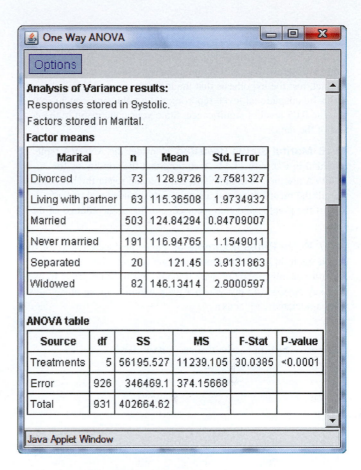

a. Write the null and alternative hypotheses for the effect of marital status on cholesterol.

b. Identify the $F$-statistic from the StatCrunch output.

c. Which marital status had the largest sample mean and which had the smallest sample mean?

d. Assuming that you did find an association between marital status and cholesterol levels, would this association mean that marital status caused different cholesterol levels? Can you think of a confounding factor?

**11.12 Marital Status and Blood Pressure** Test the hypothesis that people with different marital statuses have different mean systolic blood pressure, using a significance level of 0.05. Refer to the StatCrunch output from NHANES data.

a. Write the null and alternative hypotheses for the effect of marital status on blood pressure.

b. Identify the $F$-value from the output.

c. Which group had the largest sample mean and which had the smallest?

d. Assuming that you did find an association between marital status and systolic blood pressure, would this association show that marital status caused different systolic blood pressures? Can you think of a confounding factor?

TRY **11.13 Work Hours and Degree (Example 5)** The 2008 General Social Survey asked people how many hours of worked they worked "last week" and also categorized them in terms of educational level (less than high school, high school, junior college, bachelor's degree). Minitab output is shown. Find the missing total sum of squares.

```
One-way ANOVA: Hours Worked by Education

Source          DF      SS      MS      F      P
Factor           3     571     190   0.94  0.422
Error         1069  217076     203
Total         1072  ??????

S = 14.25    R-Sq = 0.26%    R-Sq(adj) = 0.00%
```

**11.14 Marital Status and Work Hours** In the 2008 General Social Survey, respondents who worked were asked how many hours they worked last week and what their marital status was. The Minitab output for ANOVA from the data is shown. Find the missing sum of squares.

```
One-way ANOVA: Hour Work Last versus Marital Status

Source            DF      SS     MS     F      P
Marital Status     4    1441    360  1.74  0.139
Error           1172   ?????    207
Total           1176  244011

S = 14.39 R-Sq = 0.59% R-Sq(adj) = 0.25%
```

## SECTION 11.3

**11.15 Work Hours Revisited** Consider the ANOVA table provided in Exercise 11.13. Assuming that the conditions for ANOVA are met, test the hypothesis that the mean number of hours of work varies by educational level, reporting the p-value and conclusion. Use the 0.05 level of significance. State your conclusion in the context of the data.

**11.16 Marital Status Revisited** Consider the ANOVA table provided in Exercise 11.14. Assuming that the conditions for ANOVA are met, determine whether you can reject the hypothesis that all the means are equal, using a significance level of 0.05. Report the p-value and state your conclusion in the context of the data.

**TRY 11.17 Pulse Rates (Example 6)** Pulse rates were taken for five people each, in three different situations: sitting, after meditation, and after exercise. Explain why it would not be appropriate to use one-way ANOVA to test whether the population mean pulse rates were associated with activity.

| Person | Sitting | Meditation | Exercise |
|--------|---------|------------|----------|
| A | 84 | 72 | 96 |
| B | 76 | 72 | 84 |
| C | 68 | 64 | 76 |
| D | 68 | 68 | 76 |
| E | 76 | 84 | 80 |

**11.18 Sports Television** The Nielsen organization did a study of the association between the time men spent watching certain kinds of shows and age group. The data were measured in minutes per week. Minitab output for time watching sports television shows for two groups is provided. Is the same-variance condition for ANOVA satisfied? Is ANOVA appropriate? Explain.

```
One-way ANOVA: Sports versus Group

Source   DF     SS     MS     F      P
Group     1   3050   3050   3.47   0.079
Error    18  15842    880
Total    19  18893

S = 29.67    R-Sq = 16.15%    R-Sq(adj) = 11.49%

Level       N    Mean    StDev
Men18-34   10  346.10    40.25
Men55+     10  370.80    11.83
```

**11.19 Comedy Television** The Nielsen organization did a poll to determine whether men and women in different age groups watched different amounts of comedy television. Check the Minitab output to see whether the same-variance condition for ANOVA holds. Is ANOVA appropriate? Explain your answer.

```
One-way ANOVA: Men18-34, Men55+, wom18-34, wom55+
Source  DF      SS      MS      F       P
Factor   3  226818   75606   38.38  0.000
Error   36   70915    1970
Total   39  297733

S = 44.38    R-Sq = 76.18%    R-Sq(adj) = 74.20%

Level       N    Mean    StDev
Men18-34   10  287.30   65.05
Men55+     10  171.00   40.81
wom18-34   10  353.70   20.78
wom55      10  356.90   39.37
```

**11.20 UCLA Music Survey** The figure shows side-by-side boxplots of the number of hours per week that UCLA students spent listening to music. Minitab output for ANOVA is also shown. Check whether the conditions for ANOVA hold. If not, state which ones fail and why.

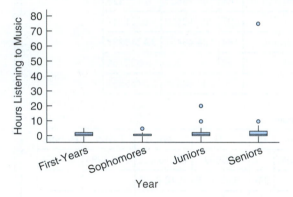

```
One-way ANOVA: Hours versus Year

Source   DF      SS     MS      F      P
YR        3   116.9   39.0   0.76   0.520
Error   117  6017.1   51.4
Total   120  6134.0

S = 7.171    R-Sq = 1.91%    R-Sq(adj) = 0.00%

Level        N    Mean    StDev
Freshman     6   1.333   1.862
Sophomore   20   0.900   1.165
Junior      47   1.596   3.295
Senior      48   3.333  10.793
```

**TRY 11.21 Baseball Player Run-Times (Example 7)** Baseball-playing statistics students John Carlson and Paul McMurray collected data on baseball player run-times. Players were required to run 50 yards and were timed in seconds (pitchers did not have to run). Assume that the distribution of each population (outfielders, infielders, catchers) is close enough to Normal to satisfy the Normal condition for using ANOVA. Test the hypothesis that different positions have different mean run-times, using a significance level of 0.05. (Do not do post-hoc tests. Exercise 11.31 will ask for the post-hoc tests.) *See page 536 for guidance.*

**11.22 Study Hours by Major** Three independent random samples of full-time college students were asked how many hours per week they studied outside of class. Their responses and their majors are shown in the table. Test the hypothesis that the mean number of hours studying varies by major by reporting the $F$-statistic, the p-value and the conclusion. Assume the conditions for ANOVA are met.

| Math | SocSci | English | Math | SocSci | English |
|------|--------|---------|------|--------|---------|
| 15 | 10 | 14 | 10 | 3 | 5 |
| 20 | 12 | 12 | 8 | 5 | 5 |
| 14 | 8 | 12 | 7 | 7 | 4 |
| 15 | 7 | 10 | 7 | 6 | 8 |
| 14 | 7 | 10 | 5 | 3 | 9 |
| 10 | 10 | 10 | 5 | 3 | 10 |
| 12 | 6 | 8 | 5 | 3 | 4 |
| 9 | 8 | 8 | 15 | 2 | 3 |
| 10 | 5 | 5 | 6 | 2 | 3 |
| 11 | 4 | 7 | 5 | 1 | 3 |
| 9 | 4 | 6 | 3 | 5 | 6 |
| 8 | 4 | 4 | 5 | 4 | 4 |
| 8 | 8 | 6 | | | |

**11.23 Ethics of Eating Meat** A survey conducted by researchers Schwitzgebel and Rust asked subjects to choose a number from 1 to 9 to explain their feelings toward eating meat in response to the prompt: "Eating meat is . . . ."

| Very Morally Bad | | Some-what Morally Bad | | Neutral | | Some-what Morally Good | | Very Morally Good |
|------|------|------|------|------|------|------|------|------|
| 1 | 2 | 3 | 4 | 5 | 6 | 7 | 8 | 9 |

Three independent random samples of 30 ethics professors (Eth), 30 philosophy professors not specializing in ethics (Phil), and 30 professors in other disciplines (Other) were taken and the data are available on the disk. Although the graphs are somewhat skewed, the sample sizes are large enough to satisfy the Normal

condition for ANOVA. Test the hypothesis that the mean response is different for different groups of professors, reporting $F$-statistic, the p-value, and the conclusion in context. Use a significance level of 0.05.

(Source: E. Schwitzgebel and J. Rust. 2009. The self-reported moral behavior of ethics professors. Unpublished manuscript. http://schwitzsplinters.blogspot.com/2009/05/do-ethicists-eat-less-meat.html)

**11.24 Meat-Eating Behavior** The survey discussed in Exercise 11.23 (Schwitzgebel and Rust 2009) also asked people how many of their meals per week included meat. The data show three groups: ethicists (Eth), other philosophy professors who are not ethicists (Phil), and professors in other departments (Other). Determine whether you can reject the hypothesis that all the population means are equal at the 0.05 level. The data were obtained from three random samples. Assume that the distribution of each population is close enough to Normal to satisfy the Normal condition of an ANOVA. Do not do post-hoc tests.

**11.25 Reaction Times for Athletes** A random sample of people were asked whether they were athletic, moderately athletic (Mod), or not athletic (NotAth). Then they were tested for reaction speed. Reaction speed was measured indirectly, through reaction distance, as follows: A vertical meter stick was dropped and they caught it. The distance (in centimeters) that the stick fell is the reaction distance, and shorter distances correspond to faster reaction times. The data are shown in the accompanying table.

| NotAth | Mod | Athletic |
|--------|------|----------|
| 16.0 | 11.7 | 24.7 |
| 25.3 | 22.3 | 15.7 |
| 19.3 | 12.7 | 17.7 |
| 19.0 | 21.3 | 12.7 |
| 14.3 | 16.0 | 21.0 |
| 34.3 | 14.0 | 21.0 |
| 17.7 | 16.3 | 19.3 |
| | 28.3 | 25.0 |
| | 30.7 | 27.0 |
| | 15.0 | 10.3 |
| | 26.7 | |
| | 26.7 | |

a. Interpret the boxplots given. Compare the medians, interquartile ranges, and shapes, and mention any potential outliers.

b. Test the hypothesis that people with different levels of athletic ability (self-described) have different mean reaction distances, reporting the *F*-statistic, p-value and conclusion. Assume that the distribution of each population is close enough to Normal to satisfy the Normal condition of an ANOVA and that the sample is randomly selected. Do not do post-hoc tests.

**11.26 Tomato Plants and Colored Light** Jennifer Brogan, a biology student who was taking a statistics class, exposed similar tomato plants to different colors of light. The average growth rates (in millimeters per week) are given in the table.

a. Interpret the boxplots given. Compare medians and interquartile ranges, and mention potential outliers.

b. Test the hypothesis that the color of light affects growth rate — in other words, that the population mean growth rates differ by color — reporting the *F*-statistic, p-value, and conclusion. Assume that conditions for ANOVA are met. Use a significance level of 0.05. Do not do post-hoc tests.

| Blue | Red | Yellow | Green |
|------|-------|--------|-------|
| 5.34 | 13.67 | 4.61 | 2.72 |
| 7.45 | 13.04 | 6.63 | 1.08 |
| 7.15 | 10.16 | 5.29 | 3.97 |
| 5.53 | 13.12 | 5.29 | 2.66 |
| 6.34 | 11.06 | 4.76 | 3.69 |
| 7.16 | 11.43 | 5.57 | 1.96 |
| 7.77 | 13.98 | 6.57 | 3.38 |
| 5.09 | 13.49 | 5.25 | 1.87 |

**11.27 Work Hours** Data were obtained from random samples of workers entering a hardware store who were asked how many hours of work they performed in the previous week. The workers were electricians (Elec), plumbers (Pb), and construction workers (Const).

a. Make side-by-side boxplots of the hours of work, and interpret the boxplots to determine whether the different professions work different numbers of hours.

b. Test the hypothesis that workers with different jobs have different mean number of hours of work, using the 0.05 level of significance. Assume that conditions for ANOVA are met. Show all four steps, numbered. (Do not do post-hoc tests.)

**11.28 Number of Televisions in Students' Homes** Random samples of students from three sister colleges were taken. In each sample, students reported the number of televisions in their homes.

| OC | VC | MC |
|----|----|----|
| 5 | 1 | 4 |
| 2 | 3 | 1 |
| 4 | 4 | 4 |
| 4 | 5 | 6 |
| 3 | 4 | 3 |
| 3 | 3 | 5 |
| 5 | 3 | 5 |
| 7 | 4 | 1 |
| 6 | 5 | 3 |
| 2 | 2 | 2 |
| 3 | 5 | 5 |
| 4 | 2 | 4 |

a. Make side-by-side boxplots, and interpret them to determine whether the typical number of televisions per student household varies among the colleges.

b. Test whether the mean number of TVs differs among these three schools. Use a 5% significance level. Assume the conditions for ANOVA are met. (Do not do post-hoc tests.)

## SECTION 11.4

For all post-hoc *t*-tests, pool variances.

**11.29 House Prices** Tukey HSD confidence intervals (with an overall significance level of 0.05) were calculated for the mean housing prices in three southern California neighborhoods: Port Hueneme, Santa Paula, and Newbury Park. The sample means and intervals are shown. Arrange the towns with smallest sample mean on the left to largest on the right, underlining pairs for which the sample means are not significantly different. (The sample mean was largest for Newbury Park and smallest for Port Hueneme.) Assume that conditions for ANOVA are met. Then write a sentence or two interpreting your results.

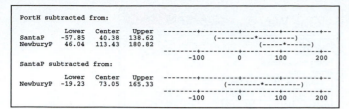

```
PortH subtracted from:

        Lower   Center  Upper   -------+---------+---------+---------+--
SantaP  -57.85   40.38  138.62               (---------*---------)
NewburyP 46.04  113.43  180.82                        (-----*------)
                                -------+---------+---------+---------+--
                                    -100        0      100      200

SantaP subtracted from:

         Lower   Center  Upper   -------+---------+---------+---------+--
NewburyP -19.23   73.05  165.33               (--------*---------)
                                 -------+---------+---------+---------+--
                                     -100        0      100      200
```

**11.30 House Prices** Tukey HSD confidence intervals (with an overall significance level of 0.05) were calculated for the mean housing prices in three southern California neighborhoods: Agoura, Ventura, and Oxnard. The sample means are shown. Arrange the towns from smallest sample mean on the left to largest sample mean on the right, underlining pairs for which the sample means are not significantly different. (The sample mean was largest for Agoura and smallest for Ventura.) Assume that conditions for ANOVA are met. Then write a sentence or two interpreting your results.

```
Agoura subtracted from:

        Lower    Center   Upper   ---+---------+---------+---------+------
Ventura -264.21 -176.78  -89.34   (----------*----------)
Oxnard  -259.09 -172.50  -85.91      (----------*----------)
                                  ---+---------+---------+---------+------
                                    -240      -160      -80       0

Ventura subtracted from:

        Lower   Center  Upper   ---+---------+---------+---------+------
Oxnard  -43.17    4.28  51.73                                (-----*----)
                                ---+---------+---------+---------+------
                                  -240      -160      -80       0
```

*For Exercises 11.31–11.36, refer to the exercise mentioned, and decide whether the p-value for ANOVA suggests that post-hoc tests should be done. For the post-hoc tests, use an overall significance level of 0.05. Report confidence intervals for the difference between all possible pairs of means, and decide for each pair whether the null hypothesis of no difference can be rejected. (Note: If your software does not compute Tukey HSD confidence intervals, you can instead find confidence intervals using the TI-83/84 by applying the Bonferroni Correction; use a confidence level 0.9833 to achieve an overall significance level of 0.05 with three comparisons (from three groups). Pool variances for your t-tests in order to be consistent with the ANOVA approach.)*

*If there is at least one difference between means, list the groups from smallest mean on the left to largest mean on the right, underlining groups for which the sample means are not significantly different and write sentences comparing means of the groups in the context of the data. If post-hoc tests are not appropriate, explain why not.*

**TRY 11.31 Baseball Run-Times (Example 8)** Refer to Exercise 11.21, which showed the overall ANOVA test. Below is output for the post-hoc tests, which are appropriate because the null hypothesis (that all means are equal) was rejected. The mean times (in seconds) for running 50 yards were as follows: Outfielders 7.02, Infielders 7.13, Catchers 7.93. Follow the instructions given above.

```
Tukey 95% Simultaneous Confidence Intervals
All Pairwise Comparisons among Levels of Position
Individual confidence level = 98.01%

Position = Catchers subtracted from:

Position        Lower    Center    Upper
Infielders     -1.3948  -0.8020  -0.2091
Outfielders    -1.4890  -0.9050  -0.3210


Position = Infielders subtracted from:

Position        Lower    Center    Upper
Outfielders    -0.5906  -0.1030   0.3846
```

**11.32 Study Hours** Refer to Exercise 11.22.

**11.33 Reaction Distances** Refer to Exercise 11.25.

**11.34 Tomato Plants** Refer to Exercise 11.26. If using TI-83/84, use 99.17% confidence intervals because there are six comparisons.

**11.35 Hours of Work** Refer to Exercise 11.27.

**11.36 Number of Televisions** Refer to Exercise 11.28.

# CHAPTER REVIEW EXERCISES

**11.37 Meat-Eating Ethics** Refer to Exercise 11.23 and test the hypothesis that professors of different subjects have different mean beliefs about the ethics of eating meat. Use a 0.05 significance level. Report the p-value and conclusion from Exercise 11.23. Perform post-hoc tests, reporting confidence intervals for all pairs if appropriate; if they are not needed, explain why not. As part of the post-hoc analysis, arrange the means from smallest on the left to largest on the right, underlining groups for which the sample means are not significantly different. Then explain the conclusion.

**11.38 Meat-Eating Behavior** Refer to Exercise 11.24 and test the hypothesis that professors of different subjects have different

mean numbers of meals containing meat in a week, using the 0.05 level of significance. Do post-hoc tests if appropriate; if they are not appropriate, explain why not.

**11.39 Marital Status and Body Mass Index** Minitab output from NHANES data, which show the association between marital status and body mass index (BMI), is given. Verify that the constant-variance condition for ANOVA is met; assume all other conditions are met. Test the hypothesis that people with different marital statuses have different mean BMIs, using the 0.05 level of significance.

```
One-way ANOVA: BMI versus Marital Status

Source    DF        SS      MS      F       P
Marital    5     137.9    27.6    0.75    0.587
Error    976   35953.2    36.8
Total    981   36091.1

S = 6.069    R-Sq = 0.38%    R-Sq(adj) = 0.00%

Level                     N     Mean    StDev
Divorced                 80   28.875    6.052
Living with partner      64   27.922    5.870
Married                 525   28.249    5.662
Never married           202   27.644    7.174
Separated                21   29.314    5.091
Widowed                  90   27.866    6.020
```

**11.40 Marital Status and Pulse Rate** Minitab output from NHANES data, which show the connection between marital status and pulse rate, is given. Verify that the constant-variance condition for ANOVA is met; assume all other conditions are met. Test the hypothesis that people with different marital statuses have different mean pulse rates, using the 0.05 level of significance.

```
One-way ANOVA: Pulse versus Marital Status

Source    DF        SS      MS      F       P
Marital    5      3039     608    3.79    0.002
Error    948    152009     160
Total    953    155048

S = 12.66    R-Sq = 1.96%    R-Sq(adj) = 1.44%

Level                     N     Mean    StDev
Divorced                 76   72.71    13.51
Living with partner      63   73.08    13.19
Married                 511   71.14    12.67
Never married           195   74.66    12.08
Separated                20   75.90    13.43
Widowed                  89   68.88    12.58
```

**11.41 Movies on Television** The Nielsen organization did a study of the association between the time spent watching certain kinds of shows on television, measured in minutes per week, and age and gender groups. StatCrunch output for time watching movies on TV for four groups is given.

a. Test the hypothesis that different groups have different mean numbers of minutes watching movies, using the 0.05 level of significance. Assume that the conditions for ANOVA are met.

b. Using the output given for Tukey post-hoc tests, determine which sample means are significantly different from which others. Report and interpret all six confidence intervals for the differences in the means. Then arrange the groups with lowest mean on the left and highest on the right, with underlines connecting groups that do not have significantly different sample means. Then write a sentence or two explaining which group means are significantly different and how they differ.

**11.42 Hours of Television by Age Group** The StatCrunch output shows the ANOVA results for testing whether there is an association between the number of hours of TV watched per week and age group: 50 and over, college students, grade school students.

a. Test the hypothesis that people in different age groups spend different amounts of time, on average, watching television. Use a significance level of 0.05. Assume that the conditions for ANOVA are met.

b. Using the output provided, determine which sample means are significantly different from each other. Report all the given confidence intervals for the difference between means, and tell what they show. Then arrange the groups with lowest mean on the left and to highest on the right, with underlines connecting groups that do not have significantly different means. Finally, write a sentence or two explaining which group means are different and how they differ.

### 11.43 Triglycerides and Gender

Using the NHANES data, we performed an ANOVA to test whether gender is associated with the level of triglycerides, a form of fat, in the blood. ANOVA and *t*-test output from a two-sample *t*-test is shown. In both cases, we

```
Source    DF       SS       MS      F      P
Sex        1    132611   132611   3.16  0.076
Error    940  39476146    41996
Total    941  39608756
```

```
Difference = mu (Female) – mu (Male)
Estimate for difference: -23.8
95% CI for difference: (-50.0, 2.5)
T-Test of difference = 0 (vs not =): T-Value = -1.78  P-Value = 0.076
```

are testing the hypothesis that the mean triglyceride levels for men and women are different. Assume that conditions for both tests have been met.

Compare the output by comparing p-values. Square the *t*-value to see what you get for comparison. Explain.

### 11.44 Cholesterol and Gender

Using NHANES data, we performed one-way ANOVA and a two-sample *t*-test. In both cases we were testing the hypothesis that the mean cholesterol values for men and women are different. Compare the output of ANOVA and the two-sample *t*-test by looking at the *t*-statistics, *F*-statistics and p-values.

| Difference | Sample Mean | Std. Err. | DF | T-Stat | P-value |
|---|---|---|---|---|---|
| $\mu_1 - \mu_2$ | 7.5490246 | 2.93151 | 940 | 2.5751317 | 0.0102 |

**ANOVA table**

| Source | df | SS | MS | F-Stat | P-value |
|---|---|---|---|---|---|
| Treatments | 1 | 13393.94 | 13393.94 | 6.631303 | 0.0102 |
| Error | 940 | 1898616.9 | 2019.8053 | | |
| Total | 941 | 1912010.9 | | | |

## gUIDED EXERCISES

**g** **11.5 Hours of Study** Three independent random samples of community college students were obtained to find out how many hours the students spent each week doing math homework outside of the classroom. The samples were made up of students enrolled in pre-algebra (PreAlg), elementary algebra (ElemAlg), and intermediate algebra (InterAlg). The numbers of hours are shown in the table. Assume that the conditions for two-sample *t*-tests are met.

| PreAlg | ElemAlg | InterAlg |
|---|---|---|
| 2 | 4 | 9 |
| 0 | 2 | 2 |
| 1 | 2 | 3 |
| 2 | 2 | 1 |
| 2 | 2 | 2 |
| 2 | 3 | 2 |
| 2.5 | 3 | 3 |
| 3 | 2 | 3 |
| 1 | 4 | 8 |
| 0 | 3 | 3 |
| 0.5 | 3 | 4 |
| | | 9 |

**QUESTION** Apply three *t*-tests, using the appropriate Bonferroni Correction with an overall significance level of 0.05, to compare all possible pairs of means. Use the following numbered steps for guidance.

**Step 1 ▶ Means and standard deviations**
Fill out the table, reporting all the sample means and standard deviations.

| | Mean | SD |
|---|---|---|
| **PreAlg** | 1.46 | 1.01 |
| **ElemAlg** | 2.73 | ___ |
| **InterAlg** | ___ | ___ |

**Step 2 ▶ List comparisons**
Finish the list of all three possible comparisons.
1. PreAlg compared to ElemAlg
2. _____
3. _____

**Step 3 ▶ Find the corrected alpha**
Find the corrected value for the significance level by dividing 0.05 by the number of comparisons.

**Step 4 ▶ Compute to compare**
We have assumed that the conditions for two-sample *t*-tests are met. For both tests, the null hypothesis is that the two population means

are the same, and the alternative hypothesis is that the two population means are different.

Complete the table below. For a significant difference, the p-value must be less than the Bonferroni-corrected value for the significance level.

|  | $t$ | p-Value | Conclusion |
|---|---|---|---|
| **PreAlg-ElemAlg** | 3.30 | 0.004 | Significantly different |
| **PreAlg-InterAlg** | 2.97 | _____ | _____ |
| **ElemAlg-InterAlg** | 1.57 | _____ | _____ |

### Conclusion

Write a clear conclusion based on what you found. Which groups have sample means that are significantly different and how do they differ?

**g** **11.21 Baseball Player Run-Times (Example 7)** Baseball-playing statistics students John Carlson and Paul McMurray collected data on baseball player run-times. Players were required to run 50 yards and were timed in seconds (pitchers did not have to run). Assume that the distribution of each population is close enough to Normal to satisfy the Normal condition for using ANOVA.

**QUESTION** Test the hypothesis that different positions have different mean run-times, using a significance level of 0.05. In other words, is run-time independent of position played? (Do not do post-hoc tests.) Follow the steps given.

| Outfielders | Infielders | Catchers |
|---|---|---|
| 6.70 | 7.62 | 7.90 |
| 6.75 | 7.20 | 7.82 |
| 7.13 | 6.37 | 7.92 |
| 7.42 | 7.12 | 7.64 |
| 7.05 | 7.04 | 8.82 |
| 6.32 | 7.15 | 7.47 |
| 7.13 | 6.89 | |
| 7.04 | 7.01 | |
| 7.47 | 8.54 | |
| 6.93 | 6.35 | |
| 7.01 | 7.10 | |
| 7.33 | | |

$\mu_{out}$ = population mean for outfielders
$\mu_{in}$ = population mean for infielders
$\mu_{cat}$ = population mean for catchers

**Step 1** ▶ **Hypothesize**

$H_o$: $\mu_{out} = \mu_{in}$ = ?
$H_a$: At least one population mean is different from another.

**Step 2** ▶ **Prepare**

Choose one-way ANOVA.
Conditions:

Random sample and observations independent of each other.

Independent groups.

Same variance: The ratio of largest standard deviation to smallest is $0.5916/0.3244 = 1.82$, which is less than 2.

Assume that the distribution of each population is close enough to Normal to satisfy the Normal condition for using ANOVA.

**Step 3** ▶ **Compute to compare**

$$F = ?$$
$$\text{p-value} = ?$$

**Step 4** ▶ **Interpret**

Reject the null hypothesis if the p-value is less than or equal to 0.05. Explain in words what the conclusion implies about the three means.

# TechTips

## General Instructions for All Technology

All the technology will use the following data set.

**EXAMPLE (ANALYSIS OF VARIANCE)** ▶ The two tables show the data we will use in two different forms. Table 11A consists of unstacked data, and Table 11B gives stacked data. Some technologies require one form and some the other form. The data were constructed for simplicity. To have context, imagine that each number represents the number of children in a random sample of families with children in fictitious countries C, E, and G.

| C | G | E |
|---|----|---|
| 1 | 11 | 2 |
| 2 | 13 | 3 |

▲ **TABLE 11A** Unstacked Data

| $n_{children}$ | Country |
|:---:|:---:|
| 1 | C |
| 2 | C |
| 11 | G |
| 13 | G |
| 2 | E |
| 3 | E |

▲ **TABLE 11B** Stacked Data

---

## TI-83/84

### Example: ANOVA

1. Press **STAT**, choose **EDIT** and enter the unstacked numbers into **L1**, **L2**, and **L3**.
2. Press **STAT**, choose **TESTS**, and scroll up to **ANOVA** and press **ENTER**.
3. See Figure 11A. When you see **ANOVA(** enter the three lists' names separated from each other by commas (the comma button is above 7). To get **L1**, for example, press **2ND** and **1**.

   Enter: **L1 ❯ L2 ❯ L3 )**

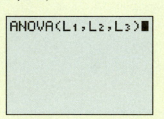

▲ **FIGURE 11A** TI-83/84 ANOVA Input

4. Press **ENTER**.

   You should get the output shown in Figure 11B. You will need to scroll down if you want to see all of the output.

▲ **FIGURE 11B** TI-83/84 ANOVA Output

### Post-Hoc Tests

To do post-hoc tests, you will need to use two sample $t$-intervals (or $t$-tests), which are shown at the end of Chapter 9. Use "Pooled Yes" for the $t$-test if you previously performed ANOVA with the data. Also, if you are using intervals and the overall significance level is 0.05:

Use 98.33% intervals for three groups, because ($0.05/3 = 0.1667$ and $1 - 0.1667 = 0.9833$).

Use 99.17% intervals for four groups, six comparisons.

Use 99.5% intervals for five groups, ten comparisons.

---

## MINITAB

### Example ANOVA

With Minitab, it is possible to use either stacked data or unstacked data, but using stacked data is the preferred convention.

1. See Figure 11C. Enter the numbers in column 1 and the category values in column 2. (It is also allowable to use numerical codes instead of words for the category.) Be consistent with the category values. If you use C (capital C) for one, do not use c (lowercase c) for the other.
2. **Stat > ANOVA > One-Way**

| ↓ | C1 | C2-T |
|---|---|---|
| | **Children** | **Country** |
| **1** | 1 | C |
| **2** | 2 | C |
| **3** | 11 | G |
| **4** | 13 | G |
| **5** | 2 | E |
| **6** | 3 | E |
| **7** | | |

▲ **FIGURE 11C** Minitab Stacked Data

3. See Figure 11D. Be sure that the variable with the numbers (here, **C1 Children**) goes in the **Response** box and that the variable with the code (here, **C2 Country**) goes in the **Factor** box.

▲ **FIGURE 11D** Minitab ANOVA Input

4. For post-hoc tests, click **Comparisons**. And choose **Tukey, family error rate 5** (for a significance level of 0.05).

5. Click **OK** and **OK** again.
   Part of the output is shown in Figure 11E.

---

**One-way ANOVA: Children versus Country**

| Source | DF | SS | MS | F | P |
|---|---|---|---|---|---|
| Country | 2 | 134.33 | 67.17 | 67.17 | 0.003 |
| Error | 3 | 3.00 | 1.00 | | |
| Total | 5 | 137.33 | | | |

▲ **FIGURE 11E** Minitab ANOVA Output

---

## EXCEL

1. See Figure 11F: Enter unstacked data with or without labels into columns A, B, and C.

| | A | B | C | D |
|---|---|---|---|---|
| **1** | C | G | E | |
| **2** | 1 | 11 | 2 | |
| **3** | 2 | 13 | 3 | |
| **4** | | | | |

▲ **FIGURE 11F** Excel Data

2. Click **Data**, **Data Analysis** and choose **Anova: Single Factor**.

3. For **Input Range**, drag the cursor over all the numbers; do not include the labels.

4. Click **OK**.
   Figure 11G shows part of the output.

| ANOVA | | | | | |
|---|---|---|---|---|---|
| *Source of Variation* | *SS* | *df* | *MS* | *F* | *P-value* |
| Between Groups | 134.3333 | 2 | 67.16667 | 67.16667 | 0.003229 |
| Within Groups | 3 | 3 | 1 | | |
| | | | | | |
| Total | 137.3333 | 5 | | | |

▲ **FIGURE 11G** Excel ANOVA Output

To do post-hoc tests, use *t*-tests, which are described in Chapter 9, but assume equal variances if you previously performed ANOVA with the data.

You may use stacked or unstacked data. We prefer stacked data.

1. See Figure 11H. Enter the numbers into one column and your code into another column. The code can be letters or numbers. Type the labels at the top, as shown.

2. **Stat > ANOVA > One Way**

3. See Figure 11I. Click **Compare values in a single column**. For **Responses in**: Click the numerical variable (here, **Children**). For **Factors in**: Click the variable with the code (here, **Country**). (If you had unstacked data, you would click **Compare selected columns**.)

▲ **FIGURE 11I** StatCrunch ANOVA Input

| Row | Children | Country | var |
|---|---|---|---|
| 1 | 1 | C | |
| 2 | 2 | C | |
| 3 | 11 | G | |
| 4 | 13 | G | |
| 5 | 2 | E | |
| 6 | 3 | E | |
| 7 | | | |

▲ **FIGURE 11H** StatCrunch Stacked Data

4. To include post-hoc tests, click **Tukey HSD with confidence level** and leave **0.95** if alpha is 0.05.

5. Click **Calculate**.

Part of the output is shown in Figure 11J.

**ANOVA table**

| Source | df | SS | MS | F-Stat | P-value |
|---|---|---|---|---|---|
| Treatments | 2 | 134.33333 | 67.166664 | 67.166664 | 0.0032 |
| Error | 3 | 3 | 1 | | |
| Total | 5 | 137.33333 | | | |

▲ **FIGURE 11J** StatCrunch ANOVA Output

# 12 Experimental Design: Controlling Variation

## THEME

**Researchers sometimes wish to detect subtle differences between experimental groups. This task can be difficult, because variation in the data may be caused by factors other than the ones that researchers are studying. We can increase the probability of seeing small differences between groups by controlling variation.**

We began our study of statistics in this textbook by talking about why and how we collect data. In Chapter 1, you saw that if the data are collected through a randomized, controlled experiment, then we can draw cause-and-effect conclusions. In Chapter 7 you learned that if the data are chosen at random from the population, then we can generalize our conclusions to the population. One lesson of Chapters 7 through 11 is that the method of data collection affects how we study and analyze the data. For example, if the data consist of paired samples and we wish to compare means, then we apply the paired *t*-test; if the data consist of two independent samples, then we apply the independent two-sample *t*-test.

In this chapter, we focus on the role that surveys and controlled experiments play in helping us understand the world by examining data. Chapter 7 introduced surveys based on simple random samples. In many realistic situations—for example, trying to determine the number of female smokers in China or the proportion of trees in a forest infected by the dangerous bark beetle—a simple random sample is very difficult to collect. In this chapter, we examine some alternative survey designs that are more practical and effective for collecting and analyzing data.

Controlled experiments can be used to determine cause-and-effect associations. Can you do better on exams by writing about your anxieties just before the exam? Does stretching before a race make you run faster? These are cause-and-effect questions that can be difficult to answer, partly because there is much variation between people when they are taking exams or running races.

Most of us will carry out very few experiments in our lifetimes, but we will read about them. The results of experiments are reported in the nightly news, on blogs, and in newspapers. Sometimes the results seem highly counterintuitive or even contradict what we heard just last week. This chapter will include some basic suggestions for reading and interpreting published scientific studies.

## CASE STUDY

## Does Stretching Improve Athletic Performance?

Most athletes take stretching seriously. Runners in particular are taught that stretching before and after running is important not just to prevent injuries but also to maximize performance. However, a growing body of research is questioning the effectiveness of stretching. For example, Nelson et al. (2005) measured the speed of 16 NCAA track athletes who ran after stretching and also after not stretching. On different days, each athlete followed these protocols: They stretched both legs, or only their forward leg positioned as at the start of a race, or only their back leg positioned as at the start of a race, or neither leg. All athletes did the same stretches for the same amount of time. The researchers found that the athletes ran the 20-meter sprint slower after stretching.

What does this study tell us about preparing competitive athletes? What does it tell us about "weekend" athletes? Is this evidence that you shouldn't stretch if you plan on running a 10K race? Are the results believable? Why was there no control group that did no stretches at all? Does it matter that all of the subjects did all of the stretching regimens? In this chapter we examine the characteristics of controlled experiments so that we can understand whether we can generalize conclusions of studies to other groups. At the end of the chapter, we will return to this case study and consider the answers to these questions in light of what we have learned.

# Variation Out of Control

How can you do your best at your next midterm? Would taking a nap (before, not during, the exam) help? How could you find out? You could just try it, of course, but then if you did really well, you wouldn't know whether it was because of the nap or because it was just an easy exam for you. You could try it a few times, sometimes napping and sometimes not, but there's a lot of variability involved here. Sometimes your naps will be longer or your sleep deeper. Sometimes the exams will be harder or easier, and your level of preparation for the material will vary. And even if napping works for you, how could you know whether it would work for someone else? This natural variability, both between people and between contexts, will make it difficult to determine just what effect, if any, napping has on test performance.

For another example, consider the need to test the effectiveness of vaccines. Beginning in mid-2009, many people began to worry about the flu, as news of the H1N1 virus (also known as the "swine flu" virus) swept the world. Eventually, a vaccine was made available, but this raised more questions. Was the vaccine really effective? Were two doses required, or was one sufficient? Could the vaccine be inhaled, or was it better to have it injected? Certainly you could get the vaccine yourself and then see whether you got the flu, but if you remained healthy, it might be just because you weren't exposed to the virus. Or if you got sick, it might be just because, even with the vaccine, your body was susceptible to this particular virus. Again, natural variability makes it difficult to determine how effective a vaccine, or a particular vaccine regimen, is at preventing the flu.

Statisticians and researchers have developed several methods for controlling variability in order to find answers to important questions like these. In Chapter 1 we explained why these questions can be answered only with controlled experiments. In this chapter, we show you how it's done.

## Review of Experimental Basics

Before getting to the heart of the chapter, let's review a few of the important concepts covered in Chapter 1.

The phrase *cause and effect* has a common English meaning, but nonetheless we need to be clear about its meaning in this context. For our purposes, we say that two variables X and Y have a cause-and-effect relationship if, whenever we change X for a person (or object), then Y typically changes in a predictable way. For example, if napping causes test scores to improve, then if you changed your test preparation from "no-nap" to "nap," you would see an increase in your test scores. If we gave you a flu vaccine, we would change your status from "not vaccinated" to "vaccinated," and we would hope to see your risk of getting the flu go down as a result.

We call the variable that we change—the "cause"—the **treatment variable**. The variable that we are interested in seeing changed—the "effect"—is called the **response variable**. The treatment variable records, for example, whether a subject received the vaccine or did not get the vaccine, and the response variable records whether or not that person later got the flu.

The trick is to determine whether changes that we see in the response variable are due to changes in the treatment variable; are due to changes in some other, unseen variables; or are due to chance. Unseen variables that might affect the response variable are called **confounding variables** (or confounders).

For example, in an evaluation of an SAT preparation company, researchers found that students who completed the SAT preparation course did better on the SAT than students who did not. The treatment variable here records whether or not a student took the preparation class, and the response variable records the student's SAT scores. One problem with the study was that the preparation course was expensive, and a student's decision to take the course depended somewhat on the student's ability to pay (or the family's ability to pay)—a possible confounding variable.

To be a confounding variable, a variable must affect *both* the treatment *and* the response. The student's ability to pay meets this criterion, because it affects both the treatment (attend or not-attend the SAT preparation course) and the response (score on SAT). A student's ability to pay for SAT preparation affects the treatment variable because, on average, students who take the SAT preparation class are better off financially than students in the group who do not take the class. And this ability to pay means that the students might, typically, have increased access to other resources that help them do better on the SAT (for example, they might not need jobs and so would have more time to prepare); for this reason, the response variable is also affected.

Another possible confounding variable is motivation. Students who are motivated enough to spend money on a prep course, to go through the trouble of finding a prep course, and to put in the extra time attending the course might also be motivated enough to study in other ways. So even if the prep course is not effective, the students' extra motivation might be the reason for their higher SAT scores (Briggs 2002).

Maybe you can think of other confounding variables as well. The presence of confounding variables means we cannot conclude that the prep course caused the higher SAT scores. To be able to conclude that the prep course caused the change, we would have to eliminate all other explanations.

To eliminate other explanations, the experimenter must make sure that the subjects in the experiment are alike in every way except that they have different values for the treatment variable. To achieve this, experimenters can take control and assign subjects to a treatment group. When this is done, we call the experiment a **controlled experiment**.

Still, making sure that the subjects in the different treatment groups are all alike is difficult, because people (and animals and even some inanimate objects) are not all alike. This variability means that if we're not careful, some differences between the groups will arise that might be confounding variables and prevent us from concluding that the treatment caused any observed changes in the response variable.

Researchers use **random assignment** to ensure that different treatment groups are as similar as possible. To take a simple example, if we have two treatment groups, one getting a vaccine and the other a placebo, then by randomly assigning subjects to the two groups, we will have roughly equal proportions of men and women in both groups. In that way, gender cannot be a confounding factor, because both groups will have about the same numbers of each gender. Random assignment ensures that other variables, even some we don't know about, will be distributed in roughly equal proportions among the groups.

**Observational studies** are quite common, in part because it is not always possible to do a controlled experiment. How could researchers decide whether using a cell phone over many years causes brain cancer in a randomized study? First, such a study would be unethical. Second, even if they ignored ethical considerations, how could researchers make sure that subjects who were assigned to the treatment group used their phones often enough over many years? And how could they be sure that subjects in the control group did *not* use cell phones for many years?

A major drawback with observational studies is that they can only establish an association between the treatment and response variables; they cannot establish a cause-and-effect relationship. The news media often ignore this fact and report the

**Looking Back**

**Observational Studies and Confounding**
Recall that in an observational study, we can never eliminate the possibility that a confounding variable exists. This means we cannot reach cause-and-effect conclusions on the basis of a single observational study.

results of observational studies as though they were controlled experiments. For example, the BBC News published this headline: "Happiness Wards Off Heart Disease, Study Suggests" (http://news.bbc.co.uk, viewed January 25, 2011, via http://jonathan .mueller.faculty.noctrl.edu). The reported study must be an observational study, because how could researchers assign people to be happy for a long period of time? Therefore, we can't conclude that happiness causes a lower risk of heart disease, despite the claims of the headline.

**KEY POINT**  Random assignment to treatment groups ensures that the subjects in all groups are as similar as possible, so that the only real difference between groups is the treatment they receive.

## EXAMPLE 1 Improving Memory

Which is better for improving memory: a short nap, caffeine, or nothing? Researchers at the Veterans Administration in San Diego randomly assigned 61 subjects to one of three groups. All subjects took a basic memory test. Depending on which group they were assigned to, subjects then took a 30-minute nap, or had a moderate dose of caffeine, or did nothing. Afterwards, subjects were again tested on their memory. Researchers found that the nappers improved more on the memory tests than either the caffeine takers or the placebo group (Mednick et al. 2008).

**QUESTION**  For this controlled experiment, identify the treatment and response variables. Restate the conclusion of the study in terms of a cause-and-effect conclusion.

**SOLUTION**  The treatment variable here records the treatment group that each subject was assigned to: subjects napped, or had caffeine, or did neither. The response variable is the change in the subjects' memory test score. The conclusion could be stated as "Napping improves memory."

**TRY THIS!**  Exercise 12.9

## Statistical Power

Random assignment helps researchers eliminate (or at least minimize) the effects of possible confounding variables. But there is always the possibility that observed changes in the response variable are simply due to chance, rather than due to the treatment the subjects received.

Because statistical tests involve chance, even when the two groups we're comparing are truly different, there is a probability that we won't see this difference. On the other hand, there is a probability that we will correctly find that the groups are truly different. This probability, called the **power**, was introduced in Chapter 8. The power is the probability that, for a given value of the parameter, we will reject the null hypothesis. In other words, power measures the probability of detecting differences that really exist between groups.

Obviously, we want this probability to be big. Many researchers like to have a power of at least 80%. Higher is even better.

The power depends primarily on three things: the sample size, the size of the true difference between the groups, and the natural variability within the population. Of these three factors, researchers and statisticians have direct control only over sample size. Still, by identifying sources of natural variability in the population, researchers can design experiments that can improve the power by controlling for the variation.

Variability that exists within a population can arise from several different sources. One source is natural variability: the subjects (or objects) being studied naturally have variability. If we are comparing the weights of different groups of people, we should expect natural variability within these groups, because everyone's weight is slightly different. Another important source of variability is measurement error. This is the variability caused by the devices or methods used to measure something. For example, if you weigh yourself twice in a row, you don't always get exactly the same weight the second time. Figure 12.1 shows the distribution of weights of two thousand 1-euro coins. You might expect coins would weigh exactly the same, because heavier coins might be perceived as more valuable. Still, whether as a consequence of measurement error or of variations in the coin-minting process, we still see variability.

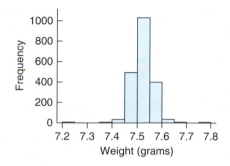

◀ FIGURE 12.1 Even though the minting of coins is very carefully controlled, there is still variability, as this distribution of the weights of two thousand 1-euro coins shows.

The best way to control for this variation when conducting a research study is to hold a variable fixed at a particular value. This is called **controlling for a variable**. If we are performing a study to see how naps affect memory, and we believe that caffeine intake affects naps, then we might control for caffeine amounts by requiring everyone to be caffeine-free. Or we might even give all of the groups the same amount of caffeine, if we were interested in seeing how our conclusions were affected by caffeine.

## Blocking

For factors that the researchers cannot control, they can apply blocking. **Blocking** is a technique in which researchers group similar objects into "blocks" and then assign treatments randomly within each block. Blocking helps achieve two purposes. It reduces bias (particularly useful when the sample sizes are small), and it increases statistical power.

In many medical studies, for example, different age groups respond differently to treatments. Researchers cannot control a person's age, but they can create blocks of subjects who are the same age (or nearly the same age) and in that way increase the statistical power of their study.

The basic idea is that once we've grouped similar subjects or objects together, we hope that subjects within a block will respond similarly to the treatment. If so, then we don't need to measure quite so many people, because there is less variability than we originally had.

**KEY POINT**  A block consists of subjects or objects that are similar on one or more variables. Objects *within* a block are then randomly assigned to treatment groups.

### Creating Blocks

To create blocks, ask whether you know of any variables that might affect the outcome and cannot be controlled. If so, create the blocks by grouping together objects that are similar to one another on those variables.

▲ **FIGURE 12.2** Each letter represents a student in the study. Students within each block are randomly assigned either to the Expressive Writing treatment (T) or to the control group (C).

▲ **FIGURE 12.3** The X's represent the locations of yucca plants. The plants near the (usually dry) riverbed have different soil types and moisture than the other plants, so it makes sense to create blocks grouping together the plants that are in a similar environment.

Once you have created the blocks, you then randomly assign objects within each block either to the control group or to the treatment group (or to the many different treatment groups, if the study is considering several different treatments). Each block should have objects in all treatment and control groups.

For example, psychologists at the University of Chicago had evidence that "expressive writing"—writing about emotions—might help prevent students from "choking" on exams (Ramirez and Beilock 2011). To test this, they recruited ninth grade biology students at a large high school. Students were evaluated about their upcoming final exam. Some students were very anxious, but others were much less anxious. Because researchers suspected that one source of variation in the effects of the treatment might be the student's anxiety level, they created two blocks. One block consisted of students who were the most anxious about the final exam. The other block consisted of the least anxious students. Ten minutes before the exam, students in each block were randomly assigned to do one of two activities. Half of the students were told to write about their anxieties concerning the test; this expressive writing constituted the treatment. The other half of the students were told to write about any topic that was not covered by the exam; these students formed the control group. Figure 12.2 illustrates this blocking design.

At the end of the study, researchers found that the high-anxiety students who performed expressive writing outperformed the high-anxiety students in the control group. On the other hand, low-anxiety students who did expressive writing did not perform any differently than their control-group counterparts. The blocking structure of the experiment helped the researcher tease out some complex relations among anxiety, the treatment, and exam performance.

Blocking is often used in agricultural and biological studies done "in the field." The reason is that when you are studying plants in the great outdoors, the amount of sunlight, water, or soil nutrients that a plant might receive can vary widely just a few feet in any direction. Figure 12.3 shows an illustration of two blocks used to study the growth of Mojave Yucca. Researchers, concerned about campers and outdoor recreationists damaging the plants, are wondering whether the plants can be made healthier by protecting some with small fences.

The illustration shows that some of the plants are near a riverbed, which fills with water after a rainstorm. Plants near the riverbed receive more moisture, so if a study is being done on these plants, then the plants near the river should be put in one block, and the other plants in another. Then, within each block, some plants are randomly chosen to be surrounded by fences, and others are randomly chosen to have no protection.

## EXAMPLE **2** Wash That Statistician Right Out of Your Hair

A cosmetics company has created a new dandruff shampoo, which it believes will work better than its older shampoo formula. Company investigators believe that the effects of the shampoo depend on hair color. For this reason, they plan to block on hair color. To do this, they create blocks that have people with blond hair, brown hair, black hair, red hair, and grey hair, respectively. They then randomly select three of the blocks to receive the new shampoo and the other two blocks to receive the old shampoo.

**QUESTION** Is this an effective design for the study? If not, describe an improvement.

**SOLUTION** No. The researchers randomly assigned entire blocks to treatment groups. Instead, they should randomize *within* blocks. If an entire block is assigned to the same treatment—for example, if all of the black-hair subjects are assigned to the old shampoo—then researchers won't be able to determine whether the new shampoo works on black hair.

To properly conduct this study, they should randomly assign half of the blond-hair group to receive the new formula and half to receive the old. Then they should randomly assign half of the brown-hair subjects to receive the new formula and half to receive the old, and so on.

**TRY THIS!** Exercise 12.17

# EXAMPLE **3** Baking Cookies

An amateur baker wants to experiment with two different cookie recipes. One recipe calls for margarine, one for butter. The problem is that the oven the baker is using is a bit old and doesn't heat up consistently. If the oven is set to 350 degrees Fahrenheit, not every region in the oven will be exactly the same temperature. Through experience, the baker has some ideas about which areas are hotter and which are cooler, and these are shown in Figure 12.4. This figure also shows where each cookie will be placed.

**QUESTION** Explain how to create blocks to test which cookie recipe tastes best.

**SOLUTION** The baker should create three blocks. The cookies within the 370°F zone (cookies 3, 4, and 8) should be in one block. The cookies within the 360°F zone (cookies 1, 2, 5, 6, 7, 9, 10, 11, 12, 15, and 16) are in another block, and the cookies in the 350°F zone (cookies 13, 14, 17, 18, 19, and 20) go in another block. Within each block, half of the spaces are randomly chosen to receive a margarine cookie, and the other half receive butter cookies. This can be done by putting the cookie numbers into three hats. One hat has the numbers for the 370°F block (numbers 3, 4, and 8), another hat has the numbers for the 360°F block, and the third hat has the numbers for the 350°F block. From each hat, half the numbers are drawn, and the selected numbers will be the margarine cookies. For blocks with an odd number of spaces, a coin-flip can be used to determine whether the last space gets margarine (heads) or butter (tails).

▲ **FIGURE 12.4** The oven has three temperature regions, and baking conditions are similar within each region.

**TRY THIS!** Exercise 12.19

Blocking is useful because if you have correctly identified a variable that leads to different outcomes in your study, then by grouping subjects together on the basis of that variable, you have decreased the variability in the study—and decreasing the variability leads to greater statistical power.

Studies with small sample sizes can be greatly improved by blocking. Suppose that you wish to replicate the study about the effects of expressive writing on students in your own class. Imagine that there are 20 students, and most—say, 15 students—are not anxious. If you do not create blocks before you randomly assign 10 students to the treatment group and the other 10 to control, then it is possible that just by chance, all 5 of the anxious students will be in the control group. If this happens, then you will not be able to see the effect of the treatment, because the people who would be best helped by the treatment will not receive it.

Instead, if you created blocks, then one block would consist of the 5 anxious students and about half (2 or 3) would be guaranteed to be assigned to the treatment. Figure 12.5 on the next page compares one possible randomization with no blocks to a blocking design.

▶ **FIGURE 12.5 (a)** If there are no blocks, it is possible that all 5 of the higher-anxiety students (HA) will be assigned to the control group. **(b)** Creating blocks based on anxiety level ensures that roughly half of the higher-anxiety students (that is, 2 or 3 students) will be assigned to receive the treatment (T).

**Looking Back**

**Paired Samples**
Paired samples are those in which each object in one group is associated with one particular object in the other group. They are also sometimes called dependent samples or dependent observations.

## Blocking and Matching

As we have just discussed, blocking combines units that are similar in order to improve the power of an experiment. What unit can be more similar than matching a person with himself or herself? Researchers can sometimes exploit the fact that measuring the same person twice (usually before and then after applying a treatment) results in positively correlated values. Studies that use this structure illustrate the **matched-pairs design**. In Chapter 9 you saw examples of such paired samples, and you learned how to apply the paired *t*-test to test hypotheses about differences between two groups of paired data. Pairs of data are, in fact, a form of blocking.

For example, in a study to detect the effects of the gastric band on weight loss (O'Brien 2010), patients had their weight measured without the band (before it was surgically put in place) and some time after the band was in place. In such a study, each patient is basically serving as his or her own control. The control group consists of the patients before the treatment. The treatment group consists of the *same* patients as in the control group, but after they've received the treatment. This sort of study design is typically more effective at measuring the effects of a treatment than one in which we have two groups of people (one group with no gastric band and the other with one) and compare the weights of the two groups.

## EXAMPLE 4 Testing Diabetes Treatments

To combat diabetes, researchers compare two treatments designed to improve kidney functioning. One treatment uses both insulin and somatostatin (IS). The other uses insulin, somatosatin, and also glucagon (ISG). Each patient receives each treatment (with some number of days between treatments so that the effects of the first can wear off). After each treatment, the amount of urea nitrogen in the urine over a 24 hour period is measured. Low levels suggest the kidneys are not properly removing waste. The data are given in Table 12.1. Summary statistics are also provided. Assume that the subjects are randomly selected from a larger population and that the distribution of urea nitrogen levels in this population is approximately Normal.

▶ **TABLE 12.1** Outcome from diabetes study. IS stands for "insulin with somatostatin"; ISG stands for "insulin with somatostatin and also glucagon." Urea nitrogen amounts are in units of grams detected in urine over a 24 hour period. (Source: Rasking and Unger, 1978.)

| Patient ID | IS (g/ 24 hr) | ISG (g/24 hr) | Difference |
|---|---|---|---|
| 1 | 14 | 17 | −3 |
| 2 | 6 | 8 | −2 |
| 3 | 7 | 11 | −4 |
| 4 | 6 | 9 | −3 |
| $\bar{x}$ | 8.25 | 11.25 | −3.0 |
| s | 3.86 | 4.03 | 0.82 |

QUESTIONS

a. Perform a two-sample *t*-test (do not treat the data as paired data) to test whether the mean urea nitrogen levels are different under IS treatment and ISG. Assume the patients are a random sample from a larger population. Use a 5% significance level.

b. Perform a paired *t*-test to test whether the mean urea nitrogen levels are different under IS treatment and under ISG treatment. Use a 5% significance level.

c. Which approach is better? Why are the conclusions different?

SOLUTIONS

a. We will quickly skip over the four steps of a hypothesis test so that we can emphasize the results of the two different procedures. Figure 12.6 shows the output from StatCrunch, which displays the null and alternative hypotheses, the observed value of the *t*-statistic, and the p-value.

◄ **FIGURE 12.6** Output from two-sample *t*-test, no pairing.

**Two sample T statistics with data**

Options

**Hypothesis test results:**

$\mu_1$ : mean of var2

$\mu_2$ : mean of var3

$\mu_1 - \mu_2$ : mean difference

$\mu_0 : \mu_1 - \mu_2 = 0$

$H_A : \mu_1 - \mu_2 \neq 0$

(without pooled variances)

| Difference | Sample Mean | Std. Err. | Df | T-Stat | P-value |
|---|---|---|---|---|---|
| $\mu_1 - \mu_2$ | −3 | 2.7913556 | 5.989039 | −1.0747466 | 0.3239 |

From this we see that the value of the *t*-statistic is −1.07 and the p-value is 0.3239. Because the p-value is bigger than 0.05, we do not reject the null hypothesis. Our conclusion is that there is no evidence of a difference in mean urea nitrogen levels between the two treatments.

b. Now we do a paired *t*-test. Figure 12.7 shows the StatCrunch output.

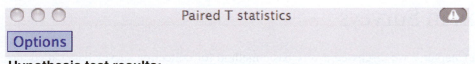

◄ **FIGURE 12.7** Output from paired *t*-test.

**Paired T statistics**

Options

**Hypothesis test results:**

$\mu_1 - \mu_2$ : mean of the paired difference between var2 and var3

$\mu_0 : \mu_1 - \mu_2 = 0$

$H_A : \mu_1 - \mu_2 \neq 0$

| Difference | Sample Diff. | Std. Err. | DF | T-Stat | P-value |
|---|---|---|---|---|---|
| $\mu_1 - \mu_2$ | −3 | 0.4082483 | 3 | −7.3484693 | 0.0052 |

This time the *t*-statistic is very different: −7.3. And the p-value is much less than 0.05. This time, we reject the null hypothesis and conclude that the treatments have different effects on the mean urea nitrogen levels.

c. The data in this sample are paired; each patient contributes two data values to the sample. Because the values within pairs are positively correlated, the paired *t*-test has greater power than the two-sample *t*-test to detect differences in the population. The paired *t*-test is the more appropriate test for this situation. One clue to why the paired *t*-test has greater power in this situation can be seen by comparing the standard error for the paired test statistic with that for the two-sample statistic. The standard error for the paired statistic is much smaller, because with this study design, correctly using the paired test statistic results in a much more precise estimator.

Note that in both part a and part b, the differences in the *sample* means are the same: $-3.0$. But our hypothesis is not about the sample, it is about the population. The paired *t*-test allowed us to see that the *population* means are different.

**TRY THIS!** Exercise 12.23

Another type of study that uses subjects as their own blocks is the **crossover design**. *Crossover design* is really just a more general term for the matched-pairs design. The difference is that sometimes there is more than one treatment, or more than one dosage of the same treatment. In a crossover design, researchers give all subjects the same set of treatments, but they usually randomly choose the order in which each subject receives the treatment.

Smoking causes many health problems, but quitting smoking can be difficult for some people. Researchers in New Zealand used a crossover design to compare three methods for helping people quit: snus (a Swedish chewing tobacco), nicotine gum, and Zonnic (which they described as an "oral nicotine sachet"). The goal was to deliver the nicotine that smokers craved in a form less harmful than cigarettes (Caldwell and Crane 2010). Patients used each treatment for two weeks. Patients were randomly assigned to the *order* in which they received each treatment. For example, some took snus first, then the gum, and then the Zonnic. Others also started with snus but then took Zonnic and ended with the gum. There were six different possible orderings to which subjects could be assigned.

During each two-week treatment, the patients recorded their desire for smoking. At the end of the trial, the researchers concluded that all three methods reduced the withdrawal symptoms created by quitting cigarettes, but that patients preferred Zonnic and snus over gum because there were fewer side effects.

## SECTION 12.2

# Controlling Variation in Surveys

In Chapter 7 we discussed a very basic method for collecting survey data called the simple random sample (SRS). In an SRS, we first create a list of all people in our population. Then we randomly choose people without replacement to participate in our study.

This basic method creates a sample that, if large enough, is representative of the population. Because it is based on a random scheme, we can compute a margin of error that measures how far away our estimate is likely to be from the population value, even when we don't know what the population value is.

In practice, however, simple random samples are difficult to carry out. In very large populations such as the United States (or any other country, for that matter), no single list of names exists for researchers to use. Also, this method can be imprecise: The margin of error can be big. In this section, we give an overview of other random sampling techniques that are more practical than simple random sampling.

# Review of Sampling Basics

Before talking about more complex sampling schemes, let's take a moment to review some of the basics we introduced in Chapter 7.

One goal of a survey is to estimate the value of a population parameter on the basis of a sample. *Parameters* are numbers that characterize populations. The proportion of Americans who bought a new car in the last two years is a parameter, as is the mean weight of all eighth graders in your town. If we collect a representative sample from the population of interest, then a statistic from that sample can be used to estimate the parameter. A *statistic* is a number based on a sample of data. The average weight of 20 eighth graders chosen from your town is a statistic, as is the proportion of 100 people selected from the United States who say they bought a new car in the last two years.

Two features that measure the quality of an estimator are bias and precision. *Bias* measures how far away an estimator tends to be from the correct population value. If, on average, the estimator tends to get the value just right, then we say that the estimator is unbiased. *Precision* measures how the estimator varies from sample to sample. Different researchers using different random samples should get similar estimates when using a precise estimator. Precision is measured by the standard error, which is just the standard deviation the estimator would have if we were to repeat the survey many times. Ideally, an estimator is unbiased and has a small standard error.

Estimators can be biased for many different reasons. One common reason is that the sample is not representative of the population. For example, if we were to survey customers at a car dealership and ask whether they bought a new car in the last two years, these people would probably not be representative of the population as a whole. (They are probably less likely than the general population to have a new car at home if they're out shopping for a car.)

One way in which a nonrepresentative sample often arises is in situations where the sample chooses itself by volunteering to participate. For example, a political website advertised a poll on health care reform, and interested readers could click on the link to answer this question: "After reading President Obama's proposal, have your views about passing health care reform changed?" About 90 percent of 131,577 respondents said "No" (FoxNews.com/opinion, viewed February 25, 2010). Is this representative of the entire U.S. population? Most probably not. People chose whether to click on the link, and this decision is quite likely to be based on the fact that they have strong opinions on the topic, which they wish to share.

This situation is similar to an observational study. People choose whether or not to participate in the poll, so we don't know whether our estimate of the population parameter is representative only of the sort of person who would answer this poll or is representative of the population as a whole. (The fact that this health care poll appeared on a politically conservative website probably accounts in part for the large proportion who don't like the idea of passing health care reform. If the poll were on a liberal news site, the results might very well go in the opposite direction.)

The only way to collect data that avoids this potential bias is to select the sample randomly. A random sample has another advantage: It gives us a way to measure our confidence in the estimate. This measure takes the form of a confidence interval. A confidence interval enables us to say we're highly confident that the true population proportion is within, say, 3 percentage points of our estimate.

The precision of a survey estimator depends on the size of the sample and on the variability in the population. The good news is that the precision can always be improved by taking a larger sample (assuming you can afford to collect more data, that is). If everyone agrees and thinks the same way—in other words, if there is no variability—then the estimator will be extremely precise. The worst-case scenario occurs when the population is evenly split. If the true proportion in the population is $p = 0.50$, then the precision will be as bad as it gets.

For example, if $p = 0.50$ and we take a simple random sample of 100, then the standard error is

$$\sqrt{\frac{p(1-p)}{100}} = \sqrt{\frac{0.5 \times 0.5}{100}} = 0.05.$$

A population with less diversity, say $p = 0.90$, would produce more precise estimators because the standard error is smaller:

$$\sqrt{\frac{p(1-p)}{100}} = \sqrt{\frac{0.9 \times 0.1}{100}} = 0.03.$$

Several approaches to collecting samples get around the difficulties posed by SRSs. These approaches either make it easier to access people to survey, or help by increasing the precision without the costs associated with a larger sample size.

## Systematic Sampling

> ⚠ **Caution**
>
> **Measurement Bias**
> Do not overlook the fact that estimates of the population proportion can also be biased if the questions are poorly worded.

**Systematic sampling** is one technique that can be used, in some situations, to obtain a random sample. In a systematic sample, objects from the population are sampled at regular intervals, such as taking every fifth person. One of the most common examples of systematic sampling occurs in exit polls held during elections.

The purpose of an exit poll is to estimate the winner of an election before the votes are counted. They are called exit polls because people are surveyed as they exit the polling place. The pollsters—the people collecting the data—are instructed to stop people at specific intervals. For example, they might stop every fifth person (or every tenth person) throughout the day.

Systematic sampling works well when you have subjects coming at you in some sort of sequence and when you have good reason to believe that the opinions will be randomly mixed during the time period when you collect that data. Exit polls would fail if, for example, pollsters collected data only first thing in the morning, or only in the middle of the afternoon. This is because it is possible that a certain type of voter goes to the polls before work, and if so, then a systematic sample taken only in the morning will be biased.

Systematic samples are used in quality control studies at manufacturing plants. Products come off an assembly line, and every tenth product might be inspected for defects. At the end of the day, this method yields a good estimate of the proportion of all products produced that day that have defects, as long as there is no particular pattern to the order in which defects arrive.

## Stratified Sampling

**Stratified sampling** is a method for collecting data with increased precision. To perform stratified sampling, researchers first identify **strata**. Strata are similar to blocks; they are collections of people who are very similar to each other. Stratified sampling is based on a very simple principle: If we know that everyone in some group will answer the question the same way, then there's no need to ask everyone.

To create strata, choose a variable that identifies people who hold similar beliefs that are associated with the variable you are studying. For example, suppose you were designing a survey to determine the proportion of the population that supports allowing gay and lesbian citizens to serve in the military. According to Gallup Polls and other research, support for this idea depends on age: Younger people are more supportive than older people. Thus, even though we could take a large sample of 1000 people to get an approximately 3-point margin of error for a 95% confidence interval, we could also stratify on age and take a smaller sample to get the same precision.

A stratified sample would divide the population into strata based on age group. One stratum (the singular form of *strata*) might consist of 18–25-year-olds, another of

26–40-year-olds, and so on. We would then randomly select people (using a simple random sample) from each stratum and ask their opinion.

Stratified sampling works because it creates "mini-populations" that have lower variability than the population as a whole. If the strata were chosen carefully, then within each stratum, there is less variability than in the population as a whole. This means that within each stratum, we can produce a more precise estimator.

The result is that we have mini-estimates from each of the mini-populations. These estimates are each very precise, because the strata were chosen to have low variability within them. We then combine the mini-estimates into one estimate for the entire population. We can show mathematically that in many cases, this combined estimator is more precise.

 **KEY POINT** Stratified sampling improves the precision of estimators by sampling from within strata. Each stratum is created in such a way that people (or objects) within the stratum are similar to one another.

Age is one variable commonly used for creating strata. Political party (Republican or Democrat) is another. Other possibilities include gender, educational level, and geographic region. For example, if we were to take an opinion poll about the pop music star Justin Beiber, a performer with a large fan base of pre-teens, we could probably improve our survey by stratifying on age. One stratum might consist of people between the ages of 8 and 13, another of people aged 14–19, another of people aged 20–25, and so on. From each of these groups, we would select a random sample for our survey.

If the variable chosen to create the strata does not actually end up creating groups with similar-thinking members, then stratification will not offer any advantage over SRS. In fact, in some very unusual cases, it can make the precision worse! For this reason, it's important to choose the right stratification plan.

## Cluster Sampling

**Cluster sampling** is a method that makes it easier to access very large populations. To perform a cluster sample, researchers divide the population into distinct groups, or **clusters**, using some sort of natural or convenient distinction. Next, they select a random sample (using SRS) of clusters and survey *everyone* within the clusters selected. For example, suppose you are interested in determining the number of children living in rental apartments in a city. Each apartment building can be a cluster, and you can then randomly select some number of apartment buildings and then survey everyone within those buildings to find out how many children are living in the apartments.

Clustering can happen over several stages. Randomly sampling, say, ten states in the United States and surveying everyone in these ten states is still a gigantic undertaking. Instead, after the states are selected, we might define a second stage of clusters consisting of cities, or maybe counties, zip codes, or the like. We would then take a random sample of these smaller clusters and survey everyone inside the clusters selected.

Foresters have used cluster sampling to estimate the extent to which bark beetles have infested forests. Bark beetles are a serious threat to forests, so being able to detect the proportion of damaged trees in areas in which there is not yet much damage is crucial to deterring their attack. By using satellite images, researchers can randomly select parcels of land to serve as clusters and then investigate every tree within those parcels (Coggins, Coops, and Wulder 2010). Cluster sampling is also used in humanitarian relief efforts to understand the extent to which people have been injured or killed by natural disasters. Researchers randomly select regions and then go door-to-door within them to collect damage reports.

Cluster sampling is sometimes confused with stratified sampling because both techniques break the population into groups. However, in some ways, cluster sampling

> **! Caution**
>
> **Stratified Sampling and Blocking**
> Stratified sampling and blocking are similar. Both put subjects into groups (strata or blocks) so that subjects are similar within the group. However, stratified sampling is about ways of selecting subjects, and blocking is about a way of randomly assigning already selected subjects to treatment groups.

is the opposite of stratified sampling. In stratified sampling, the strata contain people who are all as similar as possible (for example, all Republicans in one strata and all Democrats in another). Cluster sampling, on the other hand, works best when each cluster by itself represents the entire population. This is rarely achieved, but when it *is* achieved, it ensures that estimates are as precise as possible.

> **KEY POINT** Cluster sampling can be used to access very large populations. A cluster consists of a group of objects or people. To perform a cluster sample, we take a random sample of clusters and survey everyone within the selected clusters.

### EXAMPLE 5 Counting Young Female Smokers

Health experts are concerned about the large number of smokers among young Chinese. However, experts do not know how many smokers there are between the ages of 14 and 24. To estimate the number of young, female smokers in China, researchers in California and China conducted a survey in which they randomly selected schools in China and then randomly selected classrooms in the selected schools. They surveyed all girls and young women in the selected classrooms and concluded that about 20% of all Chinese females in this age group were smokers (Ho et al. 2009).

**QUESTION** Is this a systematic sample, a stratified sample, or a cluster sample?

**SOLUTION** This is a cluster sample. More precisely, this is a two-stage cluster sample. The clusters are schools at one stage and then classrooms at the second stage.

**TRY THIS!** Exercise 12.29

## SECTION 12.3

# Reading Research Papers

One goal of this book is to teach you enough of the basic concepts of statistics that you can critically evaluate published research. The medical literature, in particular, records many findings that can have major consequences in our lives. When we rely on the popular media to interpret these findings, we often get contradictory messages. However, you now know enough statistics so that, with a few guiding principles, you can often make sense of research results yourself.

Before discussing ways of evaluating individual research articles, we offer a few over-arching, guiding principles.

1. *Pay attention to how randomness is used.* Random sampling is used to obtain a representative sample, so that we can make inferences about a larger population. Random assignment is used to test causal associations so that we can conclude that the treatment was truly effective (or was not) on a particular sample of subjects. Many medical studies use random assignment but do not use random sampling. This means the results are not necessarily applicable to the entire population. Table 12.2 summarizes these possibilities.

2. *Don't rely solely on the conclusions of any single paper.* Research advances in small steps. A single research study, even when the conclusions are grand and ambitious, can tell us only a small part of the real story. For example, the *Los Angeles Times* reported, on the basis of a few published studies, that many people considered vitamin D to be useful for preventing cancer, cardiovascular

|  | Sample Selected Randomly | No Random Sample |
|---|---|---|
| **Random Assignment** | You can make a causal conclusion and conclude that the entire population would be affected similarly. | You can make a causal conclusion, but we do not know whether everyone would respond similarly. |
| **No Random Assignment** | You can assume that an association between the variables exists in the population, but you cannot conclude that it is a causal relationship. | You can conclude that an association exists within the sample but not in the entire population, and you cannot conclude that there is a causal relationship. |

◄ **TABLE 12.2** Four different study designs and the inferences possible.

disease, depression, and other maladies. But a panel of medical experts concluded that these beliefs were based on preliminary studies. The body of medical research, they concluded, was in fact quite mixed in its view of just how effective vitamin D really is (Healy 2011).

3. *Extraordinary claims require extraordinary evidence.* This advice was a favorite piece of wisdom of magician and professional skeptic (The Amazing) James Randi. If someone claims to have done something that had been believed impossible, don't believe it until you've seen some very compelling evidence.

4. *Be wary of conclusions based on very complex statistical or mathematical models.* Complex statistics often require complex assumptions, and one consequence of this is that the findings might be correct and true, but only in a limited set of circumstances. Also, some research studies are essentially "what if" studies: "We're studying *what* will happen *if* these assumptions hold." It is important to remember that those assumptions may not be practical—or even possible.

5. *Stick to peer-reviewed journals.* **Peer review** means that papers were read by two or three (and sometimes more) knowledgeable and experienced researchers in the same field. These reviewers can prevent papers from being published if they do not think the methods employed were sound or if they find too many mistakes. They can also demand that the authors make changes and submit the paper again. Many published papers have gone through several rounds of reviews. But be careful: Not all peer-reviewed journals are equal. The more prestigious journals have greater resources to check papers for mistakes, and they have much more careful and knowledgeable reviewers on hand to find sometimes subtle errors. Good journals have editorial boards that reflect the general practice of the community and are not restricted to people who reflect a minority point of view.

## EXAMPLE **6** Improving Tips

A sociologist wonders whether a waitress's tip is affected by whether or not she touches her customers. Below are two study designs that the sociologist might use. Read these, and then answer the questions that follow.

Design A: The sociologist chooses four large restaurants in his city and gets all waitresses at these restaurants to agree to participate in his study. On several nights during the next few weeks, the sociologist visits the restaurants and records the number of times all waitresses touch their customers on the customer's back or shoulder. He also records the total amount of tips earned by each waitress. He finds that waitresses who sometimes touched their customers earned larger tips, on average, than those who did not.

Design B: The sociologist chooses four large restaurants in his city and gets all waitresses at these restaurants to agree to participate in his study. At each restaurant, half of the waitresses are randomly assigned to either the "touch" group or the "no touch" group. The "touch" group is instructed to lightly touch each customer on the back or shoulder two or three times during the meal. The "no touch" group is instructed not to touch customers at all. At the end of the study, the sociologist finds that waitresses in the "touch" group earned larger tips, on average, than those in the "no touch" group.

**QUESTION** For each study design, state whether it is possible to generalize the results to a larger population and whether the researcher can make a cause-and-effect conclusion.

**SOLUTION** For Design A, the waitresses (or restaurants) were not randomly selected from a larger population, so we cannot generalize the results beyond the sample. This was an observational study: The waitresses themselves chose whether to be "touchers" or "no touchers." Because random assignment was not used, we cannot conclude that the difference in tips was caused by touching. A possible confounding effect is the restaurant itself, which might encourage touching and which might have a clientele that tends to tip more generously than the clientele at other restaurants.

For Design B, we also cannot generalize to a larger population. But here random assignment was used. The researcher blocks on restaurant, so he also controls for the environment of the restaurant. Because random assignment was used, we can conclude that the touching caused the increased amount of tips.

**TRY THIS!** Exercise 12.35

## Reading Abstracts

An **abstract** is a short paragraph at the beginning of a research article that describes the basic findings. If you look up science papers using Google Scholar, say, clicking on the link will usually take you to an abstract. Often, you will be able to read the abstract at no charge. (Reading the papers themselves often requires that you subscribe to the journal or that you read from the computers at a library that has purchased a subscription.) In addition, the websites of many journals display the abstracts to their published papers, even if they do not provide access to the papers themselves.

For example, an article recently published in the *New England Journal of Medicine* (Poordad, et al., 2011) reported the findings of a study on the effectiveness of a new treatment for chronic hepatitis C virus (HCV). Currently, the standard treatment, peginterferon-ribavirin, has a relatively low success rate. A success, in treating HCV, is called a "sustained virologic response." In a sustained virologic response, the virus is not eliminated from the body, but is undetectable for a long period of time. Researchers hope that adding a new medicine, boceprevir, to the standard treatment would lead to a great proportion of patients achieving a sustained virologic response. An excerpt from the abstract follows:

**Methods:** We conducted a double-blind study in which previously untreated adults with HCV genotype 1 infection were randomly assigned to one of three groups. In all three groups, peginterferon alfa-2b and ribavirin were administered for 4 weeks (the lead-in period). Subsequently, group 1 (the control group) received a placebo plus peginterferon-ribavirin for 44 weeks; group 2 received boceprevir plus peginterferon-ribavirin for 24 weeks, and those with a detectable HCV RNA level between weeks 8 and 24 received a placebo plus peginterferon-ribavirin for an additional 20 weeks; and group 3 received boceprevir plus peginterferon-ribavirin for 44 weeks. Nonblack patients and black patients were enrolled and analyzed separately.

**Results:** A total of 938 nonblack and 159 black patients were treated. In the nonblack cohort, a sustained virologic response was achieved in 125 of the 311 patients (40%) in group 1, in 211 of the 316 patients (67%) in group 2 ($P < 0.001$), and in 213 of the 311 patients (68%) in group 3 ($P < 0.001$). In the black cohort, a sustained virologic response was achieved in 12 of the 52 patients (23%) in group 1, in 22 of the 52 patients (42%) in group 2 ($P = 0.04$), and in 29 of the 55 patients (53%) in group 3 ($P = 0.004$). In group 2, a total of 44% of patients received peginterferon-ribavirin for 28 weeks. Anemia led to dose reductions in 13% of controls and 21% of boceprevir recipients, with discontinuations in 1% and 2%, respectively.

**Conclusions:** The addition of boceprevir to standard therapy with peginterferon-ribavirin, as compared with standard therapy alone, significantly increased the rates of sustained virologic response in previously untreated adults with chronic HCV genotype 1 infection. The rates were similar with 24 weeks and 44 weeks of boceprevir. (Funded by Schering-Plough [now Merck]; SPRINT-2 ClinicalTrials.gov, number NCT00705432.)

To help you evaluate this abstract, and others like it, answer these questions:

1. What is the research question that these investigators are trying to answer?

2. What is their answer to the research question?

3. What were the methods they used to collect data?

4. Is the conclusion appropriate for the methods used to collect data?

5. To what population do the conclusions apply?

6. Have the results been replicated (that is, reproduced) in other articles? Are the results consistent with what other researchers have suggested?

The sixth question is important, although it is often difficult for a layperson to answer. Research is a difficult activity, in part because of the great variability in nature. Studies are often done with samples that do not allow generalizing to populations or are subject to bias, and sometimes researchers just make mistakes. For that reason, you should not believe in the claims of a single study unless it is consistent with currently accepted theory and supported by other research. For all of these reasons, wait until there is some accumulation of knowledge before changing your lifestyle or making a major decision on the basis of scientific research.

Let's see how we would answer these questions for this abstract.

1. *What is the research question that these investigators are trying to answer?* Does the addition of boceprevir to the standard treatment lead to a higher proportion of Hepatitis C patients achieving a sustained virologic response?

2. *What is their answer to the research question?* Yes, the additional drug improved patients' responses. Researchers report an increased rate of patients achieving sustained virologic response with boceprevir plus the standard therapy compared to those receiving only standard therapy.

3. *What were the methods they used to collect data?* Patients were randomly assigned to one of three treatment groups. Assignments were double-blind. Patients were examined after 24 and 44 weeks to determine if they had achieved sustained virologic response.

4. *Is the conclusion appropriate for the methods used to collect data?* The fact that subjects were randomly assigned to treatment groups means that we can make a causal conclusion and say that the difference in virologic response rates was due to the the addition of boceprevir to the standard

treatment. We also know that the observed results were too large to be explained by chance. Note that p-values are given as, for example, "P < 0.001". For instance, we know that group 2 (nonblack cohort) had a greater rate of sustained virologic response than the treatment group, with p-value < 0.001.

5. *To what population do the conclusions apply?* The sample was not randomly selected, so although we *can* conclude that the treatment was effective, we do not know if we would see the same sized effect on other samples or in the population of all Hepatitis C patients. Because it is known that African Americans and non-African Americans have different responses to the standard treatment, these two groups were treated separately in the analysis.

6. *Have the results been replicated in other articles? Are the results consistent with what other researchers have suggested?* We cannot tell from the abstract.

Note that even though we don't know which statistical procedures were carried out (we could learn this from reading the article itself), we still have a good sense of the reliability of the study. We know that the reported success rates for standard treatment for Hepatitis C (40% in the nonblack cohort ad 23% in the black cohort) were lower than for the treatment that included boceprevir (67% for nonblack cohort after 24 weeks, 68% after three weeks; 42% for the black cohort after 24 weeks, and 53% after 44 weeks.) We know this difference cannot be plausibly assigned to chance, and we know that, because of random assignment, that boceprevir was the cause of the difference in rates.

## EXAMPLE 7 Driving While Talking on a Cell Phone

Read the abstract that follows, which concerns the hazards of driving while talking on a cell phone (Strayer et al. 2006).

**Objective:** The objective of this research was to determine the relative impairment associated with conversing on a cellular telephone while driving. **Background:** Epidemiological evidence suggests that the relative risk of being in a traffic accident while using a cell phone is similar to the hazard associated with driving with a blood alcohol level at the legal limit. The purpose of this research was to provide a direct comparison of the driving performance of a cell phone driver and a drunk driver in a controlled laboratory setting. **Method:** We used a high-fidelity driving simulator to compare the performance of cell phone drivers with drivers who were intoxicated from ethanol (i.e., blood alcohol concentration at 0.08% weight/volume). **Results:** When drivers were conversing on either a hand-held or hands-free cell phone, their braking reactions were delayed and they were involved in more traffic accidents than when they were not conversing on a cell phone. By contrast, when drivers were intoxicated from ethanol, they exhibited a more aggressive driving style, following closer to the vehicle immediately in front of them and applying more force while braking. **Conclusion:** When driving conditions and time on task were controlled for, the impairments associated with using a cell phone while driving can be as profound as those associated with driving while drunk. **Application:** This research may help to provide guidance for regulation addressing driver distraction caused by cell phone conversations.

**QUESTION** Write a short paragraph describing this research. Use the six questions above for guidance.

**SOLUTION** The study asks whether talking on a cell phone impairs driving and, if so, whether the impairment is similar to the impairment caused by driving drunk. The researchers conclude that the impairment is as great when talking on a cell phone as it is when driving drunk. To reach this conclusion, the researchers had drivers use a driving simulator. Some drivers talked on their phone during the task, and others were given sufficient alcohol to make them legally drunk. However, these techniques as described in the abstract do not lead to much faith in the conclusion.

- We do not know whether drivers were randomly assigned to treatment groups.

- We do not know how these drivers were chosen.

- We do not know how many drivers were in each group.

- Although we know that all of the drivers in the alcohol group had the same blood alcohol level, we do not know whether all of the drivers in the cell phone group were equally engaged in the task. For example, perhaps some had conversations that required more concentration.

- We also do not know whether the results are due to chance, because no statistical conclusions are provided.

- Presumably, this finding is meant to apply to all cell phone drivers, but we are not given enough information about the subjects to know how they were selected, so we don't know how well we can generalize these results to a larger population.

Despite these shortcomings, the abstract states that the study's findings are consistent with a larger body of literature that finds that cell phone driving is dangerously distracting.

**TRY THIS!**  Exercise 12.45

## Buyer Beware

Tolstoy's novel *Anna Karenina* begins, "Happy families are all alike; every unhappy family is unhappy in its own way." To (very loosely) paraphrase the great Russian novelist, there are few ways that a study can be good, but there are many ways that studies can go wrong. We've given you some tips that should help you recognize a good study, but you should also be aware of some warning signs and features that indicate that a study might not be good.

### Data Dredging

Hypothesis testing is designed to test claims that result from a theory. The theory makes predictions about what we should see in the data; for example, students who write about their anxieties will do better on an exam, so the mean exam score of students who wrote about their anxieties should be greater than the mean score of those who did not. We next collect the data and then do a hypothesis test to determine whether the theory was correct. **Data dredging** is the practice of stating our hypotheses after first looking at the data. Data dredging makes it more likely that we will mistakenly reject the null hypothesis. Even when the null hypothesis is true, our data sometimes show surprising outcomes, just because of chance variation. If we first look at the test statistic to decide what the hypotheses should be, we are rigging the system in favor of the alternative hypothesis.

The situation is analogous to betting on a horse race after the race has begun. You are supposed to place a bet before the race starts, so that everyone is on equal footing. The odds on which horse will win (which determine the payoffs) are meant to estimate probabilities of a horse winning. But if you wait until the horses are running, you have a better chance than everyone else of correctly predicting which horse will win. You have "snooped" at the data to make your decision. Your probability of winning is not the same as what everyone else believed it to be.

Theories should be based on data, but the correct procedure is to use data to formulate a theory and then to collect additional data in an independent study to test that theory. If it is too costly to collect additional data, one common approach is to randomly split the data set into two (or even more) pieces. One piece of the data is then set aside—"locked away"—and the researchers are forbidden to look. The researchers can then examine the first piece as much as they want and use these data to generate hypotheses. After they have generated hypotheses, they can test these hypotheses on the second, "locked away" data set. Another possibility is to use the data for generating hypotheses and then go out and collect *more* data.

## Publication Bias

Most scientific and medical journals prefer to publish "positive" findings. A positive finding is one in which the null hypothesis is rejected (with the result that the researcher concludes that the tested treatment is effective). Some journals prefer this sort of finding, because these are generally the results that advance science. However, suppose a pharmaceutical company produces a new drug that it claims can cure depression, and suppose this drug does not work. If many researchers are interested in studying this drug and they do statistical tests with a 5% significance level, then 5% of them will conclude that the drug is effective, even though we know it is not. If a journal favors positive findings over negative findings, then we will read only about the studies that find the drug works, even though the vast majority of researchers came to the opposite conclusion.

**Publication bias** is one reason why it is important to consider several different studies of the same drug before making decisions. A new, and somewhat controversial, form of statistical analysis called meta-analysis has been developed in recent years, in part to help with problems such as these. A **meta-analysis** considers all studies done to test a particular treatment and tries to reconcile different conclusions, attempting to determine whether other factors, such as publication bias, played a role in the reported outcomes.

Psychologists conducted a meta-analysis to conclude that violent video games are not associated with violent behavior in children, despite the fact that several studies had concluded otherwise (Ferguson and Kilburn 2010). They concluded that papers that found an association between video game violence and real-life violence were more likely to be published, so researchers who found no such associations were less likely to be published. This result is not likely to settle the controversy, but it points to a potential danger that arises when journals publish only positive findings.

## Profit Motive

Much statistical research is now paid for by corporations that hope to establish that their products make life better for people. Researchers are usually required to disclose to a journal whether they are themselves making money off the drug or product that they are researching, but this does not tell us who funded the project in the first place. There is no reason to reject the conclusions of a study simply because it was paid for by a corporation or business or some other organization with a vested interest. You should always evaluate the methods of the study used and decide whether the methods are sound.

However, you should be aware that sometimes the corporation funding the research can influence whether results get published and which results get published. For example, a researcher might be funded by a pharmaceutical company to test a new drug. If he finds the drug doesn't work, the drug company might decide that it doesn't want to publicize this fact. Or perhaps the drug works on only a small subset of people. The company might then publicize that the drug is effective but fail to mention that it is effective on only a small group.

For example, a 2007 study concluded that playing "active" video games (such as the Wii) was healthier than playing "passive" games and found that children burned more calories playing the active games (Neale 2007). The study also suggested that playing real sports was much more healthful than playing any video game, but still the researchers conclude, "Nevertheless, new generation computer games stimulated positive activity behaviors. Given the current prevalence of childhood overweight and obesity, such positive behaviors should be encouraged." A press release notes that the study was funded by Cake, the "marketing arm" of Nintendo UK, which manufactures the Wii. This does not mean that the results of the study are wrong (the children playing the active games most likely did in fact burn more calories than those who played the passive games), but it may account for the positive spin despite the finding that active video games were not nearly as effective as playing sports at promoting weight loss.

## Media

The media—newspapers, magazines, television shows, and radio broadcasts—are also profit-motivated. A good journalist strives to get to the heart of the matter. Nonetheless, scientific and medical research findings are complex, and when condensing complex ideas into easily digested sound bites (and doing so in a way that entertains those who pay for the papers and magazines), the truth sometimes gets distorted.

The media often use catchy headlines, and these headlines do not always capture the true spirit of a study. The most common problem is that headlines often suggest a cause-and-effect relationship, even though the wise statistics student will quickly recognize that such a conclusion is not supported by the data. Statistician Jonathan Mueller keeps a website of such headlines. A couple of examples: "Studies say lots of candy could lead to violence." "Texting lifts crash risk by large margin." These headlines all suggest causality: Eating candy will make you violent. Texting will increase the risk of a car accident. But you now know that such a conclusion can be made only for controlled experiments, and even then we must be cautious. Ideally, you can learn what you need to know by reading the news article. But not always. The information you need to judge whether the conclusions of a study are strong is often missing from news reports.

## Clinical Significance vs. Statistical Significance

An outcome of an experiment or study that is large enough to have a real effect on people's health or lifestyle is said to have **clinical significance**. Sometimes, researchers discover that a treatment is statistically significant (meaning that the outcome is too large to be due to chance) but too small to be meaningful (so it is not clinically significant). Studies with very large sample size have large power and so are capable of detecting even very small differences between treatment groups. This does not mean the differences matter. For example, a drug might truly lower cholesterol levels, but not enough to make a real difference in someone's health. Or playing Wii might burn more calories, but maybe not enough for it to serve as a form of exercise. Treatments that cause meaningful effects are called clinically significant. Sometimes, statistically significant results are not clinically significant.

Imagine a rare disease that only 1 person in 10 million people gets. A controlled experiment finds that a new drug "significantly" reduces your risk of getting this disease; specifically, it cuts the risk in half. Here, *significantly* means that the reduction in risk is statistically significant. But given that the disease is so rare already, is it worth taking medicine to cut your risk from 1 in 10 million to 1 in 20 million? Particularly if the drug is expensive or has side effects, most people would probably decide that this treatment is not clinically significant.

## CASE STUDY REVISITED

At the beginning of this chapter, we reported on a study that found that stretching before a race did not increase a runner's speed. Here is the abstract:

The results of previous research have shown that passive muscle stretching can diminish the peak force output of subsequent maximal isometric, concentric and stretch-shortening contractions. The aim of this study was to establish whether the deleterious effects of passive stretching seen in laboratory settings would be manifest in a performance setting. Sixteen members (11 males, 5 females) of a Division I NCAA track athletics team performed electronically timed 20 m sprints with and without prior stretching of the legs. The experiment was done as part of each athlete's Monday work-out programme. Four different stretch protocols were used, with each protocol completed on a different day. Hence, the test period lasted 4 weeks. The four stretching protocols were no-stretch of either leg (NS), both legs stretched (BS), forward leg in the starting position stretched (FS) and rear leg in the starting position stretched (RS). Three stretching exercises (hamstring stretch, quadriceps stretch, calf stretch) were used for the BS, FS, and RS protocols. Each stretching exercise was performed four times, and each time the stretch was maintained for 30 s. The BS, FS, and RS protocols induced a significant ($P < 0.05$) increase (approximately 0.04 s) in the 20 m time. Thus, it appears that pre-event stretching might negatively impact the performance of high-power short-term exercise.

We can apply the six questions to help us determine how strong these results are.

1. What is the research question that these investigators are trying to answer?
2. What is their answer to the research question?
3. What were the methods they used to collect data?
4. Is the conclusion appropriate for the methods used to collect data?
5. To what population do the conclusions apply?
6. Have the results been replicated in other articles? Are the results consistent with what other researchers have suggested?

The answers follow.

- The research question is "Does stretching improve sprinting speed (20 meters) for NCAA athletes?"

- They conclude that it does not and that, in fact, stretching makes sprinters run more slowly.

- To collect these data, the researchers did not take a random sample but instead used 16 college athletes. The researchers used a crossover design in which they blocked on athlete, and each athlete did every treatment, including the "no stretch" treatment. Although it does not say so in the abstract, the order in which an athlete did the treatments was randomly determined. This random assignment minimizes the effects of confounding factors, including the potential confounding factor caused by doing the routines in a particular order. The amount of stretching was carefully controlled so that every athlete did precisely the same stretches for the same amount of time.

- The fact that random assignment was used (to determine the order of the stretching) means that the researchers are justified in making a cause-and-effect conclusion: We can conclude that, for these 16 athletes, the slower times were caused by the stretching.

- However, the athletes were a small sample from one college and were not randomly selected. This means we should not generalize these results to other athletes, especially not to athletes of other ages or in other sports. Note that the increase in time to do the 20-meter sprint was only 0.04 second. Although this slight increase might be meaningful (clinically significant) to elite runners, for most of us this difference is undetectable. (But then, most of us don't run sprints.)

- We cannot tell from the abstract whether the results have been replicated elsewhere, nor do we know the extent to which this study is building on the previous research. But we should keep in mind that one paper is not enough for us to believe in the results. We should try to find out whether studies done at other colleges, or with different levels of athletes, find that stretching does not help performance.

# EXPLORING STATISTICS
## CLASS ACTIVITY

## Reporting on Research Abstracts

| GOALS | MATERIALS |
|---|---|
| To practice reading and interpreting scientific studies. | Six questions provided below. |

**BEFORE THE ACTIVITY**

Find a reported study (either an experiment or an observational study) on the Internet or in a magazine, journal, book, or newspaper in an area of interest to you. Print out or copy the abstract or summary. One good source for medical studies is the *New England Journal of Medicine* archives (www.nejm.org/medical-articles/research).

**ACTIVITY**

Write a report on your chosen abstract. Be sure to include the authors of the study; the title of the study; the journal, book, or magazine in which it was published; and the date of publication. In your report, answer these questions:

1. What is the research question that these investigators are trying to answer?

2. What is their answer to the research question?

3. What were the methods used to collect the data?

4. Is the conclusion appropriate for the methods? (For example, is the conclusion a cause-and-effect claim? Were the data collected in such a way as to support this claim?)

5. To what population do the conclusions apply?

6. Is any indication given that these results are consistent with what other researchers have suggested?

**AFTER THE ACTIVITY**

There is no follow-up to this activity.

## CHAPTER REVIEW

### Key Terms

Treatment variable, *542*
Response variable, *542*
Confounding variables, *542*
Controlled experiment, *543*
Random assignment, *543*
Observational study, *543*

Power, *544*
Controlling for a variable, *545*
Blocking, *545*
Matched-pairs design, *548*
Crossover design, *550*
Systematic sampling, *552*

Stratified sampling, *552*
Strata, *552*
Cluster sampling, *553*
Clusters, *553*
Peer review, *555*
Abstract, *556*

Data dredging, *559*
Publication bias, *560*
Meta-analysis, *560*
Clinical significance, *561*

### Learning Objectives

After reading this chapter and doing the assigned homework problems, you should
- Know how to form blocks of experimental units, and understand why and when these blocks are useful for comparing treatment groups.
- Understand how random assignment is used to allow cause-and-effect inference, and understand how random sampling is used to allow generalizations to a larger population.

- Be able to identify different sampling strategies and understand why they are useful.
- Be prepared to apply knowledge about collecting and analyzing data to critically evaluate abstracts in the science literature.

### Summary

Many research questions can be divided into two categories: those that ask causal questions and those that ask about associations between variables. Causal questions can only be answered with controlled experiments, whereas observational studies can answer questions about associations. Controlled experiments are difficult to design, in part because in real life, many different factors contribute to variation in the response variable. Blocking—the practice of creating groups of subjects/objects that are similar and then randomizing assignment within the blocks—is one method that helps control for known sources of variation. Blocking can help increase the power of a study, the probability of detecting true differences among the treatment groups.

Random sampling is required to conduct a valid poll or survey, but can be complicated to carry out in large, complex populations. Systematic sampling and stratified sampling are two approaches that can improve a random sample. Systematic sampling is done by

regularly selecting every *n*th (such as every 10th) object to appear in the sample, and this provides a representative sample if the objects are in a random/shuffled order or if the entire population is sampled this way. Stratified sampling creates strata (such as gender), which are groups of subjects that are similar to each other, and then randomly samples (without replacement) within each stratum. Cluster sampling is a method used to provide access to large populations. A cluster consists of objects that are all very different from one another and, ideally, a cluster is itself a representative sample. In cluster sampling, the clusters are selected randomly, without replacement, and every object in the cluster is included in the sample.

Interpreting conclusions in scientific studies is complex, because many things can go wrong in a study. Remember that extraordinary results require extraordinary evidence, and that you should trust studies that have been replicated (repeated, resulting in the same conclusion) over ones that have not.

### Sources

Briggs, D. C. 2002. Evaluating SAT coaching: Gains, effects and self-selection. In Rebecca Zwick, *Rethinking the SAT: The future of standardized testing in university admissions.* New York: RoutledgeFarmer, 2004.

Caldwell, B., and J. Crane. 2010. Randomized crossover trial of the acceptabilty of snus, nicotine gum, and Zonnic therapy for smoking reduction in heavy smokers. *Nicotine & Tobacco Research* 12(2), 179–183.

Coggins, S. B., N. C. Coops, and M. A. Wulder. 2010. Improvement of low level bark beetle damage estimates with adaptive cluster sampling. *Silva Fennica* 44(2), 289–301.

Ferguson, C. J., and J. Kilburn. 2010. Much ado about nothing: The misestimation and overinterpretation of violent video game effects in Eastern and Western nations: Comment on Anderson et al. *Psychological Bulletin* 136(2), 174–178, discussion 182–187.

Healy, M. 2010. New vitamin D recommendations: What they mean. *Los Angeles Times* December 6, www.latimes.com (accessed February 3, 2011).

Ho, M. G., S. Ma, W. Chai, W. Xia, G. Yang, and T. E. Novotny. 2009. Smoking among rural and urban young women in china. *Tob Control* 19, 13–18.

Mednick, S., D. Cai, J. Kanady, and S. Drummond, S. 2008. Comparing the benefits of caffeine, naps and placebo on verbal, motor and perceptual

memory. *Behavioural Brain Research* 193, 79–86, www.elsevier.com/locate/bbr (accessed April 22, 2010).

Neale, T. 2007. Even active video games not good enough for kids' fitness. MedPage Today, http://www.medpagetoday.com (accessed April 22, 2010).

Nelson, A., N. Driscoll, D. Landin, M. Young, and I. Schexnayder. 2005. Acute effects of passive muscle stretching on spring performance. *Journal of Sports Sciences* 23(50). 449–454.

Poordad, F., McCone, J., Bacon, B., Bruno, S., Manns, M., et al., 2011. Boceprevir for Untreated Chronis HCV Genotype 1 Infection. *New England Journal of Medicine,* v.364 (13) 1195–1206, http://www.nejm.org/doi/pdf/10.1056/NEJMoa1010494 (accessed April 26, 2011).

Ramirez, G., and S. Beilock. 2011. Writing about testing worries boosts exam performance in the classroom. *Science* 331, 211.

Raskin, P. and R. Unger, 1978. Hyperglucagonemia and its Suppression. *The New England Journal of Medicine* 299(9) 433–436.

Strayer, D. L., F. A. Drews, and D. J. Crouch. 2006. A comparison of the cell phone driver and the drunk driver. *Journal of the Human Factors and Ergonomics Society* 48(2), 381–391.

## SECTION EXERCISES

### SECTION 12.1

**12.1 Happiness and Heart Disease** These two headlines are on the same topic, but one has language that suggests a cause-and-effect relationship, and the other does not.

Headline A: "Happiness Wards Off Heart Disease, Study Says"

Headline B: "Happy People Have Healthier Hearts, Study Finds"

Which headline suggests a cause-and-effect relationship and which suggests only an association? (Source: news.bbc.co.uk, Feb 18, 2010. Accessed through Jonathan Mueller, http://jonathan.mueller.faculty.noctrl.edu/).

**12.2 Breastfeeding and IQ** These two headlines are on the same topic, but one has language that suggests a cause-and-effect relationship, and the other does not.

Headline A: "Higher IQ Found in Breastfed Babies"

Headline B: "Breastfeeding Found to Boost Baby IQ"

State which headline claims a cause-and-effect relationship and which claims only an association. (Sources: http://www.iop.kcl.ac.uk/news/ (accessed February 19, 2010) and http://health.virginmedia.com/ (accessed February 19, 2010).

**12.3 Tattoos and Hepatitis** Two doctors analyzed data from patients who were seeing a physician for back problems. They found that patients with at least one tattoo were more likely to have hepatitis than those with no tattoos. Was this a controlled experiment or an observational study? Explain. (Source: Robert W. Haley and R. Paul Fischer. 2003. *The tattooing paradox.* Archives of Internal Medicine 163, 1095–1098, http://www.menstuff.org/ (accessed August 21, 2004).

**12.4 Marital Status and Death after Breast Cancer** In a 1986 article by A. V. Neale et al., published in *Social Science and Medicine*, researchers compared 5- and 10-year survival rates for women diagnosed with and treated for breast cancer. They found that widowed patients were less likely to survive than married patients. Does this show that widowhood causes death from breast cancer? Why or why not? Can you think of a possible confounding variable? (Source: A. V. Neale, B. C. Tilley, and S. W. Vernon. 1986. Marital status, delay in seeking treatment and survival from breast cancer. *Social Science and Medicine* 23(3), http://www.sciencedirect.com (accessed March 15, 2010).)

**12.5 Tattoos and Hepatitis** Using the information from Exercise 12.3, write two headlines announcing the results of the study. Make one of the headlines state or imply causality and one not imply causality. Explain which is which. Which headline is appropriate for these data, and why?

**12.6 Marital Status and Death after Breast Cancer** Using the information from Exercise 12.4, write two headlines announcing the results of the study. Make one of the headlines state or imply causality and one not imply causality. Explain which is which. Which headline is appropriate for these data, and why?

**12.7 Premature Babies** Some premature babies have trouble breathing. Researchers conducted a study to find out whether lowering the babies' body temperature increased their chance of survival. Is the study described in the following abstract excerpt an observational study or a controlled experiment? How can you tell?

"*Methods:* We performed a randomized trial of infants who were less than 6 hours of age and had a gestational age of at least 36 weeks. . . . We compared intensive care plus cooling of the body to 33.5°C for 72 hours and intensive care alone. The primary outcome was death or severe disability at 18 months of age."

"*Conclusions:* Induction of moderate hypothermia for 72 hours in infants who had perinatal asphyxia did not significantly reduce the combined rate of death or severe disability but resulted in improved neurologic outcomes in survivors."

(Source: D. V. Azzopardi et al. 2009. Moderate hypothermia to treat perinatal asphyxial encephalopathy. *New England Journal of Medicine.* 361, 1340–1358.)

**12.8 Less Supervision for Low-Risk Offenders** The Philadelphia Adult Probation and Parole Department tried an experiment with low-risk offenders (those who are not predicted to commit a serious offence within the next two years) based on variables such as age and previous arrest record. Low-risk offenders were randomly assigned to one of two groups: the control group, for which probation and parole officers had 150 offenders to supervise, and the treatment group, for which officers had 400 offenders to supervise. The control group had an average of 2.4 visits every three months, and the experimental group had an average of 1 visit every three months. The goal was to determine whether there were more rearrests when there was less supervision to see if it might be possible to focus on supervising high-risk offenders to promote public safely. The study concluded that the arrest rates in the control and treatment groups were the same.

Is this study an observational study or a controlled experiment? Explain. (Source: Ahlman, L. C., E. M. Kurtz, and R. Malvestuto, 2010. The Philadelphia randomized controlled trial of low risk supervision: The effect of low risk supervision on rearrest. *Journal of the American Probation and Parole Association* 34(1).)

**TRY 12.9 Multiple Sclerosis Drugs (Example 1)** Multiple sclerosis is a disease of the nervous system that causes severe weakness of the arms and legs, among many other damaging symptoms. Read the following excerpt from the abstract for this controlled experiment, and then:

a. Identify the treatment and response variables.

b. Restate the conclusion of the study in terms of a cause-and-effect conclusion.

"In this 12-month, double-blind . . . study we randomly assigned 1292 patients with relapsing–remitting multiple sclerosis who had a recent history of at least one relapse to receive either oral fingolimod at a daily dose of either 1.25 or 0.5 mg or intramuscular interferon beta-1a (an established therapy for multiple sclerosis) at a weekly dose of 30 μg.

The annualized relapse rate was significantly lower in both groups receiving fingolimod—0.20 (95% confidence interval [CI], 0.16 to 0.26) in the 1.25-mg group and 0.16 (95% CI, 0.12 to 0.21) in the 0.5-mg group—than in the interferon group (0.33; 95% CI, 0.26 to 0.42; $P < 0.001$ for both comparisons).

*Conclusions*: This trial showed the superior efficacy of oral fingolimod with respect to relapse rates and MRI outcomes in patients with multiple sclerosis, as compared with intramuscular interferon beta-1a."

(Source: J. A. Cohen et al. 2010. Oral fingolimod or intramuscular interferon for relapsing multiple sclerosis. *New England Journal of Medicine* 362, 402–415.)

**12.10 Antibacterial Scrub** Read the excerpt from the following abstract from a controlled experiment, and then:

a. Identify the treatment and response variables.

b. Restate the conclusion of the study in terms of a cause-and-effect conclusion.

"We randomly assigned adults undergoing clean-contaminated surgery in six hospitals to preoperative skin preparation with either chlorhexidine–alcohol scrub or povidone–iodine scrub and paint. The primary outcome was any surgical-site infection within 30 days after surgery. Secondary outcomes included individual types of surgical-site infections.

The overall rate of surgical-site infection was significantly lower in the chlorhexidine–alcohol group than in the povidone–iodine group (9.5% vs. 16.1%; $P = 0.004$; relative risk, 0.59; 95% confidence interval, 0.41 to 0.85). Chlorhexidine–alcohol was significantly more protective than povidone–iodine against both superficial incisional infections (4.2% vs. 8.6%, $P = 0.008$) and deep incisional infections (1% vs. 3%, $P = 0.05$) but not against organ-space infections (4.4% vs. 4.5%).

*Conclusions:* Preoperative cleansing of the patient's skin with chlorhexidine–alcohol is superior to cleansing with povidone–iodine for preventing surgical-site infection after clean-contaminated surgery."

(Source: P. Daroui et al. 2010. Chlorhexidine–alcohol versus povidone–iodine for surgical-site antisepsis. *New England Journal of Medicine* 362, 18–26.)

**12.11 Mild Cognitive Impairment and Exercise** In a 2010 study, Dr. Yonas E. Geda and colleagues at Mayo Clinic in Minnesota analyzed data on 1324 individuals without dementia who completed a questionnaire about physical activity between 2006 and 2008 as part of the Mayo Clinic Study of Aging. The participants had an average age of 80 and were classified as having normal cognition ($n = 1126$) or mild cognitive impairment (MCI) ($n = 198$). Those who reported performing moderate exercise such as yoga, aerobics, strength training, swimming, or brisk walking during midlife were 39% less likely to develop MCI, and moderate exercise later in life was associated with a 32% reduction in the rate of MCI.

Was this more likely to have been a controlled experiment or an observational study? How do you know? (Source: http://newsblog.mayoclinic.org/ (accessed February 2, 2011).)

**12.12 Childhood Exposure to Tobacco Smoke and Emphysema** A March 2010 headline from esciencenews.com said, "Exposure to tobacco smoke in childhood home associated with early emphysema in adulthood."

a. Does this headline most likely refer to a controlled experiment or an observational study? Explain.

b. Can you think of a confounding factor that might cause emphysema and is not childhood exposure to smoking?

(Source: http://esciencenews.com/ (accessed March 15, 2010).)

**12.13 Body Temperatures** Two different simple random samples of body temperatures were taken from the same population of healthy adult men. One of the samples had 10 men and one had 30 men. Use the output provided to test the hypothesis that the mean temperature in the population of healthy adult men is not 98.6°F. Assume that the population distribution of temperatures is Normal and that we have random samples. Use a significance level of 0.05. The null hypothesis is that the population mean is 98.6°F, and the alternative hypothesis is that it is not 98.6°F.

a. Test the sample of 10 temperatures to decide whether you can reject the null hypothesis. Report the p-value and conclusion.

```
One-Sample T: 10men

Test of mu = 98.6 vs not = 98.6

Variable   N    Mean   StDev   SE Mean      95% CI           T      P
10men      10   98.190  0.810   0.256   (97.610, 98.770)  -1.60  0.144
```

b. Test the sample of 30 temperatures to decide whether you can reject the null hypothesis. Report the p-value and conclusion.

```
One-Sample T: 30men

Test of mu = 98.6 vs not = 98.6

Variable   N    Mean   StDev   SE Mean      95% CI           T      P
30men      30   98.097  0.726   0.132   (97.826, 98.368)  -3.80  0.001
```

c. Explain the main reason for the different p-values and the different conclusions.

(Source: www.amstat.org/publications/jse/datasets/normtemp.dat)

**12.14 Gender and Use of Turn Signals** Hector Porath, a statistics student, wanted to find out whether gender and the use of turn signals when driving are independent. He observed drivers when driving in his truck over a period of weeks, noting the drivers' gender and whether they used the turn signal when turning or changing lanes, as required by state law. Use a significance level of 0.05.

|             | Men | Women |
|-------------|-----|-------|
| **Turn Signal** | 585 | 452   |
| **No Signal**   | 351 | 155   |
|             | 936 | 607   |

a. In this sample, were men or women more likely to use their turn signals? Show both percentages.

b. Report the chi-square value and the p-value for a chi-square test of independence. Can you reject the null hypothesis that gender and use of turn signal are independent? If so, what do you conclude?

c. Now repeat your analysis using sample sizes that are about one-tenth the size of the real sample sizes:

|              | Men | Women |
|--------------|-----|-------|
| **Turn Signal** | 59  | 45    |
| **No Signal**   | 35  | 16    |
|              | 94  | 61    |

d. Compare your answers to parts b and c, and explain what happened.

**\*12.15 Diet Pill and Power** Imagine two studies of a diet pill that, manufacturers claim, will make people lose weight. The first study is based on a random sample of 100 men and women who have been taking the diet pill for 12 months. A hypothesis test is carried out to determine whether their mean weight change is negative. The second study was based on a random sample consisting solely of 100 men (no women) who had been taking the weight loss pill for 12 months. The same hypothesis test is carried out to determine whether their mean weight change is negative. Which study's hypothesis test will have greater power to detect whether the mean weight change is truly less than 0? Assume that the diet pill really does lower weight and that the effect of weight loss is larger on men than on women.

**\*12.16 SAT Prep and Power** Suppose an SAT tutoring company really can improve SAT scores by 10 points, on average. A competing company, however, uses a more intense tutoring approach and really can improve SAT scores by 15 points, on average. Suppose you've been hired by both companies to test their claims that their tutoring improves SAT scores. For both companies, you will collect a random sample of high school students to undergo tutoring. With both resulting samples, you will test the hypothesis that the mean improvement is more than 0. Suppose it is important to keep the power of both studies at 80%. Will you use the same sample size for both studies? If so, explain why you can. If not, which study would require the largest sample size, and why? Assume that both samples of students will be drawn from the same population.

**TRY 12.17 Brain Games (Example 2)** Researchers are interested in testing whether a video game that is designed to increase brain activity actually works. To test this, they plan to randomly assign subjects to either the treatment group (spend 15 minutes per day playing the game) or a control group (spend 15 minutes per day surfing the Web). At the end of the study, the researchers will administer a test of "brain teasers" to see which group has the greater mental agility. Because they suspect that age might affect the outcome, and the researchers will create blocks of ages: 18–25, 26–35, 36–45, and 46–55. To randomly assign subjects to treatment or control, they will place four tickets in a box. The tickets are labeled with the age groups. When an age group is selected, everyone in that age group will be assigned to the treatment group. They will select two age groups to go to treatment, and two to go to

control. Is this an appropriate use of blocking? If so, explain why. If not, describe a better blocking plan.

**12.18 A Smile a Day** Smiling is a sign of a good mood, but can smiling improve a bad mood? Researchers plan to assign subjects to two groups. Subjects in both groups will rate their mood at the beginning of the study. Then subjects in the treatment group will be told to smile while they are asked to recount a pleasant memory. Subjects in the control group will also be asked to recount a pleasant memory, but they will not be told to smile. Both groups will again rate their moods, and researchers will determine whether the reported moods differ between the two groups. Because the initial, baseline mood rating might affect the outcome, after the first mood rating the subjects will be broken into two groups: one group with low ratings ("bad mood") and one with higher ratings ("good mood"). Patients in each group will then be randomly assigned to either the treatment group or the control group. Is this an appropriate use of blocking? If so, explain why. If not, describe a better blocking plan.

**TRY \*12.19 Swimsuits and Racing Speeds (Example 3)** New, slick swimsuits were introduced in the 2008 summer Olympics, and some say that they gave the wearers an advantage in races. In order to test whether the suits were effective, suppose that there are 40 swimmers; 20 of them are Olympic-level swimmers, and 20 are amateur swimmers. The designers will ask the swimmers to swim 200 meters as fast as possible. It is reasonable to assume that the effects of the suits (due to dynamic forces of the water) might be different for the elite Olympic swimmers than for regular swimmers.

a. Identify the treatment variable and the response variable.

b. Describe a simple randomized design (not blocked) to test the whether the slick suits decrease race times. Explain in detail how you will assign swimmers to treatment groups. Your description should be detailed enough that a friend could carry out your instructions.

c. Describe a blocked design to test whether the slick suits decrease race times. Explain in detail how you will assign swimmers to treatment groups.

d. What advantage does the blocked design have?

e. Describe a way to use swimmers as their own controls to reduce variation.

**\*12.20 Flu Vaccines and Age** Suppose you want to compare the effectiveness of the flu vaccine in preventing the flu using one of two different forms: nasal spray vs. injection. Suppose you have 60 subjects available of different ages, and you suspect that age might have an effect on the outcome. Assume there are 20 children from ages 2 to 15, 20 people from ages 16 to 30, and 20 people from ages 31 to 49.

a. Identify the treatment variable and the response variable.

b. Describe a simple randomized design (no blocking) to test the whether the injection or the nasal spray is more effective. Explain in detail how to assign people to treatment groups.

c. Describe a blocked design (blocking by age) to test the whether the injection or the nasal spray is more effective. Explain in detail how you will assign people to treatment groups.

d. What advantage does the blocked design have?

**\*12.21 Preventing Heart Attacks with Aspirin** Suppose that you want to determine whether the use of one aspirin per day for people age 50 and older reduces the chance of heart attack. You have 200 people available for the study, 100 men and 100 women. You suspect that aspirin might affect men and women differently.

To ensure an appropriate comparison group, those who do not get aspirin should get a placebo.

   a. Identify the treatment and response variables.

   b. Describe in detail a simple randomized design (not blocked) to test whether aspirin lowers the risk of heart attack.

   c. Describe a blocked design to test whether aspirin lowers the risk of heart attack; the blocking variable is gender.

   d. Explain why researchers might prefer a blocked design.

**\*12.22 Tomato Plants and Fertilizer** Suppose you grow tomato plants in a greenhouse and sell the tomatoes by weight, so that the amount of money you make depends on plants producing a large total weight of tomatoes.

You want to determine which of two fertilizers will produce a heavier harvest of tomatoes, fertilizer A or fertilizer B. There are two distinct regions in the greenhouse, one on the southern side that gets more light and one on the northern side that gets less light. There is room for 20 tomato plants on the southern side and 20 on the northern side. Assume that all the plants are beefsteak tomato plants.

   a. Identify the treatment and response variables.

   b. Describe a simple randomized design to test whether fertilizer A is better than fertilizer B.

   c. Describe a blocked design to test which fertilizer produces a greater weight of tomatoes, blocking by southern side and northern side of the greenhouse. Explain why creating blocks based on whether plants are on the southern or northern side makes sense.

   d. Explain why researchers might prefer a blocked design.

**TRY 12.23 Comparing Navy Commissary and Kmart Prices (Example 4)** Gina Ortiz, a statistics student who lived on a naval base, looked at the prices of items at the Navy Commissary and at the prices of the same items at Kmart. Assume that these are a random sample of prices and that the distribution of prices is sufficiently close to Normal for $t$-tests. The student wanted to test the hypothesis that the mean price at the naval base was *less than* the mean price at Kmart. Assume that conditions for $t$-tests hold.

| Items | Kmart | Navy |
|---|---|---|
| Tide | 19.99 | 18.13 |
| Clorox | 4.29 | 2.57 |
| Gain | 5.39 | 5.13 |
| Swiffer | 7.99 | 5.99 |
| Pledge | 2.99 | 2.48 |
| Downy | 4.29 | 4.03 |
| Jet Dry | 3.99 | 2.99 |
| Secret | 3.99 | 2.94 |
| Dove | 5.99 | 2.79 |
| LAGel | 2.38 | 1.71 |
| Q Tips | 2.99 | 5.00 |
| Kudos | 2.69 | 2.37 |
| Spam | 2.49 | 1.30 |
| Tuna | 0.65 | 0.65 |
| Coke | 1.49 | 1.37 |

   a. Use a two-sample $t$-test, and report the value of the observed test statistic and the p-value. Report whether you can reject the null hypothesis of no difference in prices at the 0.05 level.

   b. Use a paired $t$-test, and report the observed value of the test statistic and the p-value. Report whether you can reject the null hypothesis of no difference in prices at the 0.05 level.

   c. For these data, which is the appropriate test? Explain.

**12.24 Comparing Textbook Prices** Amelia Suragairin, a statistics student, was interested in the prices of college textbooks sold at the United States and United Kingdom Amazon.com online stores. She gathered data on the prices of the same books on both websites; these data are available on the disk and website. Assuming that this is a random sample of textbooks, determine whether the prices are significantly different, using a significance level of 0.05. The prices in the UK were obtained in pounds and changed to dollars for the comparison. Assume that the distribution of prices is sufficiently close to Normal for $t$-tests.

   a. Use a two sample $t$-test; report the $t$-statistic and p-value, and report whether you can reject the null hypothesis of no difference in prices at the 0.05 level of significance.

   b. Use a paired $t$-test; report the $t$-statistic and p-value, and report whether you can reject the null hypothesis of no difference in prices at the 0.05 level of significance.

   c. Which is the appropriate test for this data set?

**12.25 Reading Colored Paper** Some people believe that it is easier to read words printed on colored paper than words printed on white paper. To test this theory, statistics student Paula Smith collected data. Subjects were timed as they read a passage printed in black ink on a sheet of salmon-colored paper and also timed for reading the same passage printed with black ink on a sheet of white paper. Each person was randomly assigned to read material printed on either the white or the salmon paper first. The times in seconds are given in the table. We are assuming that smaller times (faster reading) imply that the reading is easier. Assume that these are a random sample of times and that the distribution of times is sufficiently close to Normal for $t$-tests.

| Salmon | White | | Salmon | White |
|---|---|---|---|---|
| 79 | 72 | | 140 | 160 |
| 49 | 45 | | 64 | 57 |
| 47 | 45 | | 67 | 61 |
| 112 | 120 | | 73 | 72 |
| 65 | 63 | | 64 | 48 |
| 66 | 62 | | 126 | 122 |
| 67 | 60 | | 67 | 73 |
| 59 | 54 | | | |

a. Compare the sample means. Does your result fit with the theory given?

b. Use a two-sample *t*-test; report the *t*-statistic and p-value, and report whether you can reject the null hypothesis of no difference in times at the 0.05 level.

c. Use a paired *t*-test; report the *t*-statistic and p-value, and report whether you can reject the null hypothesis of no difference in times at the 0.05 level.

d. Which is the appropriate test for this data set?

e. Why is it essential that the researchers randomly assign the order so that some of the people read material on the salmon-colored paper first and some read material on the white first?

**12.26 The Stroop Effect** Suppose you had to identify the color of ink for a series of printed words that spelled out a color that did not match the ink color? For example, what color ink is used in the word RED? This might take longer than identifying the color when the ink and printed word matched (RED). This difference is an example of *interference* and is called the Stroop Effect, in reference to a famous study conducted by research psychologist J. R. Stroop (1897–1973).

| Same | Different |
|------|-----------|
| 32 | 66 |
| 12 | 31 |
| 17 | 40 |
| 16 | 25 |
| 25 | 36 |
| 18 | 15 |
| 18 | 39 |
| 24 | 35 |
| 20 | 32 |
| 24 | 30 |

The data in the table were collected to see whether interference was significant. Each of the 10 people studied identified colors of ink in two situations, and the time (in seconds) was recorded. There were 36 words in each trial. In column 1 (Same), the ink was the same color as the meaning of the spelled-out word. In column 2 (Different) the color of the ink was different from the meaning of the spelled-out word. Each person was randomly assigned to read the "same color" or the "different color" words first. Assume that all the conditions required for *t*-tests are satisfied.

a. Compare the sample means. Does the result fit with the theory given?

b. Use a two-sample *t*-test; report the *t*-statistic and p-value, and report whether you can reject the null hypothesis of no difference in times at the 0.05 level.

c. Use a paired *t*-test; report the *t*-statistic and p-value, and report whether you can reject the null hypothesis of no difference in times at the 0.05 level.

d. Which is the appropriate test for this data set?

e. Why is it appropriate to randomly assign the order so that some of the people read the "same color" word first and some read the "different color" word first?

**12.27 Pulse Rates** Suppose you want to test students' pulse rates in three different situations: after sitting in class, after exercise (standing up and sitting down five times), and after meditation (with eyes closed and counting silently and slowly from 1 to 5 repeatedly). To increase the power of your statistical tests, you plan to measure each student in all three situations. Why would you want to randomize the order so that some students start with sitting, some start with exercise, and some start with meditation?

**12.28 Reading Colored Paper** Suppose you want to test students' reading with three different colors of paper to see whether the color has an effect on reading speed. Assume that the same passage is written on three different colors of paper: light blue, yellow, and pink. Each student's reading speed will be tested on all three colors. Why would you want to randomize the order in which the different colors are presented?

## SECTION 12.2

TRY **12.29 Student Records (Example 5)** Suppose a person with access to student records at your college has an alphabetical list of currently enrolled students. The person looks at the records of every 10th person (starting with a randomly selected person among the first 10) to see whether they have paid their latest tuition bill. What kind of sampling does this illustrate?

**12.30 Student Records** Suppose a person with access to student records at your college has a list of currently enrolled students. The person sorts the data to create two new lists. One contains all the male names, the other all the female names. The person then uses a random number generator to select 50 men and 50 women. What kind of sampling does this illustrate?

**12.31 Social Security** Suppose you wanted to collect opinions on what to do about the Social Security program in the United States. Describe the potential benefits of stratifying on age.

**12.32 College Football** Many colleges have football programs that are expensive to run. Suppose you wanted to survey students on campus to find out whether they think that funding for the football program at your school should be expanded or reduced. Why might you want to stratify your samples by gender?

**12.33 Balancing the Books** A national retail clothing chain suspects that individual stores throughout the country are making accounting mistakes when balancing their registers. The executive office staff suspect that the problem is worse in their large-volume stores, where there are many customers per day and, consequently, many possibilities for mistakes in the accounting. They divide all of their stores into three groups: high volume, medium volume, and low volume. Within each of these groups, they take a random sample, without replacement, of 10 stores. Each store selected for the sample will have its accounts audited to look for accounting mistakes, and from this the executive office will estimate the proportion of all stores that have committed accounting errors. Is this an example of stratified random sampling, systematic sampling, or cluster sampling? Explain.

**12.34 Music Lovers** A large concert promoter that operates several hundred concert locations around the country wants to survey the managers at these locations to ask their opinions about how to improve attendance at concerts. Because the survey is rather lengthy, the promoter does not want to ask all of the managers and decides to ask a random sample of managers instead. The promoter organizes the concert locations into 20 different geographic zones, randomly select 5 zones, and surveys all of the managers in those 5 zones. Is this an example of stratified random sampling, systematic sampling, or cluster sampling?

## SECTION 12.3

TRY **12.35 Effect of Confederates on Compliance (Example 6)**
A study was done to see whether participants would ignore a sign that read, "Elevator may stick between floors. Use the stairs." The people who used the stairs were classified as compliant, those who used the elevator as noncompliant. The study was done in a university dorm on the ground floor of a building that had three floors. There were three different situations, two of which involved a person who was secretly working with the experimenter. (This person is called a confederate.) In the first situation, there was no other person using the stairs or elevator—that is, no confederate. In the second, there was a compliant confederate (one who used the stairs). In the third, there was a noncompliant confederate (one who used the elevator). Suppose that the participants (people who arrived to use the elevator at the time the experiment was going on) were randomly assigned to the three groups. There were significant differences between groups.

a. Can we generalize widely to a large group? Why or why not?

b. Can we infer causality? Why or why not?

(Source: Lary Shaffer and Matthew R. Merrens. 2001. *Research stories for introductory psychology.* Allyn and Bacon. Original Source: M. S. Wogalter, et al. 1987. Effectiveness of warnings. *Human Factors* 29, 599–612.)

**12.36 Nicotine Patch** Suppose that a new nicotine patch to help people quit smoking was developed and tested. Smokers voluntarily entered the study and were randomly assigned either the nicotine patch or a placebo patch. Suppose that a larger percentage of those using the nicotine patch were able to stop smoking.

a. Can we generalize widely to a large group? Why or why not?

b. Can we infer causality? Why or why not?

**12.37 Hospital Rooms** When patients are admitted to hospitals, they are sometimes assigned to a single room with one bed and sometimes assigned to a double room, with a roommate. (Some insurance companies will pay only for the less-expensive, double rooms.) A researcher was interested in the effect of the type of room on the length of stay in the hospital. Assume that we are not dealing with health issues that require single rooms.

Suppose that upon admission to the hospital, the names of patients who would have been assigned a double room were put onto a list and a systematic random sample was taken; every tenth patient who would have been assigned to a double room was part of the experiment. For each participant, a coin was flipped: If it landed heads up, they got a double room, and if it landed tails up, they got a single room. Then the experimenters observed how many days the patients stayed in the hospital and compared the two groups. The experiment ran for two months. Suppose those who stayed in single rooms stayed (on average) one less day, and suppose the difference was significant.

a. Can you generalize to others from this experiment? If so, to whom can you generalize and why can you do it?

b. Can you infer causality from this study? Why or why not?

**12.38 College Tours** A random sample of 50 college first-year students (out of a total of 1000 first-years) was obtained from college records using systematic sampling. Half of those students had a campus tour with a sophomore student, and half had a tour with an instructor. The tour guide was determined randomly by coin flip for each student. Suppose that those with the student guide rated their experience higher than those with the instructor guide.

a. Can you generalize to other first-year students at this college? Explain.

b. Can you infer causality from this study? Explain.

**12.39 Low-Dose Flu Vaccine** In the fall of 2004, the United States experienced a shortage of flu vaccine when vaccine used in injections from one of the pharmaceutical companies was found to be contaminated. A study was done to see whether a smaller dose of the vaccine (produced by other companies) could be used successfully; if so, a small amount of vaccine could be divided into more flu shots. The usual amount of vaccine was injected into the muscle of half of the patients (intramuscular injection). The other half of the patients had only a small amount of vaccine injected under the skin (intradermal injection), not into the muscle. Assume that patients were randomly assigned to groups. The response was measured by looking at the production of antibodies. More antibodies generally result in less risk of getting the flu. The conclusion of the study is given. Tell whether the smaller dose should be used, and explain why or why not.

*"Conclusions:* As compared with an intramuscular injection of full-dose influenza vaccine, an intradermal injection of a reduced dose resulted in similarly vigorous antibody responses among persons 18 to 60 years of age but not among those over the age of 60 years."

(Source: Robert B. Belshe et al. 2004. Serum antibody responses after intradermal vaccination against influenza. *New England Journal of Medicine* 351, 2286–2294.)

**12.40 CABG or Stent?** The following abstract is from a study to determine which is more effective for avoiding adverse health outcomes: a coronary artery bypass graft (CABG), in which arteries in one part of the body are used to replace clogged arteries, or stenting (inserting a tube into an artery to keep the artery open). The term *revascularization* means new blocking of the arteries.

a. Read the abstract, and indicate which procedure had better results, CABG or stents.

b. Does the stated conclusion make a claim for a cause-and-effect link between the treatment and response? If so, is this claim justified? If no claim of causality was made, why was it not made?

*"Methods:* We identified patients with multivessel disease who received drug-eluting stents or underwent CABG in New York State between October 1, 2003, and December 31, 2004, and we compared adverse outcomes (death, death or myocardial infarction, or repeat revascularization) through December 31, 2005, after adjustment for differences in baseline risk factors among the patients.

*Results:* In comparison with treatment with a drug-eluting stent, CABG was associated with lower 18-month rates of death and of death or myocardial infarction both for patients with three-vessel disease and for patients with two-vessel disease. . . . Patients undergoing CABG also had lower rates of repeat revascularization.

*Conclusions:* For patients with multivessel disease, CABG continues to be associated with lower mortality rates than does treatment with drug-eluting stents and is also associated with lower rates of death or myocardial infarction and repeat revascularization."

(Source: E. H. Hannan et al. 2008. Drug-eluting stents vs. coronary-artery bypass grafting in multivessel coronary disease. *New England Journal of Medicine* 358, 331–341.)

**12.41 Endovascular or Open Repair of Aortic Aneurysms?** An aortic aneurysm is the ballooning out of the aorta. If the aorta ruptures, death usually follows. There are two methods of surgery to repair the blood vessel. Conventional repair uses a larger incision in the area. Endovascular repair uses a very small incision, and the operation is done within the small incision with special instruments.

a. Read the following abstract, and write a conclusion about which treatment you would choose and why.

b. The researchers give two confidence intervals, each associated with a different outcome. Choose one of these intervals and interpret it, as you would to someone concerned about whether she or he should have conventional or endovascular repair of the aorta.

*"Methods:* We conducted a multicenter, randomized trial comparing open repair with endovascular repair in 351 patients who had received a diagnosis of abdominal aortic aneurysm of at least 5 cm in diameter and who were considered suitable candidates for both techniques. Survival after randomization was calculated with the use of Kaplan–Meier analysis and compared with the use of the log-rank test on an intention-to-treat basis.

*Results:* Two years after randomization, the cumulative survival rates were 89.6 percent for open repair and 89.7 percent for endovascular repair (difference, −0.1 percentage point; 95 percent confidence interval, −6.8 to 6.7 percentage points). The cumulative rates of aneurysm-related death were 5.7 percent for open repair and 2.1 percent for endovascular repair (difference, 3.7 percentage points; 95 percent confidence interval, −0.5 to 7.9 percentage points).

*Conclusions:* The perioperative survival advantage with endovascular repair as compared with open repair is not sustained after the first postoperative year."

(Source: Jan D. Blankensteijn et al. 2005. Two-year outcomes after conventional or endovascular repair of abdominal aortic aneurysms. *New England Journal of Medicine* 352, 2398–2405.)

**12.42 Antibiotics for Heart Patients** Read the following abstract, and state whether you think that coronary patients should be given azithromycin (an antibiotic) to reduce the risk of death or other bad outcomes. Explain your answer. Include the confidence interval for the difference in percentages (of bad outcomes), and explain what it shows and why.

*"Methods:* In this randomized, prospective trial, we assigned 4012 patients with documented stable coronary artery disease to receive either 600 mg of azithromycin or placebo weekly for one year. The participants were followed for a mean of 3.9 years at 28 clinical centers throughout the United States.

*Results:* The primary end point, a composite of death due to coronary heart disease, nonfatal myocardial infarction, coronary revascularization, or hospitalization for unstable angina, occurred in 446 of the participants who had been randomly assigned to receive azithromycin and 449 of those who had been randomly assigned to receive placebo. There was no significant risk reduction in the azithromycin group as compared with the placebo group with regard to the primary end point (risk reduction, 1 percent [95 percent confidence interval, −13 to 13 percent]). There were also no significant risk reductions with regard to any of the components of the primary end point, death from any cause, or stroke. The results did not differ when the participants were stratified according to sex, age, smoking status, presence or absence of diabetes mellitus, or *C. pneumoniae* serologic status at baseline.

*Conclusions:* A one-year course of weekly azithromycin did not alter the risk of cardiac events among patients with stable coronary artery disease."

(Source: J. Thomas Grayston et al. 2005. Azithromycin for the secondary prevention of coronary events. *New England Journal of Medicine* 352, 1637–1645.)

*****12.43 Alumni Donations** The alumni office wishes to determine whether students who attend a reception with alumni just before graduation are more likely to donate money within the next two years.

a. Describe a study based on a sample of students that would allow the alumni office to conclude that attending the reception *causes* future donations but that it is *not* possible to generalize this result to all students.

b. Describe a study based on a sample of students that does *not* allow fundraisers to conclude that attending receptions causes future donations but does allow them to generalize to all students.

c. Describe a study based on a sample of students that allows fundraisers to conclude that attending the reception causes future donations and also allows them to generalize to all students.

* **12.44 Recidivism Rates** The 3-year recidivism rate in Texas is about 30%, which means that 30% of released Texas prisoners return to prison within 3 years of release. There have been many attempts to reduce the recidivism rate. Suppose you want to determine whether electronic monitoring bracelets that track the location of the released prisoner reduce recidivism. Suppose that offenders released from prison are observed for three years to see whether they go back to prison and that the ones who wear electronic monitoring bracelets wear them for the first year only.

a. Describe a study based on a sample of released offenders that would allow the legal system to conclude that monitoring causes a reduction in recidivism but would not allow it to generalize this result to all released prisoners.

b. Describe a study based on a sample of released offenders that does *not allow* the legal system to conclude that monitoring causes a reduction in recidivism but does allow it to generalize to all released offenders.

c. Describe a study based on a sample of released prisoners that allows the legal system to conclude that monitoring causes a reduction in recidivism and also allows it to generalize to all released offenders.

TRY **12.45 Low-Dose Flu Vaccine (Example 7)** Refer to Exercise 12.39. If you cannot obtain the answer to any of the following questions from the excerpt of the abstract provided there, say so.

1. What is the research question that these investigators are trying to answer?

2. What is their answer to the research question?

3. What were the methods they used to collect data?

4. Is the conclusion appropriate for the methods used to collect data?

5. To what population do the conclusions apply?

6. Have the results been replicated in other articles? Are the results consistent with what other researchers have suggested?

**12.46 CABG or Stent?** Refer to the abstract in Exercise 12.40 to answer the questions listed in Exercise 12.45.

**12.47 Endovascular or Open Repair of Aortic Aneurysms?** Refer to the abstract in Exercise 12.41, and answer the questions listed in Exercise 12.45. The conclusion was "The perioperative survival advantage with endovascular repair as compared with open repair is not sustained after the first postoperative year."

* **12.48 Autism and MMR Vaccine** An article by Wakefield et al. in the British medical journal *Lancet* claimed that autism was caused by the measles, mumps, and rubella (MMR) vaccine. This vaccine is typically given to children twice, at about the age of 1 and again at about 4 years of age. The article reports a study of 12 children with autism who had all received the vaccines shortly before developing autism. The article was later retracted by *Lancet* because the conclusions were not justified by the design of the study.

Explain why *Lancet* might have felt that the conclusions were not justified by listing potential flaws in the study, as described above. (Source: A. J. Wakefield et al. 1998. Ileal–lymphoid-nodular hyperplasia, non-specific colitis, and pervasive developmental disorder in children. *Lancet* 351, 637–641.)

## CHAPTER REVIEW EXERCISES

* **12.49 Vaccinations for Diarrhea in Mexico** Diarrhea can kill children and is often caused by rotavirus. Read the abstract below and answer the questions that follow.

*"Methods:* We obtained data on deaths from diarrhea, regardless of cause, from January 2003 through May 2009 in Mexican children under 5 years of age. We compared diarrhea-related mortality in 2008 and during the 2008 and 2009 rotavirus seasons with the mortality at baseline (2003–2006), before the introduction of the rotavirus vaccine. Vaccine coverage was estimated from administrative data.

*Results:* Diarrhea-related mortality fell from an annual median of 18.1 deaths per 100,000 children at baseline to 11.8 per 100,000 children in 2008 (rate reduction, 35%; 95% confidence interval [CI], 29 to 39; $P < 0.001$). . . . Mortality among unvaccinated children between the ages of 24 and 59 months was not significantly reduced. The reduction in the number of diarrhea-related deaths persisted through two full rotavirus seasons (2008 and 2009).

*Conclusions:* After the introduction of a rotavirus vaccine, a significant decline in diarrhea-related deaths among Mexican children was observed, suggesting a potential benefit from rotavirus vaccination."

a. State the death rate before vaccine and the death rate after vaccine. What was the change in deaths per 100,000 children?

From the given p-value, can you reject the null hypothesis of no change in death rate?

b. Would you conclude that the vaccine was effective? Why or why not?

(Source: R. Vesta et al. 2010. Effect of rotavirus vaccination on death from childhood diarrhea in Mexico. *New England Journal of Medicine* 62, 299–305.)

**12.50 Gestational Diabetes** Comment on the following abstract, which discusses treatment of diabetes in pregnant women (gestational diabetes) and complications of childbirth.

*"[Methods:]* We conducted a randomized clinical trial to determine whether treatment of women with gestational diabetes mellitus reduced the risk of perinatal [immediately before or after childbirth] complications. We randomly assigned women between 24 and 34 weeks' gestation who had gestational diabetes to receive dietary advice, blood glucose monitoring, and insulin therapy as needed (the intervention group) or routine care. Primary outcomes included serious perinatal complications (defined as death, shoulder dystocia, bone fracture, and nerve palsy), admission to the neonatal nursery, jaundice requiring phototherapy, induction of labor, cesarean birth, and maternal anxiety, depression, and health status.

*[Results:]* The rate of serious perinatal complications was significantly lower among the infants of the 490 women in the intervention group than among the infants of the 510 women in the routine-care group (1 percent vs. 4 percent; . . . P = 0.01). . . .

*Conclusions:* Treatment of gestational diabetes reduces serious perinatal morbidity. . . ."

a. Identify the treatment and the response variables.

b. Is the stated conclusion making a claim for cause-and-effect? If so, is this claim justified? Explain.

(Source: Caroline A. Crowther et al. 2005. Effect of treatment of gestational diabetes mellitus on pregnancy outcomes. *New England Journal of Medicine* 352, 2477–2486.)

# 13 Inference without Normality

Many of the techniques you've learned for inference so far have relied on a single important mathematical idea: the Normal distribution. In many cases, when the population we're studying follows a Normal distribution, we can compute accurate confidence levels and compute p-values for our hypothesis tests. Confidence levels and p-values are important because they measure our level of uncertainty. Inference is an uncertain business and might seem completely useless if it were not for the fact that we can measure how far from the truth our estimates are likely to be.

However, not every distribution is Normal. When the population is not Normal, the Central Limit Theorem tells us that if the sample size is large enough, the distribution of the sample mean or sample proportion is still approximately Normal. But what if we are not estimating a mean or proportion? The CLT doesn't help us if we want to estimate the population median, for example.

In some contexts, such as skewed distributions, the median is a more natural measure of center than is the mean. At other times, we might have too small a sample size to rely on the CLT. For example, biologists who study animals in the wild, such as the electric rays examined in the following case study, must sometimes work with small sample sizes and so cannot rely on the Central Limit Theorem to assure them that the Normal condition is satisfied.

Several approaches to inference will work regardless of whether the population follows the Normal distribution. The formal name for such approaches is **nonparametric inference**. This term covers a variety of techniques and procedures, and this chapter focuses on hypothesis tests that do not depend on the Normal distribution. Certain conditions need to be satisfied in order for these tests to provide valid inference, but generally these conditions are less strict than those that must be satisfied for the tests presented in earlier chapters.

## CASE STUDY

# Electric Rays

Stingrays catch food by using an electric shock. Researchers studying stingrays wondered whether the rays adjust their shocks to subdue different types of prey (Lowe et al. 1994). To test this, the researchers approached Pacific electric rays (*Torpedo californica*) off the coast of southern California and offered four different types of prey. We will examine the responses of the rays to two different types of prey: live fish and freshly killed fish. When the rays attacked, researchers were able to record the electrical impulse, and one measure of this is called the *mean initial pulse rate,* or ipr, which is measured in units called *electric organ discharges per second,* or eod/s. Ten rays were offered live prey, and another six rays were offered freshly killed prey. The sample mean for rays offered live prey was 100.2 eod/s, and for those offered freshly killed prey was 78.3 eod/s.

You might think of using a two-sample *t*-test to compare these two groups of rays. However, the sample sizes are too small for us to determine whether the distributions are Normal. The *t*-test can handle non-Normal distributions if the sample size is large, but these two samples of ten and six rays are not very large. At the end of this chapter, you'll see how a nonparametric procedure— a statistical method that makes no assumptions about the distribution of the population—can be used to test this hypothesis.

# Transforming Data

One approach to dealing with data from a non-Normal population is to change the data so that they look more like they *did* come from a Normal population. Strictly speaking, this is not a nonparametric technique, because we still rely on an identifiable population distribution (the Normal, in this case) to extend our conclusions to the population. Still, this commonly used technique can be quite effective.

Suppose we want to estimate the mean income of residents in the United States on the basis of a random sample. The distribution of income is famous for being right-skewed, as you can see in Figure 13.1, which displays a histogram for a random sample of 36 U.S. residents' annual income. The fact that the sample is skewed makes us strongly suspect that the population is also skewed.

We can use technology to find a 95% confidence interval for the mean income (in dollars per year) of all U.S. citizens, based on the data in this sample. The 95% confidence interval is

$$\$19{,}374 \text{ to } \$40{,}091$$

However, one assumption behind this calculation is that either income is Normally distributed (and we know it's not) or that the sample size is "large enough." One rule of thumb for "large enough" is that the sample size should be 25 or higher. But how can we be sure this is indeed large enough?

First, we examine a tool that helps us better judge whether a distribution is Normal or non-Normal. Then we will show you how to transform the data so that the distribution is closer to a Normal distribution. We can then analyze these new, transformed data and get more precise results than we would get if we relied on the original, untransformed data.

▶ **FIGURE 13.1** The distribution of a random sample of incomes is skewed right, which is a typical feature of distributions of income.

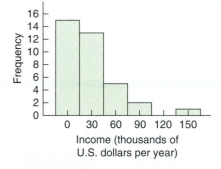

Income (thousands of U.S. dollars per year)

---

**Looking Back**

**Central Limit Theorem (CLT)**
The CLT says that the sampling distribution of the sample mean, $\bar{x}$, is approximately $N\left(\mu, \dfrac{\sigma}{\sqrt{n}}\right)$, where $\mu$ is the mean of the population and $\sigma$ is the standard deviation of the population. The larger the sample size, $n$, the better the approximation. If the population is Normal to begin with, then the sampling distribution is exactly a Normal distribution.

## Interpreting QQ Plots

A new tool that is useful here is the **QQ plot**. A QQ plot (the Q's stand for "quantile") is a graphical tool that is helpful for deciding whether a sample was drawn from a distribution that is approximately Normal. A QQ plot graphs the actual values of a variable on the *y*-axis against the values on the *x*-axis that the variable would have if, in fact, it followed a standard Normal distribution. If the distribution of the variable is a Normal distribution, then this plot is a straight line.

Several variations of the QQ plot are common, and different software packages display different variations. For example, some software uses the *y*-axis to display the values of the variable converted to standard units, rather than plotting the actual values. Some software packages display QQ plots with the *x*-axis used to plot the standard units and the *y*-axis to plot the observed values. Others show a variation of the QQ plot called a probability plot, in which the probability of seeing a value less than or equal to the displayed value, if the distribution is Normal, is plotted. No matter which variation your technology produces, the interpretation of the plot is the same. If the data are Normal, then the points on the plot fall along a straight line. If they are non-Normal, they follow a curve.

Figure 13.2 shows the QQ plot for the incomes displayed in Figure 13.1. The black line shows where the points would fall if the distribution were Normal. As we read from left to right, the points are fairly flat from –2 to –1 on the *x*-axis, and then they start to rise. This feature is commonly seen in right-skewed distributions.

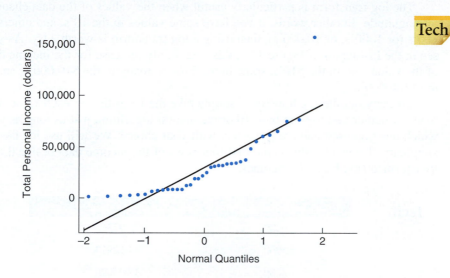

◀ **FIGURE 13.2** The QQ plot for the sample of incomes in Figure 13.1. If the data were Normally distributed, they would follow the straight line.

Reading a QQ plot takes a little practice. Even samples from true Normal distributions do not produce QQ plots that exactly follow the straight line. Indeed, samples from true Normal distributions typically have points that meander above and below the line in a QQ plot, but overall, the points follow the line.

Figure 13.3 shows a QQ plot from a true Normal distribution created using simulated data. (Because we simulated them, we know for a fact that the data come from a Normal distribution.) The points do not perfectly follow the straight line. However, unlike the points in Figure 13.2, they tend to follow a constant increasing trend.

◀ **FIGURE 13.3** The QQ plot for data simulated from a true Normal distribution (with mean 0, standard deviation 1). Note that the points "stray" a bit but for the most part follow the straight line.

▶ **TABLE 13.1** Log (base 10) transforms of the first four rows of a data set containing incomes of U.S. residents.

## The Log Transform

The **log transform** is an approach to analyzing data that can sometimes turn right-skewed distributions into something more closely resembling a Normal distribution. The log transform can be applied only to data with all positive (greater than 0) values. This approach changes—transforms—the data to new values. If all goes well, the distribution of the new values follows the Normal distribution, so the Normal condition required of so many of the techniques presented in earlier chapters will be satisfied. Then, if we want to, we can "undo" the transform once we've learned what we need from the data.

The log transform is particularly useful when the values of the data change orders of magnitude. In other words, if you have some values in the 10's, and others in the 100's (or 1000's, or 10,000's), then using a log transform is worth a try. As you can see in the histogram of Figure 13.1, this is certainly the case for the income data: Some of the values are in the $10's, some in the $100's, many in the $10,000's, and some in the $100,000's.

To carry out the log transform, simply take the logarithm of each value. It doesn't matter whether you use log base 10 or the natural logarithm, just as long as you know which one you used and are consistent with your choice. We will use log base 10 for simplicity. Table 13.1 shows the first four rows of the income data beside the log-transformed (log base 10) incomes.

**Tech**

| Total _Personal _Income | Log$_{10}$(Total _Personal _Income) |
|---|---|
| 6200 | 3.792392 |
| 65000 | 4.812913 |
| 8000 | 3.903090 |
| 13000 | 4.113943 |

As you can see from Figure 13.4, the transformed data come closer than the original data to following a Normal distribution. Compare these figures to the original distributions in Figures 13.1 and 13.3. Admittedly, it can be difficult to tell from the histogram whether the distribution of the log-transformed data are Normal, but this is exactly why the QQ plot is useful.

▲ **FIGURE 13.4** **(a)** The histogram of the transformed data is more symmetric than that of the untransformed data. Note that, with such a small sample size, it is difficult to tell from the histogram whether the data are truly Normal. **(b)** The QQ plot for the log-transformed data is much closer to the straight line than the QQ plot for the untransformed data shown in Figure 13.2. This plot shows us that the transformed data are fairly close to being Normally distributed.

## EXAMPLE 1 Examining Normality

The histogram in Figure 13.5 shows the distribution of sizes (in square feet) of houses for sale in December 2010, in Beverly Hills, California. Beverly Hills is well known for extravagant lifestyles, and this is quite apparent in the size of some of the homes represented. To put these data in perspective, the average detached home (in other words, not an apartment, townhouse, or condominium) in the United States contains about 2300 square feet. The largest house in this data set measures 28,000 square feet.

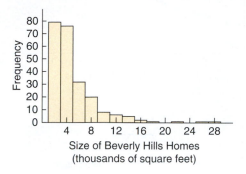

◄ **FIGURE 13.5** Distribution of the sizes (in thousands of square feet) of 232 homes for sale in Beverly Hills in December 2010.

**QUESTION**  Describe the distribution of house sizes. Why might a log transform make the distribution of house sizes more like a Normal distribution? Does it? (Refer to Figure 13.6.)

**SOLUTION**  The distribution is right-skewed, with the median home size somewhere between 2500 and 5000 square feet. (In fact, the median is 3573 square feet.) The first quartile is around 2500 square feet, and the third is just about 5000 square feet (reading off the histogram), so the interquartile range is about 2500 square feet. Figure 13.6a shows the QQ plot of the data. The very clear curve in the QQ plot tells us we have a skewed, non-Normal distribution.

Because the house sizes cover several orders of magnitude (100's, 1000's, and 10,000's), a log transform might make the data more Normally distributed. Figure 13.6b shows the QQ plot of the log-transformed data. The points in this QQ plot stay closer to the line here than in Figure 13.6a, which tells us that the distribution of the transformed data is closer to a Normal distribution. Figure 13.6c shows the histogram

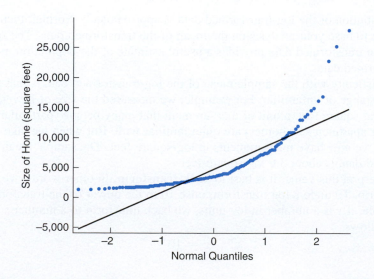

◄ **FIGURE 13.6a** A QQ plot of the original data (sizes of houses) shows that the data are highly skewed and very non-Normal.

▶ **FIGURE 13.6b** QQ plot for the log transform of the sizes of the houses: $\log_{10}$(size). The data come close to following the straight line, which suggests that they are approximately Normal but not perfectly Normal.

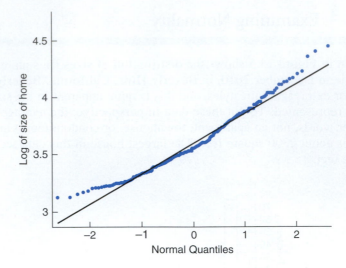

▶ **FIGURE 13.6c** A histogram of the transformed data shows that the distribution is now only slightly right-skewed.

of the transformed data with a Normal distribution superimposed. The transformed data distribution is not perfectly symmetric, but it's closer to a Normal distribution than is the distribution of the original data.

**TRY THIS!** Exercise 13.5

## Analyzing Log-Transformed Data

If the distribution of the log-transformed data is approximately Normal, then it makes good sense to base your analysis on the mean of the transformed data. The sample mean of the transformed data provides a useful estimate of the population mean of the log-transformed data.

One difficulty with the sample mean of the log-transformed data is that the units might be strange or unfamiliar. For example, we measured the size of Beverly Hills homes using square feet, a unit of measurement that many people (particularly those hunting for apartments or homes) are quite familiar with. But when we take the log transform, we now have measurements in log-square-feet. The sample mean of the transformed data is about 3.6 log-square-feet.

To interpret this value, it is best to **back transform**. In other words, we "undo" the transform. To undo a log transform, raise 10 to the power of the transformed value. For example, if $y$ is a number in log units, we back transform to a number $x$ in regular units as follows:

$$x = 10^y$$

For example, to convert the log-transformed sample mean of 3.6 log-square-feet back to square feet, we find

$$10^{3.6} = 3981 \text{ square feet}$$

To back transform, you will have to use a calculator or the calculator function of your statistical software. Figure 13.7 shows the output of a TI-83/84.

The back-transformed sample mean is called the **geometric mean**. The geometric mean is another measure of center that, like the median, is useful for summarizing the center of skewed distributions. To compute the geometric mean for a sample of data, first convert all of the values by taking the log (base 10). Next, find the sample mean of these log-transformed values. Finally, back transform the sample mean of the log-transformed values.

▲ **FIGURE 13.7** TI-83/84 Output. The button used is $10^X$, which is the second function of the Log button.

 **KEY POINT**  The geometric mean is a measure of center that, like the median, is sometimes useful for skewed distributions. The geometric mean is particularly useful when the log transform of the data results in a Normal (or nearly Normal) distribution.

To find a confidence interval for the geometric mean, first convert the original data by taking the log transform (base 10). Then find a confidence interval for the mean, as you have learned to do. Finally, back transform the left and right endpoints of the confidence interval. The next example illustrates this process.

> ! **Caution**
>
> **Round After, Not Before**
> Do not round off before performing back transformations. With no rounding, the sample mean of the transformed home-size data is 3.5970366. Using all of the digits gives a more precise value of $10^{3.5970366} = 3954$ square feet.

## EXAMPLE **2** Beverly Hills Is the Place to Be

Figure 13.6a on page 579 illustrates that the sample mean is not a good measure of the typical home size in Beverly Hills. The right-skew to the distribution "pulls" the mean up, so the mean home seems much larger than what we might think of as "typical." (Only about 1/3 of the homes are bigger than the mean size of 4858 square feet.) From Figures 13.6b and 13.6c, we know that the distribution of the log transform of these data is approximately Normal. This means that we can use the geometric mean of our sample to estimate the geometric mean of the population.

We transformed the Beverly Hills home sizes by taking the log base 10 of every value. We then used StatCrunch to calculate a 95% confidence interval for the mean of the log-transformed data. On the next page, the first few rows of the data set, including the transformed data, are shown in Figure 13.8a. The results are shown in Figure 13.8b.

**QUESTION**  Use the StatCrunch output to find a 95% confidence interval for the population geometric mean. Report the geometric mean. Interpret the confidence interval.

**SOLUTION**  To find the 95% confidence interval for the population geometric mean, we simply back transform the lower and upper limits of the 95% confidence interval for the mean of the transformed data.

$$\text{Lower limit} = 10^{3.5630028} = 3655.971$$

$$\text{Upper limit} = 10^{3.6310706} = 4276.324$$

After rounding these results to the nearest square foot, we find that a 95% confidence interval for the geometric mean of all for-sale home sizes in Beverly Hills is 3656 square feet to 4276 square feet.

To find the geometric mean, we back transform the sample mean of the transformed data:

$$\text{Geometric mean} = 10^{3.5970366} = 3953.999, \text{ or about 3954 square feet}$$

▶ **FIGURE 13.8a** The first 16 rows of data concerning homes for sale in Beverly Hills. The log-transformed data are in the fourth column, headed log10(sqft). For example, log(1500) = 3.1760912.

**beveryhills.txt**

| StatCrunch | Data | Stat | Graphics | Help |
| --- | --- | --- | --- | --- |

| Row | city | type | sqft | log10(sqft) | bed | bath |
| --- | --- | --- | --- | --- | --- | --- |
| 1 | Beverly Hills | Condo/Twh | 1500 | 3.1760912 | 2 | 2.5 |
| 2 | Beverly Hills | Condo/Twh | 1617 | 3.20871 | 2 | 2.5 |
| 3 | Beverly Hills | Condo/Twh | 1910 | 3.2810333 | 2 | 2.5 |
| 4 | Beverly Hills | Condo/Twh | 1961 | 3.2924776 | 2 | 2.5 |
| 5 | Beverly Hills | Condo/Twh | 2512 | 3.4000196 | 2 | 2.5 |
| 6 | Beverly Hills | Condo/Twh | 2526 | 3.4024334 | 2 | 2.5 |
| 7 | Beverly Hills | Condo/Twh | 2662 | 3.425208 | 2 | 2.5 |
| 8 | Beverly Hills | Condo/Twh | 2759 | 3.4407518 | 2 | 2.5 |
| 9 | Beverly Hills | Condo/Twh | 1856 | 3.268578 | 2 | 3 |
| 10 | Beverly Hills | Condo/Twh | 2210 | 3.3443923 | 2 | 3 |
| 11 | Beverly Hills | Condo/Twh | 2283 | 3.358506 | 2 | 3 |
| 12 | Beverly Hills | Condo/Twh | 1857 | 3.268812 | 2 | 3 |
| 13 | Beverly Hills | Condo/Twh | 2265 | 3.3550682 | 3 | 2.5 |
| 14 | Beverly Hills | Condo/Twh | 2265 | 3.3550682 | 3 | 2.5 |
| 15 | Beverly Hills | Condo/Twh | 2285 | 3.3588862 | 3 | 2.5 |
| 16 | Beverly Hills | Condo/Twh | 2659 | 3.4247184 | 3 | 2.5 |

▶ **FIGURE 13.8b** StatCrunch output for a 95% confidence interval for the population mean of the log-transformed house sizes. The units are log-square-feet.

One sample T statistics with data

**Options**

**95% confidence interval results:**

$\mu$ : mean of Variable

| Variable | Sample Mean | Std. Err. | DF | L. Limit | U. Limit |
| --- | --- | --- | --- | --- | --- |
| log10(sqft) | 3.5970366 | 0.017273584 | 231 | 3.5630028 | 3.6310706 |

We interpret the confidence interval to mean that we are 95% confident that the true population geometric mean lies between 3656 square feet and 4276 square feet.

**TRY THIS!** Exercise 13.11

**SNAP**SHOT    **THE GEOMETRIC MEAN**

**WHAT IS IT?** ▶ A numerical summary of a distribution.

**WHAT DOES IT DO?** ▶ Provides a measure of center for some right-skewed distributions.

**HOW DOES IT DO IT?** ▶ If the log transform of the data follows a Normal distribution, then inference based on the mean log-transformed data can be successfully back-transformed and interpreted in terms of the geometric mean.

**HOW IS IT USED?** ▶ To estimate a measure of center for skewed data when (a) the data have all positive values and (b) the distribution of the log-transformed data is approximately Normal.

In general, how does the 95% confidence interval for the geometric mean compare to the 95% confidence interval for the mean (of the untransformed distribution)? Sometimes, a confidence interval for the geometric mean will be narrower—and thus will provide a more precise estimate—than an interval for the population mean. A 95% confidence interval for the mean Beverly Hills house size is given in Figure 13.9. Note that the width of the confidence interval for the geometric mean found in Example 2 is approximately 620 square feet: 4276 − 3656 = 620. The confidence interval for the mean shown in Figure 13.9 is wider and therefore less precise: 5353 − 4364 = 989 square feet.

95% confidence interval results:

μ : mean of Variable

| Variable | Sample Mean | Std. Err. | DF | L. Limit | U. Limit |
|---|---|---|---|---|---|
| sqft | 4858.491 | 250.77806 | 231 | 4364.3867 | 5352.596 |

◀ **FIGURE 13.9** StatCrunch output showing an approximate 95% confidence interval for the population mean house size. Because the distribution of data is very right-skewed, the confidence interval could differ substantially from 95%, even though the sample size is large.

Because the distribution of house sizes is not Normal, the confidence level for the mean is also probably not really 95%. For both intervals, the 95% confidence level is only an approximation, but because the nontransformed data are right-skewed and the transformed data are roughly symmetric, the approximation is much better for the geometric mean confidence interval.

 **KEY POINT** If the distribution of the log-transformed data is approximately Normal, then confidence intervals for the geometric mean can be more precise (smaller margin of error) than confidence intervals for the mean. The confidence level can also be more accurate for confidence intervals of the geometric mean.

## Median or Geometric Mean?

The median is another useful measure of the center of a skewed distribution. If both the geometric mean and the median are measures of the center for skewed distributions, when should you use the median and when the geometric mean? There are no hard-and-fast rules here. The geometric mean is useful only when the log transform results in a Normal (or nearly Normal) distribution. Also, finding confidence intervals for the median can be complicated and may require very particular conditions; these are left for a more advanced statistics book. On the other hand, if the distribution of the log-transformed data is *not* Normal, then the confidence intervals for the geometric mean will not be reliable, and you should rely on the median. Advanced statistics courses (typically at the graduate level) discuss these approaches, so if you find yourself in this situation and do not want to take additional statistics courses, it is best to consult a statistician.

Figure 13.10 shows all three measures of center on the same histogram of Beverly Hills home sizes, marked with vertical lines. The smallest measure is the median: 3573 square feet. Next is the geometric mean, which, at 3954 square feet, is only slightly larger than the median. The mean is to the right of both and substantially larger than both: 4858 square feet. (Note that, in general, the geometric mean will never be bigger than the arithmetic mean.)

**Looking Back**

**Sample Median**
In Chapter 3 you learned that the sample median is better than the sample mean as a measure of what is "typical" for a data set if the distribution of the data is highly skewed. The sample median is the value that has (about) half of the sorted data above and half below.

◀ **FIGURE 13.10** This histogram shows the distribution of the sizes of a sample of 232 homes in Beverly Hills, California.

# The Sign Test for Paired Data

Recall that sometimes, when we are comparing two groups, the observations in the groups are paired. Paired groups occur, for example, when we are comparing subjects' responses before and after getting a drug and thus have measured each subject twice. Also, in studies of twins where one twin is in each group, the groups are dependent. In Chapter 9, we discussed using the paired *t*-test to compare the means of dependent groups.

The **sign test** is a nonparametric test that can be used in place of the paired *t*-test. The sign test is particularly useful when the Normal condition of the paired *t*-test is not satisfied. Unlike the *t*-test, the sign test is based on the median of the population, not the mean. The sign test is similar to the paired *t*-test in that both are based on examining the differences between pairs of observations. Therefore, the first step in both tests is to subtract one observation from the other in each pair.

The sign test is useful for comparing two groups when these conditions are met:

1. The two groups are paired (sometimes called matched) so that every observation in one group is coupled with one particular observation in the other.

2. The pairs are independent of other pairs.

The sign test is a nonparametric test because it makes no assumptions about the distributions of the populations. Because it works whether or not the data come from a Normal distribution, the sign test is helpful if you do not know whether your data are Normally distributed.

Interestingly, the sign test doesn't rely on the values of the differences in the pairs. As the name suggests, it relies only on the *signs* (negative or positive).

We'll begin with a brief overview and then show how to use the sign test with real data.

## Overview of the Sign Test

The paired *t*-test and the sign test share the same first step: Find the differences within each pair. The differences between the paired observations in group 1 and group 2 are found by subtracting one value from the other in each pair. Figure 13.11 shows the first few rows of data from a study examining whether the sense of smell is different when we are sitting than when we're lying down (Lundstrom et al. 2006). Each row is a person, and the person's sense of smell was measured twice: once sitting up and once lying down. The *difference* variable records the value when sitting minus the value when lying down. For the first subject, the difference score is $13.5 - 13.25 = 0.25$. (Recall from Example 9 in Chapter 9 that bigger numbers indicate a better ability to detect odors.)

### Stating Hypotheses

The null and alternative hypotheses are based on the median value of these differences.

$$H_0\text{: The median change} = 0.$$
$$H_a\text{: The median change} \neq 0.$$

Because the median is the value with half of the observations above it and half below it, if we select an observation at random, then there is a 50% chance that it will be bigger than the median and a 50% chance that it will be smaller. Therefore, the null hypothesis is saying that in the population, half of the differences are positive and half are negative.

We have shown a two-tailed alternative hypothesis, but the test can be modified easily enough to perform it with a one-tailed alternative.

| Row | Sitting Up | Lying Down | Sex | Difference |
|-----|------------|------------|-----|------------|
| 1 | 13.5 | 13.25 | woman | 0.25 |
| 2 | 13.5 | 13 | woman | 0.5 |
| 3 | 12.75 | 11.5 | woman | 1.25 |
| 4 | 12.5 | 12.5 | man | 0 |
| 5 | 9.25 | 10 | man | -0.75 |
| 6 | 12.5 | 13 | woman | -0.5 |
| 7 | 14 | 11 | woman | 3 |
| 8 | 5 | 2.75 | man | 2.25 |
| 9 | 12.75 | 11.5 | woman | 1.25 |
| 10 | 13.75 | 13.5 | woman | 0.25 |
| 11 | 8.75 | 8.75 | man | 0 |
| 12 | 12.75 | 13.75 | man | -1 |
| 13 | 9 | 10.75 | man | -1.75 |
| 14 | 10 | 3.75 | woman | 6.25 |
| 15 | 7.5 | 9 | man | -1.5 |
| 16 | 12.25 | 9.75 | woman | 2.5 |

◄ **FIGURE 13.11** Measures of the sense of smell for the first 16 people in the data table. The *difference* variable measures the difference "Sitting Up minus Lying Down."

## Calculating the Test Statistic

The test statistic is quite simple: Count the number of positive signs in the change scores. Ignore any change scores of 0. We will call this test statistic $S$ (for "sign"). For example, Figure 13.11 shows nine positive signs, five negative signs, and two values of 0 for the variable *difference*. So we find that $S = 9$. You also need to know the number of nonzero observations; we call this number $n$. Here $n = 14$.

If the null hypothesis is correct and the median is 0, then half of the values should be above 0 and half below. In other words, we should see about half of the pairs, $n/2$, with positive signs. If we see too many positive signs, or too few, then we suspect that the null hypothesis is not true.

## Finding the p-Value

After finding $S$, we need to know whether the observed value of $S$ is unusually large or unusually small. The p-value is then the probability of getting a value of $S$ as far or farther from $n/2$ as the observed value, assuming the true median is 0.

We can find an approximation of this probability through a simulation. If the null hypothesis is correct that the true median is 0, then about half of the observations will be bigger than 0 and the probability that a randomly selected observation is positive will be is 0.50. Thus we can simulate probabilities by flipping a coin. If the coin lands "Heads," we record a positive sign. We flip the coin once for each of the $n$ observations and count the number of heads. If we do this many times, say 1000, then we can get a good sense of the probability of getting a value as extreme or more extreme than our actual outcome when the null hypothesis is true.

As you learned in Section 6.3, we don't need to do a simulation. The probability distribution for the number of heads in $n$ flips of a coin is modeled by the binomial model. We can use the binomial model to find the exact p-value.

## Applying the Sign Test

Let's examine some examples to illustrate how to use the sign test.

 **Details**

**A Positive Spin**
There is nothing special about counting positive signs. You could also base the test on the number of negative signs.

**Looking Back**

**The Binomial Model**
The binomial model (Section 6.3) provides probabilities for random experiments in which you are counting the number of successes and these four characteristics hold:
1. There is a fixed number of trials, $n$ (the number of pairs for which the difference is not 0).
2. The only two possible outcomes are success and failure (positive sign and negative sign).
3. The probability of success $p$ at each trial is the same for all trials ($p = 0.50$ for the sign test).
4. The trials are independent (which means that the pairs are independent of each other).

Tech  **EXAMPLE 3 Testing Sense of Smell**

Does our sense of smell vary on the basis of whether we are sitting or lying down? Thirty-six subjects had their sense of smell measured in both positions (which position was measured first was determined randomly). The first sixteen rows of data are shown in Figure 13.11. Three of the subjects provided differences of 0, so these observations will not be used. The StatCrunch output for a sign test is provided in Figure 13.12.

▶ **FIGURE 13.12** StatCrunch output for a sign test of the data on sense of smell.

| StatCrunch | Edit | Data | Stat | Graphics | Help | | |
|---|---|---|---|---|---|---|---|
| Row | Sitting Up | Lying Down | Sex | Difference | var5 | var6 | var7 |
| 1 | 13.5 | 13.25 | woman | 0.25 | | | |
| 2 | 13.5 | 13 | woman | 0.5 | | | |
| 3 | 12.75 | 11.5 | woman | 1.25 | | | |
| 4 | 12.5 | 12.5 | man | 0 | | | |
| 5 | 9.25 | 10 | man | −0.75 | | | |
| 6 | 12.5 | 13 | woman | −0.5 | | | |
| 7 | 14 | | | | | | |
| 8 | 5 | | | | | | |
| 9 | 12.75 | | | | | | |
| 10 | 13.75 | | | | | | |
| 11 | 8.75 | | | | | | |
| 12 | 12.75 | | | | | | |
| 13 | 9 | | | | | | |
| 14 | 10 | | | | | | |
| 15 | 7.5 | | | | | | |
| 16 | 12.25 | | | | | | |

**Sign Test**

Options

**Hypothesis test results:**
Parameter : median of Variable
$H_0$ : Parameter = 0
$H_A$ : Parameter ≠ 0

| Variable | n | n for test | Sample Median | Below | Equal | Above | P-value |
|---|---|---|---|---|---|---|---|
| Difference | 36 | 33 | 1 | 10 | 3 | 23 | 0.0351 |

**QUESTION** Is one's sense of smell different when one is lying down than when one is sitting? Use the provided StatCrunch output to perform a sign test. Follow the four-step procedure for hypothesis tests.

**SOLUTION** Most of the heavy lifting was done by the software. However, it is your responsibility to make sure the test was appropriate for these data and to interpret the output in a meaningful context.

*Step 1: Hypothesis*

$H_0$: The median difference in smelling ability is 0.
$H_a$: The median difference in smelling ability is not 0.

*Step 2: Prepare*
The data clearly are paired, because each subject was measured twice. We assume that each pair is independent of every other.

*Step 3: Compute to compare*
The value of the test statistic is $S = 23$. This is found in StatCrunch under the Above column. "Above" means the number of observations that were above, or greater than, the null hypothesis value of 0. The sample size $n$ is 33. (There were 36 observations, but three had differences of 0 and so were discarded.)

If the null hypothesis is true, then about half, or 16 to 17, of the observations should be positive. We instead saw 23. Is this unusual?

The reported p-value is 0.0351. This is a two-tailed p-value, as we can see in the output in Figure 13.12, where the alternative hypothesis is given.

Figure 13.13 shows the sampling distribution for $S$. This is the binomial distribution with $n = 33$ and $p = 0.5$. The probability that $S$ will be "as extreme or more extreme" than 23 means it will be as far or farther above the mean value of $33/2 = 16.5$ than 23 is, or as far or farther below 16.5 than 10 is. These values are indicated in the figure.

◄ **FIGURE 13.13** The sampling distribution of $S$. The probability of getting 23 or more, or 10 or fewer, positive signs in 33 pairs is indicated in red. See socr.stat.ucla.edu.

*Step 4: Interpret*

Using a significance level of 5%, we reject the null hypothesis because the p-value is less than 0.05. We conclude that the sense of smell is in fact different when one is lying down than when one is sitting up.

**TRY THIS!** Exercise 13.15

---

The sign test is very useful and can even be used for one-sample tests of the median (see Exercises 13.41 and 13.42.) The sign test suffers from low power compared to the paired *t*-test. Recall that power is the probability that you will correctly reject the null hypothesis. If you know that the population distribution is Normal, or if your sample size is large, then the paired *t*-test is the best choice because you'll have a higher probability of correctly rejecting the null hypothesis. However, if you are not sure whether the sample size is large enough, or you know that it is not large enough, then the sign test is a more reliable test than the paired *t*-test.

**Tech**

**SNAP**SHOT   THE SIGN TEST

**WHAT IS IT?** ► A hypothesis test of whether the median difference among paired data is 0.

**WHAT DOES IT DO?** ► The sign test is based on the number of pairs with positive differences.

**HOW DOES IT DO IT?** ► If the null hypothesis is true, then about half of the pairs should be positive. If many more than half, or many fewer, are positive, it suggests that the null hypothesis is wrong.

**HOW IS IT USED?** ► It can be used whenever the paired *t*-test is used, but in particular when the sample size is too small to apply the paired *t*-test or when the distribution is known to be non-Normal.

# Mann-Whitney Test for Two Independent Groups

The Mann-Whitney test can be used wherever you can use a *t*-test for two independent groups (see Section 9.4). It can also be used in many situations in which you can't use the *t*-test—for example, when the population distributions are not Normal or when the sample sizes are too small for the Central Limit Theorem to provide a useful approximate result.

In order for us to use the Mann-Whitney, the following conditions must be satisfied.

1. There are two independent groups.

2. The response variable is numerical and continuous.

3. Each group is a random sample from some population.

4. The observations are independent of one another.

5. The population distribution of each group has the same shape.

In practice, the Mann-Whitney test also works for noncontinuous (that is, discrete) response variables, as long as the values can be ordered. Thus, the Mann-Whitney test is often applied to ranks—such as first place, second place, and so on—in competitions.

Even though the Mann-Whitney test requires that the two population distributions have the same shape, we don't care what this shape is. In other words, we don't care whether both groups come from a Normal distribution, a uniform distribution, or *any* particular distribution. We only care that they both come from a distribution with the same shape.

**Details**

**Mann-Whitney Test**
The Mann-Whitney test is also called the Mann-Whitney-Wilcoxon test and the Wilcoxon Rank Sum test.

## Overview of the Mann-Whitney Test

The Mann-Whitney test requires some pre-processing before you begin. Rather than using the actual data values, the Mann-Whitney is based on the **ranks** of the values. This means that the smallest value in the data set gets the rank of 1, the next smallest is ranked 2, and so on.

### Stating Hypotheses

Like the two-sample *t*-test, the Mann-Whitney test is used to compare the centers of two distributions. Unlike the *t*-test, the Mann-Whitney compares the medians of the population distributions, not the means. This is a particularly compelling approach when comparing skewed distributions; in this context, the median is a more natural measure of center than the mean.

$H_0$: The median of population 1 is equal to the median of population 2.
$H_a$: The medians are not equal.

We can also have one-tailed alternatives.

 **KEY POINT** The Mann-Whitney test is based on the ranks of the observations, not on their actual values.

If the medians of the populations are equal, then each group should have roughly the same sorts of ranks in them. If all of the low ranks fall into one group in your sample, and all of the high ranks into another, then the null hypothesis might not be true.

### Finding the Test Statistic

The first step is to compute the ranks for each observation. When doing this, you ignore which group the observations belong to. To illustrate, let's compare the weights of newborn boys and girls, using the very small data set shown in Table 13.2. (In fact, the Mann-Whitney does not work well with such very small data sets, but the small size makes it easier to understand how the test works.) The full data set consists of all babies born at a hospital in Brisbane, Australia, in one 24-hour period (Dunn 1999), and Table 13.2 shows a random sample of six of these babies. Do the weights differ by gender?

The easiest way to rank the observations is to sort them from smallest to largest, ignoring whether the weight belongs to a boy or a girl, and then assign the first the rank of 1, the second the rank of 2, and so on. Table 13.3 shows the result of this ranking of the original data.

The test statistic, represented by the letter $W$, is simply the sum of the ranks of one of the groups. Technically, it doesn't matter which group you choose, because if you know the sum of the ranks of one group, it is possible to determine the sum of the others (because the ranks must always sum to the same value). Most packages assume you used the group that produced the smallest sum, but other packages will give you results for both groups.

For example, when we add up the ranks of the girls, we get $W = 6 + 5 + 1 = 12$.

The intuition behind the Mann-Whitney $W$-statistic is this: If all of the heaviest babies were girls, then they would get the top three ranks: 4, 5, 6. In that case, $W = 4 + 5 + 6 = 15$. At the other extreme, if the lightest babies were all girls, they would get the lowest ranks: 1, 2, 3. Then $W = 1 + 2 + 3 = 6$.

However, if the null hypothesis is true and both groups are really the same, then it will be as though the ranks were randomly assigned to groups. So we would expect each group to have a mix of low and high ranks. This means that if the null hypothesis is true, then $W$ should be somewhere close to the midpoint between 6 and 15: about 10 or 11.

In this case, we observed $W = 12$, which is very slightly larger than what the null hypothesis might lead us to expect. We now ask how likely a value of 12 or larger is, if the ranks were really just distributed by chance.

## Finding the p-Value

The distribution of $W$ does not have a simple formula that allows us to compute probabilities, so we rely heavily on statistical software to compute p-values. For large sample sizes, and particularly when there are values that are tied (in other words, several values are the same and get the same rank), approximate probabilities are calculated by statistical software.

Figure 13.14 shows output from StatCrunch that gives the p-value as 0.70. Because the sample size was not terribly large and there were no ties, an exact p-value was computed. (StatCrunch, like some other statistical packages, automatically decides whether to compute an exact p-value or an approximation.)

| Gender | Weight (grams) |
|--------|---------------|
| Girl | 3866 |
| Boy | 3380 |
| Girl | 3438 |
| Boy | 3300 |
| Girl | 1745 |
| Boy | 3150 |

▲ **TABLE 13.2** Weights of six randomly selected babies all born on the same day at the same hospital.

| Gender | Weight | Rank |
|--------|--------|------|
| Girl | 3866 | 6 |
| Boy | 3380 | 4 |
| Girl | 3438 | 5 |
| Boy | 3300 | 3 |
| Girl | 1745 | 1 |
| Boy | 3150 | 2 |

▲ **TABLE 13.3** Ranks based on baby weights. The smallest is assigned the rank 1.

◀ **FIGURE 13.14** Output from the Mann-Whitney test on StatCrunch. The test statistic is $W = 12$, and the p-value for a two-tailed alternative is 0.7.

Mann–Whitney

Options

**Hypothesis test results:**

m1 = median of weight where gender="girl"

m2 = median of weight where gender="boy"

Parameter : m1 – m2

$H_0$ : Parameter = 0

$H_A$ : Parameter ≠ 0

| Difference | n1 | n2 | Diff. Est. | Test Stat. | P-value | Method |
|------------|----|----|------------|------------|---------|--------|
| m1 – m2 | 3 | 3 | 138 | 12 | 0.7 | Exact |

## Applying the Mann-Whitney Test

Many statisticians prefer to use the Mann-Whitney test, rather than the *t*-test, for many situations. This is because the Mann-Whitney is more robust (it doesn't need to know whether the populations are Normally distributed or whether the sample size is "large enough").

To see how to apply the Mann-Whitney, let's consider the full sample of 26 boys and 18 girls born in one 24-hour period at a hospital in Brisbane, Australia.

### EXAMPLE 4  Weights of Boy and Girl Babies

Figure 13.15 shows StatCrunch output from a Mann-Whitney test to determine whether boy babies tend to have different weights from girl babies. A preliminary check of histograms (not shown) for the boy and girl weights show that both distributions are roughly symmetric, although both have an outlier on the low end.

▶ **FIGURE 13.15** StatCrunch output from a Mann-Whitney test to compare the typical weight of baby boys with that of baby girls. Girls are coded with gender = 1.

**QUESTION** Do boy babies weigh a different amount, typically, than girl babies? Perform a Mann-Whitney test, using the computer output provided. Use a 5% significance level.

**SOLUTION**

*Step 1: Hypothesize*

$H_0$: The median weight of boys equals the median weight of girls.
$H_a$: The median weight of boys does not equal the median weight of girls.

*Step 2: Prepare*
The Mann-Whitney test is appropriate because:

1. The samples are independent (this might not be true if there were boy and girl twin siblings in the sample).

2. The variable measured (weight) is numerical (and continuous).

3. Although the babies are not a random sample, we shall think of the babies born this day as representative of all babies born.

4. We assume that the distributions of boys' and girls' weights have the same shape.

*Step 3: Compute to compare*
We read from the output that the test statistic has the value 365.5, with a p-value (for a two-tailed alternative) of 0.3519.

*Step 4: Interpret*
Because the p-value is larger than 0.05, we cannot reject the null hypothesis and find insufficient evidence that weights differ between baby boys and girls.

**TRY THIS!** Exercise 13.23

## Sample Size and the Mann-Whitney

The Mann-Whitney test has relatively low power. This means that even if the alternative hypothesis is true (the groups really are different), the Mann-Whitney test can be less likely than other tests to reject the null hypothesis. For example, suppose that the three heaviest weights in our small sample of six babies all belonged to the girls. In other words, the girls had the ranks 4, 5, and 6, and the boys had the three lower ranks. Even in this extreme case, the p-value for the two-tailed alternative is 0.10, and we would not reject the null hypothesis. When the sample size is very small (under ten for total number of both groups), be aware that it might not be possible to reject the null hypothesis, even if the null is not true.

For larger sample sizes, the *W*-statistic follows, approximately, a Normal distribution. For this reason, some software automatically produce an approximate p-value using the Normal distribution, rather than an exact p-value, because this makes the computation faster. The StatCrunch display in Figure 13.15 tells us that the method used was "Norm. Approx"; that is, StatCrunch made the decision to use the Normal approximation for this sample size.

## What Can Go Wrong?

Although it is true that you don't have to worry about whether the data follow the Normal distribution, you still need to check the conditions carefully. In Section 13.2 we noted that if you do not have independent groups, then you need to use the sign test. You might also have a few other things to worry about.

*Ties in the Data.* In theory, if the response variable is continuous, as in the case of baby weights, then it is impossible for two babies to weigh exactly the same amount. In practice, however, we can weigh objects only to a certain level of precision. Because we must round these numbers, it is possible to get ties—two or more babies that weigh the same. There were no ties in our data set of 44 babies, but it's not hard to imagine that there would be ties in a data set of, say, 44,000 babies.

When ties occur, it is not possible to give each observation its own rank. Instead, several observations share the same rank. One approach is to give tied observations their average rank. For example, if we saw the data

15.5, 16.0, 16.0, 16.0, 18.2

we would assign them the ranks

1, 3, 3, 3, 5

because the three values of 16.0 were assigned the ranks 2, 3, 4, and the average of these ranks is 3.

Doing this averaging leads to "conservative" p-values: p-values that tend to be too high. Most statistical software packages apply an adjustment that gives a more accurate p-value. Some software packages display a warning that says something like "adjusted for ties." If your software gives you a choice of different p-values, use the one that has been adjusted for ties.

*Different-Shaped Distributions.* The Mann-Whitney test requires that the shapes of the two population distributions be the same because if they are different (such as one being symmetric and the other left-skewed), then the Mann-Whitney test might reject the null hypothesis even when the medians are the same. In practice, any difference between the two distributions in shape, center, or spread between the two distributions can lead to rejecting the null hypothesis. Most statistical software packages treat the Mann-Whitney as a test based on the median values, and this is how we presented it. However, this interpretation is valid only when the shapes and spreads of the distributions are the same.

## SNAPSHOT THE MANN-WHITNEY TEST

**WHAT IS IT?** ▶ A hypothesis test to compare the centers of two groups of numerical variables.

**WHAT DOES IT DO?** ▶ It tests the hypothesis that the medians of two populations are different.

**HOW DOES IT DO IT?** ▶ The original values are pooled and ranked from smallest to largest and then sorted back into their original groups. If the null hypothesis is true, then the sum of the ranks in both groups should be roughly equal. If the sum of one group is much larger than the other, then it suggests that the null hypothesis might not be true.

**HOW IS IT USED?** ▶ It can be used whenever the t-test for two independent samples is used, but can also be used when the Normal condition of the t-test is not met.

## t-Test or Mann-Whitney Test?

How do you decide whether to use the two-sample *t*-test for independent samples or the Mann-Whitney test? To some extent, it is a matter of preference. There are a few things to consider, though.

Because the Mann-Whitney test can be used to compare medians, it is preferred in situations where the distributions are skewed or outliers are present.

If you don't know whether the distributions of the populations are Normal, and are not willing to assume that they are, then the Mann-Whitney is useful. If your sample sizes are large, then the Mann-Whitney and two-sample *t*-tests should lead you to the same conclusion. But when the sample sizes are too small for the Central Limit Theorem to apply (less than 25 in most cases, larger if the distributions are severely skewed), then the Mann-Whitney will produce more reliable results.

Even though one condition of the tests is that the data be continuous, the Mann-Whitney also works when the data are ordinal (can be ordered) but not necessarily continuous. The most common example is when the data are ranks to begin with. This is often the situation when data from contests are examined. For example, a panel of judges might rank 12 wines using the numbers 1 (best) to 12 (worst), and we might want to compare judges to see whether they tend to give the same ranks. The *t*-test would not be good for these data, which are very non-Normal. The Mann-Whitney would be a more suitable test.

Many statisticians feel that for most situations, the Mann-Whitney test can be used instead of the two-sample *t*-test.

SECTION 13.4

# Randomization Tests

**Randomization tests** include a wide variety of different types of tests in different types of situations. What these tests have in common is a general strategy for answering the question "Can this be due to chance?" The strategy is to use a computer (or some other simulation-based procedure) to shuffle observations together in order to simulate a world in which chance is, in fact, the only reason for differences between groups. Fisher's Exact Test, which was introduced in Chapter 10, is a randomization test used to test for associations between two categorical variables. In this section, we examine randomization tests for comparing two groups of independent, numerical observations.

Randomization tests can be used whenever the two-sample *t*-test or Mann-Whitney test can be used. The advantages of randomization tests over *t*-tests and the Mann-Whitney are that they are more versatile and that they can be used to compare statistics other than the mean and median. Randomization tests have three requirements.

1. The two groups are independent of each other, and the observations within each group are independent of other observations within the group.

2. Either the data are a random sample from some population, or the observations were assigned to groups randomly, as in a randomized controlled experiment.

3. The variability of both groups is approximately the same.

**Looking Back**

**Fisher's Exact Test**
Fisher's Exact Test is an example of a randomization test for categorical variables. In this chapter we focus on numerical variables.

**Details**

**Permutations**
Another term for the randomization tests we are presenting is *permutation tests*.

## Overview of Randomization Tests

One way to understand randomization tests is to think of them as comparing two worlds, or two alternative realities, if you prefer. One world is the real world. This is the world that produces the data we see. The other world we'll call Chance World. We create Chance World by simulation, and we do it so that every outcome depends only on chance.

Consider a well-known study on the effects of cloud seeding (Simpson et al. 1975). Cloud seeding is the practice of using airplanes to drop silver nitrate into clouds, with the intent of producing rain. In this study, 26 clouds were randomly chosen to be seeded with silver nitrate. Another 26 clouds were selected to receive a "placebo." The pilot flying the plane did not know which was which; he or she simply acted as always, not knowing what was released into the cloud. The amount of rainfall in each area was recorded in acre-feet. The boxplots in Figure 13.16 show the results.

 ◄ **FIGURE 13.16** Rainfall amounts (in acre-feet) produced by 26 clouds seeded with silver nitrate (left) and 26 clouds seeded with a placebo (right).

Figure 13.16 shows the *real-world* results, and we summarize these real-world results in Table 13.4 on the next page. We can see that the median rainfall amounts in the two groups are different. But can this difference be due to chance? Clouds were

chosen at random, and perhaps if a different subset of the clouds had been chosen for the treatment, rainfalls would have turned out differently.

▶ **TABLE 13.4** Summary statistics for rainfall from clouds, comparing seeded and unseeded clouds.

| Group | n | Mean | SD | Median | IQR |
|---|---|---|---|---|---|
| Seeded | 26 | 533.2 | 779.9 | 221.6 | 396.7 |
| Unseeded | 26 | 164.6 | 278.4 | 44.2 | 138.6 |

To create Chance World, imagine a collection of slips of paper. Each slip represents a cloud. On each slip we write a rainfall amount from our actual data. This represents the rainfall from that cloud. Now imagine two stacks. One stack represents the seeded group of clouds, and the other represents the unseeded group. Shuffle all the slips of paper (from both stacks) together very well, and randomly deal them into the two new stacks, so that each stack has 26 slips of paper. For each stack, find the median rainfall, and then calculate the difference between the medians. This is the Chance World difference.

In the real world, the differences between the two groups might have been caused by the seeding. But in the chance world, the only differences are due to chance, caused by our shuffling of the slips of paper. If we repeat this shuffling and dealing, we get another, different outcome. If we do this many times, we obtain an understanding of what the difference in medians looks like in a world in which everything is due to chance.

We now compare our real-world result to the many simulated results from Chance World. If we can't tell the difference between the real world and Chance World—in other words, if the real-world outcome is also a common outcome in Chance World—then we conclude that there is no evidence of a real difference; there is no evidence that cloud seeding results in a change in rainfall.

If, on the other hand, the real-world result is unusual in Chance World, then perhaps the outcome was not due to chance. We can use the simulated Chance World outcomes to compute the probability of getting, in our Chance World, a result as extreme as or more extreme than the real-world result. If the probability is small, less than our chosen significance level, we reject the null hypothesis.

Figure 13.17 illustrates this situation. The distribution shows the Chance World outcomes under 1000 shuffles for the cloud-seeding data. After each shuffle, the difference in median rainfalls is computed, so the histogram shows us many possible differences in Chance World. If the real-world outcome were where the green line indicates, then we would conclude that the real world looks a lot like the chance world, and we would not reject the null hypothesis. If, on the other hand, the real-world outcome fell where the red line indicates, then we would have to admit that our real-world outcome looks nothing at all like the sort of outcome seen in the chance world. We would reject the null hypothesis.

▶ **FIGURE 13.17** Histogram of a Chance World simulation. The green line represents one possible real-world outcome for which we would *not* reject the null hypothesis. The red line represents one possible real-world outcome for which we *would* reject the null hypothesis.

Of course, not all situations are as clear-cut as the red-line situation. Soon we will discuss how to find the p-value for a particular situation so that we can decide whether to reject the null hypothesis.

**KEY POINT**   Chance World is a simulated world in which every outcome is determined solely by chance. In Chance World, the null hypothesis is true. The real world may or may not be the same as Chance World. It is your job to decide whether the data in the real world look as though they could have been generated in Chance World.

The most unusual feature of the randomization test is that you get to choose which statistics you want to use to compare the two groups. You can choose either the means or the medians. Or you can choose the third quartiles. Anything you want, really. The trick is to choose a statistic that will help you answer the question that has been asked.

Because we are comparing the "typical" rainfall, we should probably choose to look at either the difference in mean rainfalls or the difference in median rainfalls. The box-plots shown in Figure 13.16 on page 593 show us that the distributions of rainfalls are skewed, so the median might be a better choice. For this reason, our test statistic will be the median rainfall of the seeded clouds minus the median rainfall of the unseeded clouds.

## Stating Hypotheses

Because we've chosen the difference in medians to compare the groups, our hypotheses look like this:

$H_0$: The median rainfall of the seeded clouds is equal to the median rainfall of the unseeded clouds.

$H_a$: The median rainfall of the seeded clouds is greater than the median rainfall of the unseeded clouds.

If you choose different statistics, you will need to reword the hypotheses accordingly.

## Finding the p-Value

To find the p-value, we need to find the sampling distribution of the median difference when the null hypothesis is true. We obtain this through simulation. We don't have time to write down the rainfalls on slips of paper, shuffle, deal, compute, and repeat this thousands of times, but we can have a computer do this. (This is easier with some software packages than with others. For guidance, see Example E on page 620.)

Each time the computer shuffles the values and then deals them into two groups, it computes the difference in median values. At the end of many repetitions, we can create a histogram of these median differences. This histogram will help us to see whether the real-world difference is unusual. We can use this histogram to find approximate p-values.

## Applying Randomization Tests

Let's carry out a randomization test to see if the cloud seeding increased rainfall. The null hypothesis is that the median rainfall is the same in both seeded and unseeded clouds. The alternative hypothesis is that the rainfall is greater in the seeded clouds.

In the real world, as Table 13.5 shows us, we saw a difference in medians of

$$221.6 - 44.2 = 177.4 \text{ acre-feet}$$

Is this unusually large? Or is this the sort of value that happens just through chance alone?

Table 13.5 on the next page shows a few outcomes from our Chance World simulation. Each row of this table represents shuffling and dealing out the slips of paper (but using the computer to do it). We actually did 1000 simulations, and we show the first four. You can see that large differences in typical rainfall can occur simply by chance.

▶ **TABLE 13.5** The first four outcomes from a chance-world simulation of rainfalls.

| Median Rainfall "Seeded" | Median Rainfall "Unseeded" | Difference |
|---|---|---|
| 57.90 | 243.40 | −185.50 |
| 93.70 | 155.40 | −61.70 |
| 133.40 | 71.15 | 62.25 |
| 101.14 | 118.65 | −17.50 |

Figure 13.18 shows the histogram of all 1000 simulations. This is the approximate sampling distribution of the difference between the medians, under the assumption that the null hypothesis is true (and therefore all differences are due to chance).

▶ **FIGURE 13.18** The distribution of the difference of medians when rainfall amounts are distributed to two groups by chance. The histogram represents 1000 randomizations. These are the sorts of rainfall differences we see in Chance World, when chance is the only mechanism that determines how much rainfall occurs.

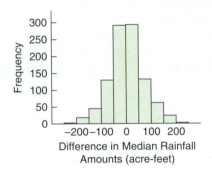

One quick look at the sampling distribution in Figure 13.18 shows us that in Chance World, a difference of 177.4 is relatively large. It's difficult to get exact values from the histogram, but it looks as though a difference like this occurred less often than about 25 out of 1000 times. If so, then a difference as extreme as or more extreme than 177.4 occurs less than 0.025, or 2.5%, of the time. This is a rather rare outcome.

In this case, we don't need to estimate the p-value by reading off the histogram. We can ask the software to actually sum up the number of times the chance-world median difference was 177.4 or bigger. From this we learn that in 13 out of 1000 simulations, the median difference was 177.4 or greater, so the approximate p-value is 0.013.

This means that, with a 5% significance level, we should conclude that this is a rare event and reject the hypothesis that the median rainfalls were the same. The observed real-world median difference is too large for us to believe it was due to chance.

Our sample of clouds was not random, so we could not extend our conclusions to all clouds. But because the clouds were randomly assigned to be seeded or unseeded, we can conclude causality; that is, we can infer that the cloud seeding caused the additional rainfall.

> **!  Caution**
>
> **Random p-Values**
>
> When you using simulations to find p-values, the p-value will be slightly different if you repeat the test. That's why it's important to do a large number of simulations (we did 1000) so that the p-value does not change too much if you do the test again.

> **↻  Looking Back**
>
> **Random Sampling and Randomized Assignment**
>
> Table 12.2 compared random sampling with randomized assignment. Random sampling is necessary in order for us to generalize to a larger population. Random assignment is required in order for us to make cause-and-effect conclusions.

## EXAMPLE 5 Slightly under Weight

The Youth Behavior Risk Study (2007) is a nationwide study of behaviors of people 12–18 years old in the United States. The purpose of the study is to assess their beliefs, attitudes, and practices concerning their health. In this example, we compare the weights of 17-year-old women who believe that they are "slightly under" the proper weight with those who believe they are "about right" in weight. How do beliefs match reality? Do those who believe they are "slightly under" actually weigh less, typically, than those who feel they are "about right"?

Figure 13.19 shows the approximate sampling distribution of the difference in mean weights. This distribution was achieved by shuffling the weights (in kilograms) and dealing them into two piles, recording the difference in means, and repeating 1000 times. Table 13.6 shows summary statistics for these two groups (55 kilograms is about 121 pounds).

Difference in Mean Weights (kilograms)

◀ **FIGURE 13.19** The approximate sampling distribution, under the condition that only chance determines any differences between groups, of difference in mean weights between 17-year-olds who reported that they are "about right" in weight and those who reported being "slightly under" weight.

| Group | n | Mean | Median | Standard Deviation | IQR |
|-------|---|------|--------|--------------------|-----|
| About right | 643 | 58.16 | 56.7 | 8.1 | 9.4 |
| Slightly under | 119 | 53.72 | 52.16 | 9.3 | 8.6 |

◀ **TABLE 13.6** Summary statistics for weights (in kilograms) of 17-year-old girls who consider themselves "about right" in weight or "slightly under" proper weight.

**QUESTION** Using the outputs in Figure 13.19 and Table 13.6, carry out a randomization test to determine whether differences in *mean* weight are due to chance. What is your conclusion? (Note that the p-value may have to be roughly estimated from the histogram.)

**SOLUTION**

*Step 0: Choose a test statistic*
We will use the difference in mean weights: the mean for the women responding "about right" minus the mean for the women responding "slightly under."

*Step 1: Hypothesize*
The hypotheses are the same as before:

$H_0$: The typical weight (measured by the mean) of the "about right" women is the same as that of the "slightly under" women.

$H_a$: The typical weight of the "about right" women is higher than that of the "slightly under" women.

*Step 2: Prepare*
The conditions are satisfied for the same reasons as given before. We should check, though, that the standard deviations and interquartile ranges (IQRs) are approximately the same for the two groups.

*Step 3: Compute to compare*
From Table 13.6, the observed value of the test statistic is 58.16 − 53.72 = 4.44 kilograms.

We compare this to the approximate sampling distribution summarized by the histogram in Figure 13.19. If chance is the only cause for differences, how often do we get values as large as 4.44 or larger? We see that we never had a value of 4.44 or larger. The largest difference in the simulations is 0.4. The p-value is approximately 0.

*Step 4: Interpret*
We reject the null hypothesis and conclude that the typical weight of the "about right" women is higher than that of the "slightly under" women, which suggests that their perceptions match reality, at least on average. The real-world outcome difference of 4.44 does not appear to be consistent with the chance-world outcome.

**TRY THIS!** Exercise 13.29

## SNAPSHOT RANDOMIZATION TESTS

**WHAT ARE THEY?** ▶ A strategy for comparing two independent groups. A chance model is created by randomly assigning the observed values to two groups, computing the test statistic, and repeating this procedure many times.

**WHAT DO THEY DO?** ▶ Randomization tests can be used to compare means, medians, or other statistics between two groups. It does not require any particular shape for the population distributions, so they can be used in a variety of situations.

**HOW DO THEY DO IT?** ▶ By repeating the randomization simulation many times, we can estimate the p-value, because we can see the distribution of the test statistic when the null hypothesis is true and there is no difference between the groups.

**HOW ARE THEY USED?** ▶ Randomization tests can be used in place of *t*-tests for two independent samples, but are much more versatile, because they can also be used with statistics other than means and also in situations where population distributions are not known.

Randomization tests can be particularly useful for randomized experiments, as in the cloud-seeding example. You probably suspect that these clouds were not a random sample of all clouds in the world, and you are correct. Note that condition 2 does not require that the data be a random sample from the population, and freedom from this restriction is useful for many real-life randomized studies. Most medical studies, for example, in which subjects are randomly assigned to treatment or control groups, do not use a random sample of subjects. It is too difficult to get a random sample of patients from a large population to visit a strange doctor to partake in a study. Instead, medical researchers recruit subjects from among those available. As long as the subjects are randomly assigned to treatment groups, researchers can still make inferences about whether the treatment under investigation caused any observed changes.

# CASE STUDY REVISITED

How can we tell whether the electric discharges of the two groups of rays (those provided live fish to eat and those provided freshly killed fish) are the same? If we carried out a two-sample t-test, we would find that the p-value associated with the test is 0.03, and we would therefore conclude that the rays do indeed respond differently to these types of prey. However, to apply the t-test, either we need large sample sizes or we must assume that the population distributions are Normal. The sample sizes in this study (ten rays received live prey and six received freshly killed prey) are small, so if we are not willing to assume that the populations are Normal (a fairly strong assumption), we can apply the randomization test.

Recall that the mean initial pulse rate (ipr) for the rays that were fed the live fish was 100.2 eod/s and that the ipr for those fed freshly killed fish was 78.3 eod/s. (The units measure the strength of the initial electric shock that the ray applies to its prey.) The difference between these sample means is $100.2 - 78.3 = 21.9$ eod/s. Figure 13.20 shows the distribution of the difference of means when the observations are "shuffled" and randomly assigned to either the "live" or the "freshly killed" group, which is consistent with the null hypothesis claim that any observed differences are due to chance. This is analogous to writing all 16 values down on slips of paper, mixing them up, randomly assigning them to two groups, and calculating the mean difference—and then repeating these steps 1000 times. Vertical lines are drawn at 21.9 eod/s and −21.9 eod/s. How often, just by chance, did the difference in means fall more than 21.9 units away from the center? Roughly 27 out of 1000 times. Our approximate p-value is therefore about 0.027, which leads us to reject the null hypothesis and conclude that the electric rays do distinguish between types of prey when delivering electric shocks. Interestingly, the p-value for the randomization test is not too different from the p-value for the t-test, which suggests that the assumption of Normality was not a bad assumption.

▶ FIGURE 13.20 Randomization distribution of the mean difference in electric discharges between eels fed live prey and eels fed freshly killed prey. The actual observed difference is 21.9, and that value and −21.9 are indicated with vertical lines. Under the null hypothesis that both groups of rays have the same distribution of electrical impulse strengths, we see that our actual outcome is fairly rare.

# EXPLORING STATISTICS
## CLASS ACTIVITY

## Balancing on One Foot

| GOALS | MATERIALS |
|---|---|
| To understand how the p-value for the sign test is calculated. | • A timer capable of timing to the nearest second.<br>• A coin for each pair of students. |

**ACTIVITY**

Pair up. Use a coin flip to determine which leg you stand on first. (Heads, start by standing on dominant foot.) With your partner, take turns recording how long each of you can stand on this foot with your eyes closed. If your partner is still balancing after 3 minutes, stop him or her and record this as 180 seconds, with an asterisk (*) to indicate that the true time was even longer. Repeat with the other foot. (Each person will have two measurements: one with the dominant foot and one with the nondominant foot.) Be sure to record which time belongs to the dominant foot and which to the nondominant foot.

Use the sign test to determine whether the students in the class do a better job (have longer times) using the dominant foot. If a student stands indeterminately long on one foot but not on the other, his or her data can still be included. If a student stands indeterminately long on *both* feet, we recommend that you exclude the observation.

**BEFORE THE ACTIVITY**

1. How should you choose which foot is used first? Does it matter?

2. When measuring the times, what do you think the shape of a histogram of the data of standing times for the dominant foot will be and why?

3. Are the time for a student's dominant foot and the time for that student's nondominant foot dependent or independent? Why?

**AFTER THE ACTIVITY**

To answer the question of whether students can balance longer on the dominant foot, gather the data for all of the students in the following table.

| Number of Students with Dominant Foot Time Greater than Nondominant | Number of Students with Dominant Foot Time Less than Nondominant | Number of Students with Both Times the Same |
|---|---|---|
|  |  |  |

Carry out a hypothesis test to test whether students can typically stand longer on their dominant foot than on their nondominant foot. Is a paired *t*-test appropriate? If yes, carry this out and then compare to the results of a sign test. If not, explain why not, and explain why a sign test is appropriate. Carry out the sign test.

## CHAPTER REVIEW

## Key Terms

Nonparametric inference, *575*
QQ plot, *576*
Log transform, *578*

Back transform, *580*
Geometric mean, *581*

Sign test, *584*
Mann-Whitney test, *588*

Rank, *588*
Randomization tests, *593*

## Learning Objectives

After reading this chapter and doing the assigned homework problems, you should

• Understand why and when a log transform is appropriate for analysis.

• Know how to apply the sign test to compare populations on the basis of dependent samples of data, and understand how the sign test compares to the paired *t*-test.

• Be able to apply the Mann-Whitney test to compare two populations on the basis of two independent groups of data, and understand how the Mann-Whitney test compares to the two-sample *t*-test.

• Know when and how to perform a randomization test to compare medians or means of two populations, and understand how this procedure compares to the two-sample *t*-test.

## Summary

The tests you have seen in this chapter—including tests on transformed data, the sign test, the Mann-Whitney test, and randomization tests—are often used when assumptions about the distribution of the population are not met or when you are unwilling to make assumptions about the shape of the sampling distribution. For example, the distributions of your sample may be strongly skewed (with a small sample size) so that *t*-tests are not appropriate, or the data may consist of ranks.

If all of the values of the variable are greater than 0, if the distribution of your data is right-skewed, and if a log transformation results in a distribution that is approximately Normal, then you can base your analysis on the geometric mean. Simply find a confidence interval for the mean of the log-transformed data, and then back transform.

When comparing two groups, if you're not willing to make assumptions about the shape of the distributions, then you can use the sign test (for paired data), the Mann-Whitney test (for two independent groups), or a randomization test.

In many situations, statisticians prefer the Mann-Whitney over the two-sample *t*-test. A randomization test can also be used, particularly if you want to use a statistic other than the mean or median to compare the groups.

| Parametric Test | Nonparametric Equivalent |
|---|---|
| Paired *t*-test | Sign test |
| Two-sample *t*-test | Mann-Whitney test, randomization test |

## Sources

Dunn, P. 1999. Time of birth, sex and birth weight of 44 babies. *Journal of Statistics Education* 7(3), http://www.amstat.org/publications/jse/jse_2001/jse_data_archive.html (viewed April 25, 2010).

Lowe, C. G., R. N. Bray, and D. R. Nelson. 1994. Feeding and associated electrical behavior of the Pacific electric ray *Torpedo californica* in the field. *Marine Biology* 120, 161–169.

Lundstrom, J., J. Boyle, and M. Jones-Gotman. 2006. Sit up and smell the roses better: Olfactory sensitivity to phenyl ethyl alcohol is dependent on body position. *Chemical Senses* 31, 249–252.

Simpson, Alsen, and Edon. 1975. A Bayesian analysis of a multiplicative treatment effect in weather modification. *Technometrics* 17, 161–166. Used in J. M. Chambers, W. S. Cleveland, B. Keliner, and P. A. Tukey. 1983. *Graphical methods for data analysis*, lib.stat.cmu.edu/DASL/Datafiles/Clouds.html (viewed April 25, 2010).

Youth Risk Behavior Study, 2007, Department of Health and Human Services, Centers for Disease Control and Prevention, www.cdc.gov/HealthyYouth/yrbs/index.htm (viewed April 26, 2010).

## SECTION EXERCISES

### SECTION 13.1

**13.1** What is the fundamental condition required for inference from a sample to a population?

**13.2** In what situations are nonparametric statistics useful?

**13.3** In addition to random samples, what other conditions are required for using the two-sample *t*-test?

**13.4** What summary statistics are best used to report the "typical" value of a data set when the distribution is strongly skewed?

TRY **13.5 QQ Plot Matching (Example 1)** Refer to the following two histograms and QQ plots of the same data.

**(B)**

Histogram 2

**(B)**

Histogram 2

**(C)**

**(C)**

**(D)**

**(D)**

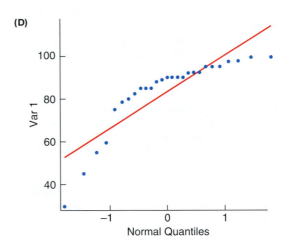

a. Match each of the histograms with the corresponding QQ plot.

Histogram 1 goes with QQ plot ____.

Histogram 2 goes with QQ plot ____.

b. Describe the shape of the histograms.

c. For which sample might a log transform be useful? (There are no zeros or negative values in either data set.)

**13.6 QQ Plot Matching** Refer to the following two histograms and QQ plots of the same data.

Histogram 1

**(A)**

a. Match each of the histograms with the corresponding QQ plot.

Histogram 1 goes with QQ plot ____.

Histogram 2 goes with QQ plot ____.

b. Describe the shape of the histograms.

c. Is a log transform likely to be useful for either sample?

**13.7 Log Transforms**

a. Find the log (base 10) of each number:

$$10, 20, 100, 555, 1000$$

b. The following numbers are in log units. Do the back transformation by finding the antilog (base 10) of each number:

$$1, 1.5, 2, 3, 3.2$$

## 13.8 Log Transforms

a. Find the log (base 10) of each number:

100, 458, 1200, 3825

b. The following numbers are in log units. Do the back transformation by finding the antilog (base 10) of each number:

2, 3, 4.5, 7.2

**13.9 Geometric Mean** Find the geometric mean for the numbers 10, 280, 600, and 15,000 by following these steps:

a. Find the log (base10) of each number: 10, 280, 600, 15,000

b. Average those four logs.

c. Find the antilog of the average by raising 10 to the number obtained in part b. The result is the geometric mean. You may round it to the nearest whole number.

d. Find the mean and median of the original four numbers and then arrange the geometric mean, the mean, and the median from smallest on the left to largest on the right.

**13.10 Geometric Mean** Find the geometric mean for the numbers 10, 200, 1200, and 26,000 by following these steps:

a. Find the log (base 10) of each number: 10, 200, 1200, and 26,000

b. Average those four logs.

c. Find the antilog by raising 10 to the power that you got in part b. The result is the geometric mean. You may round it to the nearest whole number.

d. Find the mean and median of the four original numbers. Then list the mean, median, and geometric mean from smallest on the left to largest on the right.

TRY **13.11 Circuit City (Example 2)** A statistics student who worked as a cashier at the (former) electronics superstore Circuit City wanted to estimate the amount spent by customers using either credit or debit cards. She took a random sample of customers and recorded the amount they spent. Figure A shows a histogram of the data.

**(A)**

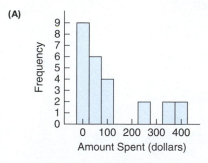

a. Figure B shows a 95% confidence interval for the mean amount spent. Interpret this confidence interval.

**(B)**

```
One-Sample T: Card


Variable   N   Mean    StDev   SE Mean        95% CI
Card       25  111.5   130.0     26.0    (57.8, 165.2)
```

b. A log transform of the data shows a more Normally distributed data set. Figure C shows a 95% confidence interval for the log (base 10) of the amounts spent). Convert the boundaries back into dollars by

raising 10 to the power given as the boundary. Report and interpret the confidence interval for the geometric mean.

**(C)**

```
One-Sample T: LogCard


Variable   N    Mean    StDev   SE Mean        95% CI
LogCard    25   1.762   0.509    0.102    (1.552, 1.972)
```

c. Which interval is narrower?

d. Which interval would you report if you wanted to tell a Circuit City manager about the typical amount of money spent via credit or debit cards? Explain.

**13.12 Morning Routine** A statistics student conducted a survey to determine how much time students at her school spent getting ready to leave the house after they got up in the morning. Figure A shows a histogram of the times for men. Assume that we have a random sample of 20 college men.

**(A)**

a. Figure B shows a 95% confidence interval for the mean time. Interpret the confidence interval, and explain why it might not be accurate.

**(B)**

```
One-Sample T: TimeMen


Variable   N    Mean    StDev   SE Mean        95% CI
Men        20   37.25   19.77     4.42    (28.00, 46.50)
```

b. After a log transform of the times is taken, a histogram of the log of the data suggests that the distribution of the transformed data is Normal. Figure C shows a 95% confidence interval for the log (base 10) of the times. Convert the boundaries back into minutes by raising 10 to the powers given as the boundary values. Interpret the confidence interval for the geometric mean.

**(C)**

```
One-Sample T: LogMen


Variable   N    Mean     StDev    SE Mean        95% CI
LogMen     20   1.5223   0.2073    0.0464    (1.4253, 1.6193)
```

c. Which interval is narrower?

d. Which interval would you report if the goal was to understand the typical amount of time spent getting ready in the morning for men at this college? Explain.

**13.13 Exercise Hours** A statistics student was interested in the amount of time that community college students exercise each week. He gathered data from a random sample of students at his community college and excluded those who did not exercise (those who reported 0 hours per week); this left 45 in the sample. All values were rounded to the nearest hour.

The table shows the data. Figure A shows a histogram of the data, and figure B shows a histogram of the log transform (base 10) of the data.

(A)

Hours of Exercise

(B)

Log (base 10) of Exercise Hours

a. Describe the distribution of the sample.

b. Find a 95% confidence interval for the mean of the number of hours of exercise per week for all students at this college.

c. Describe the distribution of the transformed data, and compare it with the distribution of the original data in part a.

d. Perform a log transform on the observations. Find the boundaries for a 95% confidence interval for the mean of the log-transformed times.

e. Convert the log interval boundaries back to units of hours. Interpret the resulting interval.

f. Which interval would you report: the interval for the population mean or for the population geometric mean? Explain.

| 1 | 2 | 3 | 5 | 8 |
|---|---|---|---|---|
| 1 | 2 | 3 | 6 | 9 |
| 1 | 2 | 4 | 6 | 9 |
| 2 | 2 | 4 | 6 | 9 |
| 2 | 3 | 4 | 6 | 10 |
| 2 | 3 | 4 | 7 | 11 |
| 2 | 3 | 5 | 7 | 11 |
| 2 | 3 | 5 | 8 | 12 |
| 2 | 3 | 5 | 8 | 20 |

**13.14 Television Viewing** A Nielsen Poll asked people the number of hours of television they watched in the last week. Assume that Nielsen obtained a random sample. We are analyzing the data for the 39 college students in the sample. The figure shows a dotplot of the distribution.

Hours of TV for College Students

a. Find a 95% confidence interval for the mean number of hours spent watching television per week.

b. Perform a log transformation on the observations, and find a 95% confidence interval for the mean of the log-transformed data.

c. Back transform the boundaries of the confidence interval into units of hours. Interpret the resulting interval.

d. Which interval is better to report and why? Consider the shape and sample size ($n = 39$).

## SECTION 13.2

TRY **13.15 Lead Exposure (Example 3)** Excessive lead levels can negatively affect brain functions; lead poisoning is particularly dangerous to children. A study was conducted to find out whether children of battery factory workers had higher levels of lead in their blood than a matched group of children. Each child in the experimental group was matched with a child in the control group of the same age living in the same neighborhood. Although these were not a random sample of children, we can test the hypothesis that the difference is too large to occur by chance if the child from the control group was randomly chosen. The figure shows a histogram of the differences in lead level.

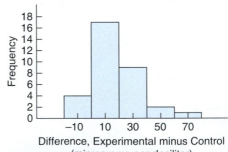
Difference, Experimental minus Control
(micrograms per deciliter)

The differences were found by calculating "factory child's lead level minus matched control child's lead level." Lead levels were concentrations in the blood, measured in micrograms per deciliter. A positive difference means the child of a factory worker has a higher lead concentration than the child in the control group. The data consisted of 1 tie (the same value for the child and the matching child), 28 pairs in which the factory worker's child had a higher level, and 4 pairs in which the factory worker's child had a lower level of lead.

Carry out a sign test to determine whether children whose parents are exposed to lead at work have a higher lead level than children whose parents are not exposed to lead at work. Use a significance level of 0.05 to see whether the experimental group had higher levels of lead.

   (Sources: Data appeared in B. Trumbo. 2001. *Learning Statistics with Real Data*, North Scituate, Massachusetts: Duxbury Press.)

**13.16 Juvenile Delinquents** Dr. Kirkland R. Gable studied 20 male juvenile delinquents who had each spent 6 months or more in a Massachusetts juvenile detention center. He wondered whether simply asking the juvenile delinquents to talk would help them stay out of jail in the future. The subjects were paid to talk into a tape recorder about anything they wanted for one hour, 3–5 days a week for 6 months; there was no therapist present. A control group was formed by matching each subject in the experimental group with a juvenile delinquent who was the same age, had the same ethnic background, grew up in the same town, had committed the same types of offenses, and had spent the same amount of time incarcerated. The control group received no treatment. The experimental and control groups were followed for three years.

| Expt | Control |
|------|---------|
| 0    | 0       |
| 12   | 3       |
| 0    | 0       |
| 8    | 0       |
| 0    | 0       |
| 0    | 11      |

The table shows *some* of the data for the number of months of incarceration in the 3-year period following the 6-month-long experiment; see the disk. The histogram shows the differences: experimental minus control. A negative difference means a subject in the experimental group spent less time in jail than did his control (which is the outcome the researcher is hoping for). Although these were not a random sample, we can test to see whether the difference is too large to attribute to chance if we assume the matched subjects in the control group were chosen at random.

   a. Summarize the months of incarceration for both groups in one or two sentences. Include appropriate numerical summaries.

   b. Perform a sign test to determine whether the typical amount of jail time after the experiment was less for the treatment group than for the control. Use a significance level of 0.05.

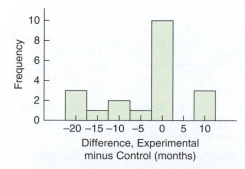

(Source: R. Schwitzgebel. 1964. *Streetcorner research: An experimental approach to the juvenile delinquent.* Cambridge, Massachusetts: Harvard University Press.)

**13.17 The Stroop Effect** Suppose you had to identify the color of ink for a series of printed words, but the printed word appeared in a color of ink that did not match the name of the color. For example, if you were shown "RED" then you should say "Blue," but you might incorrectly say "Red" because that is what the word spells. It might take longer to correctly identify a color when it was used to print a word whose meaning did not match that color than to identify the color when the ink color and printed word matched. This difference in times is an example of something psychologists call *interference* and is called the Stroop effect, after research psychologist J. R. Stroop (1897–1973).

| Same | Diff |
|------|------|
| 32   | 66   |
| 12   | 31   |
| 17   | 40   |
| 16   | 25   |
| 25   | 36   |
| 18   | 15   |
| 18   | 39   |
| 24   | 35   |
| 20   | 32   |
| 24   | 30   |

The data in the table were collected by a student conducting research on the Stroop effect. Each of 10 subjects identified colors of ink in two different situations, and the time (in seconds) was recorded. There were 36 words in each trial. In column 1, the ink was the same color as the word. In column 2, the ink and the words were different colors. Whether the subject started with the color matching the word or different was randomly determined. Treat the data as though they came from a random sample.

   a. Write a sentence comparing the median time to identify the color for the two groups. Did the task tend to take longer when the colors were different?

   b. Do a sign test to test whether those who see the ink in the "wrong" color tend to take longer to identify the color. Use a significance level of 0.05.

**13.18 Reading Material on Colored Paper** In the past, some people believed it was easier to read words printed on colored paper than on white paper, while other people believed it was easier to read words printed on white paper. To test these theories, researchers asked a sample of 15 subjects to read a passage printed in black ink twice: once on salmon-colored paper and once on white paper. The time it took to read each passage, in seconds, is given in the table. Whether the person read the salmon-colored paper first or the white paper first was determined randomly. A histogram of the differences is also shown.

| Salmon | White |
|--------|-------|
| 79 | 72 |
| 49 | 45 |
| 47 | 45 |
| 112 | 120 |
| 65 | 63 |
| 66 | 62 |
| 67 | 60 |
| 59 | 54 |
| 140 | 160 |
| 64 | 57 |
| 67 | 61 |
| 73 | 72 |
| 64 | 48 |
| 126 | 122 |
| 67 | 73 |

a. Compare the typical values for the two groups.

b. Refer to the histogram of differences. Why is the *t*-test potentially not appropriate for these data?

c. Carry out an appropriate hypothesis test that the typical reading time is not the same for words printed on salmon-colored paper (at least for these subjects and this passage). Use a significance level of 0.05 (with a two-tailed alternative) and interpret your results.

**13.19 Males' Pulse Rates** Students in a statistics class were asked to measure their resting pulse rates. After that, the instructor unexpectedly screamed and ran from one side of the class to the other. Students again measured their pulse rates. The pulse rates (in beats per minute) were recorded before and after the scream for the male students in the class. Perform a sign test to see whether the pulse rate went up significantly, using a significance level of 0.05 and treating the sample as random. The data can be found in Exercise 9.60 on page 435.

**13.20 Females' Pulse Rates** Refer to Exercise 13.19. This time, the data (beats per minute) came from female students before and after the scream. Perform a sign test to see whether the pulse

rate went up significantly, using a significance level of 0.05 and treating the sample as random. The data can be found in Exercise 9.59 on page 435.

* **13.21 Ages of Brides and Grooms** A random sample of the ages of 14 brides and their grooms showed that in 10 of the pairs the grooms were older, in 1 pair they were the same age, and in 3 pairs the bride was older. Perform a sign test with a significance level of 0.05 to test the hypothesis that grooms tend to be older than their brides.

* **13.22 Textbook Prices** A student was interested in comparing textbook prices at two universities. She matched the textbooks by subject and compared prices from the University of California at Santa Barbara (UCSB), which is on the quarter system (10 weeks) and California State University at Northridge (CSUN), which is on the semester system (16 weeks).

   a. Test the hypothesis that the books for UCSB tend to cost less than the books for CSUN, using a significance level of 0.05. For 17 of the pairs the prices were higher at CSUN, and for 7 of the pairs they were higher at UCSB. There were no ties.

   b. Why is the sign test probably a good choice for these data?

## SECTION 13.3

TRY **13.23 Meat-Eating Behavior (Example 4)** A researcher was interested in the ethics of eating meat, so he studied and compared ethicists (philosophy professors who taught ethics) with professors who taught other subjects to find out whether ethicists eat less meat. The subjects were asked how many meals they eat per week that include meat. The data provided are a random sample from the full study. Vegetarians (who eat no meat) were excluded. Assume the shapes and spreads of the distributions are the same.

   a. Refer to the output given. Compare the sample medians. What do they tell us about the research question?

   b. Refer to the output to perform a Mann-Whitney test using a significance level of 0.05. State all four steps of a significance test.

```
Mann-Whitney Test and CI
              N   Median
Ethicists    16    4.500
Control      12    6.500


Point estimate for ETA1-ETA2 is -2.000
95.2 Percent CI for ETA1-ETA2 is (-3.999, 0.001)
W = 198.0
Test of ETA1 = ETA2 vs ETA1 < ETA2 is significant at 0.0599
The test is significant at 0.0582 (adjusted for ties)
```

(Source: E. Schwitzgebel and J. Rust. 2009. The self-reported moral behavior of ethics professors. Unpublished manuscript. http://schwitzsplinters.blogspot.com)

**13.24 Credit Card Debt** A statistics student who was interested in credit card debt asked a random sample of students for the total amount of their credit card debt. We eliminated the two women and the one man who had a debt of 0, which left 18 women and 19 men.

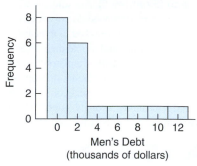

a. By looking at the histograms of the data, determine whether it would be appropriate to do a two-sample *t*-test using this data set. Explain.

b. Refer to the Minitab output. Which group had a higher median debt in this sample?

c. Refer to the Minitab output provided to test whether the typical credit card debt (as measured by the median) is different for men and women, using a significance level of 0.05.

```
Mann-Whitney Test and CI: Women, Men
          N   Median
Women    18    725.0
Men      19   1250.0

Point estimate for ETA1-ETA2 is -212.0
95.3 Percent CI for ETA1-ETA2 is (-1100.0, 700.1)
W = 322.5
The test is significant at 0.5635 level
```

**13.25 Cleaning Archeological Artifacts** Archeologists who work in a "dig" often unearth dirty objects, such as those made of ceramic, metal, glass, and other substances. The traditional method of cleaning these objects has been to brush them gently with a soft brush, which takes a long time. Therefore, archeologists have been looking for other cleaning methods that take less time, do a good job, and do not damage the artifacts. A substance called Calgon (sodium hexametaphosphate) can be used; the objects are soaked in it, and no brushing is required. It has been determined that Calgon does not damage the objects. In order to determine whether it cleans as well as brushing, investigators measured the weights of the objects (in grams) before and after cleaning. We have reported some of the differences in weights (dirty weight minus clean

weight, in grams) and also the method of treatment; b is for brushing and c is for Calgon; you should use the complete data set from the disk or website. Use a Mann-Whitney test with a significance level of 0.05 to determine whether the two methods result in significantly different changes in median weight. Assume the shapes and spreads of the two distributions are the same.

| Diff | Method |
|------|--------|
| 1.3  | b      |
| 1.9  | b      |
| 1.2  | b      |
| 0.5  | c      |
| 0.1  | c      |

(Source: Neumann & Sanford. 1998. Cleaning artifacts with Calgon. *American Antiquity* 63(1), 157–160.)

**13.26 Body Temperatures** Body temperatures for 65 healthy females and 65 healthy males were collected. The samples suggest the distributions of body temperatures are Normally distributed for both men and women. Assume that the standard deviations of the populations are roughly equal.

a. Use the following output to test the hypothesis that the means are not equal, using a significance level of 0.05. Report the p-value and conclusion or explain why the test should not be done.

**(A)**

```
Two-Sample T-Test and CI: tempF, tempM
          N     Mean   StDev   SE Mean
tempF    65   98.105   0.699     0.087
tempM    65   98.394   0.743     0.092

T-Test of difference = 0(vs not =):
T-Value = -2.29
P-Value = 0.024
```

b. Use the following output to test the hypothesis that population *medians* are the same using a significance level of 0.05.

**(B)**

```
Mann-Whitney Test and CI: tempF, tempM
          N    Median
tempF    65    98.100
tempM    65    98.400

W = 3782.0
Test of ETA1 = ETA2 vs ETA1 not =ETA2
  is significant at 0.0270
The test is significant at 0.0268
  (adjusted for ties)
```

**13.27 Cell Phone Bills** Cell phone bills (rounded to the nearest dollar) for the most recent month for random samples of college men (M) and college women (F) were studied. Histograms for the amounts of dollars for men and women (A) and output from a Mann-Whitney test (B) are given.

a. Why would it be inappropriate to compare means with a *t*-test?

b. A student feels that women talk more on their phones than men and therefore have higher bills. Test this hypothesis at a 5% significance level.

**(A)**

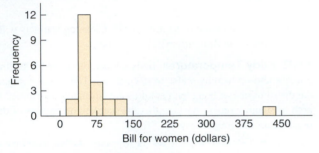

**(B)**

```
Mann-Whitney Test and CI: M, F
     N   Median
M   23    48.00
F   24    57.50

Point estimate for ETA1-ETA2 is -5.50
95.1 Percent CI for ETA1-ETA2 is (-18.00, 5.00)
W = 501.0
Test of ETA1 = ETA2 vs ETA1 < ETA2 is significant at 0.1413
The test is significant at 0.1411 (adjusted for ties)
```

**13.28 Weights of Athletes** The table shows a selection of *some* of the weights for 25 male college baseball players (B) and 25 college soccer players (S). Assume that these are random samples from all college baseball and soccer players. Refer to the Mann-Whitney output. Assume the shapes and spreads of the two distributions are the same.

| Weight | Sport |
|--------|-------|
| 190 | B |
| 200 | B |
| 187 | B |
| 181 | B |
| 165 | S |
| 190 | S |

a. Which group tends to weigh more in this sample? Compare the median weights by referring to the output.

b. Are these data paired or independent? Explain.

c. Test the hypothesis that the median weights are different. Use a 0.05 significance level, and use the Mann-Whitney test.

```
Mann-Whitney Test and CI: Baseball, Soccer
          N   Median
Baseball  25  192.00
Soccer    25  174.00

Point estimate for ETA1-ETA2 is 17.00
95.2 Percent CI for ETA1-ETA2 is (9.99, 25.00)
W = 857.5
Test of ETA1 = ETA2 vs ETA1 not = ETA2 is significant at 0.0000
```

# SECTION 13.4

**TRY** * **13.29 Sports and Extraversion (Example 5)** Are students who participate in sports more extraverted than those who do not? A random sample of students at a small university were asked to indicate whether they participated in sports (yes or no) and to rate their level of extraversion. Extraverts are outgoing, are talkative, and don't mind being the center of attention. Students were asked whether they agreed with the statement that they were extraverts, using a scale of 1–5 with 1 meaning "strongly disagree" and 5 meaning "strongly agree." There were 51 students who participated in sports and 64 who did not.

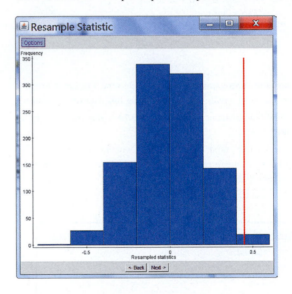

The mean extraversion score for those who participated in sports was 3.618, and the mean for those who did not participate was 3.172, a difference of 0.446 point. To determine whether the mean difference was significant, we performed a randomization test to test whether the mean extraversion level was greater for athletic students.

a. The histogram shows the results of 1000 randomizations of the data. In each randomization, we found the mean difference between two groups that were randomly selected from the combined group of data: the combined data for the sporty and nonsporty. Note that, just as you would expect under the null hypothesis, the distribution is centered at about 0. The red line shows the *observed* sample mean difference in extraversion for the sporty minus the nonsporty. From the graph, does it look like the observed mean difference is unusual for this data set? Explain.

b. The software output gives us the probability of having an observed difference of 0.446 *or more*. (See the column labeled "Proportion => Observed"). In other words, it gives us the right-tail area, which is the p-value for a one-tailed alternative that the mean extraversion score for athletes is higher than for nonathletes. State the p-value.

c. Using a significance level of 0.05, can we reject the null hypothesis that the means are equal?

d. If you did not have the computer output, explain how you would use the histogram to get an approximate p-value.

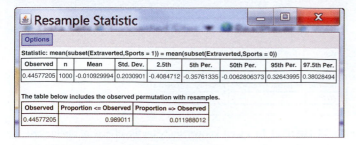

Statistic: mean(subset(Extraverted,Sports = 1)) = mean(subset(Extraverted,Sports = 0))

| Observed | n | Mean | Std. Dev. | 2.5th | 5th Per. | 50th Per. | 95th Per. | 97.5th Per. |
|---|---|---|---|---|---|---|---|---|
| 0.44577205 | 1000 | -0.010929994 | 0.2030901 | -0.4084712 | -0.35761335 | -0.0062806373 | 0.32643995 | 0.38028494 |

The table below includes the observed permutation with resamples.

| Observed | Proportion <= Observed | Proportion => Observed |
|---|---|---|
| 0.44577205 | 0.989011 | 0.011988012 |

* **13.30 Meat-Eating Ethics** A survey asked respondents about the ethics of eating meat. Survey respondents were shown the following chart and asked to choose a number that represented how they felt about the ethics of eating meat.

| Very Morally Bad | Somewhat Morally Bad | Neutral | Somewhat Morally Good | Very Morally Good |
|---|---|---|---|---|
| 1 | 2 3 | 4 5 6 | 7 | 8 9 |

The survey was given to ethics professors and to professors who did not teach ethics or philosophy. A random sample of 20 ethics professors (Eth) and 20 professors who did not teach ethics or any philosophy course (Profs) was taken. The mean for the ethics professors was 3.8, and the mean for other professors was 5.2. The observed mean difference between ethicists and nonethicists is therefore –1.4 points. (The negative value means that ethicists in this sample tended to rate eating meat as being a less ethical activity than did nonethicists.)

A randomization test was done to test whether the ethicists view meat eating as less ethical than their counterparts.

a. The histogram shows the results of randomly shuffling the labels "ethicist" and "nonethicist" between the two groups of values. This randomization was repeated 1000 times, and the mean difference was computed each time. Hence, all of the differences observed are due purely to chance. Note the red line at the observed difference of –1.4 Looking at the graph, does it appear that the observed difference is typically obtained by chance?

(A)

(B)

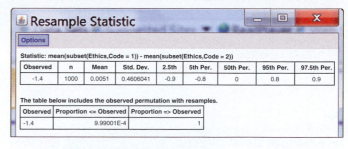

Statistic: mean(subset(Ethics,Code = 1)) - mean(subset(Ethics,Code = 2))

| Observed | n | Mean | Std. Dev. | 2.5th | 5th Per. | 50th Per. | 95th Per. | 97.5th Per. |
|---|---|---|---|---|---|---|---|---|
| -1.4 | 1000 | 0.0051 | 0.4606041 | -0.9 | -0.8 | 0 | 0.8 | 0.9 |

The table below includes the observed permutation with resamples.

| Observed | Proportion <= Observed | Proportion => Observed |
|---|---|---|
| -1.4 | 9.99001E-4 | 1 |

b. The software output gives us the probability of having an observed difference of −1.4 *or less*. (See the column labeled "Proportion <= Observed".) (The number is in exponential notation. E-4 means "10 to the negative fourth power"—that is, to the power of −4.) Convert the number to decimal notation. State the p-value. Using a significance level of 0.05, can we reject the null hypothesis that the means are equal?

(Source: E. Schwitzgebel and J. Rust. 2009. The self-reported moral behavior of ethics professors. Unpublished manuscript. http://schwitzsplinters.blogspot.com)

**13.31 College Students and Credit Card Debt** In Exercise 13.24 you compared credit card debts for college men and women using the Mann-Whitney test to compare medians. We'll use the same data again, but this time you will apply a randomization test to determine whether men and women in college have different median credit card debts. There were 19 men and 18 women in this data set. As part of the randomization test, the credit card debts were randomly assigned the label "man" or "woman," and the difference of the medians (men minus women) was calculated and recorded. This was repeated 1000 times. Figure A shows a histogram of the resulting differences of medians and figure B shows the resampling output.

(A)

a. Why are medians a better choice than means for comparing typical debts of men and women? Refer to the histograms given in Exercise 13.24.

b. The sample medians were $1250 for men and $725 for women, so the difference of men minus women was $525. This value is indicated with a red vertical line in the histogram. Is the observed difference unusually large? Explain.

c. The proportion of differences in the histogram that were greater than or equal to $525 is 0.245 as shown in figure B. Our alternative hypothesis, that the medians are different, requires a two-tailed test. However, with a significance level of 0.05, and knowing that the two-tailed p-value must be bigger than the one-tailed value of 0.245, will we reject the null hypothesis? Explain.

**(B)**

| Statistic: median(subset(Debt,Gender = 1)) = median(subset(Debt,Gender = 2)) | | | | | | | | |
|---|---|---|---|---|---|---|---|---|
| Observed | n | Mean | Std. Dev. | 2.5th | 5th Per. | 50th Per. | 95th Per. | 97.5th Per. |
| 5.25 | 1000 | -23.6625 | 635.99457 | -1150 | -1000 | -50 | 956.25 | 1075 |

The table below includes the observed permutation with resamples.

| Observed | Proportion <= Observed | Proportion => Observed |
|---|---|---|
| 525 | 0.75624377 | 0.24475524 |

### 13.32 Salaries of Men and Women

Do men typically make more than women? A random sample of salaries for men and women shows extremely skewed distributions. We'll perform a randomization test to determine whether median income is greater for men than for women.

The difference in sample medians (men's median income minus women's) is $8550, with the men having the higher sample median. (Source: SPSS) Figure A shows a histogram of the resulting differences.

**(A)**

a. State the null and alternative hypotheses for this test.

b. The histogram shows the results of 1000 "shuffles" of the labels "men" and "women" on the incomes in the data set. For each shuffle, the difference in medians (men minus women) was calculated. The red line indicates the observed difference of medians. What can we conclude from this?

c. Refer to the output shown in figure B. Can you reject the null hypothesis, using a significance level of 0.05? Explain. (E-4 means "10 to the negative fourth power"—that is, 10 to the power of –4.)

**(B)**

| Statistic: median(subset(Salary,Gender = m)) - median(subset(Salary,Gender = f)) | | | | | | | | |
|---|---|---|---|---|---|---|---|---|
| Observed | n | Mean | Std. Dev. | 2.5th | 5th Per. | 50th Per. | 95th Per. | 97.5th Per. |
| 8550 | 1000 | 50.655 | 1061.4767 | -1987.5 | -1650 | 187.5 | 1800 | 2100 |

The table below includes the observed permutation with resamples.

| Observed | Proportion <= Observed | Proportion => Observed |
|---|---|---|
| 8550 | 1 | 9.99001E-4 |

**\* 13.33 Rainfall** In a well-known study on the effects of cloud seeding to produce rainfall, experimenters randomly assigned airplanes to release either silver nitrate (which is believed to increase the amount of rainfall from a cloud) or a placebo. Fifty-two clouds were chosen at random; half were "seeded" with silver nitrate, and half were not. In the text, we compared the *median* rainfall amounts between the seeded (silver nitrate) and unseeded clouds. In this exercise, you will compare the *mean* rainfall amounts for the seeded and unseeded clouds.

In the study, the seeded clouds produced more rain, on average, by 368.9 acre-feet. To determine whether such differences could occur by chance, a statistician could have written the 52 rainfall amounts on separate slips of paper and randomly dealt them into two stacks. He or she would then have computed the mean of each stack and found the difference. This was actually done by a computer and repeated 1000 times. The results are shown in the histogram.

Carry out a hypothesis test to test whether cloud seeding increased the mean rainfall. By referring to the histogram, choose from the following possible p-values (one-tailed):

0.50, 0.25, 0.15, 0.025, less than 0.0001

Use a 5% significance level for your test.

Mean rainfall with seeding minus
Mean rainfall without

**\* 13.34 Rainfall** Refer to Exercise 13.33, which discussed a study on the effects of cloud seeding to produce rainfall. Some researchers think that cloud seeding has little effect on "low rain potential" clouds. Instead, they claim, most of the action is with clouds that would produce lots of rain even without seeding. In this scenario, clouds that would produce little rain without seeding will produce little rain with seeding. However, the clouds that would produce the most rain without seeding will produce much, much more rain with cloud seeding. To test this, researchers carried out a randomization test to test whether the third quartile of rainfall increased under cloud seeding. The table gives summary statistics.

| Treatment | Minimum | First Quartile | Median | Mean | Third Quartile | Maximum |
|---|---|---|---|---|---|---|
| Seeded | 4.10 | 98.13 | 221.60 | 533.20 | 474.30 | 2747.00 |
| Unseeded | 1.00 | 24.82 | 44.20 | 164.60 | 159.20 | 1203.00 |

a. Explain what it means to say that the third quartile of rainfall is 474.30 acre-feet.

b. Why is the third quartile an appropriate statistic to answer the researchers' question?

c. What is the observed difference in third-quartile rainfall between the seeded and unseeded clouds?

d. To determine whether such differences could occur by chance, a statistician could have written the 52 rainfall amounts on separate slips of paper and randomly dealt them into two stacks. He or she would then have computed the third quartile of each stack and found the difference. A computer actually did this 1000 times, each time finding the difference between the third quartile for the seeded clouds minus the third quartile for the unseeded clouds. The results are shown in the histogram. Referring to the histogram, carry out a hypothesis test to test whether cloud seeding increased the third-quartile rainfall. (You will have to get approximate p-values by reading the histogram.) (Remember that you need only decide whether the p-value is larger or smaller than 0.05.)

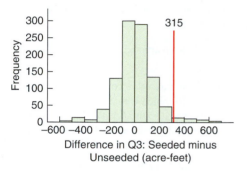

**13.35 Randomization** Exercise 13.33 describes a simulation exercise. Which of the following is the best explanation for the process described?

a. This process allows researchers to compare the actual result to what could have happened by chance (assuming the null hypothesis is true).

b. This process allows researchers to determine the percentage of time the seeded clouds would outperform the unseeded clouds for all possible scenarios.

c. This process allows researchers to determine how many times they need to replicate the experiment for valid results.

**13.36 Randomization** Exercise 13.33 describes a simulation exercise. Under which of the following assumptions were the data simulated?

a. Cloud seeding typically produces more rain than not seeding the cloud.

b. Cloud seeding typically produces the same amount of rain as not seeding.

c. Cloud seeding typically produces less rain than not seeding.

## CHAPTER REVIEW EXERCISES

*For Exercises 13.37 through 13.40, choose from the following tests, as appropriate: paired t-test, sign test, two-sample t-test, and Mann-Whitney test. There may be two acceptable choices.*

**13.37 Ages of Married Couples** You have data on the ages of both partners in a random sample of married couples and want to find out whether husbands tend to be older than their wives. There are 15 married couples in your sample. Which test(s) can you use? Answer for each circumstance.

a. Assume that the distributions of ages are Normal.

b. Assume that the distributions of ages for both men and women are not Normal, and assume that the distribution of differences in age for each couple is not Normal.

c. Assume the distributions of ages are not Normal for both men and women, but assume that the distribution of differences in age is Normal.

**13.38 Comparing GPAs** You know the GPAs of a random sample of 10 college men and a random sample of 10 college women, and you want to test the hypothesis that the typical GPAs are different. Which test(s) should you choose for each situation?

a. Suppose your preliminary investigations lead to you to conclude that both the distribution of GPAs for men and the distribution of GPAs for women are approximately Normal.

b. Suppose your preliminary investigations lead you to conclude that the distributions of both variables are not Normal.

**13.39 Voicemail** Suppose you are interested in the length of recorded outgoing voicemail messages for professors at your college. You call 10 randomly selected male professors and 10 randomly selected female professors over the weekend and time their outgoing messages. You want to see whether women tend to have longer outgoing messages than men. Which test(s) should you choose for each situation?

a. Assume that the length of messages is Normally distributed both for men and for women.

b. Assume that one or both distributions are strongly right-skewed.

**13.40 Pulse Rates** Suppose you want to determine whether meditation can cause a decrease in pulse rate. You randomly select 15 students, teach them a meditation technique, and then measure their pulse rates before and after meditation. Which test(s) should you choose for each situation?

a. Assume that your analysis shows that the differences in pulse rates are Normally distributed.

b. Assume that the differences in pulse rates are strongly skewed.

**g * 13.41 Ice Cream Cones** McDonald's claims that its ice cream cones typically weigh 3.18 ounces (converted from grams). Here are the weights, in ounces, of cones purchased on different days from different servers:

$$4.2, 3.6, 3.9, 3.4, \text{ and } 3.3$$

Carry out a sign test to test whether the median is greater than 3.18 ounces. Use a significance level of 0.05. Also, why is the sign test a more appropriate choice than the one-sample *t*-test? *See page 612 for guidance.*

**\*13.42 Average Body Temperatures** Many people believe that healthy people typically have a body temperature of 98.6°F. We took a random sample of 10 people and found the following temperatures:

$$98.4, 98.8, 98.7, 98.7, 98.6, 97.2, 98.4, 98.0, 98.3, \text{ and } 98.0$$

Use the sign test to test the hypothesis that the median is not 98.6. Refer to Exercise 13.41 for guidance.

# gUIDED EXERCISES

**g\* 13.41 Ice Cream Cones** McDonald's claims that its ice cream cones typically weigh 3.18 ounces (converted from grams). Here are the weights, in ounces, of cones purchased on different days from different servers):

$$4.2, 3.6, 3.9, 3.4, \text{ and } 3.3$$

Carry out a sign test to test whether the median is greater than 3.18 ounces. Use a significance level of 0.05. Also, why is the sign test a more appropriate choice than the one-sample *t*-test?

**Step 0 ▶** Fill in the missing numbers to find the difference between the observed value and the value we would expect if the claim that McDonald's makes is correct.

| Obs | Null Value | Difference |
|-----|-----------|-----------|
| 4.2 | 3.18 | 1.02 |
| 3.6 | 3.18 | |
| 3.9 | | 0.72 |
| 3.4 | | |
| 3.3 | 3.18 | 0.12 |

**Step 1 ▶** H$_0$: _____

H$_a$: The median weight is more than 3.18 ounces (or the median difference is more than 0.00.)

**Step 2 ▶** Choose the sign test as instructed. What assumptions should be checked?

**Step 3 ▶** Find the p-value. You may use a computer to do a sign test to test whether the median difference is more than 0. Or you may use binomial probabilities, because the p-value is equivalent to the probability of getting five out of five heads in five tosses of a fair coin.

**Step 4 ▶** Can you reject the null hypothesis that the median is 3.18 ounces?

**Why Use the Sign Test?** Why is the sign test more appropriate than the one-sample *t*-test for this data set?

## CHECK YOUR TECH

# Verifying Minitab Output for the Sign Test

The table gives data for the ages of five randomly sampled newly wed couples. We used a sign test to test the hypothesis that grooms tend to be older than their brides. The figure shows the Minitab output.

```
Sign Test for Median: Differences
Sign test of median =  0.00000 versus > 0.00000

                   N   Below   Equal   Above        P    Median
Differences        5     1       1       3     0.3125    2.000
```

▲ **FIGURE** Minitab Output for Sign Test

| Groom | Bride | Difference |
|-------|-------|------------|
| 24 | 22 | 2 |
| 38 | 28 | 10 |
| 27 | 27 | 0 |
| 24 | 25 | −1 |
| 33 | 31 | 2 |

▲ **TABLE** Ages and Differences

**QUESTION** Confirm the p-value (0.3125) by following the steps given below.

**SOLUTION**

**Step 1 ▶** The alternative hypothesis is that grooms tend to be older than their brides. After removal of the point (27, 27) in which they were the same age, what is the sample size?

**Step 2 ▶** If the null hypothesis is true, how many grooms should we expect to be older, using the sample size obtained in step 1?

Our sample shows that three out of four remaining grooms are older than their brides. To get a one-tailed p-value, we want the probability of three grooms older than their bridges or a more extreme outcome, which is four grooms older than their brides.

**Step 3 ▶** We started making a list of all the possibilities, using G for the case where the groom is older and B for the case where the bride is older. Thus **GGGB** means that in the first, second, and third couples the groom was older, but in the fourth couple the bride was older. Finish the table by putting in the missing outcomes. The outcomes are stacked so that you can see the outcomes for four grooms (out of four) older than their brides: GGGG (which can happen in only 1 way), three grooms older than their brides (which can happen in 4 ways), and so on.

| GGGG | GGGB | GGBB | BBBG | _____ |
|------|------|------|------|------|
|      | GGBG | GBGB | BBGB |      |
|      | GBGG | GBBG | BG __ |      |
|      | _____ | B__B__ | G ___ |      |
|      | BB___ |      |      |      |
|      | B___B |      |      |      |

**Step 4 ▶** Under the null hypothesis, the probability that the groom is older is 0.5, and the probability that the bride is older is also 0.5.

What is the probability of GGGG—in other words, the probability that all four grooms are older than their brides, assuming equal ages of brides and grooms in the population? Assume independence and use the multiplication rule.

**Step 5 ▶** What is the probability of each of the other listed outcomes?

**Step 6** ▶ How many different outcomes are in the list?

**Step 7** ▶ How many outcomes are there for the situation in which the groom is older in three cases out of four?

**Step 8** ▶ How many outcomes are there for the situation in which the groom is older than his bride in three cases out of four *or* in all four cases (more extreme)?

**Step 9** ▶ Finally, assuming that the probabilities of all 16 options are equal, what is the p-value, which is probability that three *or more* grooms are older than their brides? Use your answer to step 8, and divide by the total number of outcomes.

Compare your answer in Step 9 to the p-value from the Minitab table.

# TechTips

## General Instructions for All Technology

The TI-83/84 is not programmed to do nonparametric tests. A randomization test using StatCrunch is shown at the end of the steps.

**EXAMPLE A: QQ PLOTS (OR NORMALITY PLOTS)** ▶ Make a QQ plot, or normality plot, of the following ages:

18, 22, 76, 21, 25

**EXAMPLE B: TRANSFORMATIONS** ▶ Transform all of the following ages to logarithm base 10, find the mean of the logs, and back transform that mean to age to find the geometric mean.

18, 22, 76

**EXAMPLE C: SIGN TEST** ▶ Suppose that eight people go on a diet, and the following weights show what happened. Determine whether you can reject the hypothesis of no weight change.

| Person | Weight Before | Weight After | Change in Weight |
|--------|---------------|--------------|------------------|
| 1 | 188 | 182 | 6 |
| 2 | 225 | 225 | 0 |
| 3 | 203 | 200 | 3 |
| 4 | 154 | 150 | 4 |
| 5 | 140 | 142 | −2 |
| 6 | 235 | 230 | 5 |
| 7 | 372 | 365 | 7 |
| 8 | 280 | 275 | 5 |

**EXAMPLE D: MANN-WHITNEY TEST** ▶ Suppose you have a random sample of GPAs for three men and three women, and you want to use a nonparametric test (the Mann-Whitney test) to determine whether you can reject the hypothesis of no difference.

| Men | Women |
|------|-------|
| 2.30 | 3.30 |
| 3.45 | 3.10 |
| 2.96 | 3.02 |

## Example A: Normality Plot (QQ Plot)

1. Enter the five ages into L1.
2. Press **2ND STATPLOT** and **1**.
3. Turn on **Plot 1** by pressing **ENTER** when **On** is flashing. (The other plots should be **Off**.)
4. See Figure 13A: Use the arrows on the keypad to get to the last option for **Type:** (the lower right corner of the six pictures). Press **ENTER** to select this option. Be sure the **Data List** is correct (**L1**). Change the **Data Axis** to Y so that the data is on the Y axis as described in the chapter.

▲ **FIGURE 13A** TI-83/84 Input for Normality Plot

5. Press **ZOOM** and **9** to see the Normality plot. Press **Trace** to see the numbers. Figure 13B shows the Normality plot.

▲ **FIGURE 13B** TI-83/84 Normality Plot

## Example B: Transformation and Geometric Mean

1. Enter the three ages into **L1**.
2. See Figure 13C: Move the cursor up to the label for **L2**. Press the **LOG** button and press **2ND 1** (for **L1**), and **ENTER**.

▲ **FIGURE 13C** TI-83/84 Taking Logs of a List

**L2** will now contain the list of logs.

3. Find the mean of the logs by using **STAT**, **CALC**, **1-Var Stats** and pressing **2ND 2** (for **L2**), and **ENTER**.

The mean of the logs should be **1.4928**.

4. Back Transforming (finding the antilog) Press **2ND 10$^x$** (with the **LOG** button), and put in **1.4928** and press **ENTER**.

You should get 31.1 for the geometric mean.

## Example A: Normality Plot

1. Enter the five ages into the first column, **C1**.
2. **Stat** > **Basic Statistics** > **Normality Test**
3. Double click **C1** to put it in the **Variable:** box. Leave the other default choices.
4. Click **OK** to make the graph.

Figure 13D shows the Minitab Normality plot. Note that the data are on the X axis, which is different from the QQ plots explained in the chapter.

▲ **FIGURE 13D** Minitab Normality Plot

## Example B: Transformation and Geometric Mean

1. Enter the three ages in **C1**. Put a label at the top, **ages**.
2. Click **Calc** and **Calculator**.
3. See Figure 13E: Enter **c2** in the small top box. Delete any existing entry in the **Expression** box. Use the **Function** box to find and double click **log base 10**. Double click **ages** to take the log base 10 of the ages.

▲ **FIGURE 13E** Minitab Calculator

**615**

4. Click **OK**.

The logs will be in column 2.

5. Find the mean by using **Stat**, **Basic Statistics**, **Display Descriptive statistics**, choosing **C2**, and clicking **OK**.

The mean of the logs will be 1.493.

Back Transforming (finding the antilog)

6. Enter **1.493** in column 3.

7. Click **Calc** and **Calculator**. Enter **C4** in the small top box. Delete any existing entry in the **Expression** box. Choose **anti-log** from the list of functions, and double click **C3** so that you are taking the antilog of the number in column 3.

8. Click **OK**.

You will see the answer in column 4: **31.1172**.

## Example C: Sign Test

Minitab uses the column of differences for the sign test. Start with either step 1 or step 2.

1. Enter the differences into one column and go to step 3.

2. If you have entered two columns of data (labeled **Before** and **After**) and want to use Minitab to calculate the differences first, click **Calc** and **Calculator**. Figure 13F shows what to do if you want column 3 to be the differences where column 1 has the weight before and column 2 has the weight after. Click **OK**. The differences will appear in column 3.

▲ FIGURE 13F   Minitab Input for Calculating Differences

3. **Stat** > **Nonparametrics** > **1-Sample Sign**

4. See Figure 13G: Enter **c3** in the **Variables** box, if that is where the differences are. Choose **Test median** and leave the default **0.0**. If you wanted to use a one-tailed alternative expecting weight loss, you would change the **Alternative** to **greater than** because when we did the subtraction, those with a weight loss resulted in a positive number.

▲ FIGURE 13G   Minitab One-Sample Sign Input

5. Click **OK**.

The output for a two-tailed alternative is shown in Figure 13H.

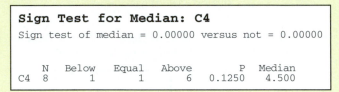

```
Sign Test for Median: C4
Sign test of median = 0.00000 versus not = 0.00000

      N   Below   Equal   Above        P   Median
C4    8       1       1       6   0.1250    4.500
```

▲ FIGURE 13H   Minitab One-Sample Sign Output

## Example D: Mann-Whitney Test

1. Enter the six GPAs unstacked (as given previously) into columns 1 and 2. Add labels (**Men** and **Women**) above the columns.

2. **Stat** > **Nonparametrics** > **Mann-Whitney**

3. Refer to Figure 13I. Double click one of the columns for the **First Sample,** and double click the other column for the **Second Sample.** You may change the **Alternative** if you want it to be one-tailed.

▲ FIGURE 13I   Minitab Mann-Whitney Input

4. Click **OK**.

The p-value comes out to be **0.6625**.

## Example A: Normality Plot (QQ Plot)

1. Enter "**Ages**" and then the five numbers in column A.

2. **Add-ins** > **XLSTAT** > **Describing data** > **Normality tests**

3. See Figure 13J. In the **Normality tests** dialog box, click in the **Data:** box to make it active; then select column of numbers including the heading (Ages) and the five numbers. Click on the **Charts** tab and select **Q-Q plots**.

4. Click **OK**, **Continue**, and **OK**.

5. Scroll down to see the plot. Note that the data are on the X axis, which is different from the QQ plots explained in the chapter.

▲ **FIGURE 13J** XLSTAT Normality Test Input

## Example B: Transformation and Geometric Mean

1. Enter the three ages in column A.

2. Put your cursor at the top of column B (click in cell **B1** to make it the active cell).

3. Click $f_x$.

4. See Figure 13K. For **category**: select **All**. Scroll down to **LOG10** and double click it to select it.

▲ **FIGURE 13K** Excel Function Input

5. See Figure 13L: For the **Number**, double click on the first entry in column A (cell A1), which contains the age of 18. The log will show up on the input screen (Figure 13K) and also at the top of column B (cell B1) when you click **OK**.

▲ **FIGURE 13L** Excel LOG10 Input

6. To get the remaining logs, put your cursor precisely in the lower right corner of the cell (B1) containing the first log until it becomes a dark cross ($+$), and drag the cross down. You will see the remaining logs.

7. Put your cursor at the bottom of the column of logs (click in cell B4). To get the average, click $f_x$ and select **Average** for the function, and then for Number1 select the columns of logs (B1-B3). Click **OK**. You will get an average of the logs of **1.492836** shown in the active cell, B4.

8. To back transform (and find the antilog), click in an empty cell (to make it active), click $f_x$, and select **Power**. For **Number** put in **10**, for the **Power** click on cell B4 or enter the average of **1.492836**. Click **OK**. You should get **31.10542**, which is the geometric mean.

## Example C: Sign Test

1. Enter "**Before**" and the list of before weights in column A. Enter "**After**" and the list of after weights in column B. You should include the labels at the top of each column.

2. **Add-ins > XLSTAT > Nonparametric tests > Comparison of two samples (Wilcoxon, Mann-Whitney, . . .)**

3. See Figure 13M. Click in the **Sample 1:** box to make it active, and then select the data in column A, including the label "Before". Click in the **Sample 2:** box to make it active, and then select the data in column B, including the label "After". For **Data Format:** select **Paired samples**. Select **Sign test**. Click **OK**.

4. Click **Continue** and then **OK**, if needed. The two tailed p-value is shown in the output to be 0.125.

▲ FIGURE 13M   XLSTAT Sign Test Input

## Example D: Mann-Whitney Test

1. Enter the GPAs into columns A and B, including the headings "Men" and "Women" at the top.

2. **Add-ins > XLSTAT > Nonparametric tests > Comparison of two samples (Wilcoxon, Mann-Whitney, . . .)**

3. Click in the **Sample 1:** box to make it active, and then select column A, including the label "**Men**". Click in the **Sample 2:** box to make it active, and then select column B, including the label "**Women**". For **Data Format**, it is *different from* Figure 13M to the left: select **One column per sample**. Select **Mann-Whitney test**. Click **OK**.

4. Click **Continue** and then **OK** if needed. The p-value (two-tailed) is shown to be 0.663.

## Example A: QQ Plot

1. Enter the five ages into the list for var1.

2. **Graphics > QQ Plot.**

3. Click **var1** and **Create Graph!**

The output is shown in Figure 13N.

▲ FIGURE 13N   StatCrunch QQ Plot

## Example B: Transformation and Geometric Mean

1. Enter the ages (18, 22, 76) into the first list, **var1**.

2. **Data > Compute expression**

3. See Figure 13O. For **Expression,** enter **log10(var1)**.

4. Click **Compute** and a list of logs will be added to the table.

▲ FIGURE 13O   StatCrunch Compute Expression Input

5. **Stat > Summary Statistics > Columns**

6. Select the column with the logs in it.

7. Click **Calculate.** You will get a table showing the mean to be **1.4928362**.

Back Transforming (finding the antilog)

8. **Data > Compute expression**

9. In the **Expression** box enter **10\*\*1.4928362**; then click **Compute** and you will get the answer 31.105 . . . in a new column.

## Example C: Sign Test

You may start with step 1 or step 2.

1. Enter the differences in one column and then go to step 5.

2. Enter in the two columns of data given in Example C (weight before and weight after).

To calculate the difference:

3. **Data > Compute expression**

4. See Figure 13P. For **Y:** choose one of the columns (such as **var1**) and for **X:** choose the other column (**var2**). Then choose

**Y-X** in the **Functions box**, and click **Set Expression** and **Compute**. There will be a column of differences added to your spreadsheet.

▲ **FIGURE 13P** StatCrunch Input for Finding Differences

5. **Stat > Nonparametrics > Sign Test**

6. Click the column of differences. Click **Next** if you want to change your alternative to be one-tailed.

7. Click **Calculate**.

Figure 13Q shows the output for a two-tailed alternative.

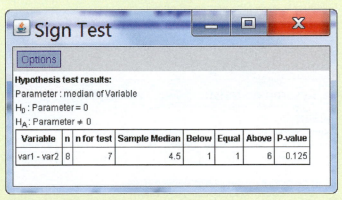

▲ **FIGURE 13Q** StatCrunch Sign Test Output

## Example D: Mann-Whitney Test (unstacked data)

1. Enter the GPAs unstacked (as shown previously) into **var1** and **var2**. You can put labels above the columns if you want.

2. **Stat > Nonparametrics > Mann-Whitney**

3. For **Group 1 in:** select one of the columns, and for **Group 2 in:** select the other column.

4. If you want to change the alternative to one-tailed, click **Next** and do it.

5. Click **Calculate**.

Figure 13R shows the output.

▲ **FIGURE 13R** StatCrunch Mann-Whitney Output

## Example D: Mann-Whitney Test (stacked data)

1. Enter the GPAs stacked (as shown in Figure 13S) including the headings (here GPA and Gender).

| Row | GPA | Gender | var3 |
|-----|------|--------|------|
| 1 | 2.3 | male | |
| 2 | 3.45 | male | |
| 3 | 2.96 | male | |
| 4 | 3.3 | female | |
| 5 | 3.1 | female | |
| 6 | 3.02 | female | |
| 7 | | | |
| 8 | | | |

▲ **FIGURE 13S** StatCrunch Stacked GPA Data

2. **Stat > Nonparametrics > Mann-Whitney**

3. See Figure 13T: For **Group 1 in** select **GPA**, and in the box for **Where:** enter **Gender="male"**. Be very careful with your typing. Typing "Male" instead of "male" will cause trouble.

4. For **Group 2 in** select **GPA**, and in the box for **Where:** enter **Gender="female"**.

5. Click **Next** to select the alternative hypothesis.

6. Click **Calculate**.

▲ **FIGURE 13T** StatCrunch Input for Mann-Whitney for Stacked Data

You will get a p-value of 0.7.

### Example E: Randomization Test

Suppose we have data on the income of 12 men (m) and 8 women (f), shown in the table.

| Income | Gender |
|---|---|
| $ 57,000.00 | m |
| $ 40,200.00 | m |
| $ 21,450.00 | f |
| $ 21,900.00 | f |
| $ 45,000.00 | m |
| $ 32,100.00 | m |
| $ 36,000.00 | m |
| $ 21,900.00 | f |
| $ 27,900.00 | f |
| $ 24,000.00 | f |
| $ 30,300.00 | f |
| $ 28,350.00 | m |
| $ 27,750.00 | m |
| $ 35,100.00 | f |
| $ 27,300.00 | m |
| $ 40,800.00 | m |
| $ 46,000.00 | m |
| $103,750.00 | m |
| $ 42,300.00 | m |
| $ 26,250.00 | f |

Use resampling to determine whether the mean difference in incomes between the men and women is significant.

1. Enter your list of numbers into one column (**var1**). Add the label **Income** in the label region at the top.
2. Enter the corresponding group labels (m or f) into a second column, **var2**. Add the label **Gender** at the top.
3. **Stat > Resample > Statistic**
4. See Figure 13U. Select **Income** for **Columns to resample**. (Do not select Gender).

▲ FIGURE 13U   StatCrunch Resampling Input

5. Select **Permutation**.
6. The most difficult part is what you enter into the **Statistic:** line. It has to be perfect. The line below shows the details. If a parenthesis is missing, if there is an extra space, or if you enter an incorrect label such as **F** instead of **f**, you will get an error message. The symbol before the second "mean" is a minus sign.

$$\text{mean}(\text{subset}(\text{Income},\text{Gender}=\text{m})) - \text{mean}(\text{subset}(\text{Income},\text{Gender}=\text{f}))$$

7. Click **Resample Statistic**.

Figure 13V shows the numerical output; yours may vary from this. The proportion => Observed is 9.99 times 10 to the negative fourth power (that is, 10 to the power −4), which is 0.000999, or approximately 0.001.

▲ FIGURE 13V   StatCrunch Resampling Output

Be aware that two people using the same data and the same steps may end up with slightly different outcomes because randomization tests are chance processes.

8. Click **Next** to see a histogram somewhat like the one shown in Figure 13W. Note the red vertical line at the observed average difference in incomes.

▲ **FIGURE 13W**  StatCrunch Resampling Histogram

*Resampling with Medians.* If you wanted to use the medians instead of the means, your line of text in step 6 would be

median(subset(Income,Gender = m)) − median(subset(Income,Gender = f))

*Vizualizing Resampling with an Applet.* If you want to see how the resampling happens with two means, enter your data, unstacked in two columns. Click **Statcrunch > Applets > Randomization test for two means**.

# 14 Inference for Regression

Using the linear model to make inferences about populations requires checking conditions carefully, as well as paying attention to how the data were collected. If the required conditions hold, then inferences can be made about the intercept and slope, as well as about predicted values of the regression line.

This chapter builds on concepts introduced in Chapter 4. There, we explained that linear regression can be used to understand associations between numerical variables, and we discussed how to use these associations to make predictions about future observations. In the meantime, you've also learned how to take knowledge about a sample and extend it to the population. In this chapter, we bring these two concepts together and show how we can use regression to make inferences about the population.

We examine two types of inference. The first is inference regarding the slope of the linear model. This is important because if the slope is 0, then there is no (linear) association between the variables. For example, suppose an oyster-processing company has a new method for measuring the size of oysters. Company executives hope that the measurements produced by the new method (which is much faster than the old) will be the same as those produced by the old method. If the slope of the regression line between old and new measurements is 0, it suggests that there is no relation between the two measurements and their new method doesn't work.

The other use for inference is to make predictions. You saw several examples of this in Chapter 4, such as predicting the end-of-year GPA of a first-year college student on the basis of her SAT score. We're now ready to present techniques that will allow you to describe the uncertainty in these predictions.

## CASE STUDY

### Building a Better Oyster Shucker

Shucking oysters—that is, removing the oysters from their shells—is a laborious activity by hand. Machines can do this, but the machines need some way to determine how large the oysters are. Lee, Lane, and Chang (2001) report on a method they developed that applied three-dimensional computer vision to estimate the volume of oysters. One important research question was how well this method works.

The researchers collected data on thirty randomly selected oysters. The volumes of the oysters were measured using traditional and accurate (but slow) methods. The oysters were measured again using the new, fast method. A regression line was fitted in order to predict the new method's measurements from the old method's. The researchers hoped that what they learned from this sample would tell them how the method works in general. By the end of this chapter, you'll see how regression can be used to estimate and measure the ability of this system to estimate oyster volume.

# The Linear Regression Model

A statistical model is a set of features that we expect our data to have. The closer our model fits reality, the better will be our inference from the model to reality. An important statistical skill is to determine how well the data fit the model. The old adage "garbage in, garbage out" applies. If we don't have good agreement between model and reality, then we won't make believable inferences about reality.

## Components of the Model

Recall from Chapter 4 that the regression line is a method used to summarize the association between two numerical variables. In Chapter 4, we required only that the association be linear. This was all that was needed for us to summarize, describe, and make predictions about associations via the regression line. Requiring linearity remains the most important condition of the model in this chapter. If the trend is not linear, then our estimates and predictions will be biased, perhaps severely.

However, we now wish to do more than simply describe an association between variables. We now want to be able to measure our uncertainty in making predictions so that we can make inferences about the population. The first step is to provide a statistical model that describes this uncertainty. With this model, we can then quantify our uncertainty by calculating margins of errors for confidence intervals and p-values for hypothesis tests.

Conceptually, the model tells us how the data were generated. This generation happens in a two-step procedure:

1. For a given value of the predictor variable, a straight-line equation determines what the value of the response variable *should* be.

2. Next, some random "noise" is added that results in the *observed* value of the response variable falling a bit above or below the line.

The first step is often called the **deterministic component** of the model—deterministic because it says, "This is the real shape that determines which observations we shall see." In this text, the deterministic component is always a straight line.

The second step is called the **random component**. The random component acknowledges that, even though the deterministic step might be the true structure of the association, randomness also appears that partially hides this structure. This randomness creates uncertainty in our inference. The random component tells us what that uncertainty looks like.

To illustrate how the model generates data, we selected four books of different lengths. We know the total number of pages in each book exactly, because we can just open the book and look at the final page number. These numbers are 203, 317, 409, and 765. We will exploit what we learned in Chapter 4 about the relation between page numbers and width of books.

1. The regression line from Chapter 4, based on a sample of 24 books, says that the width of a book (in millimeters) should be

Predicted width = 6.22 + 0.0366 (number of pages)

This is the deterministic component of our model. The deterministic component tells us that our 203-page book *should* be 6.22 + 0.0366 × 203 = 13.6498 mm wide. For our four books, the widths should be as shown in the accompanying table.

**Looking Back**

**Describing Associations**
In Chapter 4 you learned that you should examine the trend (positive, negative, nonexistent), strength (strong trends have little vertical scatter), and shape (linear or nonlinear trend) and report them in the context of the data.

| Number of Pages | Width according to Straight-line Model (mm) |
|---|---|
| 203 | 13.6 |
| 317 | 17.8 |
| 409 | 21.2 |
| 765 | 34.2 |

These points are shown in Figure 14.1a, along with the straight-line equation that tells us where other books *should* appear.

2. In statistics, as in real life, what should happen is rarely what does happen. To account for this, the random component of the model adds random noise to these widths. Instead of a book with width 13.6 mm, we see 13.6 mm plus-or-minus some random amount. We could simulate this observed width with a random number generator on a computer. We would instruct the computer to draw a number from a Normal distribution with mean 0 and standard deviation of, say, 5 mm. The value we actually observe, then, is the deterministic value (13.6) plus this random amount. If this random amount is positive, then the book we see is wider than the expected width. In this case, the observation will be above the regression line on a scatterplot. If the amount is negative, then the book is narrower than what we expected, and the observation will fall below the regression line.

In real life, where does this randomness come from? When we measured the books, randomness pushed its way into our study. Different books with the same number of pages have different widths because of the type of paper used, the way the book is bound, the type of cover on the book (hard or soft), and other unknown reasons. We also introduced some measurement error. Measurement error occurs because measurements vary from measure to measure, so if we measured the width of a book the next day, we might get a slightly different value. Thus measurements of the same object vary but are centered on the true value. The results of adding this random noise are shown in Figure 14.1b.

▲ FIGURE 14.1 (a) The first step of the model is to determine the book widths so that the points fall on the straight line that they *should* follow. (b) The second step is to add random noise, which moves each point slightly above or below the line by an amount randomly sampled from a Normal distribution.

## EXAMPLE 1 Model of Mercury Level in Fresh Water

Biologists studying the relationship between the mercury levels in freshwater lakes and the mercury levels of fish living in those lakes believe that the deterministic component of the relationship is a straight line. A scatterplot (not shown) shows that even though the general trend is linear, the points do not fall exactly on a straight line.

**QUESTION** What factors might account for the random component of this regression model?

**SOLUTION** Variability might appear in the instruments used to measure mercury levels. The size of a fish might also affect the amount of mercury in its system. Different breeds of fish might process mercury differently, and this might affect the amount of mercury found.

**TRY THIS!** Exercise 14.1

Table 14.1 lists, for each book in our sample, the width we should have seen (if the widths were completely determined by the straight line), the widths we actually did see, and the deviation of the actual widths from the straight-line widths.

▶ **TABLE 14.1** Each row represents one book, and we list the number of pages, the width the book should have if all of the points perfectly followed the linear model, the actual width, and the deviations (actual width minus straight-line width).

| Pages | Straight-line Width | Actual Width | Deviation |
|-------|---------------------|--------------|-----------|
| 203   | 13.6                | 15.0         | 1.4       |
| 317   | 17.8                | 12           | −5.8      |
| 409   | 21.2                | 27           | 5.8       |
| 765   | 34.2                | 30           | −4.2      |

The deviations from the straight line have a technical name: **residuals**. *Residual* means "what's left over after something is taken away" or "remainder." The idea here is that the residual is the excess that doesn't fit on the line. Sometimes it's referred to as the error, because it represents how far away the observed value is from what it should be (according to the straight line). The typical deviation is measured by the standard deviation, $\sigma$ (sigma). You can think of $\sigma$ as measuring the amount of random variation present.

**KEY POINT** A residual is the difference between the actual (observed) $y$-value and the predicted $y$-value that lies on the line. By examining the collection of residuals, we can understand how the real-world data differ from our straight-line predictions.

Up to now, we've described the linear model conceptually, but to be more precise, we now list the particular conditions that must hold for the linear model to be a valid description of the data.

1. Linearity: The trend is linear.

2. Normality: The errors follow a Normal distribution with mean 0 and a standard deviation represented by $\sigma$. In shorthand notation, the residuals must be $N(0, \sigma)$.

3. Constant standard deviation: The standard deviation $\sigma$ must be the same for all values of the predictor variable. This will be explained in more detail shortly. (For example, see Figure 14.9.)

4. Independence: The errors must be independent of one another. Knowing that one observation is, say, above the line, should tell us nothing about whether the next observation is above or below the line, or by how much.

## Checking the Conditions of the Model

A very useful tool for checking conditions 1 and 3 of the model is the **residual plot**. A residual plot is a scatterplot that has the residuals on the vertical axis and the original *x*-values on the horizontal axis. Residual plots are like magnifying glasses, in that they zoom in on potential problems in the model and make deviations from the linearity condition and the constant standard deviation condition easier to spot.

Most statistical software packages either produce residual plots automatically as part of the regression analysis or will produce them upon request. Figure 14.2 shows the residual plot for our four books.

◀ **FIGURE 14.2** A residual plot for the four data points in Table 14.1. The vertical axis indicates the residuals (the true width minus the width if it fell exactly on the straight line), and the horizontal axis marks the values of the predictor variable (the number of pages).

Figure 14.3a shows a scatterplot of the cost of a plane trip versus the distance traveled for five trips. The regression line is included on the scatterplot. The colored vertical lines represent the residuals. The value of the residual is the length of the vertical line, which is the distance between the point and the regression line. If the point is above the line, then the residual has a positive value. If the point is below the line, then the residual has a negative value.

▲ **FIGURE 14.3** **(a)** The original scatterplot with the regression line. **(b)** The resulting residual plot.

Figure 14.3b shows these residuals plotted on the vertical axis. It illustrates that the residual plot magnifies the residuals so that they can be seen more clearly. The colored vertical lines have the same lengths as they did in Figure 14.3a, but the scale of the plot in Figure 14.3b makes them easier to see.

## Checking Linearity

**Tech**

If the trend in the original data is linear, then the residual plot should have *no* slope. In other words, the residual plot should have a flat trend. For example, Figure 14.4a shows a scatterplot for the larger collection of books; width is plotted against length, measured in terms of number of pages. The regression line has been added to help you see that the trend is linear. Figure 14.4b shows the residual plot. The residual plot has no trend. This is good, because it indicates that the first condition of the linear model is satisfied.

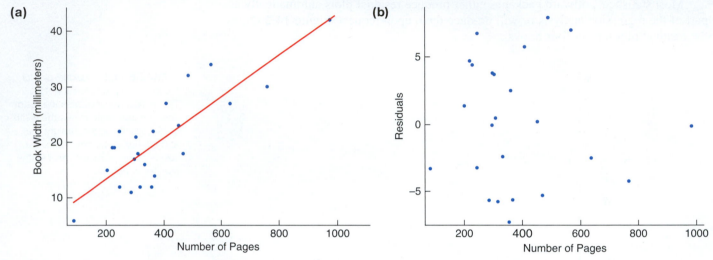

▲ **FIGURE 14.4** **(a)** The width of books versus the number of pages. The trend is linear. **(b)** The residual plot for the regression line in Figure 14.4a. The lack of a trend suggests that the first condition of the model, that *pages* and *width* have a linear trend, is satisfied.

When the trend is nonlinear, as it is for the relationship between daily ozone levels and maximum air temperature in Figure 14.5a, the residual plot will display some sort of trend, as you can see in Figure 14.5b.

▲ **FIGURE 14.5** **(a)** The association between levels of ozone (a pollutant) and maximum air temperature over a year at a site in California just outside Los Angeles. The trend is curved, and the straight best-fit line does not fit the trend. The best-fit line is too low for low temperatures and too high for moderate temperatures. **(b)** The residual plot shows a trend that decreases initially and then levels off. The presence of this trend indicates that condition 1 is not satisfied. The trend for the association between temperature and ozone is nonlinear. (Source: Hastie and Tibshirani, 1990.)

You may be wondering why we need a residual plot to tell us whether a trend is linear or nonlinear. Can't you tell just by looking at the scatterplot? It is true that in many situations, you *can* tell from the scatterplot. But the residual plot serves as a sort of microscope and makes it easier to see subtle nonlinear patterns.

Figure 14.6 shows data collected by Dan Teague of the North Carolina School of Science and Mathematics. He filled an urn with lemonade, opened the urn's spigot, and recorded the depth of the lemonade in the urn over time (seconds). At first glance, the trend looks fairly linear.

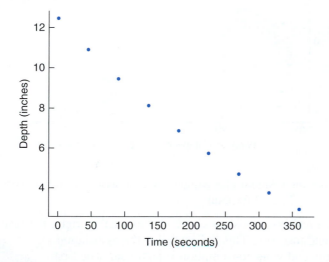

◄ **FIGURE 14.6** The depth of lemonade in an urn as the urn is drained. Does this trend look linear?

Now consider the residual plot (Figure 14.7).

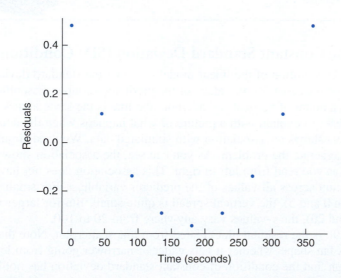

◄ **FIGURE 14.7** The residual plot for Figure 14.6. The trend in the residual plot tells us that the trend between the depth of lemonade in the urn and elapsed time is nonlinear and does not satisfy the linearity condition.

The residuals show a very strong trend. The residual plot highlights the deviations from a straight line that are hard to see without very careful scrutiny. Look back at Figure 14.6, and you can see that the points do not exactly follow a straight line. (In fact, they follow a quadratic curve.) But without the residual plot, we might not have noticed that these data do not satisfy the linearity condition.

## EXAMPLE 2 To Life!

In Chapter 4 we presented a scatterplot showing the association between mortality in a country and the wine consumption for that country in liters per person per year (see Figure 4.26). Figure 14.8 shows the residual plot that results from fitting a linear regression model.

▶ **FIGURE 14.8** Residuals from a linear regression to predict mortality for a country based on per-capita wine consumption.

**QUESTION** Does the residual plot suggest that the association between mortality and wine consumption is linear? Explain.

**SOLUTION** The residual plot shows a trend: From left to right, the residuals drop sharply at first and then rise. This indicates that the association between mortality and wine consumption is nonlinear. The first condition of the model does not hold.

**TRY THIS!** Exercise 14.3

## Checking the Constant Standard Deviation (SD) Condition

The constant SD condition of the linear model says that the standard deviation in the $y$-values must be the same for all values of the predictor variable. Essentially, this means that the amount of vertical spread about the line is the same across the entire line. This is easier to explain with a picture of what happens when the condition fails.

Figure 14.9a shows an association with simulated data. We used simulated data so that we can exaggerate the problem. As you can see, the association shows greater and greater spread as you read from left to right. This association does not have the same standard deviation across all values of the predictor variable, $x$. For small values of $x$ (say, between 0 and 5), the vertical spread is quite small. But for larger values of $x$ (between 15 and 20), the $y$-values vary anywhere from 20 to 100.

Figure 14.9b is the residual plot created from this scatterplot. Note that it has a "fan" shape. A fan shape, whether it gets wider or narrower going from left to right, is a tell-tale sign that the condition of constant standard deviation has not been satisfied. We can see a fan shape even in the original scatterplot (Figure 14.9a) for these simulated data, because we exaggerated the increasing spread. But the residual plot (Figure 14.9b) magnifies this shape and makes it easier to see. In general, a residual plot highlights the existence of nonconstant SD even when it's hard to see in the original scatterplot. A good residual plot will not show any trends or any fan shapes; it will be featureless.

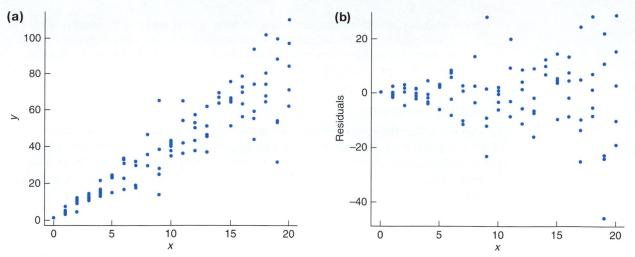

**(a)**

**(b)**

▲ **FIGURE 14.9** **(a)** Simulated data that exhibit a nonconstant standard deviation. This violates one of the conditions that must hold for the linear model to be a valid description of the data. **(b)** Residual plot for Figure 14.8a. The increasing fan shape is a sure sign that the condition requiring constant standard deviation for all values of *x* has been violated. The residual plot clearly shows little vertical variation for low values of *x*, where residuals range from about −5 to +5. However, there is quite a bit of variation for larger values of *x* (where residuals range from about −40 to 25).

# EXAMPLE 3 Predicting House Prices

What is the relation between the size of a home (in square feet) and the list price? Statistician Andrew Bray gathered information on a sample of homes for sale in Long Beach, California (a mostly middle-class city outside Los Angeles) from the website househunt.org. Then we found the regression line for these data. Figure 14.10 shows the residual plot that results from this regression line.

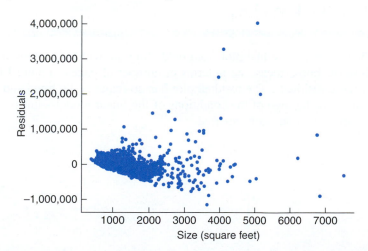

◀ **FIGURE 14.10** Residual plot for a regression line that predicts the cost of a home on the basis of the size of the home for a sample of homes in Long Beach, California.

**QUESTION** Evaluate the residual plot to determine whether the conditions of the linear regression model hold. If not, which conditions do not hold? Why?

**SOLUTION** The residual plot shows a decreasing trend. This means there is a nonlinear association between price of a house and the size of the house. The residual plot also shows a fan shape with increasing spread. This means that the condition of constant standard deviation is not satisfied.

**TRY THIS!** Exercise 14.5

> **KEY POINT** A residual plot with no structure is a good thing: It indicates that the linearity and constant SD conditions are satisfied. A fan shape in a residual plot indicates that the constant SD condition does not hold. A trend in the residual plot indicates that the linearity condition does not hold.

## Checking Normality

To check whether the errors follow a Normal distribution, examine the distribution of the residuals. You can look at either a histogram or a QQ plot, but the QQ plot is more useful in most situations. Figure 14.11 shows a QQ plot for the residuals from the home prices in Long Beach. The QQ plot is not a straight line, which suggests that these residuals are not Normal.

▶ **FIGURE 14.11** The QQ plot for the residuals shown in Figure 14.9 (from the prices of homes, given house size) indicates that these residuals do not follow a Normal distribution.

> **🔄 Looking Back**
>
> **Checking Normality**
> Recall from Chapter 13 that the QQ plot is used to check whether a sample distribution is approximately Normal. If the points follow (more or less) a straight line, then the sample distribution is approximately Normal.

💿 **EXAMPLE 4** Book Widths

 **Tech**

Figure 14.12a shows the residual plot from predicting the width of a book on the basis of the length of the book (measured in terms of number of pages). Figure 14.12b is a QQ plot of these residuals. The residual plot has no features: no trend and no fan shape. This suggests that two of the conditions of the linear model—linear trend and constant standard deviation—are satisfied.

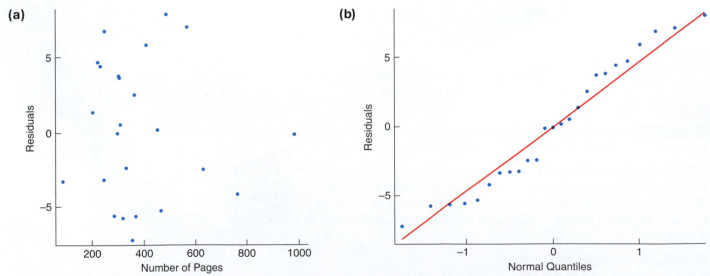

▲ **FIGURE 14.12 (a)** Residual plot for predicting the width of a book on the basis of the number of pages in the book. **(b)** The QQ plot for residuals that comes from predicting the width of a book on the basis of the number of pages in the book, using a straight line.

**QUESTION** Is the Normality condition of the linear regression model satisfied?

**SOLUTION** Yes. Because the points of the QQ plot generally follow a straight line, we can conclude that the residuals are approximately Normally distributed.

**TRY THIS!** Exercise 14.7

## Checking the Independence Condition

The independence condition usually cannot be checked by looking at the data. If you know that the data were collected in a particular order, then you might try plotting the data against the order in which they were collected. If you see a trend, it indicates that the observations might not be independent. However, you generally do not have this information. Sometimes, those who collected the data can give you enough information to determine whether this condition might be violated. For example, if you are studying the effects of a weekly music class on math scores for students in a school, you might suspect that scores of students who are in the same classroom with the same math teacher might not be independent, because if one student does very well on math, it might indicate that others in that class did very well on math, too.

In most situations, however, you will have to assume that this condition holds and be prepared to accept that your conclusions might not be sound if it is later discovered that the data are not independent.

**KEY POINT** You should always check that the conditions of the linear regression model hold before interpreting the regression line.

1. Linearity: Use a residual plot. If it has no trend, the linearity condition is satisfied.
2. Constant standard deviation: Use a residual plot. If it has no fan shape, then the constant SD condition is satisfied.
3. Normality: Make a QQ plot of the residuals. If it (mostly) follows a straight line, the normality condition is satisfied.
4. Independence: Review any information you have about how the data were collected to decide whether the independence condition is satisfied.

## SECTION 14.2

# Using the Linear Model

Marcus Vitruvius Pollio was a Roman architect and engineer who lived in the first century B.C. He proposed a theory of architecture concerning the qualities that make buildings beautiful. Vitruvius claimed that a sense of proportion was part of natural beauty. This belief led him to study the proportions of natural creatures, including humans. Vitruvius claimed, for example, that the typical human's arm span (the length from finger tip to fingertip with both arms extended to the side) was the same as his or her height. Over 1000 years later, Leonardo da Vinci was interested in this idea and made a famous drawing known as "The Vitruvian Man" (Figure 14.13 on the next page). Variations of this picture appear on the euro coin and in many other places as well.

However, is the Vitruvian Man's proportion (arm span equals height) true? To find out, we could collect data on a class of adult students, measuring their arm spans and heights. We could then plot arm span against height. If Vitruvius was correct, what would be the intercept and slope for a regression line for these data? (We must assume that the linear model conditions hold, of course.) Given the amount of variability in such data (measurement error, differences between people), we probably won't see exactly the same values as Vitruvius would have liked, even if his theory is correct. So how can we confirm or discredit his theory?

► **FIGURE 14.13** The length of the Vitruvian Man's arm span is the same as his height, as this drawing by Leonardo da Vinci illustrates.

Inference comes to the rescue. Our regression model will allow us to do hypothesis tests on the intercept and slope so that we can reject (or fail to reject) Vitruvius' theory.

## Estimators for the Intercept and Slope

We now must be careful to distinguish between the *population* values of the slope and intercept, and *estimates* of those values. The population values are the values an intercept and slope would have if we measured the entire population with perfect accuracy. The population intercept and slope are parameters, and we estimate them with our sample data.

To represent the population values of intercept and slope, we follow standard statistical practice and use Greek characters for population parameters.

$\beta_0$ represents the population intercept.

$\beta_1$ represents the population slope.

Our method for estimating the population values uses the same estimates that we used in Chapter 4. However, now we have a different notation to represent them. The "hat" over the character tells us that the symbol now represents an estimate of the population parameter.

$\hat{\beta}_0$ represents the estimator of the population intercept.

$\hat{\beta}_1$ represents the estimator of the slope.

In addition,

$\hat{y}$ represents the predicted value of $y$.

The formulas for these estimators are the same as in Chapter 4, even though the notation is different:

$$\hat{\beta}_1 = r\frac{s_y}{s_x} \quad \text{and} \quad \hat{\beta}_0 = \bar{y} - \hat{\beta}_1\bar{x}$$

We can then write the equation of the regression line:

$$\hat{y} = \hat{\beta}_0 + \hat{\beta}_1 x$$

As in Chapter 4, we expect that you will use these formulas on only rare occasions. Most of the time, your software will compute them for you. Your job is to interpret the output correctly and evaluate whether the conditions of the regression model hold sufficiently well for you to learn anything from the computer output.

---

**🔄 Looking Back**

**Regression Line Formulas**
Formulas 4.2a, 4.2b, and 4.2c in Chapter 4 represent the same regression line formulas but use a different notation:

$$b = r\frac{s_y}{s_x}$$

$$a = \bar{y} - b\bar{x}$$

Predicted $y = a + bx$

If the conditions of the linear model hold, then two useful facts about these estimators follow:

1. The sampling distributions of the estimators follow the Normal model.

2. The estimators are unbiased.

Even if the Normality condition is violated so that the errors are not Normal, the Central Limit Theorem comes to the rescue. Because of the CLT, if the sample sizes are large enough, the sampling distributions of the intercept and slope will be approximately Normal, and this approximation is better for larger sample sizes.

The fact that the estimators are unbiased means that typically our calculated values for the intercept and slope, based on our data, will be about the same as the population values. What we mean by "typical" is that if each of a large number of researchers drew a random sample of the two variables from the same population and found the regression line, then the mean of all of their intercepts and slopes would be the population values.

How close *our* estimates of the slope and intercept are likely to be to the true population values depends on how much these estimates vary from sample to sample. This variability is measured by the standard error. The estimated standard errors for the estimators will be given to you by the software package.

 **KEY POINT**    The estimators for the intercept and slope of a regression line are unbiased. If the Normality condition holds, then the sampling distributions of these estimators are Normal. If the Normality condition does not hold, then the sampling distributions are approximately Normal (and the larger the sample size, the better).

## Hypothesis Tests for Intercept and Slope

The test statistics for both the intercept and the slope have a familiar form, which was introduced in Chapter 8 for the one-proportion $z$-test and was also used in Chapter 9 for the one-sample $t$-test:

$$t = \frac{\text{estimator} - \text{null value}}{SE}$$

where $SE$ stands for the estimated standard error (the estimated standard deviation of the estimator).

For the intercept,

$$t = \frac{\hat{\beta}_0 - \text{null}}{SE_{\hat{\beta}_0}}$$

For the slope,

$$t = \frac{\hat{\beta}_1 - \text{null}}{SE_{\hat{\beta}_1}}$$

The value for the estimated standard error is given in the software output. If the conditions for the linear regression model hold, then both of these test statistics follow a $t$-distribution with $n - 2$ degrees of freedom ($n$ is the number of observations.)

Most statistical software is designed to automatically produce a test of a very specific hypothesis about the slope:

$H_0$: The slope equals 0.

$H_a$: The slope does not equal 0.

Or, in symbols,

$$H_0: \beta_1 = 0$$
$$H_a: \beta_1 \neq 0$$

 **Looking Back**

**The $t$-Test**
The test statistic for the one-sample $t$-test (Formula 9.2) is
$$t = \frac{\bar{x} - \mu_0}{SE_{EST}}$$

 **Looking Back**

**Standard Errors for Sample Means**
The standard deviation of the sampling distribution of the sample mean, which is also called the standard error for the sample mean, is
$$SE = \frac{\sigma}{\sqrt{n}}$$
where $\sigma$ is the population standard deviation. Because $n$ is bigger than 1, the standard error is always smaller than the population standard deviation.

This is a very important test. If the true slope is 0, then there is no linear association between the two variables—and no reason to do any further analyses. This means that this test could also be phrased as

$H_0$: There is no linear association between the two variables.

$H_a$: There is a linear association between the two variables.

or as

$H_0$: The correlation is 0.

$H_a$: The correlation is not 0.

The next example will show you how to find the information you need to do a test on the slope from the output provided.

 EXAMPLE 5 Mario Kart Auction

Suppose you want to buy a game on eBay, the on-line auction site. You make a bid, and if the next person to come along wants the item, he or she can bid more than you did. At the end of the scheduled time, the person who has made the highest bid gets to purchase the item. Suppose you want to buy a Mario Kart Wii game on eBay. (This is a popular video game that runs on the Wii system.) You might suspect that the more people who are bidding, the more expensive the item will be. David Diez, a statistician, collected data from a sample of auctions for Mario Kart on eBay in the fall of 2009.

QUESTION Is there an association between the number of people who bid for these items and the final cost of the item? Use Figure 14.14, which shows the results from a regression model in which the total price for a Mario Kart was predicted from the number of bids for that item. State the null and alternative hypotheses, the observed value of the test statistic, and the p-value. State your conclusion using a

▶ FIGURE 14.14 StatCrunch output from a regression model that predicts the price (totalPr), in dollars, for a Mario Kart game on an eBay auction for a given number of bids (nBids).

5% significance level. Assume that the conditions of the linear regression model are satisfied.

**SOLUTION** To test whether there is an association between price (TotalPr) and number of bids (nBids), we test whether the slope of the regression line is 0. Usually, we would begin with a scatterplot to make sure there were no unusual observations and as a preliminary check that the association between these variables is linear. However, the problem statement tells us to assume that the conditions of the linear regression model hold, so we proceed accordingly.

$H_0: \beta_1 = 0$. There is no association between price and number of bids.

$H_a: \beta_1 \neq 0$. There is an association between price and number of bids.

The $t$-statistic is given at the intersection of the column labeled T-Stat and the row labeled Slope. The value is

$$t = -0.9311$$

The p-value is given as 0.3534. Because the p-value is larger than 0.05, we fail to reject the null hypothesis.

**CONCLUSION** There is not enough evidence to conclude that the price of these items is associated with the number of bids placed on the item.

**TRY THIS!** Exercise 14.9

---

If you wish to test a value other than 0 in the null hypothesis, you need to do so by hand. The same is true of tests of the intercept; the software produces tests that the intercept is 0, and if you wish to test another value in your null hypothesis, you need to do the test by hand. The Vitruvian Man provides us with one such example.

## EXAMPLE 6 Vitruvian Students

One of the authors collected data from a statistics class. Students helped each other measure their arm spans and reported their heights (both in inches). The scatterplot is shown in Figure 14.15. Output from fitting a regression line is shown in Figure 14.16. We can use these data to test Vitruvius' theory that people's arm spans are generally equal to their height.

◄ **FIGURE 14.15** Data from college students showing arm span and height (both in inches).

▶ FIGURE 14.16 Output from the regression analysis, using height to predict arm span.

If Vitruvius was correct, then an equation of the line for the association between arm span and height (the deterministic component of the model) would be

$$\text{Predicted Armspan} = \text{Height}$$

or

$$\text{Predicted Armspan} = 0 + 1\,\text{Height}$$

In other words, if Vitruvius is correct, then the intercept should be 0, and the slope should be 1.

**QUESTION** Was Vitruvius correct? Use the regression output provided in Figure 14.16 to test Vitruvius' theory. Be sure to carefully state the null and alternative hypotheses. State your conclusion using a significance level of 5%.

**SOLUTION** From the output in Figure 14.16, we see that the equation for the regression line is

$$\text{Predicted Armspan} = -3.84 + 1.05\,\text{Height}$$

If Vitruvius was correct, then a person's arm span is equal to his or her height. This means that Vitruvius' theory says that the intercept should be 0 and the slope should be 1. We observed an intercept of −3.84 and a slope of 1.05. But are the differences between these observed values and those predicted by the Vitruvian theory simply due to chance?

First, let's examine the intercept.

$$H_0: \beta_0 = 0$$
$$H_a: \beta_0 \neq 0$$

The computer output assumes we are testing whether the intercept is equal to 0, and we can read the $t$-statistic directly from the output: $t = -0.54$. The p-value is large: p-value $= 0.594$. This is bigger than 0.05, so we do *not reject* the hypothesis that the intercept is 0. We conclude that the intercept is consistent with Vitruvius' theory. (Or, at the very least, there is not enough evidence to conclude that the data are *inconsistent* with his theory.)

As for the slope:

$$H_0: \beta_1 = 1$$
$$H_a: \beta_1 \neq 1$$

We can't simply use the output from the software, because the output is testing whether the slope is 0 or not. But we have all of the ingredients we need to find the observed value of the test statistic, and we can use a calculator to find the p-value. Recall that the test statistic is

$$t = \frac{\hat{\beta}_1 - \text{null}}{SE_{\hat{\beta}_1}}$$

Here, the null value is 1. The value for the estimated SE is found from the computer output, in the column that says "Std. Error" in the row for the slope. We find

$$t = \frac{\hat{\beta}_1 - \text{null}}{SE_{\hat{\beta}_1}} = \frac{\hat{\beta}_1 - 1}{SE_{\hat{\beta}_1}} - \frac{1.054347 - 1}{0.105442} = 0.5154$$

To find the p-value, we need the probability of getting a *t*-statistic as extreme as or more extreme than 0.5154. The output table tells us that we use a *t*-statistic with 23 degrees of freedom. We can look up this probability in a table or use a statistical calculator. Figure 14.17 shows output from StatCrunch's calculator feature. This tells us that the probability of getting a *t*-statistic equal to or greater than 0.5154 is about 0.306.

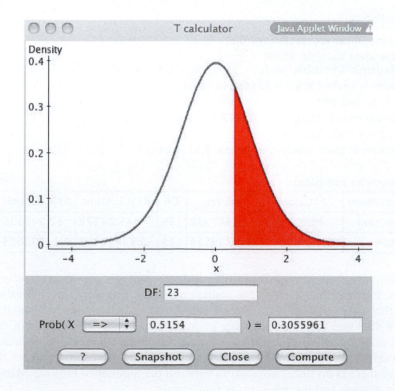

◄ **FIGURE 14.17** StatCrunch calculator showing the probability of getting a *t*-statistic greater than or equal to 0.5154 in a *t*-distribution with 23 degrees of freedom.

This is only half the story. Because our alternative hypothesis is a two-tailed hypothesis, we also need the probability of getting a $t$-statistic equal to or smaller than $-0.5154$. Because the $t$-distribution is symmetric, the p-value will simply be twice 0.306; that is, p-value $= 0.612$. We again fail to reject the null hypothesis and find insufficient evidence that the slope is anything other than 1.

**CONCLUSION** We lack evidence to refute Vitruvius' theory.

**TRY THIS!** Exercise 14.11

## Confidence Intervals for Intercept and Slope

Confidence intervals for the intercept and slope follow a familiar form:

$$\text{estimate} \pm t^*SE_{EST}$$

This is the same form for all of the confidence intervals we've presented. The constant $t^*$ is chosen from a $t$-distribution so that we get an interval with the desired confidence level. (Almost always, you'll use a 95% confidence level.) The estimate here can be either the intercept or the slope. The $SE_{EST}$ is the standard error and is obtained from the regression output.

For instance, to predict the price of a Mario Kart Wii game (totalPr) from the number of bids (nBids), we found this regression line from the output in Figure 14.14:

$$\text{Predicted totalPr} = 49.1 - 0.12 \text{ nBids}$$

The standard error for the slope is 0.134 (after rounding) and is found in the StatCrunch output column labeled Std. Err.

The best way to find the confidence intervals for the intercept and slope is to read them from the output. Most software provides confidence intervals for both parameters. Figure 14.18 shows StatCrunch output for the linear regression to predict the total price of the Mario Kart on the basis of the number of bids.

▶ **FIGURE 14.18** Regression output including confidence intervals for predicting the price of the Mario Kart on the basis of the number of bids received in an eBay auction. A 95% confidence level for the intercept is 45.2 to 52.9. A 95% confidence interval for the slope is $-0.39$ to 0.14.

**Simple Linear Regression**

**Options**

**Simple linear regression results:**
Dependent Variable: totalPr
Independent Variable: nBids
totalPr = 49.097908 – 0.12448585 nBids
Sample size: 141
R (correlation coefficient) = –0.0787
R-sq = 0.0061987373
Estimate of error standard deviation: 9.117984

**Parameter estimates:**

| Parameter | Estimate | Std. Err. | DF | 95% L. Limit | 95% U. Limit |
|---|---|---|---|---|---|
| Intercept | 49.097908 | 1.947032 | 139 | 45.24828 | 52.947536 |
| Slope | –0.12448585 | 0.13369364 | 139 | –0.38882193 | 0.13985023 |

The lower and upper bounds of the confidence intervals are taken directly from the output. For example, we are 95% confident that the true intercept (the value of the intercept that we would get if we saw all eBay auctions and not just our small sample) is between 45.2 and 52.9. Perhaps more interestingly, we are 95% confident that the true slope is between $-0.39$ and 0.14. Because this interval includes 0, we cannot rule out the possibility that the true slope is 0. In other words, we cannot rule out the possibility that there is no linear association between the number of bids the Mario Kart receives at auction and the price that is finally paid.

## Interpreting Confidence Intervals for Regression

The interpretation of the confidence intervals for the intercept and slope of a regression equation is exactly the same as the interpretation of the confidence intervals of other parameters. The confidence interval gives us a range of plausible population values. We can use this to test hypotheses, as the next example shows.

### EXAMPLE **7** Vitruvian Confidence Intervals

The output in Figure 14.19 repeats the regression for predicting arm span from height for the class of statistics students used in Example 6. This time, the software was asked to provide confidence intervals for the intercept and slope.

◀ **FIGURE 14.19** Regression output to predict arm span from height for the same data as in Example 6.

**Simple linear regression results:**
Dependent Variable: armspan
Independent Variable: Height
armspan = –3.8370333 + 1.0543467 Height
Sample size: 25
R (correlation coefficient) = 0.9017
R-sq = 0.81298727
Estimate of error standard deviation: 1.5954849

**Parameter estimates:**

| Parameter | Estimate | Std. Err. | DF | 95% L. Limit | 95% U. Limit |
|-----------|----------|-----------|-----|--------------|--------------|
| Intercept | –3.8370333 | 7.0985656 | 23 | –18.521536 | 10.847468 |
| Slope | 1.0543467 | 0.105441906 | 23 | 0.8362235 | 1.2724699 |

**QUESTION** Do these intervals support Vitruvius' theory that the intercept is 0 and the slope 1? State the 95% confidence intervals for the intercept and slope, and explain whether they support or refute Vitruvius' theory.

**SOLUTION** The 95% confidence interval for the intercept is –18.5 to 10.8. The 95% confidence interval for the slope is 0.84 to 1.27.

Because the confidence interval for the intercept includes 0, we have no reason to reject the theory that the population intercept is 0. Similarly, the 95% confidence interval for the slope includes 1.

**CONCLUSION** We cannot reject Vitruvius' theory, because the confidence interval for the intercept includes the value 0, and the confidence interval for the slope includes the value 1. This is consistent with the hypothesis tests in Example 6.

**TRY THIS!** Exercise 14.15

## SECTION 14.3

# Predicting Values and Estimating Means

The regression line is also useful for making predictions about future observations, as you saw in Chapter 4. But there are two different types of questions that we can ask of a regression line, and it is important to be able to tell these two types of questions apart.

**Looking Back**

**Predicting Observations**
Review Example 3 in Chapter 4 to see how we predicted the width of a book on the basis of the number of pages.

One type of question involves groups of individuals and is concerned with the mean value of the group. The other type of question is about an isolated individual and is concerned with an individual's value. We use the same number to answer both questions, but our *uncertainty* about this answer is different for the different questions.

Figure 14.20 shows a regression line that can be used to predict the amount of mercury found in largemouth bass (a freshwater fish) on the basis of the pH of the lake where they are caught. pH measures the acidity of the water. A value of 7 is neutral. A value of 0 represents the highest level of acidity, and a value of 14 represents the lowest level of acidity (the most basic or alkaline reading). The data are based on samples of fish caught in 53 lakes in Florida (Lange et al. 1993).

▶ **FIGURE 14.20** Mercury levels found in largemouth bass as a function of the pH of the water where they are caught. Mercury is measured in micrograms of mercury per gram of fish.

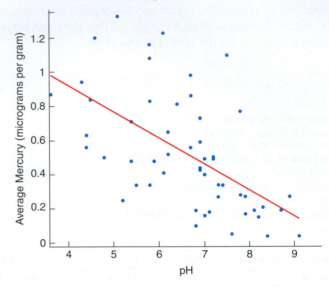

Evidence exists that humans can get mercury poisoning from eating fish with mercury in their tissue. For this reason, it is important to understand and predict the mercury content of fish that swim in these lakes.

The equation of the regression line shown is

$$\text{Predicted mercury level} = 1.53 - 0.152 \, \text{pH}$$

There are two questions we might ask:

1. If the acidity of a lake measures pH 7, what is the mean level of mercury found in all fish in the lake?

2. I've just caught a fish from a lake that measures pH 7. What mercury level will that fish have?

The answer to both questions is the same and is found with the regression line:

$$\text{Predicted mercury} = 1.53 - 0.152 \times 7 = 0.466$$

Question 2 has greater uncertainty, however, because it is asking about an individual fish. Question 1 has less uncertainty because it is asking about a collection of fish.

To express our uncertainty for questions such as question 1, we use a confidence interval. But to express uncertainty for predictions about individuals, we use something called a **prediction interval**.

For all practical purposes, prediction intervals function the same way as confidence intervals. The difference is that confidence intervals are always concerned with estimating population parameters—characteristics of a *population*. Prediction intervals are concerned with predicting values for *individuals*.

Prediction intervals are wider than confidence intervals, because there is more uncertainty in predicting an individual's value than in predicting the mean of an entire group. Individual scores have more variability than sample means. We know this because the sampling distribution for the mean has a smaller standard deviation than the population distribution. This increased uncertainty means that prediction intervals can be quite a bit wider than confidence intervals.

Figure 14.21 shows output from StatCrunch. Most statistical software either will show you both a prediction and a confidence interval, or will ask you to choose which you want. StatCrunch shows you both (and allows you to input any value for $x$ that you please). Note that the information you are looking for is given last, after the summary information about the regression model.

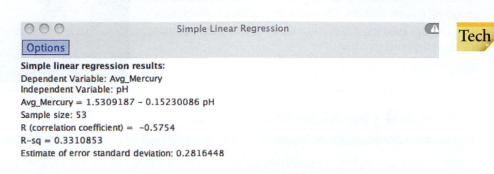

◄ **FIGURE 14.21** StatCrunch regression output for predicting mercury level in fish based on the water's pH. The last table shows a confidence interval (C.I.) for estimating the mean mercury content of fish caught in a lake with pH = 7, and the prediction interval (P.I.) for an individual fish caught in a lake with pH = 7.

**Simple linear regression results:**
Dependent Variable: Avg_Mercury
Independent Variable: pH
Avg_Mercury = 1.5309187 − 0.15230086 pH
Sample size: 53
R (correlation coefficient) = −0.5754
R-sq = 0.3310853
Estimate of error standard deviation: 0.2816448

**Parameter estimates:**

| Parameter | Estimate | Std. Err. | DF | T-Stat | P-Value |
|---|---|---|---|---|---|
| Intercept | 1.5309187 | 0.20349288 | 51 | 7.5232058 | <0.0001 |
| Slope | −0.15230086 | 0.030313263 | 51 | −5.024232 | <0.0001 |

**Analysis of variance table for regression model:**

| Source | DF | SS | MS | F-stat | P-value |
|---|---|---|---|---|---|
| Model | 1 | 2.0023627 | 2.0023627 | 25.242905 | <0.0001 |
| Error | 51 | 4.0455127 | 0.07932378 | | |
| Total | 52 | 6.0478754 | | | |

**Predicted values:**

| X value | Pred. Y | s.e.(Pred. y) | 95% C.I. | 95% P.I. |
|---|---|---|---|---|
| 7 | 0.46481267 | 0.040628992 | (0.38324657, 0.5463788) | (−0.106465735, 1.0360911) |

Note that the confidence interval is much narrower than the prediction interval. In fact, the prediction interval is so wide that it includes negative values, even though negative mercury levels are impossible. This suggests that there might be too much variability in the data for us to make a useful prediction of how much mercury an individual bass might contain, although we can get a possibly useful estimate of the mean mercury level of all bass in the lake.

**KEY POINT**    Prediction intervals are used to express the uncertainty in predicting an individual observation. They are interpreted the same way as confidence intervals. A 95% prediction interval of (−0.11, 1.04) means we are 95% confident that the amount of mercury found in a single fish caught from a lake with pH = 7 will be in this interval.

# EXAMPLE 8 Frozen Dinners

A consumer group plots the number of calories advertised by the manufacturer for frozen dinners (single portion) against the actual caloric content based on laboratory testing. The group's regression line predicts that a frozen dinner advertised to have 1000 calories actually has, on average, 1100 calories.

QUESTION Which of the two questions that follow can be answered with a confidence interval? Which can be answered with a prediction interval?

A. For all frozen dinners, what is the mean calorie content if the advertised content is 1000 calories?

B. The Frozen Man TV Dinner advertises that it contains 1000 calories. What is the actual calorie content of one dinner?

SOLUTION Question A requires a confidence interval, because we are making an estimation of the mean of a large group of frozen dinners. Question B requires a prediction interval, because we are predicting the true calorie content for an individual product.

TRY THIS! Exercise 14.19

# EXAMPLE 9 One More Book

A professor wants to buy a book from an Internet bookseller, but he has only 40 millimeters of space left on his bookshelf. If the book will not fit, he does not want to buy it. From the Internet, he knows the book has 598 pages. On the basis of statistics student Edwin Onega's sample of books, he performs a regression to predict the width of a book (in mm) from the number of pages it contains. The output is shown in Figure 14.22. Assume that the conditions of the linear regression model hold.

QUESTION How wide should the professor expect the book to be? Should he use a prediction interval or a confidence interval to decide whether the book will fit on his shelf? Should he buy the book?

▶ **FIGURE 14.22** Regression output for predicting the width of a book on the basis of its number of pages. The output is based on a sample of books.

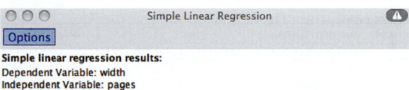

**Simple Linear Regression**

Options

**Simple linear regression results:**
Dependent Variable: width
Independent Variable: pages
width = 6.2208385 + 0.036551695 pages
Sample size: 25
R (correlation coefficient) = 0.8468
R-sq = 0.71699643
Estimate of error standard deviation: 4.707955

**Parameter estimates:**

| Parameter | Estimate | Std. Err. | DF | T–Stat | P–Value |
|-----------|----------|-----------|-----|--------|---------|
| Intercept | 6.2208385 | 2.0173435 | 23 | 3.0836785 | 0.0052 |
| Slope | 0.036551695 | 0.0047882963 | 23 | 7.6335487 | <0.0001 |

**Analysis of variance table for regression model:**

| Source | DF | SS | MS | F–stat | P–value |
|--------|-----|-----|-----|--------|---------|
| Model | 1 | 1291.5687 | 1291.5687 | 58.27107 | <0.0001 |
| Error | 23 | 509.79126 | 22.164837 | | |
| Total | 24 | 1801.36 | | | |

**Predicted values:**

| X value | Pred. Y | s.e.(Pred. y) | 95% C.I. | 95% P.I. |
|---------|---------|---------------|----------|----------|
| 598 | 28.078753 | 1.432286 | (25.115843, 31.04166) | (17.89888, 38.258625) |

**SOLUTION** The book is expected to be about 28.1 mm wide. Because the professor is predicting the width of a single book, he should use a prediction interval. This tells us that he can be highly confident that the width of the book could be as small as 17.9 mm or as great as 38.3 mm. Because he has 40 mm of space, he can be confident that the book will fit. He should buy the book.

**TRY THIS!** Exercise 14.23

# What Can Go Wrong?

What happens if the conditions are not satisfied? Sometimes, disaster. Other times, not so much.

### If the Linearity Condition Is Not Satisfied

Disaster. Estimates of slope and intercept may be biased. Predicted values and estimated means will be biased. Confidence intervals and prediction intervals are not reliable.

One thing to consider is a transform. Sometimes, taking the log of all values for either $x$ or $y$, or both, will "straighten up" a nonlinear trend. Figure 14.23a repeats the scatterplot given in Example 2. It shows a nonlinear trend between mortality in a country and the amount of wine consumed per person. Figure 14.23b shows the same data, only this time we plot the log of mortality against the log of wine consumption. Although the trend for the association of mortality and wine is not linear, the trend for the association of the log of mortality and the log of wine *is* linear.

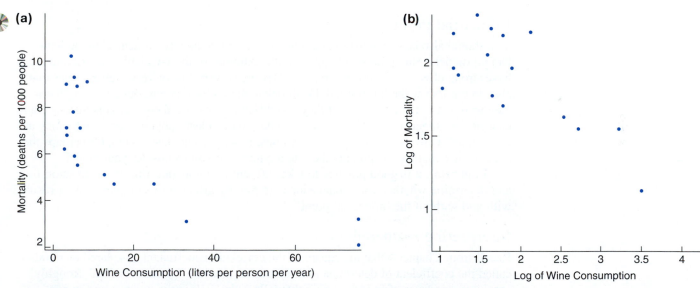

▲ FIGURE 14.23 **(a)** The association between mortality in a country and per-capita wine consumption (in liters per person per year) is nonlinear (Leger et al. 1979). **(b)** The association between the log of mortality and the log of wine consumption is relatively linear, and a regression line can be fitted and interpreted.

Transforming the variables can make interpretation of the output tricky. We will not get into details here, but we advise you to consult a statistician if you find yourself in this situation.

### If the Errors Are Not Normal

If the errors are not Normal, all is not lost. Assuming that the other conditions hold, you will still get unbiased estimates of the slope and intercept and of predicted values of $y$ (or mean values of $y$) for a given value of $x$.

Because of the Central Limit Theorem, the p-values for hypothesis tests of the slope and intercept will be approximately correct. If the sample size is large, then this approximation can be quite good. Confidence intervals for the mean value of $y$ at any given $x$-value will also be approximately correct. However, prediction intervals may not be accurate.

In some situations, the distribution of the errors is so far from Normal that the Central Limit Theorem doesn't rescue us. In many of these situations, more advanced techniques exist that can provide better p-values and confidence intervals than we can get simply by relying on the Central Limit Theorem. For example, if the response variable has only the values 0 and 1, then the linear regression model should be abandoned.

### If the Standard Deviation Is Not the Same for All Values of $x$

You will still have unbiased estimates of the slope and intercept, and unbiased estimates of means and predicted values. However, the confidence intervals, prediction intervals, and p-values will not be accurate. Statisticians have other techniques for dealing with this situation. Transformations sometimes fix this problem, but sometimes they do not.

### If the Errors Are Not Independent of Each Other

Disaster. If the errors are not independent, then the regression model is not appropriate. One common situation in which errors are not independent is when dealing with time-series data. **Time-series data** occur when the predictor variable is time. For example, we might plot daily maximum temperature against the date, intending to predict tomorrow's temperature. The linear regression model will fail for several reasons, but one is that the errors are not independent. If it is hotter than usual on one day (so the error is positive), there is more likely to be a hotter-than-usual temperature the next day. That is, if one day has a positive error, the next day is more likely to have a positive error as well.

### Influential Points

You should also be aware that regression models can be greatly influenced by outliers. Not all outliers, but outliers that appear at the extremes of the $x$-variable's distribution, can have strong effect on the estimated slope. Beware of very low or very high $x$-values that also do not follow the linear trend. Don't delete these observations, but be aware that if they were not in the data set, or if they had different $y$-values, then your conclusions could change. Investigate these points to see whether you can learn anything about why they are exceptional. Perhaps they are mistaken entries; if so, you can delete them. Or perhaps they are evidence that an exception to the rule requires that you reconsider your model.

In general, it is good practice to take influential points out, find the regression line, and determine whether your conclusions are greatly affected by comparing your results with and without the influential point.

### Interpreting $r$-squared

Recall from Chapter 4 that the square of the correlation coefficient, $r$-squared or $r^2$ (also called the coefficient of determination), is used to measure "goodness of fit." Roughly speaking, goodness of fit is a number that tells us how close the points come to our model. In order for us to interpret $r$-squared, the linearity condition of the linear regression model must be satisfied. A common mistake is to think that a high $r$-squared value means that the conditions are satisfied. In fact, you should check the conditions using the methods discussed in this chapter before you even *look* at the $r$-squared value.

The coefficient of determination ranges from 0% to 100% and represents the amount of variability in $y$ that we have "explained" with our regression line. The higher the value, the better. How do we know whether $r$-squared is high enough?

Surprisingly, this is a question that shouldn't concern you. After you've checked that the conditions of the model are satisfied, the most important thing to check in the model is whether the slope is 0 or not. Look at the hypothesis test for whether the slope is 0. If you reject the null hypothesis and conclude that the slope is not 0, then congratulations! You have discovered an association between $x$ and $y$. The value of $r$-squared is not sufficient for deciding whether the linear relationship is strong

**Looking Back**

**Coefficient of Determination**
The correlation coefficient squared ($r$-squared) is called the coefficient of determination. In Section 4.4, you learned that $r$-squared measures how much of the variation in the response variable is explained by the explanatory variable.

enough. We can decide whether the relationship is strong enough (and hence *r*-squared is high enough) by determining whether the regression line is useful for our purposes.

Recall that *r*-squared tells us how closely the points are clustered about the line. When you have a high value for *r*-squared, it means that your estimates and predictions will be more precise. But you shouldn't trust *r*-squared to tell you whether you should find prediction intervals or confidence intervals. Go ahead and compute them, and decide for yourself whether they are useful. If the intervals are too wide, then your *r*-squared is too low.

For example, suppose an admissions officer at a large college uses applicants' SAT scores to determine how successful they would be if admitted to the school. She has a large sample of data from current students that includes their SAT scores and their first-year GPAs. She carries out a regression to predict GPA on the basis of critical reading SAT score. After confirming that the conditions of the linear model are satisfied, she finds a regression line

Predicted GPA $= 1.108 + 0.00356$ (critical reading SAT score)

with *r*-squared $= 0.104$ (or 10.4%). From the output, she sees that the p-value for the hypothesis test for the slope is 0.02.

Now, this *r*-squared value seems low, but she can still learn quite a bit from the data. The output tells her to reject the null hypothesis that the slope is 0. This means she concludes that there really is a positive association between critical reading SAT score and first-year GPA. Now she's interested in what the mean GPA will be for all of the students with a 650 critical reading score.

The 95% confidence interval for the mean GPA of all students with a 650 critical reading score is (2.7, 2.9). This is useful information for her, because it tells her that these students will, on average, have a GPA of about C+ at the end of their first year. Even though the *r*-squared value was low, it still yielded useful information: (1) There is a real association between critical reading SAT score and GPA, and (2) we can find useful confidence intervals for mean GPAs for groups of students.

On the other hand, suppose a worried parent calls the college. The parent's son got a score of 650 on his critical reading SAT. What will his first-year GPA be? The same regression line provides a 95% prediction interval of 1.7 to 3.6. In other words, we are confident that this student will end his first year with a GPA somewhere between a D+ and an A. Well, we didn't really need a regression analysis to tell us that. There is too much variability in the data to make predictions for individual students.

> ### ✎ Details
>
> **r-squared**
> The coefficient of determination is actually not too useful in many situations. Knowing the value of the slope and knowing whether it is "statistically significant" (has a p-value $\leq 0.05$) provide much more useful information than knowing *r*-squared.

# CASE STUDY REVISITED

Using three-dimensional imaging equipment, engineers hope to be able to accurately predict the volume of an oyster. This information is needed in order for a machine to be able to remove the oyster from its shell. The horizontal axis of the scatterplot in Figure 14.24 on the next page shows a computer-generated measurement of the size of an oyster on the basis of the number of pixels that an image of the oyster occupies on a computer screen. The vertical axis shows the actual volume of the oyster, measured in cubic centimeters. The association looks linear. The fact that there is not much vertical scatter tells us to expect a high *r*-squared value.

The engineers hope for a positive association, because this means that they can use the number of pixels as a substitute measure for the volume of the oysters. But they also want small spread (a high *r*-squared), because this means they can get precise predictions of oyster volume given the number of pixels.

Figure 14.25a on page 649 shows the residual plot from a regression model. We see no trend to the plot and no fan shape, which tells us that the linearity condition and the constant standard deviation condition are satisfied. Figure 14.25b shows a QQ plot of the residuals, and this shows us that the errors follow a Normal distribution.

The only remaining condition is the condition of independence, and there's nothing that suggests this might be violated. Therefore, we can interpret the regression output in Figure 14.26.

From the output, we see that the equation of the line is

Predicted volume = 0.41957238 + 0.00000248 (number of pixels)

and that $r$-squared = 0.9537 (or 95%).

The p-value for the test of whether the intercept is 0 is 0.3767. This is larger than 0.05, so we fail to reject the null hypothesis. This tells us that it is plausible that the intercept is 0, which is a good sign. If there are 0 pixels on the screen, then we would predict an oyster with 0 volume.

The p-value for the slope is small enough for us to reject the null hypothesis; we conclude that the slope is not 0. The actual value for the slope is quite small (in part because the number of pixels is so large), but it is still larger than would be expected by chance. This is a good sign, too. It means that there really is an association between the number of pixels and the volume of the oyster.

Finally, we examine the 95% prediction interval for an oyster that registers 4,000,000 pixels. The regression line predicts a volume of 10.3 cubic centimeters (cc). The prediction interval says that we are confident that the oyster will be between 8.97 cc and 11.67 cc in volume. This fairly small interval tells us that we can get relatively precise predictions of the size of an oyster on the basis of the number of pixels that appear on the screen. The researchers concluded that the new method worked well, because it provided measures of oyster volume that were very close to the true values.

▶ **FIGURE 14.24** The volume of oysters (measured in cubic centimeters) plotted against the number of pixels that a picture of the oyster occupies on a screen.

**(a)**

**(b)**

▲ **FIGURE 14.25** **(a)** Residual plot for predicting oyster volume from number of pixels. The lack of a trend means that the linearity condition is satisfied. The lack of a fan shape tells us that the constant standard deviation condition is satisfied. **(b)** The QQ plot of residuals shows a Normal distribution.

**Simple Linear Regression**

**Options**

**Simple linear regression results:**
Dependent Variable: Oyster_Volume_cc
Independent Variable: Pixels_3D
Oyster_Volume_cc = 0.41957238 + 2.4752235E-6 Pixels_3D
Sample size: 30
R (correlation coefficient) =  0.9766
R–sq = 0.9537137
Estimate of error standard deviation: 0.64877564

**Parameter estimates:**

| Parameter | Estimate | Std. Err. | DF | T–Stat | P–Value |
|---|---|---|---|---|---|
| Intercept | 0.41957238 | 0.46712822 | 28 | 0.8981953 | 0.3767 |
| Slope | 2.4752235E-6 | 1.0305106E-7 | 28 | 24.01939 | <0.0001 |

**Analysis of variance table for regression model:**

| Source | DF | SS | MS | F–stat | P–value |
|---|---|---|---|---|---|
| Model | 1 | 242.83595 | 242.83595 | 576.9311 | <0.0001 |
| Error | 28 | 11.785475 | 0.4209098 | | |
| Total | 29 | 254.62143 | | | |

**Predicted values:**

| X value | Pred. Y | s.e.(Pred. y) | 95% C.I. | 95% P.I. |
|---|---|---|---|---|
| 4000000 | 10.320467 | 0.12491194 | (10.064596, 10.576337) | (8.967102, 11.673831) |

<– Back     Next –>

◀ **FIGURE 14.26** StatCrunch output from a linear regression model for oyster volume.

## Older and Slower?

| GOALS | MATERIALS |
|---|---|
| To use a collection of data to test whether two variables are associated. | Meter stick or yard stick. |

**ACTIVITY**

If you still have the chapter 3 activity data, you can use them here. The only additional information you will need is the age (in years) of the person for each recorded reaction distance.

Work in groups of two or three. One person holds the meter stick vertically, with one hand holding the top of the stick, so that the 0-centimeter mark is at the bottom. The other person then positions his or her thumb and index finger about 5 cm (2 in.) apart on opposite sides of the meter stick at the bottom. Now the first person drops the meter stick without warning; the other person catches it. Record the location of the middle of the thumb of the catcher. This is the distance that the stick traveled and is called the reaction distance, which is related to reaction time. A student who records a small distance has a fast reaction time, and a student with a larger distance has a slower reaction time. Now switch tasks. Each person should try catching the meter stick twice, and the better (shorter) distance should be reported for each person. For each trial, record the age of the catcher. Your instructor will collect your data and combine the class results.

**BEFORE THE ACTIVITY**

1. If there is no association between age and reaction distance, what will a scatterplot look like?

2. If there is an association, do you expect the association to be positive or negative? This will be your theory, which we will test with the activity.

3. What reasons other than age might explain why a catcher's reaction distances vary?

**AFTER THE ACTIVITY**

Report the regression line to predict reaction distance from age. Do the conditions of the linear regression model hold? If not, which ones fail?

Regardless of whether the conditions hold, find 95% confidence intervals to test whether your theory (from question 2 in the section Before the Activity) is supported by the data.

You can use the data in the table below, and merge your class data with these observations, if your class doesn't have much variety in ages.

| Age | ReacDist(cm) | Age | ReacDist(cm) | Age | ReacDist(cm) | Age | ReacDist(cm) | Age | ReacDist(cm) |
|---|---|---|---|---|---|---|---|---|---|
| 21 | 12 | 21 | 16 | 51 | 13 | 18 | 16 | 51 | 27 |
| 45 | 22 | 50 | 18 | 37 | 16 | 24 | 28 | 53 | 34 |
| 33 | 16 | 23 | 13 | 68 | 19 | 78 | 31 | 51 | 27 |
| 41 | 25 | 43 | 21 | 22 | 14 | 50 | 25 | 58 | 27 |
| 63 | 25 | 28 | 19 | 22 | 14 | 61 | 15 | 29 | 10 |

## CHAPTER REVIEW

### Key Terms

Deterministic component, *624*

Random component, *624*

Residuals, *626*

Residual plot, *627*

Prediction interval, *642*

Time series data, *646*

### Learning Objectives

After reading this chapter and doing the assigned homework problems, you should

- Be able to check that the conditions of the linear model hold, and understand the consequences if the conditions do not hold.

- Know how to use the output from a regression analysis to perform and interpret hypothesis tests of the slope and intercept,

and know how to interpret confidence intervals for the slope and intercept.

- Be able to use the regression line to make predictions, and understand when a situation calls for a confidence interval and when it calls for a prediction interval.

### Summary

Making inferences from the regression model requires that the linearity condition, the Normality condition, the constant standard deviation condition, and the independence condition be satisfied. A residual plot, which plots the residuals against the $x$-values, can be used to evaluate the linearity and constant standard deviation conditions. If the residual plot has no trend, the linearity condition is satisfied. If the residual plot does not have a fan shape, the constant standard deviation condition is satisfied. A QQ plot of the residuals can be used to evaluate the Normality condition. The independence condition is harder to verify, but you should be wary of time series (regressions using time as the $x$-axis), because these are often not independent.

If the conditions are satisfied, then the standard computer output for a regression analysis can be used to determine whether there is a linear association between the variables (test whether the slope is 0) or whether the intercept is 0. Confidence intervals for the intercept and slope can be found if the situation requires an estimate of the value of these parameters. A confidence interval can be used to estimate the mean of a group for any given value of $x$, and a prediction interval can be used to predict the value for an individual who has any given value of $x$. Prediction intervals are wider than confidence intervals because there is more uncertainty in predicting an individual outcome than in predicting a mean.

### Sources

Hastie, T. J. and Tibshirani, R.J., (1990), Generalized Additive Models, Chapman & Hall/CRC.

Lange T. L., H. E. Royals, and L. L. Connor. 1993. Influence of water chemistry on mercury concentration in largemouth bass from Florida lakes. Transactions of the American Fisheries Society 122, 74–84, http://wiki.stat.ucla.edu/socr/index.php/NISER_081107_ID_Data (viewed April 19, 2010).

Lee, D., R. Lane, and G. Chang. 2001. Three-dimension reconstruction for high-speed volume measurement. *Proceedings of the International*

*Society for Optical Engineering, Machine Vision and Three-Dimensional Imaging Systems for Inspection and Metrology* 4189, 258–267.

Leger, A., A. Cochrane, and F. Moore. 1979. Factors associated with cardiac mortality in developed countries with particular reference to the consumption of wine. *The Lancet* 313(8124), 1017–1020.

Onega, Edwin. 2005. Book width data. UCLA Statistics 120A class project.

## SECTION EXERCISES

### SECTION 14.1

TRY **14.1 Predicting Test Scores (Example 1)** A professor tells his class that he knows their second exam score without their having to take the test. He tells them that the second exam score can be predicted from the first with this equation:

Predicted second exam score = 5 + 0.75 (first exam score)

This tells us that the deterministic part of the regression model that predicts second exam score on the basis of first exam score is a straight line. What factors might contribute to the random component? In other words, why might a student's score not fall exactly on this line?

**14.2 Used Car Values** A student wishes to buy a used car. He finds a consumer website that says the price of a used car is determined by its age according to the following formula:

Predicted price in thousands of dollars = 17 − 0.8 (age in years)

This is the deterministic component of a regression model for predicting price on the basis of the age of the car. What factors might contribute to the random component? In other words, why might the price of the car he buys not fall exactly on this line?

TRY **14.3 Used BMWs (Example 2)** Figure A shows a scatterplot of the price and age of a random sample of used BMW cars and includes the regression line. Figure B shows a residual plot based on the regression line.

  a. Is the linear regression model appropriate for these data? Explain.

  b. How old is the car that is farthest from the regression line?

**(A)**

Scatterplot

**(B)**

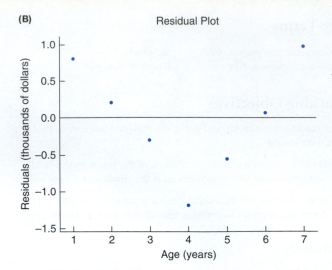
Residual Plot

**14.4 Pulse Rates** Figure A shows a scatterplot with the regression line for pulse rates for a large number of students in a college statistics class before and after running in place for two minutes. Figure B shows the residual plot of the data. Figure C shows a QQ plot of the residuals. Is the linear regression model appropriate for these data? Why or why not? Assume the observations are independently measured.

**(A)**

Scatterplot

**(B)**                Residual Plot

**(A)**                Scatterplot

**(C)**                QQ Plot

**(B)**                Residual Plot

**\* 14.6 U.S. Population** Figure A shows a scatterplot for the U.S. population (in millions) from 1850 to 1900. Figure B shows a residual plot of the same data. Give the reasons why the linear model is not appropriate for predicting the population in the year 2020. (Source: *World Almanac and Book of Facts*, 2009)

**TRY 14.5 Semesters and Credits (Example 3)** Figure A shows a scatterplot for the number of semesters that students have attended a community college and the number of credits they have accumulated. Figure B shows a residual plot of the same data. Is the linear regression model appropriate for these data? Why or why not? Assume the observations are independently measured.

**(A)**                Scatterplot

(B) Residual Plot

(C) QQ Plot

TRY **14.7 Wages of Twins (Example 4)** Figure A shows a scatterplot of wages of twins for a group of 183 pairs of twins. Figure B shows a residual plot of the same data. Figure C shows a QQ plot of these residuals. Is the linear regression model appropriate for these data? Why or why not? Assume the observations are independently measured.

**14.8 Simulated Data** Figure A shows a scatterplot of some simulated data, and figure B shows a residual plot of the same data. Is the linear regression model appropriate for these data? Why or why not? Assume the observations are independently measured.

(A) Scatterplot

(A) Scatterplot

(B) Residual Plot

(B) Residual Plot

# SECTION 14.2

TRY **14.9 Age and Weight (Example 5)** Do older college students tend to weigh more than younger college students? Our data are ages and weights for a random sample of 76 college students. A scatterplot (not shown here) shows that the association between age and weight is linear. Refer to the Minitab output given.

```
Regression Analysis: Weight versus Age
The regression equation is
Weight = 129 + 0.843 Age

76 cases used, 7 cases contain missing values

Predictor   Coef    SE Coef   T     P
Constant    129.10  15.19     8.50  0.000
Age         0.8429  0.5884    1.43  0.156

S = 37.5072   R-Sq = 2.7%   R-Sq(adj) = 1.4%
```

a. According to the equation for weight, do weights tend to be larger or smaller for older students than for younger students in this sample? Explain.

b. Use the Minitab output to test the hypothesis that the slope is zero in this population using a significance level of 0.05. Show all four steps of the hypothesis test, and explain your conclusion. Assume that all the required conditions are satisfied and that the sample is representative of the population of all college students. The statistics for the test that the slope is 0 are in the row labeled "Age."

## 14.10 Age and Weight for Baseball Players

Do older college baseball players tend to weigh more than younger baseball players? Our data are ages and weights for a random sample of 24 college baseball players. A scatterplot (not shown here) shows that the association between age and weight is linear. Refer to the Minitab output given.

a. Do weights tend to be larger or smaller for older students in this sample? Explain.

b. Use the Minitab output to test the hypothesis that the slope is zero using a significance level of 0.05. Show all four steps of the hypothesis test, and explain your conclusion. Assume that all the required conditions are satisfied. The statistics for the test that the slope is 0 are in the row labeled "Age."

```
Regression Analysis: Weight versus Age
The regression equation is
Weight = 107 + 4.58 Age

Predictor    Coef   SE Coef   T     P
Constant     107.37 21.60     4.97  0.000
Age          4.582  1.123     4.08  0.001

S = 7.98383   R-Sq = 49.5 %   R-Sq(adj) = 46.5%
```

TRY **14.11 Education of Parents (Example 6)** Each of 29 randomly sampled community college students reported the number of years of formal education for his or her mother and father. For example, a

value of 12 means that the parent completed high school but had no further education. (The numbers ranged from 0 to 18.) The regression equation shown in the output is for predicting years of education for the father on the basis of years of education for the mother. Assume that all the required conditions for regression are fulfilled.

a. Identify the intercept. Explain what it would mean if the intercept were zero.

b. Use the output given to test the hypothesis that the intercept is 0. Show all the steps, and use a significance level of 0.05. The statistics for the test that the intercept is 0 can be found in the row labeled "Constant."

```
Regression Analysis: FatherEd versus MotherEd

The regression equation is
FatherEd = 2.78 + 0.637 MotherEd

Predictor   Coef    SE Coef   T     P
Constant    2.785   1.396     2.00  0.056
MotherEd    0.6370  0.1483    4.30  0.000

S = 3.63005   R-Sq = 40.6%   R-Sq(adj) = 38.4
```

## 14.12 Education of Parents
Refer to Exercise 14.11 and the output given.

a. Identify the slope. Explain what it would mean if the slope were 0.

b. Test the hypothesis that the slope is 0 using the output. Show all the steps, and use a significance level of 0.05. The statistics for the test that the slope is 0 can be found in the row labeled "MotherEd."

c. What would it mean if the slope were 1 and the intercept 0?

* d. Test the hypothesis that the slope is 1 using a significance level of 0.05.

Use the table here to decide whether the p-value is less than 0.01, between 0.01 and 0.05 or more than 0.05 for part d. The df is 27.

| alpha | 0.05 | 0.01 |
|---|---|---|
| tails | 2 | 2 |
| df | | |
| 26 | 2.056 | 2.779 |
| 27 | 2.052 | 2.771 |
| 28 | 2.048 | 2.763 |

*14.13 Student and Parent Heights A random sample of 29 community college students were asked their heights in inches and the heights of their biological parent of the same gender. The output of a regression analysis for predicting student height from parent height is shown. Assume that all the conditions required are fulfilled, and use a significance level of 0.05 for any tests.

a. What is the slope of the regression line? Interpret the value of the slope in context.

b. Test the hypothesis that the slope is 0 (significance level is 0.05), and explain what your answer means.

c. If the slope were 1, what would that say about the association?

\* d. Test the hypothesis that the slope is 1 using a significance level of 0.05. Use the table for Exercise 14.12 to decide whether the p-value is less than 0.05. The degrees of freedom (df) is 27.

```
Regression Analysis: Student versus Parent
The regression equation is
Student = 5.63 + 0.936 Parent

Predictor   Coef  SE Coef    T     P
Constant   5.631   7.366   0.76  0.451
Parent     0.9355  0.1116  8.38  0.000

S = 1.97692  R-Sq = 72.2%  R-Sq(adj) = 71.2%
```

**14.14 Trash** The weight of trash (in pounds) produced by a household and the number of people living in the household were obtained for 13 houses. Refer to the Minitab regression output. Assume that all the conditions necessary for regression analysis are met.

```
Regression Analysis: Trash versus People
The regression equation is
Trash = 2.34 + 11.3 People

Predictor   Coef  SE Coef    T     P
Constant   2.340   6.869   0.34  0.740
People    11.300   1.867   6.05  0.000

S = 11.8519  R-Sq = 76.9%  R-Sq(adj) = 74.8%
```

a. What is the observed intercept, according to the equation?

b. What would it mean if the intercept were 0?

c. Test the hypothesis that the intercept is 0 using a significance level of 0.05.

TRY \* **14.15 Academic Performance Index (Example 7)** All California schools receive an annual Academic Performance Index (API) that is intended to measure the quality of the school. High scores indicate higher quality. The output shows the results of a regression analysis to predict a school's 2008 API score on the basis of the percentage of students enrolled at that school who receive reduced-cost meals. The regression is based on a random sample of 50 schools.

a. According to the regression line, what's the mean API score of schools in which no students receive reduced-cost meals?

b. Interpret the slope in context.

c. State the 95% confidence interval for the slope.

d. A school administrator argues that in California there is no relationship between the percentage of students who receive reduced-cost meals at a school and that school's API score. Use your confidence interval for the slope to respond to this claim.

---

○ ○ ○      Simple Linear Regression

Options

**Simple linear regression results:**

Dependent Variable: api08
Independent Variable: MEALS
api08 = 767.3472 – 2.4070053 MEALS
Sample size: 43
R (correlation coefficient) = –0.5372
R-sq = 0.2885424
Estimate of error standard deviation: 111.237785

**Parameter estimates:**

| Parameter | Estimate | Std. Err. | DF | 95% L. Limit | 95% U. Limit |
|---|---|---|---|---|---|
| Intercept | 767.3472 | 33.483723 | 41 | 699.7255 | 834.969 |
| Slope | -2.4070053 | 0.5902756 | 41 | -3.599091 | -1.2149196 |

**Analysis of variance table for regression model:**

| Source | DF | SS | MS | F-stat | P-value |
|---|---|---|---|---|---|
| Model | 1 | 205754.39 | 205754.39 | 16.628168 | 0.0002 |
| Error | 41 | 507327.66 | 12373.846 | | |
| Total | 42 | 713082.06 | | | |

\* **14.16 Father's and Mother's Education** The output shows the results of a regression analysis to predict the father's education given the mother's education for a random sample of students at a two-year college. Assume that this sample is representative of the population of all students at this two-year college.

a. A friend says she thinks love is blind and people don't pay any attention to the education level of potential partners. If this were true for parents of students at the two-year college, what value would the slope of the regression line have?

b. Report a 95% confidence interval for the slope. What does this say about the friend's theory?

**Simple linear regression results:**

Dependent Variable: FatherEd
Independent Variable: MotherEd
FatherEd = 2.7847526 + 0.63699657 MotherEd
Sample size: 29
R (correlation coefficient) = 0.6372
R-sq = 0.4059982
Estimate of error standard deviation: 3.6300526

**Parameter estimates:**

| Parameter | Estimate | Std. Err. | DF | 95% L. Limit | 95% U. Limit |
|---|---|---|---|---|---|
| Intercept | 2.7847526 | 1.3956293 | 27 | -0.07884216 | 5.6483474 |
| Slope | 0.63699657 | 0.14828151 | 27 | 0.332748 | 0.9412451 |

**14.17 Trash and Confidence Intervals** The output provided for Exercise 14.14 provides a regression line to predict the amount of trash produced by a household on the basis of the number of people living in the household. Suppose you found a 95% confidence interval for the intercept of the regression line. Would that confidence interval include 0? (Refer to the output in Exercise 14.14.)

**14.18 Baseball Players' Ages and Weights** In Exercise 14.10 you examined the association between baseball players' ages and weights. Would a 95% confidence interval for the slope, based on the same data, include 0? Explain.

## SECTION 14.3

TRY **14.19 Predicted GPA (Example 8)** A student who has been accepted by two colleges wants to estimate what GPA he might get at the two colleges. His data consist of SAT scores and GPAs from random samples of recent graduates at each college. He wishes to predict his GPA at each school, using separate linear regression models for each college. Should he use prediction intervals or confidence intervals for the prediction of his own GPA? Explain.

**14.20 Used BMWs** A used-car dealer is purchasing 50 used BMWs from one dealer in order to sell them for a profit. Working with collected data, the dealer has found a regression model to predict the selling price on the basis of the car's age. He wants to predict the total amount he will get for these 50 cars in order to make sure he does not lose money. All 50 cars are three years old (they were turned in after their leases expired). Should he use a confidence interval or a prediction interval for the mean selling price of these 50 cars? Explain.

**14.21 Loggers** A logging company has the diameter of each of a large number of trees and wants to estimate the mean number of usable cubic feet of wood the company will get if it cuts the trees down. Working with a sample of trees, company planners find the regression line that predicts the volume of lumber (in cubic feet) on the basis of the diameter of the tree. Should they use a prediction interval or a confidence interval? Explain.

**14.22 Height of Blind Date** A female college student has been set up on a blind date with the son of her father's friend. She knows the height of the blind date's father and wants to predict the height of her date. She has information on heights of fathers and sons in this community. Should she use a prediction interval or a confidence interval for her prediction? Explain.

TRY **14.23 House Prices (Example 9)** Figure A contains the selling price and area (in square feet) of 81 recently sold homes in a region where a buyer wants to purchase a home.

  a.  Use the equation to estimate the price of a home with 2500 square feet.

(A)

Price = 16209 + 101.9 Area

**(B)**

```
Predicted Values for New Observations

New
Obs   Fit  SE Fit      95% CI            95% PI
  1 271056  10085 (250975, 291138) (160581, 381532)


Values of Predictors for New Observations

New
Obs    Area
  1    2500
```

  b.  The buyer wants to know the uncertainty in the prices he might pay for a home with 2500 square feet in this region. Should he use a prediction interval or a confidence interval? Explain.

  c.  Report the correct 95% interval, confidence or prediction as determined in part b, for the predicted price. See figure B.

  d.  The buyer has prequalified for a loan up to $400,000. Is he likely to be able to afford a house in this area with 2500 square feet?

**14.24 Math SAT Score and GPA** Figure A shows information about a random sample of students' math SAT scores and GPAs at an unidentified four-year college.

  a.  Use the formula on the graph to predict the GPA for a person with a math SAT score of 600.

  b.  Figure B shows both a prediction interval and a confidence interval for a new SAT of 600; report both.

  c.  One student wants to estimate the GPA he will achieve if he attends that school. Should he use the prediction interval or the confidence interval? Explain.

  d.  Report the interval obtained for part c. Is it very useful? Explain.

**(A)**

GPA = 0.6219 + 0.003394 SAT

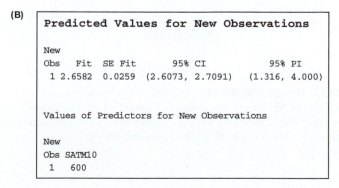

**(B)**

```
Predicted Values for New Observations

New
Obs   Fit  SE Fit      95% CI           95% PI
  1 2.6582  0.0259 (2.6073, 2.7091)  (1.316, 4.000)


Values of Predictors for New Observations

New
Obs SATM10
  1   600
```

**14.25 Height and Weight** A scatterplot of the heights and weights of 500 people was shown in Chapter 4. The accompanying scatterplot uses the same data but displays 95% prediction intervals. From the graph, estimate the boundary values for the prediction interval used for predicting the weight of someone who is 70 inches tall.

Scatterplot with Prediction Intervals

**14.26 Waist Size and Weight** A scatterplot of the waist sizes and weights of the same 500 people mentioned in Exercise 14.25 is shown. The accompanying scatterplot uses the same people and displays 95% prediction intervals. From the graph, estimate the boundary values for the prediction interval for predicting the weight of someone who has a 30-inch waist.

Scatterplot with Prediction Intervals

**14.27 Height and Weight** A scatterplot of the heights and weights of 500 people was shown in Exercise 14.25. The accompanying scatterplot shows both confidence intervals and prediction intervals. Which is which? Explain what this shows.

Scatterplot with Prediction and Confidence Intervals

**14.28 Waist Size and Weight** A scatterplot of the waist sizes and weights of 500 people was shown in Exercise 14.26. The accompanying scatterplot shows both confidence intervals and prediction intervals. Which is which? Explain how you can tell.

Scatterplot with Prediction and Confidence Intervals

**14.29 Baseball Players** Figure A shows a scatterplot with the regression line for the ages and weights of a random sample of 19 college baseball players. Figure B gives a prediction (Fit), a prediction interval, and a confidence interval for a new observation at 20 years old.

a. Identify which interval is which and explain why one of the intervals is larger than the other.

b. Interpret the confidence interval.

c. Interpret the prediction interval.

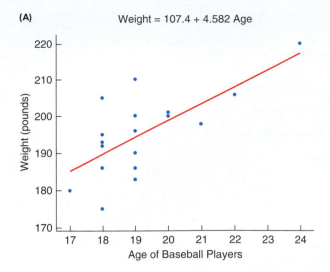

**(A)** Weight = 107.4 + 4.582 Age

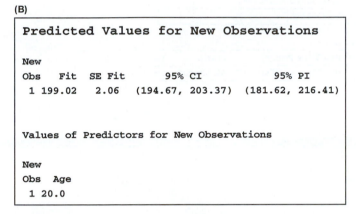

**(B)**

```
Predicted Values for New Observations

New
Obs    Fit   SE Fit      95% CI           95% PI
 1  199.02   2.06   (194.67, 203.37)  (181.62, 216.41)



Values of Predictors for New Observations

New
Obs   Age
 1   20.0
```

**(A)** FatherEd = 2.785 + 0.6370 MotherEd

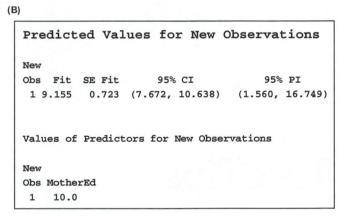

**(B)**

```
Predicted Values for New Observations

New
Obs   Fit   SE Fit      95% CI           95% PI
 1  9.155   0.723   (7.672, 10.638)  (1.560, 16.749)



Values of Predictors for New Observations

New
Obs  MotherEd
 1     10.0
```

**14.30 Predicting Education** Figure A shows a scatterplot with the regression line for predicting the father's education from the mother's education for a random sample of 29 students. Figure B shows the confidence interval and the prediction interval for the father for a new observation when the mother has 10 years of education.

a. Identify which interval is which. Explain what they show and why one interval is larger than the other.

b. Interpret the confidence interval.

c. Interpret the prediction interval.

## CHAPTER REVIEW EXERCISES

**✻ ✱ 14.31 Life Expectancy and Gestation Periods for Animals** Data were collected showing the gestation period (in days) and the average longevity (in years) for 32 animals. Assume that all conditions of the linear regression model hold. This set of data can be found in Exercise 2.28 on page 64.

a. Make a scatterplot with gestation period as the *x*-variable and longevity as the *y*-variable, and insert the correct regression line. Report the regression equation.

b. One animal lives much longer than we would expect, given its gestation period. Identify this animal.

c. From the data, find a confidence interval for the mean longevity of animals with a gestation period of 266 days.

d. Humans have a gestation period of about 266 days. Does the confidence interval for the average life span for humans seem to fit what you know about humans' life spans? Explain.

**\* 14.32 Academic Performance Index** The California Department of Education makes available the academic performance index (API) for each school. The output provided predicts the 2008 API for a random sample of high schools in California on the basis of the percentage of students receiving reduced-cost meals. Assume that all conditions of the linear regression model hold.

   a. Write the equation of the regression line for predicting API score on the basis of the percent of the students who receive reduced-cost meals.

   b. Is there an association (using a 5% significance level) between the percent receiving reduced-cost meals and the API score? Explain. If so, interpret the slope.

   c. Using a 95% confidence interval, what is the estimated mean API score of schools at which 50% of the students receive reduced-cost meals?

**Simple Linear Regression**

Options

**Simple linear regression results:**
Dependent Variable: api08
Independent Variable: MEALS
api08 = 767.3472 − 2.4070053 MEALS
Sample size: 43
R (correlation coefficient) = −0.5372
R−sq = 0.2885424
Estimate of error standard deviation: 111.237785

**Parameter estimates:**

| Parameter | Estimate | Std. Err. | DF | T−Stat | P−Value |
|---|---|---|---|---|---|
| Intercept | 767.3472 | 33.483723 | 41 | 22.917025 | <0.0001 |
| Slope | −2.4070053 | 0.5902756 | 41 | −4.077765 | 0.0002 |

**Analysis of variance table for regression model:**

| Source | DF | SS | MS | F−stat | P−value |
|---|---|---|---|---|---|
| Model | 1 | 205754.39 | 205754.39 | 16.628168 | 0.0002 |
| Error | 41 | 507327.66 | 12373.846 | | |
| Total | 42 | 713082.06 | | | |

**Predicted values:**

| X value | Pred. Y | s.e.(Pred. y) | 95% C.I. | 95% P.I. |
|---|---|---|---|---|
| 50 | 646.997 | 16.975868 | (612.71356, 681.28046) | (419.7468, 874.2472) |

# TechTips

## General Instructions for All Technology

The basic steps for finding a regression equation and making a scatterplot were shown in Chapter 4.

**EXAMPLE** ▶ The table shows the heights (in inches) and the weights (in pounds) of six women.

| Ht | Wt |
|---|---|
| 61 | 104 |
| 62 | 110 |
| 63 | 141 |
| 64 | 125 |
| 66 | 170 |
| 68 | 160 |

### TI-83/84

The only confidence interval available for regression is for the slope, and it is available only on the TI-84. To find it, press **STAT**, choose **TESTS**, and choose **LinRegTint**. For the slope of the six heights and weights given in the example, the 95% confidence interval comes out to be (2.2957, 15.763).

### Residual Plot

  1. Press **STAT** and choose **EDIT** and enter heights in **L1** and weights in **L2**.

  2. Press **STAT** and choose **CALC, 8: LinReg (a + bx)**.

  3. Press **2ND L1 , 2ND L2 ENTER.** (Remember that the predictor— here, it is height—must go in before the response and that the comma must be entered and is the button above the 7 button on the keypad.) You will see the numbers for the regression equation.

  4. Now, to create the residual plot, use Plot2 (assuming that you might have used Plot1 for the scatterplot). Choose **2ND STATPLOT**, turn **Plot1 Off**. Choose **2ND STATPLOT, Plot2, On, Type:** scatterplot (first choice, in the upper left corner), **Xlist: L1, Ylist: RESID**. (Do this by pressing **2ND LIST, RESID**. You might have to scroll down to see **RESID**.)

  5. To make the graph, press **ZOOM** and **9**.

  6. Press **Trace** to see the numbers, and use the arrows on the keypad to move to other points.

  7. Figure 14A shows a residual plot of the six points.

▲ **FIGURE 14A**   TI-83/84
Residual Plot

## Hypothesis Test

With the data set, you can test the hypothesis that the population slope, $\beta$ is 0. This is equivalent to testing that the population correlation, $\rho$ is 0. (See $\beta$ and $\rho$ in Figure 14B.)

▲ **FIGURE 14B** TI-83/84 Input for a Regression Test

1. Press **STAT**, choose **EDIT**, and enter the heights into **L1** and the weights into **L2**.
2. Press **STAT**, choose **TESTS**, and choose **LinRegTTest**.

3. See Figure 14B. Be sure the predictor (**L1** in our example) is the **XList** and that the response (**L2** in our example) is the **YList**. Here we are testing the hypothesis that the slope is greater than 0. You can use $\neq$ if you want a two-tailed hypothesis.
4. Scroll down to **Calculate** and press **ENTER**.

Figure 14C shows the output.

▲ **FIGURE 14C** TI-83/84 Output for a Regression Test

---

MINITAB

### Residual Plot

1. Enter the heights and weights into columns **C1** and **C2**. Add the labels **Ht** and **Wt** in the label area above the data.
2. **Stat > Regression > Fitted Line Plot**
3. Be sure to put the dependent variable (here, **Wt**) in the **Response (Y)** box and the independent variable (here, **Ht**) in the **Predictor (X)** box.
4. Click **Graphs.**
5. See Figure 14D. In the **Fitted Line Plot - Graphs** leave the default **Regular**, and in the **Residuals versus the variables** box put the independent variable, here **Ht**. Click **OK** and **OK** again.

▲ **FIGURE 14D** Minitab Input for a Residual Plot

You will get a scatterplot with the regression line and also a residual plot like Figure 14E. You may have to click **Window** and **3** or **4** below **Window** to see the other graph.

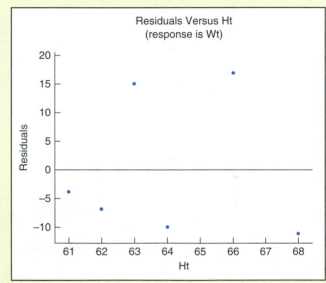

▲ **FIGURE 14E** Minitab Residual Plot

### A Scatterplot with the Regression Line with Prediction and Confidence Intervals

1. The heights should be in column **C1** and the weights in column **C2**, with the labels **Ht** and **Wt** at the top.
2. **Stat > Regression > Fitted Line Plot**
3. After entering the **Response (Wt)** and the **Predictor (Ht)**, click **Options**.
4. Click **Display confidence intervals** or **Display prediction intervals**, or both, as we have shown in Figure 14F. Click **OK**, and click **OK** again for the next screen. You may have to click **Window** and select an option below **Window** to see it.

**▲ FIGURE 14F** Minitab Input for Prediction and Confidence Intervals

Figure 14G shows the result.

**▲ FIGURE 14G** Minitab Prediction and Confidence Intervals

### Regression Numbers and Numerical Confidence and Prediction Intervals

1. Your data should be in columns **C1** and **C1**, labeled **Ht** and **Wt**.

2. **Stat > Regression > Regression**

3. In the **Response** box put Y, here **Wt**. In the **Predictors** box put X, here **Ht**.

4. Click **Options**. In the box below **Prediction intervals for new observation**, enter a height of a new woman; here we used **67** inches.

5. Click **OK** and **OK** again.

The output is shown in Figure 14H and includes the prediction and confidence intervals at the bottom. These are for prediction of ranges of weights with 95% confidence from a new woman's height of 67 inches. The output in Figure 14H can also be used to test whether the null hypothesis that the slope is 0 can be rejected; because the p-value is 0.020, you can reject the hypothesis that the slope is 0. This output can also be used to test whether the intercept is 0; because the p-value is 0.046, you can reject the hypothesis that the intercept is 0.

```
The regression equation is
Wt = -443 + 9.03 Ht


Predictor    Coef  SE Coef    T      P
Constant   -442.9    155.3  -2.85  0.046
Ht          9.029    2.425   3.72  0.020


S = 14.1419   R-Sq = 77.6%   R-Sq(adj) = 72.0%


Analysis of Variance


Source            DF    SS       MS      F      P
Regression         1  2772.0   2772.0  13.86  0.020
Residual Error     4   800.0    200.0
Total              5  3572.0


Predicted Values for New Observations


New
Obs    Fit   SE Fit      95% CI              95% PI
  1  162.09   9.29  (136.30, 187.88)  (115.11, 209.06)
```

**▲ FIGURE 14H** Minitab Output for Regression, Including Intervals

Minitab does not supply output reporting confidence intervals for the slope or intercept.

Excel can give numerical regression output and residual plots at the same time.

1. Enter the heights into column **A** and the weights into column **B**.
2. Click **Data**, click **Data Analysis**, and double click **Regression**.
3. See Figure 14I. For **Input Y Range**, highlight the data for the weights (without the label, wt), and for **Input X Range**, highlight the data for the heights (without the label, ht); then click **Residual Plots** and **OK**.

You will get numerical output like that shown in Figure 14J and a residual plot like that shown in Figure 14K. To see the numerical output you may have to click **Home**, **Format** (in the **Cells** group), and **AutoFit Column Width**. Note in Figure 14J that confidence intervals are given for the intercept ($-874.14$, $-11.6247$) and for the slope ($2.295667$, $15.76316$). These can be used to test hypotheses about the slope and intercept.

▲ **FIGURE 14I** Excel Input for Regression and Residual Plots

SUMMARY OUTPUT

| *Regression Statistics* | |
|---|---|
| Multiple R | 0.880934 |
| R Square | 0.776044 |
| Adjusted R Square | 0.720055 |
| Standard Error | 14.14188 |
| Observations | 6 |

ANOVA

| | df | SS | MS | F | Significance F |
|---|---|---|---|---|---|
| Regression | 1 | 2772.029 | 2772.029 | 13.86066 | 0.020421 |
| Residual | 4 | 799.9706 | 199.9926 | | |
| Total | 5 | 3572 | | | |

| | Coefficients | Standard Error | t Stat | P-value | Lower 95% | Upper 95% | Lower 95.0% | Upper 95.0% |
|---|---|---|---|---|---|---|---|---|
| Intercept | −442.882 | 155.3273 | −2.85129 | 0.046336 | −874.14 | −11.6247 | −874.14 | −11.6247 |
| X Variable 1 | 9.029412 | 2.425312 | 3.72299 | 0.020421 | 2.295667 | 15.76316 | 2.295667 | 15.76316 |

▲ **FIGURE 14J** Excel Regression Output

▲ **FIGURE 14K** Excel Residual Plot

1. Enter the heights and weights into **var1** and **var2**. Include labels at the top, **Ht** and **Wt**.

2. **Stat > Regression > Simple Linear**

3. For the **X-Variable** pick **Ht**, for the **Y-Variable** pick **Wt**, and click **Next.**

4. Click **Confidence intervals** and leave the default **0.95** and click **Next**. (This means you will get confidence intervals for the slope and intercept. If you wanted to do hypothesis tests on the slope and intercept, you should instead leave the default **Hypothesis tests** selected.)

5. Click **Predict Y for X=**, type in **67** (for the new height), and click **Next**.

6. You may click both **Plot the fitted line** and **Residuals vs X-values**; click **Next** and **Calculate.** The numerical output may not be entirely visible. You may have to drag the bottom edge down to see all the numbers as shown in Figure 14L. Note that there are confidence intervals given for the slope and intercept and also a confidence and a prediction interval given for the new point with height of 67 inches.

7. If you click **Next** (near the bottom of the numbers), you will see the scatterplot with the regression line, not shown.

8. If you click **Next**, you will see the residual plot shown in Figure 14M.

▲ **FIGURE 14M** StatCrunch Residual Plot

**Simple Linear Regressi...**

Options

Simple linear regression results:
Dependent Variable: Wt
Independent Variable: Ht
Wt = -442.88235 + 9.029411 Ht
Sample size: 6
R (correlation coefficient) = 0.8809
R-sq = 0.7760441
Estimate of error standard deviation: 14.141875

Parameter estimates:

| Parameter | Estimate | Std. Err. | DF | 95% L. Limit | 95% U. Limit |
|-----------|----------|-----------|-----|-------------|-------------|
| Intercept | -442.88235 | 155.32729 | 4 | -874.14 | -11.624699 |
| Slope | 9.029411 | 2.4253116 | 4 | 2.2956672 | 15.763156 |

Analysis of variance table for regression model:

| Source | DF | SS | MS | F-stat | P-value |
|--------|-----|-----|-----|--------|---------|
| Model | 1 | 2772.0293 | 2772.0293 | 13.860657 | 0.0204 |
| Error | 4 | 799.9706 | 199.99265 | | |
| Total | 5 | 3572 | | | |

Predicted values:

| X value | Pred. Y | s.e.(Pred. y) | 95% C.I. for mean | 95% P.I. for new |
|---------|---------|---------------|-------------------|------------------|
| 67 | 162.08824 | 9.288237 | (136.29996, 187.87651) | (115.112625, 209.06384) |

← Back  Next →

▲ **FIGURE 14L** StatCrunch Regression Output

# Appendix A: Tables

## Table 1: Random Numbers

| Line | | | | | | |
|------|-------|-------|-------|-------|-------|-------|
| 01 | 21033 | 32522 | 19305 | 90633 | 80873 | 19167 |
| 02 | 17516 | 69328 | 88389 | 19770 | 33197 | 27336 |
| 03 | 26427 | 40650 | 70251 | 84413 | 30896 | 21490 |
| 04 | 45506 | 44716 | 02498 | 15327 | 79149 | 28409 |
| 05 | 55185 | 74834 | 81172 | 89281 | 48134 | 71185 |
| 06 | 87964 | 43751 | 80971 | 50613 | 81441 | 30505 |
| 07 | 09106 | 73117 | 57952 | 04393 | 93402 | 50753 |
| 08 | 88797 | 07440 | 69213 | 33593 | 42134 | 24168 |
| 09 | 34685 | 46775 | 32139 | 22787 | 28783 | 39481 |
| 10 | 07104 | 43091 | 14311 | 69671 | 01536 | 02673 |
| 11 | 27583 | 01866 | 58250 | 38103 | 35825 | 94513 |
| 12 | 60801 | 04439 | 58621 | 09840 | 35119 | 60372 |
| 13 | 62708 | 04888 | 37221 | 49537 | 96024 | 24004 |
| 14 | 21169 | 14082 | 65865 | 29690 | 00280 | 35738 |
| 15 | 13893 | 00626 | 11773 | 14897 | 37119 | 29729 |
| 16 | 19872 | 41310 | 65041 | 61105 | 31028 | 80297 |
| 17 | 29331 | 36997 | 05601 | 09785 | 18100 | 44164 |
| 18 | 76846 | 74048 | 08496 | 22599 | 29379 | 11114 |
| 19 | 11848 | 80809 | 25818 | 38857 | 23811 | 80902 |
| 20 | 85757 | 33963 | 93076 | 39950 | 29658 | 07530 |
| 21 | 71141 | 00618 | 48403 | 46083 | 40368 | 33990 |
| 22 | 47371 | 36443 | 41894 | 62134 | 86876 | 18548 |
| 23 | 46633 | 10669 | 95848 | 69055 | 49044 | 75595 |
| 24 | 79118 | 21098 | 63279 | 26834 | 43443 | 38267 |
| 25 | 91874 | 87217 | 11503 | 47925 | 13289 | 42106 |
| 26 | 85337 | 08882 | 68429 | 61767 | 18930 | 37688 |
| 27 | 88513 | 05437 | 22776 | 17562 | 03820 | 44785 |
| 28 | 31498 | 85304 | 22393 | 21634 | 34560 | 77404 |
| 29 | 93074 | 27086 | 62559 | 86590 | 18420 | 33290 |
| 30 | 90549 | 53094 | 76282 | 53105 | 45531 | 90061 |
| 31 | 11373 | 96871 | 38157 | 98368 | 39536 | 08079 |
| 32 | 52022 | 59093 | 30647 | 33241 | 16027 | 70336 |
| 33 | 14709 | 93220 | 89547 | 95320 | 39134 | 07646 |
| 34 | 57584 | 28114 | 91168 | 16320 | 81609 | 60807 |
| 35 | 31867 | 85872 | 91430 | 45554 | 21567 | 15082 |
| 36 | 07033 | 75250 | 34546 | 75298 | 33893 | 64487 |
| 37 | 02779 | 72645 | 32699 | 86009 | 73729 | 44206 |
| 38 | 24512 | 01116 | 49826 | 50882 | 44086 | 87757 |
| 39 | 52463 | 30164 | 80073 | 55917 | 60995 | 38655 |
| 40 | 82588 | 59267 | 13570 | 56434 | 66413 | 99518 |
| 41 | 20999 | 05039 | 87835 | 63010 | 82980 | 66193 |
| 42 | 09084 | 98948 | 09541 | 80623 | 15915 | 71042 |

Cumulative Probability for z is The Area Under the Standard Normal Curve to The Left of z

## Table 2: Standard Normal Cumulative Probabilities

| z | .00 |
|---|---|
| −5.0 | .000000287 |
| −4.5 | .00000340 |
| −4.0 | .0000317 |
| −3.5 | .000233 |

| z | .00 | .01 | .02 | .03 | .04 | .05 | .06 | .07 | .08 | .09 |
|---|---|---|---|---|---|---|---|---|---|---|
| −3.4 | .0003 | .0003 | .0003 | .0003 | .0003 | .0003 | .0003 | .0003 | .0003 | .0002 |
| −3.3 | .0005 | .0005 | .0005 | .0004 | .0004 | .0004 | .0004 | .0004 | .0004 | .0003 |
| −3.2 | .0007 | .0007 | .0006 | .0006 | .0006 | .0006 | .0006 | .0005 | .0005 | .0005 |
| −3.1 | .0010 | .0009 | .0009 | .0009 | .0008 | .0008 | .0008 | .0008 | .0007 | .0007 |
| −3.0 | .0013 | .0013 | .0013 | .0012 | .0012 | .0011 | .0011 | .0011 | .0010 | .0010 |
| −2.9 | .0019 | .0018 | .0018 | .0017 | .0016 | .0016 | .0015 | .0015 | .0014 | .0014 |
| −2.8 | .0026 | .0025 | .0024 | .0023 | .0023 | .0022 | .0021 | .0021 | .0020 | .0019 |
| −2.7 | .0035 | .0034 | .0033 | .0032 | .0031 | .0030 | .0029 | .0028 | .0027 | .0026 |
| −2.6 | .0047 | .0045 | .0044 | .0043 | .0041 | .0040 | .0039 | .0038 | .0037 | .0036 |
| −2.5 | .0062 | .0060 | .0059 | .0057 | .0055 | .0054 | .0052 | .0051 | .0049 | .0048 |
| −2.4 | .0082 | .0080 | .0078 | .0075 | .0073 | .0071 | .0069 | .0068 | .0066 | .0064 |
| −2.3 | .0107 | .0104 | .0102 | .0099 | .0096 | .0094 | .0091 | .0089 | .0087 | .0084 |
| −2.2 | .0139 | .0136 | .0132 | .0129 | .0125 | .0122 | .0119 | .0116 | .0113 | .0110 |
| −2.1 | .0179 | .0174 | .0170 | .0166 | .0162 | .0158 | .0154 | .0150 | .0146 | .0143 |
| −2.0 | .0228 | .0222 | .0217 | .0212 | .0207 | .0202 | .0197 | .0192 | .0188 | .0183 |
| −1.9 | .0287 | .0281 | .0274 | .0268 | .0262 | .0256 | .0250 | .0244 | .0239 | .0233 |
| −1.8 | .0359 | .0351 | .0344 | .0336 | .0329 | .0322 | .0314 | .0307 | .0301 | .0294 |
| −1.7 | .0446 | .0436 | .0427 | .0418 | .0409 | .0401 | .0392 | .0384 | .0375 | .0367 |
| −1.6 | .0548 | .0537 | .0526 | .0516 | .0505 | .0495 | .0485 | .0475 | .0465 | .0455 |
| −1.5 | .0668 | .0655 | .0643 | .0630 | .0618 | .0606 | .0594 | .0582 | .0571 | .0559 |
| −1.4 | .0808 | .0793 | .0778 | .0764 | .0749 | .0735 | .0721 | .0708 | .0694 | .0681 |
| −1.3 | .0968 | .0951 | .0934 | .0918 | .0901 | .0885 | .0869 | .0853 | .0838 | .0823 |
| −1.2 | .1151 | .1131 | .1112 | .1093 | .1075 | .1056 | .1038 | .1020 | .1003 | .0985 |
| −1.1 | .1357 | .1335 | .1314 | .1292 | .1271 | .1251 | .1230 | .1210 | .1190 | .1170 |
| −1.0 | .1587 | .1562 | .1539 | .1515 | .1492 | .1469 | .1446 | .1423 | .1401 | .1379 |
| −0.9 | .1841 | .1814 | .1788 | .1762 | .1736 | .1711 | .1685 | .1660 | .1635 | .1611 |
| −0.8 | .2119 | .2090 | .2061 | .2033 | .2005 | .1977 | .1949 | .1922 | .1894 | .1867 |
| −0.7 | .2420 | .2389 | .2358 | .2327 | .2296 | .2266 | .2236 | .2206 | .2177 | .2148 |
| −0.6 | .2743 | .2709 | .2676 | .2643 | .2611 | .2578 | .2546 | .2514 | .2483 | .2451 |
| −0.5 | .3085 | .3050 | .3015 | .2981 | .2946 | .2912 | .2877 | .2843 | .2810 | .2776 |
| −0.4 | .3446 | .3409 | .3372 | .3336 | .3300 | .3264 | .3228 | .3192 | .3156 | .3121 |
| −0.3 | .3821 | .3783 | .3745 | .3707 | .3669 | .3632 | .3594 | .3557 | .3520 | .3483 |
| −0.2 | .4207 | .4168 | .4129 | .4090 | .4052 | .4013 | .3974 | .3936 | .3897 | .3859 |
| −0.1 | .4602 | .4562 | .4522 | .4483 | .4443 | .4404 | .4364 | .4325 | .4286 | .4247 |
| −0.0 | .5000 | .4960 | .4920 | .4880 | .4840 | .4801 | .4761 | .4721 | .4681 | .4641 |

Cumulative probability for *z* is the area under the standard normal curve to the left of *z*

# Standard Normal Cumulative Probabilities (*continued*)

| z | .00 | .01 | .02 | .03 | .04 | .05 | .06 | .07 | .08 | .09 |
|---|-----|-----|-----|-----|-----|-----|-----|-----|-----|-----|
| 0.0 | .5000 | .5040 | .5080 | .5120 | .5160 | .5199 | .5239 | .5279 | .5319 | .5359 |
| 0.1 | .5398 | .5438 | .5478 | .5517 | .5557 | .5596 | .5636 | .5675 | .5714 | .5753 |
| 0.2 | .5793 | .5832 | .5871 | .5910 | .5948 | .5987 | .6026 | .6064 | .6103 | .6141 |
| 0.3 | .6179 | .6217 | .6255 | .6293 | .6331 | .6368 | .6406 | .6443 | .6480 | .6517 |
| 0.4 | .6554 | .6591 | .6628 | .6664 | .6700 | .6736 | .6772 | .6808 | .6844 | .6879 |
| 0.5 | .6915 | .6950 | .6985 | .7019 | .7054 | .7088 | .7123 | .7157 | .7190 | .7224 |
| 0.6 | .7257 | .7291 | .7324 | .7357 | .7389 | .7422 | .7454 | .7486 | .7517 | .7549 |
| 0.7 | .7580 | .7611 | .7642 | .7673 | .7704 | .7734 | .7764 | .7794 | .7823 | .7852 |
| 0.8 | .7881 | .7910 | .7939 | .7967 | .7995 | .8023 | .8051 | .8078 | .8106 | .8133 |
| 0.9 | .8159 | .8186 | .8212 | .8238 | .8264 | .8289 | .8315 | .8340 | .8365 | .8389 |
| 1.0 | .8413 | .8438 | .8461 | .8485 | .8508 | .8531 | .8554 | .8577 | .8599 | .8621 |
| 1.1 | .8643 | .8665 | .8686 | .8708 | .8729 | .8749 | .8770 | .8790 | .8810 | .8830 |
| 1.2 | .8849 | .8869 | .8888 | .8907 | .8925 | .8944 | .8962 | .8980 | .8997 | .9015 |
| 1.3 | .9032 | .9049 | .9066 | .9082 | .9099 | .9115 | .9131 | .9147 | .9162 | .9177 |
| 1.4 | .9192 | .9207 | .9222 | .9236 | .9251 | .9265 | .9279 | .9292 | .9306 | .9319 |
| 1.5 | .9332 | .9345 | .9357 | .9370 | .9382 | .9394 | .9406 | .9418 | .9429 | .9441 |
| 1.6 | .9452 | .9463 | .9474 | .9484 | .9495 | .9505 | .9515 | .9525 | .9535 | .9545 |
| 1.7 | .9554 | .9564 | .9573 | .9582 | .9591 | .9599 | .9608 | .9616 | .9625 | .9633 |
| 1.8 | .9641 | .9649 | .9656 | .9664 | .9671 | .9678 | .9686 | .9693 | .9699 | .9706 |
| 1.9 | .9713 | .9719 | .9726 | .9732 | .9738 | .9744 | .9750 | .9756 | .9761 | .9767 |
| 2.0 | .9772 | .9778 | .9783 | .9788 | .9793 | .9798 | .9803 | .9808 | .9812 | .9817 |
| 2.1 | .9821 | .9826 | .9830 | .9834 | .9838 | .9842 | .9846 | .9850 | .9854 | .9857 |
| 2.2 | .9861 | .9864 | .9868 | .9871 | .9875 | .9878 | .9881 | .9884 | .9887 | .9890 |
| 2.3 | .9893 | .9896 | .9898 | .9901 | .9904 | .9906 | .9909 | .9911 | .9913 | .9916 |
| 2.4 | .9918 | .9920 | .9922 | .9925 | .9927 | .9929 | .9931 | .9932 | .9934 | .9936 |
| 2.5 | .9938 | .9940 | .9941 | .9943 | .9945 | .9946 | .9948 | .9949 | .9951 | .9952 |
| 2.6 | .9953 | .9955 | .9956 | .9957 | .9959 | .9960 | .9961 | .9962 | .9963 | .9964 |
| 2.7 | .9965 | .9966 | .9967 | .9968 | .9969 | .9970 | .9971 | .9972 | .9973 | .9974 |
| 2.8 | .9974 | .9975 | .9976 | .9977 | .9977 | .9978 | .9979 | .9979 | .9980 | .9981 |
| 2.9 | .9981 | .9982 | .9982 | .9983 | .9984 | .9984 | .9985 | .9985 | .9986 | .9986 |
| 3.0 | .9987 | .9987 | .9987 | .9988 | .9988 | .9989 | .9989 | .9989 | .9990 | .9990 |
| 3.1 | .9990 | .9991 | .9991 | .9991 | .9992 | .9992 | .9992 | .9992 | .9993 | .9993 |
| 3.2 | .9993 | .9993 | .9994 | .9994 | .9994 | .9994 | .9994 | .9995 | .9995 | .9995 |
| 3.3 | .9995 | .9995 | .9995 | .9996 | .9996 | .9996 | .9996 | .9996 | .9996 | .9997 |
| 3.4 | .9997 | .9997 | .9997 | .9997 | .9997 | .9997 | .9997 | .9997 | .9997 | .9998 |

| z | .00 |
|---|-----|
| 3.5 | .999767 |
| 4.0 | .9999683 |
| 4.5 | .9999966 |
| 5.0 | .999999713 |

# Table 3: Binomial Probabilities

| | | | | | | | | p | | | | | | | |
|---|---|---|---|---|---|---|---|---|---|---|---|---|---|---|---|
| n | x | .01 | .05 | .10 | .20 | .30 | .40 | .50 | .60 | .70 | .80 | .90 | .95 | .99 | x |
| 2 | 0 | .980 | .902 | .810 | .640 | .490 | .360 | .250 | .160 | .090 | .040 | .010 | .002 | 0+ | 0 |
| | 1 | .020 | .095 | .180 | .320 | .420 | .480 | .500 | .480 | .420 | .320 | .180 | .095 | .020 | 1. |
| | 2 | 0+ | .002 | .010 | .040 | .090 | .160 | .250 | .360 | .490 | .640 | .810 | .902 | .980 | 2 |
| 3 | 0 | .970 | .857 | .729 | .512 | .343 | .216 | .125 | .064 | .027 | .008 | .001 | 0+ | 0+ | 0 |
| | 1 | .029 | .135 | .243 | .384 | .441 | .432 | .375 | .288 | .189 | .096 | .027 | .007 | 0+ | 1 |
| | 2 | 0+ | .007 | .027 | .096 | .189 | .288 | .375 | .432 | .441 | .384 | .243 | .135 | .029 | 2 |
| | 3 | 0+ | 0+ | .001 | .008 | .027 | .064 | .125 | .216 | .343 | .512 | .729 | .857 | .970 | 3 |
| 4 | 0 | .961 | .815 | .656 | .410 | .240 | .130 | .062 | .026 | .008 | .002 | 0+ | 0+ | 0+ | 0 |
| | 1 | .039 | .171 | .292 | .410 | .412 | .346 | .250 | .154 | .076 | .026 | .004 | 0+ | 0+ | 1 |
| | 2 | .001 | .014 | .049 | .154 | .265 | .346 | .375 | .346 | .265 | .154 | .049 | .014 | .001 | 2 |
| | 3 | 0+ | 0+ | .004 | .026 | .076 | .154 | .250 | .346 | .412 | .410 | .292 | .171 | .039 | 3 |
| | 4 | 0+ | 0+ | 0+ | .002 | .008 | .026 | .062 | .130 | .240 | .410 | .656 | .815 | .961 | 4 |
| 5 | 0 | .951 | .774 | .590 | .328 | .168 | .078 | .031 | .010 | .002 | 0+ | 0+ | 0+ | 0+ | 0 |
| | 1 | .048 | .204 | .328 | .410 | .360 | .259 | .156 | .077 | .028 | .006 | 0+ | 0+ | 0+ | 1 |
| | 2 | .001 | .021 | .073 | .205 | .309 | .346 | .312 | .230 | .132 | .051 | .008 | .001 | 0+ | 2 |
| | 3 | 0+ | .001 | .008 | .051 | .132 | .230 | .312 | .346 | .309 | .205 | .073 | .021 | .001 | 3 |
| | 4 | 0+ | 0+ | 0+ | .006 | .028 | .077 | .156 | .259 | .360 | .410 | .328 | .204 | .048 | 4 |
| | 5 | 0+ | 0+ | 0+ | 0+ | .002 | .010 | .031 | .078 | .168 | .328 | .590 | .774 | .951 | 5 |
| 6 | 0 | .941 | .735 | .531 | .262 | .118 | .047 | .016 | .004 | .001 | 0+ | 0+ | 0+ | 0+ | 0 |
| | 1 | .057 | .232 | .354 | .393 | .303 | .187 | .094 | .037 | .010 | .002 | 0+ | 0+ | 0+ | 1 |
| | 2 | .001 | .031 | .098 | .246 | .324 | .311 | .234 | .138 | .060 | .015 | .001 | 0+ | 0+ | 2 |
| | 3 | 0+ | .002 | .015 | .082 | .185 | .276 | .312 | .276 | .185 | .082 | .015 | .002 | 0+ | 3 |
| | 4 | 0+ | 0+ | .001 | .015 | .060 | .138 | .234 | .311 | .324 | .246 | .098 | .031 | .001 | 4 |
| | 5 | 0+ | 0+ | 0+ | .002 | .010 | .037 | .094 | .187 | .303 | .393 | .354 | .232 | .057 | 5 |
| | 6 | 0+ | 0+ | 0+ | 0+ | .001 | .004 | .016 | .047 | .118 | .262 | .531 | .735 | .941 | 6 |
| 7 | 0 | .932 | .698 | .478 | .210 | .082 | .028 | .008 | .002 | 0+ | 0+ | 0+ | 0+ | 0+ | 0 |
| | 1 | .066 | .257 | .372 | .367 | .247 | .131 | .055 | .017 | .004 | 0+ | 0+ | 0+ | 0+ | 1 |
| | 2 | .002 | .041 | .124 | .275 | .318 | .261 | .164 | .077 | .025 | .004 | 0+ | 0+ | 0+ | 2 |
| | 3 | 0+ | .004 | .023 | .115 | .227 | .290 | .273 | .194 | .097 | .029 | .003 | 0+ | 0+ | 3 |
| | 4 | 0+ | 0+ | .003 | .029 | .097 | .194 | .273 | .290 | .227 | .115 | .023 | .004 | 0+ | 4 |
| | 5 | 0+ | 0+ | 0+ | .004 | .025 | .077 | .164 | .261 | .318 | .275 | .124 | .041 | .002 | 5 |
| | 6 | 0+ | 0+ | 0+ | 0+ | .004 | .017 | .055 | .131 | .247 | .367 | .372 | .257 | .066 | 6 |
| | 7 | 0+ | 0+ | 0+ | 0+ | 0+ | .002 | .008 | .028 | .082 | .210 | .478 | .698 | .932 | 7 |
| 8 | 0 | .923 | .663 | .430 | .168 | .058 | .017 | .004 | .001 | 0+ | 0+ | 0+ | 0+ | 0+ | 0 |
| | 1 | .075 | .279 | .383 | .336 | .198 | .090 | .031 | .008 | .001 | 0+ | 0+ | 0+ | 0+ | 1 |
| | 2 | .003 | .051 | .149 | .294 | .296 | .209 | .109 | .041 | .010 | .001 | 0+ | 0+ | 0+ | 2 |
| | 3 | 0+ | .005 | .033 | .147 | .254 | .279 | .219 | .124 | .047 | .009 | 0+ | 0+ | 0+ | 3 |
| | 4 | 0+ | 0+ | .005 | .046 | .136 | .232 | .273 | .232 | .136 | .046 | .005 | 0+ | 0+ | 4 |
| | 5 | 0+ | 0+ | 0+ | .009 | .047 | .124 | .219 | .279 | .254 | .147 | .033 | .005 | 0+ | 5 |
| | 6 | 0+ | 0+ | 0+ | .001 | .010 | .041 | .109 | .209 | .296 | .294 | .149 | .051 | .003 | 6 |
| | 7 | 0+ | 0+ | 0+ | 0+ | .001 | .008 | .031 | .090 | .198 | .336 | .383 | .279 | .075 | 7 |
| | 8 | 0+ | 0+ | 0+ | 0+ | 0+ | .001 | .004 | .017 | .058 | .168 | .430 | .663 | .923 | 8 |

NOTE: 0+ represents a positive probability less than 0.0005.

(continued)

# Binomial Probabilities (*continued*)

| | | | | | | | | p | | | | | | | |
|---|---|---|---|---|---|---|---|---|---|---|---|---|---|---|---|
| n | x | .01 | .05 | .10 | .20 | .30 | .40 | .50 | .60 | .70 | .80 | .90 | .95 | .99 | x |
| 9 | 0 | .914 | .630 | .387 | .134 | .040 | .010 | .002 | 0+ | 0+ | 0+ | 0+ | 0+ | 0+ | 0 |
| | 1 | .083 | .299 | .387 | .302 | .156 | .060 | .018 | .004 | 0+ | 0+ | 0+ | 0+ | 0+ | 1 |
| | 2 | .003 | .063 | .172 | .302 | .267 | .161 | .070 | .021 | .004 | 0+ | 0+ | 0+ | 0+ | 2 |
| | 3 | 0+ | .008 | .045 | .176 | .267 | .251 | .164 | .074 | .021 | .003 | 0+ | 0+ | 0+ | 3 |
| | 4 | 0+ | .001 | .007 | .066 | .172 | .251 | .246 | .167 | .074 | .017 | .001 | 0+ | 0+ | 4 |
| | 5 | 0+ | 0+ | .001 | .017 | .074 | .167 | .246 | .251 | .172 | .066 | .007 | .001 | 0+ | 5 |
| | 6 | 0+ | 0+ | 0+ | .003 | .021 | .074 | .164 | .251 | .267 | .176 | .045 | .008 | 0+ | 6 |
| | 7 | 0+ | 0+ | 0+ | 0+ | .004 | .021 | .070 | .161 | .267 | .302 | .172 | .063 | .003 | 7 |
| | 8 | 0+ | 0+ | 0+ | 0+ | 0+ | .004 | .018 | .060 | .156 | .302 | .387 | .299 | .083 | 8 |
| | 9 | 0+ | 0+ | 0+ | 0+ | 0+ | 0+ | .002 | .010 | .040 | .134 | .387 | .630 | .914 | 9 |
| 10 | 0 | .904 | .599 | .349 | .107 | .028 | .006 | .001 | 0+ | 0+ | 0+ | 0+ | 0+ | 0+ | 0 |
| | 1 | .091 | .315 | .387 | .268 | .121 | .040 | .010 | .002 | 0+ | 0+ | 0+ | 0+ | 0+ | 1 |
| | 2 | .004 | .075 | .194 | .302 | .233 | .121 | .044 | .011 | .001 | 0+ | 0+ | 0+ | 0+ | 2 |
| | 3 | 0+ | .010 | .057 | .201 | .267 | .215 | .117 | .042 | .009 | .001 | 0+ | 0+ | 0+ | 3 |
| | 4 | 0+ | .001 | .011 | .088 | .200 | .251 | .205 | .111 | .037 | .006 | 0+ | 0+ | 0+ | 4 |
| | 5 | 0+ | 0+ | .001 | .026 | .103 | .201 | .246 | .201 | .103 | .026 | .001 | 0+ | 0+ | 5 |
| | 6 | 0+ | 0+ | 0+ | .006 | .037 | .111 | .205 | .251 | .200 | .088 | .011 | .001 | 0+ | 6 |
| | 7 | 0+ | 0+ | 0+ | .001 | .009 | .042 | .117 | .215 | .267 | .201 | .057 | .010 | 0+ | 7 |
| | 8 | 0+ | 0+ | 0+ | 0+ | .001 | .011 | .044 | .121 | .233 | .302 | .194 | .075 | .004 | 8 |
| | 9 | 0+ | 0+ | 0+ | 0+ | 0+ | .002 | .010 | .040 | .121 | .268 | .387 | .315 | .091 | 9 |
| | 10 | 0+ | 0+ | 0+ | 0+ | 0+ | 0+ | .001 | .006 | .028 | .107 | .349 | .599 | .904 | 10 |
| 11 | 0 | .895 | .569 | .314 | .086 | .020 | .004 | 0+ | 0+ | 0+ | 0+ | 0+ | 0+ | 0+ | 0 |
| | 1 | .099 | .329 | .384 | .236 | .093 | .027 | .005 | .001 | 0+ | 0+ | 0+ | 0+ | 0+ | 1 |
| | 2 | .005 | .087 | .213 | .295 | .200 | .089 | .027 | .005 | .001 | 0+ | 0+ | 0+ | 0+ | 2 |
| | 3 | 0+ | .014 | .071 | .221 | .257 | .177 | .081 | .023 | .004 | 0+ | 0+ | 0+ | 0+ | 3 |
| | 4 | 0+ | .001 | .016 | .111 | .220 | .236 | .161 | .070 | .017 | .002 | 0+ | 0+ | 0+ | 4 |
| | 5 | 0+ | 0+ | .002 | .039 | .132 | .221 | .226 | .147 | .057 | .010 | 0+ | 0+ | 0+ | 5 |
| | 6 | 0+ | 0+ | 0+ | .010 | .057 | .147 | .226 | .221 | .132 | .039 | .002 | 0+ | 0+ | 6 |
| | 7 | 0+ | 0+ | 0+ | .002 | .017 | .070 | .161 | .236 | .220 | .111 | .016 | .001 | 0+ | 7 |
| | 8 | 0+ | 0+ | 0+ | 0+ | .004 | .023 | .081 | .177 | .257 | .221 | .071 | .014 | 0+ | 8 |
| | 9 | 0+ | 0+ | 0+ | 0+ | .001 | .005 | .027 | .089 | .200 | .295 | .213 | .087 | .005 | 9 |
| | 10 | 0+ | 0+ | 0+ | 0+ | 0+ | .001 | .005 | .027 | .093 | .236 | .384 | .329 | .099 | 10 |
| | 11 | 0+ | 0+ | 0+ | 0+ | 0+ | 0+ | 0+ | .004 | .020 | .086 | .314 | .569 | .895 | 11 |
| 12 | 0 | .886 | .540 | .282 | .069 | .014 | .002 | 0+ | 0+ | 0+ | 0+ | 0+ | 0+ | 0+ | 0 |
| | 1 | .107 | .341 | .377 | .206 | .071 | .017 | .003 | 0+ | 0+ | 0+ | 0+ | 0+ | 0+ | 1 |
| | 2 | .006 | .099 | .230 | .283 | .168 | .064 | .016 | .002 | 0+ | 0+ | 0+ | 0+ | 0+ | 2 |
| | 3 | 0+ | .017 | .085 | .236 | .240 | .142 | .054 | .012 | .001 | 0+ | 0+ | 0+ | 0+ | 3 |
| | 4 | 0+ | .002 | .021 | .133 | .231 | .213 | .121 | .042 | .008 | .001 | 0+ | 0+ | 0+ | 4 |
| | 5 | 0+ | 0+ | .004 | .053 | .158 | .227 | .193 | .101 | .029 | .003 | 0+ | 0+ | 0+ | 5 |
| | 6 | 0+ | 0+ | 0+ | .016 | .079 | .177 | .226 | .177 | .079 | .016 | 0+ | 0+ | 0+ | 6 |
| | 7 | 0+ | 0+ | 0+ | .003 | .029 | .101 | .193 | .227 | .158 | .053 | .004 | 0+ | 0+ | 7 |
| | 8 | 0+ | 0+ | 0+ | .001 | .008 | .042 | .121 | .213 | .231 | .133 | .021 | .002 | 0+ | 8 |
| | 9 | 0+ | 0+ | 0+ | 0+ | .001 | .012 | .054 | .142 | .240 | .236 | .085 | .017 | 0+ | 9 |
| | 10 | 0+ | 0+ | 0+ | 0+ | 0+ | .002 | .016 | .064 | .168 | .283 | .230 | .099 | .006 | 10 |
| | 11 | 0+ | 0+ | 0+ | 0+ | 0+ | 0+ | .003 | .017 | .071 | .206 | .377 | .341 | .107 | 11 |
| | 12 | 0+ | 0+ | 0+ | 0+ | 0+ | 0+ | 0+ | .002 | .014 | .069 | .282 | .540 | .886 | 12 |

*NOTE:* 0+ represents a positive probability less than 0.0005.

(*continued*)

## Binomial Probabilities (*continued*)

| | | | | | | | p | | | | | | | | |
|---|---|---|---|---|---|---|---|---|---|---|---|---|---|---|---|
| n | x | .01 | .05 | .10 | .20 | .30 | .40 | .50 | .60 | .70 | .80 | .90 | .95 | .99 | x |
| 13 | 0 | .878 | .513 | .254 | .055 | .010 | .001 | 0+ | 0+ | 0+ | 0+ | 0+ | 0+ | 0+ | 0 |
| | 1 | .115 | .351 | .367 | .179 | .054 | .011 | .002 | 0+ | 0+ | 0+ | 0+ | 0+ | 0+ | 1 |
| | 2 | .007 | .111 | .245 | .268 | .139 | .045 | .010 | .001 | 0+ | 0+ | 0+ | 0+ | 0+ | 2 |
| | 3 | 0+ | .021 | .100 | .246 | .218 | .111 | .035 | .006 | .001 | 0+ | 0+ | 0+ | 0+ | 3 |
| | 4 | 0+ | .003 | .028 | .154 | .234 | .184 | .087 | .024 | .003 | 0+ | 0+ | 0+ | 0+ | 4 |
| | 5 | 0+ | 0+ | .006 | .069 | .180 | .221 | .157 | .066 | .014 | .001 | 0+ | 0+ | 0+ | 5 |
| | 6 | 0+ | 0+ | .001 | .023 | .103 | .197 | .209 | .131 | .044 | .006 | 0+ | 0+ | 0+ | 6 |
| | 7 | 0+ | 0+ | 0+ | .006 | .044 | .131 | .209 | .197 | .103 | .023 | .001 | 0+ | 0+ | 7 |
| | 8 | 0+ | 0+ | 0+ | .001 | .014 | .066 | .157 | .221 | .180 | .069 | .006 | 0+ | 0+ | 8 |
| | 9 | 0+ | 0+ | 0+ | 0+ | .003 | .024 | .087 | .184 | .234 | .154 | .028 | .003 | 0+ | 9 |
| | 10 | 0+ | 0+ | 0+ | 0+ | .001 | .006 | .035 | .111 | .218 | .246 | .100 | .021 | 0+ | 10 |
| | 11 | 0+ | 0+ | 0+ | 0+ | 0+ | .001 | .010 | .045 | .139 | .268 | .245 | .111 | .007 | 11 |
| | 12 | 0+ | 0+ | 0+ | 0+ | 0+ | 0+ | .002 | .011 | .054 | .179 | .367 | .351 | .115 | 12 |
| | 13 | 0+ | 0+ | 0+ | 0+ | 0+ | 0+ | 0+ | .001 | .010 | .055 | .254 | .513 | .878 | 13 |
| 14 | 0 | .869 | .488 | .229 | .044 | .007 | .001 | 0+ | 0+ | 0+ | 0+ | 0+ | 0+ | 0+ | 0 |
| | 1 | .123 | .359 | .356 | .154 | .041 | .007 | .001 | 0+ | 0+ | 0+ | 0+ | 0+ | 0+ | 1 |
| | 2 | .008 | .123 | .257 | .250 | .113 | .032 | .006 | .001 | 0+ | 0+ | 0+ | 0+ | 0+ | 2 |
| | 3 | 0+ | .026 | .114 | .250 | .194 | .085 | .022 | .003 | 0+ | 0+ | 0+ | 0+ | 0+ | 3 |
| | 4 | 0+ | .004 | .035 | .172 | .229 | .155 | .061 | .014 | .001 | 0+ | 0+ | 0+ | 0+ | 4 |
| | 5 | 0+ | 0+ | .008 | .086 | .196 | .207 | .122 | .041 | .007 | 0+ | 0+ | 0+ | 0+ | 5 |
| | 6 | 0+ | 0+ | .001 | .032 | .126 | .207 | .183 | .092 | .023 | .002 | 0+ | 0+ | 0+ | 6 |
| | 7 | 0+ | 0+ | 0+ | .009 | .062 | .157 | .209 | .157 | .062 | .009 | 0+ | 0+ | 0+ | 7 |
| | 8 | 0+ | 0+ | 0+ | .002 | .023 | .092 | .183 | .207 | .126 | .032 | .001 | 0+ | 0+ | 8 |
| | 9 | 0+ | 0+ | 0+ | 0+ | .007 | .041 | .122 | .207 | .196 | .086 | .008 | 0+ | 0+ | 9 |
| | 10 | 0+ | 0+ | 0+ | 0+ | .001 | .014 | .061 | .155 | .229 | .172 | .035 | .004 | 0+ | 10 |
| | 11 | 0+ | 0+ | 0+ | 0+ | 0+ | .003 | .022 | .085 | .194 | .250 | .114 | .026 | 0+ | 11 |
| | 12 | 0+ | 0+ | 0+ | 0+ | 0+ | .001 | .006 | .032 | .113 | .250 | .257 | .123 | .008 | 12 |
| | 13 | 0+ | 0+ | 0+ | 0+ | 0+ | 0+ | .001 | .007 | .041 | .154 | .356 | .359 | .123 | 13 |
| | 14 | 0+ | 0+ | 0+ | 0+ | 0+ | 0+ | 0+ | .001 | .007 | .044 | .229 | .488 | .869 | 14 |
| 15 | 0 | .860 | .463 | .206 | .035 | .005 | 0+ | 0+ | 0+ | 0+ | 0+ | 0+ | 0+ | 0+ | 0 |
| | 1 | .130 | .366 | .343 | .132 | .031 | .005 | 0+ | 0+ | 0+ | 0+ | 0+ | 0+ | 0+ | 1 |
| | 2 | .009 | .135 | .267 | .231 | .092 | .022 | .003 | 0+ | 0+ | 0+ | 0+ | 0+ | 0+ | 2 |
| | 3 | 0+ | .031 | .129 | .250 | .170 | .063 | .014 | .002 | 0+ | 0+ | 0+ | 0+ | 0+ | 3 |
| | 4 | 0+ | .005 | .043 | .188 | .219 | .127 | .042 | .007 | .001 | 0+ | 0+ | 0+ | 0+ | 4 |
| | 5 | 0+ | .001 | .010 | .103 | .206 | .186 | .092 | .024 | .003 | 0+ | 0+ | 0+ | 0+ | 5 |
| | 6 | 0+ | 0+ | .002 | .043 | .147 | .207 | .153 | .061 | .012 | .001 | 0+ | 0+ | 0+ | 6 |
| | 7 | 0+ | 0+ | 0+ | .014 | .081 | .177 | .196 | .118 | .035 | .003 | 0+ | 0+ | 0+ | 7 |
| | 8 | 0+ | 0+ | 0+ | .003 | .035 | .118 | .196 | .177 | .081 | .014 | 0+ | 0+ | 0+ | 8 |
| | 9 | 0+ | 0+ | 0+ | .001 | .012 | .061 | .153 | .207 | .147 | .043 | .002 | 0+ | 0+ | 9 |
| | 10 | 0+ | 0+ | 0+ | 0+ | .003 | .024 | .092 | .186 | .206 | .103 | .010 | .001 | 0+ | 10 |
| | 11 | 0+ | 0+ | 0+ | 0+ | .001 | .007 | .042 | .127 | .219 | .188 | .043 | .005 | 0+ | 11 |
| | 12 | 0+ | 0+ | 0+ | 0+ | 0+ | .002 | .014 | .063 | .170 | .250 | .129 | .031 | 0+ | 12 |
| | 13 | 0+ | 0+ | 0+ | 0+ | 0+ | 0+ | .003 | .022 | .092 | .231 | .267 | .135 | .009 | 13 |
| | 14 | 0+ | 0+ | 0+ | 0+ | 0+ | 0+ | 0+ | .005 | .031 | .132 | .343 | .366 | .130 | 14 |
| | 15 | 0+ | 0+ | 0+ | 0+ | 0+ | 0+ | 0+ | 0+ | .005 | .035 | .206 | .463 | .860 | 15 |

*NOTE:* 0+ represents a positive probability less than 0.0005.

From Frederick C. Mosteller, Robert E. K. Rourke, and George B. Thomas, Jr., *Probability with Statistical Applications*, 2nd ed., copyright © 1970 Pearson Education. Reprinted with permission of the publisher.

Right Tail Probability

## Table 4: *t* Distribution Critical Values

| | Confidence Level | | | | | |
|---|---|---|---|---|---|---|
| | 80% | 90% | 95% | 98% | 99% | 99.8% |
| | Right-Tail Probability | | | | | |
| df | $t_{.100}$ | $t_{.050}$ | $t_{.025}$ | $t_{.010}$ | $t_{.005}$ | $t_{.001}$ |
| 1 | 3.078 | 6.314 | 12.706 | 31.821 | 63.656 | 318.289 |
| 2 | 1.886 | 2.920 | 4.303 | 6.965 | 9.925 | 22.328 |
| 3 | 1.638 | 2.353 | 3.182 | 4.541 | 5.841 | 10.214 |
| 4 | 1.533 | 2.132 | 2.776 | 3.747 | 4.604 | 7.173 |
| 5 | 1.476 | 2.015 | 2.571 | 3.365 | 4.032 | 5.894 |
| 6 | 1.440 | 1.943 | 2.447 | 3.143 | 3.707 | 5.208 |
| 7 | 1.415 | 1.895 | 2.365 | 2.998 | 3.499 | 4.785 |
| 8 | 1.397 | 1.860 | 2.306 | 2.896 | 3.355 | 4.501 |
| 9 | 1.383 | 1.833 | 2.262 | 2.821 | 3.250 | 4.297 |
| 10 | 1.372 | 1.812 | 2.228 | 2.764 | 3.169 | 4.144 |
| 11 | 1.363 | 1.796 | 2.201 | 2.718 | 3.106 | 4.025 |
| 12 | 1.356 | 1.782 | 2.179 | 2.681 | 3.055 | 3.930 |
| 13 | 1.350 | 1.771 | 2.160 | 2.650 | 3.012 | 3.852 |
| 14 | 1.345 | 1.761 | 2.145 | 2.624 | 2.977 | 3.787 |
| 15 | 1.341 | 1.753 | 2.131 | 2.602 | 2.947 | 3.733 |
| 16 | 1.337 | 1.746 | 2.120 | 2.583 | 2.921 | 3.686 |
| 17 | 1.333 | 1.740 | 2.110 | 2.567 | 2.898 | 3.646 |
| 18 | 1.330 | 1.734 | 2.101 | 2.552 | 2.878 | 3.611 |
| 19 | 1.328 | 1.729 | 2.093 | 2.539 | 2.861 | 3.579 |
| 20 | 1.325 | 1.725 | 2.086 | 2.528 | 2.845 | 3.552 |
| 21 | 1.323 | 1.721 | 2.080 | 2.518 | 2.831 | 3.527 |
| 22 | 1.321 | 1.717 | 2.074 | 2.508 | 2.819 | 3.505 |
| 23 | 1.319 | 1.714 | 2.069 | 2.500 | 2.807 | 3.485 |
| 24 | 1.318 | 1.711 | 2.064 | 2.492 | 2.797 | 3.467 |
| 25 | 1.316 | 1.708 | 2.060 | 2.485 | 2.787 | 3.450 |
| 26 | 1.315 | 1.706 | 2.056 | 2.479 | 2.779 | 3.435 |
| 27 | 1.314 | 1.703 | 2.052 | 2.473 | 2.771 | 3.421 |
| 28 | 1.313 | 1.701 | 2.048 | 2.467 | 2.763 | 3.408 |
| 29 | 1.311 | 1.699 | 2.045 | 2.462 | 2.756 | 3.396 |
| 30 | 1.310 | 1.697 | 2.042 | 2.457 | 2.750 | 3.385 |
| 40 | 1.303 | 1.684 | 2.021 | 2.423 | 2.704 | 3.307 |
| 50 | 1.299 | 1.676 | 2.009 | 2.403 | 2.678 | 3.261 |
| 60 | 1.296 | 1.671 | 2.000 | 2.390 | 2.660 | 3.232 |
| 80 | 1.292 | 1.664 | 1.990 | 2.374 | 2.639 | 3.195 |
| 100 | 1.290 | 1.660 | 1.984 | 2.364 | 2.626 | 3.174 |
| ∞ | 1.282 | 1.645 | 1.960 | 2.326 | 2.576 | 3.091 |

## Table 5: Chi-Squared Distribution for Values of Various Right Tail Probabilities

| | Right-Tail Probability | | | | | | |
|---|---|---|---|---|---|---|---|
| df | 0.250 | 0.100 | 0.050 | 0.025 | 0.010 | 0.005 | 0.001 |
| 1 | 1.32 | 2.71 | 3.84 | 5.02 | 6.63 | 7.88 | 10.83 |
| 2 | 2.77 | 4.61 | 5.99 | 7.38 | 9.21 | 10.60 | 13.82 |
| 3 | 4.11 | 6.25 | 7.81 | 9.35 | 11.34 | 12.84 | 16.27 |
| 4 | 5.39 | 7.78 | 9.49 | 11.14 | 13.28 | 14.86 | 18.47 |
| 5 | 6.63 | 9.24 | 11.07 | 12.83 | 15.09 | 16.75 | 20.52 |
| 6 | 7.84 | 10.64 | 12.59 | 14.45 | 16.81 | 18.55 | 22.46 |
| 7 | 9.04 | 12.02 | 14.07 | 16.01 | 18.48 | 20.28 | 24.32 |
| 8 | 10.22 | 13.36 | 15.51 | 17.53 | 20.09 | 21.96 | 26.12 |
| 9 | 11.39 | 14.68 | 16.92 | 19.02 | 21.67 | 23.59 | 27.88 |
| 10 | 12.55 | 15.99 | 18.31 | 20.48 | 23.21 | 25.19 | 29.59 |
| 11 | 13.70 | 17.28 | 19.68 | 21.92 | 24.72 | 26.76 | 31.26 |
| 12 | 14.85 | 18.55 | 21.03 | 23.34 | 26.22 | 28.30 | 32.91 |
| 13 | 15.98 | 19.81 | 22.36 | 24.74 | 27.69 | 29.82 | 34.53 |
| 14 | 17.12 | 21.06 | 23.68 | 26.12 | 29.14 | 31.32 | 36.12 |
| 15 | 18.25 | 22.31 | 25.00 | 27.49 | 30.58 | 32.80 | 37.70 |
| 16 | 19.37 | 23.54 | 26.30 | 28.85 | 32.00 | 34.27 | 39.25 |
| 17 | 20.49 | 24.77 | 27.59 | 30.19 | 33.41 | 35.72 | 40.79 |
| 18 | 21.60 | 25.99 | 28.87 | 31.53 | 34.81 | 37.16 | 42.31 |
| 19 | 22.72 | 27.20 | 30.14 | 32.85 | 36.19 | 38.58 | 43.82 |
| 20 | 23.83 | 28.41 | 31.41 | 34.17 | 37.57 | 40.00 | 45.32 |
| 25 | 29.34 | 34.38 | 37.65 | 40.65 | 44.31 | 46.93 | 52.62 |
| 30 | 34.80 | 40.26 | 43.77 | 46.98 | 50.89 | 53.67 | 59.70 |
| 40 | 45.62 | 51.80 | 55.76 | 59.34 | 63.69 | 66.77 | 73.40 |
| 50 | 56.33 | 63.17 | 67.50 | 71.42 | 76.15 | 79.49 | 86.66 |
| 60 | 66.98 | 74.40 | 79.08 | 83.30 | 88.38 | 91.95 | 99.61 |
| 70 | 77.58 | 85.53 | 90.53 | 95.02 | 100.43 | 104.21 | 112.32 |
| 80 | 88.13 | 96.58 | 101.88 | 106.63 | 112.33 | 116.32 | 124.84 |
| 90 | 98.65 | 107.57 | 113.15 | 118.14 | 124.12 | 128.30 | 137.21 |
| 100 | 109.14 | 118.50 | 124.34 | 129.56 | 135.81 | 140.17 | 149.45 |

# Appendix B:  Answers to Check Your Tech

## CHAPTER 3

### Finding the Standard Deviation of Vacation Days for Several Countries

1: $180/6 = 30$

2:

| $x$ | $x - \bar{x}$ | $(x - \bar{x})^2$ |
|---|---|---|
| 13 | $13 - 30 = -17$ | $(-17)^2 = 289$ |
| 25 | $25 - 30 = -5$ | $(-5)^2 = 25$ |
| 42 | $42 - 30 = 12$ | $12^2 = 144$ |
| 37 | $37 - 30 = 7$ | $7^2 = 49$ |
| 35 | $35 - 30 = 5$ | $5^2 = 25$ |
| 28 | $28 - 30 = -2$ | $(-2)^2 = 4$ |

3: $289 + 25 + 144 + 49 + 25 + 4 = 536$

4: $536/5 = 107.2$

5: $\sqrt{107.2} = 10.3537$, which rounds to 10.35. This is the same value shown in Figure A.

### Making a Boxplot of the Area of Western States

1: The axis is shown in the graph.

2: The box goes from 84.5 on the left to 134.5 on the right.

3: The median line is at 104, which is Colorado.

4: IQR $= 134.5 - 84.5 = 50$

5: Lower limit $=$ Q1 $- (1.5 \times$ IQR$) = 84.5 - (1.5 \times 50) = 84.5 - 75 = 9.5$

The smallest area is 11, which is not below 9.5, so there are no low-end potential outliers.

6: The low whisker goes down to 11 (Hawaii).

7: Upper limit $=$ Q3 $+ (1.5 \times$ IQR$) = 134.5 + (1.5 \times 50) = 134.5 + 75 = 209.5$

The only area larger than 209.5 is 656 (Alaska), which is an outlier and should have a separate mark.

8: The upper whisker goes to 164 (California), which is the largest area that is not a potential outlier.

9: One title is shown in the graph.

## CHAPTER 4

1:

| $x$ | $x - \bar{x}$ | $z_x$ | $y$ | $y - \bar{y}$ | $z_y$ | $z_x z_y$ |
|---|---|---|---|---|---|---|
| 20 | $20 - 30 = -10$ | $-10/10 = -1$ | 20 | $20 - 25 = -5$ | $-5/5 = -1$ | $(-1) \times (-1) = 1$ |
| 30 | $30 - 30 = 0$ | $0/10 = 0$ | 30 | $30 - 25 = 5$ | $5/5 = 1$ | $0 \times (1) = 0$ |
| 40 | $40 - 30 = 10$ | $10/10 = 1$ | 25 | $25 - 25 = 0$ | $0/5 = 0$ | $1 \times 0 = 0$ |

2: Add the last column to get $\sum z_x z_y = 1 + 0 + 0 = 1$.

3: Find the correlation and check it with the output,

$$r = \frac{\sum z_x z_y}{n - 1} = \frac{1}{3 - 1} = \frac{1}{2} = 0.5$$

This is the same correlation as the output.

4: Find the slope.

$$b = r \frac{s_y}{s_x} = r \times \frac{5}{10} = 0.50 \times (0.50) = 0.25$$

5: Find the $y$-intercept.

$$a = \bar{y} - b\bar{x} = 25 - 0.25 \times 30 = 25 - 7.5 = 17.5$$

6: Finally, put together the equation:

$$y = a + bx$$

$$\text{Predicted Wife} = a + b \text{ Husband}$$

$$\text{Predicted Wife} = 17.5 + 0.25 \text{ Husband}$$

The equation is the same as the Minitab output.

## CHAPTER 9

1: $73.6 - 92.5143 = -18.9143$ (which is also shown in the Minitab Output)

2: $SE_{\text{diff}} = \dfrac{S_{\text{diff}}}{\sqrt{n}} = \dfrac{15.0497}{\sqrt{35}} = \dfrac{15.0497}{5.9161} = 2.5439$

3: $t = \dfrac{\bar{x}_{\text{diff}} - 0}{SE_{\text{diff}}} = \dfrac{-18.9143}{2.5439} = -7.44$

4: a. If the difference between means were larger, that would cause $t$ to be larger, because the difference between means is in the numerator, and a larger numerator results in a larger number. b. If the standard deviation ($S_{\text{diff}}$) were larger, that would cause $t$ to be smaller, because the standard deviation is in the denominator, and a larger denominator results in a smaller number. c. If the sample size were larger, that would cause $t$ to be larger: The larger sample size would cause $SE_{\text{diff}}$ to be smaller (because the sample size is in the denominator of $SE_{\text{diff}}$), and the smaller $SE_{\text{diff}}$ would cause $t$ to be larger (because $SE_{\text{diff}}$ is in the denominator of $t$).

## CHAPTER 13

1: 4

2: Two out of four should be older.

3:

| GGGG | GGGB | GGBB | BBBG | BBBB |
|---|---|---|---|---|
| | GGBG | GBGB | BBGB | |
| | GBGG | GBBG | BGBB | |
| | BGGG | BGBG | GBBB | |
| | | BBGG | | |
| | | BGGB | | |

4: $(0.5)^4 = 0.0625$  5: The probability of each is 0.0625.

6: 16  7: 4 out of 16  8: 5 out of 16

9: $5/16 = 0.3125$, which is the same as the Minitab output.

# Appendix C:
# Answers to Odd-Numbered Exercises

## CHAPTER 1

### Section 1.2

**1.1 a.** Handedness: categorical  **b.** Age: numerical

**1.3** Male is categorical with two categories. The 1's represent males, and the 0's represent females. If you added the numbers, you would get the number of males, so it makes sense here.

**1.5 a.** Stacked  **b.** 1 means male, and 0 means female.

**c.**

| Female | Male |
|--------|------|
| 9.5 | 9.4 |
| 9.5 | 9.5 |
| 9.9 | 9.5 |
| | 9.7 |

### Section 1.3

**1.7 a.** $189/29{,}617 = 0.64\%$ changes  **b.** $124/14{,}513 = 0.85\%$ changes  **c.** The higher percentage of change was in the class told to change their answers, so the instruction did seem to have a small effect.

**1.9 a.** $15/38 = 39.5\%$ of the class were male  **b.** $0.641(234) = 149.994$, or about 150, men in the class

**c.** $0.40(x) = 20$
$20/0.4 = 50$ people in the class

**1.11** The frequency of women is 7, the proportion is $7/11$, and the percentage is **63.6%.**

**1.13** $0.202x = 88{,}547{,}000$
$\qquad x = 438{,}351{,}485$ or a rounded version of this

**1.15** Steps 1–3 are shown in the accompanying table.

| State | AIDS | Rank Cases | Population | Population (thousands) | AIDS per 1000 | Rank Rate |
|-------|------|------------|------------|-----------------------|---------------|-----------|
| New York | 75,253 | 1 | 19,297,729 | 19,298 | 3.90 | 2 |
| California | 65,582 | 2 | 36,553,215 | 36,553 | 1.79 | 4 |
| Florida | 48,059 | 3 | 18,251,243 | 18,251 | 2.63 | 3 |
| Texas | 34,940 | 4 | 23,904,380 | 23,904 | 1.46 | 6 |
| Pennsylvania | 19,236 | 5 | 12,432,792 | 12,433 | 1.55 | 5 |
| District of Columbia | 8,895 | 6 | 588,292 | 588 | 15.13 | 1 |

4: No, the ranks are not the same. The District of Columbia had the highest rate and had the lowest number of cases. (Also, the rate for Florida puts its rank above California, and the rate for Pennsylvania puts it above Texas in ranking.)

5: The District of Columbia is the place (among these six regions) where you would be most likely to meet a person living with AIDS, and Texas is the place (among these six regions) where you would be least likely to do so.

**1.17** 1990: 58.7%, 1997: 56.4%, 2000: 56.2%, 2007: 55.1%
The percentage of married people is decreasing over time (at least with these dates).

**1.19** We don't know the percentage of female students in the two classes. The larger number of women at 8 a.m. may just result from a larger number of students at 8 a.m., which may be because the class can accommodate more students because perhaps it is in a large lecture hall.

### Section 1.4

**1.21** Observational study

**1.23** Controlled experiment

**1.25** Controlled experiment

**1.27** Observational study

**1.29** This was an observational study, and from it you cannot conclude that the tutoring raises the grades. Possible confounders (answers may vary): 1. It may be the more highly motivated who attend the tutoring, and this motivation is what causes the grades to go up. 2. It could be that those with more time attend the tutoring, and it is the increased time studying that causes the grades to go up.

**1.31** It was an observational study. There was no random assignment. They simply looked at records. We cannot say that CABG causes better results, because there may have been confounding variables.

**1.33** The students should have been randomly assigned to the treatment. Half of the students should have been given a placebo. Also, ideally the person to whom the subjects report their results should not know whether they are taking vitamin C or a placebo, making the study double-blind.

**1.35** Because this was not a randomized experiment (no one assigned people to be happy), we should not infer causation. It is possible that healthier people tend to be happier, and it was the health of those people that caused them not to catch a cold. If we cannot infer causation, then you should not believe that a change in happiness would change the likelihood of catching a cold.

**1.37** Ask whether the patients were randomly assigned the full or the half dose. Without randomization there could be bias, and we cannot infer causation. With randomization we can infer causation.

**1.39** This was an observational study: vitamin C and breast milk. We cannot conclude cause and effect from observational studies.

**1.41 a.** LD: 8% tumors; LL: 28% tumors  **b.** A controlled experiment. You can tell by the random assignment.  **c.** Yes, we can conclude cause and effect because it was a controlled experiment, and random assignment will balance out potential confounding variables.

### Chapter Review Exercises

**1.43 a.** Dating: $81/440 = 18.4\%$  **b.** Cohabiting: $103/429 = 24.0\%$  **c.** Married: $147/424 = 34.7\%$  **d.** No, this was an observational study. Confounding variables may vary. Perhaps married people are likely to be older, and older people are more likely to be obese.

**1.45 a.** The two-way table follows.

| | Boy | Girl |
|--------|-----|------|
| Violent | 10 | 11 |
| Nonviolent | 19 | 4 |

**b.** For the boys, $10/29$, or 34.5%, were on probation for violent crime. For the girls, $11/15$, or 73.3%, were on probation for violent crime.  **c.** The girls were more likely to be on probation for violent crime.

**1.47** Writing: Answers will vary but should include randomization, placebo, control, and blinds.

**1.49** This was an observational study, not a controlled experiment. From observational studies you cannot conclude cause and effect. Therefore, offering to participate in clinical trials would probably have very little effect on one's chance of surviving. Possible confounders will vary. However, the people who offer to be in clinical trials may tend to follow the advice of doctors more, and it may be the practice of following advice that raises their chances of survival.

**1.51** No, we cannot conclude causation. There was no control group for comparison, and the sample size was very small.

# CHAPTER 2

## Sections 2.1 and 2.2

**2.1 a.** 11 are morbidly obese. **b.** 11/134 is about 8%, which is much more than 3%.

**2.3** New vertical axis labels: 0.04, 0.08, 0.12, 0.16, 0.20, 0.24, 0.28. Note that 0.04 comes from 1/25

**2.5 a.** 1 (or 2) have no TVs **b.** 9 TVs **c.** Between 25 and 30 **d.** Around 6 **e.** 6/90, or 0.0667

**2.7 a.** NYC **b.** Las Vegas **c.** Las Vegas

**2.9** It should be right-skewed.

**2.11** It would be bimodal because the men and women tend to have different heights and therefore different armspans.

**2.13** About 58 years (between 56 and 60)

**2.15** Riding the bus shows a larger typical value and also more variation.

**2.17 a.** Multimodal with modes at 12 years (high school), 14 years (junior college), 16 years (bachelor's degree), and 18 years (possible master's degree). It is also left-skewed with numbers as low as 0. **b.** Estimate: 300 + 50 + 100 + 40 + 50, or about 500 to 600, had 16 or more years. **c.** This is between 25% (from 500/2018) and 30% (from 600/2018) have a bachelor's degree or higher. This is very similar to the 27% given.

**2.19** Both graphs go from about 0 to about 20, but the data for years of formal education for the respondents (compared to their mothers) include more with education above 12 years. For example, the bar at 16 (college bachelor's degree) is higher for the respondents than for the mothers, which shows that the respondents tend to have a bit more education than their mothers. Also, the bar at 12 is taller for the mothers, showing that the mothers were more likely to get only a high school diploma. Furthermore, the bar graph for the mothers includes more people (taller bars) at lower numbers of years, such as 0 and 3 and 6.

**2.21 1.** B **2.** C **3.** A

**2.23 1.** See the dotplots. Histograms would also be good for visualizing the distributions. Stemplots would not work with these data sets because all the observed values have only one digit.

**2.** Full-time is a bit left-skewed, and part-time is a bit right-skewed.

**3.** Those with full-time jobs go out to eat more than those with part-time jobs.

**4.** The full-time workers have a distribution that is more spread out; full-time goes from 1 to 7, whereas part-time goes only from 1 to 5.

**5.** There are no outliers—that is, no dots detached from the main group with an empty space between.

**6.** For the full-time workers the distribution is a little left-skewed, and for the part-time workers it is a little right-skewed. The full-time workers tend to go out to eat more, and their distribution is more spread out.

**2.25** See histogram.

The histogram is bimodal with modes at about $30 and about $90.

**2.27** See histogram. The histogram is right-skewed. The typical value is around 12 (between 10 and 15) years, and there are three outliers: Asian elephant (40), African elephant (35), and hippo (41). Humans (75 years) would be way off to the right; they live much longer than other mammals.

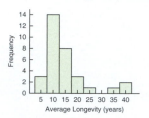

**2.29 1.** B **2.** A **3.** C

## Sections 2.3 and 2.4

**2.31** No, the largest category is Wrong to Right, which suggests that changes tend to make the answers more likely to be right.

**2.33 a.** Baidu is the mode (the most frequently used search engine) for both time periods. It becomes even more popular in 2010. **b.** There was more variability in 2009, because in 2010 the predominance of Baidu was more noticeable.

**2.35 a.** Democrat (not strong). It is easier to compare using the heights of the bars in the bar chart. **b.** Republican (not strong)

**2.37 a.** The percentage of old people is increasing, the percentage of those 25–64 is decreasing, and the percentage of those 24 and below is relatively constant. **b.** The money for Social Security normally comes from those in a working age range (which includes those 25–64), and that group is decreasing in percentage. Also, the group receiving Social Security (those 65 and older) is becoming larger. This suggests that in the future, Social Security might not get enough money from the workers to support the old people.

**2.39** A Pareto chart or pie chart would also be appropriate.

Note that the mode is Social Science and that there is substantial variation. (Of course, individual majors such as chemistry were grouped into Math and Science.)

### Chapter Review Exercises

**2.41** TV: Histograms: One for the males and one for the females would be appropriate. Dotplots or stemplots would also work for this numerical data set.

**2.43 a.** The diseases with higher rates for HRT were heart attack, stroke, pulmonary embolism, and breast cancer. The diseases with lower rates for HRT were endometrial cancer, colorectal cancer, and hip fracture. **b.** Comparing the rates makes more sense than comparing just the numbers, in case there were more women in one group than in the other.

**2.45** The vertical axis does not start at zero and exaggerates the differences. Make a graph for which the vertical axis starts at zero.

**2.47 a.** Pie chart **b.** Histogram

**2.49** The shapes are roughly Normal; the later period is warmer, but the spread is similar. This is consistent with theories on global warming. The difference is $57.9 - 56.7 = 1.2$, so the difference is only a bit more than 1 degree Fahrenheit.

**2.51** The created 10-point dotplots will vary. The dotplot for Exercise 2.51 should have skew, and the dotplot for Exercise 2.52 should not.

**2.53** Graphs will vary. Histograms, dotplots, or stemplots are all appropriate. The prices in West Los Angeles tend to be higher and more varied than the prices in Midtown.

**2.55** The data set should be right-skewed with some unusually high numbers.

Exxon Mobil is an outlier.

# CHAPTER 3

## Section 3.1

**3.1 c**

**3.3** The typical age of the CEOs is between about 56 and 60 (or any number from 56 to 60). (The distribution is symmetric, so the mean should be about in the middle.)

**3.5 a.** The typical number of vacation days is about 17.3, the mean. One could also say that number of vacation days tends to be about 17.3, on average. **b.** The standard deviation of paid vacation days for the six countries is 10.7 **c.** The United States, at 0, is farthest from the mean and contributes most to the standard deviation.

**3.7 a.** 57.8, Older: 57.8 (the mean age for the first six presidents) is more than 55.3 for the most recent six presidents. **b.** Less: 1.6 (the standard deviation of ages for the first six presidents) tends to be much less varied than the 9.3 for the most recent six presidents.

**3.9 a.** 35,520; 117,845, eastern **b.** 24,704; 124,646, western **c.** Eastern

**3.11 a.** The mean for longboards is 12.4 days which is more than the mean for shortboards, which is 9.8 days. So longboarders tend to get more surfing days. **b.** The standard deviation of 5.2 days for the longboarders was larger than the standard deviation of 4.2 days for the shortboarders. So the longboarders have more variation in days.

**3.13** The prices of the houses in Westlake (Figure B) have a larger standard deviation than the prices in Agoura, because the data from Agoura show a lot of prices near the center of the graph and the prices in Westlake show a lot of prices far from the center.

**3.15 a.** Top: $3462 + 500 = 3962$
    Bottom: $3462 - 500 = 2962$
  **b.** Yes, a birth weight of 2800 grams is more than one SD below the mean because it is less than 2962.

**3.17 a.** The standard deviation is 1.4 years. **b.** The same. The standard deviation in 20 years is still 1.4. Adding 20 to each number does not affect the standard deviation. Standard deviation does not depend on the size of the numbers, only on how far apart they are. **c.** The mean is 3 years. **d.** Larger:

The mean is 23 years of age. When 20 is added to each number, the mean increases by 20.

**3.19** SD for the 100-meter event would be less. All the runners come to the finish line within a few seconds of each other. In the marathon, the runners can be quite widely spread after running that long distance.

## Section 3.2

**3.21** Answers correspond to the guided steps.
1: 95% (See the accompanying curve.)
2: 583 is from $406 + 177$, because it is one standard deviation above the mean.
3: As shown on the curve, A is 52, B is 229, and C is 760.
4: Answer a. About 95% between 52 and 760.
5: Answer b: About 68% between 229 and 583.
6: Answer c: Most would not consider 484 unusual because it is between 229 and 583.
7: Answer d: 30 is unusually small, because it is less than 52, which means it is more than two standard deviations below the mean, and so less than 2.5% of the population have values lower than this.

**3.23** $z = \dfrac{4060 - 3331}{729} = \dfrac{729}{729} = 1$

**a.** About 68% according to the Empirical Rule, because the $z$-scores are 1 and $-1$. **b.** About 95% according to the Empirical Rule **c.** Nearly all the data should be within three standard deviations of the mean. Three standard deviations above the mean is

$$3331 + 3(729) = 5518$$

Because 9000 is quite a bit above 5518, you might think it unlikely that one state's rate would be 9000.

**3.25 a.** $-2$ **b.** 67 inches (or 5 feet 7 inches)

**3.27** The $z$-score for the SAT of 750 is 2.5, and the $z$-score for the ACT of 28 is 1.4. The score of 750 is more unusual, because its $z$-score is farther from 0.

**3.29 a.** $z = \dfrac{2500 - 3462}{500} = \dfrac{-962}{500} = -1.92$

**b.** $z = \dfrac{2500 - 2622}{500} = \dfrac{-122}{500} = -0.24$

**c.** A birth rate of 2500 grams is more common (the $z$-score is closer to 0) for babies born one month early. In other words, there are a higher percentages of babies with low birth weight among those born one month early. This makes sense because babies gain weight during gestation, and babies born one month early have had less time to gain weight.

**3.31** $70 + 1.5(10) = 85$

## Section 3.3

**3.33** Two measures of the center of data are the mean and the median. The median is preferred for data that are strongly skewed or have outliers. If the data are relatively symmetric, the mean is preferred but the median is also OK.

**3.35**

| 163 | 192 | 206 | 224 | 244 | | 246 | 256 | | 261 | 293 | 340 |
|---|---|---|---|---|---|---|---|---|---|---|---|
| | | Q1 | | | | Med | | | Q3 | | |

**a.** The median of 245 million is the typical income for the top ten grossing Pixar animated movies. **b.** $IQR = 261 - 206 = 55$ million. This is the range of the middle 50% of the sorted incomes in the top ten grossing Pixar animated movies.

**3.37**

|  | 17.3 | 22.0 | 23.0 | 24.0 | 25.0 | 27.2 | 27.7 |
|--|------|------|------|------|------|------|------|
|  |      | Q1   |      | Med  |      | Q3   |      |

**a.** 24.0 cents per gallon **b.** IQR = Q3 − Q1 = 27.2 − 22.0 = 5.2 cents per gallon

**3.39 a.** 11.5, 9.5, longboards **b.** 8, 5, longboards (Answers for interquartile range may vary with different technology.)

## Section 3.4

**3.41 a.** Outliers are observed values that are far from the main group of data. In a histogram they are separated from the others by space. If they are mistakes, they should be removed. If they are not mistakes, do the analysis twice: once with and once without outliers. **b.** The median is more resistant, which implies that it changes less than the mean (when comparing the data with and without outliers).

**3.43** The corrected value will give a different mean but not a different median. Medians are not as affected by the size of extreme scores, but the mean is affected.

**3.45 a.** The shape is right-skewed, the median is 20, and the interquartile range is 35 − 19 = 16. There is an outlier at 66. **b.** The mean is 27.3 and the median is 20. The mean is much larger because of the outlier, 66. The mean and median should be marked on the histogram of the data.

**3.47 a.** Both data sets are right-skewed and have outliers that represent large numbers of hours of study, so the medians (and interquartile ranges) should be compared. **b.** The median of 7 was larger for the women; the men's median was only 4. The interquartile range was 4.5 for the women and 3 for the men, so the IQR was larger for the women. Both data sets are right-skewed with outliers at around 15 or 20 hours. Summary: The women tended to study more and had more variation as measured by the interquartile range.

## Section 3.5

**3.49 a.** S, MW, W, NE (lowest to highest medians) **b.** The Northeast has the largest IQR. **c.** The South has the smallest IQR. **d.** The South has potential outliers. **e.** The box for the South is very small and far to the left, so the numbers in the 20's are very far from the box. The box for the Northeast is large and far to the right, so 24.5 is not a potential outlier.

**3.51** Answers will vary: Phoenix tends to be hottest, since median temperatures are high, although Honolulu and Phoenix have approximately the same medians. The most varied temperatures are in Chicago. Both Honolulu and LA have relatively little variation, as determined by the interquartile range.

The choice of favorite city will vary.

**3.53 a.** Histogram 1 goes with boxplot C.

Histogram 2 goes with boxplot B.

Histogram 3 goes with boxplot A.

Reasoning: A boxplot with a whisker (or potential outliers) on the far right corresponds to a histogram that is right-skewed. A boxplot that is centered approximately symmetrically corresponds to a histogram that is not very skewed.

**b.** Histogram 1 is strongly right-skewed, histogram 2 is left-skewed, and histogram 3 is relatively symmetric.

Histogram 1 would have a mean larger than the median.

Histogram 2 would have a mean smaller than the median.

Histogram 3 would have a mean and median that were very similar.

**3.55 a.** The median is 20. Q1 is 17.5 and Q3 is 23, so the IQR is 5.5. There is a high-end outlier of about 33, which is Wyoming.

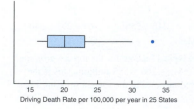

**b.** A low population relative to the number of deaths is another way of seeing a high death rate. (Wyoming has a low population.)

## Chapter Review Exercises

**3.57** Both data sets are slightly right-skewed because of the potential outliers. The men tend to be faster than the women, as shown by the lower median line for the men. The interquartile range is larger for the men, as shown by the larger box.

**3.59 a.** The median number of death row prisoners in the states in the South is 84 **b.** IQR = Q3 − Q1 = 166 − 36 = 130

The range of the middle 50% of the state data on death row prisoners in the South is 130.

**c.** The mean number of death row prisoners in the states in the South is 115.6 **d.** The mean is pulled up by the really large numbers, such as the numbers from Texas and Florida. The median is not affected by the size of these large numbers. **e.** The median is more stable with outliers.

**3.61** The answers given follow the steps in the Guided Exercises.

1:

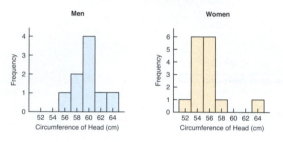

2: The men's histogram is roughly symmetric (bell-shaped). The women's histogram is bell-shaped except for the high-end outlier, so it may be called right-skewed.

3: Compare the medians and interquartile ranges because of the outlier for the women.

4: median, 59.5, median, 55, men

5: Men's IQR = 3.75. Women's IQR = 2. IQR may vary with different technology.

6: The measurement of 63 for the women was an outlier.

7: Both data sets are unimodal and roughly symmetric except for one large outlier for the women. The men tended to have larger heads with more variation in size than the women. However, there was a large outlier (63 cm) for the women.

**3.63** Summary statistics are shown below. The 5 p.m. class did better, typically; both the mean and the median are higher. Also, the spread (as reflected in both the standard deviation and the IQR) is larger for the 11 a.m. class, so the 5 p.m. class has less variation.

The visual comparison is shown by the boxplots. Both distributions are slightly left-skewed. Therefore, you can compare the means and standard deviations *or* the medians and IQRs.

```
Minitab Statistics
Variable   N     Mean    Median   StDev   Min    Max    Q1   Q3
11am      15    70.73    72.5     19.84   39     100    53   86
5pm       19    84.78    86.5     11.95   64.5   104.5  73   94
```

**3.65** The graph is bimodal, with modes around 65 inches (5 feet 5 inches) and around 69 inches (5 feet 9 inches). There are two modes because men tend to be taller than women.

**3.67 a.** The approximate mean GPA is around 3.1 or 3.2 (GPA) **b.** The approximate standard deviation of the GPAs is about 0.2, from $(3.8 - 2.6)/6$ **c.** The data set is unimodal and roughly symmetric and so all (or nearly all) of the data should be within three standard deviations of the mean. Thus one boundary is up 3SD and one is down 3SD, which is a difference of 6SD.

**3.69 and 3.71** Constructed numbers will vary.

**3.73 and 3.75** Answers will vary but should include graphs and comparison of centers and variation.

# CHAPTER 4

## Section 4.1

**4.1 a.** Acres: The number of acres has a stronger relationship with the value of the land, as shown by the fact that the points are less scattered in a vertical direction. **b.** Acreage. The association is stronger between the value of land and acreage than with the number of fireplaces because the vertical spread is less.

**4.3** Very little trend.

**4.5** The more people weigh, the more weight they tend to want to lose.

**4.7** The states with higher populations tend to have more motor vehicle fatalities.

**4.9** The trend is positive, but the trend tends to level off at higher acreages. Generally, homes on larger acreages have larger assessed values.

**4.11** Linear regression is not appropriate because the trend is not linear, it is curved.

**4.13 a.** Decreasing: States with a higher percentage of students who take the SAT generally have lower mean math SATs. **b.** Somewhat curved. **c.** No. The trend is not linear.

**4.15** The correlation between age and GPA would be near zero.

**4.17.** 0.767 A
     0.299 B
     −0980 C

**4.19 a.** 0.987 **b.** The correlation remains the same ($r = 0.987$) if you multiply a constant. **c.** The correlation remains the same ($r = 0.987$) if you add a constant.

## Section 4.2

**4.21 a.** Between 40,000 and 50,000 **b.** 44,299

**4.23 a.** Predicted Armspan = $16.8 + 2.25$ height **b.** $b = 0.948(8.10/3.41) = 2.25$ **c.** $a = 159.86 - 2.25(63.59) = 16.8$ **d.** Armspan = $16.8 + 2.25(64) = 160.8$, or about 161 cm

**4.25 a.** Refer to the scatterplot. **b.** The trend is not linear.

**4.27 a.** Predicted Armspan = $6.24 + 2.515$ Height (Rounding may vary.)
     **b.** Minitab: slope = 2.51, intercept = 6.2
         StatCrunch: slope = 2.514674, intercept = 6.2408333
         Excel: slope = 2.514674, intercept = 6.240833
         TI-83/84: slope = 2.514673913, intercept = 6.240833333

**4.29** The horizontal axis starts at 0. To zoom in on the data, the horizontal axis should have been started at about 60. This would have rescaled the graph, making it easier to interpret.

**4.31** The correlation for the women is stronger because the $r$-squared value is closer to 1.

**4.33 a.** The slope would be near 0. **b.** $r$ is about 0 **c.** Last two digits of Social Security Number is not associated with age.

**4.35** Explanations will vary.

|     | $x$    | $y$      |
| --- | ------ | -------- |
| **a.** | gas    | miles    |
| **b.** | years  | salary   |
| **c.** | weight | belt size |

**4.37 a.** The higher the percentage of smoke-free homes in a state, the lower the percentage of high school students who smoke tends to be. **b.** $56.32 - 0.4647(70) = 23.79$, or about 24%

**4.39 a.** The graph shows that young drivers and old drivers have more fatalities and that the safest drivers are between about 40 and 60 years of age. **b.** It would not be appropriate for linear regression because the trend is not linear.

**4.41.** The answers are given in the order shown in the Guidance section.

1: The regression line is shown with the scatterplot.

2: Is the linear model appropriate?

Given: In this case the answer is yes, because there is a linear trend. It is hard to see with so few points, but a strong curvature is not present and the cost tends to increase as the miles increase.

3: Predicted Cost = $162.60 + 0.07957$ Miles

4: Put in the correct regression line (shown in Step 1).

5: Interpret the slope and intercept in context.

For the slope: For every additional mile, on average, the price goes up by 0.07957 dollars (or about 8 cents).

For the intercept: A trip of 0 miles should cost about 162.60 dollars on average. Beware of extrapolation. The linear trend might not continue near 0 miles and so this prediction could be very different from truth.

$$\text{Predicted cost} = 162.60 + 0.07957 \text{ miles}$$

6: Final answer: For the hypothetical flight of 500 miles, you would pay

$$162.60 + 0.07957 \times 500 = 202.385, \text{ or about } 202 \text{ dollars}$$

**4.43 a.** Positive: The more population, the more millionaires there tend to be. **b.** See scatterplot. **c.** $r = 0.992$ **d.** For each additional hundred thousand in the population, there is an additional 1.9 thousand millionaires. **e.** Do not focus on the intercept, because it does not make sense to look for millionaires in states with no people.

**4.45  a.** The correlation should be positive. The more people living in the state, the more drivers, and the more motor vehicle fatalities there should be. **b.** See figure. **c.** $r = 0.949$ **d.** Equation and line are shown on plot. **e.** For each additional hundred thousand people in a state, there are about 9 more fatalities on average.

**Section 4.3**

**4.47  a.** An influential point is a point that changes the regression equation by a large amount. **b.** Going to church may not cause lower blood pressure; just because two variables are related does not show that one caused the other. It could be that healthy people are more likely to go to church, or there could be other confounding factors.

**4.49** Older children have larger shoes and have studied math longer. Large shoes do not cause higher grades. Both are affected by age.

**4.51** Square 0.67 and you get 0.4489, so the coefficient of determination is about 45%. Therefore, 45% of the variation in weight can be explained by the regression line.

**4.53** The cholesterol going down might be partly caused by regression toward the mean.

**4.55  a.** The salary is $2.099 thousand less for each year later that the person was hired, or $2.099 thousand more for each year earlier. **b.** The intercept ($4,255,000) would be the salary for a person who started in the year 0, which is inappropriate (and ridiculous).

**4.57  a.** See graph.

**b.** $r = 0.933$
A positive correlation suggests that larger turkeys tend to have a higher prices. **c.** Predicted Price $= -4.49 + 1.57$ Weight **d.** The line is on the graph. **e.** The slope: For each additional pound, the price goes up by $1.57. The interpretation of the intercept is inappropriate, because it is not possible to have a turkey that weighs 0 pounds.
**f.** The 30-pound free turkey changes the correlation to $-0.375$ and changes the equation to

$$\text{Predicted Cost} = 26.87 - 0.553 \text{ Weight}$$

This implies that the bigger the turkey, the less it costs! The 30-pound free turkey was an influential point, which really changed the results.

**4.59  a.** Positive correlation **b.** For each additional dollar spent on teachers' pay, the expenditure per pupil goes up about 23 cents on the average. **c.** The intercept might represent the mean cost of education excluding the cost of paying teachers. However, doing so requires extrapolation which may not be valid.

**4.61  a.** Correlation is negative. **b.** For each additional hour of work, the score tended to go down by 0.48 point. **c.** A student who did not work would expect to get about 87 on average.

**4.63  a.** *Predicted Mother's Education* $= 3.12 + 0.637(12) = 3.12 = 7.64$ $= 10.76 \, (12, 10.76)$

*Predicted Mother's Education* $= 3.12 + 0.637 \, (4.0) = 3.12 + 2.548 = 5.67 \, (4, 5.67)$
**b.** The new equation is given below.

*Predicted Father's Education* $= 2.78 + 0.637 \, (Mother's \; Education)$

*Predicted Father's Education* $= 2.78 + 0.637 \, (12) = 10.42 \, (12, 10.42)$

*Predicted Father's Education* $= 2.78 + 0.637 \, (4) = 5.33 \, (4, 5.33)$
**c.** Regression toward the mean: Values for the predictor variable that are far from the mean lead to responses that are closer to the mean.

**4.65**  1: i. Slope: $b = r\dfrac{s_y}{s_x} = 0.7\dfrac{10}{10} = 0.7$

ii. Intercept: $a = \bar{y} - b\bar{x}$

$$a = 75 - 0.7(75) = 22.5$$

iii. Equation

$$\text{Predicted final} = 22.5 + 0.7 \text{ Midterm}$$

2: Predicted final $= 22.5 + 0.7$ Midterm

$$= 22.5 + 0.7(95)$$

$$= 89$$

3: The score of 89 is lower than 95 because of regression toward the mean.

**Chapter Review Exercises**

**4.67  a.** $r = 0.941$

Predicted Weight $= -245 + 5.80$ Height

**b.**

| Height | Weight |
|--------|--------|
| $60(2.54) = 152.4$ | $105/2.205 = 47.6190$ kg |
| $66(2.54) = 167.64$ | $140/2.205 = 63.4921$ kg |
| $72(2.54) = 182.88$ | $185/2.205 = 83.9002$ kg |
| $70(2.54) = 177.8$ | $145/2.205 = 65.7596$ kg |
| $63(2.54) = 160.02$ | $120/2.205 = 54.4218$ kg |

**c.** The correlation between height and weight is 0.941. It does not matter whether you use inches and pounds or centimeters and kilograms. A change of units does not affect the correlation because it has no units.
**d.** The equations are different.

Predicted Weight Pounds $= -245 + 5.80$ Height (in inches)

Predicted Weight Kilograms $= -111 + 1.03$ Height (in centimeters)

**4.69  a.** Choose $r = +0.75$ **b.** Price $= 223.5 + 0.174$ square feet

The slope is about 0.174 thousand dollars per square foot. On average, for every additional square foot, the price is $174 more. The intercept is not appropriate to interpret because there are no houses with 0 square feet.
**c.** 1300 sq ft: about $450 thousand **d.** $0.75^2 = 56.25\%$
About 56% of the variation in price can be explained by the regression line.

**4.71 a.** See figure.

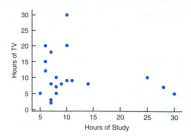

**b.** The trend is not linear. **c.** The correlation is near zero ($-0.176$).

**4.73 a.** $r$ is the square root of 0.054, which is 0.23. It is positive because the slope is positive. **b.** Predicted weight = 111.4 + 0.882(35) = 111.4 + 30.87 = 142.27, or about 142 pounds **c.** For each additional year of age, a woman weighs 0.88 pound more, on average, a little less than 1 pound. **d.** The intercept is not appropriate to interpret because no woman is near 0 years of age.

**4.75** The trend is positive. In general, if one twin has a higher-than-average level of education, so does the other twin. The point that shows one twin with 1 year of education and the other twin with 12 years is an outlier. (Another point showing one twin with 15 years and the other with 8 years is a bit unusual, as well.)

**4.77** There appears to be a positive trend. It appears that the number of hours of homework tends to increase slightly with enrollment in more units.

**4.79** Corelation table:

|        | Diameter | Height |
|--------|----------|--------|
| Height | 0.519    |        |
| Volume | 0.967    | 0.598  |

The diameter is a better predictor of volume than the height because there is a larger correlation between diameter and volume than between height and volume.

**4.81** You can see the tendency for big-budget films to gross more. One point in the upper right corner jumps out at you. *Avatar* had the largest budget (of these) and also the largest gross. *Titanic* is the other outlier.

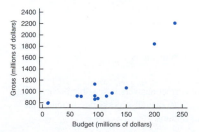

**4.83 and 4.85** Answers will vary.

# CHAPTER 5

## Section 5.1

**5.1 a.** 2 6 4 2 7      4 0 6 5 0
**b.** TTTTH      TTTHT
**c.** No. We got 2 heads.

**5.3** Theoretical probability: because it is not based on an experiment.

**5.5** Empirical probability: because it is based on an experiment.

## Section 5.2

**5.7 a.** The nine equally likely outcomes are Beasley, Blackett, Coffee, Craft, Fowlkes, Higgs, Lammey, Skahan, Ward. **b.** The outcomes that make up event A are Blackett and Skahan. **c.** 2/9, or about 22.2% **d.** Beasley, Coffee, Craft, Fowlkes, Higgs, Lammey, Ward.

**5.9** 1.3 could not be a probability because it is larger than 1. 150% could not be a probability because it is larger than 100%. $-0.50$ could not be a probability because it is negative (less than 0).

**5.11 a.** A heart: 13/52 or 1/4 **b.** A red card: 26/52 or 1/2 **c.** An ace: 4/52 or 1/13 **d.** A face card: 12/52 or 3/13 **e.** A three: 4/52 or 1/13 Answers may also be in decimal or percentage form.

**5.13 a.** P(guessing correctly) = 1/2 **b.** P(guessing incorrectly) = 1/2

**5.15**

|     | Number of Girls | Probability    |
|-----|-----------------|----------------|
| **a.** | 0            | 1/16           |
| **b.** | 1            | 4/16 = 1/4     |
| **c.** | 2            | 6/16 = 3/8     |
| **d.** | 3            | 4/16 = 1/4     |
| **e.** | 4            | 1/16           |

**5.17** The probability of being born on a Friday OR Saturday OR Sunday is 3/7, or 42.86%.

**5.19 a.** 530/1858 = 28.5% **b.** 689/1858 = 37.1%

**5.21** 104/1858 = 5.6%

**5.23** (689 + 469)/1858 = 62.3%

**5.25** The answers follow the guided steps.

1. P(liberal) = 530/1858 = 28.5%
2. P(Dem) = 689/1858 = 37.1%
3. They are not mutually exclusive because there are people who are both liberal AND Democrats.
4. P(liberal AND Dem) = 306/1858 = 16.5%
5. If you don't subtract the probability of being liberal AND a Democrat you will be counting those people twice, once when you count the liberals and once when you count the Democrats.
6. $\dfrac{530}{1858} + \dfrac{689}{1858} - \dfrac{306}{1858} = \dfrac{913}{1858} = 49.1\%$
7. The probability of being liberal OR a Democrat is 49.1%.

**5.27 a.** Mutually exclusive **b.** Not mutually exclusive

**5.29** Any two column headings or any two row headings are mutually exclusive. For example, you cannot be both a Republican AND a Democrat, so those are mutually exclusive. Likewise, you cannot be both liberal AND moderate, so those are mutually exclusive.

**5.31** You don't know what percentage of households have at least one cat AND at least one dog. You cannot simply add the percentages, because the events are not mutually exclusive and you would count households with at least one cat AND at least one dog twice.

**5.33 a.** 4/6, or 66.7% **b.** 3/6 = 1/2, or 50%

**5.35 a.** A OR B: 0.18 + 0.25 = 0.43 **b.** A OR B OR C: 0.18 + 0.25 + 0.37 = 0.80 **c.** Lower than a C: 1 − 0.80 = 0.20

**5.37 a.** Most: Category 3: married OR have children **b.** Fewest: Category 4: married AND have children

**5.39** Two coin flips
**a.** 0 heads: ¼ (Decimals and percentages are also acceptable.) **b.** 1 head: ½ (from 2/4) **c.** 2 heads: ¼ **d.** At least 1 head: ¾ **e.** Not more than 2 heads (which means 2 or fewer heads): 1 (from 4/4)

**5.41** The total is 240,032 thousand.
**a.** Widowed: 14,254/240,032, or 5.9% **b.** Divorced: 23,266/240,032, or 9.7% **c.** Divorced OR widowed: (23,266 + 14,254)/240,032, or 15.6% (You can also get this from adding the percentages.)

**5.43 a.** More than 8 mistakes: $1 - 0.48 - 0.30 = 0.22$ **b.** 3 or more mistakes: $0.30 + 0.22 = 0.52$ **c.** At most 8 mistakes: $0.48 + 0.30 = 0.78$ **d.** The events in parts a and c are complementary because "at most 8 mistakes" means from 0 mistakes up to 8 mistakes. "More than 8 mistakes" means 9, 10, up to 12 mistakes. Together, these mutually exclusive events include the entire sample space

## Section 5.3

**5.45 ii.** $D \mid C$

**5.47 a.** $306/530 = 57.7\%$ **b.** $104/593 = 17.5\%$ **c.** Liberals

**5.49** Because men are more likely to be left-handed than women, gender and handedness are associated.

**5.51 a.** Hair color and age are independent because hair color does not change in this age range. (If you included all adults, you would have to say the variables are not independent, because old people are more likely to have gray hair.) **b.** Hair color and eye color are not independent, because people with dark eyes are more likely to have dark hair.

**5.53** Gender and opinion on same-sex marriage are not independent; they are associated. The females were more likely to support same sex marriage.

**5.55** They are not independent because liberals are more likely than conservatives to be Democrats. We see that P(Dem), which is about 37.1%, does not equal P(Dem | liberal), which is about 57.7%.

**5.57** The answers follow the format for the Guided Exercises.

|       | M  | W  |     |
|-------|----|----|-----|
| Right | 18 | 42 | 60  |
| Left  | 12 | 28 | 40  |
|       | 30 | 70 | 100 |

Step 1: See table.

Step 2: 60/100, or 60%

Step 3: 18/30 = 60%

Step 4: The variables are independent because the probability of having the right thumb on top given a person is a man is equal to the probability that a person has the right thumb on top (for the whole data set).

**5.59**

|         | Agree | Don't Know | Disagree |      |
|---------|-------|------------|----------|------|
| Happy   | 345   | 261        | 656      | 1262 |
| Unhappy | 34    | 22         | 44       | 100  |
|         | 379   | 283        | 700      | 1362 |

$$P(\text{happy} \mid \text{agree}) = 345/379 = 91.0\%$$

$$P(\text{happy}) = 1262/1362 = 92.7\%$$

These percentages are very close, so you might think the factors are independent. But we said in the chapter that for this chapter, independence requires precisely equal probabilities, so they are *not independent* from that point of view.

**5.61 a.** 1/8 **b.** 1/8

**5.63** They are the same. Both probabilities are $\left(\dfrac{1}{6}\right)^5$ or $\dfrac{1}{7776}$

**5.65 a.** Neither fastened: $0.16(0.16) = 0.0256$ **b.** At least one fastened: $1.0000 - 0.0256 = 0.9744$

**5.67** P(have C AND test positive) = P(have C) P(test positive | have C)
$$= 0.00008(0.84)$$
$$= 0.000067$$

## Section 5.4

**5.69 a.** HTHTH    HHTTT    THHTH    HHHTT

**b.** 11/20

**5.71** Histogram B was for 10,000 rolls because it has nearly a flat top. In theory, there should be the same number of each outcome, and the Law of Large Numbers says that the one with the largest sample should be closest to the theory.

**5.73** The proportion should get closer to 0.5 as the number of flips increases.

**5.75 a.** You could use the numbers 1, 2, 3, and 4 to represent the outcomes and ignore 0 and 5–9, but answers to this will vary. **b.** The empirical probabilities will vary. The theoretical probability of getting a 1 is 1/4; remember that the die is four-sided.

**Chapter Review Exercises**

**5.77** $117/1300 = 9.0\%$

**5.79 a.** Gender and shoe size are associated, because men tend to wear larger shoe sizes than women. **b.** Win/loss record is independent of the number of cheerleaders. The coin does not "know" how many cheerleaders there are or have an effect on the number of cheerleaders.

**5.81 a.** Both believe: $(0.62)(0.50) = 0.31$ **b.** Neither believes: $(0.38)(0.50) = 0.19$ **c.** Same beliefs: $0.31 + 0.19 = 0.50$ **d.** Different beliefs: $1 - 0.50 = 0.50$

**5.83 a.** $0.80(0.80) = 0.64$ **b.** The married couple might have Internet access at home, in which case, if one of them has Internet access, then the other also does (unless one of them prohibits the other's use).

**5.85 a.** Both born Monday: $\dfrac{1}{7}\left(\dfrac{1}{7}\right) = \dfrac{1}{49}$

**b.** Alicia OR David were born on Monday:
P(A OR B) = P(A) + P(B) − P(A AND B)

$$= \frac{1}{7} + \frac{1}{7} - \frac{1}{7}\left(\frac{1}{7}\right)$$

$$= \frac{7}{49} + \frac{7}{49} - \frac{1}{49}$$

$$= \frac{13}{49}$$

**5.87** $0.83(5000) = 4150$

**5.89** Answers will vary. Red die is 1, blue die is 1.

**5.91** Answers will vary.

**5.93 a.** Using 0 and 1 to represent correct answers is sensible because that is 20% of the digits, and we want a 20% overall success rate to match the multiple-choice questions with five options.
**b.** 1 1 3 7 3    9 6 8 7 1
   C C I I I    I I I I C
**c.** The outcome from the first trial was 3 correct answers. **d.** Yes, the event of interest occurred, because the student got 3 correct.
**e.** 5 2 0 2 2    5 9 0 9 3
   I I C I I    I I C I I
The student got 2 correct out of 10 this time. The event of interest did not occur.
**f.** 1 4 7 0 9  9 3 2 2 0
   C I I C I  I I I I C    3 correct out of 10. The event of interest occurred.
   3 1 8 6 7  8 5 8 7 2
   I C I I I  I I I I I    1 correct out of 10. The event did not occur.
Altogether, there were 2 times out of 4 with 3 or more right, so the empirical probability of 3 or more correct is ½, or 50%.

**5.95** The smaller hospital will have more than 60% girls born more often, because, according to the Law of Large Numbers, there's more variability in proportions for small sample sizes. For the larger sample size ($n = 45$), the proportion will be more "settled" and will vary less from day to day. Over half of the subjects in Tversky and Kahneman's study said that "both hospitals will be the same." But *you* didn't, did you?

**5.97 a.** A trial consists of driving through the three intersections. **b.** Let the digits 0–5 represent a green light, and let 6–9 represent yellow or red. (Any assignment that gives 6 digits to green and 4 digits to non-green will work.) **c.** The event of interest is "get three out of three green" **d.** Digits:

275  830  186  658  250  381  033  582  594  513
608  010  443  958  621  098  403  511  960  372

Outcomes (non-green are labeled R)

GRG  RGG  GRR  RGR  GGG  GRG  GGG  GRG  GRG  GGG
RGR  GGG  GGG  RGR  RGG  GRR  GGG  GGG  RRG  GRG

Response variable:
2  2  1  1  3  2  3  2  2  3
1  3  3  1  2  1  3  3  1  2

**e.** P(all three green) is estimated as $7/20$.

## CHAPTER 6

Answers may vary slightly due to rounding or type of technology used.

### Section 6.1

**6.1 a.** Discrete **b.** Continuous

**6.3 a.** Discrete **b.** Continuous

**6.5**

| Number of Spots | 1 | 2 | 3 | 4 | 5 | 6 |
|---|---|---|---|---|---|---|
| Probability | 0.1 | 0.2 | 0.2 | 0.2 | 0.2 | 0.1 |

**The table may have a different orientation.**

**6.7 a.** See table.

| Win | Prob |
|---|---|
| 3 | 1/6 |
| 0 | 1/2 |
| −4 | 1/3 |

**b.** See graph

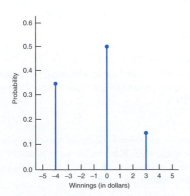

**6.9** $(6 − 3)(0.2) = 0.6$, or 60%, and the area between 3 and 6 should be shaded.

### Section 6.2

**6.11 a.** ii., 95% **b.** i., almost all **c.** iii., 68% **d.** iv., 50% **e.** ii., 13.5%

**6.13 a.** iii., 50% **b.** iii., 68% **c.** v., about 0% **d.** v., about 0% **e.** ii., 95% **f.** v., 2.5%

**6.15** Figure (B), about 0.63

**6.17 a.** 0.8461, or about 85% **b.** $1 − 0.8431 = 0.1539$, or about 15%

**6.19 a.** 0.9608, or about 96% **b.** $1 − 0.9608 = 0.0392$, or about 4% **c.** $0.1515 − 0.0968 = 0.0547$, or about 5%

**6.21 a, b,** and **c** are all 0.000. **d.** The proportion to the right of 4.00 would be the largest of the three, and the proportion to the right of 50.00 would be the smallest. **e.** below −10.00

**6.23** The answers use the steps given in the Guided Exercises.

1. $z = \dfrac{x - \mu}{\sigma} = \dfrac{675 - 500}{100} = \dfrac{175}{100} = 1.75$

2. 500 is the mean, and it belongs right below 0 because 0 is the mean of the standard Normal or the mean z-score.

3. and 4. are shown in the sketch.

5. 0.9599

6. $1 − 0.9599 = 0.0401$

7. The percentage of female college-bound seniors taking the SAT who scored 675 or more is about 4.0%.

**6.25** 0.8413, or about 84%

**6.27** 0.6827, or about 68% (Empirical Rule)

**6.29 a.** 0.2903, or about 29% **b.** 0.0487, or about 5% **c.** $0.2903(0.2903) = 0.0843$, or about 8% **d.** 29.03% of 750, or about 218

**6.31** about 0.50

**6.33 a.** 0.5000, or about 50%. **b.** 0.7881, or about 79%

**6.35** About 80% of the days in February have minimums of 32°F or less.

**6.37** Part a asks for a measurement (inverse), and part b asks for a probability.

**6.39** $z = 0.43$

**6.41 a.** $z = 0.56$ **b.** $z = −1.00$

**6.43** The answers follow the guided steps.

1. The test score will be above the mean, because 96% of students score worse, so it must be a very high grade.

2. See the figure.

3. $z = 1.75$

4. See the figure.

5. $500 + 1.75(100) = 500 + 175 = 675$

6. See the figure.

7. The SAT score at the 96th percentile is **675**.

**6.45** 99.0°F

**6.47** 62.9 inches

**6.49 a.** The Wechsler IQ at the 75th percentile is 110. **b.** The interquartile range (IQR) is $110 - 90$, or 20 **c.** The interquartile range (20) is larger than the standard deviation (15).

**6.51 a.** 84th percentile. **b.** About 73 inches, which is 6 feet 1 inch

### Section 6.3

**6.53** Conditions

Two complementary outcomes: boy or girl

Fixed number of trials: 4 children

Same probability of success on each trial: ½ probability of a boy

All trials independent: Because there are no twins, the gender of each child is independent of the gender of the others.

**6.55** There are not a fixed number of trials, because the students flip the coins *until* the class is over.

**6.57** It is not a binomial experiment, because the two people (married to each other) are not independent with respect to getting a divorce. If one gets a divorce, then the other one gets a divorce.

**6.59** $b(n, p, x)$ **a.** $b(100, 0.54, 65)$ **b.** $b(10, 0.90, 9)$ **c.** $b(10, 0.25, 6)$

**6.61** $b(6, 0.40, 2) = 0.311$

**6.63 a.** $b(10, 0.90, 7 \text{ or fewer}) = 0.070$, or about 7% **b.** $b(10, 0.90, 8 \text{ or more}) = 1 - b(10, 0.90, 7 \text{ or fewer}) = 0.930$, or about 93% **c.** $b(10, 0.90, \text{from 7 to 9}) = b(10, 0.90, 9 \text{ or fewer}) - b(10, 0.90, 6 \text{ or fewer}) = 0.6513 - 0.0128 = 0.6385$, or about 64%

**6.65 a.** $(0.5)(0.5) = 0.25$, or 25% **b.** $(0.5)(0.5)(0.5) = 0.125$, or about 13% **c.** $1 - 0.125 = 0.875$, or about 88% **d.** Yes, the more children she has, the more likely she will be to have at least one boy.

**6.67 a.** $0.8(0.8)(0.8) = 0.512$ **b.** At least one being late is the complement of all being on time:

$$1 - 0.512 = 0.488$$

**c.** The flights might not be independent, because the weather could be bad and that might cause most of the flights to be late. Thus if one flight is late, the others are more likely to be late.

**6.69 a.** $0.7(15) = 10.5$; expect 10 or 11 **b.** $b(15, 0.70, 11) = 0.219$ **c.** $b(15, 0.70, 11 \text{ or fewer}) = 0.703$

**6.71 a.** YY   YN   NY   NN
**b.** YY ¼
YN ¼
NY ¼
NN ¼

**c.** Neither has only cell: ¼ **d.** Exactly one has only cell (from ¼ + ¼): ½

**e.** Both have only cell: ¼

**6.73 a.** In 50 flips, expect 25 heads, because $np$ is 50 times ½.
**b.** $\sigma = \sqrt{np(1-p)} = \sqrt{50(0.5)(1-0.5)} = \sqrt{12.5} = 3.54$, or about 4. Therefore, you should expect between 21 and 29 heads.

**6.75 a.** Expect 60 give or take 7, because

$$\sigma = \sqrt{np(1-p)} = \sqrt{300(0.2)(0.8)} = \sqrt{48} = 6.93, \text{ or about 7}$$

**b.** 53 to 67 **c.** No, because 62 is close to the mean (within 1 SD).

### Chapter Review Exercises

**6.77** 95.2%, or about 95%

**6.79 a.** 0.7625, or about 76% **b.** 98.6°F

**6.81** Answers will vary. However, the probability that her next child will be a girl should be about 50%, because genders of children is independent unless there are identical twins (or selection of gender with in vitro fertilization).

**6.83 a.** $\mu = np = 2000(0.3) = 600$ give or take about 20, because

$$\sigma = \sqrt{np(1-p)} = \sqrt{2000(0.3)(1-0.3)} = \sqrt{420} = 20.49, \text{ or about 20}$$

**b.** 580 to about 620

**6.85 a.** **i.** $0.9^4 = 0.656$;
**ii.** $1 - 0.656 = 0.344$
**b.** **i.** $0.4^4 = 0.026$;
**ii.** $1 - 0.026 = 0.974$

**c.** They are less likely to die because of the Gleevec, so part (a) has a smaller probability of at least one dead.

# CHAPTER 7

Answers may vary slightly due to rounding or type of technology used.

### Section 7.1

**7.1** A parameter is a measure of the population, and a statistic is a measure of a sample.

**7.3 a.** $\bar{x}$ is a statistic, and $\mu$ is a parameter. **b.** $\bar{x}$

**7.5** If you know all the ages at inauguration, you should not make inferences because you have the population, not a sample from the population.

**7.7** You want to test a sample of batteries. If you tested them all until they burned out, no usable batteries would be left.

**7.9** First, all 10 cards are put in a bowl. Then one is drawn out and noted.
    "With replacement": The card that is selected is replaced in the bowl, and a second draw is done. It is possible that the same student could be picked twice.
    "Without replacement": After the first card is drawn out, it is not replaced, and the second draw must be a different card.

**7.11** Chosen: 7, 3, 5, 2

**7.13** Marco took a convenience sample. The students may not be representative of the voting population, so the proposition may not pass.

**7.15** The population is all readers of Time Magazine on the Internet. The *Time* Internet poll used a voluntary-response sample. It could be that only people who were really angry with the United States took time to submit their answers. The poll may have shown bias because of that fact.

**7.17** The administrator might dismiss the negative findings by saying the results could be biased because the small percentage who chose to return the survey might be very different from the majority who did not return the survey.

**7.19** No, the people you met would not be a random sample but a convenience sample.

**7.21**

|  | With Persuasion | Without Persuasion |
|---|---|---|
| Support Cap | 6 + 2 | 13 + 5 |
| Oppose | 9 + 8 | 2 + 5 |

Support capital punishment
**a.** With persuasion $8/25 = 32\%$
**b.** Without persuasion $18/25 = 72\%$
**c.** Yes, she spoke against it, and fewer who heard her statements against it (32%) supported capital punishment, compared with those who did not hear her persuasion (72%).

### Section 7.2

**7.23 a.** The sketch should show bullet holes consistently to the left of the target and close to each other. If the bullets go consistently to the left, then there is bias, not lack of precision. **b.** The sketch should show bullet holes that are all near the center of the target.

**7.25** The small mean might have occurred by chance.

**7.27 a.** $17/30$, or about 56.7%, odd digits **b.** $17/30$ is $\hat{p}$ (p-hat), the sample proportion. **c.** The error is $17/30$ minus $15/30$ (or $2/30$) or about 6.7%

**7.29 a.** We should expect 0.20, or 20%, orange candies.

**b.** $\sqrt{\dfrac{p(1-p)}{n}} = \sqrt{\dfrac{0.2 \times 0.8}{100}} = \sqrt{\dfrac{0.16}{100}} = \sqrt{0.0016} = 0.04$, or 4%

**c.** We expect about 20% orange candies, give or take 4%.

**7.31 a.** 50% seniors **b.** 0% seniors **c.** 50% seniors **d.** 100% seniors

| Repetition | $p$ (pop prop of seniors) | $\hat{p}$ (sample prop of seniors) | Error: $\hat{p} - p$ |
|---|---|---|---|
| 1 (from part a) | $2/5 = 0.4$ | $1/2 = 0.5$ | $0.5 - 0.4 = 0.1$ |
| 2 (from part b) | $2/5 = 0.4$ | $0/2 = 0.0$ | $0.0 - 0.4 = -0.4$ |
| 3 | $2/5 = 0.4$ | $1/2 = 0.5$ | $0.5 - 0.4 = 0.1$ |
| 4 | $2/5 = 0.4$ | $2/2 = 1.0$ | $1.0 - 0.4 = 0.6$ |

## Section 7.3

**7.33** $z = \dfrac{0.24 - 0.20}{0.04} = \dfrac{0.04}{0.04} = 1.00$

The area to the right of a $z$-score of 1.00 is 0.1587.

The probability that the percentage of orange candies will be 24% or more is about 16%.

**7.35** Because the sampling distribution for the sample proportion is approximately Normal, we know the probability of falling within two standard errors is about 0.95. Therefore, the probability of falling more than two standard errors away from the mean is about 0.05.

**7.37** Answers are given in the order shown in the Guided Exercises.

Step 1: $p = 0.65$

Step 2: $np = 200(0.65) = 130$ which is more than 10

$n(1 - p) = 200(0.35) = 70$ which is more than 10

The other assumptions were given.

Step 3:

$SE = \sqrt{\dfrac{p(1 - p)}{n}} = \sqrt{\dfrac{0.65(0.35)}{200}} = \sqrt{\dfrac{0.2275}{200}} = \sqrt{0.0011375} = 0.03373$

$z = \dfrac{\hat{p} - p}{SE} = \dfrac{0.67 - 0.65}{0.03373} = 0.59$

The area to the right of a $z$-value of 0.59 is 0.2776 (or 0.2766 with technology and no rounding of intermediate steps).

Step 4: The probability is represented by the area to the right of 0.59, because the question asks for the probability that the sample proportion will be "at least" 0.67 (134 out of 200), which translates to a $z$-score of 0.59. This means we are asked to find the probability that the $z$-score will be 0.59 or greater.

Step 5: The probability that at least 67% of 200 pass the exam is about 28%.

**7.39 a.** Expect 0.08. **b.** The approximately probability is 0.122. It should be shown in a Normal curve. **c.** 0.0099 or 0.010 and 360 chairs **d.** The answer to part c is smaller because 9% is farther out in the right tail than 8.5%, and it is the tail area that gives the probability we want.

**7.41 a.** 0.5 **b.** 0.5 **c.** 0.103 **d.** Lower, because the probability of guessing correctly on each question is lower when there are four options.

## Section 7.4

**7.43** (0.440, 0.481)

**7.45 a.** 770/1027, or 75.0%, want a stimulus bill. **b.** (72.3%, 77.6%)

**7.47 a.** I am 95% confident that the population percentage of voters supporting Candidate X is between 53% and 57%. **b.** There is no evidence that he could lose, because the interval is entirely above 50%. **c.** A sample from New York City would not be representative of the entire country and would be worthless in this context.

**7.49 a.** 17/30, or 0.5667 (or round it off to 0.567) **b.** $SE_{est}$ given as 0.090 $m = 1.96 \times 0.090 = 0.1764$ **c.** 95% CI (0.390, 0.743) **d.** $m = 1.28 \times 0.090 = 0.115$ 80% CI (0.452, 0.682) **e.** 99% CI (0.334,

0.799) **f.** The 99% confidence interval is widest, and the 80% interval is narrowest. A greater confidence level requires a larger $z^*$.

**7.51 a.** Sqrt $[(0.87 \times 0.13)/1500] = 0.0087$ or 0.87% **b.** 1.70% **c.** (85.3%, 88.7%) **d.** The interval supports this claim. Because 80% is not in the interval, and all values are above 80%, we are confident that the current percentage is higher than 80%.

**7.53 a.** 92/113, or 81.4%, reported dreaming in color. **b.** (0.742, 0.886) **c.** The interval does not capture 29.24%. This shows that the results in 2003 were quite different from the results in 1942.

**7.55 a.** He or she should find 15, because 15 is half of 30. **b.** You would expect about 4 out of 40 not to capture 50%, because with a 90% confidence interval about 10% should not capture, and 10% of 40 is 4.

**7.57** The proportion 16/100 is the population proportion, not a sample proportion. You should not find a confidence interval unless you have a sample and are making statements about the population from which the sample has been drawn.

**7.59 a.** 430/1001 = 43.0% **b.** 95% CI (0.399, 0.460) **c.** 80% CI (0.410, 0.450) **d.** The 95% interval is wider and the 80% interval is narrower. To get a higher degree of certainty, we need to widen the interval.

**7.61 a.** 595/1009 = 59.0% **b.** 95% CI (0.559, 0.620) **c.** A 99% CI would be wider.

**7.63 a.** 74/1438 = 5.15% **b.** 95% CI (0.040, 0.063) **c.** 5% is plausible because it is inside the interval.

## Section 7.5

**7.65 a.** $n = \dfrac{1}{m^2} = \dfrac{1}{0.015^2} = \dfrac{1}{0.000225} = 4444.4$

They should poll 4445 people.

**b.** A 1% margin of error would require a larger sample than a 1.5% margin of error. **c.** 90% would require a smaller sample size.

**7.67** $n = \dfrac{1}{m^2} = \dfrac{1}{0.03^2} = \dfrac{1}{0.0009} = 1111.1$: Requires 1112

$n = \dfrac{1}{m^2} = \dfrac{1}{0.015^2} = \dfrac{1}{0.000225} = 4444.4$: Requires 4445

4445/1112 is 4.

To get half the margin of error, you must multiply the sample size by 4.

## Chapter Review Exercises

**7.69 a.** 515/691, or 74.5%, had been exposed to traffic. **b.** 95% CI (0.713, 0.778) **c.** No. Without random assignment, you cannot draw conclusions about cause and effect.

**7.71** $n = 1006$

$n = \dfrac{1}{m^2}$

and so $m = \dfrac{1}{\sqrt{n}} = \dfrac{1}{\sqrt{1006}} = \dfrac{1}{31.7175} = 0.0315$, or about 3%

**7.73 a.** 78/489 is 16.0% **b.** I am 95% confident the percentage of American adults who would say yes to making smoking totally illegal is between 12.7% and 19.2%. **c.** No, the entire interval is below 50% and it is not plausible to think a majority would say smoking should be made totally illegal. **d.** The percentage is the same. If the sample size were larger the interval would be narrower.

**7.75 a.** 292/1006 is 29.0% **b.** I am 95% confident that the percentage of American adults who would say yes to banning smoking in bars is between 26.2% and 31.8%. **c.** No, the entire interval is below 50% **d.** If the sample size were larger the interval would be narrower.

**7.77** 95% confidence interval (52%, 60%) or (0.52, 0.60)

This suggests that a majority of adults believe the feeding tube should have been removed, because the interval is entirely above 50%.

## CHAPTER 8

Answers may vary slightly due to type of technology or rounding.

### Section 8.1

**8.1** population parameter

**8.3** Small.

**8.5** ii. $p = 0.50$

**8.7 a.** i. **b.** iii

**8.9** iii.

**8.11** does not, 0.05

**8.13** $z = 0.89$.

**8.15** The p-value tells us that if the true proportion of those who text while driving is 0.25, then there is only a 0.034 probability that one would get a sample proportion of 0.125 or smaller with a sample size of 40.

### Section 8.2

**8.17** In Figure (A), the shaded area could be a p-value because it includes tail areas only; it would be for a two-tailed alternative because both tails are shaded. In Figure (B) the shaded area would not be a p-value because it is the area between two z-values.

**8.19** $z = 3.00$. It is farthest from 0 and therefore has the smallest tail area.

**8.21** He has not demonstrated ESP; 10 right out of 20 is only 50% right, which you should expect from guessing.

### Section 8.3

**8.23** $H_0$: The death rate after starting hand washing is still 9.9%, or $p = 0.099$ ($p$ is the proportion of all deaths at the clinic.)
   $H_a$: The death rate after starting hand washing is less than 9.9%, or $p < 0.099$.

**8.25** $H_0$: The rate of using seat belts has not changed and is 0.71, or $p = 0.71$.
   $H_a$: The rate of using seat belts has changed and is not 0.71, or $p \neq 0.71$.

**8.27** *Step 1:* $H_0$: $p = 0.52$, $H_a$: $p \neq 0.52$ where $p$ is the population proportion favoring stricter gun control. *Step 2:* One-proportion z-test (given), sample size large (1011 times 0.48 = about 485 which is more than 10), sampling random, and population large. *Step 3:* $\hat{p} = 0.4896$, $SE = 0.01571$, $z = -1.93$, p-value = 0.053. *Step 4:* Do not reject $H_0$. Choose conclusion i.

**8.29** Random sample was mentioned.
   Large sample: $np_0 = 113(0.29) = 32.77 > 10$ and $n(1 - p_0) = 113(0.71) = 80.23 > 10$.
   Large population: There are more than 1130 people in the population of dreamers.
   So the conditions are met.

**8.31** $z = 2.56$, p-value = 0.005.

**8.33** *Step 3:* $z = 2.83$, p-value = 0.002. *Step 4:* Reject $H_0$. The probability of doing this well by chance alone is so small that we must conclude that the student is not guessing.

**8.35** Pick the graph shown in Figure (A). The observed value of 13 right out of 20 is more than half, so the z-statistic will be bigger than 0. The right tail area corresponds to the p-value.

**8.37 a.** *Step 1:* $H_a$: $p > 0.50$ where $p$ is the population proportion of people who believe that schools should ban sugary foods. *Step 2:* One-proportion z-test, the sampling is random, sample size large, and population large. *Step 3:* $z = 1.26$, $p = 0.103$ (one-tailed). *Step 4:* Do not reject the null hypothesis.
**b.** Pick ii.

**8.39 a.** $527/1014 = 52.0\%$. This is less than the 58%. **b.** *Step 1:* $H_a$: $p \neq 0.58$ where $p$ is the population proportion of people who believe there is global warming. *Step 2:* One-proportion z-test, sample size large, sampling random, population more than 10 times sample. *Step 3:* $z = -3.89$, p-value < 0.001. *Step 4:* Reject $H_0$. **c.** Choose ii.

**8.41** *Step 1:* $H_a$: $p \neq 0.20$ where $p$ is the population proportion of dangerous fish. *Step 2:* One-proportion z-test, sample size large enough, population large, assume a random sample. *Step 3:* $z = 1.58$, p-value = 0.114. *Step 4:* Do not reject $H_0$.
   We are not saying the percentage is 20%. We are only saying that we cannot reject 20%. (We might have been able to reject the value of 20% if we had had a larger sample.)

**8.43** *Step 1:* $H_a$: $p \neq 0.09$ where $p$ is the population proportion of t's in the English language. *Step 2:* One-proportion z-test, the sample is random and large enough, and population large. *Step 3:* $z = -0.86$, p-value = 0.392. *Step 4:* Do not reject $H_0$. We cannot reject 9% as the current proportion of t's.

### Section 8.4

**8.45**
*0:* Vioxx percentage heart attack = 3.5%
   Placebo percentage heart attack = 1.9%
   So the risk of heart attack in the sample was greater for those taking Vioxx.

*1:* $H_0$: $p_V = p_p$ where $p_V$ is the population proportion of heart attacks for those who could use Vioxx and $p_p$ is the population proportion of heart attacks for those who could use the placebo.
   $H_a$: $p_V > p_p$

*2:* We don't have a random sample but we do have random assignment.
   $n_1 \times \hat{p} = 1287(0.027069) = 34.84$, which is more than 10
   $n_1 \times (1 - \hat{p}) = 1287(0.972931) = 1252.16$, which is more than 10
   $n_2 \times \hat{p} = 1299(0.027069) = 35.16$, which is more than 10
   $n_2 \times (1 - \hat{p}) \geq 1299(0.027069) = 35.16$, which is more than 10

*3:* $z = 2.46$ (or $-2.46$)
   p-value = 0.007.

*4:* Reject $H_0$ and choose ii. Do not generalize because you do not have a random sample.

Causality: Yes, you can conclude cause and effect for this group of patients because of the format of the study, including random assignment to groups. Vioxx caused a significantly increased rate of heart attacks. (Vioxx was taken off the market because of this study.)

**8.47 a.** Vaccine: 1.9% got the virus. Placebo: 4.9% got the virus. So the vaccine group got the virus at a lower rate, which is what was hoped for. **b.** *Step 3:* $z = 5.85$ (or $-5.85$), p-value < 0.001. *Step 4:* Reject $H_0$. The vaccine causes a significantly lower chance of getting the virus, but don't generalize.

**8.49 a.** For nicotine gum, the proportion quitting was 0.106. For the placebo, it was 0.040. This was what was hoped for—that the drug was helpful compared to the placebo. **b.** $z = 7.23$ (or $-7.23$).

**8.51 a.** Men: 46.2% smiling. Women 51.1% smiling. **b.** *Step 1:* $H_a$: $p_{men} \neq p_{women}$. (where $p$ is the proportion smiling) *Step 2:* Two-proportion z-test, large sample size, assume random sample, population large. *Step 3:* $z = 6.13$ (or $-6.13$), p-value < 0.001. *Step 4:* Reject $H_0$. There is a significant difference in rate of smiling for men and women. **c.** The difference is significant because of the large sample size.

### Section 8.5

**8.53** No; we don't use "prove" because we cannot be 100% sure of conclusions based on chance processes.

**8.55** It is a null hypothesis.

**8.57** Interpretations b and d are valid. Interpretations a and c are both "accepting" the null hypothesis claim, which is an incorrect way of expressing the outcome.

**8.59** Far apart. Assuming the standard errors are the same, the farther apart the two proportions are, the larger the absolute value of the numerator of $z$, and therefore the larger the absolute value of $z$ and the smaller the p-value.

**8.61 a.** 336 (from 0.60 times 560) said more strict in February 1999, and 370 said more strict in late April 1999. **b.** *Step 1:* $H_a$: $p_{more\ strict\ Feb} \neq p_{more\ strict\ April}$. *Step 2:* Two-proportion z-test, samples are random, sample size large, population large. *Step 3:* $z = 2.10$ (or $-2.10$), p-value $= 0.035$. *Step 4:* Do not reject H$_0$, because alpha is 0.01. **c.** 672 out of 1120 in February vs. 739 out of 1120 in late April, 1999 *Step 1:* H$_a$: $p_{\text{more strict Feb}} \neq p_{\text{more strict April}}$. *Step 2:* Two-proportion $z$-test, samples are random, sample size is large, population large. *Step 3:* $z = 2.93$ (or $-2.93$), p-value $= 0.003$. *Step 4:* Reject H$_0$. The results of the polls were significantly different from each other. **d.** With a larger sample size (more evidence), we got a smaller p-value and were able to reject H$_0$.

**8.63** Interpretation iii.

**8.65 a.** One-proportion $z$-test. The population is all Florida voters. **b.** Two-proportion $z$-test. One population is all men at the college, and the other population is all women at that college.

**8.67 a.** $p =$ the population proportion of correct answers.

H$_0$: $p = 0.50$ (he is just guessing), H$_a$: $p > 0.50$ (he is not just guessing). One-proportion $z$-test. **b.** $p_a$ is the population proportion of athletes who can balance for at least 10 seconds. $p_n$ is the population proportion of nonathletes who can balance for at least 10 seconds.

H$_0$: $p_a = p_n$
H$_a$: $p_a \neq p_n$
Two-proportion $z$-test.

### Chapter Review Exercises

**8.69** 6 right out of 20 is less than half, so he cannot tell the difference. (Or: H$_a$: $p > 0.50$, $z = -1.79$, p-value $= 0.963$, do not reject H$_0$.)

**8.71** 0.05 (because $1 - 0.95 = 0.05$).

**8.73** 5% of 200, or 10.

**8.75** It would not be appropriate to do such a test, because the data were the entire population of people who voted. This was not a sample, so inference is not reasonable.

**8.77** *Step 1:* H$_a$: $p_{\text{money}} > p_{\text{no money}}$ where $p$ is the population proportion that successfully would quit smoking. *Step 2:* Two-proportion $z$-test, the samples large enough, not a random sample. *Step 3:* $z = 3.65$ (or $-3.65$), p-value $< 0.001$. *Step 4:* Reject H$_0$. The proportion of success was significantly higher for those who received the money. We can conclude the money caused the difference but cannot generalize to a larger group of smokers.

**8.79 a.** Aspirin: 2.39%, Placebo 2.62%. There was a slightly lower heart event rate for those using aspirin. **b.** *Step 1:* H$_a$: $p_{\text{aspirin}} < p_{\text{placebo}}$ (where $p$ is the proportion of those who had heart events). *Step 2:* Two-proportion $z$-test, sample large, random assignment but not a random sample. *Step 3:* $z = -1.44$ (or 1.44), p-value $= 0.076$. *Step 4:* You cannot reject the null hypothesis. There is not enough evidence to conclude that aspirin prevents heart events in women.

**8.81 a.** The misconduct rate was higher for those in the sample who did *not* have three strikes (30.6%) than for those in the sample who had three strikes (22.2%). This was not what was expected. **b.** *Step 1:* H$_a$: $p_{\text{three-strikers}} > p_{\text{others}}$ (where $p$ is the proportion of those with misconduct). *Step 2:* Two-proportion $z$-test, samples large enough, assume random samples. *Step 3:* $z = -4.49$ (or 4.49), p-value $> 0.999$. *Step 4:* Do not reject H$_0$. The three-strikers do not have a greater rate of misconduct than the other prisoners.

(If a two-tailed test had been done, the *p*-value would have been $< 0.001$, and we would have rejected the null hypothesis because the three-strikers had *less* misconduct.)

## CHAPTER 9

Answers may vary due to rounding or type of technology used.

### Section 9.1

**9.1 a.** Parameters: 54.4 years, 15.5 years **b.** Statistics: 50.2 years, 18.0 years

**9.3 a.** See the accompanying figure.

**b.** 16% (By the Empirical Rule, 68% of the observed values are between 60 and 80, which leaves 32% outside of those boundaries. But we want only the right half.)

**9.5** The sample mean is based on a random sample, so different samples will have different means. Having an "unbiased" estimator does not mean that the sample mean will be equal to the population mean. Instead, it means that if we took all possible samples of bats, the mean of all of the sample means would be the same as the population mean.

**9.7** The distribution of *a* sample (*one* sample).

**9.9 a.** About 128 pounds, because the sample mean is unbiased.

**b.** $\dfrac{20}{\sqrt{100}} = 2.0$ pounds

### Section 9.2

**9.11 a.** It is large enough because the sample size of 100 is larger than 25. **b.** The mean is 128 pounds.

$$SE = \frac{\sigma}{\sqrt{n}} = \frac{20}{\sqrt{100}} = \frac{20}{10} = 2$$

**c.** $1 - 0.9545 = 0.0455$, or about 5%

**9.13** A: Calls  B: Means: smaller standard deviation (narrower) and more Normal (less right-skewed)

**9.15 a.** 7.9 and 7.7 are parameters; 8.2 and 6.0 are statistics. **b.** $\mu = 7.9$, $\sigma = 7.7$, $\bar{x} = 8.2$, $s = 6.0$ **c.** Random: OK. Sample size: $40 > 25$ OK. The shape would be Normal.

### Section 9.3

**9.17 a.** i. is correct: (3.07, 3.43); ii. and iii. are both incorrect. **b.** It tells us a range of plausible values for the population mean GPA, where the population is all statistics students at the school. Yes, reject 2.81, because it is not in the interval. We are confident that the statistics students have a higher population mean than 2.81.

**9.19 a.** i. is correct: (10.125, 10.525). Both ii. and iii. are incorrect. **b.** No, it does not capture 10. Reject the claim of 10 pounds because 10 is not in the interval.

**9.21 a.** C-level; **b.** C-interval

**9.23 a.** 1.699 **b.** Wider because $t^*$ is bigger (2.045), which makes the margin of error bigger:

$$\text{Margin of error} = t^* \frac{s}{\sqrt{n}}$$

**9.25 a.** $m = t^* \dfrac{s}{\sqrt{n}} = 2.093 \dfrac{2.17}{\sqrt{20}} = 1.016$, or about 1.0

$63.3 \pm 1.0$   95% CI (62.3 to 64.3). I am 95% confident the population mean height is between 62.3 and 64.3 inches.

**b.** $m = t^* \dfrac{s}{\sqrt{n}} = 2.861 \dfrac{2.17}{\sqrt{20}} = 1.388$, or about 1.4

$63.3 \pm 1.4$   99% CI (61.9 to 64.7). I am 99% confident the population mean height is between 61.9 and 64.7 inches.

**c.** The 99% interval is wider because it has a greater confidence level, and therefore we use a bigger value of $t^*$, which creates a wider interval.

**9.27 a.** Wider **b.** Narrower **c.** Wider

**9.29** *Step 1:* $H_a$: $\mu \neq 98.6$. *Step 2:* One-sample *t*-test, random sample, not strongly skewed, use 0.05. *Step 3:* $t = -1.65$, p-value $= 0.133$. *Step 4:* Do not reject $H_0$. Choose i.

**9.31 a.** *Step 1:* $H_a$: $\mu \neq 38$. *Step 2:* One-sample *t*-test, Normal and random. *Step 3:* $t = -1.03$, p-value $= 0.319$. *Step 4:* Do not reject $H_0$. The mean for non-U.S. boys is not significantly different from 38. **b.** *Step 3:* $t = -1.46$, p-value $= 0.155$. *Step 4:* Do not reject $H_0$. The mean for non-U.S. boys is not significantly different from 38. **c.** Larger $n$, smaller standard error (narrower sampling distribution) with less area in the tails, as shown by the smaller p-value.

**9.33 a.** 3.25, higher **b.** *Step 1:* $H_a$: $\mu > 2.81$. *Step 2:* One-sample *t*-test, random and $40 > 25$. *Step 3:* $t = 4.91$, p-value $< 0.001$. *Step 4:* Reject $H_0$. The mean GPA for statistics students is significantly higher than 2.81.

**9.34 a.** Sample mean is 63.3, which is less than 64.5 inches.
**b.** *Step 1:* $H_a$: $\mu < 64.5$. *Step 2:* One sample *t*-test, random and Normal. *Step 3:* $t = -2.53$, p-value $= 0.010$. *Step 4:* Reject $H_0$. The mean height of dancers is significantly less than 64.5.

**9.35** *Step 1:* $H_0$: $\mu_{before} = \mu_{after}$ [or $\mu_{difference} = 0$]; $H_a$: $\mu_{before} > \mu_{after}$ [or $\mu_{difference} > 0$]. *Step 2:* One-sample *t*-test, assume conditions are met (given). *Step 3:* $t = 1.39$, p-value $= 0.098$. *Step 4:* Do not reject $H_0$. Meditation does not lower pulse rates significantly.

**9.37** $t = 1.20$ for all three.

**a.** $H_a$: $\mu \neq 20$, p-value $= 0.315$, do not reject $H_0$.

**b.** $H_a$: $\mu < 20$, p-value $= 0.842$, do not reject $H_0$.

**c.** $H_a$: $\mu > 20$, p-value $= 0.158$, do not reject $H_0$.

**9.39** a

**9.41** Expect $0.95(200) = 190$ to capture and 10 to miss.

**9.43** It should not be interpreted. The data are not a random sample, and so inference based on a confidence interval is not possible.

## Section 9.4

**9.45 a.** Independent; **b.** Paired (or dependent)

**9.47** *Step 1:* $H_a$: $\mu_{oc} \neq \mu_{mc}$ where $\mu$ is the population mean number of TVs. *Step 2:* Two-sample *t*-test, samples large ($n = 30$), independent, and random, use 0.05. *Step 3:* $t = 0.95$, p-value $= 0.345$. *Step 4:* Do not reject $H_0$. Choose i. Confidence Interval: $(-0.404, 1.138)$. Because the interval for the difference captures 0, we cannot reject the hypothesis that the mean difference in number of TVs is 0.

**9.49 a.** The men's sample mean triglyceride level of 139.5 was higher than the women's sample mean of 84.4. **b.** *Step 1:* $H_a$: $\mu_{men} > \mu_{women}$ where $\mu$ is the population mean trigliceride level. *Step 2:* Two-sample *t*-test, assume the conditions are met. *Step 3:* $t = 4.02$ or $-4.02$, p-value $< 0.001$. *Step 4:* Reject $H_0$. The mean triglyceride level is significantly higher for men than for women. Choose output B: Difference $= \mu_{female} - \mu_{male}$, which tests whether this difference is less than 0, which is the one-tailed hypothesis that we want.

**9.51** $(-82.5, -27.7)$; because the difference of 0 is not captured, it shows there is a significant difference. Also, the difference $\mu_{female} - \mu_{male}$ is negative, which shows that the men's mean (triglyceride level) is significantly higher than the women's mean.

**9.53** *Step 1:* $H_a$: $\mu_{men} \neq \mu_{women}$ where $\mu$ is the population mean clothing expense for one month. *Step 2:* Two-sample *t*-test, assume Normal and random. *Step 3:* $t = 1.42$ or $-1.42$, p-value $= 0.171$. *Step 4:* Do not reject $H_0$. The mean clothing expense is not significantly different for men and women.

**9.55 a.** The 95% interval would capture 0, because we could not reject the hypothesis that the mean amounts spent on clothing are the same. **b.** A 99% interval would also capture 0, because it is wider than the 95% interval and centered at the same place. **c.** $(-18.4, 97.7)$; because the interval captures 0, we cannot reject the hypothesis that the mean difference in spending on clothing is 0, which shows we cannot reject the hypothesis that the means are the same.

**9.57** *Step 1:* $H_a$: $\mu_{letter} < \mu_{no\,letter}$ where $\mu$ is the population mean rate of crime. *Step 2:* Two-sample *t*-test, large sample sizes, not random samples, so don't generalize to a larger population; random assignment allows us to assume causality. *Step 3:* $t = 1.77$ or $-1.77$, p-value $= 0.041$ (one-tailed). *Step 4:* Reject $H_0$. The police letter reduced the number of crimes reported for these properties, but we cannot generalize to other properties.

**9.59** *Step 1:* $H_a$: $\mu_{before} < \mu_{after}$ where $\mu$ is the population mean pulse rate. *Step 2:* Paired *t*-test, each woman is measured twice (repeated measures), so a measurement in the first column is coupled with a measurement of the same person in the second column, assume random and Normal, use 0.05. *Step 3:* $t = 4.90$ or $-4.90$, p-value $< 0.001$. *Step 4:* Reject $H_0$. The sample mean before was 74.8 and the sample mean after was 83.7. The pulse rates of women go up significantly after a scream.

**9.61 a.** $\bar{x}_{UCSB} = \$61.01$ and $\bar{x}_{CSUN} = \$75.55$, so the sample mean at CSUN was larger. **b.** *Step 1:* $H_a$: $\mu_{UCSB} \neq \mu_{CSUN}$ where $\mu$ is the population mean book price. *Step 2:* Paired *t*-test, matched pairs, assume random and Normal (given). *Step 3:* $t = -3.21$ or 3.21, p-value $= 0.004$. *Step 4:* You can reject $H_0$. The means are significantly different.

**9.63 a.** $\bar{x}_{groom} = 27.3$ was larger ($\bar{x}_{bride} = 25.9$).

**b.** *Step 1:* $H_a$: $\mu_{bride} \neq \mu_{groom}$ where $\mu$ is the population mean age at marriage. *Step 2:* Paired *t*-test, random, and large samples. *Step 3:* $t = 2.24$ or $-2.24$, p-value $= 0.033$. *Step 4:* Reject $H_0$. The mean ages of brides and grooms are significantly different. **c.** $H_a$: $\mu_{bride} < \mu_{groom}$. The new p-value would be half of 0.033, or about 0.017.

**9.65 a.** 95% CI $(-1.44, 0.25)$ captures 0, so the hypothesis that the means are equal cannot be rejected. **b.** *Step 1:* $H_a$: $\mu_{measured} \neq \mu_{reported(converted\,to\,cm)}$ where $\mu$ is population mean height of men. *Step 2:* Paired *t*-test, each person is the source of two numbers, assume conditions hold (given). *Step 3:* $t = 1.50$ or $-1.50$, p-value $= 0.155$. *Step 4:* Do not reject $H_0$. The mean measured and reported heights are not significantly different for men or there's not enough evidence to support the claim that the typical self-reported height differs from the typical measured height for men.

**9.67** *Step 1:* $H_a$: $\mu_{longboard} > \mu_{shortboard}$ where $\mu$ is the population mean number of surfing days last month. *Step 2:* Two-sample *t*-test, assume random, large samples. *Step 3:* $t = 2.11$, p-value $= 0.020$. *Step 4:* Reject $H_0$. Longboard users have a significantly larger mean number of days than shortboard users.

## Chapter Review Exercises

**9.69 a.** Paired *t* **b.** Two-sample *t*

**9.71 a.** *Step 1:* $H_a$: $\mu_{men} < \mu_{women}$ where $\mu$ is the population mean number of seconds for marathon runners. *Step 2:* Two-sample *t*-test, large sample, assume random. *Step 3:* $t = 9.26$, p-value $< 0.001$. *Step 4:* Reject $H_0$. The mean time for men is significantly less than the mean time for women or the typical man is quicker than the typical woman. **b.** 95% CI for difference: $(1405, 2163)$. Because the interval does not capture 0, it shows there is a significant difference. I am 95% confident that the population mean time for men is between 1405 and 2163 seconds less than the population mean time for women.

**9.73** *Step 1:* $H_a$: $\mu_{before} < \mu_{after}$ where $\mu$ is the population mean pulse rate (before and after coffee). *Step 2:* Paired *t*-test (repeated measures), assume condition hold (given). *Step 3:* $t = -2.96$, p-value $= 0.005$. *Step 4:* Reject $H_0$. Heart rates increase significantly after coffee. (The average rate before coffee was 82.4, and the average rate after coffee was 87.5.)

**9.75** The typical number of hours was a little higher for the boys and the variation was almost the same. $\bar{x}_{girls} = 9.8$ $\bar{x}_{boys} = 10.3$. $s_{girls} = 5.4$ and $s_{boys} = 5.5$. See the histograms.

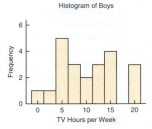

*Step 1:* H$_a$: $\mu_{girls} \neq \mu_{boys}$ where $\mu$ is the population mean number of TV viewing hours. *Step 2:* Two-sample *t*-test, random samples, assume the sample sizes of 32 girls and 22 boys is large enough that slight non-Normality is not a problem. *Step 3:* $t = -0.38$ or $0.38$, p-value $= 0.706$. *Step 4:* You cannot reject the null hypothesis. There is not enough evidence to conclude that boys and girls differ in the typical hours of TV watched.

**9.77 a.** Choose A, the two-sample t-test; the data are not paired.
*Step 1:* H$_a$: $\mu_{men} \neq \mu_{women}$ where $\mu$ is the population mean number of work hours. *Step 2:* Two sample *t*-test, random, $n = 50 > 25$ *Step 3:* $t = 4.09$, p-value $< 0.001$. *Step 4:* Reject H$_0$. The mean number of work hours for men and women are significantly different. 95% CI for difference in work hours, $\mu_{men} - \mu_{women}$: (6.18, 17.86). Because it does not capture 0, you can reject the hypothesis of no difference. Because the interval contains only positive values, we know that the mean number of work hours is greater for men than for women.

**9.79** Choose A because it is testing $\mu_1 < \mu_2$, which is what we want to test, assuming that 1 is the experimental group.
*Step 1:* H$_a$: $\mu_{drug\ tested} < \mu_{not\ drug\ tested}$ where $\mu$ is the population mean arrest rate. *Step 2:* Two-sample *t*-test, assume conditions hold (given) but not random sample, random assignment. *Step 3:* $t = 2.03$ or $-2.03$, p-value $= 0.022$. *Step 4:* Reject H$_0$. The drug testing lowered the rate of arrest significantly but do not generalize.

**9.81** The table shows the results. The average of $s^2$ in the table is 2.8889 (or about 2.89), and if you take the square root, you get about 1.6997 (or about 1.70), which is the value for sigma ($\sigma$) given in the TI-83/84 output shown

| Sample | $s$ | $s^2$ |
|--------|-----|-------|
| 1, 1 | 0 | 0 |
| 1, 2 | 0.7071 | 0.5 |
| 1, 5 | 2.8284 | 8.0 |
| 2, 1 | 0.7071 | 0.5 |
| 2, 2 | 0 | 0 |
| 2, 5 | 2.1213 | 4.5 |
| 5, 1 | 2.8284 | 8.0 |
| 5, 2 | 2.1213 | 4.5 |
| 5, 5 | 0 | 0 |
| | $26/9 = 2.8889$ | Sum 26.0 |

in the exercise. This demonstrates that $s^2$ is an unbiased estimator of $\sigma^2$, sigma squared.

**9.83** Answers will vary.

# CHAPTER 10

Answers may vary slightly due to type of technology or rounding.

## Section 10.1

**10.1 a.** Proportions are used for categorical data. **b.** Chi-square tests are used for categorical data.

**10.3**

| | Boy | Girl |
|-----------|-----|------|
| Violent | 10 | 11 |
| Nonviolent | 19 | 4 |

The table may have a different orientation. Both *Gender* and *Type of Crime* are categorical variables, as you can see from the raw data.

**10.5** *Mean GPA*: numerical and continuous. *Field of Study*: categorical.

**10.7 a.** 30/50, or 0.6 **b.** 30 **c.** 25

**10.9 a.**

| | High TV Violence | Low TV Violence | Total |
|-------------------|-----------------|----------------|-------|
| Yes, Physical Abuse | 13 | 27 | 40 |
| No Physical Abuse | 18 | 95 | 113 |
| Total | 31 | 122 | 153 |

**b.** $40/153 = 26.1\%$ **c.** $0.261438(31) = 8.10$
**d.** Expected values are shown in the table.

| | High TV Violence | Low TV Violence | Total |
|-------------------|-----------------|----------------|-------|
| Yes, Physical Abuse | 8.10 | 31.90 | 40 |
| No Physical Abuse | 22.90 | 90.10 | 113 |
| Total | 31 | 122 | 153 |

**10.11 a.** 40% of 16 is 6.4, so 6.4 should have had heart disease and 9.6 should not.

**b.** $X^2 = \dfrac{(9 - 6.4)^2}{6.4} + \dfrac{(7 - 9.6)^2}{9.6} = \dfrac{6.76}{6.4} + \dfrac{6.76}{9.6}$

$= 1.056 + 0.704 = 1.76$

**10.13 a.** One-fifth of 152 is 30.4 (expected correct), and four-fifths of 152 is 121.6 (expected incorrect).

**b.** $X^2 = \dfrac{(39 - 30.4)^2}{30.4} + \dfrac{(113 - 121.6)^2}{121.6} = \dfrac{73.96}{30.4} + \dfrac{73.96}{121.6}$

$= 2.433 + 0.608 = 3.04$

**10.15**

$X^2 = \dfrac{(13 - 8.10)^2}{8.1} + \dfrac{(27 - 31.90)^2}{31.9} + \dfrac{(18 - 22.90)^2}{22.9} + \dfrac{(95 - 90.10)^2}{90.1}$

$= 2.964 + 0.753 + 1.048 + 0.266 = 5.03$

From technology to avoid rounding: Chi-square $= 5.02$

## Section 10.2

**10.17** One categorical variable

**10.19**

*1:* H$_a$: Humans are not like random number generators and are not equally likely to pick all the integers.

*2:* If the integers are equally likely to be picked and there are 5 integers, then $38/5 = 7.6$, so the expected values counts for each value are all 7.6, which is larger than 5.

*3:* $X^2 = \frac{(3 - 7.6)^2}{7.6} + \frac{(5 - 7.6)^2}{7.6} + \frac{(14 - 7.6)^2}{7.6} + \frac{(11 - 7.6)^2}{7.6}$

$+ \frac{(5 - 7.6)^2}{7.6} = 11.47$

11.47 is the same in the output. p-value = 0.022.

*4:* Reject the null hypothesis and pick option ii.

**10.21** *Step 1:* $H_a$: $p$(heads) $\neq$ 0.50, or the coin is biased. *Step 2:* Chi-square GOF, both expected values are 25 > 5. *Step 3:* $X^2 = 8.00$, p-value = 0.005. *Step 4:* Reject $H_0$. The coin is biased.

**10.23** *Step 1:* $H_a$: The die is *not* fair and does not produce proportions of $1/6$ for each possible outcome. *Step 2:* Chi-square GOF, expected values all $1/6$ of 120 or 20 > 5, assume random sample. *Step 3:* $X^2 = 9.20$, p-value = 0.101. *Step 4:* We cannot reject $H_0$. The die has not been shown to be unfair.

**10.25 a.** *Step 1:* $H_a$: $p \neq 0.20$ where p is the population proportion of people who could correctly identify a Stradivarius violin. *Step 2:* Chi-square GOF, sample not random-testing to see whether the results could easily occur by chance, expected values are large enough. *Step 3:* Chi-square = 3.04, p-value = 0.081. *Step 4:* Because the p-value is 0.081, which is more than 0.05, we conclude that the results are not significantly different from guessing, which would produce about 20% correct identifications. **b.** *Step 1:* $H_a$: $p > 0.20$ where p is the population proportion of people who could correctly identify a Stradivarius violin. *Step 2:* One-proportion z-test, expected numbers of success and failure both more than 10, not a random sample, test whether the result could occur by chance. *Step 3:* $z = 1.74$, p-value = 0.041. *Step 4:* Because the p-value is 0.041, we can reject $H_0$ and conclude that there was a significantly higher proportion of correct identification than 20%. **c.** The one-tailed hypothesis has a p-value that is half that of the two-tailed hypothesis, and you can therefore reject $H_0$ with the one-tailed hypothesis.

## Section 10.3

**10.27** Independence: one sample.

**10.29** Homogeneity: random assignment (to four groups).

**10.31** The data are of the entire population (not a sample), and therefore there is no need for inference. The data are given as rates (percentages), not frequencies (counts), and there is not enough information for us to convert these percentages to counts.

**10.33**

*1:* $H_a$: The variables *Relationship Status* and *Obesity* are not independent (are associated).

*2:* Chi-square test of independence (given). The smallest expected value is 108.54, which is much more than 5.

*3:* $X^2 = 30.83$, p-value < 0.001.

*4:* Reject $H_0$. There is a connection between obesity and marital status; they are not independent. However, we should not generalize.
Causality? No, it is an observational study.
Percentage Obese: Dating: $81/440 = 18.4\%$ Cohabiting: $103/429 = 24.0\%$ Married: $147/424 = 34.7\%$

**10.35** *Step 1:* $H_a$: For men, watching violent TV is not independent of abusiveness. *Step 2:* Chi-square test of independence (one sample), expected values all more than 5, sample not random. *Step 3:* Chi-square = 5.02, p-value = 0.025. *Step 4:* Reject $H_0$. High TV violence as a child is associated with abusiveness as an adult in men, but don't generalize to all males and don't conclude causality.

**10.37 a.** Independence: one sample with two variables **b.** *Step 1:* $H_a$: Gender and happiness of marriage are associated (not independent). *Step 2:* Chi-square test of independence (one sample), random sample, minimum expected count is 14.6 > 5. *Step 3:* Chi-square = 1.22, p-value = 0.545. *Step 4:* You cannot reject $H_0$. Gender and happiness have not been shown to be associated. **c.** The rate of happiness in marriage has not been found to be significantly different for men and women.

**10.39 a.** HS Grad rate for no preschool: $29/64$, or 45.3%. The preschool kids had a higher graduation rate. **b.** *Step 1:* $H_a$: Graduation and preschool are not independent (they are associated). *Step 2:* Chi-square test of homogeneity, random assignment, not a random sample, the smallest expected value is 25.91 > 5. *Step 3:* $X^2 = 4.67$, p-value = 0.031. *Step 4:* Reject $H_0$. Graduation and preschool are associated: causality, yes, generalization, no.

**10.41 a.** For preschool, 50% graduated, and for no preschool, $21/39 = 53.8\%$ graduated. It is surprising to see that the boys who did not go to preschool had a bit higher graduation rate. **b.** *Step 1:* $H_a$: For the boys, graduation and preschool are associated. *Step 2:* Chi-square test for homogeneity, random assignment, not a random sample, the smallest expected value 15.32 > 5. *Step 3:* $X^2 = 0.10$, p-value = 0.747. *Step 4:* Do not reject $H_0$. For the boys, there is no evidence that attending preschool is associated with graduating from high school. c. The results do not generalize to other groups of boys and girls, but what evidence we have suggests that although preschool might be effective for girls, it may not be for boys, at least with regard to graduation from high school.

**10.43 a.** Surgery: $22/29 = 75.9\%$ free from diabetes. No surgery: $4/26 = 15.4\%$ free from diabetes. The surgery patients were much more likely to be free from diabetes. **b.** Although there is an observed value that is less than 5, there is no expected value less than 5, so you can use chi-square. **c.** *Step 1:* $H_a$: Surgery and recovery from diabetes are not independent. *Step 2:* Chi-square test of homogeneity, random assignment, not a random sample, there are no expected values less than 5. *Step 3:* $X^2 = 20.12$, p-value < 0.001. *Step 4:* Reject $H_0$. The surgery significantly affects the chance of recovery from diabetes but don't generalize.

**10.45 a.** With no confederate, $6/18$ (33.3%) followed the directions and took the stairs. With a compliant confederate, $16/18$ (88.9%) followed directions. With a noncompliant confederate, $5/18$ (27.8%) followed directions. Thus the subject was influenced to tend to do the same thing as the confederate. **b.** A p-value can never be larger than 1. The p-value is about 2.7 times 10 to the negative fourth power or 0.00027, which is less than 0.001. **c.** *Step 1:* $H_a$: Treatment and compliance are not independent. *Step 2:* Chi-square test of homogeneity, random assignment, not a random sample, expected values are all 9 > 5. *Step 3:* $X^2 = 16.44$, p-value < 0.001. *Step 4:* Reject $H_0$. There is a significant effect: causality: yes, generalization: no.

**10.47 a.** LD: $4/50 = 8\%$ tumor; LL $15/50 = 30\%$ tumor. Thus there is a higher rate of tumors in the mice that were exposed to light 24 hours a day.

**b.**

|  | LD | LL |
|---|---|---|
| Mice with Tumor(s) | 4 | 15 |
| Mice with No Tumors | 46 | 35 |

**c.** *Step 1:* $H_a$: The rate of tumors is associated with the lighting conditions. *Step 2:* Chi-square test for homogeneity, random assignment, none of the *expected values* is less than 5. *Step 3:* $X^2 = 7.86$, p-value = 0.005. *Step 4:* Reject $H_0$. Light affects tumor development in mice. **d.** Some shift workers are exposed to light at night and also during the day, because they may sleep where there is light. Perhaps they will be more likely to develop tumors.

## Section 10.4

**10.49**

|  | Under 30 | 30 and Older |
|---|---|---|
| Hospitalized: Yes | 25 | 10 |
| No | 480 | 104 |

*Step 1:* $H_a$: Age group and hospitalization are not independent. *Step 2:* Chi-square test of independence, expected values are greater than 5, random sample (given). *Step 3:* Chi-square = 2.55, p-value = 0.111. *Step 4:* Do not reject $H_0$ (at least with this age grouping). There is not enough evidence to show an association between age group and hospitalization.

**10.51 a.** The smallest expected value is 17.5 (from men and Other party), so the original table could have been used.

**b.**

|  | Male | Female |
|---|---|---|
| Democrat | 299 | 422 |
| Republican | 232 | 273 |
| Other | 395 | 389 |

**c.** Women: $422/1084 = 38.9\%$ Democrat. Men: $299/926 = 32.3\%$ Democrat. So these women are more likely to be Democrats than these men. **d.** *Step 1:* $H_a$: Gender and political party are associated (or not independent). *Step 2:* Chi-square test of independence, assume random sample, all expected values are much larger than 5. *Step 3:* Chi-square $= 12.01$, p-value $< 0.001$. *Step 4:* Reject $H_0$. Gender and political party are not independent.

**10.53 a.** You would get a smaller p-value, because it is more extreme in the direction of the antivenom working. **b.** You would get a larger p-value, because the results are less extreme. **c.** The p-value for the test in part a is 0.0002 (both one-tailed and two-tailed); yes, it is smaller. The p-value for the test in part b is 0.1002 (one-tailed) and 0.1319 (two-tailed); yes, it is larger.

### Chapter Review Exercises

**10.55 a.** 31/65, or 47.7%, of those in the control group were arrested, and 8/58, or 13.8%, of those who attended preschool were arrested. So there was a lower rate of arrest for those who went to preschool.

**b.**

|  | Preschool | No Preschool |
|---|---|---|
| Arrest | 8 | 31 |
| No Arrest | 50 | 34 |

*Step 1:* $H_a$: The treatment and arrest rate are associated. *Step 2:* Chi-square for homogeneity, random assignment, not a random sample, all the expected values are more than 5. *Step 3:* Chi-square $= 16.27$, p-value $= 0.000055$ (or $< 0.001$). *Step 4:* We can reject the hypothesis of no association at the 0.05 level. Don't generalize. We conclude that preschool attendance affects the arrest rate. **c.** Two-proportion z-test. *Step 1:* $p_{pre} < p_{nopre}$ (p is the rate of arrest). *Step 2:* Two-proportion z-test, all expected values are large. *Step 3:* $z = 4.03$, p-value $= 0.000028$ (or $< 0.001$). *Step 4:* Reject the null hypothesis; preschool lowers the rate of arrest but we cannot generalize. **d.** The z-test allows us to test the alternative hypothesis that preschool attendance *lowers* the risk of later arrest. The chi-square test allows only for testing for some sort of association, but we can't specify whether it is a positive or a negative association. Note that the p-value for the one-tailed hypothesis with the z-test is half the p-value for the two-tailed hypothesis with the chi-square test.

**10.57 a.** 11.4% of those on the drug had a heart event, and 10.0% of those on the placebo had a heart event. So those on the drug were more likely to have a heart event. **b.** *Step 1:* $H_a$: There is an association between use of the drug and heart events. *Step 2:* Chi-square test of homogeneity, random assignment, not a random sample, all expected values are large. *Step 3:* Chi-square $= 4.96$, p-value $= 0.026$. *Step 4:* Reject $H_0$. The drug affects the rate of heart events. **c.** Yes, because of the random assignment to groups and the statistically significant result.

**10.59 a.** In the experimental group the percentage arrested was 43/53, or 81.1%. In the control group the percentage was 36/53, or 67.9%. It seems that Scared Straight had a bad influence, because the control group was less likely to be arrested, but we don't yet know whether this difference was significant. **b.** *Step 1:* $H_a$: Treatment and arrests are not independent. *Step 2:* Chi-square test of homogeneity, random assignment, not random samples, no expected values less than 5. *Step 3:* Chi-square $= 2.44$, p-value $= 0.119$. *Step 4:* We cannot reject the hypothesis of independence at the 0.05 level. **c.** *Step 1:* $H_a$: $p_1 < p_2$ where $p_1$ is the population proportion of arrest for scared straight and $p_2$ is the population proportion of arrest for no treatment.

*Step 2:* Two-proportion z-test, all expected values are large enough, random assignment, not a random sample. *Step 3:* $z = 1.56$, p-value $= 0.941$ (use output A). *Step 4:* Don't reject the null hypothesis; we cannot conclude that the scared straight program lowered the arrest rate. **d.** To get the p-value for the one-tailed alternative, divide 0.119 by 2 and subtract that quotient from 1.000. The value of the chi-square test statistic is the square of the z-statistic.

**10.61 a.** *Step 1:* $H_a$: The presence or absence of robots is not independent of grouping. *Step 2:* Chi-square test for homogeneity, random (given), two expected values are 5, which is on the low side for this approximation. *Step 3:* Chi-square $= 4.32$, p-value $= 0.038$. *Step 4:* Reject $H_0$; the proportions are not the same. The robots have a significant effect. **b.** With Fisher's Exact Test, using a two-tailed alternative, the p-value is 0.080. You cannot reject $H_0$. The robots do not have a significant effect. **c.** For chi-square, the p-value was 0.038, and with Fisher's Exact Test, the p-value was 0.080. Fisher's Exact Test is accurate, and the chi-square test is a large-sample approximation. The approximation is not very good with two expected values of 5, and that is why the two p-values are so different. The p-value of 0.080 is the accurate value.

## CHAPTER 11

Correct answers may vary slightly due to rounding and type of technology used. All *t*-statistics are reported as positive values, although the sign of your *t*-statistics (positive or negative) must be consistent with which group you chose for group 1.

### Section 11.1

All calculations for this section were done without assuming equal variances.

**11.1 a.** ANOVA **b.** Two-sample *t*-test

**11.3 a.** $5(5-1)/2 = 20/2 = 10$ AB, AC, AD, AE, BC, BD, BE, CD, CE, DE **b.** $0.05/10 = 0.0050$

**11.5**

1:

|  | Mean | SD |
|---|---|---|
| PreAlg | 1.46 | 1.01 |
| ElemAlg | 2.73 | 0.79 |
| InterAlg | 4.08 | 2.88 |

2: Comparisons: 1. PreAlg-ElemAlg 2. PreAlg-InterAlg 3. ElemAlg-InterAlg
3: $0.05/3 =$ about 0.0167
4:

| | | |
|---|---|---|
| PreAlg Elem | $t = 3.30, p = 0.004$ | Significantly different |
| PreAlg Intermed | $t = 2.97, p = 0.011$ | Significantly different |
| Elem Intermed | $t = 1.57, p = 0.142$ | Not significantly different |

Conclusion: Pre-algebra students have a smaller mean number of study hours than the other two groups, which have means that are not different from each other. In other words, the population mean hours of study was smaller for Pre-algebra students than for the other two groups.

**11.7 a.**

|  | Mean |
|---|---|
| Younger Men | 287.3 |
| Older Men | 171.0 |
| Younger Women | 353.7 |
| Older Women | 356.9 |

The largest mean is for the older women, the second largest mean is for the younger women, the third largest is for the younger men, and the smallest is for the older men.

**b.** There are six pairs. Bonferroni significance level: $0.05/6 = 0.0083$.

**c.**

|  | *t* | *p* | **Conclusion** |
|---|---|---|---|
| Younger men–Older men | 4.79 | 0.000 | Different |
| Younger men–Younger women | 3.07 | 0.012 | Not different |
| Younger men–Older women | 2.89 | 0.012 | Not different |
| Older men–Younger women | 12.62 | 0.000 | Different |
| Older men–Older women | 10.37 | 0.000 | Different |
| Younger women–Older women | 0.23 | 0.824 | Not different |

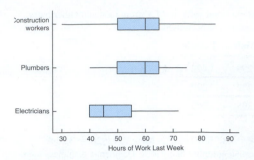

Older men have a significantly lower mean than all the others. The other differences in means are not significant. **d.** The target audience should probably *not* include the older men, because they watch significantly less comedy TV than the other groups. The other three groups should all be included as a part of the audience.

## Section 11.2

**11.9** The *F*-value of 9.38 goes with A, B, and C. And the *F*-value of 150.00 goes with L, M, and N. The reason for the difference is that the variation between groups (the separation between means) is larger for L, M, and N, relative to the variation within groups (which is the same in all groups).

$$F = \frac{\text{Variation between Groups}}{\text{Variation within Groups}}$$

**11.11 a.** $H_0$: The population means are all equal or marital status and cholesterol levels are not associated. $H_a$: At least one mean is different from another or marital status and cholesterol levels are associated. **b.** $F = 9.14$ **c.** Largest mean cholesterol: divorced. Smallest mean cholesterol: never married. **d.** This was an observational study, from which you cannot conclude causality. One possible confounder is age. For example, the "never married" may tend to be young, and youth may cause the low cholesterol.

**11.13** SST $= 571 + 217076 = 217647$

## Section 11.3

**11.15** p-value $= 0.422$. We cannot reject the null hypothesis that the population means are all the same. There's not enough evidence to conclude that there is an association between hours worked and educational level.

**11.17** The pulse rates are not in three independent groups, so the condition of independent groups fails.

**11.19** $65.05/20.78 = 3.13$, which is more than 2. The same-variance condition fails. ANOVA is not appropriate.

**11.21** *Step 1:* $H_0$: $\mu_{out} = \mu_{in} = \mu_{cat}$ $H_a$: At least one population mean is different from another. *Step 2:* Choose one-way ANOVA. Conditions are checked in the question statement. *Step 3:* $F = 8.00$, p-value $= 0.002$. *Step 4:* Reject $H_0$. The population means for the groups are not all equal. We reject the hypothesis that the mean run-time is independent of position played.

**11.23** $F = 4.85$, p-value $= 0.010$. Reject $H_0$. Feelings about the ethics of eating meat vary across the disciplines. We can reject the hypothesis that meat-eating feelings are independent of teaching discipline.

**11.25 a.** The medians and interquartile ranges are all similar, although the median for the athletes was a little larger (slower!) than the others, and the interquartile range was a bit larger for the moderate group. Also, the shapes are not strongly skewed. There are no potential outliers. **b.** $F = 0.10$, p-value $= 0.903$. The sample means were not significantly different and we conclude that we cannot reject the hypothesis that the population means are the same.

**11.27 a.** The median number of hours worked looks lowest for electricians. The medians for construction workers and plumbers are very similar. The interquartile ranges are all similar, and there are no potential outliers.

**b.** *Step 1:* $H_0$: All population means are equal. Ha: The population means are not all equal. *Step 2:* One-way ANOVA; assume conditions met (given in the problem statement). *Step 3:* $F = 7.48$, p-value $= 0.001$. *Step 4:* Reject $H_0$. The number of hours worked is associated with the type of job or the population mean numbers of hours of work are not all equal.

## Section 11.4

All calculations for post hoc tests were done with pooled variances.

**11.29** The only confidence interval that does not capture 0 is the one for NewburyP compared to PortH, so that is the only significant difference.

<u>PortH     SantaP</u>    NewburyP

The population mean price in Newbury Park is more than the mean price in Port Hueneme. The other differences between sample means are too small to conclude that the population means are different.

**11.31** Infielders minus catchers; 95% CI $(-1.39, -0.21)$: Because both boundaries are negative, we can reject the hypothesis of equality and say catchers take more time. Outfielders minus catchers; 95% CI $(-1.49 \text{ to } -0.32)$: Because both boundaries are negative, we can reject the hypothesis of equality and say catchers take more time. Outfielders minus infielders: 95% CI $(-0.59, 0.38)$: Because 0 is captured between the negative and positive endpoints, we cannot reject the null hypothesis of no difference.

<u>Outfielders     Infielders</u>    Catchers

Catchers had a significantly larger sample mean (were slower) than infielders and outfielders, and the infielders' and outfielders' sample means were not significantly different.

**11.33** The p-value for ANOVA was 0.903. We cannot reject the null hypothesis of no differences in population mean reaction distances and so you should not do post-hoc tests. In other words the sample means are not significantly different.

**11.35** Pb is plumber, Const is construction worker, and Elec is electrician For the TI-83/84, 98.33% confidence intervals were obtained, and Pooled Yes was chosen.

| **Comparison** | **Tukey CI** | **TI-83/84 CI** | **Reject $H_0$?** |
|---|---|---|---|
| Pb minus Const | $(-4.83, 8.81)$ | $(-5.39, 9.38)$ | No |
| Elec minus Const | $(-15.31, -1.67)$ | $(-16.04, -0.93)$ | Yes |
| Elec minus Pb | $(-17.37, -3.59)$ | $(-16.72, -4.24)$ | Yes |

<u>Electrician     Construction</u>    Plumber

Electricians have a mean that is less than the means for the others. There was insufficient evidence to conclude there is a difference between the mean hours of construction workers and the mean of plumbers.

## Chapter Review Exercises

**11.37** The p-value for ANOVA test was 0.01. Because overall significance level is 0.05, we can reject the null hypothesis that all the population means are the same and therefore should perform post hoc tests. For the TI-83/84 intervals, the confidence level was 98.33%, and Pooled Yes was chosen.

| Comparison | Tukey CIs | TI-83/84 CI | Reject $H_0$? |
|---|---|---|---|
| Phil minus Eth | $(-1.12, 0.26)$ | $(-1.21, 0.34)$ | No |
| Other minus Eth | $(-0.22, 1.16)$ | $(-0.13, 1.06)$ | No |
| Other minus Phil | $(0.21, 1.59)$ | $(0.15, 1.66)$ | Yes |

Philosophy Professors     Ethicists     Other professors

Remember that smaller numbers correspond to a worse opinion about the ethics of eating meat. The mean opinion of eating meat is worse for philosophy professors than for other professors. The other two differences are too small to be significant.

**11.39** The ratio of the smallest standard deviation to the largest is 7.174/5.091, or about 1.41, which is less than 2, so the same-variance condition is satisfied. The hypothesis that the population mean BMI (for different marital statuses) are all the same cannot be rejected, because the p-value is 0.587.

**11.41 a.** There are significant differences. **b.** OM is older men, YM is younger men, OW is older women, and YW is younger women.

| Comparison | CI | Reject Ho? |
|---|---|---|
| OM minus YM | $(-19.70, 43.10)$ | No |
| YW minus YM | $(19.00, 81.80)$ | Yes |
| OW minus YM | $(153.80, 216.60)$ | Yes |
| YW minus OM | $(7.30, 70.10)$ | Yes |
| OW minus OM | $(142.10, 204.90)$ | Yes |
| OW minus YW | $(103.40, 166.20)$ | Yes |

Younger men     Older men     Younger women     Older women

The population mean for older women is larger than all the others. The population mean for younger women is larger than those for both older men and younger men. There is insufficient evidence to conclude that the population mean for younger men and the mean for older men are different.

**11.43** The p-values are the same. The F-statistic is the square of the t-statistic. You get the same conclusion either way: Do not reject the null hypothesis of equal population mean triglyceride levels for men and women.

# CHAPTER 12

## Section 12.1

**12.1** Headline A suggests a cause-and-effect relationship, because it tells us that happiness prevents ("wards off") heart diseases. Headline B merely claims an association: People who are happier tend to have less heart disease.

**12.3** It was an observational study. Subjects in the study themselves chose which treatment group to belong to (tattoo or no tattoo).

**12.5** Answers may vary but "Tattoos Cause Hepatitis" is incorrect because the study is observational, not a randomized experiment. "Tattoos Linked with Hepatitis" and "Tattoos Associated with Hepatitis" are OK because they do not imply causality.

**12.7** It is a controlled experiment. We can tell because it was a "randomized" trial and therefore researchers assigned subjects to treatment groups.

**12.9 a.** The treatment variable is the drug administered: fingolimod or interferon. The response variable is whether or not the patient relapses. **b.** Conclusion: The drug fingolimod causes a decrease in the relapse rate of multiple sclerosis compared to interferon.

**12.11** It was probably an observational study, because subjects themselves chose their level of exercise.

**12.13** Both use the one-sample t-test. **a.** $p = 0.144$, do not reject $H_0$. **b.** $p = 0.001$, reject $H_0$. **c.** The larger sample size results in a smaller standard error, which, in turn, results in a t-statistic value farther from 0 and a p-value which is smaller. (The sample means are very similar, and the standard deviations are also very similar.)

**12.15** The second study will have greater power. The fact that the effect of the pill differs for men and women means that there is more variability in the population that combines men and women (study 1) than in the population with only men (study 2). Thus, although the results of the studies are generalized to different populations, the second study is drawing a sample from a population with less variability (and a larger weight loss) and therefore more power.

**12.17** This is not an appropriate use of blocking. In this design, the researchers randomized the blocks, not the subjects. Randomization should happen within blocks. For this study, each patient in a block should be assigned a number, and these numbers should be put into a bowl, mixed up. There will be four bowls, one for each age group. From each bowl (each block), half of the subjects' numbers are chosen, and these people will receive the treatment.

**12.19 a.** The treatment variable is the type of swimsuit. The response variable is the amount of time to swim the 200 meters. **b.** Randomly assign each person to wear a slick suit or a nonslick suit, and record the times for the races. To do this, place in a bag 20 tickets that say "slick" and 20 that say "nonslick." Each swimmer chooses a ticket and will use that type of suit. Record the swimming times. **c.** Block on whether the swimmer is an Olympic swimmer or not. Randomly assign half of the amateur swimmers to the slick suits: Place in a bag 10 tickets marked "Slick" and 10 marked "Nonslick"; each amateur swimmer chooses a ticket and will use that suit. For the Olympic swimmers, use a separate bag and place 10 "Slick" and 10 "Nonslick" tickets in it; each Olympic swimmer chooses a ticket and will use that type of suit. Record the swimming times. **d.** The blocked design will prevent having uneven groups with more Olympic swimmers in one group than in another. This blocked design will increase the power of the analysis by reducing variation and make it easier for designers to see whether the new design really improves swimming speed. **e.** Have each swimmer race with a slick suit and also race with a regular suit and use a paired t-test for comparison. This will reduce variation and make you more likely to see an advantage to the new suits if, in fact, such an advantage exists. To do this, randomly assign half of all swimmers to wear the slick suit first and the other half to wear a regular suit first to avoid the effects of order. (Otherwise, some people might swim more slowly in their second race because they were tired from the first race.)

**12.21 a.** The treatment variable records whether subjects get aspirin or placebo. The response variable is whether the person has a heart attack. **b.** You could put 100 slips of paper that marked A and 100 slips of paper marked B in a bag. Each person would draw out a slip of paper. Those who got A would get the aspirin, and those who got B would get the placebo. Then observe the subjects for a given time interval to see whether they have a heart attack, and compare the percentages with heart attacks for the two groups. **c.** Randomly assign half of the men and half of the women to use the aspirin and the rest to use a placebo. You could use two separate bags, one for the men and one for the women. Each bag would have 100 slips of paper: 50 marked A and 50 marked B. Each woman draws randomly from the women's bag, and each man draws randomly from the men's bag. Then observe the subjects for a given time interval to see whether they have a heart attack, and determine the percentage of aspirin-taking men who had a heart attack, the percentage of placebo-taking men who had a heart attack, the percentage of aspirin-taking women who had a heart attack, and the percentage of placebo-taking women who had a heart attack. **d.** The blocked design improves statistical power, in part by preventing an uneven distribution of men and women in the two groups. With the blocked design, we have a higher probability of determining whether aspirin reduces the risk of heart attack, if it actually does so.

**12.23 a.** t-value = 0.50, p-value = 0.309 (one-tailed). Do not reject $H_0$. **b.** t-value = 2.67, p-value = 0.009 (one-tailed). Reject $H_0$. **c.** The paired t-test is appropriate because the items are matched.

**12.25 a.** The mean for white paper was smaller (the means were 74.3 and 76.3). This contradicts the idea that reading material written on colored paper is easier to read. **b.** Two sample $t$-test: t-value = 0.19, p-value = 0.854. Do not reject $H_0$. **c.** Paired $t$-test: t-value = 0.97, p-value = 0.349. Do not reject $H_0$. **d.** The paired $t$-test is appropriate because each person is tested twice, so the numbers are coupled. **e.** People might get faster if they had read the passage previously, and you don't want the order of reading to affect the answer.

**12.27** The order might have an effect. For example, after exercise the pulse might stay elevated for a while, and if there was not a long waiting time between the two sets of observations, the higher pulse after exercise might affect the next situation. Randomizing the order ensures that effects such as these (which are sometimes called carryover effects) are balanced out in the total analysis.

### Section 12.2

**12.29** Systematic sampling

**12.31** The opinions on Social Security might vary widely from age group to age group, but they might be fairly similar within an age group. Possibly the older people will want it continued because they are receiving money from the system, and the younger people may want it abolished because they are paying into it and are afraid the system will go bankrupt before they get any money back from the system. Stratified sampling would allow us to compare answers in different age groups and improve the precision of our estimates.

**12.33** This is an example of stratified sampling. The strata consist of high-volume, medium-volume, and low-volume stores. These strata were chosen because stores within a stratum are seen as similar. A random sample is then taken from within each stratum.

### Section 12.3

**12.35 a.** No, we cannot generalize, because this was not a random sample. **b.** Yes, we can infer causality because of random assignment.

**12.37 a.** You can generalize to other people admitted to this hospital who would have been assigned a double room because of the random sampling from that group. **b.** Yes, you can infer causality because of the random assignment.

**12.39** Because a random sample was not used (or we do not know whether one was used), we cannot generalize to the larger public. Because patients were randomly assigned to treatment groups (full-dose or reduced dose), we can conclude cause-and-effect. However, it might appear that both full-dose and reduced-dose have the same effect only because researchers did not have a large enough sample to detect a difference. We *can* say that among persons 60 years of age or older, the intradermal injection causes a less vigorous antibody response.

**12.41 a.** It does not matter which treatment is chosen; the treatments are not significantly different. **b.** Report one interval: The first interval given, (−6.8% to 6.7%) represents the difference in survival rates for patients receiving the two treatments. The fact that the interval includes 0 means we cannot rule out the possibility that the survival rate is the same for both treatments. The other interval (−0.5, 7.90) also does not show a significant difference because it captures 0.

**12.43 a.** Take a nonrandom sample of students and randomly assign some to the reception and some to attend a "control group" meeting where they do something else (such as learn the history of the college). **b.** Take a random sample of students and offer them the choice of attending the reception or attending a "control group" meeting where they do something else (such as learn the history of the college). **c.** Take a random sample of students. Then randomly assign some of the students in this sample to the reception and some to the "control group" meeting.

**12.45**

1. They want to know whether low-dose vaccine is as effective as the regular dose.

2. It appears to be effective in people aged 18 to 60 but not for those above 60.

3. You cannot tell whether random assignment was used, because it was not mentioned in this excerpt.

4. Because we do not know the methods, we do not know whether the conclusion was appropriate.

5. Because there was no random sampling and we cannot tell whether there is random assignment, we cannot generalize beyond these subjects.

6. No other studies are mentioned, so this might be the first study of its kind.

**12.47**

1. Which is the better procedure for aneurysms, open repair or endovascular repair?

2. There is no difference between them in terms of the outcomes observed.

3. They used a randomized design.

4. They used random assignment and found no difference, so the conclusion was appropriate. However, we cannot eliminate the possibility that the effect is subtle and the sample size wasn't large enough to detect it.

5. Because there was not a random sample, we cannot generalize. But the random assignment would have allowed us to make cause-and-effect conclusions, had we rejected the null hypothesis.

6. We cannot tell whether previous research was done or not from the excerpt.

### Chapter Review Exercises

**12.49 a.** The death rate before the vaccine was 18.1 deaths per 100,000 children. After the vaccine, the death rate fell to 11.8 deaths per 100,000 children. The difference is 6.3 fewer deaths per 100,000 children after the vaccine was introduced. The small p-value (less than 0.001) means we can reject the null hypothesis that the death rate was unchanged and conclude that the death rate decreased. **b.** Although there are many indications that the vaccine is successful, this was not a randomized study. We cannot rule out the possibility that a confounding variable, not the vaccine, caused the decrease in death rates. (For example, because the comparison was done using different years, there might have been a difference in weather which contributed to the difference in disease rates.)

## CHAPTER 13

### Section 13.1

**13.1** You should have a random sample from the population.

**13.3** The data should be drawn from Normal distributions, or the sample sizes should be large (typically at least 25 from each population). Observations must be independent of each other. The groups must be independent of each other.

**13.5 a.** Histogram 1 goes with D, and 2 goes with C. **b.** 1 is roughly bell-shaped (Normal), and 2 is right-skewed. **c.** Use a log transform on the data that are shown in histogram 2.

**13.7 a.** 1, 1.30, 2, 2.74, 3 **b.** 10 (because $10^1 = 10$), 31.62, 100, 1000, 1584.89

**13.9** Don't worry if your numbers are a bit different because of rounding. **a.** 1, 2.447, 2.778, 4.176 **b.** 2.600 **c.** 398 **d.** Mean is 3972.5, median is 440.

| Geometric Mean | Median | Mean |
|---|---|---|
| 398 | 440 | 3972.5 |

**13.11 a.** I am 95% confident the population mean amount spent is between $57.80 and $165.20. **b.** (35.65, 93.76); I am 95% confident that the population geometric mean is between $35.65 and $93.76.

**c.** $165.2 - 57.8 = 107.40$
$93.76 - 35.65 = 58.11$

The confidence interval for the geometric mean (from the logs) is narrower. **d.** The confidence interval for the geometric mean is more appropriate, because the distribution of the log-transformed data is more symmetric, and so the confidence level for the geometric mean will be more accurate. Also, we can get a more precise (smaller margin of error) estimate for the geometric mean than for the mean, based on this sample.

**13.13 a.** Right-skewed **b.** (4.04, 6.31) **c.** It is closer to Normal than the distribution of untransformed data. **d.** (0.515, 0.7025) **e.** (3.27, 5.04); We are 95% confident that the population geometric mean number of hours of exercise per week for all students at this college is between 3.27 hours and 5.04 hours. **f.** Answers may vary. The interval for the geometric mean is more precise (smaller margin of error), which suggests that the geometric mean might be a better measure of center. On the other hand, the sample size is large, so both confidence intervals are valid and the population mean could also be used. (If the sample size were under 25, the reported confidence level for the resulting confidence interval of the population mean would not be accurate.)

### Section 13.2

**13.15** *Step 1:* $H_0$: The median levels of lead in the blood are the same. $H_a$: The median level of lead in the blood is larger for the experimental group (one-tailed). *Step 2:* Sign test (paired). *Step 3:* $S = 28$ or $S = 4$, p-value $< 0.001$. *Step 4:* Reject $H_0$. The median level of lead in the blood was larger for the children in the experimental group. For a p-value from a TI-83/84: $b(32, 0.5, 4$ or fewer$) = 9.7 \times 10^{-6}$, or about 0.00001

**13.17 a.** The medians were 19 (for same) and 33.50 (for different). Yes, it tended to take longer for the different colors. **b.** *Step 1:* $H_a$: Median time for same $<$ median time for different. *Step 2:* Sign test (paired). *Step 3:* $S = 1$ or 9, p-value $= 0.0107$. *Step 4:* Reject $H_0$. The difference is significant.

**13.19** *Step 1:* $H_a$: The median pulse rate goes up (one-tailed). *Step 2:* Sign test (paired). *Step 3:* $S = 4$ or 5, p-value $= 0.500$. *Step 4:* Do not reject $H_0$. The median pulse rate did not go up significantly for men.

**13.21** *Step 1:* $H_a$: Median age for grooms $>$ median age for brides. *Step 2:* Sign test (paired). *Step 3:* $S = 10$ or 3, p-value $= 0.046$ from $b(13, 0.5, 3$ or fewer$)$. *Step 4:* Reject $H_0$. The median age for the grooms is significantly more than the median age for the brides.

### Section 13.3

**13.23 a.** The median for the ethicists was 4.5 meals per week, which is lower than the median for the control group (6.5). Typically, ethicists eat fewer meals with meat per week than nonethicists. **b.** *Step 1:* $H_a$: Median meals of meat per week of control $>$ median of ethicists. *Step 2:* Mann-Whitney test (independent). *Step 3:* $W = 198$, p-value $= 0.058$. *Step 4:* Do not reject $H_0$. We have not found a significant difference in behavior.

**13.25** *Step 1:* $H_a$: Median weight change for brushing $\neq$ median weight change for Calgon. *Step 2:* Mann-Whitney test (independent). *Step 3:* $W = 45$, p-value $= 0.540$. *Step 4:* The difference is not significant. We have not shown a difference between brushing and Calgon.

**13.27 a.** We don't compare means because there is an outlier of around $400 for the women, and the samples are not large. **b.** *Step 1:* $H_a$: Median for the men $<$ median for the women. *Step 2:* Mann-Whitney test (independent). *Step 3:* $W = 501.0$, p-value $= 0.141$. *Step 4:* Do not reject $H_0$. The medians for men and women with regard to cell phone bills are not significantly different.

### Section 13.4

**13.29 a.** The red line looks like it is pretty far out in the tail of the data and suggests visually that there is a real difference in extraversion between the sporty and the nonsporty students. **b.** The p-value is 0.011988012, or 0.012. **c.** We reject the null hypothesis and conclude that "sporty" students typically have a higher level of extraversion than "nonsporty" students. **d.** We

could get an approximate p-value using the histogram, by finding the (approximate) proportion of observations to the right of the red vertical line.

**13.31 a.** The distributions are strongly right-skewed, and the median is often a better choice for skewed data than the mean. **b.** 525 is not far out in the tail, so it is not unusually large. **c.** The one-tailed p-value is larger than 0.05, so the two-tailed p-value will be even larger. Do not reject $H_0$. The population median for the men has not been shown to be different from the population median for the women.

**13.33** Choose 0.025 by looking at the graph, knowing the observed number is 368.9 acre-feet. Reject the null hypothesis. The mean rainfall is greater for the seeded clouds.

**13.35** Explanation a

### Chapter Review Exercises

**13.37 a.** Paired *t*-test or sign test **b.** Sign test **c.** Paired *t*-test or sign test

**13.39 a.** Two-sample *t*-test or Mann-Whitney test **b.** Mann-Whitney test

**13.41** Step 0:

| Obs | Null Value | Difference |
|-----|-----------|------------|
| 4.2 | 3.18 | 1.02 |
| 3.6 | 3.18 | 0.42 |
| 3.9 | 3.18 | 0.72 |
| 3.4 | 3.18 | 0.22 |
| 3.3 | 3.18 | 0.12 |

1: $H_0$: The median weight of the cones is 3.18 ounces, as advertised (or the median difference is 0). $H_a$: The median weight is more than 3.18 ounces (or the median difference is more than 0).

2: Use the one-sample sign test. The cones are independent because they are purchased from different servers, and although we don't have a truly random sample, we hope we have a representative sample.

3: $p = 0.031$

4: Reject $H_0$ and conclude that the population median is more than 3.18 ounces (or the population median difference is more than 0.00).

**Why?**: We have a small sample and do not know whether the distribution is Normal or not.

## CHAPTER 14

### Section 14.1

**14.1** Answers will vary. Some possibilities: the amount of time the student could study, the amount of sleep the student got the night before, the particular choice of questions on the exam, the noise level in the room, the health of the student.

**14.3** The residual plot shows that the trend is not a straight line, so the linear condition fails. The linear model is not appropriate. The car that is 4 years old is the farthest from the regression line because its residual is the biggest (in absolute value).

**14.5** The residual plot is fan-shaped, showing more variation in the number of units for students who have attended for many semesters and less variation for those who have attended for few semesters. This shows that the constant standard deviation condition does not hold, so inference (including confidence intervals or prediction intervals) would not be appropriate.

**14.7** The residual plot shows an increasing trend, and the QQ plot does not follow a straight line. Linear regression is inappropriate for this data set, because the linearity condition and the Normality condition fail.

## Section 14.2

**14.9 a.** The slope (0.843) is positive, showing that the older people *in the sample* tended to weigh a bit more. **b.** *Step 1:* $H_0$: Slope $= 0$ $H_a$: Slope $\neq 0$. *Step 2:* t-test for slope. *Step 3:* $t = 1.43$, p-value $= 0.156$. *Step 4:* Do not reject $H_0$. We do not have enough evidence to reject the hypothesis that the slope is 0. We have not shown an association between age and weight.

**14.11 a.** The intercept is 2.78. If the intercept were 0, it would mean that if a mother had 0 years of education, then the father would be predicted also to have 0 years of education. **b.** *Step 1:* $H_0$: Intercept $= 0$, $H_a$: Intercept $\neq 0$. *Step 2:* t-test for intercept. The conditions are assumed met. *Step 3:* $t = 2.00$, p-value $= 0.056$. *Step 4:* We cannot reject $H_0$. We don't have enough evidence to reject an intercept of 0.

**14.13 a.** The slope is 0.936. On the average, for each inch taller a parent is, the child is about 0.94 inch taller, *in the sample*. **b.** *Step 1:* $H_0$: Slope $= 0$ $H_a$: Slope $\neq 0$. *Step 2:* t-test for slope. *Step 3:* $t = 8.38$, p-value $< 0.001$. *Step 4:* Reject $H_0$. There is a significant relationship between parent height and student height. **c.** If the slope were 1, it would mean that on the average, for every inch taller the parent was, the student would be 1 inch taller also. **d.** *Step 1:* $H_0$: Slope $= 1$, $H_a$: Slope $\neq 1$. *Step 2:* t-test for slope. *Step 3:* $t = -0.578$, p-value $=$ larger than 0.05 (because 0.578 is less than 2.052). *Step 4:* Do not reject $H_0$.

$$t = \frac{\hat{\beta}_1 - \text{null}}{SE_{\hat{\beta}_1}} = \frac{0.9355 - 1}{0.1116} = -0.578$$

**14.15 a.** 767.3 **b.** The slope tells us that schools that have one percentage point more students receiving reduced-cost meals score, on average, 2.4 points lower on API than schools with a lower percentage of students receiving reduced-cost meals. **c.** $(-3.6, -1.2)$ **d.** The administrator is arguing that if we could see all schools in California, we would find a slope of 0. But 0 is not in the confidence interval, which suggests that the administrator is wrong.

**14.17 a.** Yes. The output shows that we cannot reject the null hypothesis that the intercept is 0 (because the p-value is 0.740, which is larger than 0.05). This means that the confidence interval will include 0 pounds of trash.

## Section 14.3

**14.19** Use a prediction interval. This concerns prediction of a single value, not a mean.

**14.21** Use a confidence interval. This concerns prediction of a mean, not a single value.

**14.23 a.** 270,959 **b.** Prediction interval, because we are predicting the value for one house, not for the mean of a group of houses. **c.** ($160,581, $381,532) **d.** Yes, he can afford a house because he has access to enough money to pay for the price at the top of the 95% interval.

**14.25** (About 125, about 200)

**14.27** The green lines are prediction intervals (for individuals), and the pink lines are confidence intervals (for means). Means tend to be more stable than individual measurements and give more precise results, which is why their intervals are narrower.

**14.29 a.** PI is (182, 216); CI is (195, 203). The confidence interval is narrower because it is estimating a population mean, and there is less uncertainty in this than in the prediction interval, which is predicting the weight of an individual. **b.** We are 95% confident that the mean weight of all baseball players who are 20 years old is between 195 pounds and 203 pounds. **c.** We are 95% confident that one 20-year-old baseball player will weigh between 182 and 216 pounds.

## Chapter Review Exercises

**14.31 a.** See graph. The equation is

Longevity $= 7.84 + 0.0327$ Gestation

**b.** The hippopotamus is predicted to live about 10 years but lives over 40 years, on average.

**c.** CI is 13.92 to 19.18 years

```
Predicted Values for New Observations

New
Obs   Fit  SE Fit    95% CI          95% PI
  1 16.55   1.29  (13.92, 19.18)  (1.75, 31.35)
```

**d.** Humans do not fit into this pattern and are not included in the data set. The average human life expectancy is much more than the top of the confidence interval, which is 19.18 years.

# Index